Tissue Engineering

Tissue Engineering

Edited by
John P. Fisher
Antonios G. Mikos
Joseph D. Bronzino

CRC Press
Taylor & Francis Group
Boca Raton London New York

CRC Press is an imprint of the
Taylor & Francis Group, an **informa** business

CRC Press
Taylor & Francis Group
6000 Broken Sound Parkway NW, Suite 300
Boca Raton, FL 33487-2742

First issued in paperback 2019

© 2007 by Taylor & Francis Group, LLC
CRC Press is an imprint of Taylor & Francis Group, an Informa business

No claim to original U.S. Government works

ISBN-13: 978-0-8493-9026-5 (hbk)
ISBN-13: 978-0-367-38905-5 (pbk)

Library of Congress Cataloging-in-Publication Data

Tissue engineering / editors, John P. Fisher, Antonios G. Mikos, and Joseph D. Bronzino.
    p. ; cm.
    "A CRC title."
    Includes bibliographical references and index.
    ISBN-13: 978-0-8493-9026-5 (hardcover : alk. paper)
    ISBN-10: 0-8493-9026-5 (hardcover : alk. paper)
    1. Tissue engineering. I. Fisher, John P. II. Mikos, Antonios G. III. Bronzino, Joseph D., 1937-
    [DNLM: 1. Tissue Engineering. QT 37 T61545 2007]

R857.T55T548 2007
610.28--dc22   2007007401

Visit the Taylor & Francis Web site at
http://www.taylorandfrancis.com

and the CRC Press Web site at
http://www.crcpress.com

Preface

Tissue engineering is increasingly viewed as the future of medicine. As evidenced in both the scientific and popular press, there exists considerable excitement surrounding the strategy of regenerative medicine. In an effort to put the numerous advances in the field into a broad context, this book is devoted to the dissemination of current thoughts on the development of engineered tissues. Three main topics are considered and form the basis for the three sections of the text: Fundamentals of Tissue Engineering, Enabling Technologies, and Tissue Engineering Applications. Fundamentals of Tissue Engineering examines the properties of stem cells, primary cells, growth factors, and extracellular matrix as well as their impact on the development of tissue-engineered devices. Enabling Technologies focuses upon those strategies typically incorporated into tissue-engineered devices or utilized in their development, including scaffolds, nanocomposites, bioreactors, drug delivery systems, and gene therapy techniques. Tissue Engineering Applications presents synthetic tissues and organs that are currently under development for regenerative medicine applications. The contributing authors are a diverse group with backgrounds in academia, clinical medicine, and industry. Furthermore, the text includes contributions from Europe, Asia, and North America, helping to broaden the views on the development and application of tissue-engineered devices.

The text is largely derived from the Advances in Tissue Engineering Short Course, a pioneering forum on regenerative medicine held at Rice University since 1993. The ATE Short Course has educated researchers, students, clinicians, and engineers on both the fundamentals of tissue engineering and recent advances in many of the most prominent tissue engineering laboratories around the world. For many of the contributors, the chapter included in this text presents findings that have been recently discussed at the Advances in Tissue Engineering Short Course.

The target audience for this text includes not only researchers, but also advanced students and industrial investigators. The text should be a useful reference for courses devoted to tissue engineering fundamentals and those laboratories developing tissue-engineered devices for regenerative medicine therapy.

John P. Fisher
University of Maryland

Antonios G. Mikos
Rice University

Editors

John P. Fisher is an assistant professor in the Fischell Department of Bioengineering at the University of Maryland. Dr. Fisher completed his BS (1995) at The Johns Hopkins University in chemical engineering, MS (1998) at the University of Cincinnati, his PhD (2003) at Rice University in bioengineering, and postdoctoral fellowship (2003) at the University of California, Davis in cartilage biology and engineering. At Maryland, Dr. Fisher directs the Biomaterial Laboratory which is involved in the development of biomaterials for engineered tissues, especially bone and cartilage. The lab focuses on the development of novel materials that can support the growth of both adult progenitor and adult stem cells, and is particularly interested in how the supporting biomaterials affect communication among cell populations. Dr. Fisher has received a NSF CAREER Award (2005), an Arthritis Foundation Investigator Award (2006), and the University of Maryland Invention of the Year Award (2006). Dr. Fisher has served as editor of several works, and is currently the reviews editor of the journal *Tissue Engineering*.

Antonios G. Mikos is the J.W. Cox professor of bioengineering and professor of chemical and biomolecular engineering at Rice University. He received his Dipl.Eng. (1983) from the Aristotle University of Thessaloniki, Greece, and his PhD (1988) in chemical engineering from Purdue University under the direction of Professor Nicholas A. Peppas. He was a postdoctoral researcher at the Massachusetts Institute of Technology and the Harvard Medical School working with Professors Robert Langer and Joseph Vacanti before joining the Rice Faculty in 1992 as an assistant professor.

Mikos' research focuses on the synthesis, processing, and evaluation of new biomaterials for use as scaffolds for tissue engineering, as carriers for controlled drug delivery, and as non-viral vectors for gene therapy. His work has led to the development of novel orthopaedic, dental, cardiovascular, neurologic, and ophthalmologic biomaterials. He is the author of over 310 publications and 22 patents. He is the editor of nine books and the author of one textbook.

Mikos is a fellow of the International Union of Societies for Biomaterials Science and Engineering and a fellow of the American Institute for Medical and Biological Engineering. He has been recognized by various awards including the Distinguished Lecturer Award of the Biomedical Engineering Society, the Edith and Peter O'Donnell Award in Engineering of The Academy of Medicine, Engineering and Science of Texas, the Marshall R. Urist Award for Excellence in Tissue Regeneration Research of the Orthopaedic Research Society, and the Clemson Award for Contributions to the Literature of the Society for Biomaterials.

Mikos is a founding editor of the journal *Tissue Engineering* and a member of the editorial boards of the journals *Advanced Drug Delivery Reviews, Biomaterials, Cell Transplantation, Journal of Biomaterials Science Polymer Edition, Journal of Biomedical Materials Research (Part A and B)*, and *Journal of Controlled Release*. He is the organizer of the continuing education course Advances in Tissue Engineering offered annually at Rice University since 1993.

Joseph D. Bronzino received the B.S.E.E. degree from Worcester Polytechnic Institute, Worcester, MA, in 1959, the M.S.E.E. degree from the Naval Postgraduate School, Monterey, CA, in 1961, and the Ph.D. degree in electrical engineering from Worcester Polytechnic Institute in 1968. He is presently the Vernon

Roosa Professor of Applied Science, an endowed chair at Trinity College, Hartford, CT and President of the Biomedical Engineering Alliance and Consortium (BEACON), which is a nonprofit organization consisting of academic and medical institutions as well as corporations dedicated to the development and commercialization of new medical technologies (for details visit www.beaconalliance.org).

He is the author of over 200 articles and 11 books including the following: *Technology for Patient Care* (C.V. Mosby, 1977), *Computer Applications for Patient Care* (Addison-Wesley, 1982), *Biomedical Engineering: Basic Concepts and Instrumentation* (PWS Publishing Co., 1986), *Expert Systems: Basic Concepts* (Research Foundation of State University of New York, 1989), *Medical Technology and Society: An Interdisciplinary Perspective* (MIT Press and McGraw-Hill, 1990), *Management of Medical Technology* (Butterworth/Heinemann, 1992), *The Biomedical Engineering Handbook* (CRC Press, 1st ed., 1995; 2nd ed., 2000; Taylor & Francis, 3rd ed., 2005), *Introduction to Biomedical Engineering* (Academic Press, 1st ed., 1999; 2nd ed., 2005).

Dr. Bronzino is a fellow of IEEE and the American Institute of Medical and Biological Engineering (AIMBE), an honorary member of the Italian Society of Experimental Biology, past chairman of the Biomedical Engineering Division of the American Society for Engineering Education (ASEE), a charter member and presently vice president of the Connecticut Academy of Science and Engineering (CASE), a charter member of the American College of Clinical Engineering (ACCE) and the Association for the Advancement of Medical Instrumentation (AAMI), past president of the IEEE-Engineering in Medicine and Biology Society (EMBS), past chairman of the IEEE Health Care Engineering Policy Committee (HCEPC), past chairman of the IEEE Technical Policy Council in Washington, DC, and presently editor-in-chief of Elsevier's BME Book Series and Taylor & Francis' *Biomedical Engineering Handbook*.

Dr. Bronzino is also the recipient of the Millennium Award from IEEE/EMBS in 2000 and the Goddard Award from Worcester Polytechnic Institute for Professional Achievement in June 2004.

Contributors

Alptekin Aksan
University of Minnesota
Center for Engineering in
   Medicine/Surgical Services
Massachusetts General Hospital
Harvard Medical School
Shriners Hospital for Children
Boston, Massachusetts

Kyle D. Allen
Department of Oral and
   Maxillofacial Surgery
Department of Bioengineering
University of Texas Dental Branch
Rice University
Houston, Texas

Jose F. Alvarez-Barreto
School of Chemical Engineering
   and Materials Science
Bioengineering Center
University of Oklahoma
Norman, Oklahoma

James M. Anderson
Department of Pathology
Macromolecular Science and
   Biomedical Engineering
Case Western Reserve University
Cleveland, Ohio

Anthony Atala
Department of Urology
Wake Forest University Baptist
   Medical Center
Winston-Salem, North Carolina

Kyriacos A. Athanasiou
Department of Oral and
   Maxillofacial Surgery
Department of Bioengineering
University of Texas Dental Branch
Rice University
Houston, Texas

C. Becker
Department of Pharmaceutical
   Technology
University of Regensburg
Regensburg, Germany

B.L. Beckstead
Department of Bioengineering
University of Washington
Seattle, Washington

François Berthiaume
Center for Engineering in
   Medicine/Surgical Services
Massachusetts General Hospital
Harvard Medical School
Shriners Hospital for Children
Boston, Massachusetts

Steven T. Boyce
Department of Surgery
University of Cincinnati
College of Medicine
Cincinnati, Ohio

Nenad Bursac
Department of Biomedical
   Engineering
Duke University
Durham, North Carolina

Baohong Cao
Department of Orthopaedic
   Surgery
University of Pittsburgh
Pittsburgh, Pennsylvania

Arnold I. Caplan
Skeletal Research Center
Case Western Reserve University
Cleveland, Ohio

Ayse B. Celil
Bone Tissue Engineering Center
Department of Biological
   Sciences
Carnegie Mellon University
Pittsburgh, Pennsylvania

Christina Chan
Center for Engineering in
   Medicine/Surgical Services
Massachusetts General Hospital
Harvard Medical School
Shriners Hospital for Children
Boston, Massachusetts
Michigan State University
Chemical Engineering and
   Materials Science
East Lansing, Michigan

Gang Cheng
Department of Chemical
   Engineering
Institute of Biosciences and
   Bioengineering
Rice University
Houston, Texas

K.S. Chian
School of Mechanical and
   Production Engineering
Nanyang Technological University
Singapore

Rena N. D'Souza
Department of Orthodontics
University of Texas Health Science
   Center at Houston-Dental
   Branch
Rice University
Houston, Texas

Brian Dunham
Department of
  Otolaryngology/Head and Neck
  Surgery
Department of Biomedical
  Engineering
Johns Hopkins School of Medicine
Baltimore, Maryland

Wafa M. Elbjeirami
Department of Biochemistry and
  Cell Biology
Rice University
Houston, Texas

Jennifer H. Elisseeff
Department of Biomedical
  Engineering
Johns Hopkins University
Baltimore, Maryland

Mary C. Farach-Carson
Department of Biological Sciences
University of Delaware
Newark, Delaware

John P. Fisher
University of Maryland
College Park, Maryland

William H. Fissell
Department of Internal Medicine
University of Michigan
Ann Arbor, Michigan

Paul Flint
Department of
  Otolaryngology/Head and Neck
  Surgery
Johns Hopkins School of Medicine
Baltimore, Maryland

Andrés J. García
Woodruff School of Mechanical
  Engineering
Petit Institute for Bioengineering
  and Bioscience
Georgia Institute of Technology
Atlanta, Georgia

Andrea S. Gobin
Department of Bioengineering
Rice University
Houston, Texas

W.T. Godbey
Laboratory for Gene Therapy and
  Cellular Engineering
Department of Chemical and
  Biomolecular Engineering
Tulane University
New Orleans, Louisiana

Aaron S. Goldstein
Department of Chemical
  Engineering
Virginia Polytechnic Institute and
  State University
Blacksburg, Virginia

A. Göpferich
Department of Pharmaceutical
  Technology
University of Regensburg
Regensburg, Germany

K. Jane Grande-Allen
Department of Bioengineering
Rice University
Houston, Texas

Scott Guelcher
Bone Tissue Engineering
  Center
Department of Biological
  Sciences
Carnegie Mellon University
Pittsburgh, Pennsylvania

Kiki B. Hellman
The Hellman Group, LLC
Clarksburg, Maryland

Jeffrey O. Hollinger
Bone Tissue Engineering Center
Department of Biological
  Sciences
Carnegie Mellon University
Pittsburgh, Pennsylvania

Johnny Huard
Department of Orthopaedic
  Surgery
University of Pittsburgh
Pittsburgh, Pennsylvania

H. David Humes
Departments of Internal
  Medicine
University of Michigan and
  Ann Arbor Veteran's Affairs
  Medical Center
Ann Arbor, Michigan

Esmaiel Jabbari
Department of Orthopedic
  Surgery
Department of Physiology and
  Biomedical Engineering
Mayo Clinic College of Medicine
Rochester, Minnesota

John A. Jansen
Department of Biomaterials
University Medical Center
  St. Radboud
Nijmegen, The Netherlands

Kristi L. Kiick
Department of Materials Science
  and Engineering
University of Delaware
Newark, Delaware

Catherine Le Visage
Department of Biomedical
  Engineering
Johns Hopkins School of
  Medicine
Baltimore, Maryland

Kam Leong
Department of Biomedical
  Engineering
Johns Hopkins School of
  Medicine
Baltimore, Maryland

Yong Li
Department of Orthopaedic
  Surgery
University of Pittsburgh
Pittsburgh, Pennysylvania

Lichun Lu
Department of Orthopedic
  Surgery
Department of Physiology and
  Biomedical Engineering
Mayo Clinic College of Medicine
Rochester, Minnesota

Surya K. Mallapragada
Department of Chemical
  Engineering
Iowa State University
Ames, Iowa

Antonios G. Mikos
Department of Bioengineering
Rice University
Houston, Texas

Michael Miller
Department of Plastic Surgery
University of Texas
Houston, Texas

Amit S. Mistry
Department of Bioengineering
Rice University
Houston, Texas

David J. Mooney
Departments of Chemical
  Engineering, Biomedical
  Engineering
Biologic and Materials Sciences
University of Michigan
Ann Arbor, Michigan
Division of Engineering and
  Applied Sciences
Harvard University
Cambridge, Massachusetts

Michael J. Moore
Department of Orthopedic
  Surgery
Department of Physiology and
  Biomedical Engineering
Mayo Clinic College of Medicine
Rochester, Minnesota

J.M. Munson
Laboratory for Gene Therapy and
  Cellular Engineering
Department of Chemical and
  Biomolecular Engineering
Tulane University
New Orleans, Louisiana

Hairong Peng
Department of Orthopaedic
  Surgery
University of Pittsburgh
Pittsburgh, Pennsylvania

J. Daniell Rackley
Department of Urology
Wake Forest University Baptist
  Medical Center
Winston-Salem, North Carolina

B.D. Ratner
Department of Bioengineering
University of Washington
Seattle, Washington

Jennifer B. Recknor
Department of Chemical
  Engineering
Iowa State University
Ames, Iowa

A. Hari Reddi
Center for Tissue Regeneration
  and Repair
Department of Orthopaedic
  Surgery
University of California
Davis School of Medicine
Sacramento, California

A.C. Ritchie
School of Mechanical and
  Production Engineering
Nanyang Technological University
Singapore

P. Quinten Ruhé
Department of Biomaterials
University Medical Center
  St. Radboud
Nijmegen, The Netherlands

Rachael H. Schmedlen
Department of Bioengineering
Rice University
Houston, Texas

Songtao Shi
National Institutes of Health
National Institute of Dental and
  Craniofacial Research
Craniofacial and Skeletal Diseases
  Branch
Bethesda, Maryland

Xinfeng Shi
Department of Bioengineering
Rice University
Houston, Texas

Yan-Ting Shiu
Department of Bioengineering
University of Utah
Salt Lake City, Utah

Vassilios I. Sikavitsas
School of Chemical Engineering
  and Materials Science
Bioengineering Center
University of Oklahoma
Norman, Oklahoma

Sunil Singhal
Department of General Surgery
Johns Hopkins School of Medicine
Baltimore, Maryland

David Smith
Teregenics, LLC
Pittsburgh, Pennsylvania

Paul H.M. Spauwen
Department of Plastic and
  Recontructive Surgery
University Medical Center
  St. Radboud
Nijmegen, The Netherlands

Dorothy M. Supp
Shriners Hospital for Children
Cincinnati Burns Hospital
Cincinnati, Ohio

Arno W. Tilles
Center for Engineering in
  Medicine/Surgical Services
Massachusetts General Hospital
Harvard Medical School
Shriners Hospitals for Children
Boston, Massachusetts

Mehmet Toner
Center for Engineering in
  Medicine/Surgical Services
Massachusetts General Hospital
Harvard Medical School
Shriners Hospital for Children
Boston, Massachusetts

Roger C. Wagner
Department of Biological
  Sciences
University of Delaware
Newark, Delaware

Jennifer L. West
Department of Bioengineering
Rice University
Houston, Texas

Joop G.C. Wolke
Department of Biomaterials
University Medical Center
  St. Radboud
Nijmegen, The Netherlands

Mark E.K. Wong
Department of Oral and
  Maxillofacial Surgery
Department of
  Bioengineering
University of Texas Dental
  Branch
Rice University
Houston, Texas

Fan Yang
Department of Biomedical
  Engineering
Johns Hopkins University
Baltimore, Maryland

Martin L. Yarmush
Center for Engineering in
  Medicine/Surgical Services
Massachusetts General
  Hospital
Shriners Burns Hospital for
  Children
Harvard Medical School
Boston, Massachusetts

Michael J. Yaszemski
Department of Orthopedic
  Surgery
Department of Physiology and
  Biomedical Engineering
Mayo Clinic College of Medicine
Rochester, Minnesota

Diana M. Yoon
Department of Chemical
  Engineering
University of Maryland
College Park, Maryland

Yu Ching Yung
Department of Chemical
  Engineering
University of Michigan
Ann Arbor, Michigan

Kyriacos Zygourakis
Department of Chemical
  Engineering
Institute of Biosciences and
  Bioengineering
Rice University
Houston, Texas

Contents

SECTION III   Tissue Engineering Applications

I

Fundamentals of Tissue Engineering

1

Fundamentals of Stem Cell Tissue Engineering

Arnold I. Caplan
Case Western Reserve University

1.1 Introduction

In adults, stem cells are fundamental cell units within every tissue that function as a renewal source of highly specialized, terminally differentiated cells. The cell renewal serves to compensate for the normal cell turnover (cell death) or serves to provide reparative cells for the repair of minor defects. The stem cells can be thought of as the rejuvenation potential of the organism (high during the young or growth phase). Unfortunately, this renewal capacity decreases with age; even amphibians that are able to perfectly regenerate an entire limb lose this capacity with age [1]. Thus, one of the long-term goals of Tissue Engineering is to learn how to control and regulate this natural regeneration potential, so that tissue performance can be enhanced or massive defects can be repaired via an intrinsic regenerative pathway.

As a scientific discipline, Tissue Engineering is very young, and thus, is quite distant from its long-term goal. To approach this goal, a series of sequential technological advancements must be made. Our earliest achievements, material-assisted repair of various tissues, have been accomplished in a variety of preclinical models and, in the case of skin, with clinical success in humans [2–5]. In all these cases, scaffolds, cells and growth factors/cytokines, or a combination of these have been surgically implanted. Like the first crude cardiac pacemakers, their initial successful implantation served as a catalyst for their improvement and perfection, a process that is ongoing.

THE MESENGENIC PROCESS

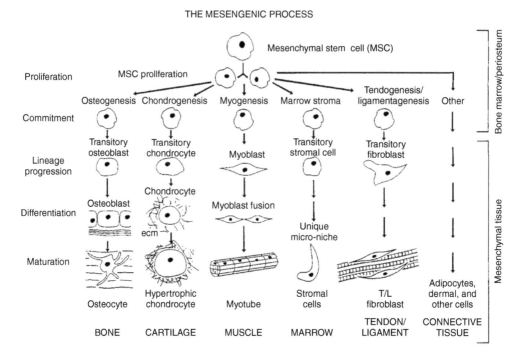

FIGURE 1.1   The mesengenic process involves the replication of MSCs and their entrances along multistep lineage pathways to produce differentiated cells that fabricate specific tissues such as bone, cartilage, and so on. We know most about the lineages on the left and least about those on the right of the diagram.

It follows that if we are to learn how to manage the various intrinsic organ stem cells to reconstruct or repair specific tissues, we must first obtain a deep understanding of these unique stem cells; we must understand what makes these cells divide, differentiate, grow old, and expire. We must learn how to position these stem cells in defects, how to coordinate the integration of blood vessels and nerves, and how to integrate the host tissue with the *neo*-tissue. Lastly, as Tissue Engineers, we must recognize that each individual has a genetically controlled variation, even between close family members, that will affect the fine-tuning of every repair logic.

It would be impossible to review the fundamental characteristics of every stem cell/organ system in the body in this chapter. Thus, I will focus on only one stem cell system, the mesenchymal stem cells (MSCs), that has already proven to be a versatile source of reparative cells for Tissue Engineering applications.

1.2   Mesenchymal Stem Cells

I suggested long ago that bone marrow contained a stem cell capable of differentiating into a number of mesenchymal tissues; I call this cell a mesenchymal stem cell (MSC) and the lineage sequences Mesengenesis, as pictured in Figure 1.1 [6–11]. This suggestion was based on my familiarity with embryonic mesenchymal progenitors [12–14] and partially on the early studies of Freidenstein [15] and, in particular, of Owen. Indeed, it was Owen [16] who drew me into the adult MSC realm by her scholarly treatise. In the late 1980s, Stephen Haynesworth and I [10,17,18] embarked in the task of isolating these rare MSCs from human bone marrow; the key to our success was a selected batch of fetal bovine serum that worked quite well with embryonic chick limb bud mesenchymal progenitor cells [19]. Subsequently, my collaborators have shown that marrow-derived MSCs are capable of differentiating into cartilage [20,21], bone [22,23], muscle [24,25], bone marrow stroma (hematopoietic support tissue) [26–28], fat [29], tendon [30,31], and other connective tissues. Other laboratories have provided evidence that MSCs can differentiate into neural cells [32,33], cardiac myocytes [34,35], vascular support cells (pericytes, smooth muscle cells) [36,37], and perhaps other tissues [38,39]. The multipotential of culture expanded MSCs provides the stimulus to consider them as candidates for various Tissue Engineering strategies. Such strategies, by necessity, require both scaffolds and various growth factors/cytokines to manage the proliferation and differentiation of MSCs to form specific tissues *in vivo*. In some cases, the tissue defect, itself, and its microenvironment provide instructional support; in other cases, pretreatment of the cells or placing the cells within unique scaffolds provides the instructional cues [40]. The details of these experiments provide the experimental basis for improving these early Tissue Engineering logics in preparation for their clinical use.

1.3   Fundamental Principles

When in doubt of how to manage the multiple parameters of tissue repair/regeneration, Mother Nature should be asked to reveal her secrets as a guide. Philosophically, we believe that many of the fundamental principles of Tissue Engineering involve the recapitulation of specific aspects of embryonic tissue formation [41,42]. For example, the embryonic mesenchyme that will form the cartilage anlagen of long bones has a high ratio of undifferentiated progenitor cells to extracellular matrix (ECM). This embryonic mesenchymal ECM is composed of type I collagen, hyaluronan, fibronectin, and water. In experiments with chick limb bud mesenchymal progenitor cells in culture, we showed that the molecular weight of hyaluronan is instructive to these progenitor cells, in that high molecular weight chondro-inductive or chondro-permissive [43,44]. Others have shown that high molecular weight hyaluronan is antiangiogenic. Indeed, the exclusion of blood vessels is also chondrogenic. Thus, a scaffold of hyaluronan coated with type I collagen or fibronectin to bind the MSCs would be expected to be chondrogenic; indeed, we have shown it to be just that [45]. The scaffold must be quite porous to allow the newly differentiated chondrocytes to fabricate their unique and voluminous ECM that controls the cushioning properties of the cartilaginous tissue. By mimicking the cell density and ECM of embryonic progenitor mesenchyme, cartilaginous tissue forms in large, full thickness defects in adult rabbit knees [41].

In the absence of hyaluronan, the key physical characteristic of prechondrocytes is their close proximity to their neighbors and maintenance of the cells in a rounded shape [20,21]. One way to achieve this condition is to place adult MSCs in a type I collagen lattice. The cells will bind to the lattice fibrils and rapidly contract the lattice to bring the cells into a high density configuration [46]. In culture, in the presence of TGF-β [20,21] or *in vivo* by the contracted network excluding blood vessels, the MSCs will form cartilage tissue. Thus, mimicking the embryonic microenvironment both chemically and physically can result in the specific differentiation of MSCs.

In contrast, the rules for bone formation are quite different from those governing cartilage formation [13,47,48]. In this case, we again studied embryonic chick limb development and observed that vasculature is the driver for bone formation. In the context of Tissue Engineering scaffolds, bone formation requires rapid invasion of blood vessels into the pores of the scaffold. For example, porous calcium phosphate ceramics coated with fibronectin to bind MSCs provide an inductive microenvironment for bone formation in subcutaneous or orthotopic sites [52]; it may be that the calcium phosphate, itself, is informational, but more likely it binds osteogenic growth factors that stimulate the MSCs. The MSCs bind to the walls of the pores in the ceramic where they divide and, as vasculature invades from the host tissue at the implantation site, the cells differentiate into sheets of osteoblasts and fabricate the lamellae of bone [49]. The vasculature does not go to the walls of dead-ends of the ceramic and in this location, the MSCs divide and pile-up on one another and form compact areas of cartilage. Thus, in the two different microenvironments (vascular and avascular), the MSCs form two very different tissues (bone and cartilage). This bone/cartilage forming capacity has been quantified and has become our gold standard for judging the quality of MSC preparations [19,50].

Again, for emphasis, mimicking Mother Nature, especially her very efficient embryological events, is, for us, a fundamental rule of engineering tissue repair or regeneration in adults.

1.3.1 *In Vitro* Assays for the Osteogenic and Chondrogenic Lineages

We have established an *in vitro* assay for the differentiation of human and animal MSCs into bone and cartilage phenotypes [20–23]. For osteogenesis, the inclusion of dexamethasone (*dex*), ascorbate, and eventually β-glycerophosphate (a phosphate donor) causes the MSCs to enter the osteogenic lineage, upregulate alkaline phosphatase activity (two- to ten-fold), secrete various bone proteins, and eventually, organize calcium apatite deposition within type I collagen fibrils comparable to that observed in *in vivo* osteoid.

Likewise, we have established the *in vitro* conditions for causing MSCs to differentiate into chondrocytes [21,22]. Simply, MSCs are pelleted to the bottom of 15-ml conical plastic tubes, and the medium is changed to a chemically defined medium containing TGF-β at a concentration of 2–10 ng/ml. The pelleted cells come off the bottom of the tube and form a sphere of cells that differentiate into chondrocytes that fabricate type II collagen- and aggrecan-rich ECM. Both of these *in vitro* assays and the porous ceramic assay *in vivo* have been tested by diluting the human or rat marrow-derived MSCs with dermal fibroblasts. In these experiments, the human dermal fibroblasts were shown to *not* differentiate into bone or cartilage phenotypes. The results of these assays are that MSC preparations can have 25–50% non-MSCs without experiencing a major loss of osteo- or chondrogenic tissue formation [51]. In the context of Tissue Engineering logics, these observations could mean that the implanted MSCs could be "contaminated" by host-derived fibroblasts, but would still retain their capacity to regenerate osteochondral tissue.

1.4 MSCs and Hematopoietic Support

The marrow stroma is a highly specialized connective tissue that provides different microenvironmental niches for the different lineage pathways of hematopoiesis. Each of the different lineage pathways of hematopoiesis requires a different combination of cytokines and growth factors. The marrow stromacytes not only fabricate these unique physical niches of connective tissue, but they secrete specific bioactive molecules for the initiation and control of these lineage pathways. The MSC differentiation sequence into these different hematopoietic lineage support phenotypes has yet to be described. However, just as complex multicellular Dexter cultures are able to support *in vitro* hematopoiesis, so can homogeneous populations of MSCs incubated in Dexter medium provide support for hematopoiesis. Because of these very specific hematopoietic-support functions, the term "marrow stromal cells" should be reserved for these highly differentiated marrow connective tissue cells that facilitate hematopoietic differentiation.

It is important to stress that the differentiation of MSCs into marrow stroma or osteoblasts involves a time-dependent sequence of differentiation steps (the lineage) with each step involving the up- and down-regulation of many genes. For example, we have compared the cytokine secretion into the medium in 24 h by human MSCs cultured in growth medium, in osteogenic medium (*dex*, ascorbate), and in stroma-generating medium (Dexter conditions or in the presence of IL-1α). The cytokine quantity is very different for these three culture conditions [28]. Also of considerable importance was the realization that the constitutive levels of these cytokines are different for each donor batch of cells under standard growth or differentiation conditions. These measured differences emphasize the influence of the genotype on the observed differentiation/lineage pathways and multigenic expression profiles of each MSC donor or recipient or both.

1.4.1 Muscle, Tendon, and Fat

Both *in vitro* and *in vivo*, microenvironments control the differentiation and expressional profile of the MSCs and their differentiated descendants. Without going into all of the details, MSCs have been shown to differentiate into skeletal muscle, into Achilles or patella tendon tissue, and into adipocytes as reviewed earlier in this chapter. The cell culture conditions or *in vivo* tissue sites were different for each of the phenotypes observed. Importantly, the introduction of MSCs into specific tissue locations was followed by using markers or molecular probes [53,54]. For example, normal rodent marrow-derived MSCs form

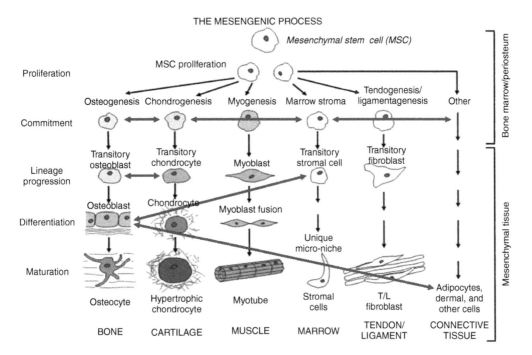

FIGURE 1.2 The mesengenic process in which transdifferentiation is depicted by horizontal arrows. For example, adipocytes can transdifferentiate into osteoblasts.

dystrophin-positive skeletal muscle in the limbs of *mdx* mice that are dystrophin-negative [25]. Likewise, reparative *neo*-tissue formed in rabbit tendon defects when autologous MSCs were introduced in suitable scaffolds [30,31].

The isolation of MSCs from fat provides an intriguing question of the origin of these progenitor cells [34,55]: Are these MSCs an intrinsic occupant of fat, are they associated with the blood vessels, or are they dedifferentiated adipocytes? Both marrow- and fat-derived MSCs differentiate into adipocytes and the other phenotypes of the Mesengenic Lineage (Figure 1.1). These observations raise the question of phenotypic *plasticity* and whether differentiated cells dedifferentiate into a stem cell and then redifferentiate into an alternate phenotype or whether the differentiated cells *transdifferentiate* directly into a different mesenchymal phenotype (as pictured in Figure 1.2). The current data indicate that cells can transdifferentiate along certain permissible routes without backing up to the stem cell status.

1.4.2 A New Fundamental Role for MSCs

Previously, all of the preclinical models used MSCs to provide the cells that fabricate the specialized, differentiated tissues. I propose that MSCs could be used in a different clinical context: that MSCs could provide a trophic influence to structure a reparative microenvironment. The precedent for cells secreting cytokines/growth factors that have regulator effects is well established in regenerative biology. For example, Singer long ago established the trophic effect of nerves on amphibian limb regeneration [56]. He showed that a certain quantum of nerves in the limb stump is required for successful limb regeneration; the nerves delivered neural trophic factors and the electrical or neural transmitter release was not associated with this trophic effect as evidenced by switching the ratios of different nerves into the limb stump.

In this regard, the bone marrow stroma derived from MSCs provides very specific instructional niches for various lineage pathways of hematopoiesis as discussed earlier. The question arises, "Can MSCs structure reparative microenvironments *without* differentiating into specific differentiated cells or tissue?" The answer appears to be "yes!" In the cases of cardiac infarct (ischemia) and stroke (brain ischemia), bone marrow-derived MSCs when injected into the ischemic zones do not massively differentiate into cardiac myocytes or neural cells. Rather, the MSCs appear to establish a trophic influence that is antifibrotic (antiscar tissue) and angiogenic [57–60]. The lack of the formation of extensive fibrotic, scar tissue and the rapid revascularization of the affected tissue zones result in clinically improved tissues.

These new findings suggest to me a new Tissue Engineering paradigm: the site-specific delivery of trophic cells (cells that secrete specific cytokines/growth factors) that structure reparative microenvironments. Already well established is the concept of genetically engineering the insertion of a specific gene, for example, transfection with a BMP-containing insert, to be synthesized by an implanted cell to stimulate bone regeneration at that implantation site [61,62]. In this context, why not implant to a specific location a normal cell that has a powerful trophic activity that will cause the implantation site to support restorative therapy? Likewise, the trophic effect may not stimulate repair, but rather, it may inhibit a debilitating effect, such as fibrosis, so that the slower, more natural host-mediated regenerative events can take place. In some cases, as mentioned earlier, MSCs can provide such a trophic effect.

1.4.3 The Use of MSCs Today and Tomorrow

The current use of MSCs is facilitated by their capacity to be expanded in culture. The introduction of these cells back *in vivo* is facilitated by site-specific scaffolds. In some cases, the MSCs can be jump-started down a lineage pathway by exposing them in culture to a specific bioactive agent (*dex* for bone [23,49], TGF-β for cartilage [20,21], etc.). In the long-term, there will be a way to mobilize the body's own MSCs, direct them to a site, expand them at the site, and signal them into a lineage pathway including modulation of the phenotypic expression of the cells to produce the appropriate tissues for that repair site (articular cartilage at the knee and auricular cartilage in the ear). There are at least two unknowns in this process. First, we do not know what mobilizes MSCs from marrow or other depots, although there are data that suggest such mobilization takes place following injury [34,35,63]. Second, we do not know how such mobilized MSCs dock into specific tissues. Consistent with this lack of knowledge is our inability to obtain high engraftment percentages for systemically administered, culture-expanded (marked) MSCs [64]. At best, only 2 to 3% of systemically administered MSCs home back to bone marrow with most of the infused cells lodging in the lung (probably a size issue) and liver (probably a fibronectin issue).

1.5   Cell Targeting

To systematically learn more about control of engraftment of systemically or locally introduced reparative cells, we have developed protocols which facilitate the targeting of cells to specific molecular docking sites. The approach is to insert specific targeting molecules into the cell membranes. The first approach involved palmitoylating protein G and painting this "primer" coat on cells, since the hydrophobic fatty acid tail of palmitic acid will quantitatively insert into the cell's plasma membrane lipid bilayer core. This positions the protein to orient out of the cell's surface. Protein G binds strongly to the Fc-region of antibodies. Thus, we can put different paints (antibodies) on the primer coat on the cell's surface. Moreover, if we had specific addressing or docking peptides, we could construct a fusion protein with such addresses in series with the Fc-region of antibodies in a molecularly reconstructed fusion protein as pictured in Figure 1.3.

In a proof of this technology concept, Dennis et al. [65] have created a deep cartilage defect in the condyle of rabbits. A library of antibodies exists against various epitopes in the ECM of deep articular cartilage as opposed to the surface of the condylar cartilage. With these antibodies, cultured rabbit or human chondrocytes were shown to home to the defect and subsequently fabricate cartilage ECM in an organ culture test system. The painting moieties were shown to not affect the viability or replication of the painted cells nor the differentiation capacity of painted MSCs. The long-term goal is the injection of painted chondrogenic cells into the synovial space to direct their docking to cartilage defects and, then, to facilitate the repair/regeneration of the cartilage.

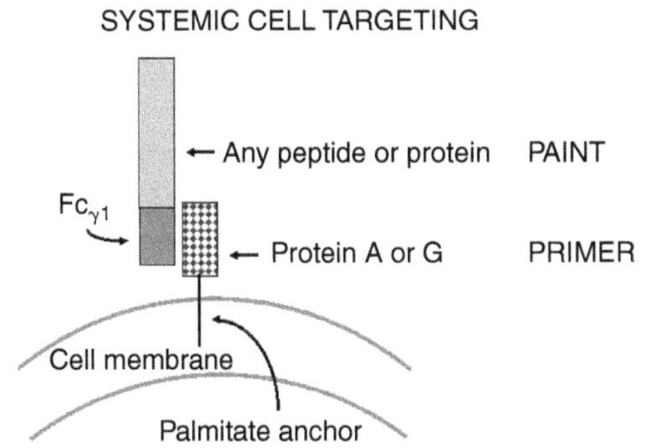

SYSTEMIC CELL TARGETING

FIGURE 1.3 Cell targeting technology involves palmitic acid that is derivatized with Protein A or G (PRIMER). The fatty acid hydrophobic tail quantitatively inserts into the lipid bilayer of the cell membrane. An antibody or fusion protein with the Fc domain at one end is referred to as the Paint.

I could imagine using the painting technology to insert tissue-specific address or targeting molecules to implanted scaffold docking sites as a basis for various systemically or locally supplied reparative cells, for example, stem cells for a variety of tissues. The docking of these cells could improve tissue performance or repair localized defects. Enhanced engraftment of stem cells, either by mobilizing intrinsic populations or by injecting extrinsically expanded cells, should provide useful Tissue Engineering strategies for tomorrow. Clearly, a lot of work needs to be done before this long-term goal will be realized.

Acknowledgments

The research was supported by grants from NIH, DOE, and the State of Ohio. I thank my colleagues at Case Western Reserve University and elsewhere for providing the experimental basis for this chapter.

References

[1] Nye, H.L. et al., Regeneration of the urodele limb: a review, *Dev. Dyn.*, 226, 280–294, 2003.
[2] Bell, E., Living tissue formed *in vitro* and accepted as skin-equivalent tissue of full thickness, *Science*, 211, 1052–1054, 1981.
[3] Boyce, S.T., Cultured skin for wound closure, in *Skin Substitute Production by Tissue Engineering: Clinical and Fundamental Applications*, chap. 4, Rouabhila, M. (Ed.), Landes Bioscience, Austin, 1997.
[4] Hardin-Young, J. et al., Approaches to transplanting engineered cells and tissues, in *Principles of Tissue Engineering*, chap. 23, Guilak, F., Butler, D.L., Goldstein, S.A., and Mooney, D.J. (Eds.), 2nd ed. Academic Press, 2000.
[5] Brisco, D.M. et al., The allogeneic response to cultured human skin equivalent in the hu-PBL-SCID mouse model of skin rejection, *Transplantation*, 67, 1590–1599, 1999.
[6] Caplan, A.I., Biomaterials and bone repair, *BIOMAT*, 87, 15–24, 1988.
[7] Caplan, A.I., Cell delivery and tissue regeneration, *J. Control. Release*, 11, 157–165, 1989.
[8] Caplan, A.I., Mesenchymal stem cells, *J. Orthop. Res.*, 9, 641–650, 1991.
[9] Caplan, A.I. et al., Cell-based technologies for cartilage repair, in *Biology and Biomechanics of the Traumatized Synovial Joint: The Knee as a Model*, Finerman, G.A.M. and Noyes, F.R. (Eds.), American Academy of Orthopaedic Surgeons, Rosemont, pp. 111–122, 1992.
[10] Haynesworth, S.E., Baber, M.A., and Caplan, A.I., Cell surface antigens on human marrow-derived mesenchymal cells are detected by monoclonal antibodies, *Bone*, 13, 69–80, 1992.
[11] Caplan, A.I., The mesengenic process, *Clin. Plast. Surg.*, 21, 429–435, 1994.

[12] Jargiello, D.M. and Caplan, A.I., The fluid flow dynamics in the developing chick wing, in *Limb Development and Regeneration, Part A*, Fallon, J.F. and Caplan, A.I. (Eds.), Alan R. Liss, Inc., New York, pp. 143–154, 1983.

[13] Caplan, A.I., Cartilage, *Sci. Am.*, 251, 84–94, 1984.

[14] Caplan, A.I., The molecular control of muscle and cartilage development, in *39th Annual Symposium of the Society for Developmental Biology*, Subtelney, S. and Abbott, U. (Eds.), Alan R. Liss, Inc., New York, pp. 37–68, 1981.

[15] Friedenstein, A.J., Petrakova, K.V., Kurolesova, A.I., and Frolova, G.P., Heterotopic of bone marrow. Analysis of precursor cells for osteogenic and hematopoietic tissues, *Transplantation*, 6, 230–247, 1968.

[16] Owen, M., Lineage of osteogenic cells and their relationship to the stromal system, in *Bone and Mineral Research*, chap. 1, Peck, W.A. (Ed.), Elsevier Science, Oxford, 1985.

[17] Haynesworth, S.E. et al., Characterization of cells with osteogenic potential from human marrow, *Bone*, 13, 81–88, 1992.

[18] Haynesworth, S.E., Goldberg, V.M., and Caplan, A.I., Diminution of the number of mesenchymal stem cells as a cause for skeletal aging, in *Musculoskeletal Soft-Tissue Aging: Impact on Mobility, Section 1*, chap. 7, Buckwalter, J.A., Goldberg, V.M., and Woo, S.L.-Y. (Eds.), American Academy of Orthopaedic Surgeons, Publishers, pp. 79–87, 1994.

[19] Lennon D.L. et al., Human and animal mesenchymal progenitor cells from bone marrow: identification of serum for optimal selection and proliferation, *In Vitro Cell Dev. Biol.*, 32, 602–611, 1996.

[20] Johnstone, B. et al., *In vitro* chondrogenesis of bone marrow-derived mesenchymal progenitor cells, *Exp. Cell Res.*, 238, 265–272, 1998.

[21] Yoo, J.U. et al., The chondrogenic potential of human bone-marrow-derived mesenchymal progenitor cells, *J. Bone Joint Surg.*, 80, 1745–1757, 1998.

[22] Dennis, J.E. and Caplan, A.I., Differentiation potential of conditionally immortalized mesenchymal progenitor cells from adult marrow of a H-2Kb-tsA58 transgenic mouse, *J. Cell. Physiol.*, 167, 523–538, 1996.

[23] Jaiswal, N., Haynesworth, S.E., Caplan, A.I., and Bruder, S.P., Osteogenic differentiation of purified, culture-expanded human mesenchymal stem cells *in vitro*, *J. Cell Biochem.*, 64, 295–312, 1997.

[24] Wakitani, S., Saito, T., and Caplan, A.I., Myogenic cells derived from rat bone marrow mesenchymal stem cells exposed to 5-azacytidine, *Muscle Nerve*, 18, 1417–1426, 1995.

[25] Saito, T. et al., Myogenic expression of mesenchymal stem cells within myotubes of mdx mice *in vitro* and *in vivo*, *Tissue Eng.*, 1, 327–344, 1996.

[26] Majumdar, M. et al., Human marrow-derived mesenchymal stem cells (MSCs) express hematopoietic cytokines and support long-term hematopoiesis when differentiated toward stromal and osteogenic lineages. *J. Hematother. Stem Cell Res.*, 9, 841–848, 2001.

[27] Dennis, J.E., Carbillet, J.-P., Caplan, A.I., and Charbord, P., The STRO-1+ marrow cell population is multi-potential, *Cells Tissues Organs*, 170, 73–82, 2002.

[28] Haynesworth, S.E., Baber, M.A., and Caplan, A.I., Cytokine expression by human marrow-derived mesenchymal progenitor cells *in vitro*: effects of dexamethasone and IL-1α, *J. Cell Physiol.*, 166, 585–592, 1996.

[29] Dennis, J.E. et al., A quadripotential mesenchymal progenitor cell isolated from the marrow of an adult mouse, *J. Bone Mineral. Res.*, 14, 1–10, 1999.

[30] Young, R.G. et al., The use of mesenchymal stem cells in Achilles tendon repair, *J. Orthop. Res.*, 16, 406–413, 1998.

[31] Awad, H. et al., Autologous mesenchymal stem cell-mediated repair of tendon, *Tissue Eng.*, 5, 267–277, 1999.

[32] Weimann, J.M. et al., Contribution of transplanted bone marrow cells to Purkinje neurons in human adult brains, *Proc. Natl Acad. Sci. USA*, 100, 2088–2093, 2003.

[33] Prockop, D.J., Marrow stromal cells as stem cells for nonhematopoetic tissues, *Science*, 276, 71–74, 1997.
[34] Laflamme, M.A., Myerson, D., Saffitz, J.E., and Murry, C.E., Evidence for cardiomyocyte repopulation by extracardiac progenitors in transplanted human hearts, *Circ. Res.*, 90, 634–640, 2002.
[35] Orlic, D. et al., Mobilized bone marrow cells repair the infracted heart, improving function and survival, *Proc. Natl Acad. Sci. USA*, 98, 10344–10349, 2001.
[36] Jackson, K.A. et al., Regeneration of ischemic cardiac muscle and vascular endothelium by adult stem cells, *J. Clin. Invest.*, 107, 1395–1402, 2001.
[37] Shi, S. and Gronthos, S., Pervascular niche of postnatal mesenchymal stem cells in human bone marrow dental pulp, *J. Bone Mineral Res.*, 18, 696–704, 2003.
[38] Jiang, Y. et al., Pluripotency of mesenchymal stem cells derived from adult marrow, *Nature*, 418, 41–49, 2002.
[39] Reyes, M. et al., Purification and *ex vivo* expansion of postnatal human marrow mesodermal progenitor cells, *Blood*, 98, 2615–2625, 2001.
[40] Caplan, A.I., Tissue engineering strategies for the use of mesenchymal stem cells to regenerate skeletal tissues, in *Functional Tissue Engineering*, chap. 10, Guilak F., Butler, D.L., Goldstein, S.A., and Mooney, D.J. (Eds.), Springer-verlag, New York, 2003.
[41] Caplan, A.I. et al., The principles of cartilage repair/regeneration, *Clin. Orthop. Relat. Res.*, 342, 254–269, 1997.
[42] Caplan, A.I., Embryonic development and the principles of tissue engineering, in *Novartis Foundation: Tissue Engineering of Cartilage and Bone*, John Wiley & Sons, London, pp. 17–33, 2003.
[43] Kujawa, M.J. and Caplan, A.I., Hyaluronic acid bonded to cell culture surfaces stimulates chondrogenesis in stage 24 limb mesenchyme cell cultures, *Dev. Biol.*, 114, 504–518, 1986.
[44] Kujawa, M.J., Carrino, D.A., and Caplan, A.I., Substrate-bonded hyaluronic acid exhibits a size-dependent stimulation of chondrogenic differentiation of stage 24 limb mesenchymal cells in culture, *Dev. Biol.*, 114, 519–528, 1986.
[45] Caplan, A.I., Tissue engineering designs for the future: new logics, old molecules, *Tissue Eng.*, 6, 1–8, 2000.
[46] Ponticiello, M.S. et al., Gelatin-based resorbable sponge as a carrier matrix for human mesenchymal stem cells in cartilage regeneration therapy, in *Osiris Therapeutics*, John Wiley & Sons, 2000.
[47] Caplan, A.I. and Pechak, D.G., The cellular and molecular embryology of bone formation, in *Bone and Mineral Research*, Vol. 5, Peck, W.A. (Ed.), Elsevier, New York, pp. 117–184, 1987.
[48] Caplan, A.I., Bone development, in *Cell and Molecular Biology of Vertebrate Hard Tissues*, CIBA Foundation Symposium 136, Wiley, Chichester, pp. 3–21, 1988.
[49] Ohgushi, H. and Caplan, A.I., Stem cell technology and bioceramics: from cell to gene engineering, *J. Biomed. Mater. Res.*, 48, 1–15, 1999.
[50] Dennis, J.E., Konstantakos, E.K., Arm, D., and Caplan, A.I., *In vivo* osteogenesis assay: a rapid method for quantitative analysis, *Biomaterials*, 19, 1323–1328, 1998.
[51] Lennon, D.P. et al., Dilution of human mesenchymal stem cells with dermal fibroblasts and the effects on *in vitro* and *in vivo* osteogenesis, *Dev. Dynam.*, 219, 50–62, 2000.
[52] Goshima, J., Goldberg, V.M., and Caplan, A.I., The osteogenic potential of culture-expanded rat marrow mesenchymal cells assayed *in vivo* in calcium phosphate ceramic blocks, *Clin. Orthop. Relat. Res.*, 262, 298–311, 1991.
[53] Allay, J.A. et al., LacZ and IL-3 expression *in vivo* after retroviral transduction of marrow-derived human osteogenic mesenchymal progenitors, *Hum. Gene Ther.*, 8, 1417–1427, 1997.
[54] Gimble, J.M. et al., The function of adipocytes in the bone marrow stroma: an update, *Bone*, 19, 421–428, 1996.
[55] Dragoo, J.L. et al., Tissue-engineered cartilage and bone using stem cells from human infrapatellar fat pads, *J. Bone Joint Surg.*, 85, 740–747, 2003.

[56] Singer, M., Neurothophic control of limb regeneration in the newt, *Ann. NY Acad. Sci.*, 228, 308–322, 1974.

[57] Stamm, C. et al., Autologous bone-marrow stem-cell transplantation for myocardial regeneration, *Lancet,* 361, 45–46, 2003.

[58] Shake, J.G. et al., *In-vivo* mesenchymal stem cell grafting in a swine myocardial infarct model: molecular and physiologic consequences, *Ann. Thorac. Surg.*, 73, 1919–1926, 2002.

[59] Chen, J. et al., Therapeutic benefit of intracerebral transplantation of bone marrow stromal cells after cerebral ischemia in rats, *J. Neurol. Sci.*, 189, 49–57, 2001.

[60] Chen, J. et al., Intravenous bone marrow stromal cell therapy reduces apoptosis and promotes endogenous cell proliferation after stroke in female rat, *J. Neurosci. Res.*, 73, 778–786, 2003.

[61] Lieberman, J.R. et al., The effect of regional gene therapy with bone morphogenetic protein-2 producing bone-marrow cells on the repair of segmental femoral defects in rats, *J. Bone Joint Surg. Am.*, 81, 905–917, 1999.

[62] Evans, C.H. and Robbins, P.D., Possible orthopaedic applications of gene therapy, *J. Bone Joint Surg. Am.*, 77, 1103–1114, 1995.

[63] Mahmood, A. et al., Treatment of traumatic brain injury in adult rats with intravenous administration of human bone marrow stromal cells, *Neurosurgery,* 53, 697–703, 2003.

[64] Gao, J. et al., The dynamic *in vivo* distribution of bone marrow-derived mesenchymal stem cells after infusion, *Cells Tissues Organs,* 169, 12–20, 2001.

[65] Dennis, J.E., Cohen, N., Goldberg, V.M., and Caplan, A.I., Targeted delivery of progenitor cells for cartilage repair, *J. Orthop. Res.*, 22, 735–741, 2004.

2

Growth Factors and Morphogens: Signals for Tissue Engineering

A. Hari Reddi
University of California

2.1 Introduction

Tissue engineering is the exciting discipline of design and construction of spare parts for the human body to restore function based on biology and biomedical engineering. The basis of tissue engineering is the triad of signals for tissue induction, responding stem cells, and the scaffolding of extracellular matrix. Among the many tissues in the human body bone has the highest power of regeneration and therefore is a prototype model for tissue engineering based on morphogenesis. Morphogenesis is the developmental cascade of pattern formation, body plan establishment, and culmination of the adult body form. The cascade of bone morphogenesis in the embryo is recapitulated by demineralized bone matrix-induced bone formation. The inductive signals for bone morphogenesis, the bone morphogenetic proteins (BMPs) were isolated from demineralized bone matrix. BMPs and related cartilage-derived morphogenetic proteins (CDMPs) initiate cartilage and bone formation. The promotion and maintenance of the initiated skeleton is regulated by several growth factors. Tissue engineering is the symbiosis of signals (growth factors and morphogens), stem cells, and scaffolds (extracellular matrix). The rules of architecture for tissue engineering are a true imitation of principles of developmental biology and morphogenesis.

2.2 Tissue Engineering and Morphogenesis

An understanding of the molecular principles of development and morphogenesis, is a prerequisite for tissue engineering. We define tissue engineering as the science of design and manufacture of new tissues for functional restoration of the impaired organs and replacement of lost parts due to disease, trauma, and tumors [1]. Tissue engineering is based on principles of developmental biology and morphogenesis, biomedical engineering, and biomechanics.

Morphogenesis is initiated by morphogens. The promotion and maintenance of morphogenesis is achieved by a variety of growth factors. Generally, morphogens are first identified in fly and frog embryos by genetic approaches, differential displays, and subtractive hybridization expression cloning. An alternate biochemical approach of "grind and find" from adult mammalian bone led to the isolation of BMPs, the premier signals for bone morphogenesis. We now discuss the identification, isolation, and molecular cloning of BMPs from a natural biomaterial, the demineralized bone matrix.

2.3   The Bone Morphogenetic Proteins

Bone grafts have been used to aid the healing of recalcitrant fractures. Demineralized bone matrix induced new bone formation. Bone induction by demineralized bone matrix is a sequential cascade [2–4]. The key steps in this cascade are chemotaxis of progenitor cells, proliferation of progenitor cells, and finally differentiation first into cartilage and then bone. The demineralized bone matrix is devoid of any living cells and is a biomaterial that elicits new bone formation. The insoluble collagenous bone matrix binds plasma fibronectin [3] and promotes the proliferation of cells. Proliferation was maximal on day 3, chondroblast differentiation was evident on day 5, and chondrocytes were abundant on day 7. The cartilage hypotrophied on day 9 with concominant vascular invasion and osteogenesis. On days 10 to 12 maximal alkaline phosphatase activity, a marker of bone formation, was observed. Hematopoietic differentiation was observed in the ossicle on day 21. The sequential bone development cascade is reminiscent of bone morphogenesis in limb.

A systematic study of the biochemical basis of bone induction was initiated. A bioassay for bone induction was established *in vivo* in rats. The insoluble demineralized bone matrix was extracted in 4 M guanidine hydrochloride, a dissociative extractant. About 3% of the proteins were solubilized and the rest was the insoluble residue. The extract and the residue alone were unable to induce bone formation. However, reconstitution of the extract to residue yielded new bone morphogenesis. Thus there is collaboration between soluble signals and the insoluble matrix scaffold to yield new bone formation [5,6]. This key experiment predates the term tissue engineering and demonstrates the collaboration of soluble signals and insoluble scaffolding as a critical concept in practical tissue engineering. Collagen appear to be an optimal scaffold [7]. The bone induction is dependent on the hormonal status including vitamin D [8,9]. Irradiation of the recipient blocked the cellular cascade of osteogenesis [10].

This bioassay was a critical development in the quest for the purification of the bioactive bone morphogens, the bone morphogenetic proteins [11–15]. There are nearly 15 members of BMPs in the human genome (Table 2.1). BMPs are dimeric molecules with a single disulfide bond. The mature monomer consists of seven canonical cysteines contributing to three interchain disulfides and one interchain disulfide bond.

BMPs stimulate chondrogenesis in limb bud mesodermal cells [16]. BMP 2 stimulates osteoblast maturation [17]. BMPs are chemotactic for human monocytes [18]. In addition to initiating chondrogenesis BMPs maintain proteoglycan biosynthesis in bovine articular cartilage explants [19,20]. Recombinant human growth/differentiation factors (GDF-5) stimulate chondrogenesis during limb development [21].

BMPs interact with BMP receptors I and II on the cell surface/membrane [1,22]. BMP receptors are serine/threonine kinases. The intracellular substrates for these kinases called Smads function as relays to activate the transcriptional machinery [23]. The three functional classes are (1) receptor-regulated Smads, namely, Smads 1, 5, and 8, (2) the common partner Smad — 4, and (3) the inhibitory Smads 6 and 7. There are Smad-dependent and independent pathways for activation of BMP signaling including new bone formation.

TABLE 2.1   The BMP Family in Mammals[a]

BMP subfamily	Generic name	BMP designation
BMP 2/4	BMP-2A	BMP-2
	BMP-2B	BMP-4
BMP-3	Osteogenin	BMP-3
	Growth/differentiation factor-10 (GDF-10)	BMP-3B
OP-1/ BMP-7	BMP-5	BMP-5
	Vegetal related-1 (Vgr-1)	BMP-6
	Osteogenic protein-1 (OP-1)	BMP-7
	Osteogenic protein-1 (OP-2)	BMP-8
	Osteogenic protein-1 (OP-3)	BMP-8B
	Growth/differentiation factor-2 (GDF-2)	BMP-9
	BMP-10	BMP-10
	Growth/differentiation factor-11 (GDF-11)	BMP-11
GDF-5,6,7	Growth/differentiation factor-7 (GDF-7) or cartilage-derived morphogenetic protein-3 (CDMP-3)	BMP-12
	Growth/differentiation factor-6 (GDF-6) or cartilage-derived morphogenetic protein-2 (CDMP-2)	BMP-13
	Growth/differentiation factor-5 (GDF-5) or cartilage-derived morphogenetic protein-1 (CDMP-1)	BMP-14
	BMP-15	BMP-15

[a] BMP-1 is not a BMP family member with seven canonical cysteines. It is a procollagen-C proteinase related to *Drosophila* Tolloid.

2.4   Growth Factors

Growth factors are proteins with profound influence on proliferation and growth of cells. Growth factors stimulate the differentiation of progenitor/stem cells. The growth factors include many subgroups and such as Insulin like growth factors (IGFs), fibroblast growth factors (FGFs), and platelet-derived growth factors (PDGFs).

Insulin like growth factors are polypeptides related to Insulin. There are two members, IGF-I and IGF-II. Liver is the predominant site of IGF-I synthesis, and is stimulated by pituitary growth hormone. IGF biological activity is modulated by IGF-binding proteins (IGFBP). There are six different IGFBPs. IGFs promote extracellular matrix biosynthesis by osteoblasts.

Fibroblast growth factors (FGFs) are proteins with multiple members. FGFs are mitogens for endothelial cells. Along with Vascular Endothelial Growth Factors (VEGFs), FGFs are critical for bone formation. It is well known that vascular invasion is a prerequisite for endochodral bone formation.

Platelet-derived growth factors come in three isoforms, namely, AA, AB, and BB. They are primarily produced by platelets in blood. Various isoforms stimulate bone formation.

2.5   BMPs Bind to Extracellular Matrix

The critical role of extracellular matrix in morphogenesis of many tissues during development is well known. The extracellular matrix is a supramolecular assembly of collagens, proteoglycans, and glycoproteins. The collagens are tissue specific and the proteoglycans include chondroitin sulfate, dermatan sulfate, heparan sulfate/heparin, and keratan sufate. Recombinant BMP 4 and BMP 7 bind to heparan sulfate/heparin, collagen IV of the basement membrane [24]. The binding of a soluble morphogen to insoluble extracellular matrix renders the morphogen to act locally and protects it from proteolytic degradation and therefore extends its biological half-life. Thus, extracellular matrix scaffolding is an efficient delivery system for tissue engineering. Growth factors such as FGFs bind to heparan sulfate. An emerging concept for tissue engineering is the tethering of signals to scaffolds to restrict their diffusion.

2.6   Clinical Applications

Recombinant BMP 2 has been approved by the Food and Drug administration for spine fusion and open fractures of tibia due to orthopaedic trauma. There have been several clinical applications of BMPs in orthopaedic surgery [25–29]. The developing experience of BMPs will be of immense utility to the nascent field of tissue engineering, the science of design-based manufacture of spare parts for human skeleton based on signals, stem cells, and scaffolding [1,30] in medicine and dentistry. A prototype paradigm has validated the proof of principle for tissue engineering based on tissue transformation by BMPs and scaffolding [31].

2.7   Challenges and Opportunities

Despite the exciting advances in clinical applications of BMPs there remain many challenges. Foremost among them is the need for developing synthetic scaffolds to deliver recombinant BMPs for skeletal tissue engineering. The development of synthetic scaffolds with an ability to respond to biomechanical influences that are known to be critical for musculoskeletal structures will lead to a quantum improvement of current tissue engineering approaches to bone, cartilage, and meniscus. The remaining challenges make the field of morphogen-based tissue engineering an exciting frontier with unlimited opportunities.

Acknowledgments

The research in the Center for Tissue Regeneration is supported by the Lawrence Ellision Chair in Musculoskeletal Molecular Biology and grants from NIH and DOD. I thank Danielle Neff for the outstanding help in completion of this article.

References

[1] Reddi, A.H., Role of morphogenetic proteins in skeletal tissue engineering and regeneration, *Nat. Biotechnol.*, 16, 247, 1998.

[2] Reddi, A.H. and Anderson, W.A., Collagenous bone matrix-induced endochondral ossification hemopoiesis, *J. Cell Biol.*, 69, 557, 1976.

[3] Weiss, R.E. and Reddi, A.H., Synthesis and localization of fibronectin during collagenous matrix-mesenchymal cell interaction and differentiation of cartilage and bone *in vivo*, *Proc. Natl Acad. Sci. USA*, 77, 2074, 1980.

[4] Reddi, A.H., Cell biology and biochemistry of endochondral bone development, *Coll. Relat. Res.*, 1, 209, 1981.

[5] Sampath, T.K. and Reddi, A.H., Dissociative extraction and reconstitution of extracellular matrix components involved in local bone differentiation, *Proc. Natl Acad. Sci. USA*, 78, 7599, 1981.

[6] Sampath, T.K. and Reddi, A.H., Homology of bone-inductive proteins from human, monkey, bovine, and rat extracellular matrix, *Proc. Natl Acad. Sci. USA*, 80, 6591, 1983.

[7] Ma, S., Chen, G., and Reddi, A.H., Collaboration between collagenous matrix and osteogenin is required for bone induction, *Ann. NY Acad. Sci.*, 580, 524, 1990.

[8] Reddi, A.H., Extracellular matrix and development, in *Extracellular Matrix Biochemistry*, Piez, K.A. and Reddi, A.H. Eds., Elsevier, New York, 1984, p. 247.

[9] Sampath, T.K., Wientroub, S., and Reddi, A.H., Extracellular matrix proteins involved in bone induction are vitamin D dependent, *Biochem. Biophys. Res. Commun.*, 124, 829, 1984.

[10] Wientroub, S. and Reddi, A.H., Influence of irradiation on the osteoinductive potential of demineralized bone matrix, *Calcif. Tissue Int.*, 42, 255, 1988.

[11] Wozney, J.M. et al., Novel regulators of bone formation: molecular clones and activities, *Science*, 242, 1528, 1988.

[12] Luyten, F.P. et al., Purification and partial amino acid sequence of osteogenin, a protein initiating bone differentiation, *J. Biol. Chem.*, 264, 13377, 1989.

[13] Celeste A.J. et al., Identification of transforming growth factor beta family members present in bone-inductive protein purified from bovine bone, *Proc. Natl Acad. Sci. USA*, 87, 9843, 1990.

[14] Ozkaynak, E. et al., OP-1 cDNA encodes an osteogenic protein in the TGF-beta family, *EMBO J.*, 9, 2085, 1990.

[15] Wang, E.A. et al., Recombinant human bone morphogenetic protein induces bone formation, *Proc. Natl Acad. Sci. USA*, 87, 2220, 1990.

[16] Chen, P. et al., Stimulation of chondrogenesis in limb bud mesoderm cells by recombinant human bone morphogenetic protein 2B (BMP-2B) and modulation by transforming growth factor beta 1 and beta 2, *Exp. Cell Res.*, 195, 509, 1991.

[17] Yamaguchi, A. et al., Recombinant human bone morphogenetic protein-2 stimulates osteoblastic maturation and inhibits myogenic differentiation *in vitro*, *J. Cell Biol.*, 113, 681, 1991.

[18] Cunningham, N.S., Paralkar, V., and Reddi, A.H., Osteogenin and recombinant bone morphogenetic protein 2B are chemotactic for human monocytes and stimulate transforming growth factor beta 1 mRNA expression, *Proc. Natl Acad. Sci. USA*, 89, 11740, 1992.

[19] Luyten, F.P. et al., Natural bovine osteogenin and recombinant human bone morphogenetic protein-2B are equipotent in the maintenance of proteoglycans in bovine articular cartilage explant cultures, *J. Biol. Chem.*, 267, 3691, 1992.

[20] Lietman, S.A. et al., Stimulation of proteoglycan synthesis in explants of porcine articular cartilage by recombinant osteogenic protein-1 (bone morphogenetic protein-7), *J. Bone Joint Surg. Am.*, 79, 1132, 1997.

[21] Khouri, R.K., Koudsi, B., and Reddi, A.H. M Tissue transformation into bone *invivo* a potential practical application, *JAMA*, 266, 1953, 1991.

[22] ten Dijke, P. et al., Identification of type I receptors for osteogenic protein-1 and bone morphogenetic protein-4, *J. Biol. Chem.*, 269, 16985, 1994.

[23] Imamura T. et al., Smad6 inhibits signalling by the TGF-beta superfamily (see comments), *Nature*, 389, 622, 1997.

[24] Paralkar, V.M. et al., Interaction of osteogenin, a heparin binding bone morphogenetic protein, with type IV collagen, *J. Biol. Chem.*, 265, 17281, 1990.

[25] Einhorn, T.A., Clinical applications of recombinant BMPs: early experience and future development, *J. Bone Joint Surg.*, 85A, 82, 2003.

[26] Li, R.H. and Wozney, J.M. Delivering on the promise of bone morphogenetic proteins, *Trends Biotechnol.*, 19, 255, 2001.

[27] Ripamonti, U., Ma, S., and Reddi A.H. The critical role of geometry of porous hydroxyapatite delivery system in induction of bone by osteogenin, a bone morphogenetic protein, *Matrix*, 12, 202, 1992.

[28] Geesink, R.G., Hoefnagels, N.H., and Bulstra, S.K. Osteogenic activity of OP1, bone morphogenetic protein 7 (BMP 7) in a human fibular defect, *J. Bone Joint Surg.*, 81, 710, 1999.

[29] Friedlaender, G.E. et al., Osteogenic protein 1 (bone morphogenic protein 7) in the treatment of tibial non-unions, *J. Bone Joint Surg. Am.*, 83A, S151, 2001.

[30] Govender et al., Recombinant human bone morphogenetic protein 2 for treatment of open tibiol fractures: a prospective, controlled, randomized study of four hundred and fifty patients. *J. Bone Joint Surg. Am.*, 84A, 2123, 2002.

[31] Nakashima, M. and Reddi, A.H. The application of bone morphogenetic protein to dental tissue engineering, *Nat. Biotechnol.*, 21, 1025, 2003.

[32] Khouri, R.K., Koudsi, B., and Reddi, A.H. Tissue transformation into bone *in vivo*: a potential practical application. *JAMA*, 266, 1953, 1991.

3

Extracellular Matrix: Structure, Function, and Applications to Tissue Engineering

Mary C. Farach-Carson
Roger C. Wagner
Kristi L. Kiick
University of Delaware

3.1 Introduction

A major goal of tissue engineering is to employ the principles of rational design to recreate appropriate signals to cells that promote biological processes leading to production of new tissues or repair of damaged ones. A key modulator of cell behavior is the ECM that provides individual cells with architectural cues of time and space, modulates bioavailability of soluble growth and differentiation factors, and organizes multicellular tissue development. This chapter will focus on key ECM components and their functions, emphasizing relationships between natural matrices present in both hard and soft tissues and cell function. The concept of "mining" the natural matrix for active motifs that are useful in tissue engineering applications also is introduced. Finally, the utility of translating knowledge gained from study of native ECM to controlled delivery of growth factors and deliberate modulation of cell and tissue phenotype for engineering purposes is discussed.

3.2 ECM and Functional Integration of Implanted Materials

Cells sense and respond to a variety of signals that include those that are soluble such as growth factors, differentiation factors, cytokines, and ion gradients. In addition, cell behavior and phenotype is governed by responses to other types of signals that include mechanical forces, electrical stimuli, and various physical cues. Immobilized protein matrices that generally are fixed in space also regulate cell function. The general term that has come to denote the complex mixture of proteins on the outside of cells that governs their behavior is ECM. Evolution has provided cells with surface receptors to ECM components that enable them to recognize and decipher the signals that they encounter from the ECM and which influence cell growth, division, and differentiation [1].

For descriptive purposes, cell adhesion is classified into categories of cell–substratum and cell–cell attachment. Cell–cell interactions may occur between like cells (homotypic events) or between dissimilar cells (heterotypic). Homotypic interactions stabilize epithelia, which typically lie upon a sheet of specialized ECM called the basement membrane, discussed in Section 3.2. Heterotypic interactions govern many normal cellular phenomena including embryo–uterine implantation, immune surveillance, cell migration during embryogenesis, and neurotransmission. They also characterize the pathological states of cancer metastasis, rejection of transplanted tissues and organs, and inflammation [2,3]. In the context of tissue engineering, both types of cellular interactions must be understood, and thus hopefully manipulated, to ensure successful integration of transplanted materials.

There exists in tissues an exquisite balance between the anabolic process of ECM production and the catabolic process of ECM turnover. Introduction of foreign materials inevitably disrupts this natural homeostasis. A goal of tissue engineering is to successfully introduce replacement tissues that will, through stimulation of anabolic processes, lead to ECM production and acceptance of engineered materials rather than their immediate or eventual destruction via activation of catabolic pathways. To facilitate biointegration, both degradable and nondegradable foreign materials can be modified with native ECM or motifs derived from proteins in the ECM. Achievement of this goal requires a thorough understanding of the structural relationships among molecules in the ECM, their molecular interactions directing cell adhesion events, their biosynthesis and turnover, and their natural functions.

3.3 Basement Membranes and Focal Adhesions

Basement membranes, also called basal lamina, are sheets of highly organized ECM that are associated with the basal membrane of epithelial cells, around muscle cells, below endothelial cells in blood vessels, supporting fat cells, and associated with Schwann cells surrounding peripheral nerve axons [4,5]. In addition to physically separating the cell layers comprising epithelia and stroma, they provide a diversity of functions including structural support, selective filtration, and serve as barriers to invasion. They also establish cell polarity, influence cell metabolism, induce cell differentiation, direct cell migration, and organize membrane receptors. Figure 3.1 depicts scanning and transmission (SEM and TEM, respectively) electron micrographs of natural basement membranes illustrating their typical functions and structural roles in separating cells in tissues. Note that cells contact the basement membrane on a single face that establishes cell polarity and creates the basal surface of attaching cell. The basement membrane exists as a complex meshwork comprised of the proteins collagen IV, laminin, proteoglycans including perlecan, agrin, and collagens XV and XVIII, fibulin, BM-40 (SPARC), and nidogen/entactin [4]. The spatial relationships among these molecules were examined using atomic force microscopy where thin, threadlike strands of heparan sulfate attached to certain proteoglycans were seen to protrude from the core meshwork and proposed to function as an "entropic brush" that could filter proteins by their constant thermal

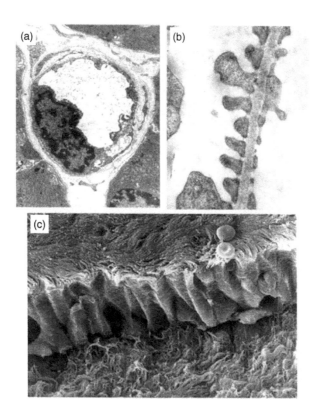

FIGURE 3.1 (a) Transmission electron micrograph of a transverse section through a blood capillary consisting of a single endothelial cell wrapped around to join upon itself to create a lumen. The endothelial cell is (as are all epithelia) bordered by a basement membrane on its external aspect. The basement membrane is shared by pericapillary cells called "pericytes" that are represented by the small bits of seemingly isolated cell material at the capillary periphery. (b) High magnification TEM of the glomerular basement membrane separating the capillaries of the kidney glomerulus and the "podocytes" on their surface. The capillary endothelium (on the right) contains apparent gaps which in three dimensions represent "fenestrae" or "little windows" in the capillary wall. The podocyte "pedicles" on the left actually are part of larger cell processes out of the plane of section. This panel highlights the function of the basement membrane as part of a filtration barrier which sieves large molecules and those with cationic charges. (c) A SEM of the ciliated respiratory epithelium of the trachea. Part of the epithelium has been denuded during preparation (bottom) to reveal meshwork collagen IV fibers of the underlying basement membrane and connective tissue underneath it.

motion within the basement membrane [5]. Collagen IV and laminin within the basement membrane form two overlapping polymeric networks [6,7] with which the other components interact in a highly organized fashion. The basement membrane provides an anchor for cells, which adhere to it using specific surface receptors called integrins [1]. Focal adhesions are the sites of cell–ECM attachment that were first identified ultrastructurally as cell surface electron dense regions near sites of cell attachment to substrata and then recognized as interfaces with the cytoskeleton [8]. Focal adhesions belong to the contractile class of matrix contacts, distinct from protrusive contacts or those that provide mechanical support [9]. Figure 3.2 diagrams a cartoon of a focal adhesion consisting of both integrin and nonintegrin receptors in the plasma membrane attached to ECM in the basement membrane. The processes of anchoring and spreading of cells on substrata occur as multistep processes that involve, among other things, clustering of surface receptors at sites of focal adhesions. Protein complexes link the cytoplasmic tails of surface receptors to the cytoskeleton, facilitating cell adhesion and transmitting signals through the intracellular network that ultimately signal to the nucleus to inform the cell that it has attached. This form of signaling has come to be called "outside in" signaling and is a key component of engineering functional interfaces between materials and living cells. It is important to note that cells seldom rely on one adhesion system

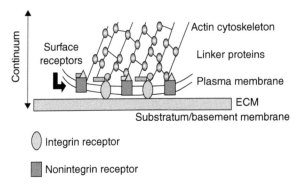

FIGURE 3.2 Focal adhesion schematic. The diagram illustrates the components of the focal adhesion attaching an adherent cell to an ECM substratum. Note the continuum of functional connections between the ECM and the cytoskeleton, bridged by both integrin and nonintegrin transmembrane receptors.

to support cell attachment, a concept thought to confer some degree of protection to cells against single mutations that would completely abolish adhesion-competence. Nonetheless, mutations in key adhesion molecules found in focal adhesions frequently have extreme and adverse consequences on tissue and organ development [10]. Guided tissue regeneration with artificial two-dimensional matrices such as GORE-TEX® (expanded polytetrafluoroethylene or ePTFE) that resemble basement membranes has been widely used for treatment of periodontal and bone defects and relies upon the membrane for physical separation of cell layers during healing [11].

3.4 Focal Adhesions as Signaling Complexes

Attachment of cells to substrata provides signals for spreading, migration, survival, and proliferation. As mentioned above, cell surface receptors recognize ECM components in a manner that is both specific and reversible. The initial attraction and attachment often involves nonintegrin adhesion events that may be related solely to charge or hydrophobic interactions between surfaces [12,13]. Examples include recognition of polyanionic polymers such as those presented by the heparan sulfate or chondroitin sulfate glycosaminoglycans (GAGS) attached to surface proteoglycans. Depending on the nature of the molecules involved, these adhesive events may or may not be dependent on the presence of divalent cations such as Ca^{2+} [14–16]. In any case, these early events tend to rely on multiple, low affinity interactions similar to the way that Velcro® functions as a two-sided hook and loop fastener. Once cells adhere, higher affinity interactions such as those involving integrin receptors are stabilized during the spreading phase of adhesion. A very interesting feature of integrin interactions with ECM proteins is that they display a wide range of ligand affinities ranging from 10^{-6} to 10^{-9} l/M under different conditions [17]. It is this quality that ensures reversibility of integrin-mediated cell adhesion and allows cells to *both* attach and detach from biological substrata at sites of focal adhesions. In general, cells attach to differentiate and carry out specialized cell functions such as secretion, absorption, or signal transmission. They detach in order to migrate and proliferate. Interestingly, many epithelia undergo a form of programmed cell death known as anoikis if forcibly detached from their substrata [18]. Elegant studies of migrating cells have shown simultaneous formation of focal adhesions with substrata on the leading edge of the cell, and dissolution of attachment sites on the trailing end [19].

Receptor clustering during adhesion triggers cascades of events involving protein phosphorylation and dephosphorylation that carry intracellular signals through the cell from surface to nucleus, ultimately leading to changes in gene transcription. Frequently, changes in cell shape mediated by the cytoskeleton accompany and promote these signals. An emerging paradigm for signal transduction involves the formation of protein complexes or "signalosomes" held together by specific interactions of regulatory molecules with scaffolding proteins that, like the electron transport chain of the mitochondrion, reduce diffusion and speed up signal transmission. The interested reader is referred to one of several recent reviews on this exciting subject [20–22].

Once activated, feedback loops within cells act quickly to attenuate signals, preventing desensitization to extracellular signals and in some cases preserving cell viability. For example, prolonged Ca^{2+} signals are associated with activation of apoptotic pathways and cell death [23]. These loops also allow cells to return to a responsive state in which they can respond to subsequent signals that they may encounter from the ECM. Changes that occur within cells and that feed back to modulate the activity of surface receptors have come to be called "inside out" signals and modulate activity of both integrin and nonintegrin receptors in focal adhesions.

3.5 ECM and Skeletal Tissues

The ECM in cartilage is frequently described as part of a pericellular or "territorial" matrix that is both avascular and noninnervated. Figure 3.3 depicts several views illustrating the architecture of cartilage at the cellular level. Note that, unlike cells interacting with basement membranes, cells in mesenchyme are surrounded on all sides by the territorial matrix. This is particularly evident in Figure 3.3c showing a single chondrocyte in its lacunae. The chondrocyte has a very unique relationship with its microenvironment that has been well described [24]. The cartilage matrix consists of type II collagen and various noncollagenous

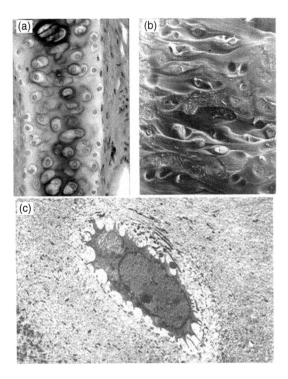

FIGURE 3.3 (a) Section through hyaline cartilage of the trachea. Dark staining regions surrounding cells (chondrocytes) are pericellular or "territorial" matrix and represent most recently-deposited cartilage matrix. (b) SEM of hyaline cartilage showing lacunae or "little lakes" in which chondrocytes reside. In this preparation some are occupied by cells and others are empty because the cells have fallen out during cutting of the cartilage. The fibrous nature of the matrix (due to the presence of type II collagen) is evident. (c) TEM of a chondrocyte within a lacuna. Collagen fibers can be seen in the cartilage matrix surrounding the lacunae. These fibers are not seen in light microscopic preparations such as those in panel a since they have the same refractive index as the collagen matrix material.

ECM molecules that differ depending on the type and differentiated state of cartilage. In articular cartilage, type II, IX, and XI collagens form a meshwork that provides form and tensile stiffness and strength [25]. Type VI collagen is present in the matrix surrounding the chondrocyte; aggrecan provides resilience to compressive forces, and small proteoglycans such as fibromodulin help to stabilize the matrix. Other noncollagenous proteins that are present include cartilage oligomeric matrix protein (COMP), perlecan, anchorin II, and tenascin. It has been proposed that the presence of tenascin-C in the articular cartilage matrix, a protein which is absent from growth plate cartilage, may help articular chondrocytes avoid endochondral ossification [26]. The role of the cartilage ECM in tissue engineering was reviewed recently [27]. In the growth plate, a progressive series of events occur to produce a cartilaginous matrix that gives way to bone matrix [28–30]. Early stages are marked by high rates of cell proliferation and ECM production. The final stage of development of growth plate cartilage is hypertrophy, marked by turnover of matrix, mineralization, invasion of marrow vasculature, and chondrocyte apoptosis. Type X collagen is the classical marker for hypertrophic cartilage ECM, and is absent from articular cartilage [31].

The ECM of bone is distinct from that of cartilage. Bone tissue (Figure 3.4a) is greater than 95% type I collagen based and includes noncollagenous proteins such as the acidic calcium binding proteins osteocalcin, bone sialoprotein, osteopontin, and osteonectin/SPARC. Of these, osteocalcin and bone sialoprotein can be considered specific to bone. It has been proposed that compact lamellar bone and woven trabecular bone have distinct matrix protein compositions and ratios [32]. The marrow compartment (Figure 3.4b) can be considered as structurally distinct from proper bone, although it houses cellular precursors for bone cells. As shown in Figure 3.4c,d, osteocytes, like chondrocytes, are surrounded by and embedded in matrix, but in this case the matrix normally remains fully mineralized. The processes

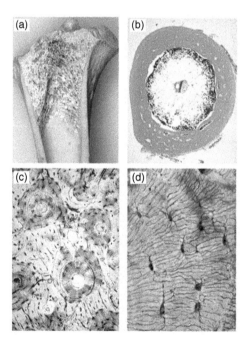

FIGURE 3.4 (a) Transverse cut through the metaphysis of a calf femur. The bone shaft is compact bone and spongy bone with bony trabeculae lining the interior. (b) Cross-section through a decalcified preparation of a fetal femur. Compact bone encloses the marrow cavity. (c) Preparation of compact bone. Concentric lamellae of Haversian systems surround blood vessels. Interstitial lamellae represent older Haversian systems replaced during bone remodeling. (d) High magnification image of a compact bone preparation. India ink fills lacunae in which embedded bone cells (osteocytes) are normally found. Channels connecting lacunae are called "canaliculi." These are occupied by cell processes and provide a means whereby osteocytes communicate with one another.

of bone remodeling that maintain structural integrity of bone and include a finely tuned balance between osteoclastic bone resorption and osteoblastic bone formation have been reviewed recently [33]. ECM is one of many influences on bone cell function that also include growth factors, physical stimuli, metabolic demands, and structural responsibilities [34].

3.6 Sources of ECM for Tissue Engineering Applications

It is critical to choose an ECM that will support cell adhesion, and also ensure survival, growth, and appropriate cell differentiation following adhesion. A number of alternatives exist for the tissue engineer including commercial mixed matrices based upon the composition of basement membranes (collagen IV, laminin, perlecan, nidogen/entactin). One example of such a commercial matrix is Matrigel™ that is available in both growth factor replete and depleted forms. This matrix is extracted from Engelbreth–Holm–Swam (EHS) mouse sarcoma and provides a material similar to the mammalian basement membrane. Available from the global nonprofit ATCC (http://www.atcc.org/) is an ECM solution also representing a solubilized basement membrane solution (30-2501) similar to Matrigel. This preparation contains tissue plasminogen activator as well as a number of growth factors including transforming growth factor β, epidermal growth factor, insulin-like growth factor 1, fibroblast growth factor 2, and platelet derived growth factor. Both these products form a gel at room temperature that will support growth and differentiation of cells grown on (or in) it as a two- or three-dimensional matrix.

Serum also provides a variety of adhesion factors including fibronectin and vitronectin, along with a rich selection of soluble, circulating growth factors. The response of cells to serum components is highly variable, as is the composition of serum itself. Because of the uncertainty in knowing the exact composition of serum for use in tissue engineering applications, batch testing of serum for long-term applications is encouraged when its use is necessitated. Purified ECM molecules provide a more reliable and predictable source for engineering applications. Smaller molecules can be expressed as recombinant proteins in either bacteria, insect, or mammalian cells in bioreactors [35–37]. Each of these expression systems provides a set of advantages and disadvantages that have been reviewed [38]. For ECM molecules of large size or complexity (collagens, many proteoglycans, noncollagenous glycoproteins), problems can occur related to size, microheterogeneity, yield, and scalability. For most of the molecules in this class, function requires posttranslational modifications that can only be acquired when expressed by mammalian cells. In this case, some luck has been achieved by expressing recombinant constructs in cell lines such as the HEK-293 cell line or even in tissue systems [37,39]. For very large matrix molecules, such as perlecan, that are too large to be expressed as full length recombinant proteins, it has been possible to create smaller expression constructs that retain functions of individual protein domains [37]. Of interest, there appears to be considerable cell type diversity in the manner and extent of posttranslational modification affecting the fine structure of the carbohydrate chains. For example, domain I of perlecan that contains consensus sites for both heparan and chondroitin sulfate chain addition can be produced in both active and inactive forms depending on the cells producing it [40].

An alternative to the use of recombinant ECM proteins is the use of synthetic peptides or proteolytic fragments of native molecules that retain selected functional properties of their parent molecules. Many ECM molecules contain a consensus sequence recognized by integrin receptors that consists of the triplet peptide, arginine–glycine–aspartate (RGD) sequence. The context in which this peptide is presented by the ECM molecule containing it determines the strength and the specificity of the interaction. Examples of common ECM molecules containing the RGD sequence include fibronectin, osteopontin, laminin, vitronectin, some collagens, and tenascin. Current effort is focused on identifying other active motifs present within ECM molecules that are small enough to be prepared as synthetic proteins. One promising example of this is the laminin-derived cell binding YIGSR sequence that has been proposed for derivatization of scaffolds for breast and soft tissue reconstruction [41]. Another is the tetrapeptide sequence REDV which simulates an active motif in the CS5 domain of fibronectin [42]. This subject is discussed further in Section 3.7 devoted to mining of the ECM.

3.7   Properties of ECM

Extracellular matrix in all tissues possesses certain properties that allow it to support tissue cohesion and provide microstructure. The most salient of these properties is that of *self-assembly* [43]. Protein and carbohydrate components of secreted ECM will self-associate in a highly ordered and predictable fashion that is tissue and cell type specific. Associations are based upon the presence of highly conserved motifs and independently folded domains whose structures are increasingly appreciated [44]. Interactions occur based upon complementary secondary and tertiary structural features including charge properties, ion and metal bridging, hydrophobic domains, redox interactions, and covalent bonding. The structures that form may produce either two-dimensional networks such as the meshwork of the basement membrane, or three-dimensional structures in space such as that of the territorial matrix. While most of the assembly is thought to occur extracellularly, there is evidence that some degree of assembly may be initiated during biosynthesis and take place within intracellular secretory vesicles creating a sort of "pre-fabricated" scaffold to promote rapid assembly once secretion into the extracellular compartment occurs [45]. In the sea urchin model system, it has been proposed that distinct intracellular vesicles form and are directionally released during exocytosis; these have been termed "basal laminar vesicles" and "apical vesicles" [46].

3.8   Mining the ECM for Functional Motifs

In the new age of the fruition of the genome projects, it is exciting to consider the possibility that bioinformatic approaches can be used to identify structure–function relationships in natural ECM molecules for use in translational biology including tissue engineering. This process, which has come to be termed "data mining," can identify motifs in larger molecules that can be used to manipulate behavior of cells and tissues *in vitro* and *in vivo*. Motifs can be included in small synthetic peptides, recombinant domains expressed in viral, bacterial, or mammalian expression systems, or rationally designed chemically synthesized mimics. The latter, in particular, frame the enormous potential of rational drug design when combined with computer modeling of protein interactions. The following sections will briefly describe the properties of some individual ECM molecules which provide the raw material for mining the ECM for functional motifs. In each case, the Swiss-Protein (http://au.expasy.org/sprot/) ID number for the *human* protein is provided as a link to the electronic database for that protein. The complete domain structure of each molecule may be accessed using Pfam (http://www.sanger.ac.uk/Software/Pfam/) and typing in the accession number in Protein Search.

3.8.1   Collagen

Members of the collagen family are diverse; they make up some one-third of all the protein in the body and model the framework of connective tissues [47,48]. Collagens form what can be thought of as "functional aggregates" with noncollagenous molecules to form macrostructures including fibrils, basement membranes, filaments, canals, and sheets. Fibril forming collagens are synthesized in precursor forms that are sequentially processed during or following secretion into the ECM [45]. Collagens and proteins with collagen-like domains now number at least 27 types with 42 distinct polypeptide chains; there are 20 additional proteins with collagen-like domains and 20 isoenzymes that modify collagen structure [49]. Examples of fibrillar collagens include type I collagen [collagen I α1 (PO2452), α2 (P08123)] found in connective tissues, and type II collagen [collagen II α1 (PO2458)] found in cartilage. Each of these assembles as a triple helix that once incorporated into rod-like fibrils can also assemble with other collagenous and noncollagenous proteins [48]. Figure 3.5 shows several views of collagen type I fibrils. Figure 3.5c depicts the structural relationship of a fibroblast and its ECM, including collagen fibrils seen in cross-section. The dimensional comparison between the cell and the collagen is also evident in this photograph (bar in the figure is 1 μm). The inset clearly shows the banding pattern characteristic of collagen type I when viewed by transmission electron microscope (TEM). Fibril associated collagens with interrupted triple helices (FACIT) collagens associate with surfaces of collagen fibrils in many tissues including skin,

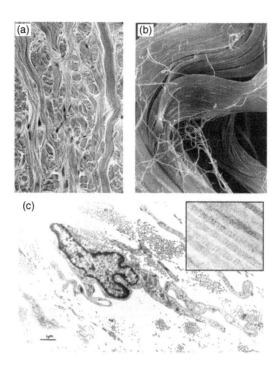

FIGURE 3.5 (a) Light micrograph of collagen fiber bundles in dense, irregular connective tissue such as the dermis of the skin. Nuclei of fibroblasts (cells that secrete procollagen which is processed and assembled into collagen extracellularly) can be seen closely associated with the fibers. (b) SEM of bands of type I collagen fiber bundles. Individual fibers can be seen independent of the fiber bundles. (c) TEM of a fibroblast in close association with bundles of collagen fibers sectioned transversely, obliquely and longitudinally. The inset shows individual collagen fibers exhibiting the characteristic periodic banding pattern.

tendon, and cartilage. Meshwork or basement membrane collagens include basement membrane-localized collagen IV [α1 (PO2462) and α2 (P08572)], hypertrophic cartilage-specific collagen X [α1 (Q03692)] and endothelial collagen VIII [α1 (P27658)]. Collagen VII [α1 (Q99715)] is also a member of this family and forms anchoring fibrils that connect skin and mucosa to underlying tissue.

3.8.2 Fibronectin

Fibronectin (PO2751) is a glycoprotein found in soluble form in plasma and in insoluble form in loose connective tissue and basement membranes. Fibronectin binds cell surfaces as well as various other ECM molecules including collagen, heparan sulfate proteoglycans, and fibrin. Fibronectins are involved in diverse functions including cell migration, wound healing, cell proliferation, blood coagulation, and maintenance of cell cytoskeleton. Fibronectin is a multidomain protein that contains three types of domains (FN1, 2, 3) that are repeated multiple times depending on the isoform. Fibronectin was the first noncollagenous component of the ECM that was thoroughly studied as a ligand for its integrin receptor, now termed $\alpha_5\beta_1$, but originally studied as the "fibronectin receptor" [50]. Fibronectin served as the prototype for the development of the RGD-peptides now widely used in modification of biomaterials for the purpose of tissue engineering [51,52].

3.8.3 Laminin

Laminin is a common ECM component found in basement membranes and used as a substratum for cell migration by many cell types. It has a clear role in cell migration and tissue morphogenesis during embryonic development [4]. The classical laminin-1 is a cruciform shaped molecule composed of three chains [α1 (P25391), β1 (PO7942), and γ1 (P11047)]. There are at least 15 heterotrimers that can form using various α, β, and γ chains [see α2 (P24043), α3 (Q16787), α4 (Q16363), α5 (O15230), β2 (P55268), and β3 (Q13751), γ2 (Q13753)]. Laminin self-assembles into polygonal lattices *in vitro* [53]. It is a favorite ECM-based substrate for cells in the neural system [54].

3.8.4   Tenascin, Thrombospondin, and Osteonectin/SPARC/BM-40

Tenascin (P24821), thrombospondin-1 (P07996) and osteonectin/SPARC (P09486) are all modulators of cell adhesion, migration, and growth. Expression changes in these molecules are associated with neoplastic transformation and acquisition of migratory metastatic states such as those that occur during wound healing [55–57]. This regulation may be attributable to the ability of these molecules to trigger "de-adhesion," a process that has been speculated to represent an intermediate adaptive condition that facilitates expression of specific genes that are involved in repair and adaptation of cells and tissues [58]. In this regard, these molecules and motifs therein might be of particular importance in processes related to tissue engineering, although this has not yet been exploited.

3.8.5   Proteoglycans and Glycosaminoglycans

Perlecan (P98160), glypican (1-P35052) and syndecan (1-P18827, 2-P34741, 3-O75056, 4-P31431) are heparan sulfate proteoglycans (HSPGs) involved in diverse functions and essential to embryogenesis. Both the core proteins and the HS side chains are thought to play important functional roles. HS chains consist of a repeating disaccharide structure (glycosaminoglycan or GAG) that is regionally modified by enzymes that produce epimerization, vary sulfation patterns, and alter chain length [59]. The GAG chains create polycationic binding sites for attachment of proteins, primarily heparin-binding growth factors (HBGFs). Perlecan is secreted entirely into the matrix [60], syndecan possesses a transmembrane domain and remains as an integral component of the plasma membrane [61], and glypican is lipid-linked [62].

Chondroitin sulfate proteoglycans (CSPGs) such as the large cartilage matrix protein aggrecan (P16112) have very large hydration spheres. Aggrecan is secreted into the pericellular matrix of cartilage where it self-assembles into a superstructure that can have as many as 50 monomers bound to a central filament composed of hyaluronic acid [63]. This structure is often described as resembling a "bottle brush." Aggrecan provides the osmotic resistance that cartilage needs to resist large compressive loads; its destruction during aging and in pathologic states contributes to skeletal erosion. The enzymes responsible for this destruction are matrix metalloproteinases (MMPs) called "aggrecanases" that are members of the ADAMTS (A Disintegrin And Metalloproteinase with Thrombospondin Motifs) gene family [64].

3.8.6   Osteopontin

Osteopontin (P10451) is a secreted matrix molecule that regulates cell responses through several integrin receptors and also is recognized by the CD44 receptor, through which it acts as a chemoattractant [65]. Osteopontin is thus a multifunctional molecule able to support both adhesion and migration, distinct properties that may be dependent on the degree and sites of phosphorylation [66]. The cell attachment site in osteopontin was mined and used to generate peptides for incorporation into peptide modified hydrogels that recently were used for modifying the function of marrow stromal osteoblasts [52].

3.9   Summary of Functions of ECM Molecules

The properties of ECM molecules make them ideal for cell and tissue engineering. ECM-derived molecules can be used to coat implants, modify surfaces, direct cell growth and differentiation, and engineer cell phenotype and behavior. Their multifunctional nature makes them ideal for promotion of cell specific adhesion via integrins and other surface receptors. Conversely, they can be used to release cells from adhesion ("counteradhesion") and thus migrate, for example, to cover a scaffold. They serve as efficient co-receptors for sequestration and delivery of growth factors, establishing morphogenic gradients recognized by cellular receptors during development, wound healing, and tissue repair [67,68]. For each of these functional roles, the context of the presentation of the ECM motif is critical in determination of outcome. For example, RGD peptides when presented to integrin-bearing cells *inhibit* adhesion and may lead to anoikis, but when immobilized these same motifs *support* adhesion, focal adhesion formation, and activate survival and proliferation pathways. These properties, if well understood, allow the ECM to serve as a rich source for mining and future development of novel tissue engineering applications.

3.10 Polymeric Materials and their Surface Modification

The rich biochemical and mechanical properties of the ECM have long been used as a model for tissue engineering constructs for the production of tissues such as cartilage, bone, nerve, and skin, as well as for the continued efforts toward the production of more complicated organs. Historically, the materials used for tissue engineering applications have relied primarily on readily available polymeric materials, both naturally derived and chemically synthesized. Polymers for specific applications are chosen on the basis of their aggregate mechanical properties, ease of processing, degradation profiles, and biochemical activity, with the latter becoming more prominent in the recent design of tissue engineering materials that elicit a specific and desired biological response. Among the most commonly utilized materials are natural ECM-based polymers such as collagen, fibrin glues, hyaluronic acid, and alginate. Although the biological activities and biocompatibility of these materials are useful, the lack of control over desired mechanical, degradation, and processing properties has motivated the use of synthetic polymers such as poly(glycolic acid), poly(lactic acid), poly(glycolic-*co*-lactic acid), poly(ethylene glycol) hydrogels, poly(N-isopropylacrylamide), poly(hydroxyethylmethacrylate), poly(anhydrides), and poly(*ortho*-esters). There are many good reviews on these and other polymeric systems and their use in tissue engineering [69–73], and descriptions of these materials are not included here.

Recent research effort in tissue engineering has focused on the incorporation of biologically active motifs derived from ECM proteins into matrices to integrate essential molecular elements of the ECM into the biomaterial and to promote a desired biological response. The choice of biologically active motifs incorporated into engineered matrices depend upon the ultimate end use of the matrix, but rely heavily on the coating of scaffolds with ECM-derived proteins, peptides, and glycosaminoglycans. Among the most straightforward of strategies is the simple coating of polymeric scaffolds (fibers, meshes, sponges, foams) with solutions of ECM protein(s) of interest based on adsorption via noncovalent interactions. While this method has demonstrated improvements in cell adhesion, proliferation, and secretion of ECM for a variety of cell types, it requires the adsorption of a high density of protein, since the protein can denature on the surface or adsorb in a suboptimal orientation, which significantly reduces affinity for given cell types. Alternatively, ECM proteins can be modified and then immobilized onto surfaces. In one recent example, fibronectin modified with Pluronic™ F108 could be easily attached to both poly(styrene) surfaces and poly(propylene) filaments with demonstrated improvements in neuronal cell attachment and neurite outgrowth [74].

Short, bioactive peptides from ECM proteins and growth factors, such as the RGD, YIGSR, IKVAV, REDV, heparin binding, and other sequences, also have been attached via straightforward chemical methods. The attachment of peptides (vs. proteins) is advantageous because the surface density and orientation of the peptide can be more easily controlled, allowing more quantitative understanding and manipulation of the surface modification process. Further, select peptides or combinations of peptides can yield materials that better mimic desirable signals present in the ECM. For example, the peptide REDV can be immobilized on surfaces instead of RGD to mediate the selective adhesion of endothelial cells over smooth muscle cells, fibroblasts, and platelets. Additionally, the combination of both the RGD and the heparin-binding motifs from bone sialoprotein (FHRRIKA) can synergistically improve cell adhesion and mineralization in osteoblast culture [75], and RGD modified polymers can impart desirable osteoblast adhesion and mineralization on titanium implants [76]. Peptides also can be incorporated into materials to reduce biodegradation, for example, the peptide aprotinin, which inhibits several serine proteases, has been used to derivatize the commercially available Tissucol® fibrin gels so that they maintain their bioactivity while exhibiting a reduced degradation rate (Immuno AG, Vienna, Austria). Numerous experiments have been conducted with surfaces and tissue engineering matrices covalently modified with ECM-derived peptides, and they have confirmed the bioactivity of these peptides and their utility in many tissue engineering applications. The most widely used methods for peptide immobilization, with just a few relevant examples, are summarized.

A very commonly employed strategy for peptide immobilization is the reaction of the thiol-terminated peptide with surfaces and polymers that are functionalized with maleimide, thiol, and vinyl sulfone groups. The reaction of thiol groups with vinyl sulfone modified polymers and surfaces has been employed in many recent investigations owing to its high selectivity over reactions with amines and its high reaction efficiency at physiological temperature and near physiological pH [77–79]. Reactions of amine-terminated side chains and carboxylic acid side chains also can be used for certain amino acid sequences in which these side chains are not required for biological activity. A variety of chemical reaction strategies can be used [80], one of the most common being the carbodiimide-activated coupling of amines with carboxylic acids. The advantage of the latter reactions is that they are generally applicable to a variety of proteins, synthetic polymers, and plasma-treated surfaces. The overall surface density of the peptides also can be controlled easily via these chemical modification strategies, which can have a large impact on cell proliferation and differentiation, particularly when coupled with manipulation of the chemical composition of the matrix material [81,82].

Another strategy for the incorporation of biologically active peptide motifs onto tissue engineering scaffolds involves the covalent coupling of the peptide into the matrix during matrix formation. Acrylated peptides such as RGD and plasmin substrate sequences have been incorporated into a variety of poly(ethylene glycol) (PEG) hydrogel materials in this manner and have imparted desirable cell adhesion and cell-mediated degradation [83,84]. Essentially, any peptide can be incorporated into hydrogels via these strategies, and recently, GRGDS or an osteopontin derived peptide (ODP), DVDVPDGRGDSLAYG, were incorporated in oligo(poly(ethylene glycol) fumarate) hydrogels to determine the impact on osteoblast migration. Osteoblasts migrate both faster and for longer distances on the ODP-modified hydrogels vs. the RGD modified hydrogels [52].

Enzymatic attachment of a peptide that carries both a cell attachment or signaling sequence and a domain that is a substrate for enzymatic coupling by Factor XIIIa (-NQEQVSP-) also has been used to covalently incorporate ECM-derived peptide sequences into tissue engineering materials. This strategy has been recently developed for the incorporation of a variety of cell adhesion peptides (derived from fibronectin, laminin, and N-cadherin), heparin-binding peptides, and growth factors into fibrin matrices, via the action of Factor XIIIa, to demonstrate improvements in cell adhesion, neurite outgrowth, and axonal regeneration [85–87].

Strategies that involve assembly and adsorption also have been useful for mediating the presentation of peptides at surfaces. For example, a very general strategy for surface modification of both polymeric and inorganic matrices involves the incorporation of mussel-adhesive-protein-derived dihydroxylphenylalanine (DOPA) residues at the termini of polymers [88]. Interaction of the DOPA residues with surfaces results in attachment, and although this method has currently been applied to the modification of surfaces with ethylene glycol oligomers, it should be equally applicable to the simple attachment of bioactive peptides to a variety of surfaces. Strategies for presenting labile ligands at an interface via the use of RGD-modified nanoparticles also have been developed as a strategy to improve cell migration on biomaterials surfaces [89]. Another emerging strategy is the use of peptide-modified self-assembling materials for the multivalent presentation of biologically active peptides. Peptide–amphiphiles, decorated with either phosphate-functionalized amino acids or sequences such as IKVAV, assemble into stable fibrous hydrogel materials and demonstrate promising activities such as mineralization and selective differentiation of neural progenitor cells *in vitro* [90,91].

A more experimentally complicated, but useful, method for the incorporation of biologically active peptide motifs is the inclusion of these peptide sequences directly in the backbone of recombinant artificial proteins and in protein/polymer conjugates. Cell adhesion sequences from fibronectin (RGD and REDV) have been incorporated into silk-like, collagen-like, and elastin-like artificial proteins. The mechanical properties of the silk, collagen, or elastin provide excellent adhesion and elasticity, and the ECM-derived proteins mimic some essential features of the ECM. The fibronectin-derived cell-binding sequences result in marked improvements in cell adhesion to films of these proteins [92–95]. The silk-fibronectin proteins form autoclave-stable coatings on polystyrene culture plates and are sold commercially as the Pronectins®, originally by Protein Polymer Technologies, Inc. The mechanical properties of the elastin-like protein polymers can be engineered via chemical cross-linking strategies to closely match those of native elastin [96,97], which aid in compliance matching of small-diameter vascular graft replacement materials with host tissues.

Investigations to engineer biologically active domains of other proteins also have been initiated. Given the prominent developmental and physiological functions of collagen, and its widespread use as a tissue engineering material, collagen proteins have received significant attention. Individual domains of collagen II have been isolated to determine their biological activity, and the amino acid region 704-938 in collagen II was identified in these studies as critical for the spreading of chondrocytes [98]. These studies may direct the production of repetitive proteins containing that amino acid sequence for improved chondrocyte adhesion and spreading. Chimeric proteins of collagen III and EGF also have been produced, and retain the fibril-forming properties of the collagen domain and the activity of EGF; these materials may find application in cell culture, wound healing, and tissue engineering applications [99]. Similarly, biologically active motifs can be genetically engineered and then conjugated to synthetic polymers to produce biologically active protein–polymer block copolymers. Specifically, an artificial protein equipped with an RGD sequence and two plasmin degradation sites from fibrinogen, and an ATIII-derived heparin binding site, was grafted to PEG-acrylates and incorporated into hydrogels. These hydrogels supported three-dimensional outgrowth of human fibroblasts mediated by adhesion to the RGD sequences [100].

3.11 Formation of Gradient Structures

While the coating of tissue engineering matrix surfaces with bioactive molecules has improved cell adhesion and biological properties of matrices, the native temporal and spatial presentation of molecules in the ECM also has been increasingly mimicked as a strategy to control materials response. The establishment of functional morphogen gradients and patterns in materials controls cell signaling and proliferation, and can be achieved by a variety of methods, including temporally controlled deposition, microfluidic approaches, photoinitiated polymerizations, photolithography, soft lithography, block copolymer assembly, and printing strategies. One area in which the deposition of functional gradients of ECM-derived proteins has had large impact is in the generation of materials for guiding neurite outgrowth. For example, the formation of simple concentration gradients via controlled photopolymerization of nerve growth factor (NGF) in poly(2-hydroxyethyl methacrylate) microporous gels has guided the growth of PC12 cell neurites up the concentration gradient [101]. Microfluidic methods also have been used to control deposition of proteins in channels; in one example, the deposition of laminin was controlled, and the growth of hippocampal neurons toward the regions of highest laminin deposition was observed [102]. Protein gradients also have been produced on surfaces via the temporally controlled deposition of protein-coated gold nanoparticles onto poly(lysine) derivatized surfaces [103]. Proteins immobilized in this manner retain their bioactivity, and the presence of the particles is not detrimental to the growth of hippocampal neurons. Furthermore, this method is generally applicable to the production of protein gradients on the surfaces of a variety of two-dimensional and three-dimensional scaffolds.

In addition to the production of concentration gradients in hydrogel materials and on surfaces, there has been increasing interest in controlling cell placement for studying and manipulating cellular responses to materials. Lithographic methods have emerged as a viable strategy for achieving this patterning. Photolithographic strategies permit the production of patterns of proteins on the micron length scale, via selective activation of photosensitive groups on a surface followed by coupling to a protein of interest. These methods have been used to create many patterned surfaces including those for controlling neuronal patterning [104,105]. Perhaps the most popular lithographic method for patterning surfaces is the soft lithographic approach of microcontact printing, in which an elastomeric stamp is used to directly "print" chemically or biologically active peptides and proteins onto surfaces of both hard and soft materials [106–108]. The methods have been used to print laminin, collagen, and a variety of other ECM-derived proteins and peptides, and provide a useful means to control cell adsorption and response to materials [109–112]. These methods have been used to pattern surfaces primarily at the micron length scale, but can also be used to pattern proteins at surfaces at length scales smaller than 1 μm.

Direct printing of ECM proteins and peptides also has been demonstrated. Ink jet printing can modify and fabricate scaffolds for tissue engineering applications, including the printing of active proteins [113], and the fabrication and surface modification of three-dimensional scaffolds [114]. In one recent example, automated ink jet printing was used to produce 350 μm features of collagen in lines, circles, arrays of dots, and gradients. Smooth muscle cells and dorsal root ganglia neurons adhere to these patterned regions and achieve confluency in the shape of the pattern in as few as 4 h [115]. Dip-pen nanolithography (DPN), in which an atomic force microscope (AFM) tip is "inked" with a solution of interest and then is used to deposit a pattern, also has been applied to produce protein patterns with feature sizes below 200 nm. Dots and lines of antibodies and ECM-derived proteins have been printed [116–118]. Collagen has been printed to feature sizes of 30 to 50 nm [116], and Retronectin, a recombinant fragment of fibronectin comprising the central cell-binding domain, the heparin binding domain II, and the CS1 site, was printed onto gold surfaces via DPN at feature sizes of approximately 200 nm [117]. Fibroblasts adhere selectively to the areas printed with Retronectin, but with morphologies different than those on unpatterned surfaces modified with Retronectin. The use of these methods may therefore permit the study and manipulation of cell adhesion and signaling phenomena via patterning at lengthscales relevant to individual receptors on cell surfaces.

The directed assembly of polymeric materials also can be used to pattern bioactive elements on the nanometer length scale. For example, the spatial distribution of RGD-containing peptides has been controlled by the presentation of the peptide on star-shaped polymers. By controlling the number of peptides per star polymer, as well as by controlling the density of peptide-functionalized star polymers on a surface, these methods have been useful for clustering peptides and increasing fibroblast adhesion strength and actin stress fiber formation [119]. The clustering of the ligands also has altered fibroblast adhesion behavior under centrifugal detachment forces, with clustered ligands increasing adhesion with increasing detachment force for forces from 70 to 150 pN/cell [120].

3.12   Delivery of Growth Factors

The development of polymeric controlled release systems has facilitated the development of tissue engineering scaffolds capable of the controlled release of growth factors, and has enabled the control of cellular processes for prolonged periods of times via a sustained release of appropriate growth factors directly into the cell microenvironment (for a recent review see Reference 121). Simple injection of growth factors has not reached its potential therapeutically owing to the very short half-lives of the growth factor *in vivo* (a few minutes), which has necessitated studies of their controlled release. Growth factors have been both noncovalently and covalently incorporated into a wide variety of materials over the past decade, which has resulted in numerous reports of their benefit in bone, cartilage, and nerve tissue engineering investigations. Growth factors can be simply encapsulated into matrices comprised of collagen, alginate, hyaluronic acid, or other synthetic hydrogels. In many cases the half-life of the growth factor is significantly increased vs. that of the free growth factor in solution. For example, in alginate matrices, electrostatic interactions stabilize the growth factor in the matrix and control its slow release [122]. Interest in extending the release time of the growth factors, improving their bioactivity, and allowing for delivery of multiple growth factors have led to additional strategies for incorporating growth factors into materials and onto surfaces [123,124]. There have been a myriad of studies on the passive encapsulation of growth factors into particles, fibers, and hydrogels; the discussion below focuses on recently described methods involving immobilization of growth factors via interaction with heparinized materials or via covalent attachment.

Heparin has been widely applied to surfaces and encapsulated into hydrogels as a strategy for immobilization of growth factors. The interaction of heparin and growth factors permits immobilization and stabilization of the growth factor via sequence-specific electrostatic interactions with heparin, which mimic the natural mechanisms by which growth factors are protected in the ECM via interactions with heparan sulfate. Accordingly, hydrogels in which growth factor and heparin are co-encapsulated show improvements in growth factor delivery and half-life. Heparin has been covalently attached to polymers to reduce heparin diffusion from the matrix and further increase retention and half-life of growth factors. For example, when bFGF is immobilized in a heparinized collagen matrix, bFGF retains its bioactivity and promotes endothelial cell growth [125]. A variety of other materials have been covalently modified with heparin using other chemical strategies to enable growth factor binding, and these achieve desirable cellular responses [126–128]. In a recent example with a slightly different approach, PEG star polymers were modified with heparin-binding peptides (hbp) and then cross-linked via noncovalent interactions with high molecular weight heparin. The mechanical properties and delivery rates of the hydrogels can be mediated by the affinity and kinetics of the heparin–hbp interactions [78]. Likewise, hydrogels have also been assembled via noncovalent interactions between PEG–hbp star copolymers and PEG star polymers modified with low molecular weight heparin. These networks have demonstrated release profiles for bFGF that correlate with erosion of the noncovalent network [129]. It is envisioned that growth factors noncovalently bound to the heparin will be released at rates that correlate with the heparin–hbp affinities and that can be tuned for a desired tissue engineering or drug delivery application.

Another recently developed approach for growth factor delivery is the development of fibrin matrices in which hbp have been covalently incorporated into fibrin via the action of Factor XIIIa [86]. Both heparin and growth factors are then incorporated into these matrices, and the passive release of the heparin-associated growth factor is minimized via the binding between the heparin and the covalently bound hbp. Prudent choice of the ratio of growth factor: heparin:hbp permits growth factor release to be controlled by cell-mediated degradation of the fibrin matrix, rather than by passive diffusion. When bFGF was delivered from this growth factor delivery system *in vitro*, the results demonstrated a nearly 100% enhancement in neurite extension over an unmodified fibrin control [86]. Additionally, NGF has been immobilized in similar matrices. Despite the fact that NGF does not bind heparin with high affinity, the electrostatic interactions between basic regions of the NGF and heparin immobilize NGF in the fibrin matrices. These materials also enhance neurite extension from chick dorsal root ganglia by up to 100% relative to unmodified fibrin at 48 h [130], and more recent investigations demonstrate that these materials perform similarly to isografts when tested in 13-mm rat sciatic nerve defects [131]. Another recently employed strategy for growth factory delivery takes advantage of the native heparan sulfate chains of the ECM protein, perlecan, which binds HBGFS including TGFB, BMP, and bFGF. A recombinant domain of perlecan was used to sequester and release bFGF from native collagen matrices [124].

While the rates of release of bioactive growth factors from hydrogels of all the above kinds can be controlled via control of matrix pore size and heparin loading, the release mechanism is passive. There have been several strategies developed for the localized delivery of growth factors, and the covalent attachment of bioactive growth factors, such as EGF and TGF-β1, to different surfaces is a useful strategy for producing bioactive biomaterials [132,133]. Although the growth factors are immobilized by chemical methods, the bioactivity of both EGF and TGF-β1 are well preserved in these systems. The EGF, immobilized to a glass surface using a PEG linker, retained its activity as assessed by mitogenic and morphological assays of primary rat hepatocytes, while physisorbed EGF demonstrated no bioactivity [132]. The incorporation of TGF-β1 into PEG hydrogels increases ECM production by smooth muscle cells grown in these gels [133]. The incorporation of collagenase or elastase sensitive amino acid sequences in cross-linking regions of the gels also permits the migration of the cells into the matrix during remodeling [134]. The combination of the increased ECM production stimulated by TGF-β1 and the cell-mediated degradation of the matrix may provide materials that can provide sufficient mechanical stability throughout all stages of cell remodeling of the implant material.

Growth factors such as VEGF and β-NGF also have been covalently incorporated into fibrin matrices via the enzymatic action of Factor XIIIa [87,135,136]. In these examples, the growth factors were genetically engineered to carry both a Factor XIIIa substrate domain (for cross-linking) as well as an MMP substrate domain, so that they could be covalently incorporated into fibrin gels, and liberated upon cell migration into the gel via enzymatic cleavage of the MMP substrate domain. The β-NGF containing fibrin matrices enhanced neurite extension from embryonic chick dorsal root ganglia by 50% relative to soluble β-NGF and by 350% relative to a negative control without β-NGF [87]. Subcutaneous implantation of the VEGF-containing fibrin materials (5.4 mm diameter, 2 mm thickness) in rats demonstrated that these tissue engineering materials, after 2 weeks, were completely remodeled into native, vascularized tissue [136]. These results suggest the promise of such approaches to the development of useful tissue engineering materials that are responsive to cellular demand.

3.13   Summary and Conclusions

This chapter has summarized the present status of the science of tissue engineering using the ECM as a rich source of biologically relevant motifs. No doubt the next decade will provide a wealth of knowledge both about the native ECM and its components and the application of this knowledge to engineering purposes. The deliberate modulation of cell behavior using rational design and ECM-based biomaterials will serve as a major growth area in future biotechnology.

Acknowledgments

The authors thank the editors for the opportunity to compile the information in this chapter. We gratefully acknowledge the contributions to Figure 3.1a,c, Figure 3.2b,c, and Figure 3.3b,c by Dr. Fred Hossler, Department of Anatomy and Cell Biology at East Tennessee State University, J.H. Quillen College of Medicine. The work in the authors laboratories is supported by NIH R01 DE13542 (DDC and MCFC) and NIH R01 EB003172 (to KLK).

References

[1] Danen, E.H. and Sonnenberg, A., Integrins in regulation of tissue development and function, *J. Pathol.* 201, 632–41, 2003.

[2] Mareel, M. and Leroy, A., Clinical, cellular, and molecular aspects of cancer invasion, *Physiol. Rev.* 83, 337–76, 2003.

[3] Mekori, Y.A. and Baram, D., Heterotypic adhesion-induced mast cell activation: biologic relevance in the inflammatory context, *Mol. Immunol.* 38, 1363–7, 2002.

[4] Sasaki, T., Fassler, R., and Hohenester, E., Laminin: the crux of basement membrane assembly, *J. Cell Biol.* 164, 959–63, 2004.

[5] Chen, C.H. and Hansma, H.G., Basement membrane macromolecules: insights from atomic force microscopy, *J. Struct. Biol.* 131, 44–55, 2000.

[6] Yurchenco, P.D., Cheng, Y.S., and Colognato, H., Laminin forms an independent network in basement membranes, *J. Cell Biol.* 117, 1119–33, 1992.

[7] Timpl, R. and Brown, J.C., Supramolecular assembly of basement membranes, *Bioessays* 18, 123–32, 1996.

[8] Otey, C.A. and Burridge, K., Patterning of the membrane cytoskeleton by the extracellular matrix, *Semin. Cell Biol.* 1, 391–9, 1990.

[9] Adams, J.C., Regulation of protrusive and contractile cell-matrix contacts, *J. Cell Sci.* 115, 257–65, 2002.

[10] Klinghoffer, R.A. et al., Src family kinases are required for integrin but not PDGFR signal transduction, *EMBO J.* 18, 2459–71, 1999.

[11] Murphy, K.G. and Gunsolley, J.C., Guided tissue regeneration for the treatment of periodontal intrabony and furcation defects. A systematic review, *Ann. Periodontol.* 8, 266–302, 2003.

[12] Krasteva, N. et al., The role of surface wettability on hepatocyte adhesive interactions and function, *J. Biomater. Sci. Polym. Ed.* 12, 613–27, 2001.

[13] Siczkowski, M., Clarke, D., and Gordon, M.Y., Binding of primitive hematopoietic progenitor cells to marrow stromal cells involves heparan sulfate, *Blood* 80, 912–9, 1992.

[14] Kan, M. et al., Divalent cations and heparin/heparan sulfate cooperate to control assembly and activity of the fibroblast growth factor receptor complex, *J. Biol. Chem.* 271, 26143–8, 1996.

[15] Dais, P., Peng, Q.J., and Perlin, A.S., A relationship between 13C-chemical-shift displacements and counterion-condensation theory, in the binding of calcium ion by heparin, *Carbohydr. Res.* 168, 163–79, 1987.

[16] Kirchhofer, D., Grzesiak, J., and Pierschbacher, M.D., Calcium as a potential physiological regulator of integrin-mediated cell adhesion, *J. Biol. Chem.* 266, 4471–7, 1991.

[17] Buck, C.A. and Horwitz, A.F., Cell surface receptors for extracellular matrix molecules, *Annu. Rev. Cell Biol.* 3, 179–205, 1987.

[18] Frisch, S.M. and Screaton, R.A., Anoikis mechanisms, *Curr. Opin. Cell Biol.* 13, 555–62, 2001.

[19] Wehrle-Haller, B. and Imhof, B.A., Actin, microtubules and focal adhesion dynamics during cell migration, *Int. J. Biochem. Cell Biol.* 35, 39–50, 2003.

[20] Farach-Carson, M.C. and Davis, P.J., Steroid hormone interactions with target cells: cross talk between membrane and nuclear pathways, *J. Pharmacol. Exp. Ther.* 307, 839–45, 2003.

[21] Werlen, G. and Palmer, E., The T-cell receptor signalosome: a dynamic structure with expanding complexity, *Curr. Opin. Immunol.* 14, 299–305, 2002.

[22] Elion, E.A., The Ste5p scaffold, *J. Cell Sci.* 114, 3967–78, 2001.

[23] Fawthrop, D.J., Boobis, A.R., and Davies, D.S., Mechanisms of cell death, *Arch. Toxicol.* 65, 437–44, 1991.

[24] Archer, C.W. and Francis-West, P., The chondrocyte, *Int. J. Biochem. Cell Biol.* 35, 401–4, 2003.

[25] Buckwalter, J.A. and Mankin, H.J., Articular cartilage: tissue design and chondrocyte matrix interactions, *Instr. Course Lect.* 47, 477–86, 1998.

[26] Pacifici, M., Tenascin-C and the development of articular cartilage, *Matrix Biol.* 14, 689–98, 1995.

[27] Goessler, U.R., Hormann, K., and Riedel, F., Tissue engineering with chondrocytes and function of the extracellular matrix (review), *Int. J. Mol. Med.* 13, 505–13, 2004.

[28] de Crombrugghe, B. et al., Transcriptional mechanisms of chondrocyte differentiation, *Matrix Biol.* 19, 389–94, 2000.

[29] van der Eerden, B.C., Karperien, M., and Wit, J.M., Systemic and local regulation of the growth plate, *Endocr. Rev.* 24, 782–801, 2003.

[30] Kronenberg, H.M., Developmental regulation of the growth plate, *Nature* 423, 332–6, 2003.

[31] Linsenmayer, T.F., Eavey, R.D., and Schmid, T.M., Type X collagen: a hypertrophic cartilage-specific molecule, *Pathol. Immunopathol. Res.* 7, 14–19, 1988.

[32] Gorski, J.P., Is all bone the same? Distinctive distributions and properties of non-collagenous matrix proteins in lamellar vs. woven bone imply the existence of different underlying osteogenic mechanisms, *Crit. Rev. Oral Biol. Med.* 9, 201–23, 1998.

[33] Jilka, R.L., Biology of the basic multicellular unit and the pathophysiology of osteoporosis, *Med. Pediatr. Oncol.* 41, 182–5, 2003.

[34] Sommerfeldt, D.W. and Rubin, C.T., Biology of bone and how it orchestrates the form and function of the skeleton, *Eur. Spine J.* 10, S86–95, 2001.

[35] Panda, A.K., Bioprocessing of therapeutic proteins from the inclusion bodies of *Escherichia coli*, *Adv. Biochem. Eng. Biotechnol.* 85, 43–93, 2003.

[36] Ikonomou, L., Schneider, Y.J., and Agathos, S.N., Insect cell culture for industrial production of recombinant proteins, *Appl. Microbiol. Biotechnol.* 62, 1–20, 2003.

[37] French, M.M. et al., Chondrogenic activity of the heparan sulfate proteoglycan perlecan maps to the N-terminal domain I, *J. Bone Miner. Res.* 17, 48–55, 2002.

[38] Sodoyer, R., Expression systems for the production of recombinant pharmaceuticals, *BioDrugs* 18, 51–62, 2004.

[39] Bulleid, N.J., John, D.C., and Kadler, K.E., Recombinant expression systems for the production of collagen, *Biochem. Soc. Trans.* 28, 350–3, 2000.

[40] Knox, S., Melrose, J., and Whitelock, J., Electrophoretic, biosensor, and bioactivity analyses of perlecans of different cellular origins, *Proteomics* 1, 1534–41, 2001.

[41] Patrick, C.W., Jr., Adipose tissue engineering: the future of breast and soft tissue reconstruction following tumor resection, *Semin. Surg. Oncol.* 19, 302–11, 2000.

[42] Heilshorn, S.C. et al., Endothelial cell adhesion to the fibronectin CS5 domain in artificial extracellular matrix proteins, *Biomaterials* 24, 4245–52, 2003.

[43] Yurchenco, P.D., Smirnov, S., and Mathus, T., Analysis of basement membrane self-assembly and cellular interactions with native and recombinant glycoproteins, *Meth. Cell Biol.* 69, 111–44, 2002.

[44] Hohenester, E. and Engel, J., Domain structure and organisation in extracellular matrix proteins, *Matrix Biol.* 21, 115–28, 2002.

[45] Hulmes, D.J., Building collagen molecules, fibrils, and suprafibrillar structures, *J. Struct. Biol.* 137, 2–10, 2002.

[46] Matese, J.C., Black, S., and McClay, D.R., Regulated exocytosis and sequential construction of the extracellular matrix surrounding the sea urchin zygote, *Dev. Biol.* 186, 16–26, 1997.

[47] Ottani, V. et al., Hierarchical structures in fibrillar collagens, *Micron* 33, 587–96, 2002.

[48] Gelse, K., Poschl, E., and Aigner, T., Collagens — structure, function, and biosynthesis, *Adv. Drug Deliv. Rev.* 55, 1531–46, 2003.

[49] Myllyharju, J. and Kivirikko, K.I., Collagens, modifying enzymes and their mutations in humans, flies and worms, *Trends Genet.* 20, 33–43, 2004.

[50] Brown, P.J. and Juliano, R.L., Expression and function of a putative cell surface receptor for fibronectin in hamster and human cell lines, *J. Cell Biol.* 103, 1595–603, 1986.

[51] Hersel, U., Dahmen, C., and Kessler, H., RGD modified polymers: biomaterials for stimulated cell adhesion and beyond, *Biomaterials* 24, 4385–415, 2003.

[52] Shin, H. et al., Attachment, proliferation, and migration of marrow stromal osteoblasts cultured in biomimetic hydrogels modified with an osteopontin-derived peptide, *Biomaterials* 25, 895–906, 2004.

[53] Colognato, H., Winkelmann, D.A., and Yurchenco, P.D., Laminin polymerization induces a receptor-cytoskeleton network, *J. Cell Biol.* 145, 619–31, 1999.

[54] Grimpe, B. and Silver, J., The extracellular matrix in axon regeneration, *Prog. Brain Res.* 137, 333–49, 2002.

[55] Mukaratirwa, S. and Nederbragt, H., Tenascin and proteoglycans: the role of tenascin and proteoglycans in canine tumours, *Res. Vet. Sci.* 73, 1–8, 2002.

[56] Sage, E.H., Regulation of interactions between cells and extracellular matrix: a command performance on several stages, *J. Clin. Invest.* 107, 781–3, 2001.

[57] Sid, B. et al., Thrombospondin 1: a multifunctional protein implicated in the regulation of tumor growth, *Crit. Rev. Oncol. Hematol.* 49, 245–58, 2004.

[58] Murphy-Ullrich, J.E., The de-adhesive activity of matricellular proteins: is intermediate cell adhesion an adaptive state? *J. Clin. Invest.* 107, 785–90, 2001.

[59] Esko, J.D. and Selleck, S.B., Order out of chaos: assembly of ligand binding sites in heparan sulfate, *Annu. Rev. Biochem.* 71, 435–71, 2002.

[60] Iozzo, R.V., Perlecan: a gem of a proteoglycan, *Matrix Biol.* 14, 203–8, 1994.

[61] Bernfield, M. et al., Biology of the syndecans: a family of transmembrane heparan sulfate proteoglycans, *Annu. Rev. Cell Biol.* 8, 365–93, 1992.

[62] Fransson, L.A., Glypicans, *Int. J. Biochem. Cell Biol.* 35, 125–9, 2003.

[63] Knudson, C.B. and Knudson, W., Cartilage proteoglycans, *Semin. Cell Dev. Biol.* 12, 69–78, 2001.

[64] Roberts, S. et al., Matrix metalloproteinases and aggrecanase: their role in disorders of the human intervertebral disc, *Spine* 25, 3005–13, 2000.

[65] Sodek, J. et al., Novel functions of the matricellular proteins osteopontin and osteonectin/SPARC, *Connect. Tissue Res.* 43, 308–19, 2002.

[66] Denhardt, D.T., The third international conference on osteopontin and related proteins, San Antonio, Texas, May 10–12, 2002, *Calcif. Tissue Int.* 74, 213–19, 2004.

[67] Gritli-Linde, A. et al., The whereabouts of a morphogen: direct evidence for short- and graded long-range activity of hedgehog signaling peptides, *Dev. Biol.* 236, 364–86, 2001.

[68] Cadigan, K.M., Regulating morphogen gradients in the *Drosophila* wing, *Semin. Cell Dev. Biol.* 13, 83–90, 2002.

[69] Hirano, Y. and Mooney, D.J., Peptide and protein presenting materials for tissue engineering, *Adv. Mater.* 16, 17–25, 2004.

[70] Drury, J.L. and Mooney, D.J., Hydrogels for tissue engineering: scaffold design variables and applications, *Biomaterials* 24, 4337–51, 2003.

[71] Shin, H., Jo, S., and Mikos, A.G., Biomimetic materials for tissue engineeering, *Biomaterials* 24, 4353–64, 2003.

[72] Sakiyama-Elbert, S.E. and Hubbell, J.A., Functional biomaterials: design of novel biomaterials, *Ann. Rev. Mater. Res.* 31, 183–201, 2001.

[73] Seal, B.L., Otero, T.C., and Panitch, A., Polymeric biomaterials for tissue and organ regeneration, *Mat. Sci. Eng. Res. Rep.* 34, 147–230, 2001.

[74] Biran, R. et al., Surfactant-immobilized fibronectin enhances bioactivty and regulates sensory neurite outgrowth, *J. Biomed. Mater. Res.* 55, 1–12, 2001.

[75] Healy, K.E., Rezania, A., and Stile, R.A., Designing biomaterials to direct biological responses, *Ann. NY Acad. Sci.* 875, 24–35, 1999.

[76] Barber, T.A. et al., Peptide-modified p(AAm-co-EG/AAc)IPNs grafted to bulk titanium modulate osteoblast behavior *in vitro*, *J. Biomed. Mater. Res. A.* 64A, 38–47, 2003.

[77] Lutolf, M.P. and Hubbell, J.A., Synthesis and physicochemical characterization of end-linked poly(ethylene glycol)-co-peptide hydrogels formed by Michael-type addition, *Biomacromolecules* 4, 713–2, 2003.

[78] Seal, B.L. and Panitch, A., Physical polymer matrices based on affinity interactions between peptides and polysaccharides, *Biomacromolecules* 4, 1572–82, 2003.

[79] Heggli, M. et al., Michael-type addition as a tool for surface functionalization, *Bioconj. Chem.* 14, 967–73, 2003.

[80] Hermanson, G.T., *Bioconjugate Techniques*. Academic Press, New York, 1996.

[81] Rezania, A. and Healy, K.E., The effect of peptide surface density on mineralization of a matrix deposited by osteogenic cells, *J. Biomed. Mater. Res.* 52, 595–600, 2000.

[82] Rowley, J.A. and Mooney, D.J., Alginate type and RGD density control myoblast phenotype, *J. Biomed. Mater. Res.* 60, 217–23, 2002.

[83] Hern, D.L. and Hubbell, J.A., Incorporation of adhesion peptides into nonadhesive hydrogels useful for tissue resurfacing, *J. Biomed. Mater. Res.* 39, 266–76, 1998.

[84] West, J.L. and Hubbell, J.A., Polymeric biomaterials with degradation sites for proteases involved in cell migration, *Macromolecules* 32, 241–4, 1999.

[85] Schense, J.C. et al., Enzymatic incorporation of bioactive peptides into fibrin matrices enhances neurite extension, *Nat. Biotechnol.* 18, 415–9, 2000.

[86] Sakiyama-Elbert, S.E. and Hubbell, J.A., Development of fibrin derivatives for controlled release of heparin-binding growth factors, *J. Control. Release* 65, 389–402, 2000a.

[87] Sakiyama-Elbert, S.E., Panitch, A., and Hubbell, J.A., Development of growth factor fusion proteins for cell-triggered delivery, *FASEB J.* 15, 1300–2, 2001.

[88] Dalsin, J.L. et al., Mussel adhesive protein mimetic polymers for the preparation of nonfouling surfaces, *J. Am. Chem. Soc.* 125, 4253–8, 2003.

[89] Tjia, J.S. and Moghe, P.V., "Cell-internalizable" ligand microinterfaces on biomaterials, in *Biomimetic Materials and Design*, Dillow, A.K. and Lowman, A.M. (Eds.), Marcel Dekker, Inc., New York, 2002, pp. 335–73.

[90] Hartgerink, J.D., Beniash, E., and Stupp, S.I., Self-assembly and mineralization of peptide-amphiphile nanofibers, *Science* 294, 1684–8, 2001.

[91] Silva, G.A. et al., Selective differentiation of neural progenitor cells by high-epitope density nanofibers, *Science* 303, 1352–5, 2004.

[92] Ferrari, F.A. and Cappello, J., Biosynthesis of protein polymers, in *Protein-Based Materials*, McGrath, K. and Kaplan, D. (Eds.), Birkhauser, Boston, 1997, pp. 37–60.

[93] Urry, D.W. et al., Transductional elastic and plastic protein-based polymers as potential medical devices, in *Drug Targeting and Delivery, Handbook of Biodegradable Polymers*, Domb, A.J., Kost, J., and Wiseman, D.M. (Eds.), Harwood Academic Publ., Amsterdam, 1997, pp. 367–86.

[94] Panitch, A. et al., Design and biosynthesis of elastin-like extracellular matrix proteins containing periodically spaced fibronectin CS5 domains, *Macromolecules* 32, 1701–3, 1999.

[95] Liu, J.C., Heilshorn, S.C., and Tirrell, D.A., Comparative cell response to artificial extracellular matrix proteins containing the RGD and CS5 cell-binding domains, *Biomacromolecules* 5, 497–504, 2004.

[96] Welsh, E.R. and Tirrell, D.A., Engineering the extracellular matrix: a novel approach to polymeric biomaterials. I. Control of the physical properties of artificial protein matrices designed to support adhesion of vascular endothelial cells, *Biomacromolecules* 1, 23–30, 2000.

[97] Di Zio, K. and Tirrell, D.A., Mechanical properties of artificial protein matrices engineered for control of cell and tissue behavior, *Macromolecules* 36, 1553–8, 2003.

[98] Fertala, A., Han, W.B., and Ko, F.K., Mapping critical sites in collagen II for rational design of gene-engineered proteins for cell-supporting materials, *J. Biomed. Mater. Res.* 57, 48–58, 2001.

[99] Hayashi, M., Tomita, M., and Yoshizato, K., Production of EGF-collagen chimeric protein which shows the mitogenic activity, *Biochim. Biophys. Acta* 1528, 187–95, 2001.

[100] Halstenberg, S. et al., Biologically engineered protein-graft-poly(ethylene glycol) hydrogels: a cell adhesive and plasmin-degradable biosynthetic material for tissue repair, *Biomacromolecules* 3, 710–23, 2002.

[101] Kapur, T.A. and Shoichet, M.S., Immobilized concentration gradients of nerve growth factor guide neurite outgrowth, *J. Biomed. Mater. Res. A* 68A, 235–43, 2004.

[102] Dertinger, S.K.W. et al., Gradients of substrate-bound laminin orient axonal specification of neurons, *Proc. Natl Acad. Sci. USA* 99, 12542–7, 2002.

[103] Kramer, S. et al., Preparation of protein gradients through the controlled deposition of protein-nanoparticle conjugates onto functionalized surfaces, *J. Am. Chem. Soc.* 126, 5388–95, 2004.

[104] Ravenscroft, M.S. et al., Developmental neurobiology implications from fabrication and analysis of hippocampal neuronal networks on patterned silane-modified surfaces, *J. Am. Chem. Soc.* 120, 12169–77, 1998.

[105] Chang, J.C., Brewer, G.J., and Wheller, B.C., Modulation of neural network activity by patterning, *Biosensors Bioelectron.* 16, 527–33, 2001.

[106] Mrksich, M. et al., Controlling cell attachment on contoured surfaces with self-assembled monolayers of alkanethiolates on gold, *Proc. Natl Acad. Sci. USA* 93, 10775–8, 1996.

[107] Kane, R.S. et al., Patterning proteins and cells using soft lithography, *Biomaterials* 20, 2363–76, 1999.

[108] Bernard, A. et al., Microcontact printing of proteins, *Adv. Mater.* 12, 1067–70, 2000.

[109] Patel, N. et al., Printing patterns of biospecifically adsorbed protein, *J. Biomater. Sci. Polym. Ed.* 11, 319–31, 2000.

[110] McDevitt, T.C. et al., Spatially organized layers of cardiomyocytes on biodegradable polyurethane films for myocardial repair, *J. Biomed. Mater. Res. A* 66A, 586–95, 2003.

[111] Lee, K.B. et al., Pattern generation of biological ligands on a biodegradable poly(glycolic acid) film, *Langmuir* 20, 2531–5, 2004.

[112] Schmalenberg, K.E., Buettner, H.M., and Uhrich, K.E., Microcontact printing of proteins on oxygen plasma-activated poly(methyl methacrylate), *Biomaterials* 25, 1851–7, 2004.

[113] Roda, A. et al., Protein microdeposition using a conventional ink-jet printer, *Biotechniques* 28, 492–6, 2000.

[114] Park, A., Wu, B., and Griffith, L.G., Integration of surface modification and 3D fabrication techniques to prepare patterned poly(L-lactide) substrates allowing regionally selective cell adhesion, *J. Biomater. Sci. Polym. Ed.* 9, 89–110, 1998.

[115] Roth, E.A. et al., Inkjet printing for high-throughput cell patterning, *Biomaterials* 25, 3707–15, 2004.

[116] Wilson, D.L. et al., Surface organization and nanopatterning of collagen by dip-pen nanolithography, *Proc. Natl Acad. Sci. USA* 98, 13660–4, 2001.

[117] Lee, K.B. et al., Protein nanoarrays generated by dip-pen nanolithography, *Science* 295, 1702–5, 2002.

[118] Lee, K.B., Lim, J.H., and Mirkin, C.A., Protein nanostructures formed via direct-write dip-pen nanolithography, *J. Am. Chem. Soc.* 125, 5588–9, 2003.

[119] Maheshwari, G. et al., Cell adhesion and motility depend on nanoscale RGD clustering, *J. Cell Sci.* 113, 1677–86, 2000.

[120] Koo, L.Y. et al., Co-regulation of cell adhesion by nanoscale RGD organization and mechanical stimulus, *J. Cell Sci.* 115, 1423–33, 2002.

[121] Lee, K.Y. and Mooney, D.J., Controlled growth factor delivery for tissue engineering, in *Advances in Controlled Drug Delivery: Science, Technology, and Products*, American Chemical Society, New York, 2003, pp. 73–83.

[122] Peters, M.C. et al., Release from alginate enhances the biological activity of vascular endothelial growth factor, *J. Biomater. Sci. Polym. Ed.* 9, 1267–78, 1998.

[123] Richardson, T.P. et al., Polymeric system for dual growth factor delivery, *Nat. Biotechnol.* 19, 1029–34, 2001.

[124] Yang, W.D. et al., Perlecan domain I promotes FGF-2 delivery in collagen I fibril scaffolds, *Tissue Eng.*, 11, 76–89, 2005.

[125] Wissink, M.J.B. et al., Improved endothelialization of vascular grafts by local release of growth factor from heparinized collagen matrices, *J. Control. Release* 64, 103–14, 2000.

[126] Liu, L.S. et al., Hyaluronate-heparin conjugate gels for the delivery of basic fibroblast growth factor (FGF-2), *J. Biomed. Mater. Res.* 62, 128–35, 2002.

[127] Matsuda, T. and Magoshi, T., Preparation of vinylated polysaccharides and photofabrication of tubular scaffolds for potential use in tissue engineering, *Biomacromolecules* 3, 942–50, 2002.

[128] Chinen, N. et al., Action of microparticles of heparin and alginate crosslinked gel when used as injectable artificial matrices to stablilize basic fibroblast growth factor and induce angiogenesis by controlling its release, *J. Biomed. Mater. Res. A* 67A, 61–8, 2003.

[129] Yamaguchi, N. and Kiick, K.L., Polysaccharide-poly(ethylene glycol) star copolymers as scaffolds for the production of bioactive hydrogels, *Biomacromolecules* 6, 1921–30, 2005.

[130] Sakiyama-Elbert, S.E. and Hubbell, J.A., Controlled release of nerve growth factor from a heparin-containing fibrin-based cell ingrowth matrix, *J. Control. Release* 69, 149–58, 2000b.

[131] Lee, A.C. et al., Controlled release of nerve growth factor enhances sciatic nerve regeneration, *Exp. Neurol.* 184, 295–303, 2003.

[132] Kuhl, P.R. and Griffith-Cima, L.G., Tethered epidermal growth factor as a paradigm for growth factor-induced stimulation from the solid phase, *Nat. Med.* 2, 1022–7, 1996.

[133] Mann, B.K., Schmedlen, R.H., and West, J.L., Tethered-TGF-beta increases extracellular matrix production of vascular smooth muscle cells, *Biomaterials* 22, 439–44, 2001a.

[134] Mann, B.K. et al., Smooth muscle cell growth in photopolymerized hydrogels with cell adhesive and proteolytically degradable domains: synthetic ECM analogs for tissue engineering, *Biomaterials* 22, 3045–51, 2001b.

[135] Zisch, A.H. et al., Covalently conjugated VEGF-fibrin matrices for endothelialization, *J. Control. Release* 72, 101–13, 2001.

[136] Zisch, A.H. et al., Cell-demanded release of VEGF from synthetic, biointeractive cell-ingrowth matrices for vascularized tissue growth, *FASEB J.* 17, 2260–2, 2003.

4

Mechanical Forces on Cells

Yan-Ting Shiu
University of Utah

4.1 Introduction

All cells in the body are subjected to mechanical forces that are either self-generated or originate from the environment. Depending on their location within the body, cells may be selectively exposed to various forces such as pressure, fluid shear stress, stretch, and compression. These externally applied mechanical forces play a significant role in normal tissue homeostasis and remodeling. For example, gravitational compressive forces control bone deposition, mechanical loads on skeletal muscle determine muscle mass, and blood flow-associated mechanical forces regulate the homeostasis of vascular walls [1–3]. All external forces that impinge on cells are imposed on a dynamic backdrop of various internally generated forces necessary for carrying out fundamental cellular events (e.g., cell division and migration). When cells sense a change in their net external loading, they actively alter their internal forces to counteract external forces. There is growing recognition that the balance between internally generated forces and externally applied forces is a key determinant of cell fate [1,4].

The importance of mechanical forces in tissue engineering applications is clear. The main goal of tissue engineering is the fabrication of artificial tissues for replacing damaged body structures. To produce functional tissues outside the body, it is necessary to create an *in vitro* culture environment that embodies the basic parameters of a physiological setting. Enormous strides have been made to understand the biochemical aspects of the *in vivo* microenvironment. However, the same level of understanding does not exist for the mechanical contributions. The mechanical environment of mammalian cells in the body is defined by an intricate balance between external loading and intracellular tension. The goal of this chapter is to examine the characteristics of each component within the mechanical microenvironment and highlight their implications in tissue engineering applications. The discussion will focus on anchorage-dependent, nonsensory cells.

4.2 The Role of Cytoskeletal Tension in Anchorage-Dependent Cells

For anchorage-dependent cells, the ability to apply cytoskeletal forces against the extracellular matrix (ECM) through integrin receptors is essential for cell survival and proliferation [4–6]. When a cell resides on an ECM scaffold, its contractile bundles of actin and myosin filaments (i.e., stress fibers) pull on an array of well-established connections between the cell and ECM known as focal adhesions (FAs). FAs consist of clustered integrins that span the plasma membrane, interacting with specific ECM ligands on the outside and with bundles of actin filaments via cytoskeletal-associated proteins (e.g., paxillin, α-actinin, and vinculin) on the inside [7–9]. In this way, cytoskeletal forces are transmitted via integrins to the underlying ECM, which acts as an external support for anchoring the cell and balancing the forces that maintain cell shape (Figure 4.1a) [7,8]. Thus, the adherent cell is under tension due to the ECM's resistance to deformation. The tension residing in the cytoskeleton of a resting adherent cell (often referred to as initial tension, resting tension, or prestress) is a major determinant of cell shape and functions such as proliferation, differentiation, deformability, migration, signal transduction, and ECM remodeling (see References 1,6, and 10–12 for review). Cytoskeletal tension is dynamic and can change without external stimuli during specific fundamental cellular events such as cell division and migration. Cytoskeletal tension also changes when the cell receives and responds to externally applied mechanical stimuli. Importantly, cellular responses to an external load may differ depending on the level of the initial tension (or, the "mechanical tone") in the cell [4,10–12].

The amount of the initial tension in an adherent cell is collectively controlled by its own actomyosin contractile machinery and its interaction with the ECM. Actomyosin contraction is driven by the motor protein myosin II and is triggered by the phosphorylation of the myosin light chain (MLC) in nonmuscle and smooth muscle cells [13]. The Rho GTPase (a member of the Rho family of GTP binding proteins) and its effector Rho kinase (ROCK) are important regulators of myosin activity. Blocking cell contractility by inactivating Rho or ROCK inhibits the formation of tension-dependent structures such as stress fibers and FAs [14–17]. The intracellular contractile force exerted on the ECM (also called the traction force) is essential for the assembly of fibronectin fibrils [18]. When the cell is cultured on a silicone rubber membrane, traction forces distort the substrate, forming wrinkles that can be visualized using phase-contrast microscopy [19].

Several techniques have been developed to characterize traction forces by measuring the deformation of elastic substrates (see References 13, and 20–22 for review). These methods can be combined with fluorescence imaging of FA proteins in living cells (e.g., green fluorescent protein [GFP]-tagged vinculin) to examine the relationship between the size/shape of FAs and the forces transmitted through them over time. Many studies have demonstrated conclusively that FAs transmit cytoskeletal forces in the range of several nanonewtons per square micrometers to the underlying substrate [13]. In addition, it was found that in stationary fibroblasts expressing GFP-tagged vinculin, the size of FAs is proportional to the local transmitted force (Figure 4.1b) and the orientation of FAs is parallel to the direction of the force applied at each FA (Figure 4.1c); the relaxation of the forces (induced by contractility inhibitors) and the disassembly of FAs occurred simultaneously [23]. It is important to note that the traction force is directly related, but not identical, to the intracellular contractile force, part of which could be dissipated

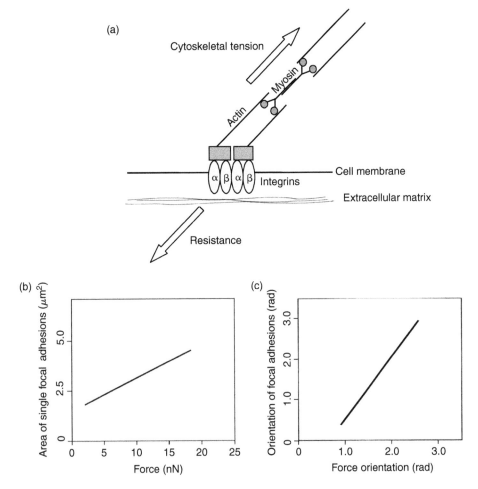

FIGURE 4.1 Forces on focal adhesions (FAs). (a) FAs are specialized sites of adhesion that form between cultured cells and a substrate. They contain clustered integrin receptors (heterodimeric transmembrane glycoproteins) whose extracellular domain binds to specific ECM ligands and intracellular domain interacts with bundles of actin filaments via cytoskeletal-associated proteins (the boxes between integrins and actin filaments). Myosin II-driven contractile forces applied to a cluster of integrins can lead to the development of tension if the underlying ECM is sufficiently rigid. (Adapted from Burridge, K. and Chrzanowska-Wodnika, M., *Annu. Rev. Cell Dev. Biol.*,12, 463, 1996; Geiger, B. and Bershadsky, A., *Cell*, 110, 139, 2002. With permission.) (b) and (c) Fibroblasts expressing GFP-tagged vinculin were cultured on silicone elastomers imprinted with micropatterns of dots. The traction force applied by the cell on the substrate was calculated to the precision of a single adhesion site based on the displacements of dots (markers) and the locations of the FAs. In stationary fibroblasts, mature FAs were elongated structures; the size of vinculin-containing FAs was proportional to the local transmitted force (b) and the main axis of this elongation is parallel to the direction of the force applied at each FA (c). (Adapted from Balaban, N.Q., et al., *Nat. Cell Biol.*, 3, 466, 2001. With permission.)

due to cell deformation or other cellular processes [23,24]. Furthermore, the magnitude and direction of traction forces vary among different regions of a cell [13,23,25,26]. Experimental evidence shows that the average traction force magnitude over the entire cell area correlates with the state of cell contraction. Therefore, quantification of traction forces provides insightful information on the state of cellular tension.

Finally, in addition to the ECM, other structures exist that may provide mechanical support for the tensed actin network in the cytoplasm. Candidates include microtubules and cell–cell contacts. Several studies have reported that microtubules are under compression in living cells, and that compressive loads could be transferred from microtubules to the ECM based on the observation of an increase in traction forces upon the disruption of microtubules [27–30]. Cells can also transmit forces to their neighbors through cell–cell junctions. Cells in a confluent monolayer generally form fewer and smaller FAs with the ECM than subconfluent cells [31,32], suggesting a decreased tension at the interface between the ECM and a cell monolayer. The interactions between groups of cells and the ECM define "the resting stress field" within a tissue and are essential for guiding tissue development, remodeling, repair, and maintaining tissue homeostasis [6,33].

4.3   The Role of ECM Scaffolds in Regulating Cellular Tension

The effect of ECM molecules on cells is primarily mediated through integrins. It is well recognized that the chemical composition of the ECM influences integrin-mediated signaling pathways. However, a number of observations have shown that adhesion to the ECM (i.e., ligand occupation) alone is not sufficient to elicit a complete integrin-mediated response unless the matrix proteins are immobilized and can physically resist tension [34,35]. *In vitro* studies have demonstrated that although many integrin signaling events can be induced in suspended cells by allowing the cells to bind to ECM-coated microbeads, these cells never enter S phase and may even undergo apoptosis [36–38]. The tension-dependent control of cell growth is attributed to ensure that only anchored cells can grow. Loss of this control (i.e., anchorage independence) is a hallmark of cancerous cells [37].

While the chemical composition of the ECM determines whether a cell can bind to it or not, once ligation is established, the development of tension is influenced by the physicality of the ECM.

4.3.1   Effects of the Compliance of ECM Scaffolds

Because the mechanical properties of the ECM determine its deformation under compressive loads, they affect the level of tension that a cell can develop; a rigid surface can resist higher tension than a softer surface and thus allow cells to carry more tension in the cytoskeleton [1,37]. Experimental observations have confirmed this notion. Wang et al. [39,40] have developed ECM-coated polyacrylamide substrates that allow the compliance to be varied while maintaining a constant chemical environment. When compared with rigid substrates, fibroblasts grown on soft substrates exert smaller traction forces, indicating a decrease in their intracellular tension (Figure 4.2a). This response to a soft substrate is accompanied by a decrease in the cell spreading area, a decrease in the rate of DNA synthesis, and an increase in the rate of apoptosis [40]. Similar phenomena have also been observed in three-dimensional cell cultures using stabilized and freely floating collagen gels (i.e., stressed vs. relaxed gels). Fibroblasts grown in stabilized collagen gels generate isometric stresses within the gels while those cultured on freely floating gels do not [41]. The implication of these results for tissue engineering applications is clear; the compliance of the scaffolds for cells is an important regulator of cell behavior through its influence on cell tension.

It is worth noting that when cells are grown on a substrate containing a stiffness gradient, cells move preferentially toward the rigid side (a phenomenon known as durotaxis) [42]. This finding indicates that cells not only respond to but also actively probe substrate flexibility, most likely by applying contractile forces to the substrate via adhesion sites and then responding to the feedback (i.e., counter-forces from the substrate) via the same sites [39,40]. Hence, cell-substrate adhesion sites may act as mechanoprobing devices, translating "external" mechanical input into intracellular signals [39,40]. Several lines of experimental evidence strongly support a pivotal role for integrin-mediated adhesions in the mechanosensing process (see References 5,8,9,13,43, and 44 for review). FAs are multimolecular complexes consisting of more than 50 different proteins that link ECM-attached integrins to the actin cytoskeleton [7,8]. Enrichment of signaling and structural proteins at FAs could facilitate intracellular signaling by bringing enzymes and their substrates into close proximity, thereby enhancing rate and opportunity of the reaction [45]. It is hypothesized that external forces received by integrins may physically change the structure of specific FA molecules and rearrange the relative positions of FA components, thereby affecting the function of

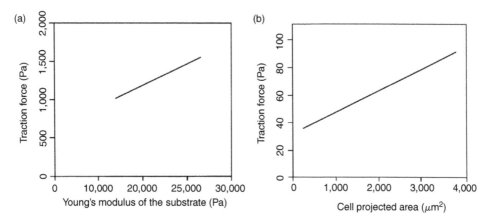

FIGURE 4.2 Effects of the physicality of ECM substrates on traction forces. Cells were cultured on polyacrylamide gels embedded with 0.2 μm-diameter fluorescent beads. Cellular traction forces were estimated based on the displacements of the beads (which reflected the deformation of the gel). The average traction force magnitude over the entire cell area was reported here. (a) Fibroblasts were cultured on collagen-coated polyacrylamide gels with different stiffness. Cells exerted larger traction forces on stiffer than softer gels. (Adapted from Wang, H.B., Dembo, M., and Wang,Y.L., *Am. J. Physiol. Cell Physiol.*, 279, C1345, 2000. With permission.) (b) Human airway smooth muscle cells were cultured on various micro-sized adhesive islands on the surface of polyacrylamide gels. The Young's modulus of the gels was $\approx$1,300 Pa. Promoting cell spreading resulted in increased traction forces. (Adapted from Wang, N. et al., *Cell Motil. Cytoskeleton*, 52, 97, 2002. With permission.)

associated signaling molecules and triggering a cascade of signaling events [8,13]. Alternatively, forces distributed along noncovalent bonds in a multimolecular FA complex may alter bond formation and dissociation kinetics, thereby altering signal transduction events [3,46].

4.3.2 Effects of the Spatial Distribution of ECM Ligands

An apparent effect of the ECM distribution on a cell is the cell shape (or projected area in a planar culture). It is well recognized that there is a close relationship between cell area and cell growth. Cells that are forced to spread over a large surface area survive better and proliferate faster than cells that are more confined [38,47]. Cell area may also affect the amount of cytoskeletal tension; a larger area underneath the cell body may resist greater levels of traction, thereby increasing isometric tension inside the cell [26,37,48]. A common and simple way to control cell area is to control the density of the ECM molecules coated on otherwise nonadhesive cell culture dishes. A higher ECM coating density allows cells to spread better and form more FAs than a lower coating density. However, because ECM coating density also affects integrin activation, it remained controversial whether the effect of increased cell area was due to the ECM density or was separate from it. This issue has been recently resolved by advances in micropatterning techniques, which allow the synthesis of surfaces on which different micron-sized islands are coated with the same ECM density and surrounded by nonadhesive regions to constrain cell spreading. When ECM islands are created on elastic substrates, traction forces can be estimated based on substrate deformation [26]. It was found that cultured cells spread to take the size and shape of the islands, and that traction forces increased as cell spreading was promoted (Figure 4.2b) [26]. These results indicate that larger cells carry greater cytoskeletal tension, and demonstrate that it is the extent of cell spreading, rather than ECM density, that influences cell tension. Furthermore, blocking the generation of actomyosin-based tension in well-spread cells (with an inhibitor that does not alter cell shape) was found to inhibit cell growth; thus, cytoskeletal tension is required for shape-dependent growth control [49].

It has been shown that myosin II-driven tension promotes cell spreading, and cell spreading stimulates MLC phosphorylation, thereby further increasing cytoskeletal tension generation [26,50]. Hence, there is an intimate crosstalk between the generation of cytoskeletal tension and the extent of cell distortion, the latter being restricted by the ECM area that is available for cell attachment. In the context of tissue engineering applications, the spatial distribution of ECM ligands may be a powerful means for controlling cell behavior. Using the micropatterned substrate mentioned above, a recent study showed that human mesenchymal stem cells (MSCs) undergo osteogenesis when allowed to spread out (cell area $\approx$ 10,000 μm^2) but become adipocytes when confined with a small area ($\approx$ 1,000 μm^2); this shape-dependent control of lineage commitment is mediated by the Rho GTPase, specifically via its effect on ROCK-mediated cytoskeletal tension [51].

Finally, the studies summarized above address the spatial effect of the ECM on the extent of cell distortion and tension in a two-dimensional cell culture system. With few exceptions, the mechanical interaction between a cell and the ECM occurs around the whole cell surface in a true physiological setting. While there is not yet a direct measurement method to correlate tractions and the spatial distribution of ECM ligands in a three-dimensional system, it is likely that the correlation is similar to that in a planar culture system.

4.3.3 Physicality of ECM Scaffolds in Tissue Engineering

The control of cell function by the ECM is a subject under vigorous investigation in the field of cell biology and has great potential in tissue engineering applications. While most studies have previously focused on the chemical composition of the ECM, it is becoming clear that the physicality of the ECM scaffold is equally important as it has a profound impact on the cellular tension. Thus, design of future artificial ECMs for tissue engineering applications should take into consideration both the chemical and physical properties of ECM scaffolds. It is important to note that the mechanical influence between cells and ECM scaffolds is mutual. Cells exert strong traction forces that deform and rearrange their surrounding matrix proteins [18,41]. Hence, traction forces may affect the structure of the scaffolds (e.g., porosity, pore size, etc.), and changes in material properties will feedback to affect cell functions. Finally, as discussed below, cells in the body exist in a dynamic mechanical environment. Cell-scaffold constructs that are formed *in situ* will inevitably experience mechanical loading. Therefore, it is important to understand how the material properties of the scaffold will be affected by mechanical forces, and how cells embedded within the scaffold will sense and respond to these physiologic loads transmitted through the scaffold.

4.4 The Role of Externally Applied Mechanical Forces in Cell Function

Throughout their lifetime, cells are exposed to various kinds of mechanical forces generated during common physiological processes. The ability to sense these external forces is not a unique property of cells in specialized sensory organs, but instead is shared by most, if not all, cell types. Furthermore, many nonsensory cells rely on appropriate mechanical inputs for regulating cell growth, function, and remodeling. For example, gravitational compressive forces control the deposition of bone, and mechanical load on skeletal muscle determines muscle mass; under conditions of diminished mechanical loading, such as prolonged bed rest or microgravity, bone and muscle mass is quickly lost [1–3]. From an engineering perspective, under normal conditions cells appear to convert "increases in the net external force acting on a tissue" into "internal tension" by increasing tissue mass or other responses [33,52]. Such a response is not always beneficial and may contribute to pathological states such as cardiac and vascular hypertrophy that is caused by chronic high blood pressure. It is generally agreed that forces beyond the physiological range, both over- and underloading, could lead to adverse consequences [11,45,52].

In order to influence cell growth and function, mechanical forces may trigger the same signaling pathways that are activated by conventional chemical stimuli (e.g., growth factors and hormones). A vast amount of data has shown that many of the biochemical events generated by cells in response to forces are similar to those that occur following recognition of chemical stimuli (see References 3,33,45, and 52–56

TABLE 4.1   A Summary of Devices Used for Mechanical Stimulation of Cells *In Vitro*

Name of device/method of developing forces	Primary force	Fluid flow in and out of device	Comments
Cone-and-plate	Shear stress	No	Possible presence of secondary flows
Parallel-plate flow chamber	Shear stress	Yes	Possible presence of not fully developed laminar flow where fluid enters device Pressure varies linearly as a function of position, though this variation is often ignored as it is very small
Uniaxial stretch	Tensional stress	No	May also generate shear stress due to motion of cells relative to fluid
Biaxial stretch	Tensional stress	No	May also generate shear stress due to motion of cells relative to fluid
Compression of gas phase by addition of an inert gas	Pressure	No	Change in concentration of dissolved gas due to nonideal behavior
Direct compression of liquid phase	Pressure	No	No gas phase present

Source: Adapted from Gooch, K.J. and Tennant, C.J., *Mechanical Forces: Their Effects on Cells and Tissues*, Springer-Verlag, New York, 1997. With permission.

for review). The exact mechanisms by which cells transduce mechanical stimuli into biochemical signaling events (mechanochemical transduction) are not yet clear and are an active area of research.

4.4.1 Devices and Methodology Used for Mechanical Stimulation of Cells *In Vitro*

In complex *in vivo* environments, it is difficult to clarify the exact effect of a specific force or to delineate the role of a specific signaling pathway in mechanotransduction processes due to the interference of myriad chemical factors and the presence of other mechanical forces. Therefore, investigations on cellular responses to mechanical stimulation have relied heavily on the use of *in vitro* systems. Table 4.1 summarizes devices that are commonly used to subject cultured cells to flow, stretch, and pressure [52]. These devices expose a large number of cells to well-defined mechanical stimuli that replicate physiological loading. Techniques for applying localized forces to individual cells have been developed. The reader is referred to other resources on the subject [2].

4.4.1.1 Shear Stress

It is possible that most cell types are exposed to fluid shearing due to interstitial flow. However, the effects of shear stress on cell behavior have been studied most extensively in vascular endothelial cells (ECs), as they are constantly exposed to blood flow. Two common flow systems used for *in vitro* studies of shear effects on cells are the cone-and-plate viscometer and the parallel-plate flow chamber. In a cone-and-plate device (Figure 4.3a), cells are placed on a plate that remains stationary. Medium is filled in the space between the plate and cone and fluid flow is achieved by rotating the cone. In a parallel-plate flow chamber (Figure 4.3b), cells are usually cultured on a glass slide that forms the floor of a rectangular flow channel. Flow is driven to pass over the cell surface by an imposed pressure gradient. Both devices generate macroscopically uniform shear stress acting on the surface of cells. The derivations for describing the velocity profile and shear stress for these two devices from hydrodynamic theory are described elsewhere [57,58]. Note that the estimated shear stress is usually nominal because only macro (or bulk) fluid dynamics is considered. At a subcellular length scale, the magnitude and gradient of shear stresses may vary significantly with the local cell surface topography [59–65]. In addition, the fluid is usually assumed to be an incompressible Newtonian fluid. The assumption of incompressibility is valid if

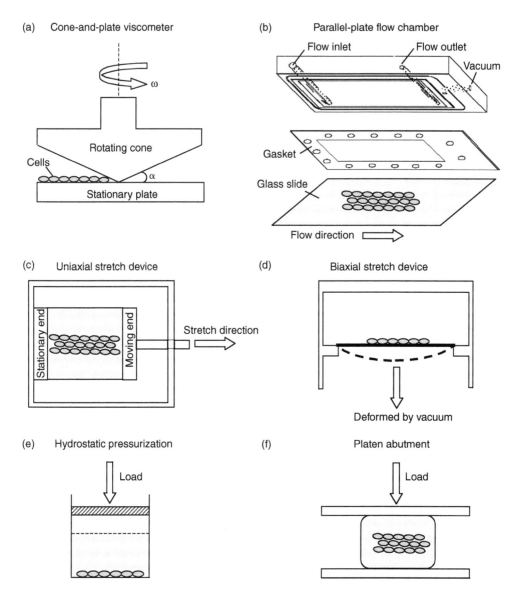

FIGURE 4.3 Schematic diagrams of devices used for mechanical stimulation of cells *in vitro*. (a) *The cone-and-plate viscometer*: cells are placed on the stationary plate and fluid flow is achieved by rotating the cone. For small cone angles ($\alpha \ll 1$) and small rotational rates ω, shear stress is uniform throughout the fluid phase between the cone and the plate. Although outside the scope of this chapter, it is important to note that this device is often used for studying the effect of bulk shear stress on suspended cells such as platelets and leukocytes. (Adapted from Tran-Son-Tay, R., *Physical Forces and the Mammalian Cell.*, Frangos, J.A. (Ed.) Academic Press, San Diego, 1993; Konstantopoulos, K., Kukreti,S., and McIntire, L.V., *Adv. Drug. Deliv. Rev.*, 33, 141, 1998. With permission.) (b) *The parallel-plate flow chamber*: the flow channel is formed by a cutout in a silicon gasket. The top plate of the flow channel is the surface of a polycarbonate base with flow inlet and outlet. The bottom plate is a glass slide where cells are cultured. The apparatus is held together by vacuum (shown here), clamps, or torqued screws. (c) *The uniaxial device*: cells are cultured on an elastic membrane, which is subjected to elongational stretch along one axis of the membrane in plane. (d) *The biaxial device*: cells are cultured on an elastic membrane, which is deformed by an applied vacuum. Solid and dashed lines are the positions of the membrane before and under deformation, respectively. (e, f) Devices for compressive loading. (Adapted from Brown, T.D., *J. Biomech.*, 33, 3, 2000. With permission.) See text for details.

water is the dominant component of the biological fluid under investigation. Other biological fluids such as blood, however, are non-Newtonian, and thereby limit the accuracy of the predictions [52,66].

4.4.1.2 Stretch

Many different cell types are constantly exposed to stretch during normal physiological processes. For example, cells on the compliant aortic wall are subjected to cyclic stretch due to pulsatile blood flow, and cells in the musculoskeletal system are subjected to cyclic stretch due to movement or gravity or both. A common method for *in vitro* studies of stretch effects on cells involves culturing cells on a flexible substrate, such as a silicone membrane, and then stretching the membrane with a controlled strain magnitude, frequency, and duration. Cells can be exposed to uniaxial or biaxial stretch (Figure 4.3c,d). This type of cell deformation device has several inherent problems. One problem is that the load delivered to cells is dependent on the state of adhesion between the cells and their substrate. Cells not fully adhered to the substrate do not experience the same amount of stretching as those fully adhered, and the subsequent data could be misleading [52]. Another problem is that fluid shear stress is often present concurrently with the imposed substrate deformation, due to coupled motion of the nutrient media [52,67]. Therefore, experimental design and data interpretation need to account for the relative contribution of stretch and shear stress.

4.4.1.3 Pressure/Compression

Hydrostatic pressurization is commonly used for investigating the effect of elevated pressure on cell functions (Figure 4.3e) [67]. This approach is simple and can deliver static and cyclic loading. However, the ensuing increase in the concentration of dissolved gases due to elevated pressure of the gas phase may affect cell functions, making it difficult to determine whether the effects observed in this system result from the pressure or from the change in gas concentration [52]. Two methods have been developed to circumvent this problem. One is to directly compress the fluid in the absence of a gas phase, and the other is to increase pressure by the addition of an inert gas (e.g., helium) while maintaining the partial pressure of biologically active gases (e.g., oxygen and carbon dioxide) [52,67].

An alternative approach to achieving compressive stress is to use direct platen abutment (Figure 4.3f) [67]. In this approach, a three-dimensional specimen (e.g., cells that are seeded in a matrix) is placed between two flat plates; the bottom plate remains stationary, whereas the top plate is pushed downward to deliver unconfined uniaxial compression (shown in Figure 4.3f; no peripheral support of the specimen so that it can freely expand laterally) or confined compression. This device can deliver static and cyclic compressive stresses. Direct platen abutment has an inherent problem that is similar to stretch devices: the loading delivered to cells is strongly dependent on the cell–matrix adhesion. Additionally, several variables arise concurrently with the compressive strain (e.g., flow, tensile strains, changes in specimen volume, etc.) [45,67]. When using them for experiments, it is important to include proper controls so that the effect due to compression can be distinguished from those due to other variables.

Finally, it should be noted that in all of the apparatuses mentioned earlier, cells usually are abruptly exposed to an imposed mechanical stimuli from static conditions. The sudden onset of mechanical inputs has been found to have a profound effect on some force-induced cell responses [68–72]. Detailed numerical analysis of the stress in the device may provide information for characterizing the time history of the mechanical stress under investigation, as well as for describing its spatial distribution and identifying the presence of unwanted inputs [52,67].

4.4.2 Responses of Cells to Mechanical Stimulation *In Vitro*

There is a rapidly growing list of reports detailing force-induced cellular responses by various experimental models as described earlier. Several excellent reviews have addressed the effects of particular forces on specific cell types *in vitro* [33,45,52–56,73–80]. The intention here is to avoid repetition with other papers and to address the general features of cellular responses to mechanical stimulation. The following discussion uses the force-induced responses of vascular ECs, vascular smooth muscle cells (SMCs), osteoblasts, and articular chondrocytes to illustrate the main points. These cell types are chosen because of the large body of data that are available.

Generally speaking, the responses of cells to a change in their external mechanical loads can be separated into rapid responses that occur within seconds to minutes and delayed responses that develop over many hours [45,75]. The rapid responses are due to the activation of a variety of intracellular signaling events, including potassium channel activation, elevated intracellular free calcium concentration, inositol trisphosphate generation, adelylate cyclase activation, G protein activation, phosphorylation of protein kinases, and, eventually activation of transcription factors [45,56,73,75]. Delayed responses usually require the modulation of gene expression [45,56,75]. Advances in the DNA microarray technology have allowed wide-scale screening of the genes affected by mechanical stimulation of cells, providing significant insights into the mechanical regulation of cell functions and homeostasis. For example, it was found that several genes related to inflammation and EC proliferation were downregulated after exposure of ECs to an arterial level of laminar shear stress (12 dyn/cm^2) for 24 h [81]. This finding suggests that long-term exposure to laminar flow may keep ECs in a relatively noninflammatory and nonproliferate state, which is consistent with the clinical observation that ECs at the straight part of arteries are mostly in a quiescent state and are relatively protected from the development of atherosclerotic plaques [81–85]. Reports concerning the genomic programming of other cell types in response to mechanical stimulation are available [86–89].

It is interesting to note that although force-induced delayed responses (and adaptive changes) vary at the cellular level among cell types, different mechanical stimuli induce several similar biochemical events in a variety of cells during rapid responses. For example, shear stress increases the concentration of intracellular free calcium and enhances the production of nitric oxide and at least one type of prostaglandin in osteoblasts, chondrocytes, SMCs, and ECs; in ECs, the above biochemical events can be induced by either shear stress or stretch [66,90–109]. The similarity of force-induced early biochemical events suggests that the way cells perceive different forces may be similar [52,73]. A cell may sense an imposed external force by detecting a deformation or resistance to deformation under stress. Different forces cause different kinds of cell deformation, thereby differentially inducing the common biochemical events and subsequently leading to diverse long-term changes. Indeed, it has been hypothesized that similar mechanisms might have evolved for cell recognition and response to external forces, since many cell types are subjected to external forces in the body, and even those thought not to be subjected to large forces *in vivo* respond to mechanical forces *in vitro* [45,52]. However, cell type-specific mechanosensing processes and the intrinsic heterogeneity among different cells cannot be excluded, but they are beyond the scope of this chapter.

4.4.3 Mechanosensing of Cultured Cells to Externally Applied Mechanical Forces

4.4.3.1 Direct Mechanosensing

Although intracellular signaling events triggered by external forces have been elucidated in many cell types, the primary mechanosensor for transducing mechanical input into biochemical signals remains elusive. It is hypothesized that forces may physically alter the molecular structure or displace the position of a sensor, thereby altering/triggering chemical signal transduction events. In conjunction, mechanosensors should be located at a site where the force acts directly or can be transmitted to efficiently [52,56,80]. Since most forces first act directly on the plasma membrane, the majority of the mechanotransensors that have been proposed are structures on the plasma membrane. Membrane structures that have been implicated in the role of mechanosensors in several cell types include ion channels, G protein-linked receptors, tyrosine kinase receptors, and integrins. Alternatively, because forces applied to the plasma membrane are transferred to the cytoskeleton, it too could act as a mechanosensor. Any reader interested in further details regarding the structures and signaling pathways associated with the proposed mechanosensors is referred to other resources [5,45,52–56,73,75,79,110–116].

Of the proposed mechanosensors, cytoskeleton and integrins have been the most extensively studied. Evidence suggests that they regulate mechanotransduction via mechanical and chemical mechanisms. The chemical nature of their functions is referred to the publications listed above. The mechanical nature of their actions is briefly considered here. As discussed in Section 4.2, an anchorage-dependent cell exists in a state of tension that is maintained by its cytoskeleton and balanced by the surrounding matrix via integrin-mediated adhesion sites (i.e., FAs). When an external force is loaded on the cell, the internal cellular tension changes to equalize the external force by actively rearranging its cytoskeleton and adhesion sites. The resulting change in cytoskeletal tension may convey a regulatory signal to the cell and subsequently alter its functional state. Dynamic changes in cytoskeleton organization, integrin-ECM binding, FAs, and traction forces may thus play a critical role in regulating mechanotransduction. These changes have been observed in EC responses to shear stress [117–121].

Finally, although each of the candidates mentioned above has been proposed to be a primary mechano-sensor, it should be noted that they have a high degree of association with one another [3,56,75,80]. Considering the vast array of signaling pathways induced by forces, it is likely that several mechanosensors are induced simultaneously. Hence, forces may be transduced to biological signals through interactions of activated mechanoreceptors (Figure 4.4). Such a "decentralized model" was first proposed by Davies to describe EC responses to mechanical stresses but is applicable to mechanically induced responses in other cell types [56,80,122]. In this model, forces acting on one region of the cell surface are also transmitted by the cytoskeleton to other locations where signaling can occur, such as FAs at the cell–ECM interface, cell–cell junctions, the nuclear membrane, and, in case of two-dimensional culture, membrane proteins at the apical surface; the cytoskeleton itself is also a mechanosensor. This model predicts mechanotrans-duction as an integrated response of multiple signaling networks that are spatially organized in the cell. There is an increasing amount of data supporting the decentralization model (see References 13,56,75, and 122–124 for review).

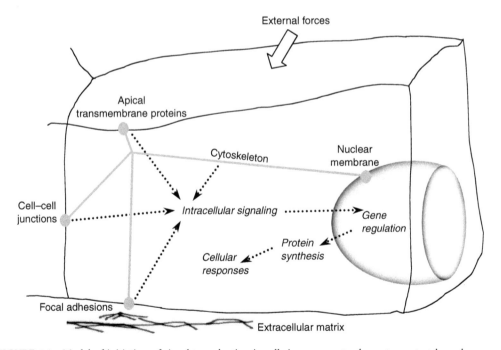

FIGURE 4.4 Model of initiation of signal transduction in cells in response to shear stress, stretch, and pressure. Forces may directly activate individual mechanosensors or may be transmitted by the cytoskeleton to intracellular locations where signaling can occur. In either case, cascades of intracellular signaling events are initiated, leading to altered gene expression and cell behavior. (Adapted from Davies, P.F., *Physiol. Rev.*, 75, 519, 1995. With permission.) See text for details.

4.4.3.2 Indirect Mechanosensing

It is important to note that mechanical forces may be accompanied by modifications of the chemical environment. For example, fluid flow influences the transport of agonists and other compounds to and from the cell surface (or, the local concentration of chemicals at the cell surface). If these agonists are degraded at the cell surface rapidly (as ATP is), flow will cause an increase in their concentration at the cell surface, and consequently modulate the cellular response [52,56,122]. Therefore, flow-induced responses may be mediated by increases in agonist concentrations (i.e., "indirect" mechanosensing), rather than physical alterations of mechanosensors (i.e., "direct" mechanosensing) by shear stress. There is *in vitro* experimental evidence that supports this mass transport hypothesis [125–132]. Indirect and direct mechanosensing are not mutually exclusive. It is likely that mechanical forces are incorporated into the biological signaling network by both physical alterations of mechanosensors and modifications of local chemical environments.

4.4.4 Applications of Externally Applied Mechanical Forces in Tissue Engineering

Conventional cell culture techniques grow cells under static conditions; in large-scale bioreactors (e.g., fluidized bed reactors, spinner flasks, rotating vessels, and perfused vessels), flow and mixing patterns are introduced merely to enhance spatially uniform cell distributions on three-dimensional scaffolds and provide efficient mass transfer to the growing tissues. As the significance of externally applied mechanical forces in maintaining appropriate cell physiology has come into the light, tissue engineers have incorporated mechanical stresses into bioreactor design and found that physiological loading has positive effects on growing cells/tissues *in vitro*. For example, increasing fluid shear forces significantly increases the mineral deposition by rat marrow stromal osteoblasts in a three-dimensional titanium fiber mesh scaffold [133,134], and the application of cyclic stretch to vascular SMCs cultured in collagen gels can help maintain the contractile phenotype of SMCs, align them in the correct physiological orientation, and improve the mechanical properties of cell–gel composites [135–137]. In the context of tissue-engineered cartilage, researchers have found that artificial cartilage grown under cyclic compressive loading has superior biochemical compositions and material properties than those grown statically [138–140]. Furthermore, cyclic compression can promote the chondrogenesis of rabbit bone-marrow MSCs by inducing the synthesis of transforming growth factor (TGF-β1), which then stimulates the MSCs to differentiate into chondrocytes [141]. These results show that appropriate mechanical stimulation may be a determining factor of tissue development *in vitro* and may improve the performance of engineered tissues in the body.

4.5 Concluding Remarks

This chapter has discussed three critical elements that define the mechanical microenvironment of a cell: self-generated forces, counter forces from the ECM, and externally applied forces in the body. Extensive work still needs to be done to create a coherent theory of mechanotransduction processes. At the cellular level, the nature of the primary mechanosensor(s) remains to be a central question. In this regard, investigations are limited by techniques that allow us to observe the force-induced changes in potential mechanosensors at a subcellular length scale and in a miniature time-frame. At the tissue level, tissue-scale responses to forces result from a dynamic and orchestrated interaction between different cell types in a three-dimensional matrix environment. In most of the *in vitro* studies described here, mechanical stimuli are imposed on monolayer cell cultures made of a single cell type. Although cultured cells sense and respond to mechanical stimulation in this setting, it is not yet clear whether the same responses or sensing mechanisms occur *in vivo*, but the wealth of *in vitro* data can guide *in vivo* experiments. Finally, in the context of tissue engineering, it is important to understand how the physicality of scaffolds affects cells, how the structure, composition, and mechanical properties of scaffolds may change as a result of traction forces from the cells and the external forces from the body, and how scaffolds affect cells to sense the external forces. These are just a few of the challenges for the future.

Acknowledgments

This work was supported by a Biomedical Engineering Research Grant from the Whitaker Foundation (RG-02-0133). The author thanks Matthew Iwamoto, Cole Quam, In Suk Joung, and Deepa Mishra for their help in manuscript preparation.

References

[1] Zhu, C., Bao, G., and Wang, N., Cell mechanics: mechanical responses, cell adhesion, and molecular deformation, *Annu. Rev. Biomed. Eng.*, 2, 189, 2000.

[2] Chen, C.S., Tan, J.L., and Tien, J., Mechanotransduction at cell–matrix and cell–cell contacts, *Annu. Rev. Biomed. Eng.*, 6, 3.1, 2004.

[3] Asthagiri, A.R. and Lauffenburger, D.A., Bioengineering models of cell signaling, *Annu. Rev. Biomed. Eng.*, 2, 31, 2000.

[4] Chicurel, M.E., Chen, C.S., and Ingber, D.E., Cellular control lies in the balance of forces, *Curr. Opin. Cell Biol.*, 10, 232, 1998.

[5] Katsumi, A. et al., Integrins in mechanotransduction, *J. Biol. Chem.*, 279, 12001, 2004.

[6] Galbraith, C.G. and Sheetz, M.P., Forces on adhesive contacts affect cell function, *Curr. Opin. Cell Biol.*, 10, 566, 1998.

[7] Burridge, K. and Chrzanowska-Wodnicka, M., Focal adhesions, contractility, and signaling, *Annu. Rev. Cell Dev. Biol.*, 12, 463, 1996.

[8] Geiger, B. and Bershadsky, A., Exploring the neighborhood: adhesion-coupled cell mechano-sensors, *Cell*, 110, 139, 2002.

[9] Geiger, B. and Bershadsky, A., Assembly and mechanosensory function of focal contacts, *Curr. Opin. Cell Biol.*, 13, 584, 2001.

[10] Ingber, D.E., Tensegrity I. Cell structure and hierarchical systems biology, *J. Cell Sci.*, 116, 1157, 2003.

[11] Ingber, D.E., Mechanobiology and diseases of mechanotransduction, *Ann. Med.*, 35, 564, 2003.

[12] Ingber, D.E., Tensegrity II. How structural networks influence cellular information processing networks, *J. Cell Sci.*, 116, 1397, 2003.

[13] Bershadsky, A.D., Balaban, N.Q., and Geiger, B., Adhesion-dependent cell mechanosensitivity, *Annu. Rev. Cell Dev. Biol.*, 19, 677, 2003.

[14] Narumiya, S., Ishizaki, T., and Watanabe, N., Rho effectors and reorganization of actin cytoskeleton, *FEBS Lett.*, 410, 68, 1997.

[15] Amano, M. et al., Phosphorylation and activation of myosin by Rho-associated kinase (Rho-kinase), *J. Biol. Chem.*, 271, 20246, 1996.

[16] Chrzanowska-Wodnicka, M. and Burridge, K., Rho-stimulated contractility drives the formation of stress fibers and focal adhesions, *J. Cell Biol.*, 133, 1403, 1996.

[17] Ridley, A.J. and Hall, A., The small GTP-binding protein rho regulates the assembly of focal adhesions and actin stress fibers in response to growth factors, *Cell*, 70, 389, 1992.

[18] Zhong, C. et al., Rho-mediated contractility exposes a cryptic site in fibronectin and induces fibronectin matrix assembly, *J. Cell Biol.*, 141, 539, 1998.

[19] Harris, A.K., Wild, P., and Stopak, D., Silicone rubber substrata: a new wrinkle in the study of cell locomotion, *Science*, 208, 177, 1980.

[20] Beningo, K.A., Lo, C.M., and Wang, Y.L., Flexible polyacrylamide substrata for the analysis of mechanical interactions at cell–substratum adhesions, *Meth. Cell Biol.*, 69, 325, 2002.

[21] Beningo, K.A. and Wang, Y.L., Flexible substrata for the detection of cellular traction forces, *Trends Cell Biol.*, 12, 79, 2002.

[22] Marganski, W.A., Dembo, M., and Wang, Y.L., Measurements of cell-generated deformations on flexible substrata using correlation-based optical flow, *Meth. Enzymol.*, 361, 197, 2003.

[23] Balaban, N.Q. et al., Force and focal adhesion assembly: a close relationship studied using elastic micropatterned substrates, *Nat. Cell Biol.*, 3, 466, 2001.

[24] Lauffenburger, D.A. and Horwitz, A.F., Cell migration: a physically integrated molecular process, *Cell*, 84, 359, 1996.

[25] Beningo, K.A. et al., Nascent focal adhesions are responsible for the generation of strong propulsive forces in migrating fibroblasts, *J. Cell Biol.*, 153, 881, 2001.

[26] Wang, N. et al., Micropatterning tractional forces in living cells, *Cell Motil. Cytoskeleton*, 52, 97, 2002.

[27] Wang, N. et al., Mechanical behavior in living cells consistent with the tensegrity model, *Proc. Natl Acad. Sci. USA*, 98, 7765, 2001.

[28] Stamenovic, D. et al., Effect of the cytoskeletal prestress on the mechanical impedance of cultured airway smooth muscle cells, *J. Appl. Physiol.*, 92, 1443, 2002.

[29] Brown, R.A. et al., Balanced mechanical forces and microtubule contribution to fibroblast contraction, *J. Cell Physiol.*, 169, 439, 1996.

[30] Dennerll, T.J. et al., Tension and compression in the cytoskeleton of PC-12 neurites. II: quantitative measurements, *J. Cell Biol.*, 107, 665, 1988.

[31] Ryan, P.L. et al., Tissue spreading on implantable substrates is a competitive outcome of cell–cell vs. cell–substratum adhesivity, *Proc. Natl Acad. Sci. USA*, 98, 4323, 2001.

[32] Nelson, C.M. et al., Vascular endothelial-cadherin regulates cytoskeletal tension, cell spreading, and focal adhesions by stimulating RhoA, *Mol. Biol. Cell*, 15, 2943, 2004.

[33] Silver, F.H. and Siperko, L.M., Mechanosensing and mechanochemical transduction: how is mechanical energy sensed and converted into chemical energy in an extracellular matrix? *Crit. Rev. Biomed. Eng.*, 31, 255, 2003.

[34] Miyamoto, S. et al., Integrin function: molecular hierarchies of cytoskeletal and signaling molecules, *J. Cell Biol.*, 131, 791, 1995.

[35] Miyamoto, S., Akiyama, S.K., and Yamada, K.M., Synergistic roles for receptor occupancy and aggregation in integrin transmembrane function, *Science*, 267, 883, 1995.

[36] Ingber, D.E., Fibronectin controls capillary endothelial cell growth by modulating cell shape, *Proc. Natl Acad. Sci. USA*, 87, 3579, 1990.

[37] Huang, S. and Ingber, D.E., The structural and mechanical complexity of cell-growth control, *Nat. Cell Biol.*, 1, E131, 1999.

[38] Chen, C.S. et al., Geometric control of cell life and death, *Science*, 276, 1425, 1997.

[39] Pelham, R.J., Jr. and Wang, Y., Cell locomotion and focal adhesions are regulated by substrate flexibility, *Proc. Natl Acad. Sci. USA*, 94, 13661, 1997.

[40] Wang, H.B., Dembo, M., and Wang, Y.L., Substrate flexibility regulates growth and apoptosis of normal but not transformed cells, *Am. J. Physiol. Cell Physiol.*, 279, C1345, 2000.

[41] Halliday, N.L. and Tomasek, J.J., Mechanical properties of the extracellular matrix influence fibronectin fibril assembly *in vitro*, *Exp. Cell Res.*, 217, 109, 1995.

[42] Lo, C.M. et al., Cell movement is guided by the rigidity of the substrate, *Biophys. J.*, 79, 144, 2000.

[43] Ingber, D.E., Mechanical signaling and the cellular response to extracellular matrix in angiogenesis and cardiovascular physiology, *Circ. Res.*, 91, 877, 2002.

[44] Geiger, B. et al., Transmembrane crosstalk between the extracellular matrix — cytoskeleton crosstalk, *Nat. Rev. Mol. Cell Biol.*, 2, 793, 2001.

[45] Millward-Sadler, S.J. and Salter, D.M., Integrin-dependent signal cascades in chondrocyte mechanotransduction, *Ann. Biomed. Eng.*, 32, 435, 2004.

[46] Bell, G.I., Models for the specific adhesion of cells to cells, *Science*, 200, 618, 1978.

[47] Ingber, D.E. and Folkman, J., Mechanochemical switching between growth and differentiation during fibroblast growth factor-stimulated angiogenesis *in vitro*: role of extracellular matrix, *J. Cell Biol.*, 109, 317, 1989.

[48] Huang, S. and Ingber, D.E., Shape-dependent control of cell growth, differentiation, and apoptosis: switching between attractors in cell regulatory networks, *Exp. Cell Res.*, 261, 91, 2000.

[49] Huang, S., Chen, C.S., and Ingber, D.E., Control of cyclin D1, p27(Kip1), and cell cycle progression in human capillary endothelial cells by cell shape and cytoskeletal tension, *Mol. Biol. Cell*, 9, 3179, 1998.

[50] Polte, T.R. et al., Extracellular matrix controls myosin light chain phosphorylation and cell contractility through modulation of cell shape and cytoskeletal prestress, *Am. J. Physiol. Cell Physiol.*, 286, C518, 2004.

[51] McBeath, R. et al., Cell shape, cytoskeletal tension, and RhoA regulate stem cell lineage commitment, *Dev. Cell*, 6, 483, 2004.

[52] Gooch, K.J. and Tennant, C.J., *Mechanical Forces: Their Effects on Cells and Tissues*, Springer-Verlag, New York, 1997.

[53] Hamill, O.P. and Martinac, B., Molecular basis of mechanotransduction in living cells, *Physiol. Rev.*, 81, 685, 2001.

[54] Apodaca, G., Modulation of membrane traffic by mechanical stimuli, *Am. J. Physiol. Renal Physiol.*, 282, F179, 2002.

[55] Shieh, A.C. and Athanasiou, K.A., Principles of cell mechanics for cartilage tissue engineering, *Ann. Biomed. Eng.*, 31, 1, 2003.

[56] Davies, P.F., Flow-mediated endothelial mechanotransduction, *Physiol. Rev.*, 75, 519, 1995.

[57] Bird, R.B., Stewart, W.E., and Lightfoot, E.N., *Transport Phenomena*, John Wiley & Sons, New York, 1960.

[58] Schlichting, H., *Boundary Layer Theory*, McGraw-Hill, New York, 1979.

[59] Liu, S.Q., Yen, M., and Fung, Y.C., On measuring the third dimension of cultured endothelial cells in shear flow, *Proc. Natl Acad. Sci. USA*, 91, 8782, 1994.

[60] Barbee, K.A., Davies, P.F., and Lal, R., Shear stress-induced reorganization of the surface topography of living endothelial cells imaged by atomic force microscopy, *Circ. Res.*, 74, 163, 1994.

[61] Barbee, K.A. et al., Subcellular distribution of shear stress at the surface of flow-aligned and nonaligned endothelial monolayers, *Am. J. Physiol.*, 268, H1765, 1995.

[62] Dewey, C.F., Jr. and DePaola, N. Exploring flow-cell interactions using computational fluid dynamics, in *Tissue Engineering*, Woo, S.-L.Y. and Seguchi, Y. (Eds.), ASME, New York, 1989.

[63] Satcher, R.L., Jr. et al., The distribution of fluid forces on model arterial endothelium using computational fluid dynamics, *J. Biomech. Eng.*, 114, 309, 1992.

[64] Sakurai, A. et al., A computational fluid mechanical study of flow over cultured endothelial cells, *Adv. Bioeng.*, 20, 299, 1991.

[65] Yamaguchi, T.H. et al., Shear stress distribution over confluently cultured endothelial cells studied by computational fluid dynamics, *Adv. Bioeng.*, 20, 167, 1993.

[66] Papadaki, M. and McIntire, L.V. Quantitative measurement of shear-stress effects on endothelial cells, in *Methods in Molecular Medicine, Vol. 18: Tissue Engineering Methods and Protocols.*, Morgan, J.R. and Yarmush, M.L. (Eds.), Humana Press, Totowa, NJ, 1998.

[67] Brown, T.D., Techniques for mechanical stimulation of cells *in vitro*: a review, *J. Biomech.*, 33, 3, 2000.

[68] Haidekker, M.A., White, C.R., and Frangos, J.A., Analysis of temporal shear stress gradients during the onset phase of flow over a backward-facing step, *J. Biomech. Eng.*, 123, 455, 2001.

[69] Frangos, J.A., Huang, T.Y., and Clark, C.B., Steady shear and step changes in shear stimulate endothelium via independent mechanisms — superposition of transient and sustained nitric oxide production, *Biochem. Biophys. Res. Commun.*, 224, 660, 1996.

[70] McKnight, N.L. and Frangos, J.A., Strain rate mechanotransduction in aligned human vascular smooth muscle cells, *Ann. Biomed. Eng.*, 31, 239, 2003.

[71] Clark, C.B., McKnight, N.L., and Frangos, J.A., Strain and strain rate activation of G proteins in human endothelial cells, *Biochem. Biophys. Res. Commun.*, 299, 258, 2002.

[72] Gudi, S.R. et al., Equibiaxial strain and strain rate stimulate early activation of G proteins in cardiac fibroblasts, *Am. J. Physiol.*, 274, C1424, 1998.

[73] Duncan, R.L. and Turner, C.H., Mechanotransduction and the functional response of bone to mechanical strain, *Calcif. Tissue Int.*, 57, 344, 1995.

[74] McIntire, L.V., 1992 ALZA Distinguished lecture: bioengineering and vascular biology, *Ann. Biomed. Eng.*, 22, 2, 1994.

[75] Papadaki, M. and Eskin, S.G., Effects of fluid shear stress on gene regulation of vascular cells, *Biotechnol. Prog.*, 13, 209, 1997.

[76] Patrick, C.W., Jr. and McIntire, L.V., Shear stress and cyclic strain modulation of gene expression in vascular endothelial cells, *Blood Purif.*, 13, 112, 1995.

[77] Skalak, T.C. and Price, R.J., The role of mechanical stresses in microvascular remodeling, *Microcirculation*, 3, 143, 1996.

[78] Liu, S.Q., Biomechanical basis of vascular tissue engineering, *Crit. Rev. Biomed. Eng.*, 27, 75, 1999.

[79] Chien, S., Li, S., and Shyy, Y.J., Effects of mechanical forces on signal transduction and gene expression in endothelial cells, *Hypertension*, 31, 162, 1998.

[80] Davies, P.F. and Tripathi, S.C., Mechanical stress mechanisms and the cell. An endothelial paradigm, *Circ. Res.*, 72, 239, 1993.

[81] Chen, B.P. et al., DNA microarray analysis of gene expression in endothelial cells in response to 24-h shear stress, *Physiol. Genom.*, 7, 55, 2001.

[82] Davies, P.F. et al., The convergence of haemodynamics, genomics, and endothelial structure in studies of the focal origin of atherosclerosis, *Biorheology*, 39, 299, 2002.

[83] Garcia-Cardena, G. et al., Biomechanical activation of vascular endothelium as a determinant of its functional phenotype, *Proc. Natl Acad. Sci. USA*, 98, 4478, 2001.

[84] McCormick, S.M. et al., DNA microarray reveals changes in gene expression of shear stressed human umbilical vein endothelial cells, *Proc. Natl Acad. Sci. USA*, 98, 8955, 2001.

[85] McCormick, S.M. et al., Microarray analysis of shear stressed endothelial cells, *Biorheology*, 40, 5, 2003.

[86] Segev, O. et al., CMF608-a novel mechanical strain-induced bone-specific protein expressed in early osteochondroprogenitor cells, *Bone*, 34, 246, 2004.

[87] Lee, R.T. et al., Mechanical strain induces specific changes in the synthesis and organization of proteoglycans by vascular smooth muscle cells, *J. Biol. Chem.*, 276, 13847, 2001.

[88] Karjalainen, H.M. et al., Gene expression profiles in chondrosarcoma cells subjected to cyclic stretching and hydrostatic pressure. A cDNA array study, *Biorheology*, 40, 93, 2003.

[89] Carinci, F. et al., Titanium–cell interaction: analysis of gene expression profiling, *J. Biomed. Mater. Res.*, 66B, 341, 2003.

[90] Schwarz, G. et al., Shear stress-induced calcium transients in endothelial cells from human umbilical cord veins, *J. Physiol.*, 458, 527, 1992.

[91] Segurola, R.J., Jr. et al., Cyclic strain is a weak inducer of prostacyclin synthase expression in bovine aortic endothelial cells, *J. Surg. Res.*, 69, 135, 1997.

[92] Rosales, O.R. et al., Exposure of endothelial cells to cyclic strain induces elevations of cytosolic Ca^2+ concentration through mobilization of intracellular and extracellular pools, *Biochem. J.*, 326, 385, 1997.

[93] Reich, K.M. and Frangos, J.A., Effect of flow on prostaglandin E2 and inositol trisphosphate levels in osteoblasts, *Am. J. Physiol.*, 261, C428, 1991.

[94] Geiger, R.V. et al., Flow-induced calcium transients in single endothelial cells: spatial and temporal analysis, *Am. J. Physiol.*, 262, C1411, 1992.

[95] Shen, J. et al., Regulation of adenine nucleotide concentration at endothelium–fluid interface by viscous shear flow, *Biophys. J.*, 64, 1323, 1993.

[96] Kuchan, M.J. and Frangos, J.A., Role of calcium and calmodulin in flow-induced nitric oxide production in endothelial cells, *Am. J. Physiol.*, 266, C628, 1994.

[97] Frangos, J.A. et al., Flow effects on prostacyclin production by cultured human endothelial cells, *Science*, 227, 1477, 1985.

[98] Donahue, S.W., Donahue, H.J., and Jacobs, C.R., Osteoblastic cells have refractory periods for fluid-flow-induced intracellular calcium oscillations for short bouts of flow and display multiple low-magnitude oscillations during long-term flow, *J. Biomech.*, 36, 35, 2003.

[99] Hung, C.T. et al., Real-time calcium response of cultured bone cells to fluid flow, *Clin. Orthop.*, 256, 1995.

[100] Lee, M.S. et al., Effects of shear stress on nitric oxide and matrix protein gene expression in human osteoarthritic chondrocytes *in vitro*, *J. Orthop. Res.*, 20, 556, 2002.

[101] Osanai, T. et al., Cross talk between prostacyclin and nitric oxide under shear in smooth muscle cell: role in monocyte adhesion, *Am. J. Physiol. Heart Circ. Physiol.*, 281, H177, 2001.

[102] Sharma, R. et al., Intracellular calcium changes in rat aortic smooth muscle cells in response to fluid flow, *Ann. Biomed. Eng.*, 30, 371, 2002.

[103] Yellowley, C.E. et al., Effects of fluid flow on intracellular calcium in bovine articular chondrocytes, *Am. J. Physiol.*, 273, C30, 1997.

[104] Smalt, R. et al., Induction of NO and prostaglandin E2 in osteoblasts by wall-shear stress but not mechanical strain, *Am. J. Physiol.*, 273, E751, 1997.

[105] Smalt, R. et al., Mechanotransduction in bone cells: induction of nitric oxide and prostaglandin synthesis by fluid shear stress, but not by mechanical strain, *Adv. Exp. Med. Biol.*, 433, 311, 1997.

[106] McAllister, T.N. and Frangos, J.A., Steady and transient fluid shear stress stimulate NO release in osteoblasts through distinct biochemical pathways, *J. Bone Miner. Res.*, 14, 930, 1999.

[107] Johnson, D.L., McAllister, T.N., and Frangos, J.A., Fluid flow stimulates rapid and continuous release of nitric oxide in osteoblasts, *Am. J. Physiol.*, 271, E205, 1996.

[108] Abulencia, J.P. et al., Shear-induced cyclooxygenase-2 via a JNK2/c-Jun-dependent pathway regulates prostaglandin receptor expression in chondrocytic cells, *J. Biol. Chem.*, 278, 28388, 2003.

[109] Alshihabi, S.N. et al., Shear stress-induced release of PGE2 and PGI2 by vascular smooth muscle cells, *Biochem. Biophys. Res. Commun.*, 224, 808, 1996.

[110] Williams, B., Mechanical influences on vascular smooth muscle cell function, *J. Hypertens.*, 16, 1921, 1998.

[111] Lee, T. and Sumpio, B.E., Cell signalling in vascular cells exposed to cyclic strain: the emerging role of protein phosphatases, *Biotechnol. Appl. Biochem.*, 39, 129, 2004.

[112] Shyy, J.Y. and Chien, S., Role of integrins in cellular responses to mechanical stress and adhesion, *Curr. Opin. Cell Biol.*, 9, 707, 1997.

[113] Shyy, J.Y. and Chien, S., Role of integrins in endothelial mechanosensing of shear stress, *Circ. Res.*, 91, 769, 2002.

[114] Davidson, R.M., Membrane stretch activates a high-conductance K+ channel in G292 osteoblastic-like cells, *J. Membr. Biol.*, 131, 81, 1993.

[115] Berk, B.C. et al., Protein kinases as mediators of fluid shear stress stimulated signal transduction in endothelial cells: a hypothesis for calcium-dependent and calcium-independent events activated by flow, *J. Biomech.*, 28, 1439, 1995.

[116] Berk, B.C. et al., Atheroprotective mechanisms activated by fluid shear stress in endothelial cells, *Drug News Perspect.*, 15, 133, 2002.

[117] Helmke, B.P. et al., Spatiotemporal analysis of flow-induced intermediate filament displacement in living endothelial cells, *Biophys. J.*, 80, 184, 2001.

[118] Jalali, S. et al., Integrin-mediated mechanotransduction requires its dynamic interaction with specific extracellular matrix (ECM) ligands, *Proc. Natl Acad. Sci. USA*, 98, 1042, 2001.

[119] Li, S. et al., The role of the dynamics of focal adhesion kinase in the mechanotaxis of endothelial cells, *Proc. Natl Acad. Sci. USA*, 99, 3546, 2002.

[120] Shiu, Y.T. et al., Rho mediates the shear-enhancement of endothelial cell migration and traction force generation, *Biophys. J.*, 86, 2558, 2004.

[121] Stamatas, G.N. and McIntire, L.V., Rapid flow-induced responses in endothelial cells, *Biotechnol. Prog.*, 17, 383, 2001.

[122] Davies, P.F., Zilberberg, J., and Helmke, B.P., Spatial microstimuli in endothelial mechanosignaling, *Circ. Res.*, 92, 359, 2003.

[123] Helmke, B.P. and Davies, P.F., The cytoskeleton under external fluid mechanical forces: hemodynamic forces acting on the endothelium, *Ann. Biomed. Eng.*, 30, 284, 2002.

[124] Davies, P.F. et al., Spatial relationships in early signaling events of flow-mediated endothelial mechanotransduction, *Annu. Rev. Physiol.*, 59, 527, 1997.

[125] Ando, J. et al., Wall shear stress rather than shear rate regulates cytoplasmic $Ca^{+}+$ responses to flow in vascular endothelial cells, *Biochem. Biophys. Res. Commun.*, 190, 716, 1993.

[126] Ando, J. et al., Flow-induced calcium transients and release of endothelium-derived relaxing factor in cultured vascular endothelial cells, *Front. Med. Biol. Eng.*, 5, 17, 1993.

[127] Koller, A., Sun, D., and Kaley, G., Role of shear stress and endothelial prostaglandins in flow- and viscosity-induced dilation of arterioles *in vitro*, *Circ. Res.*, 72, 1276, 1993.

[128] Ando, J., Komatsuda, T., and Kamiya, A., Cytoplasmic calcium response to fluid shear stress in cultured vascular endothelial cells, *In Vitro Cell Dev. Biol.*, 24, 871, 1988.

[129] Dull, R.O. and Davies, P.F., Flow modulation of agonist (ATP)-response ($Ca^{2}+$) coupling in vascular endothelial cells, *Am. J. Physiol.*, 261, H149, 1991.

[130] Dull, R.O., Tarbell, J.M., and Davies, P.F., Mechanisms of flow-mediated signal transduction in endothelial cells: kinetics of ATP surface concentrations, *J. Vasc. Res.*, 29, 410, 1992.

[131] Mo, M., Eskin, S.G., and Schilling, W.P., Flow-induced changes in $Ca^{2}+$ signaling of vascular endothelial cells: effect of shear stress and ATP, *Am. J. Physiol.*, 260, H1698, 1991.

[132] Nollert, M.U. and McIntire, L.V., Convective mass transfer effects on the intracellular calcium response of endothelial cells, *J. Biomech. Eng.*, 114, 321, 1992.

[133] Bancroft, G.N. et al., Fluid flow increases mineralized matrix deposition in 3D perfusion culture of marrow stromal osteoblasts in a dose-dependent manner, *Proc. Natl Acad. Sci. USA*, 99, 12600, 2002.

[134] Sikavitsas, V.I. et al., Mineralized matrix deposition by marrow stromal osteoblasts in 3D perfusion culture increases with increasing fluid shear forces, *Proc. Natl Acad. Sci. USA*, 100, 14683, 2003.

[135] Nerem, R.M., Role of mechanics in vascular tissue engineering, *Biorheology*, 40, 281, 2003.

[136] Kanda, K. and Matsuda, T., Mechanical stress-induced orientation and ultrastructural change of smooth muscle cells cultured in three-dimensional collagen lattices, *Cell Transplant.*, 3, 481, 1994.

[137] Seliktar, D., Nerem, R.M., and Galis, Z.S., Mechanical strain-stimulated remodeling of tissue-engineered blood vessel constructs, *Tissue Eng.*, 9, 657, 2003.

[138] Davisson, T. et al., Static and dynamic compression modulate matrix metabolism in tissue engineered cartilage, *J. Orthop. Res.*, 20, 842, 2002.

[139] Park, S., Hung, C.T., and Ateshian, G.A., Mechanical response of bovine articular cartilage under dynamic unconfined compression loading at physiological stress levels, *Osteoarthr. Cartil.*, 12, 65, 2004.

[140] Hung, C.T. et al., A paradigm for functional tissue engineering of articular cartilage via applied physiologic deformational loading, *Ann. Biomed. Eng.*, 32, 35, 2004.

[141] Huang, C.Y. et al., Effects of cyclic compressive loading on chondrogenesis of rabbit bone-marrow derived mesenchymal stem cells, *Stem Cells*, 22, 313, 2004.

[142] Tran-Son-Tay, R. Techniques for studying the effects of physical forces on mammalian cells and measuring cell mechanical properties, in *Physical Forces and the Mammalian Cell*, Frangos, J.A. (Ed.), Academic Press, San Diego, 1993.

[143] Konstantopoulos, K., Kukreti, S., and McIntire, L.V., Biomechanics of cell interactions in shear fields, *Adv. Drug Deliv. Rev.*, 33, 141, 1998.

5

Cell Adhesion

Aaron S. Goldstein
Virginia Polytechnic Institute and State University

5.1 Introduction

Adhesion plays a critical role in the normal function of mammalian cells by regulating proliferation, differentiation, and phenotypic behavior. Although the molecular mechanism of adhesion involves several families of transmembrane receptors as well as numerous structural and regulatory proteins, they can be divided phenomenologically into two groups: those that mediate adhesive contacts to neighboring cells, and those that mediate adhesive contacts with structural extracellular matrix (ECM) proteins. In development, specific adhesive interactions between adjacent cells lead to the assembly of three-dimensional tissue structures, while cell adhesion to ECM proteins (e.g., collagens, elastin, and laminin) establishes cellular orientation and spatial organization of cells into tissues and organs. Consequently, both types are necessary for proper function of mammalian cells and tissues.

Initial efforts to design biomaterials — either to replace damaged tissues, or as a temporary scaffolds for the manufacture of engineered tissues and organs — have focused on achieving robust mechanical interactions between the biomaterial and adjacent cells (i.e., integration). However, evidence is emerging that cell/biomaterial interactions can affect cell/cell interactions, and consequently impact tissue development and function. Therefore, successful outcomes of tissue engineering efforts may require the development of advanced materials that can facilitate both cell/cell and cell/biomaterial adhesion.

This chapter discusses the phenomenon of cell adhesion both from a tissue perspective and from a biomaterial perspective. Section 5.2 describes the cell biology of adhesion receptors and their relevance to particular tissue functions. Section 5.3 addresses cell adhesion to biomaterials with specific focus on how the chemistry, biochemistry, and topography of biomaterial interfaces affect cell adhesion. Section 5.4 describes quantitative techniques to measure cell adhesion to biomaterials. Finally, Section 5.5 describes how cell spreading, proliferation, and migration depend on cell adhesion to biomaterials.

5.2   Adhesion Receptors in Tissue Structures

Mammalian cells establish and maintain adhesive contacts with extracellular structures and adjacent cells through several types of adhesion receptors that are displayed on the cell surface. The three main families of these receptors are integrins, cadherins, and members of the immunoglobulin superfamily [1–3]. Although they are involved in different adhesion-related tasks, they share several common features. First, when associated with extracellular ligands (e.g., ECM proteins, receptors on adjacent cells) they frequently assemble into clusters. Second, these transmembrane receptors are often linked to cytoskeletal structures (e.g., actin filaments, microtubules), and thus both anchor the cellular skeleton to extracellular structures and mediate the transmission of mechanical forces. Third, regulatory proteins (e.g., protein kinases) are associated with receptor clusters, and are responsible for controlling cluster stability and initiating signaling events that regulate cell proliferation and function.

5.2.1   Integrins

Integrins are a family of receptors that mediate cell adhesion to ECM proteins. They are found as heterodimers, consisting of α and β subunits, and to date at least 8 β and 18 α subunits and at least 24 different dimer combinations have been reported [4]. A short list of known integrin heterodimers and their respective ligands (Table 5.1) reveals that individual integrins can have multiple ligands (e.g., $\alpha_3\beta_1$ binds collagen, fibronectin, and laminin) and multiple integrins may bind the same ligand (e.g., laminin binds integrins $\alpha_1\beta_1$, $\alpha_2\beta_1$, $\alpha_3\beta_1$, $\alpha_6\beta_1$, $\alpha_7\beta_1$) [1,5]. In addition, most cell types express multiple integrin

TABLE 5.1   Integrin Dimers and Their Known Ligands

β subunit	α subunit	Ligand/counterreceptor	Binding site
β_1	α_1	LM (COL)	
	α_2	COL (LM)	DGEA (in COL I)
	α_3	FN, LM, COL	RGD?
	α_4	V-CAM (FN-alt)	REDV (in FN-alt)
	α_5	FN	RGD + PHSRN
	α_6	LM	RGD
	α_7	LM	
	α_8	?	
	α_V	FN	RGD
β_2	α_L	ICAM-1,2	
	α_M	FB, C3bi	
	α_X	FB, C3bi	GPRP
β_3	α_{IIb}	FB, FN, VN, VWF	RGD (+PHSRN in FN)
	α_V	VN, FB, VWF, TSP	RGD
β_5	α_V	VN, FN	RGD
β_6	α_V	FN	RGD

Source: Adapted from Albelda, S.M. and Buck, C.A., *FASEB J.* 4, 2868, 1990. With permission and Hynes, R.O. and Lander, A.D., *Cell* 68, 303, 1992. With permission.

Notes: LM laminin, COL collagens, FN fibronectin, VN vitronectin, FB fibrinogen, VWF von Willebrand Factor, TSP thrombospondin, C3bi breakdown product of third component of complement, FN-alt alternatively spliced fibronectin. DGEA, RGD, REDV, PHSRN, and GPRP are amino acid sequences.

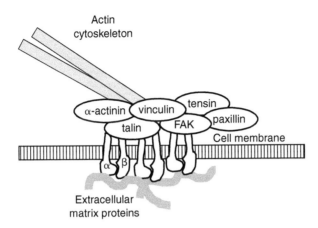

FIGURE 5.1   Illustration of a focal adhesion complex. A cluster of transmembrane integrin receptor heterodimers is shown interacting with extracellular matrix proteins and cytoplasmic proteins vinculin, α-actinin, talin, and focal adhesion kinase (FAK). This complex interacts with the actin cytoskeleton. (Adapted from Petit, V. and Thiery, J.-P., *Biol. Cell.* 92, 477, 2000. With permission; Yamada, K.M. and Miyamoto, S., *Curr. Opin. Cell Biol.* 7, 681, 1995. With permission.)

subunits. For example, the pluripotent mesenchymal progenitor bone marrow stromal cells have been shown to express integrin subunits α_1, α_5, α_6, and β_1 [6], while aortic smooth muscle cells express α_2, α_5, α_v, β_1, and β_5 [7]. It should be noted that integrins do not exclusively mediate cell/ECM adhesion: $\alpha_4\beta_1$ and $\alpha_M\beta_2$, presented on activated leukocytes and neutrophils mediate cell/cell adhesion by binding adhesion receptors of the immunoglobulin superfamily (e.g., vascular cell adhesion molecules [VCAM], intracellular adhesion molecules [ICAMs]) on activated endothelial cells [1,8].

Binding of integrin receptors to their extracellular ligands initiates a process of receptor clustering and the assembly of focal adhesion complexes [4,9–11] (Figure 5.1). Conserved sequences in the cytoplasmic tail of β subunit are responsible for recruiting integrins into clusters (except with β_4 and β_5 which do not aggregate), while the cytoplasmic tail of the α subunit plays a regulatory role to ensure aggregation of only ligand-bound receptors. The cytoplasmic proteins that associate with integrins in focal adhesion complexes can be divided into the categories of structural and regulatory proteins. The structural proteins — which include talin, α-actinin, filamin, and vinculin — stabilize the receptor aggregates and mechanically link the integrins to actin filaments, thus anchoring the cell cytoskeleton to extracellular structures. In particular, talin and α-actinin possess binding domains for β integrins, while vinculin has bonding domains for talin and actin. Regulatory proteins include paxillin, focal adhesion kinase (FAK), Src, and members of the Rho and Ras families [4]. FAK and Src are kinases that regulate focal adhesion complex assembly by phosphorylating structural proteins. In contrast, Rho and Ras are GTPases involved in initiating intracellular signaling. Rho is thought to act through a number of secondary messengers to stimulate focal adhesion assembly, actin stress fiber formation, and contraction, while Ras is thought to regulate integrin activation [12]. In addition, there is evidence that Ras activates mitogen-activated protein (MAP) kinase signaling cascades that are involved in regulating cell proliferation, apoptosis, migration, and development [10].

In addition to initiating signaling cascades, integrins also can stimulate signaling events initiated by cytokine receptors and growth factor receptors [12]. For example, interleukin (IL)-1 activation of MAP kinases requires integrin-mediated cell adhesion [13]. In addition, insulin and platelet-derived growth factor (PDGF) receptors — when bound to their respective ligands — physically interact with the $\alpha_v\beta_3$ integrin, and this association enhances MAP kinase signaling and cell proliferation [14]. Integrin binding also can modulate effects of other adhesion receptors. For example, the association of $\alpha_4\beta_1$ integrins with fibronectin has been shown to block the upregulation of metalloproteinase by $\alpha_5\beta_1$ binding to fibronectin

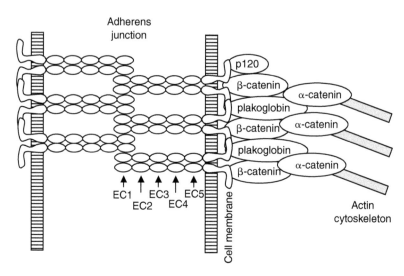

FIGURE 5.2   Illustration of cadherin receptors associated with actin cytoskeleton. Cadherin dimers are shown in a linear zipper structure with antiparallel binding of the EC1 domains. Cytoplasmic tails of cadherins interact with actin filaments through structural proteins β-catenin, p120[cas] catenin, plakoglobin, and α-catenin.

[15]. Similarly, integrin-meditated adhesion to the ECM downregulates cadherin-mediated intercellular adhesion and cell clustering [16]. This last example illustrates a competition between integrin-mediated adhesion to the ECM and cadherin-mediated intercellular adhesion [17].

5.2.2   Cadherins

Cadherins are a family of calcium-dependent transmembrane glycoproteins that mediate cell/cell adhesion. Classical cadherins (e.g., E-, N-, P-, and VE-cad) possess five extracellular (EC) cadherin repeats, assemble into dimers, and mediate primarily homotypic binding between adjacent cells (Figure 5.2) [18–20]. Although the outermost repeat (EC1) is thought to be responsible for intercellular binding, it has been shown that cadherins can interact in at least three antiparallel alignments [21]. In addition, crystallography has revealed that cadherins may assemble into a linear zipper structure [22]. The cytoplasmic tails of classical cadherins interact with structural proteins β-catenin, p120[cas] catenin, plakoglobin, and through them α-catenin and the actin cytoskeleton. The cytoplasmic tail of endothelial cell-specific receptor VE-cad also can interact vimentin intermediate filaments [18,20]. The protein p120[cas] — a target of the Src kinase — is thought to be involved in regulating lateral clustering of cadherins, which has been shown to increase adhesive strength [23]. Thus, Src may be involved in regulating robust cadherin contacts.

Although cadherins do not exclusively cluster, the formation of adherens junctions is important for several tissue functions [18]. In epithelial tissues cadherins mediate zonula adherens junctions, which partition the cell membrane into apical and basolateral surfaces. In neuronal cells they mechanically stabilize synaptic junctions by mediating adhesion between the presynaptic and postsynaptic cells. In cardiac myocytes they establish intercalated discs, which both stabilize gap junctional coupling and transmit contractile forces.

In addition to classical cadherins, several other cadherin types have been identified [18,20]. Among these, desmocollins and desmogleins are expressed in cells that experience mechanical stress (e.g., epidermis, myocardium) and are the transmembrane proteins present in desmosomes [18,20]. Like classical cadherins, they have five extracellular cadherin repeats (except desmoglein 1, which has four repeats), but they form desmoglein/desmocollin heterodimers and are thought to mediate intercellular adhesion through heterotypic desmoglein/desmocollin interactions between adjacent cells. The cytoplasmic tails of these receptors interact, through structural proteins desmoglobin and desmoplakin, with the keratin intermediate filaments. To date three desmocollins and desmogleins have been identified. Desmoglein 2 and desmocollin 2 are expressed in all epithelial tissues that form desmosomes, while desmocollin 1 and desmoglein 1 are expressed predominantly in epidermal tissue.

In addition to regulating organized tissue functions, E-, N-, and VE-cadherin-mediated intercellular contacts have been shown to suppress cell proliferation [24–26]. This phenomenon is thought to be a consequence of cadherin-induced alterations in the cytoskeleton structure [27], and to contribute to the abrogation of cell proliferation of confluent tissue cultures.

5.2.3 Immunoglobulins

Members of the diverse immunoglobulin superfamily are characterized by the immunoglobulin (Ig) fold motif, a 70–110 amino acid structure consisting of two sheets of antiparallel β-strands stabilized by a disulfide bridge and numerous hydrophobic side chains [28,29]. Because the amine and carboxy termini extend from opposite ends of this structure, members of the immunoglobulin family frequently possess strings of Ig folds. Immunoglobulin proteins are most noted for their role in antigen presentation and recognition functions of the immune system, but at least ten members of the superfamily are involved in cell/cell adhesion. ICAM-1, ICAM-2, and VCAM, for example, are presented on the surface of activated endothelial cells and mediate heterotypic binding to integrins on activated leukocytes and neutrophils (Table 5.1) [8]. Neural cell adhesion molecules (NCAMs) and L1 — present in neurons and glia — primarily mediated homotypic binding and are involved in axon guidance, neuronal migration, neurite outgrowth, synapse formation, and maintenance of the integrity of myelinated fibers [29,30]. Finally, junctional adhesion molecules (JAMs) stabilize tight junctions, which regulate paracellular ion transport across epithelial and endothelial layers [31].

5.3   Cell Adhesion to Biomaterials

The efficacy of biomaterials — either as replacements for damaged tissues, or as a temporary scaffolds for the manufacture of engineered tissues and organs — relies on the ability of the material surface to regulate cell adhesion, and strategies to engineer this cell/biomaterial interaction have evolved over the past few decades through three basic generations [32]. The first generation is exemplified by bulk materials such as titanium, Dacron, and ultrahigh molecular weight polyethylene, which are durable and exhibit good mechanical properties, but are nondegradable and essentially bioinert (i.e., they elicit minimal effects *in vivo*). Second generation biomaterials are those that — when brought into contact with bodily fluids and cells — are modified chemically or biochemically to achieve more favorable biologic properties. Biochemical modifications include the adsorption of ECM proteins to the biomaterial surface in conformations that enhance cell and tissue behavior (e.g., adhesion [33], phenotype [34]), while examples of chemical modifications include the ion-exchange reactions of bioactive glasses, dissolution and reorganization of amorphous calcium phosphate, and hydrolytic degradation of polyurethanes and polyesters. Finally, third generation biomaterials are those that incorporate biomimetic moieties (e.g., peptides sequences) that are designed to interact with target proteins (e.g., adhesion receptors, growth factor receptors, matrix metalloproteinases) and elicit specific cell responses. Thus, third generation biomaterials are often referred to as bioactive materials because they act on cells and tissues.

Cell adhesion to biomaterials may be characterized in terms of specific and nonspecific interactions. Specific interactions entail cell receptor recognition and binding to proteins or biomimetic moieties on the biomaterial surface, whereas nonspecific interactions encompass noncovalent attractive forces (e.g., electrostatic interactions, hydrogen bonds, van der Waals forces) between the cell and biomaterial that are not associated with a specific receptor. In general, both specific and nonspecific interactions contribute to cell adhesion to biomaterials, and can be regulated through design of substratum chemistry and biochemistry. The following subsections address (a) the role of substratum chemistry on nonspecific adhesion, the roles of (b) proteins and (c) biomimetic moieties on specific adhesion, and (d) the effect of substratum topography on cell adhesion.

5.3.1 The Role of Interfacial Chemistry in Cell Adhesion

Although cells express receptors for ECM proteins and adjacent cells, they have been shown to adhere through nonspecific interactions to a variety of biomaterials. For example, a comprehensive study by Schakenraad et al. [35] demonstrated cell adhesion and spreading on 13 commercially available polymers. In particular, this study showed that cell spreading was markedly greater on materials with surface free energies greater than 50 erg/sec (hydrophilic surfaces) relative to low energy (hydrophobic) surfaces [35]. With the development of self-assembling methods (e.g., alkanethiolates on gold and alkylsilanes on glass), chemically well-defined model surfaces have been constructed and used to characterize cell adhesion [36–41]. These model studies indicate that terminal amine (NH_2) and carboxylic acid (COOH) groups produce superior adhesion relative to hydrophobic methyl-(CH_3)-terminated surfaces [36,38–40]. In contrast, surfaces presenting poly(ethylene glycol) subunits have been shown to inhibit cell adhesion [42]. In addition to organic polymers, alloys [43], and ceramics [44] have also been studied for their capacity to support cell adhesion. These materials have proven more difficult to characterize as the chemistries are complex, and the grain size appears to play an important role [44,45]. Nevertheless, a consistent theme that emerges from these studies is that cell adhesion increases with the material's capacity to promote adsorption of ECM proteins [33,38,40,45].

5.3.2 The Role of Interfacial Biochemistry in Cell Adhesion

A simple method to enhance cell adhesion to first and second generation biomaterials is to adsorb ECM proteins to the biomaterial surface from protein solutions (e.g., collagens [46], fibronectin [46–48]), serum, or serum-containing culture medium. Although albumin is the predominant protein in serum, vitronectin and fibronectin have been shown to deposit readily from serum onto both organic [49] and ceramic [50] biomaterial surfaces. The choice of ECM protein can be important to regulating cell behavior; cell adhesion and function have been shown to vary with the adhesive protein [46,49]. In addition, the chemistry of the biomaterial surface — to which the ECM proteins are adsorbed — can also affect cell adhesion and function [33,34,51] by altering protein orientation and inducing conformational changes (i.e., denaturation) [51–56] that undermine receptor binding. Conformational changes have been inferred from an increase in the ratio of β-turn to β-sheet structures, and indicate a higher degree of protein denaturation on hydrophobic methyl-terminated surfaces than on more hydrophilic carboxyl-terminated surfaces [54]. In addition, protein denaturation is inversely related to the rate of protein adsorption [57], and decreases as the protein concentration in solution is increased.

5.3.3 Biomimetic Approaches to Regulate Cell Adhesion

An attractive alternative to immobilization of entire ECM proteins at biomaterial interfaces is the use of short peptide sequences that are specifically recognized by integrin adhesion receptors [58–60]. Usually these peptides are immobilized in a random, spatially uniform manner across the biomaterial surface using bioconjugate techniques [61]. However, enhanced cell adhesion and migration has been reported when peptides are immobilized to surfaces in clusters that permit integrin clustering [62]. Particular peptide sequences that have been studied include argenine–glysine–aspartic acid (RGD), tyrosine–isoleucine–glycine–serine–aspartic acid (YIGSR), isoleucine–lysine–valine–alanine–valine (IKVAV), and arginine-glutamic acid–aspartic acid–valine (REDV). The RGD sequence — found in collagens, fibronectin, vitronectin, fibrinogen, von Willebrand factor, osteopontin, and bone sialo-protein [63–65] — is recognized by a large number of integrin receptors (Table 5.1), and consequently has been exploited extensively to promote integrin-mediated adhesion to biomaterials. Interestingly, the sequence proline–histidine–serine–arginine–asparagine (PHSRN) on fibronectin has been shown to act synergistically with RGD to enhance integrin-mediated adhesion and cell signaling via integrins $\alpha_5\beta_1$ and

$\alpha_{11b}\beta_3$ [66,67]. REDV is an integrin binding sequence found in the alternatively spliced type III connecting segment region of fibronectin and is recognized by $\alpha_4\beta_1$ integrin (Table 5.1) [68]. It has been studied for its ability to support endothelial cell adhesion for development of engineered blood vessels [69–71]. Lastly, YIGSR [72] and IKVAV [73] are integrin binding sequences found in laminin. Because laminin is found in peripheral nerve tissue and the basal lamina of endothelial tissues, these peptide sequences hold promise for mediating adhesion and orientation of engineered epithelial [74] and neural tissues [75,76].

5.3.4 The Role of Interfacial Topography in Cell Adhesion

Cell adhesion to biomaterials is also influenced by topographical features, which restrict sites for cell adhesion. Studies performed nearly a century ago using spiderwebs revealed that adherent cells would spread with an orientation that could be dictated by the substratum topography [77]. This phenomenon, referred to as contact guidance, holds promise as a means to enhance engineered tissue function by inducing cell alignment [78,79]. Using microfabrication techniques (e.g., micromachining, photolithography) model substrates have been constructed to characterize the effect of topographical surface features on cell adhesion and function. For example, surfaces textured with parallel grooves can restrict focal adhesions to the raised edges of the substrate [80], while cytoskeletal elements — including microtubules, vimentin intermediate filaments, and actin filaments — are oriented parallel to the grooves [81]. Further, these effects are limited to feature sizes of 0.3 [81] to 2 μm [80]. When spacing between parallel grooves becomes too large, cells are able to attach to the depressed regions of the substratum and orientation is lost. Substrates exhibiting random topographies (e.g., acid-etched, sand blasted) also affect the behavior of adherent cells. Such substrates have been shown to enhance cell adhesion [82], synthesis of ECM proteins [82,83], and phenotypic behavior [83]. However, the effect of roughness on cell proliferation remains ambiguous: increasing roughness has been shown to decrease proliferation of osteoblasts [83], but may increase proliferation of endothelial and smooth muscle cells [84,85].

5.4 Measurement of Cell Adhesion to Biomaterials

Over the past two decades several devices have been developed to measure and characterize the adhesive interaction of cells with biomaterials, particles, and other cells. These devices share the common experimental strategy that nonadherent spherical cells are allowed to establish adhesive contacts to the test material under quiescent conditions and then are subjected to a well-defined distractive force. From the examination of large numbers of cells over a range of distractive forces, a probability distribution for cell adhesion as a function of distractive force can be constructed (Figure 5.3). From this distribution, an adhesion characteristic (e.g., τ_{50}, the shear stress necessary to detach 50% of the cells [47,86–88] may be determined. The primary differences between the various devices for measuring cell adhesion is the type of distractive force that is applied (e.g., membrane tension, buoyancy, and hydrodynamic shear stress) and the direction of the force relative to the plane of cell/surface contacts.

5.4.1 Micropipette Aspiration

Micropipette aspiration is a sensitive technique capable of measuring the strength of individual receptor/ligand adhesion complexes [89,90]. The advantage of this technique is its ability to probe cell contacts and adhesive structures with a range of forces (10^{-3} to 10^2 pN) that are capable of deforming the cell membrane (<0.1 pN) and extracting lipid-anchored receptors (~20 pN), but not breaking actin filaments (>100 pN) [90].

In this approach a cell is held at the end of micropipette by a small amount of suction. The cell is then brought into contact with a test surface, adhesive contacts are allowed to form, and then suction is applied to break this contact (Figure 5.4). The underlying principle is that suction, ΔP, used to draw a spherical cell of radius R_0 into the end of micropipette tip of radius R_p creates a membrane tension, T_m, at the

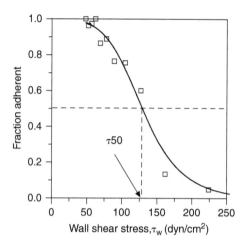

FIGURE 5.3   Probability distribution for cell adhesion. NIH 3T3 fibroblasts were detached from glass using a radial-flow chamber. Squares correspond to the fraction of cells that resisted detachment as a function of the applied shear stress. The curve is a best fit of the integral of a normal distribution function to the data. The characteristic measure of cell adhesion, τ_{50}, is 129 dyn/cm^2 for this test.

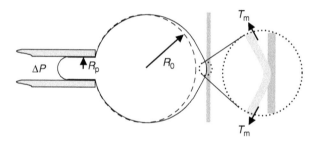

FIGURE 5.4   Micropipet aspiration. As suction, ΔP, is applied the cell is drawn into the pipet and membrane tension, T_m, is exerted at the point of cell/material contact. (Adapted from Evans, E., Berk, D., and Leung, A., *Biophys. J.* 59, 838, 1991. With permission.)

receptor/ligand contact [89]

$$T_m \approx \Delta P R_p / 2[1 - R_p/R_0] \qquad (5.1)$$

The tension disrupts adhesive contacts in a manner that may be modeled as either a peeling process [91] or the failure of discrete cross-bridges [92], and can be used to predict receptor/ligand dissociation kinetics under mechanical strains [93].

5.4.2   Centrifugation

The centrifugation approach exploits the difference in density between cells, $\rho_c = 1.07$ [62], and culture medium, $\rho_m = 1.00$, to exert a supergravitational body force,

$$f = (\rho_c - \rho_m)V_c a \qquad (5.2)$$

on the cell. Here V_c and a are the volume of the cell and the acceleration generated by the centrifuge, respectively. Distractive vectors both tangential [94] and normal [95,96] to the adhesive substrate have been used experimentally, although computational modeling predicts that tangential forces can be 20 to 56 times more disruptive to cell adhesion [97].

The experimental approach of Chu et al. [95] involves seeding cells onto substrates within 96 well plates, placing the plates into the 96 well-plate buckets of a table top centrifuge, and then applying accelerations of $a = 50, 100, 200$, or 300 g. Images are then collected to calculate the percent of cells that remain attached as a function of applied force, f. In contrast, the approach of Thoumine et al. [94] involves seeding cells onto plastic slides that are then fit into centrifuge tubes and placed in a swinging bucket centrifuge. Cells are then exposed to accelerations of 9,000 to 70,000 g for 5 to 60 min. Thoumine et al. also showed that the distractive force can be increased further by allowing the cells to phagocytose 1.5 μm gelatin-coated glass beads. As with Chu et al., images are analyzed to determine the percent of adherent cells as a function of time and distractive force.

5.4.3 Laminar Flow Chambers

Laminar flow devices are commonly used to characterize cell adhesion, and include rotating disc, parallel-plate (and variants), and radial-flow devices. These devices all employ tangential fluid flow to exert hydrodynamic drag and torque on adherent cells [98]. Among the different devices, the parallel-plate flow chamber (Figure 5.5a) is described most extensively, and has been used primarily to characterize cell and tissue phenomena under a well-defined hydrodynamic shear stress, τ. Here, τ can be predicted from fluid mechanics,

$$\tau = \frac{6\mu Q}{wh^2} \tag{5.3}$$

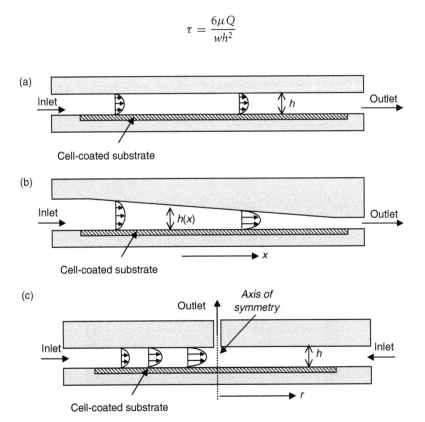

FIGURE 5.5 Various laminar flow chambers geometries. (a) Parallel-plate flow chamber uses linear laminar flow through a rectangular conduit with constant width, b, and height, h, where $b \gg h$. (b) Variable height laminar flow chamber uses flow through a tapered conduit, $h(x)$, of constant width to produce a flow profile that depends on position, x. (c) Radial-flow chamber uses cylindrical geometry to produce a flow profile that depends on position, r.

where Q is volumetric flow rate, h and w are the height and width of the flow path, respectively, and μ is the fluid viscosity [99]. The advantage of this device is that it permits nondestructive *in situ* visualization of cell attachment [100], detachment [101], and rolling processes [102]. However, because a single shear stress is generated across the cell-seeded substrate, multiple substrates must be tested at different shear stresses in order to construct shear-dependent patterns of cell adhesion [100,101].

Minor modifications have been made to the parallel-plate flow chamber geometry in order to produce a spatially dependent range of shear stresses across a single substrate. For example, a variable gap height device has been constructed in which h is a linear function of the distance from the inlet, x (Figure 5.5b) [103]. In another device the width of the flow path varies inversely with distance from the inlet, $w \propto 1/x$, to produce a Hele–Shaw flow pattern and a hydrodynamic shear stress that increases linearly with distance, $\tau \propto x$ [104,105].

An alternative device, the radial-flow chamber uses axisymmetric flow between parallel surfaces to generate a spatially dependent range of hydrodynamic shear stresses [65,106] (Figure 5.5c). Here, the magnitude of the shear stress is predicted by the equation

$$\tau = \left| \frac{3\mu Q}{\pi h^2 r} - \frac{3\rho Q^2}{70\pi^2 h r^3} \right| \tag{5.4}$$

where r is radial distance from the axis of symmetry, and ρ is fluid density. The first term of this equation is the creeping flow solution (analogous to Equation 5.3) and the second term is a first-order correction to account for inertial effects at high Reynolds numbers or low radial positions. It is interesting to note that the contribution of the second term depends on the direction of flow (i.e., the sign of Q) [106].

5.4.4 Rotating Disc

The rotating disc differs from laminar flow chambers in that a motor is used to put the cell-seeded substrate in motion, rather than a pressure gradient to put the fluid in motion [47,107,108] (Figure 5.6). An advantage of this device is that generates a shear stress, τ, that varies linearly with radial position, r, by the equation [107]

$$\tau = 0.800r\sqrt{\rho\mu\omega^3} \tag{5.5}$$

Here, ρ, μ, and ω are fluid density, fluid viscosity, and angular velocity, respectively. However, a disadvantage is that cell adhesion cannot be examined *in situ*. Instead, the cell-seeded substrate must be removed from the apparatus in order to collect images of adherent cells.

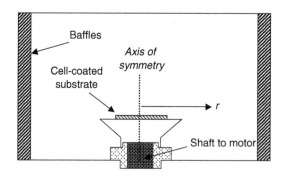

FIGURE 5.6 Rotating disc apparatus. The rapid rotation of the cell-seeded substrate generates a hydrodynamic shear that is proportional to radial position, r. (Adapted from García, A.J., Huber, F., and Boettiger, D., *J. Biol. Chem.* 273, 10988, 1998. With permission.)

5.4.5 Interpretation of Adhesion Data

Proper interpretation of adhesion data remains a technical challenge because quantitative measures are sensitive to cell history. As spherical, nonadherent cells are brought into contact with biomaterial interfaces, receptor-mediated contact initiates, followed by the clustering of receptors, cell spreading, and reorganization of the actin cytoskeleton. This dynamic process involves changes in the number and organization of adhesive contacts, cell shape, and the viscoelastic properties of the cell body — all of which influence the measured strength of adhesion. For example, quantitative measurements have demonstrated a linear dependence of the strength of cell adhesion on the density of adhesive ligands in the absence of cooperative binding [47] that is consistent with kinetic theory [109], but the strength of adhesion is increased by receptor activation [107] and modulated by receptor clustering [62]. In addition, cell adhesion has been shown to increase with the duration of cell/surface incubation [101], but it is difficult to deconvolve the individual effects of cell/surface contact area, focal adhesion organization, and filamentous actin content. Further, detachment assays that use hydrodynamic shear are sensitive to the hydrodynamic profile of the cell [110] and its deformability [111]. Consequently cells that are more spread or deform rapidly at the onset of shearing flow are more likely to resist detachment [48].

5.5 Effect of Biomaterial on Physiological Behavior

Establishment of adhesive cellular contacts with a biomaterial surface is critical for anchorage-dependent cell functions, such as spreading, proliferation, and migration. Studies of cell spreading indicate systematic increases in the rate and the extent of spreading with increasing density of adhesive ligands [112] or peptides [113]. In addition, microtubule and filamentous actin content increase during the initial spreading process [112]. Cell shape is also affected by the extent of cell spreading. For example, the aspect ratio of fibroblastic cells — which typically exhibit a spindle-shaped morphology — is maximal at an intermediate projected cell area [113]. Cell viability and proliferation also depend on projected cell area. Using microcontact printing to restrict the extent of cell spreading, Chen et al. [114] showed that when cells are able to adhere but not spread, apoptosis (programmed cell death) is significant. However, as the area for cell spreading is increased apoptosis declines and cell proliferation occurs.

Migration is another adhesion-dependent phenomenon, and requires polarized cells to execute coordinated steps that include pseudopodal extension, firm adhesion, contraction, and uropodal release [115]. Because both the pseudopodal attachment and uropodal release are adhesive processes, the capacity of the substratum to mediate cell adhesion directly affects the speed of cell migration. Manipulation of the strength of cell adhesion by changing the density, and type of ECM proteins has been shown to affect migration, with maximal speed occurring at an intermediate substratum adhesiveness [46]. In addition, migration rate can be modulated by depositing adhesive moieties as clusters [116] or by introducing topographical features [117].

5.6 Summary

In mammalian tissues and organs cell/cell and cell/ECM interactions are mediated through several families of adhesion receptors, which include integrins, cadherins, and members of the immunoglobulin superfamily. Biological studies have revealed that these adhesive contacts mechanically link the cell cytoskeleton to extracellular structures, initiate intercellular signaling cascades, and regulate cell viability, proliferation, organization of tissue structures, and cell function. Existing biomaterials — intended to replace tissue function or to serve as temporary scaffolds for engineered tissues — are capable of supporting cell adhesion, viability, and proliferation through a combination of nonspecific and integrin-mediated interactions. The current challenge is to develop bioactive materials that can act through different types of cell receptors to regulate cell and tissue behavior.

References

[1] Albelda, S.M. and Buck, C.A., Integrins and other cell adhesion molecules, *FASEB J.* 4, 2868, 1990.

[2] Hynes, R.O. and Lander, A.D., Contact and adhesive specificities in the associations, migrations, and targeting of cells and axons, *Cell* 68, 303, 1992.

[3] Juliano, R.L., Signal transduction by cell adhesion receptors and the cytoskeleton: functions of integrins, cadherins, selectins, and immunoglobulin-superfamily members, *Ann. Rev. Pharmacol. Toxicol.* 42, 283, 2002.

[4] Petit, V. and Thiery, J.-P., Focal adhesions: structure and dynamics, *Biol. Cell.* 92, 477, 2000.

[5] Hynes, R.O., Integrins: versatility, modulation, and signaling in cell adhesion, *Cell* 69, 11, 1992.

[6] ter Brugge, P.T., Torensma, R., de Ruijter, J.E. et al., Modulation of integrin expression on rat bone marrow cells by substrates with different characteristics, *Tissue Eng.* 8, 615, 2002.

[7] Nikolovski, J. and Mooney, D.J., Smooth muscle cell adhesion to tissue engineering scaffolds, *Biomaterials* 21, 2025, 2000.

[8] Springer, T.A., Adhesion receptors of the immune system, *Nature* 346, 425, 1990.

[9] Sastry, S.K. and Burridge, K., Focal adhesions: a nexus for intercellular signalling and cytoskeletal dynamics, *Exp. Cell Res.* 261, 25, 2000.

[10] Giancotti, F.G. and Ruoslahti, E., Integrin signaling, *Science* 285, 1028, 1999.

[11] Yamada, K.M. and Miyamoto, S., Integrin transmembrane signaling and cytoskeletal control, *Curr. Opin. Cell Biol.* 7, 681, 1995.

[12] Howe, A., Alpin, A.E., Alahari, S.K. et al., Integrin signaling and cell growth control, *Curr. Opin. Cell Biol.* 10, 220, 1998.

[13] Lo, Y.Y.C., Luo, L., McCulloch, C.A. et al., Requirements of focal adhesions and calcium fluxes for interleukin-1-induced Erk kinase activation and c-fos expression in fibroblasts, *J. Biol. Chem.* 273, 7059, 1998.

[14] Schneller, M., Vuori, K., and Ruoslahti, E., $\alpha v\beta 3$ integrin associates with activated insulin and PDGFβ receptors and potentiates the biological activity of PDGF, *EMBO J.* 16, 5600, 1997.

[15] Huhtala, P., Humphries, M.J., McCarthy, J.B. et al., Cooperative signaling by $\alpha_5\beta_1$ and $\alpha_4\beta_1$ integrins regulates metalloproteinase gene expression in fibroblasts adhering to fibronectin, *J. Cell Biol.* 129, 867, 1995.

[16] Monier-Gavelle, F. and Duband, J.L., Cross talk between adhesion molecules: control of N-cadherin activity by intercellular signals elicited by beta1 and beta3 integrins in migrating neural crest cells, *J. Cell Biol.* 137, 1663, 1997.

[17] Lauffenburger, D.A. and Griffith, L.G., Who's got pull around here? Cell organization in development and tissue engineering, *Proc. Natl. Acad. Sci. USA* 98, 4282, 2001.

[18] Wheelock, M.J. and Johnson, K.R., Cadherins as modulators of cellular phenotype, *Ann. Rev. Cell Dev. Biol.* 19, 207, 2003.

[19] Yap, A.S., Brieher, W.M., and Gumbiner, B.M., Molecular and functional analysis of cadherin-based adherins junctions, *Ann. Rev. Cell Dev. Biol.* 13, 119, 1997.

[20] Angst, B.D., Marcozzi, C., and Magee, A.I., The cadherin superfamily: diversity in form and function, *J. Cell Sci.* 114, 629, 2001.

[21] Zhu, B., Chappuis-Flament, S., Wong, E. et al., Functional analysis of the structural basis of homophilic cadherin adhesion, *Biophys. J.* 84, 4033, 2003.

[22] Shapiro, L., Fannon, A.M., Kwong, P.D. et al., Structural basis of cell–cell adhesion by cadherins, *Nature* 374, 327, 1995.

[23] Yap, A.S., Brieher, W.M., Pruschy, M. et al., Lateral clustering of the adhesive ectodomain: a fundamental determinant of cadherin function, *Curr. Biol.* 7, 308, 1997.

[24] St Croix, B., Sheehan, C., Rak, J.W. et al., E-cadherin-dependent growth suppression is mediated by the cyclin-dependent kinase inhibitor p27(KIP1), *J. Cell Biol.* 142, 557, 1998.

[25] Levenberg, S., Yarden, A., Kam, Z. et al., p27 is involved in N-cadherin-mediated contact inhibition of cell growth and S-phase entry, *Oncogene* 18, 869, 1999.

[26] Caveda, L., Martin-Padura, I., Navarro, P. et al., Inhibition of cultured cell growth by vascular endothelial cadherin (cadherin-5/VE-cadherin), *J. Clin. Invest.* 98, 886, 1996.

[27] Nelson, C.M. and Chen, C.S., VE-cadherin simultaneously stimulates and inhibits cell proliferation by altering cytoskeletal structure and tension, *J. Cell Sci.* 116, 3571, 2003.

[28] Stryer, L., *Biochemistry*, 3rd ed., W.H. Freeman and Company, New York, 1988, p. 901.

[29] Walsh, F.S. and Doherty, P., Neural cell adhesion molecules of the immunoglobulin superfamily: role in axon growth and guidance, *Ann. Rev. Cell Dev. Biol.* 13, 425, 1997.

[30] Panicker, A.K., Buhusi, M., Thelen, K. et al., Cellular signalling mechanism of neural cell adhesion molecules, *Front. Biosci.* 8, 900, 2003.

[31] González-Mariscal, L., Betanzos, A., Nava, P. et al., Tight junction proteins, *Prog. Biophys. Mol. Biol.* 81, 1, 2003.

[32] Hench, L.L. and Polak, J.M., Third-generation biomedical materials, *Science* 295, 1014, 2002.

[33] García A.J., Ducheyne, P., and Boettiger, D., Effect of surface reaction stage on fibronectin-mediated adhesion of ostoblast-like cells to bioactive glass, *J. Biomed. Mater. Res.* 40, 48, 1998.

[34] Stephansson, S.N., Byers, B.A., and García A.J., Enhanced expression of the osteoblastic phenotype on substrates that modulate fibronectin conformation and integrin receptor binding, *Biomaterials* 23, 2527, 2002.

[35] Schakenraad, J.M., Busscher, H.J., Wildevuur, C.R.H. et al., The influence of substratum free energy on growth and spreading of human fibroblasts in the presence and absence of serum proteins, *J. Biomed. Mater. Res.* 20, 773, 1986.

[36] Tidwell, C.D., Ertel, S.I., Ratner, B.D. et al., Endothelial cell growth and protein adsorption on terminally functionalized, self-assembled monolayers of alkanethiolates on gold, *Langmuir* 13, 3404, 1997.

[37] Tegoulia, V.A. and Cooper, S.L., Leukocyte adhesion on model surfaces under flow: effects of surface chemistry, protein adsorption, and shear rate, *J. Biomed. Mater. Res.* 50, 291, 2000.

[38] McClary, K.B., Ugarova, T., and Grainger, D.W., Modulating fibroblast adhesion, spreading, and proliferation using self-assembled monolayer films of alkylthiolates on gold, *J. Biomed. Mater. Res.* 50, 428, 2000.

[39] Scotchford, C.A., Cooper, E., Leggett, G.J. et al., Growth of human osteoblast-like cells on alkane-thiol on gold self-assembled monolayers: the effect of surface chemistry, *J. Biomed. Mater. Res.* 41, 431, 1998.

[40] Faucheux, N., Schweiss, R., Lützow K. et al., Self-assembled monolaers with different terminating groups as model substrates for cell adhesion studies, *Biomaterials* 25, 2721, 2004.

[41] Webb, K., Hlady, V., and Tresco, P.A., Relationships among cell attachment, spreading, cytoskeletal organization, and migration rate for anchorage-dependent cells on model surfaces, *J. Biomed. Mater. Res.* 49, 362, 2000.

[42] López G.P., Albers, M.W., Schreiber, S.L. et al., Convenient methods for patterning the adhesion of mammalian cells to surfaces using self-assembled monolayers of alkanethiolates on gold, *J. Am. Chem. Soc.* 115, 5877, 1993.

[43] Puleo, D.A., Holleran, L.A., Doremus, R.H. et al., Osteoblast responses to orthopedic implant materials *in vitro*, *J. Biomed. Mater. Res.* 25, 711, 1991.

[44] Webster, T.J., Siegel, R.W., and Bizios, R., Osteoblast adhesion on nanophase ceramics, *Biomaterials* 20, 1221, 1999.

[45] Webster, T.J., Schadler, L.S., Siegel, R.W. et al., Mechanisms of enhanced osteoblast adhesion on nanophase alumina involve vitronectin, *Tissue Eng.* 7, 291, 2001.

[46] DiMilla, P.A., Stone, J.A., Quinn, J.A. et al., Maximal migration of human smooth muscle cells on fibronectin and type IV collagen occurs at an intermediate attachment strength, *J. Cell Biol.* 122, 729, 1993.

[47] García A.J., Ducheyne, P., and Boettiger, D., Cell adhesion strength increases linearly with adsorbed fibronectin surface density, *Tissue Eng.* 3, 197, 1997.

[48] Goldstein, A.S. and DiMilla, P.A., Effect of adsorbed fibronectin concentration on cell adhesion and deformation under shear on hydrophobic surfaces, *J. Biomed. Mater. Res.* 59, 665, 2001.

[49] Thomas, C.H., McFarland, C.D., Jenkins, M.L. et al., The role of vitronectin in the attachment and spatial distribution of bone-derived cells on materials with patterned surface chemistry, *J. Biomed. Mater. Res.* 37, 81, 1997.

[50] Kilpadi, K.L., Chang, P.L., and Bellis, S.L., Hydroxylapatite binds more serum proteins, purified integrins, and osteoblast precursor cells than titanium or steel, *J. Biomed. Mater. Res.* 57, 258, 2001.

[51] Iuliano, D.J., Saavedra, S.S., and Truskey, G.A., Effect of the conformation and orientation of adsorbed fibronectin on endothelial cell spreading and the strength of adhesion, *J. Biomed. Mater. Res.* 27, 1103, 1993.

[52] Yu, J.-L., Johansson, S., and Ljungh, Å., Fibronectin exposes different domains after adsorption to a heparinized and an unheparinized poly(vinyl chloride) surface, *Biomaterials* 18, 421, 1997.

[53] Giroux, T.A. and Cooper, S.L., FTIR/ATR studies of human fibronectin adsorption onto plasma derivatized polystyrene, *J. Colloid Interface Sci.* 139, 352, 1990.

[54] Cheng, S.-S., Chittur, K.K., Sukenik, C.N. et al., The conformation of fibronectin on self-assembled monolayers with different surface composition: an FTIR/ATR study, *J. Colloid Interface Sci.* 162, 135, 1994.

[55] Narasimhan, C. and Lai, C.-S., Conformational changes of plasma fibronectin detected upon adsorption to solid substrates, *Biochemistry* 28, 5041, 1989.

[56] Lhoest, J.-B., Detrait, E., van den Bosch de Aguilar, P. et al., Fibronectin adsorption, conformation, and orientation on polystyrene substrates studied by radiolabeling, XPS, and ToF SIMS, *J. Biomed. Mater. Res.* 41, 95, 1998.

[57] Wertz, C.F. and Santore, M.M., Effect of surface hydrophobicity on adsorption and relaxation kinetics of albumin and fibrinogen: single-species and competitive behavior, *Langmuir* 17, 3006, 2001.

[58] Mann, B.K., Gobin, A.S., Tsai, A.T. et al., Smooth muscle cell growth in photopolymerized hydrogels with cell adhesive and proteolytically degradable domains: synthetic ECM analogs for tissue engineering, *Biomaterials* 22, 3045, 2001.

[59] Massia, S.P. and Stark, J., Immobilized RGD peptides on surface-grafted dextran promote biospecific cell attachment, *J. Biomed. Mater. Res.* 56, 390, 2001.

[60] Alsberg, E., Anderson, K.W., Albeiruti, A. et al., Cell-interactive alginate hydrogels for bone tissue engineering, *J. Dental Res.* 80, 2025, 2001.

[61] Hermanson, G.T., Ed., *Bioconjugate Techniques*, Academic Press, San Diego, 1996.

[62] Koo, L.Y., Irvine, D.J., Mayes, A.M. et al., Co-regulation of cell adhesion by nanoscale RGD organization and mechanical stimulus, *J. Cell Sci.* 115, 1423, 2002.

[63] Pierschbacher, M.D. and Ruoslahti, E., Cell attachment activity of fibronectin can be duplicated by small synthetic fragments of the molecule, *Nature* 309, 30, 1984.

[64] Ruoslahti, E. and Pierschbacher, M.D., New perspectives in cell adhesion: RGD and integrins, *Science* 238, 491, 1987.

[65] Rezania, A., Thomas, C.H., Branger, A.B. et al., The detachment strength and morphology of bone cells contacting materials modified with a peptide sequence found within bone sialoprotein, *J. Biomed. Mater. Res.* 37, 9, 1997.

[66] Bowditch, R.D., Hariharan, M., Tominna, E.F. et al., Identification of a novel integrin binding site in fibronectin: differential utilization by $\beta3$ integrins, *J. Biol. Chem.* 269, 10856, 1994.

[67] Aota, S., Nomizu, M., and Yamada, K.M., The short amino acid sequence Pro–His–Ser–Arg–Asn in human fibronectin enhances cell-adhesive function, *J. Biol. Chem.* 269, 24756, 1994.

[68] Mould, A.P., Komoriya, A., Yamada, K.M. et al., The CS5 peptide is a second site in the IIICS region of fibronectin recognized by the integrin alpha 4 beta 1. Inhibition of alpha 4 beta 1 function by RGD peptide homologues, *J. Biol. Chem.* 266, 3579, 1991.

[69] Hodde, J., Record, R., Tullius, R. et al., Fibronectin peptides mediate HMEC adhesion to porcine-derived extracellular matrix, *Biomaterials* 23, 1841, 2002.

[70] Heilshorn, S.C., DiZio, K.A., Welsh, E.R. et al., Endothelial cell adhesion to the fibronectin CS5 domain in artificial extracellular matrix proteins, *Biomaterials* 24, 4245, 2003.

[71] Massia, S.P. and Hubbell, J.A., Vascular endothelial cell adhesion and spreading promoted by the peptide REDV of the IIICS region of plasma fibronectin is mediated by integrin a4b1, *J. Biol. Chem.* 267, 14019, 1992.

[72] Iwamoto, Y., Robey, F.A., Graf, J. et al., YIGSR, a synthetic laminin pentapeptide, inhibits experimental metastasis formation, *Science* 238, 1132, 1987.

[73] Tashiro, K., Sephel, G.C., Weeks, B. et al., A synthetic peptide containing the IKVAV sequence from the A chain of laminin mediates cell attachment, migration, and neurite outgrowth, *J. Biol. Chem.* 264, 16174, 1989.

[74] Li, F., Carlsson, D., Lohmann, C. et al., Cellular and nerve regeneration within a biosynthetic extracellular matrix for corneal transplantation, *Proc. Natl. Acad. Sci. USA* 100, 15346, 2003.

[75] Massia, S.P., Holecko, M.M., and Ehteshami, G.R., *In vitro* assessment of bioactive coatings for neural implant applications, *J. Biomed. Mater. Res.* 68A, 177, 2004.

[76] Shaw, D. and Shoichet, M.S., Toward spinal cord injury repair strategies: peptide surface modification of expanded poly(tetrafluoroethylene) fibers for guided neurite outgrowth *in vitro*, *J. Craniofac. Surg.* 14, 308, 2003.

[77] Harrison, R.G., The cultivation of tissues in extraneous media as a method of morphogenetic study, *Anat. Rec.* 6, 181, 1912.

[78] Singhvi, R., Stephanopoulos, G., and Wang, D.I.C., Review: effects of substratum morphology on cell physiology, *Biotechnol. Bioeng.* 43, 764, 1994.

[79] Flemming, R.G., Murphy, C.J., Abrams, G.A. et al., Effects of synthetic micro- and nano-structured surfaces on cell behavior, *Biomaterials* 20, 573, 1999.

[80] den Braber, E.T., Jansen, H.V., de Boer, H.J. et al., Scanning electron microscopic, transmission electron microscopic, and confocal laser scanning microscopic observation of fibroblasts cultured on microgrooved surfaces of bulk titanium substrata, *J. Biomed. Mater. Res.* 40, 425, 1998.

[81] Manwaring, M.E., Walsh, J.F., and Tresco, P.A., Contact guidance induced orientation of extracellular matrix, *Biomaterials* 25, 3631, 2004.

[82] Lampin, M., Warocquier-Clérout R., Legris, C. et al., Correlation between substratum roughness and wettability, cell adhesion, and cell migration, *J. Biomed. Mater. Res.* 36, 99, 1997.

[83] Martin, J.Y., Schwartz, Z., Hummert, T.W. et al., Effect of titatium surface roughness on proliferation, differentiation, and protein synthesis of human osteoblast-like cells (MG63), *J. Biomed. Mater. Res.* 29, 389, 1995.

[84] Miller, D.C., Thapa, A., Haberstroh, K.M. et al., Endothelial and vascular smooth muscle cell function on poly(lactic-co-glycolic acid) with nano-structured surface features, *Biomaterials* 25, 53, 2004.

[85] Chung, T.W., Liu, D.Z., Wang, S.Y. et al., Enhancement of the growth of human endothelial cells by surface roughness at nanometer scale, *Biomaterials* 24, 4655, 2003.

[86] Goldstein, A.S. and DiMilla, P.A., Application of fluid mechanic and kinetic models to characterize mammalian cell detachment in a radial-flow chamber, *Biotechnol. Bioeng.* 55, 616, 1997.

[87] Cozens-Roberts, C., Quinn, J.A., and Lauffenburger, D.A., Receptor-mediated adhesion phenomena: model studies with the radial-flow detachment assay, *Biophys. J.* 58, 107, 1990.

[88] Truskey, G.A. and Proulx, T.L., Relationship between 3T3 cell spreading and the strength of adhesion on glass and silane surfaces, *Biomaterials* 14, 243, 1993.

[89] Evans, E., Berk, D., and Leung, A., Detachment of agglutinin-bonded red blood cells, I. Forces to rupture molecular-point attachments, *Biophys. J.* 59, 838, 1991.

[90] Evans, E., Ritchie, K., and Merkel, R., Sensitive force technique to probe molecular adhesion and structural linkages at biological interfaces, *Biophys. J.* 68, 2580, 1995.

[91] Evans, E., Detailed mechanics of membrane–membrane adhesion and separation. I. Continuum of molecular cross-bridges, *Biophys. J.* 48, 175, 1985.

[92] Evans, E., Detailed mechanics of membrane-membrane adhesion and separation. II. Discrete, kinetically trapped molecular cross-bridges, *Biophys. J.* 48, 185, 1985.

[93] Chelsa, S.E., Selvaraj, P., and Zhu, C., Measuring two-dimensional receptor–ligand binding kinetics by micropipette, *Biophys. J.* 75, 1553, 1998.

[94] Thoumine, O., Ott, A., and Louvard, D., Critical centrifugal forces induce adhesion rupture or structural reorganization in cultured cells, *Cell Motil. Cytoskeleton* 33, 276, 1996.

[95] Chu, L., Tempelman, L.A., Miller, C. et al., Centrifugation assay of IgE-mediated rat basophilic leukemia cell adhesion to antigen-coated polyacrilimide gels, *AIChE J.* 40, 692, 1994.

[96] Piper, J.W., Swerlick, R.A., and Zhu, C., Determining force dependence of two-dimensional receptor-ligand binding affinity by centrifugation, *Biophys. J.* 74, 492, 1998.

[97] Chang, K.-C. and Hammer, D., Influence of direction and type of applied force on the detachment of macromolecularly-bound particles from surfaces, *Langmuir* 12, 2271, 1996.

[98] Hammer, D.A. and Lauffenburger, D.A., A dynamical model for receptor-mediated cell adhesion to surfaces, *Biophys. J.* 52, 475, 1987.

[99] Frangos, J.A., McIntire, L.V., and Eskin, S.G., Shear stress induced stimulation of mammalian cell metabolism, *Biotechnol. Bioeng.* 32, 1053, 1988.

[100] Tempelman, L.A. and Hammer, D.A., Receptor-mediated binding of IgE-sensitized rat basophilic leukemia cells to antigen-coated substrates under hydrodynamic flow, *Biophys. J.* 66, 1231, 1994.

[101] Truskey, G.A. and Pirone, J.S., The effect of fluid shear sress upon cell adhesion to fibronectin-treated surfaces, *J. Biomed. Mater. Res.* 24, 1333, 1990.

[102] Rinker, K.D., Prabhakar, V., and Truskey, G.A., Effect of contact time and force on monocyte adhesion to vascular endothelium, *Biophys. J.* 80, 1722, 2001.

[103] Xiao, Y. and Truskey, G.A., Effect of receptor–ligand affinity on the strength of endothelial cell adhesion, *Biophys. J.* 71, 2869, 1996.

[104] Usami, S., Chen, H.-H., Zhao, Y. et al., Design and construction of a linear shear stress flow chamber, *Ann. Biomed. Eng.* 21, 77, 1993.

[105] Powers, M.J., Rodriguez, R.E., and Griffith, L.G., Cell–substratum adhesive strength as a determinant of hepatocyte aggregate morphology, *Biotechnol. Bioeng.* 53, 415, 1997.

[106] Goldstein, A.S. and DiMilla, P.A., Comparison of converging and diverging radial flow for measuring cell adhesion, *AIChE J.* 44, 465, 1998.

[107] García, A.J., Huber, F., and Boettiger, D., Force required to break $\alpha_5\beta_1$ integrin–fibronectin bonds in intact adherent cells is sensitive to integrin activation state, *J. Biol. Chem.* 273, 10988, 1998.

[108] Reutelingsperger, C.P.M., van Gool, R.J.G., Heinjen, V. et al., The rotating disc as a device to study the adhesive properties of endothelial cells under differential shear stresses, *J. Mater. Sci., Mater. Med.* 5, 361, 1994.

[109] Cozens-Roberts, C., Lauffenburger, D.A., and Quinn, J.A., Receptor-mediated cell attachment and detachment kinetics: I. Probabilistic model and analysis, *Biophys. J.* 59, 841, 1990.

[110] Olivier, L.A. and Truskey, G.A., A numerical analysis of forces exerted by laminar flow on spreading cells in a parallel plate flow chamber assay, *Biotechnol. Bioeng.* 42, 963, 1993.

[111] Dong, C., Struble, E.J., and Lipowsky, H.H., Mechanics of leukocyte deformation and adhesion to the endothelium in shear flow, *Ann. Biomed. Eng.* 27, 298, 1999.

[112] Mooney, D.J., Langer, R., and Ingber, D.E., Cytoskeletal filament assembly and the control of cell spreading and function by extracellular matrix, *J. Cell Sci.* 108, 2311, 1995.

[113] Neff, J.A., Tresco, P.A., and Caldwell, K.D., Surface modification for controlled studies of cell–ligand interactions, *Biomaterials* 20, 2377, 1999.

[114] Chen, C.S., Mrksich, M., Huang, S. et al., Geometric control of cell life and death, *Science* 276, 1425, 1997.

[115] Lauffenburger, D.A. and Horwitz, A.F., Cell migration: a physically integrated molecular process, *Cell* 84, 359, 1996.

[116] Maheshwari, G., Brown, G., Lauffenburger, D.A. et al., Cell adhesion and motility depend on nanoscale RGD clustering, *J. Cell Sci.* 113, 1677, 2000.

[117] Tan, J., Shen, H., and Saltzman, W.M., Micron-scale positioning of features influences the rate of polymorphonuclear leukocyte migration, *Biophys. J.* 81, 2569, 2001.

6
Cell Migration

Gang Cheng
Kyriacos Zygourakis
Rice University

6.1 Introduction

Cell migration is an essential component of normal development, inflammation, tissue repair, angiogenesis, and tumor invasion. After conception, selected cells of the developing mammalian zygote invade the uterine wall to establish the placenta, while the intricately programmed migration of other cells within the embryo shapes the complex form of the emerging organism [1,2]. The nervous system is another example of large-scale cell migration during fetal development. The growth of axons and dendrites is preceded by a phase of cell migration in which immature neurons (or neuroblasts) move from their birthplace to settle in some other location in order to make the right connections [3]. Certain kinds of white blood cells are able to migrate through the walls of blood vessels and into the surrounding tissues, actively seeking and engulfing sources of decay [4]. Migrating fibroblastic and epithelial cells heal wounds, and osteoclasts and osteoblasts are in constant movement as they remodel bone [5–7]. Tumor cell motility is also required for invasion and metastasis. The crawling malignant tumor cells that invade and disrupt tissue architecture account as much or more for the lethality of cancer as does uncontrolled growth [8].

Cell migration also plays a key role in determining the structure and growth rate of bioartificial tissues built on scaffolds made from suitable biomaterials [9]. In recent years, a lot of attention has been focused on the development of *biomimetic* materials capable of promoting cell functions, including migration, by biomolecular recognition [10]. Such recognition can be achieved by surface or bulk modification of the material with bioactive molecules such as extracellular matrix (ECM) proteins or short peptide sequences that can induce specific interactions with cell receptors. In order to design biomimetic scaffolds with optimal properties for each application, however, we must thoroughly understand not only the mechanism of cell migration, but also the many factors that modulate this important process. We must also develop assays to accurately characterize cell movement on various biomaterials and, ultimately, build theoretical models that can quantitatively predict the effect of system parameters (scaffold properties, nutrient or growth factor concentrations, pH, etc.) on tissue development. Because tissues are highly *heterogeneous systems* exhibiting complicated cell population dynamics, such theoretical models must be able to accurately describe cell–cell and cell–substrate interactions.

This review attempts to address some of these issues for anchorage-dependent mammalian cells. After a brief description of the mechanism of cell movement, we outline the role that growth factors, substrate-adhesion molecules, and other environmental factors play in modulating cell migration. Since accurate measurements are essential for elucidating the effect of a specific stimulus on cell migration, we discuss the application of several assays that may be used to characterize cell motility. Finally, the use of theoretical models for analyzing cell population dynamics and predicting tissue growth rates is discussed.

6.2 Characteristics of Mammalian Cell Migration

6.2.1 Cell Movement Cycle

The movement of a mammalian cell on a substrate is an intricate process requiring at least three structural elements: an ECM ligand on the substrate, its cell surface receptor, and the intracellular cytoskeleton [6]. The receptors that play key roles for cell movement belong to a large family of transmembrane proteins called *integrins* [11]. Migration can be considered as a continual cycle consisting of four essential steps [4] (1) extension of the cell's leading margin over the substratum to form a lamellipod (i.e., a thin piece of membrane and cytoplasm at the front of the cell); (2) attachment to the substrate; (3) pulling or contraction using the newly formed points of adhesion as anchorage; and (4) release or detachment of adhesions at the rear of the cell. These four steps are orchestrated by the interaction of various extracellular and intracellular molecules. Extension of the leading margin of the cell is caused by the polymerization of actin filaments at the lamellipodia and crucial factors involved in this process are the Arp2/3 complex, gelsolin, and capping protein [4,12,13]. At the base of the cortical actin meshwork, cofilin promotes the disassembly of filaments [4,14]. Integrins anchor the cell to its substratum by binding both to ECM molecules on the outside of the cell and to the actin cytoskeleton on the inside. Cortical contractions due to myosin molecules pull structures toward the center of the cell, causing the uropod (lamellipod's counterpart at the rear of the cell) to retract and unattached structures on the dorsal surface to move backward [15–17]. The cycle is completed by a forward movement of actin and other constituents through the cytoplasm to the leading margin of the cell [4].

6.2.2 Persistent Random Walk

Migration of individual mammalian cells in isotropic environments can be described as a persistent random walk [18,19]. Over short time periods, cells follow a relatively straight path, showing persistence of movement (see Figure 6.1a). If long time intervals are used to observe the cell position, however, cell movement appears similar to Brownian motion with frequent direction changes (Figure 6.1b). At least two parameters are needed to describe persistent random walk [20]. The first one is the *speed S* that is intuitively defined as the displacement of the cell centroid per unit time. The second one is the *persistence time* (usually denoted by P) that is a measure of the average time between "significant" direction changes. The magnitude of P and S depend both on the type of the cell and on its microenvironment. Reported values of migration speed and persistence time range from 0.5 μm/min and 4 to 5 h, respectively for human microvessel endothelial cells and smooth muscle cells [20,21] to 20 μm/min and 4 min for rabbit neutrophils [22]. Lauffenburger and coworkers [23] pointed out that there exists a rough inverse relationship between S and P and that this could be understood by considering the product of these two parameters as the cell's analog to a "mean free-path length." Rigorous definition of P and S necessary for mathematical modeling and the assays to measure them will be described in Section 6.4.

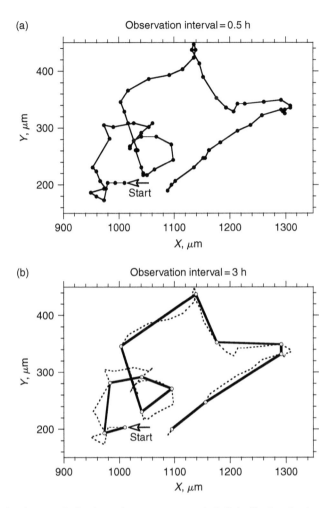

FIGURE 6.1 Typical trajectory of a bovine pulmonary artery endothelial cell migrating in a uniform environment. Symbols represent the position of the centroid of the same cell recorded at 30 min intervals (top panel) and 3 h intervals (bottom panel). When the observation interval is short (top panel), the cell clearly exhibits persistence in movement direction. If a long observation interval (bottom panel) is chosen, however, the movement of the same cell appears to be a random walk with frequent direction changes. (Adapted from Lee, Y., Mcintire, L.V., and Zygourakis, K., *Biochem. Cell Biol.*, 1995, **73**: 461–472. With permission.)

6.2.3 Cell–Cell Contacts

In a population of migrating cells, the persistent random walk of individual cells can be interrupted by contacts with other cells. Since cells usually stop and change the direction of their movement after such contacts, this phenomenon is often called *contact inhibition of locomotion* [6]. When two fibroblasts collide with each other, for example, ruffling of the membrane near the contact point stops to form a quiescent region, while ruffling continues at other regions of the membrane. After about 25 min, the cells break the adhesion and move away in new directions [6]. Cell–cell contacts have an even more profound effect on the migration of epithelial cells. Following a collision, the leading lamellae of the epithelial cells are gradually lost and an adhesion is formed between two cells. Sequential collisions with other cells result in the formation of small colonies and eventually a sheet of contiguous cells [6,4]. This process is essential for the coverage of the wounded area in wound healing [24,25]. It should be noted that cell–cell contacts also affect cell division and the effect is also often referred to as "contact inhibition," which actually means *contact inhibition of division*. As is shown later, the mechanisms of these two contact inhibitions are different, even though they are related in a rather complicated way.

6.3   Regulation of Cell Movement

6.3.1   Soluble Factors Modulate Cell Movement

Many polypeptide growth factors can upregulate cell motility. Sato and Rifkin [26] found that when a confluent monolayer of bovine aortic endothelial cells (BAECs) was wounded with a razor blade, cells at the edge of the wound released basic fibroblast growth factor (bFGF), which stimulated the rapid movement of nearby cells into the denuded area. Addition of anti-bFGF IgG slowed down considerably the cell movement and this inhibition was dose dependent. Sato and Rifkin also found that bFGF regulated the basal level of synthesis of the protease plasminogen activator (PA) and the basal level of DNA synthesis. These findings are consistent with earlier work demonstrating that plasmin contributed to cell migration [27,28]. Sato and Rifkin [26] also offer the alternative explanation that bFGF acts as a motility factor via its adhesive interactions described by Baird and coworkers [29]. Platelet-derived growth factor-BB (PDGF-BB) has been found to be the major motility enhancing factor in human serum for human dermal fibroblasts migrating on type I collagen, even though it is not needed for the initiation of cell movement [30]. Transforming growth factor-α (TGF-α) enhances locomotion of cultured human keratinocytes [31]. Stimulation of MTLn3 cells (a metastatic carcinoma cell line) with epidermal growth factor (EGF) causes rapid and transient lamellipod protrusion along with an increase in actin polymerization at the leading edge [32]. EGF was also found to greatly enhance random dispersion of fibroblasts by increasing the frequency of direction changes and at the same time slightly increasing the path length [33]. When multiple types of growth factors are present, they may affect cell motility synergistically with certain extent of specificity. For example, TGF-β1 has been found to synergistically enhance EGF-stimulated hepatocyte motility responses on collagen-containing extracellular matrices. However, the same effect was not achieved when TGF-β1 was added together with hepatocyte growth factor (HGF) [34]. Placental growth factor (PlGF) can augment the migration of endothelial cell in response to vascular endothelial growth factor (VEGF) in pathological angiogenesis [35].

Recent studies have revealed additional soluble motility-stimulating proteins, which include (a) The scatter factor or SF (also known as HGF), a mesenchymal cell-derived protein, which causes contiguous sheets of epithelium to separate into individual cells and stimulates the migration of epithelial as well as vascular endothelial cells [36–38]; (b) the autocrine motility factor (AMF), a tumor cell-derived protein, which stimulates migration of the producer cells [39–41]; and (c) the migration-stimulating factor (MSF), a protein produced by fetal and cancer patient fibroblasts, which stimulates penetration of three-dimensional collagen gels by nonproducing adult fibroblasts [42–44].

The direction of cell movement can be affected by the concentration gradient of certain growth factors, a response called *chemotaxis*. Chemotaxis is very important for processes such as normal development, inflammation, angiogenesis, and wound healing in which directed migration of specific cell types is essential. It has been found that when a wound is caused in the human body, large amounts of PDGF, EGF, and TGF-β are secreted at different times by cells around the wound to coordinate the influx of neutrophils, macrophages, fibroblasts, smooth muscle cells, and endothelial cells for fast and complete healing of the wound [45–48]. Lack of these chemotactic growth factors may impair the healing process, while overproduction can cause excessive repair or scarring [49]. A chemotactic response is typically a function of both the absolute concentration of the attractant and the steepness of its concentration gradient. Directional orientation bias increases with concentration gradient steepness, asymptotically approaching a maximal level as steepness increases. The dependence on attractant concentration is biphasic, increasing at low concentration to reach a maximum, then decreasing as concentration increases further [50–52].

Several hypotheses have been proposed to explain the underlying mechanisms of these phenomena. Lackie [6] suggested that cells decide their movement direction by receptor-mediated comparison of the spatial difference in attractant concentration across the cell dimension. This explains the biphasic dependence of directional orientation bias on attractant concentration for constant concentration gradient. At low-attractant concentrations, very few receptors are bound and, thus, only a small orientation bias results. At high-attractant concentrations almost all receptors are bound and, again, only a small bias is observed. Maximum bias is found for an intermediate attractant concentration, where the number of bound receptors is significant and most sensitive to differences in local attractant concentration. It should be noted that during the course of directed cell movement in chemotaxis, there are periods during which the cell may randomly stray toward the lower concentration or any other direction. Mechanistic models aimed at simulating chemotaxis must account for these random fluctuations in the cell's orientation even at the presence of the concentration gradient of a chemoattractant. Tranquillo and coworkers [53,54] hypothesized that this is caused primarily by the probabilistic kinetics of receptor/attractant binding. Their model divides a cell into two compartments along its polarization axis and assumes that the instantaneous numbers of receptor/attractant complexes at the compartment surfaces are governed by a stochastic differential equation with both a deterministic and a probabilistic part, which accounts for the random fluctuations in the binding process. The numbers of receptor/attractant complexes are then used to calculate the concentration of motile effectors in each intracellular compartment. Finally, the model postulates that the cell changes direction with an angular rate proportional to the imbalance between the levels of motile effectors in the two compartments. Model results were shown to provide good prediction for the chemotaxis of neutrophil leukocytes [55]. A recent extension of this model [56] relaxes several simplifying assumptions regarding receptor dynamics in the original model using newly obtained knowledge on transient G-protein signaling, cytoskeletal association, and receptor internalization and recycling, including statistical fluctuations in the numbers of receptors among the various states.

6.3.2 ECM Proteins and Cell–Substrate Interactions Regulate Cell Movement

As described earlier, each of the four phases in cell-movement cycle involves the interaction of cell-surface receptors with the ECM components on substratum surface. The ECM is a molecular complex whose components include collagens, glycoproteins, hyaluronic acid, proteoglycans, glycosaminoglycans, and elastins [57,58]. In addition, ECM harbors molecules such as growth factors, cytokines, matrix-degrading enzymes, and their inhibitors [57]. The distribution of these molecules varies from tissue to tissue and changes with time during tissue development, making the ECM a highly dynamic system [59,60]. The binding of ECM molecules to the extracellular domains of integrins triggers the receptor/ligand binding, trafficking and signaling cascade, activation of transcription factor and expression of target genes, and eventually results in the regulation of specific cell functions, which may include adhesion, migration, proliferation, and differentiation.

Integrin receptors are heterodimeric proteins composed of α and β subunits [11]. At least 15 α and β subunits have been identified so far and they pair with each other in a variety of combinations, giving rise to specific recognition on the ECM molecules with different selectivity. These combinations include the $\alpha_5\beta_1$ fibronectin receptor, $\alpha_2\beta_1$, $\alpha_3\beta_1$, and the vitronectin receptor $\alpha_v\beta_3$ [61–63]. While the $\alpha_5\beta_1$ integrin binds exclusively to fibronectin, $\alpha_1\beta_1$ can bind either to laminin or to collagen-IV and the $\alpha_v\beta_3$ receptor recognizes fibrinogen, vitronectin, and probably fibronectin. The $\alpha_4\beta_1$ integrin has also been found to mediate cell motility on fibronectin and vascular adhesion molecule-1 (VCAM-1) independently of the $\alpha_5\beta_1$ [64].

The ability of integrin receptors to recognize and bind the short peptide sequences corresponding to the adhesive domains of ECM proteins stimulated a lot of interest in developing *biomimetic* materials. Several studies by Hubbell and coworkers [65–71] have shown that covalent immobilization of adhesive peptides like RGD or YIGSR (which are the adhesive domains of fibronectin and laminin respectively) on the surface of glass or polymeric substrates can promote adhesion of endothelial cells. Kouvroukoglou and coworkers [72] found that such surface modifications significantly enhanced the migration of endothelial cells, a fact that might lead to higher endothelization rates of the surfaces of implantable biomaterials. Yang and coworkers [73] found that surface modification of poly(lactic acid) (PLA) films and poly(lactic-co-/glycolic acid) (PLGA) porous structures with RGD peptides or fibronectin greatly promoted human osteoprogenitor adhesion, migration, growth, and differentiation. Shin and coworkers [74] found that bulk modification of oligo(poly[ethylene glycol] fumarate) (OPF) hydrogel with a rat osteopontin-derived peptide (ODP) or a RGD peptide could increase the motility of marrow stromal osteoblasts and accelerate the expansion of megacolonies of these cells on the surface of the hydrogel.

The sudden introduction of a migratory ECM ligand into the environment of a cell may provide a key signal for the initiation of cell migration [75]. Similarly, the biosynthetic induction of such a migration protein or of its receptors might also promote migration [64,75–79]. The driving mechanism for migration appears to be provided by the physical interactions between specific sequences in adhesion proteins and their receptors [75], and by intracellular contractile proteins [4,17]. The signals involved in the termination of cell migration include the reacquisition of cell–cell adhesion molecules (e.g., N-cadherin), often accompanied by differentiation into the final tissue type [75].

The speed of cell movement on ECM proteins is regulated not only by the type of receptor/ligand interactions, but also by the ligand density on the substrate and the ligand affinity for cell adhesion receptors [80–83]. The complexity of these interactions leads to a *biphasic dependence* of cell speed on substrate adhesiveness. Goodman and coworkers [84] found that the migration speed of murine myoblasts on substrates covered with laminin or laminin fragment E8 depended in a biphasic fashion on the density of surface ligand. Maximum cell speeds were observed for ligand densities ranging from one third to one tenth of those required for strongest cell attachment. Low cell speeds were measured when the cells adhered very strongly or very weakly to the surface. A slow, monotonic increase of cell speed was observed, however, when the density of fibronectin ligands was increased. DiMilla and coworkers [22] measured the speed of human smooth muscle cells on fibronectin and collagen-IV. They found that cell speed varied in a biphasic fashion with increasing cell adhesion strengths for both ligands. Similar results are seen by varying the number of receptors on the cell surface. Keely and coworkers [85] found that the strength of adhesion of human breast carcinoma T47D cells to collagen I and IV decreased with decreasing levels of expression of the $\alpha_2\beta_1$ integrin (a collagen and laminin receptor). T47D clones that exhibited intermediate levels of adhesion to collagen had the highest motility across collagen-coated filters, suggesting that an intermediate density of cell-surface $\alpha_2\beta_1$ integrin optimally supports cell motility.

The reasons for this contrasting behavior were revealed by an elegant mathematical analysis of the cell-migration cycle by Lauffenburger [86] and DiMilla and coworkers [87]. The first key system parameter for explaining the experimental data is the *substratum adhesiveness* that is proportional to the surface ligand density and inversely proportional to the receptor/ligand equilibrium dissociation constant. Substrate adhesiveness, however, may not only depend on the strength of receptor/ligand association. A later analysis of Ward and Hammer [88] showed that the formation of focal contacts (which were modeled as cytoplasmic nucleation centers binding adhesion receptors) and the elastic rigidity of cytoskeletal connections may significantly affect cellular adhesive strength. The second parameter describes the *asymmetry in bond affinity* as the ratio of the dissociation rate constants between the front and the rear of the migrating cell. The model correctly predicted a biphasic dependence of cell speed on substrate adhesiveness, with cell locomotion occurring over an intermediate and (in many cases) limited range of adhesiveness. The size of this range was primarily governed by the asymmetry in adhesiveness between the front and the rear of the cell. Several mechanisms can provide a front-to-rear asymmetry in cell/substrate adhesiveness including dynamic integrin/cytoskeletal interactions [89] and polarized distributions of integrins maintained by receptor trafficking mechanisms [90].

The substratum adhesiveness can also affect the direction of cell movement, a response called *hapto-taxis*. For example, cells cultured on a surface will move onto tracks of artificial adhesive material, such as polylysine or silicon oxide [4]. When fibroblasts are plated on a surface coated with a uniformly increasing gradient of a charged substance, they turn and move in the direction of increasing adhesiveness [91]. Dickinson and Tranquillo [92] have developed a stochastic mathematical model based on receptor/ligand binding and trafficking mechanism to provide a mechanistic understanding of how the magnitude and distribution of adhesion ligands in the substratum influence cell movement. Additional sources for directionality in movement may be simple population pressure, tissue or physical barriers [4,93], or the three-dimensional structure of the matrix. In a process known as *contact guidance*, matrix fibers can be spatially arranged as to facilitate cell movement in a preferred direction [94]. Micromachined grooves cut into the substratum have similar effect [95].

While properties of the substratum can greatly influence cell movement, cells can also direct their own migration by modifying the physical properties of the substratum. For example, pioneering cells within a population of migrating neural crest cells may be laying down extracellular cues that trailing cells recognize [96]. When a suspension of human keratinocytes is plated on a fibrin matrix, single cells invade the matrix and progress through it by dissolving the fibrin and thereby creating tunnels [97].

6.3.3 Electrical Fields Direct Cell Movement

The migration of many mammalian cells is affected by the presence of electric fields, a phenomenon called *galvanotaxis*. Nerve cells can detect electric field as weak as 10 mV/mm and turn their growth cones to move in the direction of the negative pole [4]. Fibroblasts and cells from the neural crest also move toward the negative pole in a steady electric field [98,99]. During wound healing in vertebrates, a steady lateral electric field of 40 to 200 mV/mm is generated in the disrupted epithelia layers to coordinate the directed migration of epithelial cells from the nearby regions [100]. Some experiments indicate that when the electric field is removed, the wound healing rate is 25% slower, while nearly every clinical trial using electric fields to stimulate healing in mammalian wounds reports a significant increase in the rate of healing from 13 to 50% [101].

Cell's response to electrical fields typically requires Ca^{2+} influx [102], the presence of specific growth factors [100] and intracellular kinase activity. Protein kinase C is required by neural crest cells [103] and cAMP-dependent protein kinase is used in keratinocytes [104], while mitogen-activated protein kinase is required by corneal epithelial cells [105]. Specifically, Zhao and coworkers [100] discovered that corneal epithelial cells cultured in serum-free medium showed no reorientation in an electric field until 250 mV/mm and addition of EGF, bFGF, or TGF-β 1 singly or in combination significantly restored the cathodal reorientation response at low field strengths. Interestingly, however, the directed migration of two fibroblastic cells, NIH 3T3 and SV101, was found to be calcium independent and was, instead, related to the lateral redistribution of plasma membrane glycoproteins involved in cell–substratum adhesion [98].

6.4 Cell Migration Assays

6.4.1 Cell-Population Assays

The methods used to assess the locomotory capabilities and characteristics of cells fall into two major categories. The first category includes techniques that monitor large populations of migrating cells and analyze the number density profiles of the population after a given time period of migration. The Boyden chamber is the most popular assay in this category [106–111]. According to this technique, cells are placed on the upper surface of a micropore filter installed in the chamber and incubated for a sufficient period of time to allow the cells to migrate to the lower surface of the filter. Cell motility is then quantified either by counting the number of cells that have migrated through the filter or by measuring the distance traveled into the filter by several of the fastest moving cells. A comparison of the number of cells that passed through the filter or of the migration distance measured for various experimental conditions reveal if, for example, a test substance is chemotactic for the cells under study. With appropriate modifications, the Boyden chamber can also be used to study cell migration under flow conditions or temporal chemotactic gradients [112,113].

Another cell-population assay measures the migration distance of endothelial cells or osteoblasts released from growth arrest using a silicon or steel template compartmentalization technique [74,114].

This assay system employs tissue culture dishes that are subdivided (using, for example, stainless steel annular rings) into two separate compartments: an inner circular core and an annular outer ring. Cells are seeded into the inner compartment and grown to confluence. The inner ring is then removed, and cells (released from growth arrest) start migrating from a sharp starting line. Cell motility is quantified by measuring the distance of the migrating front from the starting line. By using a growth-arrested monolayer as a starting cell population for the quantification of migration, this assay system produces few or no wounded cells at the migration front. It also allows us to evaluate the migratory response to other cell types by coculturing them with the test cells after having seeded the effector cells into the outer compartment of the assay system. Since migration of endothelial cells is one of the critical features of wound repair, several researchers have also studied the movement of cells from a wound edge into a denuded area formed by scraping a confluent monolayer with a razor blade [26,115,116]. Migration is then quantified by counting the number of cells in successive 125-μm sections from the wound edge.

A third popular technique is the under-agarose assay where cells are allowed to migrate under a layer of agarose gel deposited on a glass or plastic surface [117–120]. For random motility experiments, a migration stimulus is incorporated at uniform concentration into the gel and the cell well medium. For chemotaxis experiments, the migration stimulus is placed in a separate well from which it forms a concentration gradient by diffusing through the gel. Typically measured quantities are again the location of the leading front or the total number of cells migrating away from the well.

However, intrinsic cell locomotory properties are not the only factors influencing the measurements obtained by cell-population assays. Parameters like the assay chamber geometry and size, initial cell number, or attractant diffusivity can significantly affect the population dispersal and complicate comparison of different sets of experimental data. To alleviate these problems, a phenomenological model has been proposed by Keller and Segel [121] and reformulated by Rivero and coworkers [122]. This mathematical model takes the form of a *partial differential equation* describing the temporal evolution of the cell number density profile and having two key parameters (a) the random motility coefficient μ and (b) a chemotaxis coefficient χ. By comparing model predictions for cell number density profiles to experimental profiles measured with the linear under-agarose assay [123] or the filter assay [124], the random motility μ and the chemotaxis coefficient χ can be determined. A different approach was recently followed by Cheng and coworkers [125] to analyze the differential effect of cell migration and proliferation on the expansion of megacolonies of marrow stromal cells cultured on biomimetic hydrogels modified with RGD or osteopontin-derived peptides [74]. This study used a *discrete model* based on the Markov chain approach to simulate the cell-population dynamics of expanding megacolonies. A comparison of model predictions to experimental data showed that surface modifications enhance the expansion rates of cell megacolonies by upregulating the speeds of cell migration.

6.4.2 Individual-Cell Assays

Cell-population assays cannot provide detailed information on how cells move. They cannot directly quantify important locomotory parameters (like cell speed, persistence, turn angles, etc.) or evaluate the effects of external stimuli on these parameters. An accurate characterization of cell locomotion can only be obtained by continuously monitoring a sufficiently large population of migrating cells using a video microscopy system with digital or analog time-lapse capabilities. In most applications, cells migrate on two-dimensional surfaces beneath liquid culture media or agarose gels [21,72,126–129]. Procedures allowing tracking in three-dimensional collagen matrices have also been reported [50,130–136].

By analyzing a sequence of images obtained at fixed time intervals, the actual cell positions at the corresponding times can be identified to reconstruct the individual cell trajectories (see Figure 6.1a). The mean square *displacement* $\langle D^2 \rangle$ of the cells from their original positions can then be calculated and plotted vs. time. If cell movement were a random walk, the mean square *displacement* $\langle D^2 \rangle$ would vary with time according to

$$\langle D^2 \rangle = 2n\mu t \qquad (6.1)$$

where μ is the *random motility coefficient* (formally equivalent to a *diffusion coefficient*) and n is a constant depending on the dimensionality of the random walk. For two-dimensional walks $n = 2$, while for three-dimensional walks $n = 3$ [18,137,138]. Equation 6.1 implies that the average distance traveled by a cell is proportional to the square root of the elapsed time.

As we mentioned in Section 6.2.2, however, mammalian cells migrating in isotropic environments execute *persistent random walks*. Dunn [19] and Othmer and coworkers [139] developed the following mathematical model to describe persistent random walks:

$$\langle D^2 \rangle = nS^2[Pt - P^2(1 - e^{-t/P})] \tag{6.2}$$

where $\langle D^2 \rangle$ is the *mean square displacement* of the tracked cells, S is the *root-mean-square cell speed*, P is the *persistence time*, and n is the constant that gives the dimensionality of the persistent random walk (n is 2 or 3 for 2D or 3D walks, respectively). This model assumes that S and P are time invariant. The root-mean-square cell speed S and persistence time P can be computed by fitting the experimental $\langle D^2 \rangle$ vs. time data with the persistent random walk model of Equation 6.2. Nonlinear parameter estimation algorithms [123] must be used, since graphical techniques [19] lead to results dependent on the size of the time interval. Note that for long times ($t \gg P$), Equation 6.2 reduces to the much simpler expression:

$$\langle D^2 \rangle = nS^2 Pt \tag{6.3}$$

Equation 6.3 implies *random walk* behavior and allows us to compute the *random motility coefficient* μ as follows:

$$\mu = \tfrac{1}{2}S^2 P \tag{6.4}$$

Many investigators have successfully used the persistent random walk model to quantify cell migration [72,140–144]. The chemotactic motion of cells in response to a chemical concentration gradient can also be modeled by adding a directional bias to the persistent random walk process [21]. The magnitude of the directional bias is characterized by the *chemotactic responsiveness* κ. By using experimentally measured values of S, P, and κ, Stokes and Lauffenburger generated computer simulations of theoretical individual cell paths, which were useful in elucidating the role of cell migration in physiological processes [21].

Although the persistent random walk model has been very successful in assaying and comparing the motility of cells under various conditions, it cannot provide all the parameters that may be necessary for an accurate quantification of the cell-migration process. Experimental observations with endothelial cells [127] have revealed that cells slow down or even stop migrating when they divide or when they collide with other cells. Migrating cells may also stop for a while before changing their direction of movement [127]. A detailed description of cell movement (suitable, e.g., for the computer implementation of migration-proliferation models of tissue growth [145]) requires the following information (1) the speed of cell locomotion (swimming speed); (2) the expected duration of cell movement in any given direction; (3) the probability distribution of turn angles that will decide the next direction of cell movement; (4) the frequency of cell stops; and (5) the duration of cell stops. The ultimate direction of cell movement should also be obtained to check for *spatial heterogeneities* or *chemotactic phenomena*.

When such a detailed description of the migration process is desired, models based on *Markov chain* concepts must be used to analyze the cell trajectory data [146–148]. At any time t, a migrating cell can exist in either a *directional state* (if it moves in a certain direction) or the *stationary state* (if it has stopped moving). Clearly, there is a *change of state* every time the migrating cell changes direction and when it starts or stops moving. The central assumption here is that state changes are random and do not depend either on the past history of the cell or on the length of time the cell spent in its current state. Under these assumptions, the sequence of states is a stochastic *Markov sequence*. Details for this analysis may be found in the monograph of Noble and Levine [147]. One usually allows only a finite number of *directional states* by partitioning the set of all possible directions of movement. Four or eight directional states are typically considered for two-dimensional walks (in addition to the stationary state) [147]. The reconstructed cell trajectories can then be modeled as Markov chains [146,149–151] and the following cell locomotory parameters can be extracted from the cell trajectory data using the well-known Markov chain theory [152] (a) *transition-state probabilities* that quantify the frequency with which the cells (individually or collectively) move from one state to another and, thus, characterize the turning and stopping behavior of the cells; (b) *waiting times* or the average time cells spend in a certain state (i.e., moving in a certain direction); and (c) *steady-state probabilities* that provide a measure of the directionality of cell movement (*chemotaxis*). The average speed of locomotion can also be estimated from the reconstructed cell trajectories using the formula:

$$v = \sum_{i=1}^{N} d_i / N \cdot \Delta t \qquad (6.5)$$

where d_i is the displacement of the centroid of a cell between two successive observations, and N is the numbers of observations made at a fixed time interval Δt.

The Markov chain model can be used to analyze *chemotaxis* experiments. If there is a preferred direction for cell movement, its steady-state probability will be significantly higher than the steady-state probabilities of the other directions. The Markov chain approach can also provide a more detailed description of cell migration since it accounts for stops in cell movement and uses more than one descriptor (e.g., transition state probabilities and average waiting times) to define persistence [148].

A detailed comparison of the random walk and the Markov chain models using experimental data for migrating endothelial cells has shown [148] that all five locomotory parameters defined above affect the speed S and persistence time P estimated by the persistent random walk model. The persistent random walk model can provide a measure of the speed of locomotion S, appropriately weighted to account for the frequency and duration of cell stops (or slow-downs). The persistence time P, however, is a composite measure of all five locomotory parameters mentioned above. Hence, it may not be always possible to extract all the information necessary to describe cell locomotion from the two parameters of the persistent random walk model. The Markov chain approach, however, also has its drawbacks. Perhaps the most serious drawback is that the values of cell speeds computed according to Equation 6.5 depend on the time interval Δt between cell observations. Accurate estimations of locomotion speeds with Equation 6.5 are possible only when (a) cell trajectories are not very tortuous and (b) the cell positions are observed at short time intervals Δt. This time interval must be carefully chosen using Richardson plots to minimize errors in the computation of cell speed [148].

6.5 Mathematical Models for Cell Migration and Tissue Growth

Cell migration influences both the structure and the growth rate of bioartificial tissues built on biomimetic scaffolds. As we have seen in the previous sections, however, cell migration is a process modulated by a multitude of system parameters that include the local scaffold properties and nutrient or growth factor concentrations. Despite the complexity of this process, however, a purely empirical approach is still used to design scaffolds and select bioreactors appropriate for each application. Noting the limitations of this approach, Ratcliffe and Niklason [153] proposed an interdisciplinary effort to improve our fundamental understanding of the dynamic processes characterizing tissue growth and bioreactor operation. This goal can be achieved with the help of comprehensive computer models that allow for systematic analysis of tissue regeneration processes. Simulations can shorten the development stage by allowing tissue engineers to rapidly screen many alternatives on the computer and choose the most promising ones for laboratory experimentation.

The key challenge here involves the development of computationally efficient algorithms to describe the dynamics of large cell populations that evolve in a continuously changing environment. Early studies

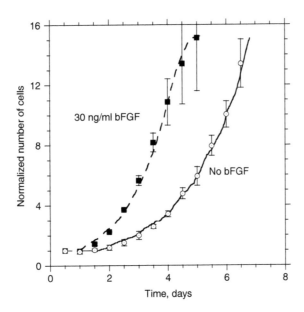

FIGURE 6.2 Normalized cell counts from experiments (symbols) and simulations (lines) for bovine pulmonary artery endothelial (BPAE) cells seeded uniformly on tissue culture plates and cultures without and with 30 ng/ml of bFGF. Standard errors are used as error bars. (Adapted from Lee, Y., Mcintire, L.V., and Zygourakis, K., *Biochem. Cell Biol.*, 1995, **69**: 1284–1298. With permission.)

utilized either discrete [154–156] or continuous models [157–161] to treat tissue growth problems. Using an approach that combined (a) modeling based on cellular automata, (b) computer simulations, and (c) experimentation (migration and proliferation assays), our group initially focused on the growth of two-dimensional tissues like vascular endothelium or keratinocyte colonies [145,155,156]. Using experimental data obtained from long-term tracking and analysis of cell locomotion [162], we developed a class of discrete models that described the population dynamics of cells migrating and proliferating on the surface of biomaterials. These models simulated persistent random walks on 2-D square lattices and accurately accounted for cell–cell collisions. Figure 6.2 shows that model predictions agreed well with experimentally measured proliferation rates of bovine pulmonary artery endothelial (BPAE) cells cultured with and without bFGF, a growth factor that upregulates cell motility. Results also showed that the seeding cell density, the population-average speed of locomotion, and the spatial distribution of the seed cells are crucial parameters in determining the temporal evolution of cell proliferation rates. More recently, models were developed to solve 2-D problems involving the aggregation and self-organization of the cellular slime mold *Dictyostelium discoideum* [163–166] and to study the interactions between extracellular matrix and fibroblasts [167]. To account for the effect of transport limitations that restrict our ability to grow bioartificial tissues [168,169], our group has also developed a *hybrid multi-scale model* that combines (a) partial differential equations modeling the simultaneous reaction–convection–diffusion problems for nutrients and growth factors, (b) discrete models describing the cell-population dynamics, and (c) single-cell models describing intracellular processes modulating cell function.

To demonstrate the potential of computational models for tissue engineering, we will briefly discuss the application of a 3-D model we have developed [170] to study a problem in *biomimetics*. Our objective was to evaluate the effectiveness of surface modifications in enhancing tissue-growth rates through increased cell-migration speeds. Such modifications may involve, for example, the immobilization of signaling or adhesive peptides on the surface of biomaterials [171–173]. To isolate the effect of population dynamics on tissue growth, simulations on 3-D scaffolds were performed by assuming that nutrient and growth factor concentrations remained constant throughout the tissue regeneration process. We considered cell-migration speeds varying across the entire range of observed values (from 0 to 50 μm/h) and used two different cell *seeding modes*. In the first mode, cells were seeded uniformly and randomly throughout the

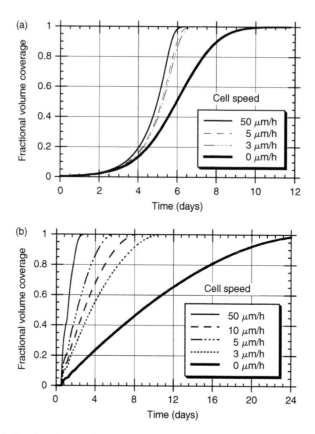

FIGURE 6.3 Effect of cell-migration speed on tissue-growth rates for two different cell-seeding modes. All simulations were carried out on a cubic domain with side equal to 1.9 mm (128 × 128 × 128 computational sites). (a) Cells were seeded uniformly and randomly with initial density equal to 0.5%. (b) Seeding simulated wound healing. The "wound" was a cylinder 0.95 mm in diameter and 1.9 mm in height. (Adapted from Belgacem, B.Y., Markenscoff, P., and Zygourakis, K. *Proceedings of the 3rd Chemical Engineering Symposium*, Athens, Greece, Vol. 2, pp. 1133–1136, 2001. With permission.)

scaffold. The second seeding mode was designed to simulate a "wound-healing" process. We assumed that a cylindrical wound was created in the center of a 3-D tissue. The wound was then filled with a biomaterial that allowed cells from the surrounding tissue to move into the wound and proliferate to heal it [170].

For uniform cell seeding, Figure 6.3a shows that increasing migration speeds initially enhanced tissue-growth rates. As cell speeds increased above 5 μm/h, however, this beneficial effect diminished and disappeared completely for large-migration speeds. For the wound-healing model, however, the simulations predicted significant enhancements of tissue-growth rates with increasing cell-migration speeds (Figure 6.3b). A careful analysis of the simulation results for these two cases revealed that the uniform cell seeding resulted in many more cell–cell collisions that slowed down the migration of cells. These results point out that the *cell-population dynamics*, the *geometry* of the problem, and the *initial conditions* can have a profound effect on tissue regeneration rates. The simulations also provide us with invaluable guidance for the design of experiments [174] that can test the efficacy of surface modifications designed to enhance cell-migration speeds.

References

[1] Le Douarin, N.M., Cell migrations in embryos. *Cell*, 1984, **38**: 353–360.
[2] Trinkaus, J.P., *Cells into Organs. The forces that Shape the Embryo.* Englewood Cliffs, NJ: Prentice-Hall, 1984.

[3] Condeelis, J., Life at the leading edge: the formation of cell protrusions. *Annu. Rev. Cell Biol.*, 1993, **9**: 411–444.

[4] Bray, D., *Cell Movements.* 2nd ed. New York: Garland Publishing, Inc., 2001.

[5] Devreotes, P.N. and Zigmond, S.H., Chemotaxis in eukaryotic cells: a focus on leukocytes and Dictyostelium. *Annu. Rev. Cell Biol.*, 1988, **4**: 649–686.

[6] Lackie, J.M., *Cell Movement and Cell Behaviour.* London: Allen and Unwin, 1986, pp. 253–275.

[7] Singer, S.J. and Kupfer, A., The directed migration of eukaryotic cells. *Annu. Rev. Cell Biol.*, 1986, **2**: 337–365.

[8] Liotta, L.A., Stracke, M.L., Aznavoorian, S.A., Beckner, M.E., and Schiffmann, E., Tumor cell motility. *Semin. Cancer Biol.*, 1991, **2**: 111–114.

[9] Langer, R. and Vacanti, J.P., Tissue engineering. *Science*, 1993, **260**: 920–926.

[10] Shin, H., Jo, S., and Mikos, A.G., Biomimetic materials for tissue engineering. *Biomaterials*, 2003, **24**: 4353–4364.

[11] Hynes, R.O., Integrins: a family of cell surface receptors. *Cell*, 1987, **48**: 549–554.

[12] Egelhoff, T.T. and Spudich, J.A., Molecular genetics of cell migration: dictyostelium as a model system. *Trends Genet.*, 1991, **7**: 161–166.

[13] Schafer, D.A., Welch, M.D., Machesky, L.M., Bridgman, P.C., Meyer, S.M. et al., Visualization and molecular analysis of actin assembly in living cells. *J. Cell Biol.*, 1998, **143**: 1919–1930.

[14] Carlier, M.F. and Pantaloni, D., Control of actin dynamics in cell motility. *J. Mol. Biol.*, 1997, **269**: 459–467.

[15] Novak, K.D. and Titus, M.A., Myosin I overexpression impairs cell migration. *J. Cell Biol.*, 1997, **136**: 633–647.

[16] Pollard, T.D. and Ostap, E.M., The chemical mechanism of myosin—I: implications for actin-based motility and the evolution of the myosin family of motor proteins. *Cell Struct. Funct.*, 1996, **21**: 351–356.

[17] Stossel, T.P., On the crawling of animal cells. *Science*, 1993, **260**: 1086–1094.

[18] Gail, M.H. and Boone, C.W., The locomotion of mouse fibroblasts in tissue culture. *Biophys. J.*, 1970, **10**: 980–993.

[19] Dunn, G.A., Characterizing a kinesis response: time averaged measures of cell speed and directional persistence. *Agents Actions* (Suppl), 1983, **12**: 14–33.

[20] Stokes, C.L., Lauffenburger, D.A., and William, S.K., Migration of individual microvessel endothelial cells: stochastic model and parameter measurement. *J. Cell Sci.*, 1991, **99**: 419–430.

[21] Dimilla, P., Stone, J., Quinn, J., Albelda, S., and Lauffenburger, D.A., Maximal migration of smooth muscle cells on fibronectin and collagen-IV occurs at an intermediate attachment strength. *J. Cell Biol.*, 1993, **122**: 729–737.

[22] Zigmond, S.H., Klausner, R., Tranquillo, R.T., and Lauffenburger, D.A., Analysis of the requirements for time-averaging of receport occupancy for gradient detection by polymorphonuclear leukocytes, in *Membrane Receptors and Cellular Regulation*, Kahn, C.R., Ed. New York,: Alan R. Liss pp. 347–356, 1985.

[23] Lauffenburger, D.A. and Linderman, J.J., *Receptors: Models for Binding, Trafficking, and Signaling.* New York, NY: Oxford University Press, 1993.

[24] Chan, K.Y., Patton, D.L., and Cosgrove, Y.T., Time-lapse videomicroscopic study of *in vitro* wound closure in rabbit corneal cells. *Invest. Ophthalmol. Visual Sci.*, 1989. **30**: 2488–2498.

[25] Brewitt, H., Sliding of epithelium in experimental corneal wounds. A scanning electron microscopic study. *Acta Ophthalmol.*, 1979, **57**: 945–958.

[26] Sato, Y. and Rifkin, D.B., Autocrine activities of basic fibroblast growth factor: regulation of endothelial cell movement, plasminogen activator synthesis, and DNA synthesis. *J. Cell Biol.*, 1988, **107**: 1199–1205.

[27] Morioka, S., Lazarus, G.S., Baird, J.L., and Jensen, P.J., Migrating keratinocytes express urokinase-type plasminogen activator. *J. Invest. Dermatol.*, 1987, **88**: 418–423.

[28] Ossowski, L., Quigley, J.P., Kellerman, G.M., and Reich, E., Fibrinolysis associated with onco-genic transformation requirement for plasminogen for correlated changes in cellular morphology, colony formation in agar and cell migration. *J. Exp. Med.*, 1973, **138**: 1056–1064.

[29] Baird, A., Schubert, D., Ling, N., and Guillemin, R., Receptor- and heparin-binding domains of basic fibroblast growth factor. *Proc. Natl Acad. Sci. USA* 1988, **70**: 369–373.

[30] Li, W., Fan, J., Chen, M., Guan, S., Sawcer, D. et al., Mechanism of human dermal fibroblast migration driven by type I collagen and platelet-derived growth factor-BB. *Mol. Biol. Cell*, 2004, **15**: 294–309.

[31] Ju, W.D., Schiller, J.T., Kazempour, M.K., and Lowy, D.R., TGF alpha enhances locomotion of cultured human keratinocytes. *J. Invest. Dermatol.*, 1993, **100**: 628–632.

[32] Lorenz, M., Desmarais, V., Macaluso, F., Singer, R.H., and Condeelis, J., Measurement of barbed ends, actin polymerization, and motility in live carcinoma cells after growth factor stimulation. *Cell Motil. Cytoskeleton*, 2004, **57**: 207–217.

[33] Ware, M.F., Wells, A., and Lauffenburger, D.A., Epidermal growth factor alters fibroblast migration speed and directional persistence reciprocally and in a matrix-dependent manner. *J. Cell Sci.*, 1998, **111**: 2423–2432.

[34] Stolz, D.B. and Michalopoulos, G.K., Synergistic enhancement of EGF, but not HGF, stimulated hepatocyte motility by TGF-beta 1 *in vitro*. *J. Cell. Physiol.*, 1997, **170**: 57–68.

[35] Gabhann, F.M. and Popel, A.S., Model of competitive binding of vascular endothelial growth factor and placental growth factor to VEGF receptors on endothelial cells. *Am. J. Physiol. Heart Circ. Physiol.*, 2004, **286**: H153–H164.

[36] Stoker, M., Gherardi, E., Perryman, M., and Gray, J., Scatter factor is a fibroblast-derived modulator of epithelial cell mobility. *Nature*, 1987, **327**: 239–242.

[37] Nusrat, A., Parkos, C.A., Bacarra, A.E., Godowski, P.J., Delp-Archer, C. et al., Hepatocyte growth factor/scatter factor effects on epithelia. Regulation of intercellular junctions in transformed and nontransformed cell lines, basolateral polarization of c-met receptor in transformed and natural intestinal epithelia, and induction of rapid wound repair in a transformed model epithelium. *J. Clin. Invest.*, 1994, **93**: 2056–2065.

[38] Poomsawat, S., Whawell, S.A., Morgan, M.J., Thomas, G.J., and Speight, P.M., Scatter factor regulation of integrin expression and function on oral epithelial cells. *Arch. Dermatol. Res.*, 2003, **295**: 63–70.

[39] Funasaka, T., Haga, A., Raz, A., and Nagase, H., Tumor autocrine motility factor induces hyper-permeability of endothelial and mesothelial cells leading to accumulation of ascites fluid. *Biochem. Biophys. Res. Commun.*, 2002, **293**: 192–200.

[40] Silletti, S., Paku, S., and Raz, A., Autocrine motility factor and the extracellular matrix. I. Coordinate regulation of melanoma cell adhesion, spreading and migration involves focal contact reorganization. *Int. J. Cancer*, 1998, **76**: 120–128.

[41] Silletti, S., Paku, S., and Raz, A., Autocrine motility factor and the extracellular matrix. II. Degradation or remodeling of substratum components directs the motile response of tumor cells. *Int. J. Cancer*, 1998, **76**: 129–135.

[42] Schor, S.L., Ellis, I.R., Jones, S.J., Baillie, R., Seneviratne, K. et al., Migration-stimulating factor: a genetically truncated onco-fetal fibronectin isoform expressed by carcinoma and tumor-associated stromal cells. *Cancer Res.*, 2003, **63**: 8827–8836.

[43] Hamada, J., Cavanaugh, P.G., Miki, K., and Nicolson, G.L., A paracrine migration-stimulating factor for metastatic tumor cells secreted by mouse hepatic sinusoidal endothelial cells: identification as complement component C3b. *Cancer Res.*, 1993, **53**: 4418–4423.

[44] Schor, S.L. and Schor, A.M., Characterization of migration-stimulating factor (MSF): evidence for its role in cancer pathogenesis. *Cancer Invest.*, 1990, **8**: 665–667.

[45] Postlethwaite, A.E., Keski-Oja, J., Moses, H.L., and Kang, A.H., Stimulation of the chemotactic migration of human fibroblasts by transforming growth factor β. *J. Exp. Med.*, 1987, **165**: 251–256.

[46] Seppä, H., Grotendorst, G., Seppä, S., Schiffmann, E., and Martin, G.R., Platelet-derived growth factor is chemotactic for fibroblasts. *J. Cell Biol.*, 1982, **142**: 1119–1130.

[47] Grotendorst, G.R., Soma, Y., and Takehara, K., EGF and TGF-alpha are potent chemoattractants for endothelial cells and EGF-like peptides are present at sites of tissue regeneration. *J. Cell Physiol.*, 1989, **139**: 617–623.

[48] Wahl, S.M., Hunt, D.A., Wakefield, L.M., Mccart-Ney-Francis, N., Wahl, L.M. et al., Transforming growth factor type β induces monocyte chemotaxis and growth factor production. *Proc. Natl Acad. Sci. USA*, 1987, **84**: 5788–5792.

[49] Grotendorst, G.R., Chemoattractants and growth factors, in *Wound Healing: Biochemical and Clinical Aspects*, Lindblad, W.J., Ed. Philadelphia, Saunders: pp. 237–246, 1992.

[50] Knapp, D.M., Helou, E.F., and Tranquillo, R.T., A fibrin or collagen gel assay for tissue cell chemotaxis: assessment of fibroblast chemotaxis to GRGDSP. *Exp. Cell Res.*, 1999, **247**: 543–553.

[51] Fisher, P.R., Merkl, R., and Gerisch, G., Quantitative analysis of cell motility and chemotaxis in Dictyostelium discoideum by using an image processing system and a novel chemotaxis chamber providing stationary chemical gradients. *J. Cell Biol.*, 1989, **108**: 973–984.

[52] Zigmond, S.H., Ability of polymorphonuclear leukocytes to orient in gradients of chemotactic factors. *J. Cell Biol.*, 1977, **75 Pt 1**: 606–616.

[53] Tranquillo, R.T. and Lauffenburger, D.A., Stochastic model of leukocyte chemosensory movement. *J. Math. Biol. (IYS)*, 1987, **25**: 229–262.

[54] Tranquillo, R.T., Theories and models of gradient perception, in *Biology of the Chemotactic Response*, Lackie, J.M., Ed. Cambridge: Cambridge University Press, pp. 35–75, 1990.

[55] Tranquillo, R.T., Lauffenburger, D.A., and Zigmond, S.H., A stochastic model for leukocyte random motility and chemotaxis based on receptor binding fluctuations. *J. Cell Biol.*, 1988, **106**: 303–309.

[56] Moghe, P.V. and Tranquillo, R.T., Stochastic model of chemoattractant receptor dynamics in leukocyte chemosensory movement. *Bull. Math. Biol.*, 1994, **56**: 1041–1093.

[57] Ruoslahti, E. and Engvall, E., Eds., *Extracellular Matrix Components*. San Diego: Academic Press, 1994.

[58] Bosman, F.T. and Stamenkovic, I., Functional structure and composition of the extracellular matrix. *J. Pathol.*, 2003, **200**: 423–428.

[59] Rodgers, R.J., Irving-Rodgers, H.F., and Russell, D.L., Extracellular matrix of the developing ovarian follicle. *Reproduction*, 2003, **126**: 415–424.

[60] Kanwar, Y.S., Wada, J., Lin, S., Danesh, F.R., Chugh, S.S. et al., Update of extracellular matrix, its receptors, and cell adhesion molecules in mammalian nephrogenesis. *Am. J. Physiol. Renal Physiol.*, 2004, **286**: F202–F215.

[61] Albelda, S.M., Daise, M., Levine, E.M., and Buck, C.A., Identification and characterization of cell–substratum adhesion receptors on cultured human endothelial cells. *J. Clin. Invest.*, 1989, **83**: 1992–2002.

[62] Albelda, S.M. and Buck, C.A., Integrins and other cell adhesion molecules. *FASEB J.*, 1990, **4**: 2868–2880.

[63] Languino, L.R., Gehlsen, K.R., Wayner, E., Carter, W.G., Engvall, E. et al., Endothelial cells use $\alpha_2\beta_1$ integrin as a laminin receptor. *J. Cell Biol.*, 1989, **109**: 2455–2462.

[64] Wu, C., Fields, A.J., Kapteijn, B.A., and Mcdonald, J.A., The role of $\alpha_4\beta_1$ integrin in cell motility and fibronectin matrix assembly. *J. Cell Sci.*, 1995, **108**: 821–829.

[65] Massia, S.P. and Hubbell, J.A., Covalent surface immobilization of Arg–Gly–Asp- and Tyr–Ile–Gly–Ser–Arg-containing peptides to obtain well-defined cell-adhesive substrates. *Anal. Biochem.*, 1990, **187**: 292–301.

[66] Massia, S.P. and Hubbell, J.A., Covalently attached GRGD on polymer surfaces promotes biospecific adhesion of mammalian cells. *Ann. NY Acad. Sci.*, 1990, **589**: 261–270.

[67] Hubbell, J.A., Massia, S.P., Desai, N.P., and Drumheller, P.D., Endothelial cell-selective materials for tissue engineering in the vascular graft via a new receptor. *Biotechnology*, 1991, **9**: 568–572.

[68] Massia, S.P. and Hubbell, J.A., An RGD spacing of 440 nm is sufficient for integrin $\alpha_v\beta_3$-mediated fibroblast spreading and 140 nm for focal contact and stress fiber formation. *J. Cell Biol.*, 1991, **114**: 1089–1100.

[69] Massia, S.P. and Hubbell, J.A., Human endothelial cell interactions with surface-coupled adhesion peptides on a nonadhesive glass substrate and two polymeric biomaterials. *J. Biomed. Mater. Res.*, 1991, **25**: 223–242.

[70] Hubbell, J.A., Massia, S.P., and Drumheller, P.D., Surface-grafted cell-binding peptides in tissue engineering of the vascular graft. *Ann. NY Acad. Sci.*, 1992, **665**: 253.

[71] Massia, S.P., Rao, S.S., and Hubbell, J.A., Covalently immobilized laminin peptide Tyr–Ile–Gly–Ser–Arg (YIGSR) supports cell spreading and co-localization of the 67-kDa laminin receptor with alpha-actinin and vinculin. *J. Biol. Chem.*, 1993, **268**: 8053–8059.

[72] Kouvroukoglou, S., Dee, K.C., Bizios, R., Mcintire, L.V., and Zygourakis, K., Endothelial cell migration on surfaces modified with immobilized adhesive peptides. *Biomaterials*, 2000, **21**: 1725–1733.

[73] Yang, X.B., Roach, H.I., Clarke, N.M., Howdle, S.M., Quirk, R. et al., Human osteoprogenitor growth and differentiation on synthetic biodegradable structures after surface modification. *Bone*, 2001, **29**: 523–531.

[74] Shin, H., Zygourakis, K., Farach-Carson, M.C., Yaszemski, M.J., and Mikos, A.G., Attachment, proliferation, and migration of marrow stromal osteoblasts cultured on biomimetic hydrogels modified with an osteopontin-derived peptide. *Biomaterials*, 2004, **25**: 895–906.

[75] Humphries, M.J., Mould, A.P., and Yamada, K.M., Matrix receptors in cell migration, in *Receptors for Extracellular Matrix*, Mecham, R.P., Ed. San Diego, CA: Academic Press, pp. 195–253, 1991.

[76] Grossi, F.S., Keizer, D.M., Saracino, G.A., Erkens, S., and Romijn, J.C., Flow cytometric analysis and motility response to laminin and fibronectin of four new metastatic variants of the human renal cell carcinoma line RC43. *Prog. Clin. Biol. Res.*, 1992, **378**: 195–205.

[77] Mooradian, D.L., Mccarthy, J.B., Skubitz, A.P., Cameron, J.D., and Furcht, L.T., Characterization of FN-C/H-V, a novel synthetic peptide from fibronectin that promotes rabbit corneal epithelial cell adhesion, spreading, and motility. *Invest. Ophthalmol. Vis. Sci.*, 1993, **34**: 153–164.

[78] Savarese, D.M., Russell, J.T., Fatatis, A., and Liotta, L.A., Type IV collaged stimulates an increase in intracellular calcium. Potential role in tumor cell motility. *J. Biol. Chem.*, 1992, **267**: 21928–21935.

[79] Lin, M.L. and Bertics, P.J., Laminin responsiveness is associated with changes in fibroblast morphology, motility and anchorage-independent growth: cell system for examining the interaction between laminin and EGF signalling pathways. *J. Cell Physiol.*, 1995, **164**: 593–604.

[80] Palecek, S.P., Loftus, J.C., Ginsberg, M.H., Lauffenburger, D.A., and Horwitz, A.F., Integrin–ligand binding properties govern cell migration speed through cell–substratum adhesiveness. *Nature*, 1997, **385**: 537–540.

[81] Schense, J.C. and Hubbell, J.A., Three-dimensional migration of neurites is mediated by adhesion site density and affinity. *J. Biol. Chem.*, 2000, **275**: 6813–6818.

[82] Gobin, A.S. and West, J.L., Cell migration through defined, synthetic ECM analogs. *FASEB J. Exp. Biol.*, 2002, **16**: 751–753.

[83] Holub, A., Byrnes, J., Anderson, S., Dzaidzio, L., Hogg, N. et al., Ligand density modulates eosinophil signaling and migration. *J. Leukocyte Biol.*, 2003, **73**: 657–664.

[84] Goodman, S.L., Risse, G., and Von Der Mark, K., The E8 subfragment of laminin promotes locomotion of myoblasts over extracellular matrix. *J. Cell Biol.*, 1989, **109**: 799–809.

[85] Keely, P.J., Fong, A.M., Zutter, M.M., and Santoro, S.A., Alteration of collagen-dependent adhesion, motility, and morphogenesis by the expression of antisense alpha 2 integrin mRNA in mammary cells. *J. Cell Sci.*, 1995, **108**: 595–607.

[86] Lauffenburger, D.A., A simple model for the effects of receptor-mediated cell–substratuum adhesion on cell migration. *Chem. Eng. Sci.*, 1989, **44**: 1903–1914.

[87] Dimilla, P., Barbee, K., and Lauffenburger, D.A., Mathematical model for the effects of adhesion and mechanics on cell migration speed. *Biophys. J.*, 1991, **60**: 15–37.

[88] Ward, M.D. and Hammer, D.A., A theoretical analysis for the effect of focal contact formation on cell-substrate attachment strength. *Biophys. J.*, 1993, **64**: 936–959.

[89] Schmidt, C.E., Horwitz, A.F., Lauffenburger, D.A., and Sheetz, M.P., Integrin-cytoskeletal interactions in migrating fibroblasts are dynamic, asymmetric, and regulated. *J. Cell Biol.*, 1993, **123**: 977–991.

[90] Lawson, M.A. and Maxfield, F.R., Ca(2+)- and calcineurin-dependent recycling of an integrin to the front of migrating neutrophils. *Nature*, 1995, **376**: 75–79.

[91] Carter, S.B., Principles of cell motility: the direction of cell movement and cancer invasion. *Nature*, 1965, **208**: 1183–1187.

[92] Dickinson, R.B. and Tranquillo, R.T., A stochastic model for adhesion-mediated cell random motility and haptotaxis. *J. Math. Biol.*, 1993, **31**: 563–600.

[93] Bray, D., *Cell Movements*. New York, NY Garland Publishing, Inc., 1992.

[94] Nakatsuji, N. and Johnson, K.E., Experimental manipulation of a contact guidance system in amphibian gastrulation by mechanical tension. *Nature*, 1984, **307**: 453–455.

[95] Oakley, C., Jaeger, N.A., and Brunette, D.M., Sensitivity of fibroblasts and their cytoskeletons to substratum topographies: topographic guidance and topographic compensation by micromachined grooves of different dimensions. *Exp. Cell Res.*, 1997, **234**: 413–424.

[96] Newgreen, D. and Thiery, J.P., Fibronectin in early avian embryos: synthesis and distribution along the migration pathways of neural crest cells. *Cell Tissue Res.*, 1980, **211**: 269–291.

[97] Ronfard, V. and Barrandon, Y., Migration of keratinocytes through tunnels of digested fibrin. *Proc. Natl Acad. Sci. USA*, 2001, **98**: 4504–4509.

[98] Brown, M.J. and Loew, L.M., Electric field-directed fibroblast locomotion involves cell surface molecular reorganization and is calcium independent. *J. Cell Biol.*, 1994, **127**: 117–128.

[99] Gruler, H. and Nuccitelli, R., Neural crest cell galvanotaxis: new data and a novel approach to the analysis of both galvanotaxis and chemotaxis. *Cell Motil. Cytoskeleton*, 1991, **19**: 121–133.

[100] Zhao, M., Agius-Fernandez, A., Forrester, J.V., and Mccaig, C.D., Orientation and directed migration of cultured corneal epithelial cells in small electric fields are serum dependent. *J. Cell Sci.*, 1996, **109**: 1405–1414.

[101] Nuccitelli, R., A role for endogenous electric fields in wound healing. *Curr. Top. Develop. Biol.*, 2003, **58**: 1–26.

[102] Fang, K.S., Farboud, B., Nuccitelli, R., and Isseroff, R.R., Migration of human keratinocytes in electric fields requires growth factors and extracellular calcium. *J. Invest. Dermatol.*, 1998, **111**: 751–756.

[103] Nuccitelli, R., Smart, T., and Ferguson, J., Protein kinases are required for embryonic neural crest cell galvanotaxis. *Cell Motil. Cytoskeleton*, 1993, **24**: 54–66.

[104] Pullar, C.E., Isseroff, R.R., and Nuccitelli, R., Cyclic AMP-dependent protein kinase A plays a role in the directed migration of human keratinocytes in a DC electric field. *Cell Motil. Cytoskeleton*, 2001, **50**: 207–217.

[105] Mcbain, V.A., Forrester, J.V., and Mccaig, C.D., HGF, MAPK, and a small physiological electric field interact during corneal epithelial cell migration. *Invest. Ophthalmol. Visual Sci.*, 2003, **44**: 540–547.

[106] Boyden, S.B., The chemotactic effect of mixtures of antibody and antigen on polymorphonuclear leukocytes. *J. Exp. Med.*, 1962, **115**: 453–466.

[107] Keller, H.U., Borel, J.F., Wilkinson, P.C., Hess, M., and Cottier, H., Reassessment of Boyden's technique for measuring chemotaxis. *J. Immunol. Meth.*, 1972, **1**: 165–168.

[108] Nind, A.F., Neutrophil chemotaxis: technical problems with nitrocellulose filters in Boyden-type chambers. *J. Immunol. Meth.*, 1981, **49**: 39–52.

[109] Glaser, B.M., D'amore, P.A., Seppa, H., Seppa, S., and Schiffmann, E., Adult tissues contain chemoattractants for vascular endothelial cells. *Nature*, 1980, **288**: 483–484.

[110] Pelz, G., Schettler, A., and Tschesche, H., Granulocyte chemotaxis measured in a Boyden chamber assay by quantification of neutrophil elastase. *Eur. J. Clin. Chem. Clin. Biochem.: J.* 1993, **31**: 651–656.

[111] Kohyama, T., Ertl, R.F., Valenti, V., Spurzem, J., Kawamoto, M. et al., Prostaglandin E(2) inhibits fibroblast chemotaxis. *Am. J. Physiol. Lung Cell Mol. Physiol.*, 2001, **281**: L1257–L1263.

[112] Ebrahimzadeh, P.R., Högfors, C., and Braide, M., Neutrophil chemotaxis in moving gradients of fMLP. *J. Leukocyte Biol.*, 2000, **67**: 651–661.

[113] Slattery, M.J. and Dong, C., Neutrophils influence melanoma adhesion and migration under flow conditions. *Int. J. Cancer*, 2003, **106**: 713–722.

[114] Augustin-Voss, H.G. and Pauli, B.U., Quantitative analysis of autocrine-regulated, matrix-induced, and tumor cell-stimulated endothelial cell migration using a silicon template compartmentalization technique. *Exp. Cell Res.*, 1992, **198**: 221–227.

[115] Mignatti, P., Tsuboi, R., Robbins, E., and Rifkin, D.B., *In vitro* angiogenesis on the human amniotic membrane: requirement for basic fibroblast growth factor-induced proteinases. *J. Cell Biol.*, 1989, **108**: 671–682.

[116] Tsuboi, R., Sato, Y., and Rifkin, D.B., Correlation of cell migration, cell invasion, receptor number, proteinase production, and basic fibroblast growth factor levels in endothelial cells. *J. Cell Biol.*, 1990, **110**: 511–517.

[117] Rothman, C. and Lauffenburger, D.A., Analysis of the linear under-agarose leukocyte chemotaxis assay. *Ann. Biomed. Eng.*, 1983, **11**: 451–460.

[118] Rupnick, M.A., Stokes, C.L., Williams, S.K., and Lauffenburger, D.A., Quantitative analysis of random motility of human microvessel endothelial cells using a linear under-agarose assay. *Lab. Invest.*, 1988, **59**: 363–372.

[119] Newton-Nash, D.K., Tonellato, P., Swiersz, M., and Abramoff, P., Assessment of chemokinetic behavior of inflammatory lung macrophages in a linear under-agarose assay. *J. Leukocyte Biol.*, 1990, **48**: 297–305.

[120] Heit, B. and Kubes, P., Measuring chemotaxis and chemokinesis: the under-agarose cell migration assay. Science's STKE [electronic resource]: signal transduction knowledge environment. 2003, **2003**: PL5.

[121] Keller, E.F. and Segel, L.A., Model for chemotaxis. *J. Theor. Biol.*, 1971, **30**: 225–234.

[122] Rivero, M.A., Tranquillo, R.T., Buettner, H.M., and Lauffenburger, D.A., Transport models for chemotactic cell populations based on individual cell behavior. *Chem. Eng. Sci.*, 1989, **44**: 2881–2897.

[123] Farrell, B.E., Daniele, R.P., and Lauffenburger, D.A., Quantitative relationships between single-cell and cell-population model parameters for chemosensory migration responses of alveolar macrophages to C5a. *Cell Motil. Cytoskeleton*, 1990, **16**: 279–293.

[124] Buettner, H.M., Lauffenburger, D.A., and Zigmond, S.H., Measurement of leukocyte motility and chemotaxis parameters with the Millipore filter assay. *J. Immunol. Meth.*, 1989, **123**: 25–37.

[125] Cheng, G., Shin, H., Mikos, A.G., and Zygourakis, K., Expansion of marrow stromal osteoblast megacolonies on biomimetic hydrogels: interpreting and evaluating the assay data. *AIChE Annual Meeting* (paper 107cy), 2003.

[126] Dimilla, P.A., Stone, J.A., Quinn, J.A., Albelda, S.M., and Lauffenburger, D.A., Maximal migration of human smooth muscle cells on fibronectin and type IV collagen occurs at an intermediate attachment strength. *J. Cell Biol.*, 1993, **122**: 729–737.

[127] Lee, Y., Mcintire, L.V., and Zygourakis, K., Analysis of endothelial cell locomotion: differential effects of motility and contact inhibition. *Biotechnol. Bioeng.*, 1994, **43**: 622–634.

[128] Demuth, T., Hopf, N.J., Kempski, O., Sauner, D., Herr, M. et al., Migratory activity of human glioma cell lines *in vitro* assessed by continuous single cell observation. *Clin. Exp. Metastasis*, 2000, **18**: 589–597.

[129] Krooshoop, D.J., Torensma, R., Van Den Bosch, G.J., Nelissen, J.M., Figdor, C.G. et al., An automated multi well cell track system to study leukocyte migration. *J. Immunol. Meth.*, 2003, **280**: 89–102.

[130] Dickinson, R.B., Mccarthy, J.B., and Tranquillo, R.T., Quantitative characterization of cell invasion *in vitro*: formulation and validation of a mathematical model of the collagen gel invasion assay. *Ann. Biomed. Eng.*, 1993, **21**: 679–697.

[131] Parkhurst, M.R. and Saltzmann, W.M., Quantification of human neutrophil motility in three-dimensional collagen gels: effect of collagen concentration. *Biophys. J.*, 1992, **61**: 306–315.

[132] Shields, E.D. and Noble, P.B., Methodology for detection of heterogeneity of cell locomotory phenotypes in three-dimensional gels. *Exp. Cell Biol.*, 1987, **55**: 250–256.

[133] Burgess, B.T., Myles, J.L., and Dickinson, R.B., Quantitative analysis of adhesion-mediated cell migration in three-dimensional gels of RGD-grafted collagen. *Ann. Biomed. Eng.*, 2000, **28**: 110–118.

[134] Niggemann, B., Maaser, K., Lü, H., Kroczek, R., Zänker, K.S. et al., Locomotory phenotypes of human tumor cell lines and T lymphocytes in a three-dimensional collagen lattice. *Cancer Lett.*, 1997, **118**: 173–180.

[135] Friedl, P., Noble, P.B., and Zänker, K.S., Lymphocyte locomotion in three-dimensional collagen gels. Comparison of three quantitative methods for analysing cell trajectories. *J. Immunol. Meth.*, 1993, **165**: 157–165.

[136] Friedl, P. and Bröcker, E.B., Reconstructing leukocyte migration in 3D extracellular matrix by time-lapse videomicroscopy and computer-assisted tracking. *Meth. Mol. Biol.*, 2004, **239**: 77–90.

[137] Berg, H.C., *Random Walks in Biology*. Princeton: Princeton University Press, pp. 5–16, 1983.

[138] Peterson, S.C. and Noble, P.B., A two-dimensional random-walk analysis of human granulocyte movement. *Biophys. J.*, 1972, **12**: 1048–1055.

[139] Othmer, H.G., Dunbar, S.R., and Alt, W., Models of dispersal in biological systems. *J. Math. Biol.*, 1988, **26**: 263–298.

[140] Dickinson, R.B., McCarthy, J.B., and Tranquillo, R.T., Quantitative characterization of cell invasion *in vitro*: formulation and validation of a mathematical model of the collagen gel invasion assay. *Ann. Biomed. Eng.*, 1993, **21**: 679–697.

[141] Saltzman, W.M., Parsons-Wingerter, P., Leong, K.W., and Shin, L., Fibroblast and hepatocyte behavior on synthetic polymer surfaces. *J. Biomed. Mater. Res.*, 1991, **25**: 741–759.

[142] Dimilla, P.A., Quinn, J.A., Albelda, S.M., and Lauffenburger, D.A., Measurement of individual cell migration parameters for human tissue cells. *AIChE J.*, 1992, **38**: 1092–1104.

[143] Glasgow, J.E., Farrell, B.E., Fisher, E.S., Lauffenburger, D.A., and Daniele, R.P., The motile response of alveolar macrophages. An experimental study using single-cell and cell population approaches. *Am. Rev. Respir. Dis.*, 1989, **139**: 320–329.

[144] Tan, J., Shen, H., and Saltzman, W.M., Micron-scale positioning of features influences the rate of polymorphonuclear leukocyte migration. *Biophys. J.*, 2001, **81**: 2569–2579.

[145] Lee, Y., Kouvroukoglou, S., Mcintire, L.V., and Zygourakis, K., A cellular automaton model for the proliferation of migrating contact-inhibited cells. *Biophys. J.*, 1995, **69**: 1284–1298.

[146] Boyarsky, A., A Markov chain model for human granulocyte movement. *J. Math. Biol.*, 1975, **2**: 69–78.

[147] Noble, P.B. and Levine, M.D., *Computer-Assisted Analysis of Cell Locomotion and Chemotaxis*. Boca Raton, FL: CRC Press, Inc. pp. 17–48, 1986.

[148] Lee, Y., Mcintire, L.V., and Zygourakis, K., Characterization of endothelial cell locomotion using a Markov chain model. *Biochem. Cell Biol.*, 1995, **73**: 461–472.

[149] Boyarsky, A. and Noble, P.B., A Markov chain characterization of human neutrophil locomotion under neutral and chemotactic conditions. *Can. J. Physiol. Pharmacol.*, 1977, **55**: 1–6.

[150] Noble, P.B. and Bentley, K.C., Locomotory characteristics of human lymphocytes undergoing negative chemotaxis to oral carcinomas. *Exp. Cell Res.*, 1981, **133**: 457–461.

[151] Noble, P.B., Boyarsky, A., and Bentley, K.C., Human lymphocyte migration *in vitro*: characterization and quantitation of locomotory parameters. *Can. J. Physiol. Pharmacol.*, 1979, **57**: 108–112.

[152] Papoulis, A., *Probability, Random Variables and Stochastic Processes*. New York, NY: McGraw-Hill, pp. 532–551, 1965.

[153] Ratcliffe, A. and Niklason, L.E., Bioreactors and bioprocessing for tissue engineering. *Ann. NY Acad. Sci.*, 2002, **961**: 210–215.

[154] Lim, J.H.F. and Davies, G.A., A stochastic model to simulate the growth of anchorage dependent cells on flat surfaces. *Biotechnol. Bioeng.*, 1990, **36**: 547.

[155] Zygourakis, K., Bizios, R., and Markenscoff, P., Proliferation of anchorage dependent contact-inhibited cells. I. Development of theoretical models based on cellular automata. *Biotechnol. Bioeng.*, 1991, **38**: 459–470.

[156] Zygourakis, K., Markenscoff, P., and Bizios, R., Proliferation of anchorage dependent contact-inhibited cells. II. Experimental results and comparison to theoretical model predictions. *Biotechnol. Bioeng.*, 1991, **38**: 471–479.

[157] Frame, K.K. and Hu, W.S., A model for density-dependent growth of anchorage-dependent mammalian cells. *Biotechnol. Bioeng.*, 1988, **32**: 1061.

[158] Cherry, R.S. and Papoutsakis, E.T., Modeling of contact-inhibited animal cell growth on flat surfaces and spheres. *Biotechnol. Bioeng.*, 1989, **33**: 300.

[159] Sherratt, J.A., Martin, P., Murray, J.D., and Lewis, J., Mathematical models of wound healing in embryonic and adult epidermis. *IMA J. Math. Appl. Med. Biol.*, 1992, **9**: 177–196.

[160] Sherratt, J.A. and Murray, J.D., Epidermal wound healing: the clinical implications of a simple mathematical model. *Cell Transplant*, 1992, **1**: 365–371.

[161] Dale, P.D., Maini, P.K., and Sherratt, J.A., Mathematical modeling of corneal epithelial wound healing. *Math. Biosci.*, 1994, **124**: 127–147.

[162] Lee, Y., Mcintire, L.V., and Zygourakis, K., Analysis of endothelial cell locomotion: differential effects of motility and contact inhibition. *Biotechnol. Bioeng.*, 1994, **43**: 622–634.

[163] Dallon, J.C. and Othmer, H.G., A discrete cell model with adaptive signalling for aggregation of *Dictyostelium discoideum*. *Phil. Trans. R. Soc. Lond. B Biol. Sci.*, 1997, **352**: 391–417.

[164] Hogeweg, P., Evolving mechanisms of morphogenesis: on the interplay between differential adhesion and cell differentiation. *J. Theor. Biol.*, 2000, **203**: 317–333.

[165] Marée, A.F. and Hogeweg, P., How amoeboids self-organize into a fruiting body: multicellular coordination in Dictyostelium discoideum. *Proc. Natl Acad. Sci. USA*, 2001, **98**: 3879–3883.

[166] Palsson, E. and Othmer, H.G., A model for individual and collective cell movement in Dictyostelium discoideum. *Proc. Natl Acad. Sci. USA*, 2000, **97**: 10448–10453.

[167] Dallon, J.C., Sherratt, J.A., and Maini, P.K., Mathematical modelling of extracellular matrix dynamics using discrete cells: fiber orientation and tissue regeneration. *J. Theor. Biol.*, 1999, **199**: 449–471.

[168] Sikavitsas, V.I., Bancroft, G.N., and Mikos, A.G., Formation of three-dimensional cell/polymer constructs for bone tissue engineering in a spinner flask and a rotating wall vessel bioreactor. *J. Biomed. Mater. Res.*, 2002, **62**: 136–148.

[169] Bancroft, G.N., Sikavitsas, V.I., Van Den Dolder, J., Sheffield, T.L., Ambrose, C.G. et al., Fluid flow increases mineralized matrix deposition in 3D perfusion culture of marrow stromal osteoblasts in a dose-dependent manner. *PNAS*, 2002, **99**: 12600–12605.

[170] Belgacem, B.Y., Markenscoff, P., and Zygourakis, K., A computational model for tissue regeneration and wound healing. *Proceedings of the 3rd Chemical Engineering Symposium*, Athens, Greece, Vol. 2, pp. 1133–1136, 2001.

[171] Huttenlocher, A.F., Ginsberg, M.H., and Horwitz, A.F., Modulation of cell migration by integrin-mediated cytoskeletal linkages and ligand-binding affinity. *J. Cell Biol.*, 1996, **134**: 1551–1562.

[172] Palecek, S.P., Loftus, J.C., Ginsberg, M.H., Lauffenburger, D.A., and Horwitz, A.F., Integrin–ligand binding properties govern cell migration speed through cell–substratum adhesiveness. *Nature*, 1997, **385**: 537–540.

[173] Shin, H., Zygourakis, K., Farach-Carson, M.C., Yaszemski, M.J., and Mikos, A.G., Attachment, proliferation, and migration of marrow stromal osteoblasts cultured on biomimetic hydrogels modified with an osteopontin-derived peptide. *Biomaterials*, 2004, **25**: 895–906.

[174] Gosiewska, A., Rezania, A., Dhanaraj, S., Vyakarnam, M., Zhou, J. et al., Development of a three-dimensional transmigration assay for testing cell — polymer interactions for tissue engineering applications. *Tissue Eng.*, 2001, **7**: 267–277.

7

Inflammatory and Immune Responses to Tissue Engineered Devices

James M. Anderson
Case Western Reserve University

7.1   Introduction

Tissue-engineered devices are biologic–biomaterial combinations in which some component of tissue has been combined with a biomaterial to create a device for the restoration or modification of tissue or organ function. Four significant goals must be achieved if these devices are to function adequately and appropriately in the host environment. These four goals are (1) restoration of the target tissue with its appropriate function and cellular phenotypic expression; (2) inhibition of the macrophage and foreign body giant cell foreign body response that may degrade or adversely modify device function; (3) inhibition of scar and fibrous capsule formation that may be deleterious to the function of the device; and (4) inhibition of immune responses that may inhibit the proposed function of the device and ultimately lead to the destruction of the tissue component of the tissue-engineered device. The range of types of tissue-engineered devices is large, yet each device is considered to be unique in its combination of tissue component and biomaterial, thus requiring a unique set of tests to ensure that the four goals are achieved for the lifetime of the device in its *in vivo* environment.

The implantation of a tissue-engineered device activates the host defense systems that include the inflammatory and immune responses. The purpose of this chapter is to provide an overview and fundamental understanding of the inflammatory and immune responses that may be responsive following the *in vivo* implantation of a tissue-engineered device. In general, tissue-engineered devices contain a biomaterials component for which the evaluation of the inflammatory and foreign body reaction is of importance. Tissue-engineered devices also contain an active biological component, that is, proteins and cells, for which evaluation of the immune responses is of importance to the overall safety and efficacy of the tissue-engineered device. In addition, tissue-engineered devices are also considered as combination devices

TABLE 7.1   The Mononuclear Phagocytic System

Tissues	Cells
Implant sites	Inflammatory macrophages, foreign body giant cells
Liver	Kupffer cells
Lung	Alveolar macrophages
Connective tissue	Histiocytes
Bone marrow	Macrophages
Spleen and lymph nodes	Fixed and free macrophages
Serous cavities	Pleural and peritoneal macrophages
Nervous system	Microglial cells
Bone	Osteoclasts
Skin	Langerhans' cells, dendritic cells
Lymphoid tissue	Dendritic cells

and the interaction between the synthetic and biologic components and their interactive inflammatory and immune responses must be considered and evaluated. For clarification, it should be noted that the inflammatory responses are also known as the innate immune system and immune responses are generally considered to be the acquired or adaptive immune system.

The inflammatory and immune systems overlap considerably through the activity and phenotypic expression of macrophages that are derived from blood-borne monocytes. Monocytes and macrophages belong to the mononuclear phagocytic system (MPS) (Table 7.1). Cells in the MPS may be considered as resident macrophages in the respective tissues that take on specialized functions that are dependent on their tissue environment. From this perspective, the host defense system may be seen as blood-borne or circulating inflammatory and immune cells as well as mononuclear phagocytic cells that reside in specific tissues with specialized functions. As will be seen in the overview of the inflammatory and immune responses, the macrophage plays a pivotal role in both the induction and effector phases of these responses.

7.2   Inflammatory Responses

The process of implantation of a biomaterial, prosthesis, medical device, or tissue-engineered device results in injury to tissues or organs and the subsequent perturbation of homeostatic mechanisms that lead to the cellular cascades of wound healing [1–6]. The response to injury is dependent on multiple factors including the extent of injury, the loss of basement membrane structures, blood–material interactions, provisional matrix formation, the extent or degree of cellular necrosis, and the extent of the inflammatory response. These events, in turn, may affect the extent or degree of granulation tissue formation, foreign body reaction, and fibrosis or fibrous capsule development. These host reactions are considered to be tissue-, organ-, and species-dependent. In addition, it is important to recognize that these reactions occur very early, that is, within 2 to 3 weeks of the time of implantation, for biocompatible materials or devices in the normal resolution of the inflammatory and wound healing responses.

Table 7.2 identifies the events in the inflammatory responses and indicates the predominant cell type found in these responses. These characteristic cell types are utilized in histological studies to identify the phase or event in the inflammatory and wound healing sequence of events.

To better appreciate the sequence of inflammatory responses that occur within an implant site, Figure 7.1 illustrates the sequence of events that occur at the tissue/biomaterial interface, that is, foreign body reaction, and the events that occur adjacent to the interfacial foreign body reaction and within the surrounding tissue of the implant site.

Inflammation is generally defined as the reaction of vascularized living tissue to local injury, that is, implantation of a biomaterial, prosthesis, medical device, or tissue-engineered device. Immediately following injury, blood–material interactions occur and a provisional matrix is formed that consists

TABLE 7.2 Principal Cell Types in Inflammatory Responses

Response	Cell type
Acute inflammation	Polymorphonuclear leukocyte (neutrophil)
Chronic inflammation	Monocytes and lymphocytes
Granulation tissue	Fibroblasts and endothelial cells (capillaries)
Foreign body reaction	Macrophages and foreign body giant cells
Fibrous encapsulation	Fibroblasts

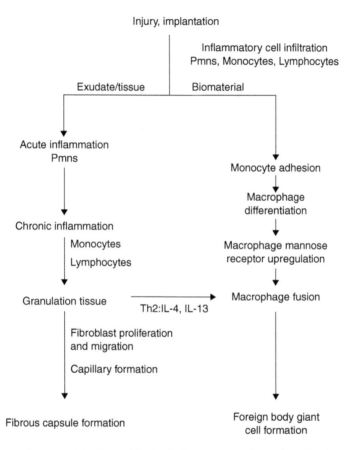

FIGURE 7.1 Inflammation, wound healing and foreign body responses at the implant site. The biomaterial pathway occurs at the surface of the biomaterial, whereas the exudate/tissue pathway occurs in the space surrounding the biomaterial. Both pathways can occur in a simultaneous manner and time frame.

of a platelet/fibrin thrombus/blood clot at the implant surface. As previously indicated, the predominant cell type (Table 7.2) present in the inflammatory response varies with the age of injury. In general, neutrophils (polymorphonuclear leukocytes) predominate during the first several days following injury and then are replaced by monocytes as the predominant cell type. Three factors occur for this change in cell type: neutrophils are short-lived and disintegrate and disappear after 24 to 48 h; neutrophil immigration is of short duration; and chemotactic factors for neutrophil migration are activated early in the inflammatory response. Following immigration from the vasculature, monocytes differentiate into macrophages and these cells are very long-lived (up to months). Monocyte immigration may continue for days to weeks dependent on the injury and implanted device and chemotactic factors for monocytes are activated over longer periods of time. The size, shape, and chemical and physical properties of the biomaterial or device may be responsible for variations in the intensity and duration of the inflammatory and wound-healing processes. Thus, intensity and time duration of the inflammatory reaction may characterize the biocompatibility of the biomaterial or device. Chemical mediators, released from plasma, cells, and injured tissue mediate these responses. These mediators include vasoactive amines, plasma proteases, arachidonic acid metabolites, lysosomal proteases, oxygen-derived free radicals, platelet activating factors, cytokines, and growth factors.

Acute inflammation is dependent on the extent of injury and may be of relatively short duration, lasting from minutes to days. As seen in Figure 7.1, its main characteristics are the formation of a fluid exudate and immigration of polymorphonuclear leukocytes across the endothelial lining of blood vessels and into tissue and the injury (implant) site. Adhesion molecules and receptors present on leukocyte and endothelial cells facilitate this process that is controlled, in part, by chemotaxis. A wide variety of exogenous and endogenous substances have been identified as chemotactic agents. Following localization of leukocytes at the injury (implant) site, phagocytosis and the release of enzymes occur following activation of neutrophils, monocytes, and macrophages. The major role of the polymorphonuclear leukocytes in acute inflammation is to phagocytose microorganisms and foreign materials. Phagocytosis is a three-step process in which the injurious agent undergoes leukocyte attachment, engulfment, and killing or degradation. In regard to biomaterials, phagocytosis and degradation may not occur, depending on the properties of the biomaterial. Biomaterials are not generally phagocytosed by neutrophils or macrophages because of the disparity in size, that is, the surface of the biomaterial is greater than the size of the cell. In general, particles, microcapsules, microspheres, or liposomes less than 10 μm in greatest dimension may undergo phagocytosis. The process of recognition and attachment is expedited when the biomaterial has adsorbed plasma-derived proteins such as immunoglobulin G (IgG) and the complement-activated fragment, C3b. Fibrinogen has also been identified as an adhesion molecule to facilitate leukocyte adhesion to biomaterial surfaces. IgG and C3b are the two major opsonins. These opsonins are naturally occurring serum factors that facilitate inflammatory cell adhesion. This may be significant with tissue-engineered devices in which the synthetic scaffold material or cell encapsulating synthetic membrane is in direct contact with tissue at the time of implantation. Thus, depending on the characteristics of the tissue-engineered device, protein adsorption and cellular adhesion in the inflammatory, wound healing and foreign body responses may be important factors in biocompatibility of the tissue-engineered device as well as its function.

The disparity in size between the biomaterial surface and the adherent cell generally leads to frustrated phagocytosis, which is the release of cellular enzymes, acid and reactive oxygen and nitrogen intermediates by either direct extrusion or exocytosis from the cell [7]. These agents may play significant roles in the biodegradation of biodegradable scaffold materials.

Following resolution of the acute inflammatory response, chronic inflammation with the presence of monocytes and lymphocytes is predominant. Chronic inflammation is characterized by the presence of monocytes, lymphocytes, and macrophages with the proliferation of blood vessels and connective tissue at the implant site. With biocompatible materials, the chronic inflammatory phase is of short duration and usually lasts several days and is seen within the first week to two weeks following implantation. Persistent inflammatory stimuli and motion of the implant may lead to focal chronic inflammation with extended time periods.

Whereas macrophages and lymphocytes play key roles in immune responses, their presence in the early inflammatory response is generally not considered to be an immune reaction. As will be seen later under Immune Responses, specific events in which the macrophages, lymphocytes, and plasma cells participate can lead to immune responses. Macrophages process and present antigens (foreign materials) to immunocompetent cells and thus are key mediators in the development of immune reactions.

In the inflammatory responses, the macrophage is probably the most important cell in chronic inflammation due to the large number of biologically active products that it produces and releases. Important classes of products produced and secreted by macrophages include neutral proteases, chemotactic factors, arachidonic acid metabolites, reactive oxygen metabolites, complement components, coagulation factors, growth-promoting factors, and cytokines. Chemotactic factors, cytokines and growth factors are important in the development of the next phase of the inflammatory and wound healing responses, which is the formation of granulation tissue. Granulation tissue is generally defined as the proliferation of new small blood vessels and the immigration of fibroblasts into the injury site. Depending on the extent of injury, granulation tissue may be seen as early as 3 to 5 days following implantation. As seen in Figure 7.1, at the same time that granulation tissue is being formed, biomaterial adherent macrophages, derived from monocytes, are fusing to form multinucleated foreign body giant cells on the surface of the biomaterial [8,9].

The form and topography of the surface of the biomaterial determines the composition of the foreign body reaction. With biocompatible materials, the composition of the foreign body reaction in the implant site may be controlled by the surface properties of the biomaterial, the form of the implant, and the relationship between the surface area of the biomaterial and the volume of the implant. Porous scaffold materials have high surface-to-volume ratios and can be expected to display large numbers of macrophages and foreign body giant cells. The foreign body reaction consisting mainly of macrophages and foreign body giant cells may persist at the tissue/implant interface for the lifetime of the implant.

Macrophages and foreign body giant cells are capable of releasing acid, enzymes, and reactive oxygen and nitrogen intermediates that can degrade and modify the surfaces to which they are adherent. The foreign body response with macrophages and foreign body giant cells has been identified as the principal cell types responsible for polyurethane biodegradation in clinical devices such as pacemaker leads. It can be anticipated that macrophages and foreign body giant cells at the surfaces of tissue-engineered devices can lead to destruction of the device and its components. For these reasons, mitigation and more preferably, total inhibition, of the foreign body response with macrophages and foreign body giant cells is desirable for tissue-engineered devices. Although the foreign body response may be inhibited through the use of pharmacologic agents, that is, dexamethasone, these agents are broad in their action and may adversely influence other types of cells and events in the normal wound healing response. Genetic engineering approaches to modulate macrophage and foreign body giant cell behavior are scientifically interesting but may provide tortuous and time-consuming regulatory constraints in their development for human application. Approaches targeting macrophage adhesion and activation may be helpful in developing viable tissue-engineered devices. Material surface chemistry may control monocyte adhesion that, of course, would significantly affect subsequent macrophage formation. Also, material surface chemistry may control adherent macrophage apoptosis, that is, programmed cell death, that renders potentially harmful macrophages nonfunctional, while the surrounding environment of the implant remains unaffected [10]. The level of adherent macrophage apoptosis appears to be inversely related to the surface's ability to promote fusion of macrophages into foreign body giant cells. This appears to be a mechanism by which adherent macrophages escape apoptosis.

The end-stage healing response to devices is generally fibrosis or fibrous encapsulation. This, of course, is the replacement of normal or injured tissue by scar or fibrous tissue formation. The replacement of normal or injured tissue by connective tissue that constitutes the fibrous capsule may be deleterious to the function of the tissue-engineered device [11,12]. Well-formed fibrous capsules are both acellular and avascular. The lack of vascularity within the fibrous capsule would certainly indicate that the fibrous encapsulated tissue-engineered device would not be vascularized and cells in the tissue-engineered device would eventually undergo ischemic cell death.

It is clear that the end-stage healing response with fibrous encapsulation of the tissue-engineered device and the presence of the foreign body reaction with macrophages and foreign body giant cells at the interface between the fibrous capsule and the tissue-engineered device would ultimately lead to failure of the tissue-engineered device. To achieve viable and functional tissue-engineered devices, at least until the target tissue or organ has been restored, control of the adverse aspects of the inflammatory and wound healing responses must be achieved. This continues as a challenge for the development of tissue-engineered devices for human application.

The inflammatory (innate) and immune (adaptive) responses have common components. It is possible to have inflammatory responses only with no adaptive immune response. In this situation, both humoral and cellular components that are shared by both types of responses may only participate in the inflammatory response. Table 7.3 indicates the common components to the inflammatory (innate) and immune

TABLE 7.3 Common Components in the Inflammatory (Innate) and Immune (Adaptive) Responses

Humoral components
    Complement cascade components
    Immunoglobulins
Cellular components
    Macrophages
    NK (natural killer) cells
    Dendritic cells
    Cells with dual phagocytic and antigen presenting capabilities

TABLE 7.4 Cell Types and Function in the Adaptive Immune System

Cell type	Function
Macrophages (APC)	Process and present antigen to immunocompetent T cells phagocytosis Activated by cytokines, that is, IFN-γ, from other immune cells
T cells	Interact with antigen presenting cells (APCs) and are activated through two required cell membrane interactions Facilitate target cell apoptosis Participate in transplant rejection (Type IV hypersensitivity)
B cells	Form plasma cells that secrete immunoglobulins (IgG, IgA, and IgE) Participate in antigen–antibody complex mediated tissue damage (Type III hypersensitivity)
Dendritic cells (APC)	Process and present antigen to immunocompetent T cells Utilize Fc receptors for IgG to trap antigen–antibody complexes
NK (natural killer) cells (Non-T, Non-B lymphocytes)	Innate ability to lyse tumor, virus infected, and other cells without previous sensitization Mediates T and B cell function by secretion of IFN-γ

(adaptive) responses. Macrophages are known as professional antigen presenting cells responsible for the initiation of the adaptive immune response.

7.3 Immune Responses

The acquired or adaptive immune system acts to protect the host from foreign agents or materials and is usually initiated through specific recognition mechanisms and the ability of humoral and cellular components to recognize the foreign agent or material as being "nonself" [11–15]. Generally, the adaptive immune system may be considered as having two components: humoral or cellular. Humoral components include antibodies, complement components, cytokines, chemokines, growth factors, and other soluble mediators. These humoral components are synthesized by cells of the immune response and, in turn, function to regulate the activity of these same cells and provide for communication between different cells in the cellular component of the adaptive immune response. Cells of the immune system arise from stem cells in the bone marrow (B lymphocytes) or the thymus (T lymphocytes) and differ from each other in morphology, function and the expression of cell surface antigens (Table 7.4). They share the common features of maintaining cell surface receptors that assist in the recognition and elimination of foreign materials. Regarding tissue-engineered devices, the adaptive immune response may recognize the biological components, modifications of the biological components, or degradation products of the biological components, commonly known as antigens, and initiate immune response through humoral or cellular mechanisms.

TABLE 7.5   Effector T Lymphocytes in Adaptive Immunity

TH1 helper cells	CD4+ Proinflammatory Activation of macrophages Produces IL-2, interferon-γ (IFN-γ), IL-3, tumor necrosis factor-α, GM-CSF, macrophage chemotactic factor (MCF), migration inhibitor factor (MIF) induce IgG2a
TH2 helper cells	CD4+ Anti-inflammatory Activation of B cells to make antibodies Produces IL-4, IL-5, IL-6, IL-10, IL-3, GM-CSF, and IL-13 Induce IgG1
Cytotoxic T cells (CTL)	CD8+ Induce apoptosis of target cells Produce IFN-γ, TNF-β, and TNF-α Release cytotoxic proteins

Components of the humoral immune system play important roles in the inflammatory responses to foreign materials. Antibodies and complement components C3b and C3bi adhere to foreign materials, act as opsonins and facilitate phagocytosis of the foreign materials by neutrophils and macrophages that have cell surface receptors for C3b. Complement component C5a is a chemotactic agent for neutrophils, monocytes, and other inflammatory cells and facilitates the immigration of these cells to the implant site. The complement system is composed of classic and alternative pathways that eventuate in a common pathway to produce the membrane attack complex (MAC), which is capable of lysing microbial agents. The complement system, that is, complement cascade, is closely controlled by protein inhibitors in the host cell membrane that may prevent damage to host cells. This inhibitory mechanism may not function when nonhost cells are used in tissue-engineered devices.

The T (thymus-derived) lymphocytes are significant cells in the cell-mediated adaptive immune response and their cell-adhesion molecules play a significant role in lymphocyte migration, activation and effector function. The specific interaction of cell membrane adhesion molecules, sometimes also called ligands or antigens, with antigen-presenting cells produce specific types of lymphocytes with specific functions. Table 7.4 indicates cell types and function in the adaptive immune response. Obviously, the functions of these cells are more numerous than indicated in Table 7.4 but the major function of these cells is provided to indicate similarities and differences in the interaction and responsiveness of these cells. Effector T cells (Table 7.5) are produced when their antigen-specific receptors and either the CD4 or the CD8 co-receptors bind to peptide-MHC (major histocompatibility complex) complexes. A second, co-stimulatory signal is also required and this is provided by the interaction of the CD28 receptor on the T cell and the B7.1 and B7.2 glycoproteins of the immunoglobulin superfamily present on antigen-presenting cells. B lymphocytes bind soluble antigens through their cell-surface immunoglobulin and thus can function as professional antigen-presenting cells by internalizing the soluble antigens and presenting peptide fragments of these antigens as MHC:peptide complexes. Once activated, T cells can synthesize the T cell growth factor interleukin-2 and its receptor. Thus, activated T cells secrete and respond to interleukin-2 to promote T cell growth in an autocrine fashion.

Cytokines are the messenger molecules of the immune system. Most cytokines have a wide spectrum of effects, reacting with many different cell types, and some are produced by several different cell types. Table 7.6 presents common categories of cytokines and lists some of their general properties. It should be noted that while cytokines can be subdivided into functional groups, many cytokines such as IL-1, TNF-α, and IFN-γ are pleotropic in their effects and regulate, mediate, and activate numerous responses by numerous cells.

Cytokines produce their effects in three ways. The first type of effect is the autocrine effect in which the cytokine acts on the same cell that produced it. An example is when IL-2 produced by activated T cells promotes T-cell growth. The second way is when a cytokine affects other cells in its vicinity. This

TABLE 7.6   Cytokines and Their Effects

Cytokine	Effect
IL-1, TNF-α, INF-γ, IL-6	Mediate natural immunity
IL-1, TNF-α, IL-6	Initiate nonspecific inflammatory responses
IL-2, IL-4, IL-5, IL-12, IL-15, and TGF-β	Regulate lymphocyte growth, activation, and differentiation
IL-2 and IL-4	Promote lymphocyte growth and differentiation
IL-10 and TGF-β	Down-regulate immune responses
IL-1, INF-γ, TNF-α, and MIF	Activate inflammatory cells
IL-8	Produced by activated macrophages and endothelial cells
	Chemoattractant for neutrophils
MCP-1, MIP-α, and RANTES	Chemoattractant for monocytes and lymphocytes
GM-CSF and G-CSF	Stimulate hematopoiesis
IL-4 and IL-13	Promote macrophage fusion and foreign body giant cell formation

TABLE 7.7   Mechanisms of Immune Mediated Responses

Type	Immune mechanism
TYPE I Anaphylactic	IgE antibodies, produced by B cells, affect the immediate release of basophilic amines and other mediators from basophils and mast cells followed by the recruitment of other inflammatory cells
TYPE II Cytotoxic	Formation and binding of IgG and IgM to antigens on target cell surfaces that facilitate phagocytosis of the target cell or lysis of the target cell by activated complement components
TYPE III Immune complex	Circulating antigen–antibody complexes activate complement whose components are chemotactic for neutrophils that release enzymes and other toxic mediators leading to cellular and tissue injury
TYPE IV Cell-mediated (delayed)	Sensitized T lymphocytes release cytokines and other mediators that lead to cellular and tissue injury

is the paracrine effect and an example is when IL-7 produced by bone marrow stromal cells promotes the differentiation of B-cell progenitors in the bone marrow. The third way is when the cytokine affects many cells systemically. This is the endocrine effect and an example is when circulating IL-1 and TNF-α produce the acute-phase response during inflammation.

This brief and very limited overview of the humoral and cellular mediated immune responses due to space limitations, now provides a basis for understanding mechanisms of immunologic tissue injury, which is also called hypersensitivity reactions. Hypersensitivity reactions can be initiated either by the interaction of antigen with humoral antibody or by cell-mediated immune mechanisms. Table 7.7 lists the four types of hypersensitivity reactions together with a very brief description of their immune mechanism. Type I hypersensitivity (anaphylactic) is generally defined as a rapidly developing immunologic reaction occurring within minutes after the combination of an antigen with antibody bound to mast cells or basophils in individuals previously sensitized to the antigen. Type II hypersensitivity (cytotoxic) is mediated by antibodies directed toward antigens present on the surface of cells or other tissue components. Three different antibody-dependent mechanisms may be involved in this type of reaction: complement-dependent reactions, antibody-dependent cell-mediated cytotoxicity, or antibody-mediated cellular dysfunction. Type III hypersensitivity (immune complex mediated) reaction is induced by antigen-antibody complexes that produce tissue damage as a result of their capacity to activate the complement system. Type IV hypersensitivity (cell-mediated) reactions are initiated by specifically sensitized T lymphocytes. This reaction includes the classic delayed-type hypersensitivity reaction initiated by CD^4+ T cells and direct cell cytotoxicity mediated by CD^8+ T cells. Immunologic reactions that occur with organ transplant rejection also offer insight into potential immune responses to tissue-engineered devices. Mechanisms involved in organ transplant rejection include T cell-mediated reactions by direct and indirect pathways and antibody-mediated reactions. Immune responses may be avoided or diminished by using autologous or isogeneic cells in cell/polymer scaffold constructs. The use of allogeneic or xenogenic cells incorporated into the device require prevention of immune rejection by immune suppression of the host, induction of tolerance in the host, or immunomodulation of the tissue-engineered construct. The development of tissue-engineered constructs by immunoisolation using polymer membranes and the use of nonhost cells have been compromised by immune responses. In this concept, a polymer membrane is used to encapsulate nonhost cells or tissues thus separating them from the host immune system. However, antigens shed by encapsulated cells were released from the device and initiated immune responses [12,16,17].

Although exceptionally minimal and superficial in its presentation, the previously discussed humoral and cell-mediated immune responses demonstrate the possibility that any known tissue-engineered construct may undergo immunologic tissue injury. To date, our understanding of immune mechanisms and their interactions with tissue-engineered constructs is markedly limited. One of the obvious problems is that preliminary studies are generally carried out with nonhuman tissues and immune reactions result when tissue-engineered constructs from one species are used in testing the device in another species. Ideally, tissue-engineered constructs would be prepared from cells and tissues of a given species and subsequently tested in that species. Whereas this approach does not guarantee that immune responses will not be present, the probability of immune responses in this type of situation is markedly decreased.

The following examples provide perspective to these issues. They further demonstrate the detailed and in-depth approach that must be taken to appropriately and adequately evaluate tissue-engineered constructs or devices and their potential adverse responses.

Babensee et al. [18,19] have tested the hypothesis that the biomaterial component of a medical device, by promoting an inflammatory response can recruit antigen-presenting cells (APCs, e.g., macrophages and dendritic cells) and induce their activation, thus acting as an adjuvant in the immune response to foreign antigens originating from the histological component of the device. Utilizing polystyrene and polylactic-glycolic acid microparticles and polylatic-glycolic scaffolds together with their model antigen, ovalbumin, in a mouse model for 18 weeks, Babensee et al. demonstrated that a persistent humoral immune response that was Th2 helper T cell dependent, as determined by the IgG1, was present. These findings indicated that activation of CD^4+ T cells and the proliferation and isotype switching of B-cells had occurred. A Th1 immune response, characterized by the presence of IgG2a was not identified. Moreover, the humoral immune responses for all three types of microparticles were similar indicating that the production of antigen-specific antibodies was not material chemistry-dependent in this model. Babensee suggests that the presence of the biomaterial functions as an adjuvant for initiation and promotion of the immune response and augments the phagocytosis of the antigen with expression of MHC class II and co-stimulatory molecules on APCs with the presentation of antigen to CD^4+ T cells.

Cleland et al. [20] have shown the significance of the use of appropriate animal models in characterizing protein/biodegradable polymer constructs. Utilizing biodegradable PLGA microspheres containing recombinant human growth hormone (rhGH), they used Rhesus monkeys, transgenic mice expressing hGH and normal control (Balb/C) mice in their in vivo studies. The Rhesus monkeys were used for serum assays in determining growth hormone release as well as tissue responses to the injected microcapsule formulations. Placebo injection sites were also used and a comparison of the injection sites from rhGH PLGA microspheres and placebo PLGA microspheres demonstrated a normal inflammatory and wound healing response with a normal focal foreign body reaction. To further examine the tissue response and potential for immune reactions, transgenic mice were used to assess the immunogenicity of the rhGH PLGA formulation. Transgenic mice expressing a heterologous protein have been previously used for assessing the immunogenicity of sequence or structural mutant proteins [21]. With the transgenic animals, no detectable antibody response to the rhGH was found. In contrast, the Balb/C control mice had a rapid onset of high titer antibody response to the rhGH PLGA formulation.

Immunotoxicity is any adverse effect on the function or structure of the immune system or other systems as a result of an immune system dysfunction. Adverse or immunotoxic effects occur when humoral or cellular immunity needed by the host to defend itself against infections or neoplastic diseases (immunosuppression) or unnecessary tissue damage (chronic inflammation, hypersensitivity or autoimmunity) is compromised. *In vivo* responses indicating immunotoxicity include histopathological changes, humoral responses, host resistance, clinical symptoms, and cellular responses (T cells, natural killer cells, macrophages, and granulocytes). The inflammatory response considered to be immunotoxic is persistent chronic inflammation. It is this persistent chronic inflammation that is of concern as immune granuloma formation and other serious immunological reactions such as autoimmune disease may occur. In biological response evaluation, it is important to discriminate between the short-lived chronic inflammation that is a component of the normal inflammatory and healing responses vs. long-term, persistent chronic inflammation that may indicate an adverse immunological response.

Immunosuppression may occur when antibody and T cell responses (adaptive immune response) are inhibited. A potentially significant consequence of this type of response is more frequent in serious infections resulting from reduced host defense. Immunostimulation may occur when unintended or inappropriate antigen-specific or nonspecific activation of the immune system is present. From a tissue-engineering perspective, antibody and cellular immune responses to a foreign protein may lead to unintended immunogenicity. Enhancement of the immune response to an antigen by a biomaterial with which it is mixed *ex vivo* or *in situ* may lead to adjuvancy that is a form of immunostimulation. This effect must be considered when biodegradable controlled release systems are designed and developed for use as vaccines. Autoimmunity is the immune response to the body's own constituents that are considered in this response to be autoantigens. An autoimmune response, indicated by the presence of autoantibodies or T lymphocytes that are reactive with host tissue or cellular antigens may, but not necessarily, result in autoimmune disease with chronic, debilitating and sometimes life-threatening tissue and organ injury.

Direct measures of immune system activity by functional assays are the most important types of tests for immunotoxicity [22–25]. Functional assays are generally more important than tests for soluble mediators, which are more important than phenotyping. Functional assays include skin testing, immunoassays (e.g., ELISA), lymphocyte proliferation, plaque-forming cells, local lymph node assay, mixed lymphocyte reaction, tumor cytotoxicity, antigen presentation, and phagocytosis. As with any type of test for biological response evaluation, immunotoxicity tests should be valid and have been shown to provide accurate, reproducible results that are indicative of the effect being studied and are useful in a statistical analysis. This implies that appropriate control groups are also included in the study design.

References

[1] Anderson, J.M., Biological responses to materials, *Ann. Rev. Mater. Res.*, 31, 81, 2001.
[2] Anderson, J.M., Mechanisms of inflammation and infection with implanted devices, *Cardiovasc. Pathol.*, 2, 33S, 1993.
[3] Cotran, R.Z., Kumar, V., and Robbins, S.L., Eds., Inflammation and repair in *Pathologic Basis of Disease*, 6th ed., W.B. Saunders, Philadelphia, 1999, p. 50.
[4] Gallin, J.L. and Snyderman, R., Eds., *Inflammation: Basic Principles and Clinical Correlates*, 3rd ed., Raven Press, New York, 1999.
[5] Babensee, J.E. et al., Host responses to tissue engineered devices, *Adv. Drug Del. Rev.*, 33, 111, 1998.
[6] Anderson, J.M., Multinucleated giant cells, *Curr. Opin. Hematol.*, 7, 40, 2000.
[7] Henson, P.M., The immunologic release of constituents from neutrophil leukocytes: II. Mechanisms of release during phagocytosis, and adherence to nonphagocytosable surfaces, *J. Immunol.*, 107, 1547, 1971.
[8] McNally, A.K. and Anderson, J.M., Beta1 and beta2 integrins mediate adhesion during macrophage fusion and multinucleated foreign body giant cell formation, *Am. J. Pathol.*, 160, 621, 2002.
[9] McNally, A. and Anderson, J.M., Interleukin-4 induces foreign body giant cells from human monocytes/macrophages. Differential lymphokine regulation of macrophage fusion leads to morphological variants of multinucleated giant cells, *Am. J. Pathol.*, 147, 1487, 1995.
[10] Brodbeck, W.G. et al., Influence of biomaterials surface chemistry on apoptosis of adherent cells, *J. Biomed. Mater. Res.*, 55, 661, 2001.

[11] Janeway, C.A. et al., *Immunobiology 5: The Immune System in Health and Disease*, Garland Publishing Inc., New York, 2001, p. 1.

[12] Babensee, J.E. et al., Host response to tissue engineered devices, in *Advanced Drug Delivery Reviews*, Vol. 33, Elsevier, Amsterdam, 1998, p. 111.

[13] Lefell, M.S., Donnenberg, A.D., and Rose, N.R., Eds., *Immunologic Diagnosis of Autoimmune Disease*, CRC Press, Boca Raton, FL, 1997, p. 111.

[14] Cohen, I.R. and Miller, A., Eds., *Autoimmune Disease Models, A Guidebook*, Academic Press, New York, 1994.

[15] Anderson, J.M. and Langone, J.J., Issues and perspectives on the biocompatibility and immunotoxicity evaluation of implanted controlled release systems, *J. Control. Release*, 57, 107, 1999.

[16] Brauker, J.H. et al., Neovascularization of synthetic membranes directed by membrane microarchitecture, *J. Biomed. Mater. Res.*, 29, 1517, 1995.

[17] Brauker, J. et al., Neovascularization of immuno-isolation membranes: the effect of membrane architecture and encapsulated tissue, *Cell Transplant.*, 1, 163, 1992.

[18] Matzelle, M.M. and Babensee, J.E., Humoral immune responses to model antigen co-delivered with biomaterials used in tissue engineering, *Biomaterials*, 25, 295, 2004.

[19] Babensee, J.E., Stein, M.M., and Moore, L.K., Interconnections between inflammatory and immune responses in tissue engineering, *Ann. NY Acad. Sci.*, 961, 360, 2002.

[20] Cleland, J.L. et al., Recombinant human growth hormone poly(lactic-co-glycolic acid) (PLGA) microspheres provide a long lasting effect, *J. Control. Release*, 49, 193, 1997.

[21] Stewart, T.A. et al., Transgenic mice as a model to test the immunogenicity of proteins altered by site-specific mutagenesis, *Mol. Biol. Med.*, 6, 275, 1989.

[22] Rose, N.R., de Mecario, E.C., Folds, J.D., Lane, H.C., and Nakamura, R.M., Eds., *Manual of Clinical Laboratory Immunology*, ASM Press, Washington, DC, 1997.

[23] Smialowicz, R.J. and Holsapple, M.P., Eds., *Experimental Immunotoxicology*, CRC Press, Boca Raton, FL, 1996.

[24] Burleson, G.R., Dean, J.H., and Munson, A.E., Eds., *Methods in Immunotoxicology*, Wiley-Liss, New York, 1995.

[25] Coligan, J.E., Kruisbeek, A.M., Margulies, D.H., Shevach, E.M., and Strober, R., Eds., *Current Protocols in Immunology*, Greene Publishing Associates and Wiley Interscience, New York, 1992.

II

Enabling Technologies

8

Polymeric Scaffolds for Tissue Engineering Applications

Diana M. Yoon
John P. Fisher
University of Maryland

8.1 Introduction

The incorporation of polymeric scaffolds in tissue regeneration occurred in the early 1980s, and it continues to play a vital role in tissue engineering [1–3]. The function of a degradable scaffold is to act as a temporary support matrix for transplanted or host cells so as to restore, maintain, or improve tissue. Scaffolds may be created from various types of materials, including polymers. There are two main classes of polymers, based upon their source: natural or synthetic. Polymeric scaffolds may be used to support a variety of cells for numerous tissues within the body. The design of a polymeric scaffold plays a significant role in proper cell growth. Therefore, several important properties must be considered: fabrication, structure, biocompatibility, biodegradability, and mechanical strength. This review will discuss natural and synthetic polymers, as well as the properties that scaffolds exhibit.

8.2 Natural Polymers for Scaffold Fabrication

There are two major classes of natural polymers used as scaffolds: polypeptides and polysaccharides (Table 8.1, Figure 8.1). Natural polymers are typically biocompatible and enzymatically biodegradable. The main advantage for using natural polymers is that they contain bio-functional molecules that aid the attachment, proliferation, and differentiation of cells. However, disadvantages of natural polymers do exist. Depending upon the application, the previously mentioned enzymatic degradation may inhibit

TABLE 8.1 Properties of Degradable Natural Homopolymers That Have Been Utilized in the Fabrication of Tissue Engineering Scaffolds

	Natural polymer	Curing method	Primary degradation method	Primary degradation products	References
A	Agarose	Entanglement	Enzymatic agarases	Oligosaccharides	[4,8]
B	Alginate	Cross-linking	Alginate lyases	Mannuronic acid and guluronic acid	[4,10–13]
C	Hyaluronic acid	Entanglement	Enzymatic hyaluronidase	$\beta(1$–$4)$ linked glucuronic acid and glucosamine	[4,8,21,22]
D	Chitosan	Cross-linking	Enzymatic chitosanase	Glucosamines	[4,8,22,25,26]
E	Collagen	Cross-linking	Enzymatic collagenase or lysozyme	Various peptides depending on sequence	[8,31–33]
F	Gelatin	Entanglement	Enzymatic collagenases	Various peptides depending on sequence	[4,8,13,22]
G	Silk	Entanglement	Proteolytic proteases	Various peptides depending on sequence	[43]

function. Further, the rate of this degradation may not be easily controlled. Since the enzymatic activity varies between hosts, so will the degradation rate. Therefore it may be difficult to determine the lifespan of natural polymers *in vivo*. Additionally, natural polymers are often weak in terms of mechanical strength but cross-linking these polymers have shown to enhance their structural stability.

8.2.1 Polysaccharides

8.2.1.1 Agarose

Agarose is a polysaccharide polymer extracted from algae (Table 8.1, Figure 8.1). Its molecular structure contains an alternating copolymer linkage of 1,4-linked 3,6 anhydro-α-L-galactose and 1,3-linked β-D-galactose [4]. Agarose is water-soluble due to the presence of hydroxyl groups. Two agarose chains interact with each other through hydrogen bonding to form a double helix. At low temperatures the agarose chains form a thermally reversible gel. Thus, at high temperatures agarose gels may be made water-soluble. Many of the properties of agarose gels, particularly strength and permeability, can be adjusted by altering the concentration of agarose [4]. For example, low concentrations of agarose lead to a highly porous entangled structure with limited mechanical strength. Agarose in its native state is enzymatically degradable by agarases.

For tissue culture systems, agarose hydrogels are widely investigated because they permit growth of cells and tissues in a three-dimensional suspension [5]. Agarose gels have been found to maintain chondrocytes, the predominant cell type in cartilage, in culture for 2 to 6 weeks [6]. Furthermore, agarose hydrogels embedded with chondrocytes allow the expression of type II collagen and proteoglycans [5,7]. These results show promise for using agarose gels in cartilage repair. Additionally, agarose gels have been investigated as a matrix for nerve regeneration [8,9]. Dorsal root ganglia (DRG), a neural cell, has been encapsulated in agarose hydrogels. The influence of increasing porosity and incorporating charged biopolymers (e.g., chitosan) produced an increase in neurite extension of DRG isolated from chicks [9].

8.2.1.2 Alginate

Alginate is a naturally derived water-soluble polysaccharide obtained from algae (Table 8.1, Figure 8.1) [4]. It is a polyanion composed of two repeating monomer units: β-D-mannuronate and α-L-guluronate [10,11]. Alginate is readily degraded at the $\beta(1$–$4)$ linkage site, located in between the monomer units, by alginate lyases [10]. The physical and mechanical properties of alginate are dependent on the length and proportion of the guluronate block within the chain. Due to alginate's polyelectrolytic nature, it readily forms into an ionotropic cross-linked hydrogel when exposed to divalent ions, the most common being calcium ions [12,13]. Furthermore, alginate gels may be solubilized when these cations are removed

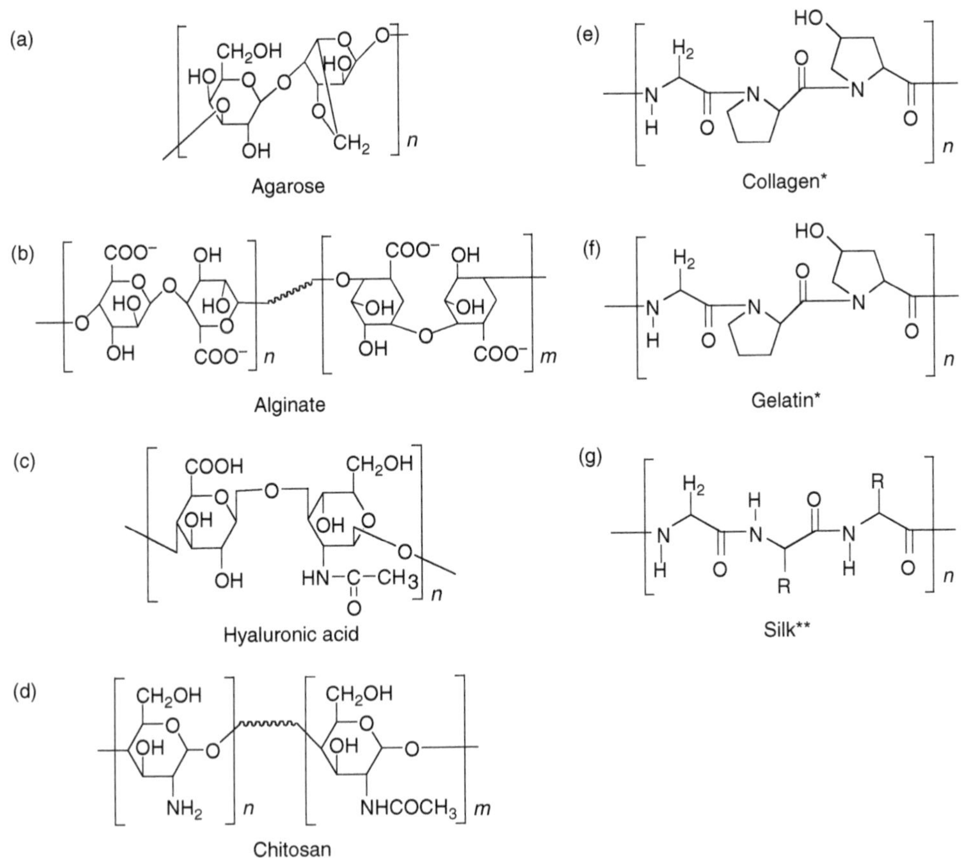

FIGURE 8.1 Repeating structural unit of degradable natural homopolymers under investigation as scaffold materials for tissue engineering applications. (*Collagen and gelatin have variable sequences, but are predominately glycine, proline, and hydroxyproline. **Silks have a glycine rich sequence, with glycine repeating every second or third amino acid residue. The variable residues are primarily alanine or serine.)

or sequestered. This ease in gel reversibility allows alginate to be especially useful for studies requiring the retrieval of encapsulated cells. Alginate does not contain cellular recognition proteins, limiting cell attachment to the natural polymer. By cross-linking alginate with poly(ethylene glycol)-diamine the mechanical properties of alginate can be controlled and even improved [4].

A variety of different cells have been found to maintain their morphology in alginate scaffolds [10,14–17]. The adhesion of fibroblasts in alginate sponges was found not to be affected by the porosity of the scaffold [14]. Chondrocytes embedded in alginate proliferated and expressed type II collagen [18]. Hepatocytes, the functional cell in liver, were also found to grow in alginate scaffolds. When hepatocytes were seeded in three-dimensional porous alginate, albumin was secreted, indicating proper cell function [19]. Other types of cells studied in alginate include cardiomyocytes, rat marrow cells, and Schwann cells [10,17,20].

8.2.1.3 Hyaluronic Acid

Hyaluronic acid (hyaluronan, HA) is a naturally occurring polysaccharide with a $\beta(1\text{–}3)$ linkage of two sugar units: D-glucuronate and N-acetyl-D-glucosamine (Table 8.1, Figure 8.1). Hyaluronic acid is an abundant glycosaminoglycan within the extracellular matrix of tissues that promotes early inflammation critical for wound healing [21]. Furthermore, hyaluronic acid is nonimmunogenic and nonadhesive, making it an attractive option for biomedical applications. In dilute concentrations hyaluronic acid has a random coil structure that becomes entangled as its concentration increases [22]. Some disadvantages of hyaluronic acid are that it is highly water-soluble and it degrades rapidly by enzymes, such as hyaluronidase. However, when fabricated into a hydrogel, hyaluronic acid is often more stable [4,8,21].

Currently, an ester derivative of hyaluronic acid has been used as a tissue engineering scaffold [8,23,24]. Adipose precursor cells proliferated and differentiated in HA sponges [23,24]. Additionally, HA has been used for osteochondral repair by incorporating mesenchymal progenitor cells, which differentiate into osteoblasts and chondrocytes [8]. Furthermore, the interaction of chondrocytes embedded in HA has resulted in the expression of collagen type II, forming a tissue similar to native cartilage [23].

8.2.1.4 Chitosan

Chitosan is a partially deacetylated derivative of chitin, found in the exoskeleton of crustaceans and insects (Table 8.1, Figure 8.1). It is a linear polysaccharide composed of $\beta(1-4)$ linked D-glucosamine with randomly dispersed N-acteyl-D-glycosamine groups [4,22,25,26]. Less than 40% of chitosan is composed of the N-acteyl-D-glycosamino group, which makes its molecular structure similar to glycosaminoglycans [4,8]. Chitosan is easily degraded by enzymes such as chitosanase and lysozyme. Chitosan is catonic with semi-crystalline properties of a high charge density. These attributes allow chitosan to be insoluble above pH 7 and fully soluble below pH 5. At high pH solutions, chitosan can be gelled into strong fibers [26]. Furthermore, hydrogels can also be formed by either ionic bonding or covalent cross-linking, using cross-linking agents such as glutaraldehyde [4].

Chitosan is a well-defined matrix and as a porous scaffold, hepatocytes were found to maintain a rounded morphology when cultured within chitosan scaffolds [27]. Results showed an increase in albumin secretion as well as urea synthesis, vital metabolic activities of liver cells [27]. Additionally, cross-linking chitosan gels with glutaraldehyde showed increased urea formation by hepatocytes [28]. Furthermore, chitosan has shown promise as an orthopaedic scaffold. When a sponge form of chitosan was seeded with rat osteoblasts, alkaline phosphatase production was detected, as well as bone spicules, indicating initial bone formation [29]. Also, chitosan scaffolds were found to support chondrocyte attachment and expression of extracellular matrix proteins [30].

8.2.2 Polypeptides

8.2.2.1 Collagen

Collagen is a major component of structural mammalian tissues (Table 8.1, Figure 8.1). Currently, 19 different types of collagen have been found, with the most abundant being type I collagen [31]. Collagen is composed of three polypeptide chains that intertwine with one another to form a triple helix. The polypeptide chains have a repeating sequence of glycine-X–Y, where X and Y are most often found to be proline and hydroxyproline. The chains are held together through hydrogen bonding of the peptide bond in glycine and an adjacent peptide carbonyl group [32]. The presence of adhesion domains allows collagen to display an attractive surface for cell attachment [31]. Some drawbacks in collagen are its high variability due to the numerous forms. Collagen is enzymatically biodegradable, and has a tendency to degrade quickly, limiting its mechanical properties. Cross-linking collagen with glutaraldehyde decreases its degradation rate [33].

Collagen naturally promotes healing wounds by promoting blood coagulation [31]. These attributes allow collagen to be used as a scaffold for cells within the epithelium, such as fibroblasts and keratinocytes [34–36]. Type II collagen is suitable for attaching chondrocytes and aiding in their proliferation and differentiation [37]. Additionally, rat hepatocytes attached onto collagen–chitosan scaffolds secreted albumin, which confirms cell function [38]. Furthermore, corneal keratocytes cultured on collagen sponges synthesized proteoglycans, indicating corneal extracellular matrix formation [39].

8.2.2.2 Gelatin

Gelatin is a collagen derivative acquired by denaturing the triple-helix structure of collagen into single-strand molecules (Table 8.1, Figure 8.1). It is water-soluble, and entangles easily to form into a gel through changes in temperature [4,13]. Gelatin is broken down enzymatically by various collagenases. The primary form of a gelatin scaffold is a hydrogel. Stabilization of gelatin hydrogels occurs through chemical cross-linking with agents such as glutaraldehyde [22].

Gelatin has been used to support cells for orthopaedic applications. Rat marrow stromal osteoblasts encapsulated in gelatin microparticles retained their phenotype [40]. Also, porous gelatin scaffolds supported human adipose derived stem cell attachment and differentiation into a variety of cell lineages. These constructs expressed a chondrogenic phenotype, indicated by the expression of hydroxyproline and glycosaminoglycans, both of which are present in the extracellular matrix of cartilage [41]. Furthermore, nonorthopaedic cell types, such as respiratory epithelial cells and cardiocyocytes, have been shown to attach and form rounded morphologies in gelatin scaffolds [17,42].

8.2.2.3 Silk

Silks are fibers that have a protein polymer basis (Table 8.1, Figure 8.1) [43]. There are two main sources of silks: spiders and silkworms. The primary structure of silks is a repetitive sequence of proteins that are glycine-rich. The secondary structure of silks is a β-sheet, which allows silks to exhibit enormous mechanical strength and is the primary reason for its strong interest as a scaffold. Most types of silks undergo slow proteolytic degradation by proteases, such as chymotrypsin [43]. A drawback for using silks is the potential formation of a granuloma as well as an allergic response. There has been a positive attachment and growth response of fibroblasts on fibroin silks. And recently, osteoblasts seeded on silks have displayed bone growth characteristics [44,45].

8.3 Synthetic Polymers for Scaffold Fabrication

Polymers that are chemically synthesized offer several notable advantages over natural-origin polymers. A major advantage of synthetic polymers is that they can be tailored to suit specific functions and thus exhibit controllable properties. Furthermore, since many synthetic polymers undergo hydrolytic degradation, a scaffold's degradation rate should not vary significantly between hosts. The most common synthetic polymers are polyesters. Other types include polyanhydrides, polycarbonates, and polyphosphazenes (Table 8.2, Figure 8.2). A significant disadvantage for using synthetic polymers is that some degrade into unfavorable products, often acids. At high concentrations of these degradation products, local acidity may increase, resulting in adverse responses such as inflammation or fibrous encapsulation [8,46,47].

8.3.1 Polyesters

8.3.1.1 Poly(glycolic) Acid

Poly(glycolic) acid (PGA) is a linear aliphatic polyester in the family of poly(α-hydroxy esters) (Table 8.2, Figure 8.2). It is formed by ring-opening polymerization of cyclic diesters of glycolide [46]. Due to the crystallinity of glycolide, PGA is a highly crystalline polymer that has a high melting point (185 to 225°C) and low solubility in organic solvents [47,48]. Further, PGA is hydrophilic and undergoes bulk degradation, often leading to a sudden loss in mechanical strength. PGA degrades hydrolytically into glycolic acid, which is metabolized in the body. An unfortunate attribute of glycolic acid is that at high concentrations it lowers the pH of the surrounding tissue, causing inflammation and possible tissue damage [46].

Poly(glycolic) acid polymers have been investigated as a scaffold to support various types of cell growth. Bovine chondrocytes seeded on PGA scaffolds *in vitro* expressed sulfated glycosaminoglycans 25% more than native cartilage [49]. *In vivo* results using a mouse model were comparable to these *in vitro* results. Other studies seeded myofibroblasts and endothelial cells on a PGA fiber matrix. These cells persisted and expressed elastic filaments and collagen, demonstrating the potential use of PGA scaffolds for heart valve engineering [50]. Hepatocytes have also attached to PGA and initiated albumin synthesis [51]. Some

TABLE 8.2 Properties of Degradable Synthetic Homopolymers That Have Been Utilized in the Fabrication of Tissue Engineering Scaffolds

	Synthetic polymer	Curing method	Primary degradation method	Primary degradation products	References
A	Poly(glycolic acid)	Entanglement	Ester hydrolysis	Glycolic acid	[8,46–48]
B	Poly(L-lactic acid)	Entanglement	Ester hydrolysis	Lactic acid	[8,46–48,53]
C	Poly(D,L-lactic acid-co-glycolic acid)	Entanglement	Ester hydrolysis	Lactic acid and glycolic acid	[8,48,57]
D	Poly(caprolactone)	Entanglement	Ester hydrolysis	Caproic acid	[8,22,48,62–64]
E	Poly(propylene fumarate)	Cross-linking	Ester hydrolysis	Fumaric acid and propylene glycol	[8,67–70]
F	Polyorthoester	Entanglement	Ester hydrolysis	Various acids depending upon R group	[8,22,57,71,72]
G	Polyanhydride	Entanglement	Anhydride hydrolysis	Various acids depending upon R group	[8,75,76]
H	Polyphosphazene	Entanglement	Hydrolysis	Phosphate and ammonia	[8,79]
I	Polycarbonate (tyrosine derived)	Entanglement	Ester and carbonate hydrolysis	Alkyl alcohol and desaminoyrosyl-tyrosine	[53,83,84]
J	Poly(ethylene glycol)/ Polyethylene oxide	Entanglement	Nondegradable	Not applicable	[4,8,32,86–91]
K	Polyurethane	Cross-linking	Ester, urethane, or urea hydrolysis	Diisocyanate and diols	[47,94–96]

investigations have coated PGA with other poly(α-hydroxy esters). Results with these scaffolds have shown that smooth muscle and endothelial cell growth may be supported [52].

8.3.1.2 Poly(L-lactic) Acid

Poly(L-lactic) acid (PLLA) is one of two isomeric forms of poly(lactic acid): D(−) and D,L(−) (Table 8.2, Figure 8.2). Similar to PGA, PLLA is classified as a linear poly(α-hydroxy acid) that is formed by ring-opening polymerization of L-lactide [53]. PLLA is structurally similar to PGA, with the addition of a pendant methyl group. This group increases the hydrophobicity and lowers the melting temperature to 170 to 180°C for PLLA [48]. Additionally, the methyl group hinders the ester bond cleavage of PLLA, and thus decreasing the degradation rate [46]. PLLA typically undergoes bulk, hydrolytic ester-linkage degradation, decomposing into lactic acid. The body is thought to excrete lactic acid in the form of water and carbon dioxide via the respiratory system [47,53]. The hydrophobic nature of PLLA allows for protein absorption and cell adhesion making them suitable as scaffolds.

In recent investigations, results have shown that the hydrophobic surface of PLLA has resulted in decreased adhesion of cells compared to other types of polymers, such as PGA [54]. However, PLLA has been found to support various cell types. Nerve stem cells seeded in PLLA fibers differentiated and expressed positive cues for neurite growth for potential regeneration of neurons [55]. Furthermore, human bladder smooth muscle cells seeded onto PLLA scaffolds also exhibited normal metabolic function and cell growth [16]. There has been similar work done with chondrocytes. The tissue engineered cartilage showed collagen, glycosaminoglycan, and elastin expression, vital for extracellular matrix formation [56].

8.3.1.3 Poly(D,L-lactic acid-co-glycolic acid)

Poly(glycolic acid) and poly(lactic acid) can be copolymerized to form poly(D,L-lactic acid-co-glycolic acid) (PLGA) (Table 8.2, Figure 8.2). This copolymer is amorphous, which decreases the degradation rate and mechanical strength of PLGA compared to PLLA. PLGA typically undergoes bulk degradation by ester hydrolysis. The degradation rate of PLGA can be controlled by adjusting the ratio of PLLA/PGA.

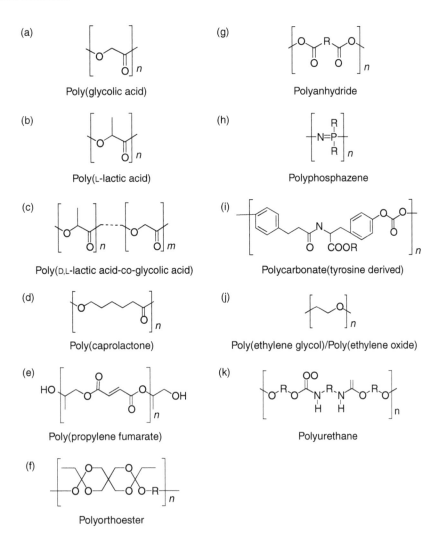

(a) Poly(glycolic acid)

(b) Poly(L-lactic acid)

(c) Poly(D,L-lactic acid-co-glycolic acid)

(d) Poly(caprolactone)

(e) Poly(propylene fumarate)

(f) Polyorthoester

(g) Polyanhydride

(h) Polyphosphazene

(i) Polycarbonate(tyrosine derived)

(j) Poly(ethylene glycol)/Poly(ethylene oxide)

(k) Polyurethane

FIGURE 8.2 Repeating structural unit of degradable synthetic homopolymers under investigation as scaffold materials for tissue engineering applications.

However, it is important to note that the copolymer composition is not linearly related to the mechanical and degradation properties of PLGA. For example, a 50 : 50% ratio of PGA and PLLA typically degrades faster than either homopolymer [48,57].

PLGA has been extensively researched as a tissue engineering scaffold and shown to support the growth of a variety of cell types. For instance, PLGA can facilitate human foreskin fibroblasts to regenerate an extracellular matrix [8]. Similar studies has been conducted with chondrocytes and osteoblastic cells [54,58]. Osteoblastic cells seeded on PLGA scaffolds expressed collagen, fibronectin, laminin, and a variety of other extracellular matrix proteins, indicating proper cell function. Furthermore, PLGA promoted smooth muscle cell growth as well as the production of collagen and elastin [59]. Similar works showed enhancement of axon regeneration by murine neural cells seeded on PLGA scaffolds [60]. Additionally, an epithelial layer has been formed on PGLA by seeded enteroctyes derived from intestinal epithelial cells [61].

8.3.1.4 Poly(ε-caprolactone)

Poly(ε-caprolactone) (PCL) is a aliphatic polyester with semi-crystalline properties (Table 8.2, Figure 8.2). PCL has repeating unit of one ester group and five methylene groups. It is highly water-soluble with a melting temperature of 58 to 63°C [22,48]. PCL is formed through ring-opening polymerization, forming a degradable ester linkage. The degradation of PCL occurs by bulk or surface hydrolysis of these ester linkages, resulting in a byproduct of caproic acid. PCL degrades at a slow rate with results showing that it can persist *in vivo* for up to 2 years [48]. To increase the degradation rate, PCL has been copolymerized with collagen, poly(glycolic acid), and poly(lactic acid), polyethylene oxide [62–64]. The ease with which PCL can be copolymerized with a variety of polymers has made it an attractive component of polymeric scaffolds.

PCL has been investigated as a stabilizing polymer on PGA scaffolds to aid in the formation of spherical aggregates by human biliary epithelial cells (hBEC). These hBECs proliferated on the synthetic scaffold and expressed phenotypic proteins, indicating cell stability [63]. Further studies of PCL as a homopolymer has been demonstrated to support human osteoblast and dermal fibroblast cell viability [65,66]. The results with osteoblasts on porous PCL has shown a production of alkaline phosphatase, a marker of bone mineralization [65]. Also, dermal fibroblasts on mechanically stretched PCL films proliferated and maintained their rounded morphology [66].

8.3.1.5 Poly(propylene fumarate)

Poly(propylene fumarate) (PPF) is a linear polyester with a repeating unit containing two ester groups and one unsaturated carbon–carbon double bond (Table 8.2, Figure 8.2). The hydrolysis of the ester bonds in PPF, forms fumaric acid and propylene glycol as the two primary degradation products. Covalent cross-linking of PPF through its double bond allows cured PPF to possess significant compressive and tensile strength. Furthermore, this cross-linking aids in the formation of a synthetic scaffold *in situ* [67,68]. Cross-linked PPF is a potential orthopaedic, especially bone, engineering material due to the strength that it exhibits [67]. Additional research has also been conducted with PPF copolymerized with PEG (PPF-co-PEG). This hydrophilic copolymer has been shown to support endothelial cell attachment and proliferation [69,70].

8.3.1.6 Polyorthoester

A family of surface eroding polyorthoesters (POE) has been synthesized and studied for tissue engineering applications (Table 8.2, Figure 8.2). One particular form of POE has been considered as a scaffold. This type of POE is created by reacting ketene acetals with diols [71]. The hydrophobic nature of POE's surface allows this polymer to undergo primarily surface degradation [8,72]. By incorporating short acid groups, such as glycolic or lactic acid, the degradation rate of the copolymer can be controlled [72]. Furthermore, the hydrolysis of the orthoester linkages in POE is mostly sensitive to acids and is stable in bases [22,57].

The surface eroding characteristics of POE has led to the research of this polymer for bone reconstruction [8,73]. Surface degradation allows a scaffold to sustain mechanical support for the surrounding tissue since the bulk of the material remains structurally intact. In other studies, hepatocytes were grafted onto polyorthoesters and adhered to the surface of the synthetic scaffold [74].

8.3.2 Other Synthetic Polymers

8.3.2.1 Polyanhydride

Polyanhydrides are a class of hydrophobic polymers containing anhydride bonds, which are highly water-reactive (Table 8.2, Figure 8.2) [75]. Polyanhydrides are also crystalline polymers, with a melting temperature of approximately 100°C [76]. Aliphatic polyanhydrides are synthesized by a dehydration reaction between diacids. The instability of the anhydride bond allows polyanhydrides to degrade within a period of weeks. The degradation rate is predictable for polyanhydrides and can be altered through changes in the hydrophobicity of the diacid building blocks [76]. Polyanhydrides are generally thought to degrade following a surface degradation mechanism.

Similar to polyorthoesters, polyanhydrides have been investigated primarily for orthopaedic applications. However, the modest Young's Modulus for entangled polyanhydride networks has limited their application in weight-bearing environments [47]. This limitation resulted in further studies involving the formation of cross-linked polyanhydrides networks with incorporated imides [53,77]. Scaffolds formed from these networks possess significant mechanical properties, such as compressive strength (30–60 MPa), which are in the intermediate range of cortical and trabecular human bone (5–150 MPa) [47,76–78].

8.3.2.2 Polyphosphazene

Polyphosphazenes are a class of polymers formed from an inorganic backbone containing nitrogen and phosphorous atoms (Table 8.2, Figure 8.2). Polyphosphazene is hydrophobic and typically undergoes hydrolytic surface degradation into phosphate and ammonium salt byproducts. There are a variety of polyphosphazenes that may be synthesized, due to the numerous types of hydrolytically labile substituents that can be can be added to the phosphorous atoms [79]. These viable substituents allow the properties of polyphosphazenes to be extremely controllable. Nevertheless, most polyphosphazenes display a slow degradation rate *in vivo* [79].

A number of different polyphosphazenes have been synthesized for use as tissue engineering scaffolds. Due to its high strength and surface degradation properties, it has been particularly investigated as an orthopaedic biomaterial [80]. Osteoblast cells that support skeletal tissue formation has been seeded and proliferated on a three-dimensional polyphosphazene scaffold [81]. Additionally, poly(organo)phosphazenes has been studied as scaffolds for synthetic nerve grafts in peripheral nerve regeneration [82].

8.3.2.3 Polycarbonate

Tyrosine-derived polycarbonate (P(DTR carbonate)) is an amorphous polycarbonate widely studied for biomedical applications (Table 8.2, Figure 8.2). The linear chain of P(DTR carbonate) contains a pendant R group, allowing it to be easily modified [83]. Further, P(DTR carbonate) has three bonds susceptible to hydrolytic degradation: amide, carbonate, and ester [53,83]. The carbonate and ester bonds readily degrade, with the former typically having a faster degradation rate. Nevertheless, the overall degradation time for P(DTR carbonate) can be months to years [84]. The ester bond degrades into carboxylic acid and alcohol, while the carbonate bond releases two alcohols and carbon dioxide [83]. The amide has been found to be relatively stable *in vitro* [83]. Overall, P(DTR carbonate) undergoes bulk degradation and its mechanical properties are determined by the pendant group [53,83].

The slow degradation rate of polycarbonates has led to its investigation in orthopaedic tissue engineering applications. Additionally, P(DTR carbonate) elicits a response for bone ingrowth at the bone-polymer implant interface, supporting the use of P(DTR carbonate) as a bone scaffold [84]. Recent studies have shown that osteoblast cells do attach to the surface of P(DTR carbonate). Results indicate that these osteoblasts maintained their phenotype and rounded cell morphology [85].

8.3.2.4 Poly(ethylene glycol)/Poly(ethylene oxide)

Poly(ethylene glycol) (PEG) is generally a linear-chained polymer consisting of an ethylene oxide repeating unit ($-O-CH_2-CH_2-)_n$ (Table 8.2, Figure 8.2). Poly(ethylene oxide) (PEO) has the same backbone as PEG, but an longer chain length and thus a higher molecular weight [86,87]. PEG is hydrophilic and synthesized by a anionic/cationic polymerization [4,8]. The ability of PEG to act as a swelling polymer has primarily led to its function as a hydrogel, and in some instances as an injectable hydrogel. Unfortunately, the linear chain form of PEG leads to rapid diffusion and low mechanical stability [8]. PEG networks may be created by attaching functional groups to the ends of the PEG chain and then initiating covalent cross-linking [8,32,87–90]. PEG is naturally nondegradable. However, PEG may be made degradable by copolymerization with hydrolytically or enzymatically degradable polymers [91].

The ease with which PEG may be modified, whether with cross-linkable groups for network formation or degradable groups for resorbable applications, has probably led to the widespread interest in PEG for tissue engineering or other biomedical applications. For example, PEG has been copolymerized with PLGA as well as alginate to form hydrogels. Results have shown that DNA content of chondrocytes proliferation on copolymerized PLGA-PEG increases as the percentage of degradable components becomes higher [92]. Similarly, islets of Langerhans in alginate–PEG hydrogels retained their viability and expressed function through insulin excretion [93].

8.3.2.5 Polyurethane

Recent investigations have developed polyurethane polymers as scaffold materials. Polyurethane is an elastomeric polymer that is typically nondegradable (Table 8.2, Figure 8.2). Positive attributes, such as flexible mechanical strength and biocompatibility, has led to the synthesis of degradable polyurethanes with nontoxic diisocyanate derivatives [47,94–96]. Studies have shown that polyurethane scaffolds support cell attachment with chondrocytes, bone marrow stromal cells, and cardiomyocytes [96–98].

8.4 Scaffold Design Properties

8.4.1 Fabrication

There are two main considerations in the assembly of a polymeric tissue engineering scaffold. First is the curing method, or the method by which the polymer is assembled into a bulk material. Second is the fabrication strategy. The chemical nature of the polymer often determines both the curing method and fabrication strategy. There are two main curing methods: polymer entanglement and polymer cross-linking. Entanglement usually involves intertwining long, linear polymer chains to form a loosely bound polymer network. An advantage of this type of method is that it is simple, allowing the polymer to be molded into a bulk material using heat, pressure, or both. However, a disadvantage of this process is that the bulk material sometimes lacks significant mechanical strength. The second curing method is cross-linking, which involves the formation of covalent or ionic bonds between individual polymer chains. Typically either a radical or ion is needed to promote cross-linking along with an initiator, such as heat, light, chemical accelerant, or time. The advantages of cross-linked polymers are that they often have significant mechanical properties. Furthermore, cross-linked polymers may be injected into a tissue defect and cured *in situ*. However, one major disadvantage is the possible cytotoxicity issues with cross-linking systems. The multiple components used by cross-linking polymers as well as the necessary chemical reaction may lead to the use of cytotoxic components or the formation of cytotoxic reaction byproducts.

There are two basic strategies of polymeric scaffold fabrication: prefabrication and *in situ*. Prefabrication structures are cured before implantation. This strategy is often preferred since it is formed outside the body, allowing the removal of cytotoxic and nonbiocompatible components prior to implantation. However, the notable disadvantage of prefabricated scaffolds is that it may not properly fit in a tissue defect site, causing gaps between the engineered graft and the host tissue. These void spaces may lead to undesirable results, including fibrous tissue formation. Therefore, the deficiencies of prefabricated scaffolds have led to investigations toward *in situ* fabrication of a scaffold, which involves curing of a polymeric matrix within the tissue defect itself. The deformability of an *in situ* fabricated matrix creates an interface between the scaffold and the surrounding tissue, facilitating tissue integration [99]. Furthermore, the implantation of an *in situ* fabricated scaffold may require as little as a narrow path for injection of the liquid scaffold, allowing minimally invasive surgery techniques to be utilized. A disadvantage of *in situ* fabrication of scaffolds is that the chemical or thermal means of curing the polymer can significantly affect the host tissue as well as any biological component of the engineered graft.

8.4.2 Micro-Structure

Tissue regeneration through cell implantation in scaffolds is dependent on the micro-structure of the scaffold. Since most cells are anchorage dependent, the scaffold should possess properties that aid cell growth and facilitate attachment of a large cell population [46,100]. To this end, a large scaffold surface area is typically favorable. Furthermore, highly porous scaffolds allow for an abundant number of cells to infiltrate the scaffold's void space. Similarly, the continuity of the scaffold's pores is important for proper transport of nutrients and cell migration. It is also generally advantageous for scaffolds to possess a large surface area to volume ratio. This ratio promotes the use of small diameter pores that are larger than the diameter of a cell. However, high porosities compromise the mechanical integrity of the polymeric scaffold. Clearly, balancing a scaffold's surface area, void space, and mechanical integrity is a necessary challenge that must be overcome in the construction of viable tissue engineering scaffolds.

Additionally, a polymer's effectiveness as a scaffolding material is dependent on its interaction with transplanted or host cells. Thus the polymer's surface properties should facilitate their attachment, proliferation, and (possible) differentiation. A strong cell adhesion favors the proliferation of cells, while a rounded morphology promotes their differentiation [46]. The hydrophilic nature of some polymers promotes a highly wettable surface and allows cells to be encapsulated by capillary action [101]. Furthermore, cellular attachment and function on polymeric scaffolds may be enhanced by providing a biomimetic surface through the incorporation of proteins and ligands.

8.4.3 Macro-Structure

A scaffold can be formed into a number of different macro-structures including, fiber meshes, hydrogels, and foams. Fiber meshes are formed by weaving or knitting individual polymeric fibers into a three-dimensional structure and are attractive for tissue engineering as they provide a large surface area for cell attachment [48]. Furthermore, a fiber mesh scaffold structure mimics the properties of the extracellular matrix, allowing the diffusion of nutrients to cells and waste products from cells. A disadvantage to this form is the lack of structural stability. However, fiber bonding, which involves bonding at fiber cross points, has been introduced to create a more stable structure [46,48,102].

Hydrophilic polymers are often formed into hydrogels by physical polymer entanglements, secondary forces, or chemical cross-linking [13,87]. The significant property of hydrogels is their ability to absorb a tremendous amount of water, up to a thousand times their own dry weight [13]. This aqueous environment, ideal for cell encapsulation and drug loading, has encouraged many investigations into hydrogels for biomedical applications. Hydrogels are generally easy to inject or mold, and therefore are often incorporated into strategies involving minimally invasive surgical techniques [32]. Some drawbacks to hydrogels are that they lack strong mechanical properties and they are difficult to sterilize [13,32].

Foam and sponge scaffolds provide a macro-structural template for the cells to form into a three-dimensional tissue structure. There are several different processing techniques for porous constructs, including phase separation, emulsion freeze drying, gas foaming, and solvent casting/particulate leaching [46,103–105]. A recent technology that has become of interest in tissue engineering is rapid prototyping or solid freeform fabrication. These techniques are quite exciting since they allow for the precise construction of scaffold architecture, often based upon a computer-aided design model [106]. The most appropriate scaffold macro-structure is dependent on the type of the polymer utilized and the application (tissue) of interest.

8.4.4 Biocompatibility

One of the most essential properties of an engineered construct is its biocompatibility, or ability to not elicit a significant or prolonged inflammatory response. It is important to know that any injury will elicit some inflammatory response, and this is certainly true with the implantation of a polymeric scaffold. However, this response should be limited and not prolonged.

When a polymeric scaffold is implanted in a tissue, there is typically a three-phase tissue response [107–109]. Phase I incorporates both acute and chronic inflammation, which occurs in a short period of time (1 to 2 weeks). Phase II involves the granulation of tissue, a foreign body reaction, and fibrosis. The rate of phase II is primarily dependent on the degradation rate of the polymeric scaffold. Finally, the fastest step, phase III is when the bulk of the scaffold is lost. These phases are directly influenced by the properties of the polymeric scaffold. Therefore, it is important to evaluate a variety of scaffold properties including synthesis components, fabrication, micro-structure, and macro-structure. It is a general thought that as the complexity of a scaffold system increases, the more likely it will result in a significant response by the body [73].

8.4.5 Biodegradability

Most polymeric scaffolds are designed to provide temporary support and, therefore to be biodegradable. Therefore, the degraded products of the scaffold must have a safe route for removal from the host. The rate of degradation is often affected by the properties of the polymeric scaffold including synthesis components, fabrication, micro-structure, and macro-structure. Most investigations intend for the scaffold degradation rate to closely follow the rate of tissue repair [22,46]. However, this intention may be difficult to implement. An alternative approach is to design the scaffold's degradation rate so that it is quicker than the rate of degradation product removal, in an effort to minimize negative host responses.

There are two types of degradation: surface and bulk [110]. Surface degradation of a scaffold is characterized by a gradual decrease in the dimensions of the scaffold with no change in its mechanical attributes. At a critical point during surface degradation, both size and mechanical properties of the scaffold decrease rapidly. Bulk degradation is characterized by loss of material throughout the scaffold's volume during degradation. Thus, mechanical strength decreases during bulk degradation and is dependent on the degradation rate. Since bulk degradation maintains an intact surface, it may facilitate cell adhesion to a greater extent than surface eroding scaffolds. Natural polymers mainly undergo surface degradation, since enzymes are generally too large to penetrate into the bulk of the scaffold. However, synthetic polymers can degrade by surface, bulk, or both, depending on its composition.

8.4.6 Mechanical Strength

Since many tissues undergo mechanical stresses and strains, the mechanical properties of a scaffold should be considered. This is especially true for the engineering of weight-bearing orthopaedic tissues. In these instances, the scaffold must be able to provide support to the forces applied to both it and the surrounding tissues. Furthermore, the mechanical properties of the polymeric scaffold should be retained until the regenerated tissue can assume its structural role [111]. In cases where mechanical forces are thought to be required for cell growth and phenotype expression, a scaffold which displays surface eroding properties may be preferred. Alternatively, hydrophobic polymers tend to resist water absorption and thus sustain their strength longer than hydrophilic polymers.

8.5 Summary

Tissue engineering has emerged as a growing field for restoring, repairing, and regenerating tissue. Polymeric scaffolds especially have a profound impact on the possibilities of tissue engineering by providing structure for the attachment, proliferation, and differentiation of transplanted or host cells. These constructs have led to exciting and novel clinical options for the repair of tissues, including but not limited to skin, bone, cartilage, kidney, liver, nerve, and smooth muscle. To this end, scaffold properties must be carefully considered for the intended application. Most importantly, the polymeric scaffold must allow invading cells to express proper functionality of the tissue being formed. Thus, tissue engineering research is continuously trying to enhance polymeric scaffold properties to form tissue that closely resembles the native tissue.

References

[1] Langer, R., Tissue engineering, *Mol. Ther.* 2000, 1(1), 12–15.

[2] Bonassar, L.J. and Vacanti, C.A., Tissue engineering: the first decade and beyond, *J. Cell. Biochem. Suppl.* 1998, 30–31, 297–303.

[3] Hutmacher, D.W., Scaffold design and fabrication technologies for engineering tissues — state of the art and future perspectives, *J. Biomater. Sci. Polym. Ed.* 2001, 12(1), 107–124.

[4] Lee, K.Y. and Mooney, D.J., Hydrogels for tissue engineering, *Chem. Rev.* 2001, 101(7), 1869–1879.

[5] Vinall, R.L., Lo, S.H., and Reddi, A.H., Regulation of articular chondrocyte phenotype by bone morphogenetic protein 7, interleukin 1, and cellular context is dependent on the cytoskeleton, *Exp. Cell Res.* 2002, 272(1), 32–44.

[6] Hung, C.T., Lima, E.G., Mauck, R.L., Taki, E., LeRoux, M.A., Lu, H.H., Stark, R.G., Guo, X.E., and Ateshian, G.A., Anatomically shaped osteochondral constructs for articular cartilage repair, *J. Biomech.* 2003, 36(12), 1853–1864.

[7] Rahfoth, B., Weisser, J., Sternkopf, F., Aigner, T., von der Mark, K., and Brauer, R., Transplantation of allograft chondrocytes embedded in agarose gel into cartilage defects of rabbits, *Osteoarthr. Cartil.* 1998, 6(1), 50–65.

[8] Seal, B.L., Otero, T.C., and Panitch, A., Polymeric biomaterials for tissue and organ regeneration, *Mater. Sci. Eng. R-Rep.* 2001, 34(4–5), 147–230.

[9] Dillon, G.P., Yu, X., Sridharan, A., Ranieri, J.P., and Bellamkonda, R.V., The influence of physical structure and charge on neurite extension in a 3D hydrogel scaffold, *J. Biomater. Sci. Polym. Ed.* 1998, 9(10), 1049–1069.

[10] Wang, L., Shelton, R.M., Cooper, P.R., Lawson, M., Triffitt, J.T., and Barralet, J.E., Evaluation of sodium alginate for bone marrow cell tissue engineering, *Biomaterials* 2003, 24(20), 3475–3481.

[11] Davis, T.A., Volesky, B., and Mucci, A., A review of the biochemistry of heavy metal biosorption by brown algae, *Water Res.* 2003, 37(18), 4311–4330.

[12] Rowley, J.A., Madlambayan, G., and Mooney, D.J., Alginate hydrogels as synthetic extracellular matrix materials, *Biomaterials* 1999, 20(1), 45–53.

[13] Hoffman, A.S., Hydrogels for biomedical applications, *Adv. Drug Deliv. Rev.* 2002, 54(1), 3–12.

[14] Shapiro, L. and Cohen, S., Novel alginate sponges for cell culture and transplantation, *Biomaterials* 1997, 18(8), 583–590.

[15] Guo, J.F., Jourdian, G.W., and MacCallum, D.K., Culture and growth characteristics of chondrocytes encapsulated in alginate beads, *Conn. Tissue Res.* 1989, 19(2–4), 277–297.

[16] Pariente, J.L., Kim, B.S., and Atala, A., *In vitro* biocompatibility evaluation of naturally derived and synthetic biomaterials using normal human bladder smooth muscle cells, *J. Urol.* 2002, 167(4), 1867–1871.

[17] Shimizu, T., Yamato, M., Kikuchi, A., and Okano, T., Cell sheet engineering for myocardial tissue reconstruction, *Biomaterials* 2003, 24(13), 2309–2316.

[18] Liu, H., Lee, Y.W., and Dean, M.F., Re-expression of differentiated proteoglycan phenotype by dedifferentiated human chondrocytes during culture in alginate beads, *Biochim. Biophys. Acta* 1998, 1425(3), 505–515.

[19] Glicklis, R., Shapiro, L., Agbaria, R., Merchuk, J.C., and Cohen, S., Hepatocyte behavior within three-dimensional porous alginate scaffolds, *Biotechnol. Bioeng.* 2000, 67(3), 344–353.

[20] Mosahebi, A., Simon, M., Wiberg, M., and Terenghi, G., A novel use of alginate hydrogel as Schwann cell matrix, *Tissue Eng.* 2001, 7(5), 525–534.

[21] Leach, J.B., Bivens, K.A., Patrick, C.W., and Schmidt, C.E., Photocrosslinked hyaluronic acid hydrogels: natural, biodegradable tissue engineering scaffolds, *Biotechnol. Bioeng.* 2003, 82(5), 578–589.

[22] Hayashi, T., Biodegradable polymers for biomedical uses, *Prog. Polym. Sci.* 1994, 19(4), 663–702.

[23] Aigner, J., Tegeler, J., Hutzler, P., Campoccia, D., Pavesio, A., Hammer, C., Kastenbauer, E., and Naumann, A., Cartilage tissue engineering with novel nonwoven structured biomaterial based on hyaluronic acid benzyl ester, *J. Biomed. Mater. Res.* 1998, 42(2), 172–181.

[24] Halbleib, M., Skurk, T., de Luca, C., von Heimburg, D., and Hauner, H., Tissue engineering of white adipose tissue using hyaluronic acid-based scaffolds. I: *in vitro* differentiation of human adipocyte precursor cells on scaffolds, *Biomaterials* 2003, 24(18), 3125–3132.

[25] VandeVord, P.J., Matthew, H.W.T., DeSilva, S.P., Mayton, L., Wu, B., and Wooley, P.H., Evaluation of the biocompatibility of a chitosan scaffold in mice, *J. Biomed. Mater. Res.* 2002, 59(3), 585–590.

[26] Suh, J.K.F. and Matthew, H.W.T., Application of chitosan-based polysaccharide biomaterials in cartilage tissue engineering: a review, *Biomaterials* 2000, 21(24), 2589–2598.

[27] Li, J., Pan, J., Zhang, L., Guo, X., and Yu, Y., Culture of primary rat hepatocytes within porous chitosan scaffolds, *J. Biomed. Mater. Res.* 2003, 67A(3) 938–943.

[28] Kawase, M., Michibayashi, N., Nakashima, Y., Kurikawa, N., Yagi, K., and Mizoguchi, T., Application of glutaraldehyde-crosslinked chitosan as a scaffold for hepatocyte attachment, *Biol. Pharm. Bull.* 1997, 20(6), 708–710.

[29] Seol, Y.J., Lee, J.Y., Park, Y.J., Lee, Y.M., Ku, Y., Rhyu, I.C., Lee, S.J., Han, S.B., and Chung, C.P., Chitosan sponges as tissue engineering scaffolds for bone formation, *Biotechnol. Lett.* 2004, 26(13), 1037–1041.

[30] Nettles, D.L., Elder, S.H., and Gilbert, J.A., Potential use of chitosan as a cell scaffold material for cartilage tissue engineering, *Tissue Eng.* 2002, 8(6), 1009–1016.

[31] Lee, C.H., Singla, A., and Lee, Y., Biomedical applications of collagen, *Int. J. Pharm.* 2001, 221(1–2), 1–22.

[32] Drury, J.L. and Mooney, D.J., Hydrogels for tissue engineering: scaffold design variables and applications, *Biomaterials* 2003, 24(24), 4337–4351.

[33] Ma, L., Gao, C., Mao, Z., Shen, J., Hu, X., and Han, C., Thermal dehydration treatment and glutaraldehyde cross-linking to increase the biostability of collagen–chitosan porous scaffolds used as dermal equivalent, *J. Biomater. Sci. Polym. Ed.* 2003, 14(8), 861–874.

[34] Langer, R. and Vacanti, J.P., Tissue engineering, *Science* 1993, 260(5110), 920–926.

[35] Falanga, V., Margolis, D., Alvarez, O., Auletta, M., Maggiacomo, F., Altman, M., Jensen, J., Sabolinski, M., and Hardin-Young, J., Rapid healing of venous ulcers and lack of clinical rejection with an allogeneic cultured human skin equivalent. Human Skin Equivalent Investigators Group, *Arch. Dermatol.* 1998, 134(3), 293–300.

[36] Noah, E.M., Chen, J., Jiao, X., Heschel, I., and Pallua, N., Impact of sterilization on the porous design and cell behavior in collagen sponges prepared for tissue engineering, *Biomaterials* 2002, 23(14), 2855–2861.

[37] Nehrer, S., Breinan, H.A., Ramappa, A., Hsu, H.P., Minas, T., Shortkroff, S., Sledge, C.B., Yannas, I.V., and Spector, M., Chondrocyte-seeded collagen matrices implanted in a chondral defect in a canine model, *Biomaterials* 1998, 19(24), 2313–2328.

[38] Risbud, M.V., Karamuk, E., Schlosser, V., and Mayer, J., Hydrogel-coated textile scaffolds as candidate in liver tissue engineering: II. Evaluation of spheroid formation and viability of hepatocytes, *J. Biomater. Sci. Polym. Ed.* 2003, 14(7), 719–731.

[39] Orwin, E.J. and Hubel, A., *In vitro* culture characteristics of corneal epithelial, endothelial, and keratocyte cells in a native collagen matrix, *Tissue Eng.* 2000, 6(4), 307–319.

[40] Payne, R.G., Yaszemski, M.J., Yasko, A.W., and Mikos, A.G., Development of an injectable, *in situ* crosslinkable, degradable polymeric carrier for osteogenic cell populations. Part 1. Encapsulation of marrow stromal osteoblasts in surface crosslinked gelatin microparticles, *Biomaterials* 2002, 23(22), 4359–4371.

[41] Awad, H.A., Wickham, M.Q., Leddy, H.A., Gimble, J.M., and Guilak, F., Chondrogenic differentiation of adipose-derived adult stem cells in agarose, alginate, and gelatin scaffolds, *Biomaterials* 2004, 25(16), 3211–3222.

[42] Risbud, M., Endres, M., Ringe, J., Bhonde, R., and Sittinger, M., Biocompatible hydrogel supports the growth of respiratory epithelial cells: Possibilities in tracheal tissue engineering, *J. Biomed. Mater. Res.* 2001, 56(1), 120–127.

[43] Altman, G.H., Diaz, F., Jakuba, C., Calabro, T., Horan, R.L., Chen, J., Lu, H., Richmond, J., and Kaplan, D.L., Silk-based biomaterials, *Biomaterials* 2003, 24(3), 401–416.

[44] Minoura, N., Aiba, S., Gotoh, Y., Tsukada, M., and Imai, Y., Attachment and growth of cultured fibroblast cells on silk protein matrices, *J. Biomed. Mater. Res.* 1995, 29(10), 1215–1221.

[45] Sofia, S., McCarthy, M.B., Gronowicz, G., and Kaplan, D.L., Functionalized silk-based biomaterials for bone formation, *J. Biomed. Mater. Res.* 2001, 54(1), 139–148.

[46] Thomson, R.C., Wake, M.C., Yaszemski, M.J., and Mikos, A.G., Biodegradable polymer scaffolds to regenerate organs, *Biopolymers Ii* 1995, 122, 245–274.

[47] Gunatillake, P.A. and Adhikari, R., Biodegradable synthetic polymers for tissue engineering, *Eur. Cell. Mater.* 2003, 5, 1–16.

[48] Yang, S., Leong, K.F., Du, Z., and Chua, C.K., The design of scaffolds for use in tissue engineering. Part I. Traditional factors, *Tissue Eng.* 2001, 7(6), 679–689.

[49] Freed, L.E., Marquis, J.C., Nohria, A., Emmanual, J., Mikos, A.G., and Langer, R., Neocartilage formation *in vitro* and *in vivo* using cells cultured on synthetic biodegradable polymers, *J. Biomed. Mater. Res.* 1993, 27(1), 11–23.

[50] Shinoka, T., Ma, P.X., Shum-Tim, D., Breuer, C.K., Cusick, R.A., Zund, G., Langer, R., Vacanti, J.P., and Mayer, J.E., Jr., Tissue-engineered heart valves. Autologous valve leaflet replacement study in a lamb model, *Circulation* 1996, 94(9Suppl.), II164–II168.

[51] Kaihara, S., Kim, S., Kim, B.S., Mooney, D.J., Tanaka, K., and Vacanti, J.P., Survival and function of rat hepatocytes cocultured with nonparenchymal cells or sinusoidal endothelial cells on biodegradable polymers under flow conditions, *J. Pediatr. Surg.* 2000, 35(9), 1287–1290.

[52] Mooney, D.J., Mazzoni, C.L., Breuer, C., McNamara, K., Hern, D., Vacanti, J.P., and Langer, R., Stabilized polyglycolic acid fibre-based tubes for tissue engineering, *Biomaterials* 1996, 17(2), 115–124.

[53] Agrawal, C.M. and Ray, R.B., Biodegradable polymeric scaffolds for musculoskeletal tissue engineering, *J. Biomed. Mater. Res.* 2001, 55(2), 141–150.

[54] Ishaug-Riley, S.L., Okun, L.E., Prado, G., Applegate, M.A., and Ratcliffe, A., Human articular chondrocyte adhesion and proliferation on synthetic biodegradable polymer films, *Biomaterials* 1999, 20(23–24), 2245–2256.

[55] Yang, F., Murugan, R., Ramakrishna, S., Wang, X., Ma, Y.X., and Wang, S., Fabrication of nano-structured porous PLLA scaffold intended for nerve tissue engineering, *Biomaterials* 2004, 25(10), 1891–1900.

[56] Park, S.S., Jin, H.R., Chi, D.H., and Taylor, R.S., Characteristics of tissue-engineered cartilage from human auricular chondrocytes, *Biomaterials* 2004, 25(12), 2363–2369.

[57] Middleton, J.C. and Tipton, A.J., Synthetic biodegradable polymers as orthopedic devices, *Biomaterials* 2000, 21(23), 2335–2346.

[58] El-Amin, S.F., Lu, H.H., Khan, Y., Burems, J., Mitchell, J., Tuan, R.S., and Laurencin, C.T., Extra-cellular matrix production by human osteoblasts cultured on biodegradable polymers applicable for tissue engineering, *Biomaterials* 2003, 24(7), 1213–1221.

[59] Kim, B.S., Nikolovski, J., Bonadio, J., Smiley, E., and Mooney, D.J., Engineered smooth muscle tissues: regulating cell phenotype with the scaffold, *Exp. Cell. Res.* 1999, 251(2), 318–328.

[60] Teng, Y.D., Lavik, E.B., Qu, X., Park, K.I., Ourednik, J., Zurakowski, D., Langer, R., and Snyder, E.Y., Functional recovery following traumatic spinal cord injury mediated by a unique polymer scaffold seeded with neural stem cells, *Proc. Natl Acad. Sci. USA* 2002, 99(5), 3024–3029.

[61] Mooney, D.J., Organ, G., Vacanti, J.P., and Langer, R., Design and fabrication of biodegradable polymer devices to engineer tubular tissues, *Cell Transplant* 1994, 3(2), 203–210.

[62] Dai, N.T., Williamson, M.R., Khammo, N., Adams, E.F., and Coombes, A.G., Composite cell support membranes based on collagen and polycaprolactone for tissue engineering of skin, *Biomaterials* 2004, 25(18), 4263–4271.

[63] Barralet, J.E., Wallace, L.L., and Strain, A.J., Tissue engineering of human biliary epithelial cells on polyglycolic acid/polycaprolactone scaffolds maintains long-term phenotypic stability, *Tissue Eng.* 2003, 9(5), 1037–1045.

[64] Park, Y.J., Lee, J.Y., Chang, Y.S., Jeong, J.M., Chung, J.K., Lee, M.C., Park, K.B., and Lee, S.J., Radioisotope carrying polyethylene oxide–polycaprolactone copolymer micelles for targetable bone imaging, *Biomaterials* 2002, 23(3), 873–879.

[65] Ciapetti, G., Ambrosio, L., Savarino, L., Granchi, D., Cenni, E., Baldini, N., Pagani, S., Guizzardi, S., Causa, F., and Giunti, A., Osteoblast growth and function in porous poly epsilon-caprolactone matrices for bone repair: a preliminary study, *Biomaterials* 2003, 24(21), 3815–3824.

[66] Ng, K.W., Hutmacher, D.W., Schantz, J.T., Ng, C.S., Too, H.P., Lim, T.C., Phan, T.T., and Teoh, S.H., Evaluation of ultra-thin poly(epsilon-caprolactone) films for tissue-engineered skin, *Tissue Eng.* 2001, 7(4), 441–455.

[67] Fisher, J.P., Holland, T.A., Dean, D., and Mikos, A.G., Photoinitiated cross-linking of the bio-degradable polyester poly(propylene fumarate). Part II. *In vitro* degradation, *Biomacromolecules* 2003, 4(5), 1335–1342.

[68] Peter, S.J., Miller, M.J., Yasko, A.W., Yaszemski, M.J., and Mikos, A.G., Polymer concepts in tissue engineering, *J. Biomed. Mater. Res.* 1998, 43(4), 422–427.

[69] Suggs, L.J. and Mikos, A.G., Development of poly(propylene fumarate-co-ethylene glycol) as an injectable carrier for endothelial cells, *Cell Transplant* 1999, 8(4), 345–350.

[70] Shung, A.K., Behravesh, E., Jo, S., and Mikos, A.G., Crosslinking characteristics of and cell adhesion to an injectable poly(propylene fumarate-co-ethylene glycol) hydrogel using a water-soluble crosslinking system, *Tissue Eng.* 2003, 9(2), 243–254.

[71] Davis, K.A. and Anseth, K.S., Controlled release from crosslinked degradable networks, *Crit. Rev. Ther. Drug Carrier Syst.* 2002, 19(4–5), 385–423.

[72] Heller, J., Barr, J., Ng, S.Y., Abdellauoi, K.S., and Gurny, R., Poly(ortho esters): synthesis, characterization, properties and uses, *Adv. Drug Deliv. Rev.* 2002, 54(7), 1015–1039.

[73] Andriano, K.P., Tabata, Y., Ikada, Y., and Heller, J., *In vitro* and *in vivo* comparison of bulk and surface hydrolysis in absorbable polymer scaffolds for tissue engineering, *J. Biomed. Mater. Res.* 1999, 48(5), 602–612.

[74] Vacanti, J.P., Morse, M.A., Saltzman, W.M., Domb, A.J., Perez-Atayde, A., and Langer, R., Selective cell transplantation using bioabsorbable artificial polymers as matrices, *J. Pediatr. Surg.* 1988, 23(1pt 2) 3–9.

[75] Langer, R., Biomaterials in drug delivery and tissue engineering: one laboratory's experience, *Acc. Chem. Res.* 2000, 33(2), 94–101.

[76] Kumar, N., Langer, R.S., and Domb, A.J., Polyanhydrides: an overview, *Adv. Drug Deliv. Rev.* 2002, 54(7), 889–910.

[77] Uhrich, K.E., Gupta, A., Thomas, T.T., Laurencin, C.T., and Langer, R., Synthesis and characterization of degradable poly(anhydride-co-imides), *Macromolecules* 1995, 28(7), 2184–2193.

[78] Muggli, D.S., Burkoth, A.K., and Anseth, K.S., Crosslinked polyanhydrides for use in ortho-pedic applications: degradation behavior and mechanics, *J. Biomed. Mater. Res.* 1999, 46(2), 271–278.

[79] Qiu, L.Y. and Zhu, K.J., Novel biodegradable polyphosphazenes containing glycine ethyl ester and benzyl ester of amino acethydroxamic acid as cosubstituents: syntheses, characterization, and degradation properties, *J. Appl. Polym. Sci.* 2000, 77(13), 2987–2995.

[80] Laurencin, C.T., Norman, M.E., Elgendy, H.M., el-Amin, S.F., Allcock, H.R., Pucher, S.R., and Ambrosio, A.A., Use of polyphosphazenes for skeletal tissue regeneration, *J. Biomed. Mater. Res.* 1993, 27(7), 963–973.

[81] Laurencin, C.T., El-Amin, S.F., Ibim, S.E., Willoughby, D.A., Attawia, M., Allcock, H.R., and Ambrosio, A.A., A highly porous 3-dimensional polyphosphazene polymer matrix for skeletal tissue regeneration, *J. Biomed. Mater. Res.* 1996, 30(2), 133–138.

[82] Langone, F., Lora, S., Veronese, F.M., Caliceti, P., Parnigotto, P.P., Valenti, F., and Palma, G., Peripheral nerve repair using a poly(organo)phosphazene tubular prosthesis, *Biomaterials* 1995, 16(5), 347–353.

[83] Tangpasuthadol, V., Pendharkar, S.M., and Kohn, J., Hydrolytic degradation of tyrosine-derived polycarbonates, a class of new biomaterials. Part I: study of model compounds, *Biomaterials* 2000, 21(23), 2371–2378.

[84] Choueka, J., Charvet, J.L., Koval, K.J., Alexander, H., James, K.S., Hooper, K.A., and Kohn, J., Canine bone response to tyrosine-derived polycarbonates and poly(L-lactic acid), *J. Biomed. Mater. Res.* 1996, 31(1), 35–41.

[85] Lee, S.J., Choi, J.S., Park, K.S., Khang, G., Lee, Y.M., and Lee, H.B., Response of MG63 osteoblast-like cells onto polycarbonate membrane surfaces with different micropore sizes, *Biomaterials* 2004, 25(19), 4699–4707.

[86] Cai, J., Bo, S., Cheng, R., Jiang, L., and Yang, Y., Analysis of interfacial phenomena of aqueous solutions of polyethylene oxide and polyethylene glycol flowing in hydrophilic and hydrophobic capillary viscometers, *J. Colloid. Interface Sci.* 2004, 276(1), 174–181.

[87] Gutowska, A., Jeong, B., and Jasionowski, M., Injectable gels for tissue engineering, *Anat. Rec.* 2001, 263(4), 342–349.

[88] Novikova, L.N., Novikov, L.N., and Kellerth, J.O., Biopolymers and biodegradable smart implants for tissue regeneration after spinal cord injury, *Curr. Opin. Neurol.* 2003, 16(6), 711–715.

[89] Sawhney, A.S., Pathak, C.P., and Hubbell, J.A., Bioerodible hydrogels based on photopolymerized poly(ethylene glycol)-co-poly(alpha-hydroxy acid) diacrylate macromers, *Macromolecules* 1993, 26(4), 581–587.

[90] Sims, C.D., Butler, P.E., Casanova, R., Lee, B.T., Randolph, M.A., Lee, W.P., Vacanti, C.A., and Yaremchuk, M.J., Injectable cartilage using polyethylene oxide polymer substrates, *Plast. Reconstr. Surg.* 1996, 98(5), 843–850.

[91] Bourke, S.L. and Kohn, J., Polymers derived from the amino acid L-tyrosine: polycarbonates, polyarylates and copolymers with poly(ethylene glycol), *Adv. Drug Deliv. Rev.* 2003, 55(4), 447–466.

[92] Bryant, S.J. and Anseth, K.S., Controlling the spatial distribution of ECM components in degradable PEG hydrogels for tissue engineering cartilage, *J. Biomed. Mater. Res.* 2003, 64A(1), 70–79.

[93] Desai, N.P., Sojomihardjo, A., Yao, Z., Ron, N., and Soon-Shiong, P., Interpenetrating polymer networks of alginate and polyethylene glycol for encapsulation of islets of Langerhans, *J. Microencapsul.* 2000, 17(6), 677–690.

[94] Ganta, S.R., Piesco, N.P., Long, P., Gassner, R., Motta, L.F., Papworth, G.D., Stolz, D.B., Watkins, S.C., and Agarwal, S., Vascularization and tissue infiltration of a biodegradable polyurethane matrix, *J. Biomed. Mater. Res.* 2003, 64A(2), 242–248.

[95] Gorna, K. and Gogolewski, S., Preparation, degradation, and calcification of biodegradable polyurethane foams for bone graft substitutes, *J. Biomed. Mater. Res.* 2003, 67A(3) 813–827.

[96] Zhang, J., Doll, B.A., Beckman, E.J., and Hollinger, J.O., A biodegradable polyurethane–ascorbic acid scaffold for bone tissue engineering, *J. Biomed. Mater. Res.* 2003, 67A(2) 389–400.

[97] Grad, S., Kupcsik, L., Gorna, K., Gogolewski, S., and Alini, M., The use of biodegradable polyurethane scaffolds for cartilage tissue engineering: potential and limitations, *Biomaterials* 2003, 24(28), 5163–5171.

[98] McDevitt, T.C., Woodhouse, K.A., Hauschka, S.D., Murry, C.E., and Stayton, P.S., Spatially organized layers of cardiomyocytes on biodegradable polyurethane films for myocardial repair, *J. Biomed. Mater. Res.* 2003, 66A(3), 586–595.

[99] Anseth, K.S., Metters, A.T., Bryant, S.J., Martens, P.J., Elisseeff, J.H., and Bowman, C.N., *In situ* forming degradable networks and their application in tissue engineering and drug delivery, *J. Control. Release* 2002, 78(1-3), 199–209.

[100] Kim, B.S., Baez, C.E., and Atala, A., Biomaterials for tissue engineering, *World J. Urol.* 2000, 18(1), 2–9.

[101] Dar, A., Shachar, M., Leor, J., and Cohen, S., Cardiac tissue engineering — optimization of cardiac cell seeding and distribution in 3D porous alginate scaffolds, *Biotechnol. Bioeng.* 2002, 80(3), 305–312.

[102] Sachlos, E. and Czernuszka, J.T., Making tissue engineering scaffolds work. Review: the application of solid freeform fabrication technology to the production of tissue engineering scaffolds, *Eur. Cell. Mater.* 2003, 5, 29–39.

[103] Lu, L., Zhu, X., Valenzuela, R.G., Currier, B.L., and Yaszemski, M.J., Biodegradable polymer scaffolds for cartilage tissue engineering, *Clin. Orthop.* 2001, (391 Suppl), S251–S270.

[104] Chen, G.P., Ushida, T., and Tateishi, T., Development of biodegradable porous scaffolds for tissue engineering, *Mater. Sci. Eng. C-Biomim. Supramol. Syst.* 2001, 17(1-2), 63–69.

[105] Griffith, L.G., Polymeric biomaterials, *Acta Materialia* 2000, 48(1), 263–277.

[106] Yang, S., Leong, K.F., Du, Z., and Chua, C.K., The design of scaffolds for use in tissue engineering. Part II. Rapid prototyping techniques, *Tissue Eng.* 2002, 8(1), 1–11.

[107] Shive, M.S. and Anderson, J.M., Biodegradation and biocompatibility of PLA and PLGA microspheres, *Adv. Drug Deliv. Rev.* 1997, 28(1), 5–24.

[108] Ziats, N.P., Miller, K.M., and Anderson, J.M., *In vitro* and *in vivo* interactions of cells with biomaterials, *Biomaterials* 1988, 9(1), 5–13.

[109] Anderson, J.M. and Miller, K.M., Biomaterial biocompatibility and the macrophage, *Biomaterials* 1984, 5(1), 5–10.

[110] von Burkersroda, F., Schedl, L., and Gopferich, A., Why degradable polymers undergo surface erosion or bulk erosion, *Biomaterials* 2002, 23(21), 4221–4231.

[111] Woodfield, T.B., Bezemer, J.M., Pieper, J.S., van Blitterswijk, C.A., and Riesle, J., Scaffolds for tissue engineering of cartilage, *Crit. Rev. Eukaryot. Gene Expr.* 2002, 12(3), 209–236.

9

Calcium Phosphate Ceramics for Bone Tissue Engineering

P. Quinten Ruhé
Joop G.C. Wolke
Paul H.M. Spauwen
John A. Jansen
University Medical Center
St. Radboud

9.1 Introduction

In reconstructive surgery, repair and regeneration of large bone defects is a major challenge. The use of autologous bone is still the gold standard, although concomitant problems as donor site morbidity and limited supply have resulted in worldwide endeavors for the development of bone graft substitutes. Each potential substitute, including tissue engineering approaches with delivery of osteogenic cells or osteoinductive macromolecules, or both is based on an appropriate scaffold biomaterial that is biocompatible, allows bone ingrowth and shows subsequent degradation of the material. In view of this, a biomaterial used for tissue replacement shows by preference a resemblance with the inorganic or organic components of the tissue, or both, to be substituted.

Bone tissue is a living organ composed of an organic and an inorganic component. The organic substance (approx. $\frac{1}{4}$ to $\frac{1}{3}$ of total dry bone weight) consists of more than 90% of collagen and only for a small fraction of cells (2%) and noncollagenous proteins (5%). The inorganic bone mineral (approx. $\frac{2}{3}$ to $\frac{3}{4}$ of total dry bone weight) is composed of specific phases of calcium phosphate (Ca-P), especially carbonate rich hydroxyapatite (HA). In biologically carbonated HA, PO_4 is substituted by CO_3 (so-called type B carbonation). Biological HA also contains other impurity ions as Cl, Mg, Na, K, and F and

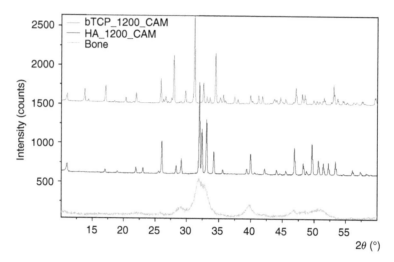

FIGURE 9.1   X-ray diffraction (XRD) pattern of bone, HA and β-tricalcium phosphate, clearly indicating the similarity between HA and bone.

trace elements like Sr and Zn [1]. The apatite in bone mineral is composed of small platelet-like crystals of just 2 to 4 nm in thickness, 25 nm in width, and 50 nm in length [2].

Together with the — organic — collagenous materials, Ca-P materials are among the few biomaterials that show high similarity to natural tissue. Ca-P biomaterials resemble the mineral phase of bone to a higher or lesser extent, depending or their stoichiometry and crystallinity (Figure 9.1). In general, Ca-P biomaterials are crystalline ceramics characterized by a high biocompatibility, the ability of direct bone bonding and osteoconduction, and a variable resorbability. Since the early seventies, Ca-P ceramics are clinically used in dentistry, followed by surgical specialties in the eighties. The various available Ca-P materials differ in their origin (naturally derived or synthetic), chemico-physical composition (calcium to phosphate ratio, crystallinity, [micro]porosity) and preparation process (prefabricated blocks and granules, coatings on other biomaterials, self setting cements and composite materials). In this review, the essential Ca-P characteristics, differences between various Ca-P biomaterials, and proceedings in Ca-P ceramic bone tissue engineering research will be discussed.

9.2 Chemico-Physical Properties of Calcium Phosphate Ceramics

Calcium phosphate (Ca-P) compounds exist in several phases. Discern can be made regarding (1). The crystallinity of various compounds (amorphous vs. various crystalline Ca-P phases), (2). eventual heat treatment of the materials (sintering), and (3). the calcium to phosphate ratio.

9.2.1 Crystallinity

Amorphous Ca-P (ACP) is the first solid phase to appear in solution containing high concentrations of calcium and phosphate. ACP lacks the orderly internal structure of crystalline Ca-P compounds and typically has a spherical morphology [3]. Chemically, it has a Ca : P ratio of around 1.5 and is characterized by the absence of diffraction peaks on x-ray diffraction (XRD) patterns. ACP is unstable in aqueous fluids and transforms into crystalline phases such as octacalciumphosphate (OCP) and apatite. Heating of ACP also results in conversion to poorly crystallized apatite (600°C), β-TCP (800°C, dry heat), or HA (800°C, humid heat) [4]. Therefore, ACP should be considered as a precursor for crystalline Ca-P compounds. There has been, and still is, considerable debate whether or not ACP is substantially present in skeletal tissue. For example, Transmission Electron Microscope (TEM) analysis has never revealed conclusive evidence for the presence of ACP in bone mineral [3].

Other phases of Ca-P than ACP reveal a crystalline structure with characteristic peaks on XRD analysis. There is a broad range in crystal morphology depending on composition and preparation characteristics such as temperature, pH, impurity, and the presence of macromolecules. Impurities, as commonly occur in bone mineral, greatly influence crystallinity (reflecting crystal size and crystal strain) but depend on the type of substitution. For example, type B carbonated apatite (CO_3 for PO_4 substitution) has a lower crystallinity and increased solubility, whereas F substitution (F for OH) give the opposite effects due to a better fit of the F^- ion in the apatite crystal structure.

9.2.2 Sintering

Sintering is the key step in processing of the majority of ceramic materials and conventionally consists of two separate phases: compacting the initial powder and heating the compacted powder up to temperatures only a little lower than their melting points. Thereby, atoms and molecules are set in rapid motion, and the particles coalesce. Fusion of the crystals reduces the porosity and increases the strength and density of the final ceramic product. Currently, several sintering techniques are available, such as continuous hot pressing, microwave sintering, pressureless sintering, and plasma sintering. Obviously, chemical composition of initial powders as well as variation of pressure and temperature influence final structure and composition of the sintered product.

9.2.3 Stoichiometry

The quantitative relationship (stoichiometry) of calcium to phosphate in Ca-P salts is essential for several material characteristics as strength, solubility, and crystal structure. Table 9.1 lists the main Ca-P compounds for biomedical applications. From a crystallographic point of view, the most stable form of Ca-P has a Ca : P ratio of 1.67 (hydroxyapatite). With a decrease of the Ca to P ratio, other Ca-deficient phases occur like tricalcium phosphate (TCP, Ca : P ratio 1.50), octacalcium phosphate (OCP, Ca : P ratio 1.33), dicalcium phosphate anhydrous (DCPA or Monetite, Ca : P ratio 1.00), and dicalcium phosphate dihydrate (DCPD or Brushite, Ca : P ratio 1.00). Most of these calcium deficient compounds are used as raw material for sintering procedures for ceramics, or as ingredients for example, Ca-P cements. TCP and HA will be discussed more in detail, since these materials are most used for biomedical applications.

9.2.3.1 TCP ($Ca_3(PO_4)_2$)

Tricalcium phosphate (TCP) has a Ca : P ratio of 1.5, similar to the amorphous biologic precursors of bone [5]. It can be prepared by sintering Ca deficient apatite (Ca : P ratio 1.5). TCP is a polymorph ceramic and exhibits two phases (α- and β-whitlockite), known as α- and β-TCP. Variation in sintering temperature and humidity determine, which phase is being formed; α-TCP occurs at dry heat >1300°C and subsequent quenching in water [4]. Solubility and resorbability of both forms is much higher compared to HA. However, α-TCP is unstable in water and reacts to produce HA. α-TCP is used mainly as a compound

TABLE 9.1 Ca-P Compounds, Names, Formulas, and Ca : P Ratios

Ca/P	Formula	Name/mineral	Abbreviation
0.5	$Ca(H_2PO_4)_2 \cdot H_2O$	Monocalcium phosphate monohydrate	MCPM
1.0	$CaHPO_4 \cdot 2H_2O$	Hydrated calcium phosphate/brushite	DCP
1.0	$CaHPO_4$	Anhydrous calcium phosphate/Monetite	ADCP
1.33	$Ca_8H_2(PO_4)_6 \cdot 5H_2O$	Octacalcium phosphate	OCP
1.5	$Ca_3(PO_4)_2$	Tricalcium phosphate/whitlockite	TCP
1.67	$Ca_{10}(PO_4)_6(F)_2$	Fluorapatite	FA
1.67	$Ca_{10}(PO_4)_6(OH)_2$	Hydroxylapatite	HA
2.0	$CaO \cdot Ca_3(PO_4)_2$	Tetracalcium phospate/hilgenstockite	TTCP

of some Ca-P cements. β-TCP is less soluble and less reactive than α-TCP. It is used pure as bone substitute, and in combination with HA (biphasic Ca-P ceramic). Due to its solubility, dissolution and reprecipitation occur *in vivo*. This results in gradual phase transition into carbonated apatite, and resorption by macrophages, giant cells, and osteoclasts.

9.2.3.2 Hydroxyapatite ($Ca_{10}(PO_4)_6(OH)_2$)

Hydroxyapatite (HA) is the term commonly used for calcium hydroxyapatite $Ca_{10}(PO_4)_6(OH)_2$, the most stable form of Ca-P under physiological pH and body temperature. With fluorapatite and chlorapatite, hydroxyapatite forms the group of apatite minerals that share a similar hexagonal monoclinic crystal structure and the general formula $A_{10}(PO_4)_6(OH,F,Cl)_2$. The A cation can be several metal ions besides calcium, such as barium, strontium, and magnesium. The phosphate anion can be substituted by a carbonate anion to a limited extent.

HA has a Ca:P ratio of 1.67 which is similar to bone mineral. It can be prepared by sintering of precipitated Ca-P salts in a Ca:P ratio of 1.67 at temperatures above 1000°C. Pure HA is hardly soluble under physiological conditions. Impurities like substitution of Ca^{2+} by other metal ions cause variation in solubility and crystal size due to the differences in ionic radius. With respect to mechanical strength, pure HA materials are superior to other Ca-P materials. *In vivo*, however, pure HA hardly shows any cellular resorption by macrophages, giant cells or osteoclasts unless the particle size is small enough for phagocytosis. As a consequence, HA should be considered as nonresorbable whereas other compositions such as β-TCP show substantial dissolution and resorption.

9.2.4 Strength

Biomechanical properties are a considerable concern in the use of Ca-P ceramics. Compressive strength of cortical and trabecular bone varies, depending on the bone density, from 130 to 180 MPa and 5 to 50 MPa respectively. For tensile strength these values fluctuate from 60 to 160 MPa and 3 to 15 MPa respectively. Dense sintered Ca-P ceramics may reach compressive strengths much higher than cortical bone (300–900 MPa), whereas tensile strengths similar and higher than cortical bone (40–300 MPa) have been reported [6]. Problem, however, is that ceramics do not exhibit substantial elastic or plastic deformation before fracturing. Due to the lack of ductility, virtually all Ca-P ceramics are brittle. The brittleness can be explained by the atomic structure of ceramics. Elasticity is manifested as small reversible changes in the interatomic spacing and stretching of interatomic bonds. Plastic deformation corresponds to breaking of existing bonds and reforming of bonds with new neighboring atoms. In crystalline materials, this occurs in planes by means of atomic slip. As a consequence of the electrically charged nature of the ions, both elastic and plastic deformations in ceramics are limited. The brittleness of ceramics is further enhanced by very small and omnipresent flaws in the material. These microcracks result in local amplification of applied tensile stresses and cause a relatively low fracture strength.

Due to the inferior biomechanical properties, Ca-P ceramics are less suitable for clinical application under weight-bearing conditions compared to, for example, metallic or polymeric biomaterials. Obviously, mechanical properties of porous ceramics deteriorate even further with an increasing porosity. Nevertheless, compressive strengths similar to trabecular bone have been reported [7].

9.2.5 Porosity

Cortical bone has pores ranging from 1 to 100 μm (volumetric porosity 5 to 10%), whereas trabecular bone has pores of 200 to 400 μm (volumetric porosity 70 to 90%). Porosity in bone provides space for nutrients supply in cortical bone and marrow cavity in trabecular bone. As mentioned in the previous paragraph, porosity is devastating for mechanical properties of Ca-P ceramics. On the other hand, it is known that porosity enhances degradation of Ca-P ceramics and determines the nature of ingrowing tissue. Consequently, this delicate equilibrium depends on more factors than mechanical material properties alone, that is, application site (vascularization, weight loading, defect dimensions), biological material properties (biocompatibility, osteoconductivity, and degradation) and other aspects as pore geometry and the use of tissue engineering strategies (addition of osteogenic cells or bioactive factors). Therefore, no general optimal pore size and architecture for all applications and biomaterials can be given.

9.2.5.1 Microporosity

Microporosity covers pores sizes smaller than — an arbitrary — 5 μm: too small for penetration an ingrowth of cells, but sufficient for penetration of fluids. Crystalline Ca-P materials intrinsically exhibit, depending on crystal size and structure, a nano- or microporous structure. However, microporosity of ceramics is highly decreased or virtually absent due to high pressure compaction and (partial) fusion of crystals after sintering. Microporous Ca-P structures have an increased surface area, which influences the biological behavior and the Ca-P dissolution/precipitation characteristics. It has even been reported that microstructure plays a crucial role in osteoinductive properties of ceramics. Yuan et al. [8] observed heterotopic bone formation in macro- and microporous ceramics implanted in the dorsal muscle in dogs, whereas similar implants without microporous structure did not reveal bone formation. They suggested that the increased surface area, resultant of the microporous structure, could have caused accumulation of adsorbed proteins (e.g., BMPs) which could have triggered osteogenesis.

9.2.5.2 Macroporosity

Pores larger than 10 μm can be considered as macropores. Various methods can be used to induce macroporosity in Ca-P material. For example, porous ceramics can be obtained by merging the Ca-P slurry with a polymer sponge-like mold or polymer beads before sintering. During the sintering, the polymer is completely burnt out, which results in a porous ceramic structure. In this way, architecture of porous ceramics can be controlled and a completely interconnected structure with any desirable pore size and pore geometry is feasible. Obviously, interconnection of pores is essential for tissue growth throughout the scaffold. Tsuruga and Kuboki have investigated the influences of pore size and geometry in induced bone formation in ceramics. They reported that for porous sintered HA blocks with pore sizes ranging from 106–212 μm to 500–600 μm, the highest amount of bone was produced in implants with pore sizes of 300 to 400 μm [9]. In another study, Kuboki et al. [10] observed that an optimal vascular ingrowth was achieved in pores with a diameter of 350 μm, and claimed that this was the key factor for effective bone formation. Macroporous dimensions are also reported to play a role in osteoinductive behavior of Ca-P ceramics. In a series of experiments in primates, Ripamonti et al. [11,12] reported that implant geometry is of critical importance for cell shape, cell locomotion and cell differentiation. They hypothesized that concavities in sintered HA mediate intrinsic osteoinduction.

9.3 Ca-P Products

Numerous Ca-P products are currently marketed for bone regenerative purposes. Most frequently used Ca-P products in dentistry and surgery are Ca-P coatings applied to prostheses. Several techniques, such as plasma spraying and RF magnetron sputtering can provide a thin bio-active Ca-P coating on the inert metal prosthesis surface [13]. Consequently, the superior mechanical metal properties can be combined with favorable bone bonding Ca-P properties. Coatings applied usually consist of HA to prevent resorption of the coating in time.

Other Ca-P products are marketed as bone filler and are available in various forms as granules, blocks, and injectable cements. Both granules and blocks are available with a dense or porous structure and several resorption rates. Most synthetic products are composed of β-TCP, HA or a combination of both (so called bi-phasic ceramics [Figure 9.2]). Besides synthetic ceramics, several naturally derived ceramics are commercially available. These and injectable Ca-P cements are being discussed in detail in the next paragraphs.

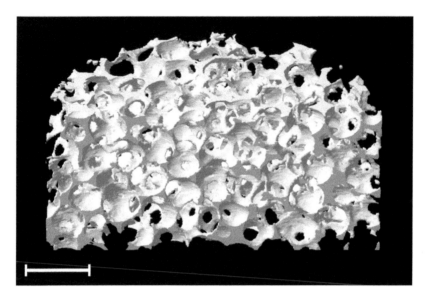

FIGURE 9.2 μCT representation of a highly porous biphasic ceramic implant material (Camceram®, Cam Implants, Leiden, The Netherlands) with a HA : TCP ratio of 60 : 40, porosity of 90%, and macropores ranging from 300 to 500 μm. Bar represents 1 mm.

9.3.1 Natural Ca-P Ceramics

Natural Ca-P biomaterials can be prepared from the inorganic bone mineral matrix of bone. Chemical, or high temperature heat treatment, or both, eliminate organic substances responsible for immunologic response and potential disease transmission (e.g., prions). Chemically treated materials derived from human spongiosa (Puros Tutoplast®, Zimmer Dental, Carslbad, CA, USA) or bovine spongiosa (Ortoss® and Bio-oss®, Geistlich Biomaterials, Geistlich, Switzerland) maintain the macroporous structure of spongious bone as well as the natural small carbonatehydroxyapatite crystal structure. Sintered bovine spongious bone products, such as Osteograf®-N (Dentsply Ceramed, Denver Co, USA) and Endobon® (Biomet Merck, Darmstadt, Germany), also maintain its macroporous structure, but due to the high sintering temperatures (>1100°C) the natural carbonate HA crystals convert into larger HA crystals without CO_3, resembling synthetic HA.

Besides bovine origin, natural Ca-P biomaterials can also be derived from special species of coral (genus Porites and Goniopora). Hydrothermal "replamiform" treatment (260°C; 15.000PSI) of the calcium carbonate (calcite) exoskeletal microstructure of these corals results in conversion into hydroxyapatite [14]. Like natural bone, this hydroxyapatite contains minor elements of Mg, Sr, F, and CO_3. The porous coralline hydroxyapatite is completely interconnected and available in various porosities (Interpore™, Pro Osteon™, Interpore Cross International Inc, Irvine, CA, USA). Different hydrothermal treatment of coral used for Pro Osteon-R results in only partial conversion of the calcium carbonate to HA [5,15]. As a result, the composite of HA/$CaCO_3$ is being resorbed faster than pure HA. Another coral derived bone substitute, which has not been converted into HA at all and therefore cannot be considered as a Ca-P ceramic, shows even faster degradation. This $CaCO_3$ material (Biocoral®, Inoteb, St. Gonnery, France) currently does not have FDA approval, but is available for clinical use in Europe.

9.3.2 Injectable Ca-P Cements

Difficulties with the clinical applicability of preformed ceramic blocks and granules have led to the development of injectable ceramic bone graft substitutes. In the early eighties, Brown and Chow were the first to describe the principles of a self setting Ca-P paste [16,17]. Currently, several different Ca-P cements are commercially available like BoneSource® (Stryker Leibinger, Kalamazoo, MI, USA), Norian® (Synthes, Oberdorf, Switzerland), α-BSM/Biobon® (Etex Corporation, Cambridge, MA, USA), and Calcibon® (Biomet Merck, Darmstadt, Germany). Each cement has a (slightly) different chemical composition, which results in differences in handling properties, setting time, strength and resorbability [18]. The major advantage of Ca-P cements is the injectability of the cement paste. Obviously, this results in favorable moldability of the bone graft, but also it results in immediate, seamless contact of cement to the surrounding bone [19]. Moreover, Ca-P cements have been found to exhibit excellent bone biocompatibility [20,21].

Roughly, the different Ca-P cements can be divided into four distinct groups [17]:

1. Cements composed of a solid Ca-P powder compound with an aqueous liquid component with or without Ca or P ions. After mixing, hardening of the cement is the result of the formation of one or more Ca-P compounds.
2. Cements with a powder component similar to those in (1), but an organic acid as liquid component. Hardening of the cement is the result of complex formation of calcium and the organic acid.
3. Cements similar to those in (2), except that the liquid component is a polymer solution. Cement hardening can be as described in (1) or complex formation of calcium and the polymer solution or both.
4. Cement composites on a polymer basis. Hardening of the cement is based on polymerization of the monomers. The Ca-P compounds present in these materials act as fillers and do not play a significant role in the mechanism of cement hardening.

In this review, we will focus only on the first group, as these cements currently play the most important role in research and clinical applications. The chemico-physical reaction that occurs during and after mixing of the powder and the liquid components is complex. After mixing, soluble compounds dissolve in the aqueous environment and subsequently precipitate, which results in formation and growth of different forms of Ca-P crystals. This crystallization finally results in entanglement of the insoluble Ca-P compounds and hardening of the cement. The setting time and phase transformation depend on the solubility of the various solid compounds, as well as of external factors as pH and temperature [22]. The setting reaction can be accelerated when the liquid phase contains phosphate ions. Therefore, setting time is different for each Ca-P cement formulation and may vary from minutes to hours. After setting, Ca-P cements may have an amorphous appearance under low magnification, but high magnification (100,000×) reveals extremely small crystals. Crystal size and interparticle micropores were observed to grow in time [17]. The final result is a microporous structure of nanometer sized crystals composed of more or less apatitic compounds, resembling the small crystal size of biological apatite.

After setting of the cement, gradual phase transformation of the Ca-P compounds occurs into more apatitic phases. Due to the nonsintered nature and the higher surface area of Ca-P cements, dissolution of calcium and phosphate ions in physiological conditions occurs to a higher extent than in sintered ceramics. However, the relatively large surface area in combination with supersaturated body fluids also results in high Ca-P precipitation. Ishikawa et al. [23] reported that the total precipitation of (B-type) carbonate HA at the surface of Ca-P cement increased the total surface dissolution in simulated blood plasma *in vitro* for a period of 20 weeks. *In vivo*, formation of carbonated HA on cement surface has been reported to occur within 12 h of implantation [24]. This phenomenon could be one of the essential aspects of the exceptionally positive osteoconductive behavior of Ca-P cements.

9.4   *In Vivo* Interactions and Osteoinductivity

In general, Ca-P compounds are found to be biocompatible and favorable for the final bone response [4]. The ability of Ca-P ceramics to bond to bone is a rather unique property. For example, it has been a consistent finding that the larger the ceramic solubility rate, the more pronounced bone ingrowth is observed [25]. Ducheyne et al. hypothesized that the dissolution of Ca-P and precipitation of a carbonated HA layer is essential in the bioactive behavior of ceramics. On the other hand, Davies claimed that microtopography played a crucial role as bone formation at the implant surface was achieved by micromechanical interdigitation of the cement line with the material surface [26].

Protein–biomaterial interactions also play a key role in the biomaterial behavior *in vivo*. Directly after implantation, the surface of a biomaterial is being covered by serum proteins, which determine crucial reactions such as complement activation, thrombogenicity, and cell adhesion. Protein binding capacity depends on the specific ceramic characteristics as well as on the protein involved [27,28]. Ceramic materials exhibit a high binding affinity for proteins. Interestingly, several authors report that the combination of a material with high protein affinity and an appropriate (micro)architecture, can exhibit intrinsic osteoinductive behavior [8,11,12]. In a study in dogs, Yuan et al. observed ectopic bone formation after 3 and 6 months in micro- and macroporous HA, whereas macroporous HA without microporous structure did not reveal bone formation. It was hypothesized that due to the microporosity, total surface area and subsequent protein adsorption was increased, which could have induced bone formation. On the other hand, Ripamonti et al. hypothesized that concave macroporosity was the driving force for intrinsic osteoinduction in ceramics. The findings of intrinsic osteoinductive behavior in ceramic are interesting. However, it must be emphasized that regarding the numerous studies using similar porous ceramics without evidence for intrinsic osteoinduction, these observations currently are a rarity without confirmation of reproducibility.

Protein adsorption on ceramics not only plays an important role in biological processes, it also influences the ceramic itself with respect to dissolution and precipitation of calcium and phosphate ions. Radin et al. [29] reported that in an *in vitro* environment, serum proteins slowed down the dissolution (OHA and β-TCP) and precipitation (carbonated HA) of crystals at the surface of a crystalline multiphasic ceramic coating. A similar observation was done by Ong et al. [30], who reported an initial decrease in dissolution of phosphate molecules from Ca-P coatings in the presence of proteins with a high binding affinity. Combes et al. [31] investigated the interference of protein adsorption and HA maturation in detail and found that, depending on the concentration, serum albumin can exhibit a dual role and act as a promotor or inhibitor of nucleation and growth of (octa)Ca-P crystals. The energy needed for the first step in the precipitation process, nucleation, was lower when adsorbing molecules were present. Therefore, multiplication of nuclei occurred as a consequence of protein adsorption. The subsequent steps, growth to critical nuclei and formation of Ca-P crystals, were slowed down by a protein coating. However, at low serum albumin concentrations, the mineral surface could not be efficiently coated and crystal growth occurred faster than in the absence of albumin due to the larger number of nuclei and their greater reactivity. At high — physiological — serum albumin concentration, the mineral surface was efficiently coated slowing down the ionic diffusion to the crystal surface and inhibiting crystal growth [31].

9.5 Calcium Phosphate Ceramics for Bone Tissue Engineering

For reconstruction of large bone defects, implantation of osteoconductive porous scaffold materials alone is not sufficient for complete bone regeneration. Therefore, osteogenesis needs to be induced throughout the scaffold material. In the tissue engineering paradigm, osteogenic cells or osteoinductive macromolecules or both are being delivered to a suitable scaffold material. Many prerequisites have been postulated for the ideal scaffold material but none of the current materials meet all the demands. Porous ceramics are potential scaffold materials for bone tissue engineering due to their outstanding osteocompatible properties, especially with respect to osteoconduction and gradual resorption. In the recent years, a variety of researchers have investigated Ca-P ceramic scaffold materials for cellular as well as growth factor based tissue engineering.

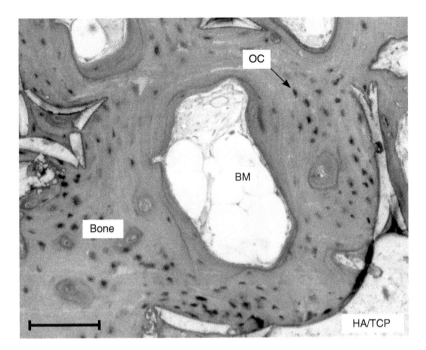

FIGURE 9.3 Bone formation in rat bone marrow stromal cell loaded porous HA/TCP implants (Camceram®), implanted in muscular flaps in rats for 8 weeks. Direct apposition of bone on the ceramic surface and bone marrow formation in the remaning porosity is observed. OC = osteocyte, BM = bone marrow. Bar represents 100 μm. Light microscopical undecalcified section stained with methylene blue and basic fuchsin (original magnification 10×). (Printed with permission from Hartman et al. Unpublished work, 2004.)

9.5.1 Ca-P Ceramics and Osteogenic Cells

In 1988, Maniatopoulos et al. published a study on harvesting and culturing of rat bone marrow stromal cells (BMSCs) *in vitro* [32]. Through this method, the small number of undifferentiated progenitor cells present in bone marrow could be isolated, expanded, and differentiated into the osteoblastic lineage. Currently, the essence of this technique is still in use to administer osteoprogenitor cells to various scaffold materials for cell-based bone engineering. Many studies in rodent animal studies have shown the validity of this concept in conjunction with ceramic scaffold materials as coralline HA, porous HA, porous HA/β-TCP and porous β-TCP [33–38] (Figure 9.3). Less studies, however, have investigated the bone forming capacities of large BMSC loaded ceramic implants in higher mammals. This is in particular challenging, as the living constructs are transposed from ideal cell culture conditions to an *in vivo* environment where oxygen supply prior to capillary ingrowth is only provided by (limited) diffusion. Vitality of the BMSCs within ceramic scaffold material was shown to be essential for osteogenicity [39]. Despite the potential jeopardy of cell death, bone formation was observed in the studies investigating ectopic implants in dogs and goats [40,41], or critical size defects in dogs, goats, and sheep [41–46]. In most of these studies, bone formation was also observed in the center of the implants where the unloaded controls did not show osteogenesis. This indicates that possibly even a small number of surviving progenitor cells may be sufficient to result in significant bone formation. Also, it emphasizes that future tissue engineering research should also focus on multicomponent constructs, in which BMSCs are being combined with bone- and vasculature inducing growth factors.

9.5.2 Ca-P Ceramics and Osteoinductive Growth Factors

During the last decade, several strategies have been developed to deliver growth factors to the desired site. Basically, the proteins can be delivered directly via a carrier/delivery vehicle, or by gene therapy

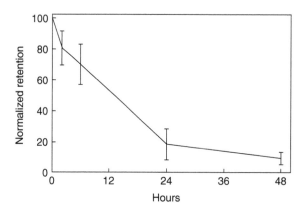

FIGURE 9.4   Retention of ^{131}I rhBMP-2 injected subcutaneously in rats. Over 80% of rhBMP-2 was cleared from the application site within 24 h, underlining the short residence time of rhBMP-2 after release from a carrier material. Error bars represent means $\pm$ standard deviation for $n = 6$.

techniques, in which a growth factor encoding gene is delivered to host cells via a viral factor or plasmid. Through this method, transfected cells start secreting growth factor. Cells can be transfected *in vitro*, prior to transplantation to the desired site, or directly *in vivo*. In this review, only direct delivery of proteins will be discussed as this approach has developed furthest and recently has led to FDA approval for specific clinical applications [47].

The growth factor loaded ceramic has a twofold function; first, it serves as a scaffold material, which enables and stimulates bone and blood vessel ingrowth; second, it serves as a carrier or delivery vehicle for the inductive factor, which should result in most favorable biopresentation and optimal activity of the protein. The role of ceramics on the bioactivity of growth factors is not completely clear. The ceramic might potentiate the activity of BMP by binding the protein and presenting it to target cells in a "bound" form [48,49]. On the other hand, the ceramic can act as vehicle for factor delivery to the surrounding tissues. Released protein may provide a supra-physiological concentration of free protein in the vicinity of the implant, what may attract target cells to the implant-site by chemotaxis [50]. It is hypothesized that for optimal bioactivity, initial burst release chemotactically recruits osteoprogenitor cells, and following sustained release — or retained factor — differentiates these cells toward the desired phenotype [51]. To our judgment, however, no convincing evidence provides further specification of this partition. Moreover, optimal pharmacokinetics are presumably specific for each surgical site.

9.5.2.1   Growth Factor Release from Ca-P Ceramics

The correlation between factor bioactivity and release or retention is a delicate issue. Since the protein needs to induce bone locally and not systemically, the released protein only has a short time to attract cells before it is being cleared from the application site. For example, a depot of 5 μg rhBMP-2 injected subcutaneously in the back in rats is cleared from the application site for more than 80% within 24 h (Figure 9.4). After clearance, the proteins are being metabolized by the body and excreted through the urinary tract [52]. The systemic exposure to released rhBMP-2 is reported to be neglectible [52].

The specific parameters resulting in protein release or retention are the driving factors behind the final clinical efficacy of a carrier system. Unfortunately, only few researchers have investigated these phenomena for ceramics. In a series of experiments in a rat ectopic model, Uludag et al. [53,54] have shown that growth factor retention depend on carrier- as well as on factor characteristics. Proteins with a higher isoelectric point (pI) show higher implant retention and higher retention of BMP yields higher osteoinductive activity. Various rhBMP-2 loaded ceramic implants (TCP cylinders and coral derived HA cylinders) show initial burst release of approximately 70% within three hours after implantation, which is followed by a slower second phase release and a final retention of approximately 6% of the initial dose after 14 days. In another study carried out by Louis-Ugbo et al. [52], initial burst release from HA/TCP granules (60 : 40) was less than 20% within the first day in a rabbit spinal fusion model. After 2 weeks $27 \pm 9\%$ of rhBMP-2 was still retained in this model. Li et al. [55] determined the release kinetics of rhBMP-2 mixed through Ca-P cement (α-BSM) and injected in a rabbit ulnar defect. Within the first day, approximately 40% of the initial factor was released. After 14 days, $15 \pm 7\%$ of the initial factor was still present in the cement. The differences in release and retention rates between these studies could be explained by variations in surgical site, dose, adsorption procedures, material surface area and chemico-physical interactions between material and growth factor. Overall, ceramics exhibit multiphasic release kinetics *in vivo* with a decrease of release in time and a certain retention for several weeks. It has been hypothesized, that ceramic surface area plays the key role in protein binding [56].

As previously mentioned, Ca-P ceramic materials show a high binding affinity for proteins. Conformational changes of proteins upon adsorption on Ca-P surface have been observed [30]. Protein conformation is of great importance for the accessibility of the specific binding sites and subsequent protein-cell interaction [57,58]. As a result, protein-biomaterial bonding may directly effect the protein's intrinsic function due to orientation or conformational changes. To evaluate the bioactivity of retained proteins, cells can be cultured on the surface of protein loaded ceramics *in vitro* [59]. A complicating factor is that no material exhibits a 100% retention [54]. For a true differentiation between activity expressed by retained protein vs. presently released protein, control groups have to correct for the released fraction bioactivity. In an *in vitro* model, Santos et al. [59] investigated the bioactivity of rhBMP-2 adsorbed onto a Ca-P layer on bioactive glass-gel (Xerogel S70) in comparison with the activity of "free" rhBMP-2 present in cell culture medium. The osteogenic response of osteoblast like cells seeded on the Ca-P surface was found significantly higher in the adsorbed rhBMP-2 group. The authors suggested that the biomaterial could generate, maintain, or even concentrate an effective level of growth factor.

9.5.2.2 Growth Factor Loading in Ca-P Ceramics

In view of the previous paragraph, discern has to be made between proteins/growth factors adsorbed at the ceramic surface and proteins/growth factors mixed through the material during preparation. For sintered ceramic materials, the latter is impossible as sintering temperatures would result in complete protein destruction. For the self setting Ca-P cements though, mixing of proteins during setting reaction has been described [60–62]. Due to physical entrapment by cement particles, the majority of the protein binding sites will be unavailable until the protein is released from the cement. This situation is quite similar to bioactive macromolecules entrapped in polymeric drug delivery vehicles, where protein release is the main focus [63–68]. However, a frequently neglected issue is that mixing of proteins through ceramics can influence the crystallization process during the setting reaction. This can modify the *in vivo* dissolution and/or degradation properties of the formed material.

For the growth factors entrapped within a Ca-P ceramic material, bioactivity is likewise expected only after release from its carrier. Bioactivity of released protein can be demonstrated through *in vitro* models where sensible cells are being exposed to the released protein [55,64,68]. To evaluate loss of activity, the induced biological response should be compared to the response to similar protein directly suspended in the culture medium [55]. *In vivo*, however, this comparison is difficult and most authors just examine whether or not the protein expresses (part of its) bioactivity.

9.5.2.3 Osteoinductive Capacity of Growth Factor Loaded Ca-P Ceramics

Osteogenic differentiation induced by BMPs is dependent on the regulating growth factor as well as on the carrier specifications. Murata et al. [69] observed that in a synthetic HA carrier, BMP induced ossification was predominantly intramembranous, whereas enchondral bone formation was observed in a nonceramic carrier. Direct bone formation on BMP loaded Ca-P ceramics was also observed by others [70–73].

However, the osteogenic behavior of growth factor loaded Ca-P ceramics is not as straightforward as suggested, because a wide variety of material compositions, animal models, surgical sites and factor dosage have been used in various studies. Another complicating factor is that physiological bone regeneration is induced by a cascade of growth factors, whereas most tissue engineered concepts investigate osteoinduction by single factors only. Consequently, most research in the recent years has focused on BMPs, like rhBMP-2

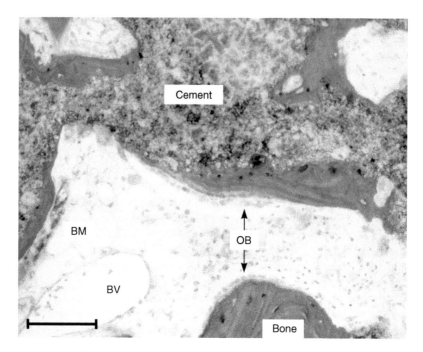

FIGURE 9.5   Bone formation in rhBMP-2 loaded porous Ca-P cement implants, implanted subcutaneously in rabbits for 10 weeks. Macroporosity in the cement (Calcibon®, Biomet Merck, Darmstadt, Germany) was induced by formation of CO_2 during setting of the cement, after which a dosage of 5 μg rhBMP-2 was applied. OB = osteoblasts, BM = bone marrow, BV = blood vessel. Bar represents 100 μm. Light microscopical undecalcified section stained with methylene blue and basic fuchsin (original magnification 10×). (From Kroese-Deutman, H.C., Ruhe, P.Q., Spauwen, P.H., and Jansen, J.A. *Biomaterials*, in press: 2004. With permission.)

and rhBMP-7 (OP-1). Due to the numerous parameters, there is no such thing as one ideal dosage or scaffold composition. Most animal studies investigating growth factor loaded ceramics have been carried out in smaller animals as rats [9,10,53,54,69,70,72,74–78] and rabbits [52,55,73,79–85] (Figure 9.5) whereas only few studies were performed in larger animals as minipigs [86,87], dogs [71,88,89], and primates [90–92]. Generally speaking for all carrier materials, the relative dosage of growth factor needed for bone induction raises with the animal size [51]. However, the dosages applied to ceramics vary substantially within similar animal models as well as between different species. For rhBMP-2, dosages applied for effective osteoinduction varied from 0.03 to 0.46 mg/ml implants in rats [9,54,75] to 0.004 to 2.1 mg/ml implant in rabbits [55,73,80,81,83,85], whereas in canine and primate models dosages have been used from 0.15 to 1.2 mg/ml [71] and 1.35 to 2.7 mg/ml [92] respectively. A dose–response relation for BMP loaded ceramics has been reported in some studies in rodents [75,80,83], but was absent or even reciprocal in studies with larger animal models [71,91,92]. Yet, it seems likely that in the dose–response relation, there is first a certain threshold dosage to achieve consistent results [92], followed by a linear dose–response interval and finalized by a certain saturation effect.

Besides BMPs, other growth factors like transforming growth factor-β (TGF-β) and insulin-like growth factor-I (IGF-I) have been investigated. TGF-β may potentiate the osteoinductivity of BMPs, but its overall osteoinductive capacity when applied alone is weak compared to BMPs [93]. IGF-I has also been applied to TCP ceramic carriers, which resulted in increased osteogenesis in an orthotopic implantation site four weeks after implantation [94]. Similarly, Damien et al. [95] reported that IGF-I loaded porous hydroxyapatite accelerated orthotopic bone ingrowth during an implantation period up to three weeks.

The synergism of various growth factors, as present in the physiological bone regeneration cascade, has only been investigated in few studies with Ca-P ceramics. Ono et al. investigated the role of prostaglandin E_1 (PGE$_1$) and basic fibroblast growth factor (bFGF) loaded on porous hydroxyapatite and observed that PGE$_1$ as well as bFGF acted synergistically with rhBMP-2 [80,83,96]. Meraw et al. prepared a true growth factor cocktail and reported that orthotopic bone formation in a canine model was enhanced by a combination of extremely low dosages of TGF-β, bFGF, BMP-2, and PDGF mixed through a calcium phosphate cement [89].

9.6 Conclusion and Future Perspective

Like all scaffolds materials currently under investigation for bone tissue engineering applications, Ca-P ceramics reveal both advantages and disadvantages. Intrinsic brittleness of Ca-P ceramics unquestionably is the greatest shortcoming of these materials, whereas their biocompatibility, osteoconductivity and resorption potential are virtues of an ideal scaffold material. Future research should further clarify the importance of release and retention of growth factors, optimize their delivery and investigate dose–response relation for the various factors and applications. Furthermore, the potentials of multicomponent constructs, in which various bone- and vasculature inducing growth factors or cultured bone marrow stromal cells or both are being combined, should be investigated to come to a successful alternative for autologous bone grafting.

References

[1] LeGeros, R.Z. Properties of osteoconductive biomaterials: calcium phosphates. *Clin. Orthop.* 81–98, 2002.

[2] Dorozhkin, S.V. and Epple, M. Biological and medical significance of calcium phosphates. *Angew. Chem. Int. Ed. Engl.* 41: 3130–3146, 2002.

[3] Eanes, E.D. Amorphous calcium phosphate. *Monogr. Oral Sci.* 18: 130–147, 2001.

[4] LeGeros, R.Z. Calcium phosphates in oral biology and medicine. *Monogr. Oral Sci.* 15: 1–201, 1991.

[5] Vaccaro, A.R. The role of the osteoconductive scaffold in synthetic bone graft. *Orthopedics* 25: s571–s578, 2002.

[6] Ravaglioli, A. and Krajewski, A. *Bioceramics.* London: Chapman & Hall, 1992.

[7] Le Huec, J.C., Schaeverbeke, T., Clement, D., Faber, J., and Le Rebeller, A. Influence of porosity on the mechanical resistance of hydroxyapatite ceramics under compressive stress. *Biomaterials* 16: 113–118, 1995.

[8] Yuan, H., Kurashina, K., de Bruijn, J.D., Li, Y., de Groot, K., and Zhang, X. A preliminary study on osteoinduction of two kinds of calcium phosphate ceramics. *Biomaterials* 20: 1799–1806, 1999.

[9] Tsuruga, E., Takita, H., Itoh, H., Wakisaka, Y., and Kuboki, Y. Pore size of porous hydroxyapatite as the cell-substratum controls BMP-induced osteogenesis. *J. Biochem. (Tokyo)* 121: 317–324, 1997.

[10] Kuboki, Y., Jin, Q., and Takita, H. Geometry of carriers controlling phenotypic expression in BMP-induced osteogenesis and chondrogenesis. *J. Bone Joint Surg. Am.* 83-A (Suppl 1): 105–115, 2001.

[11] Ripamonti, U., Crooks, J., and Kirkbride, A.N. Sintered porous hydroxyapatites with intrinsic osteoinductive activity: geometric induction of bone formation. *S. Afr. J. Sci.* 95: 335–343, 1999.

[12] Ripamonti, U. Smart biomaterials with intrinsic osteoinductivity: Geometric control of bone differentiation. In *Bone Engineering*, ed. Davies, J.E., Toronto: EM Squared Incorporated, 2000, pp. 215–222.

[13] Jansen, J.A., Wolke, J.G., Swann, S., van der Waerden, J.P., and de Groot, K. Application of magnetron sputtering for producing ceramic coatings on implant materials. *Clin. Oral. Implants Res.* 4: 28–34, 1993.

[14] Roy, D.M. and Linnehan, S.K. Hydroxyapatite formed from coral skeletal carbonate by hydrothermal exchange. *Nature* 247: 220–222, 1974.

[15] Truumees, E. and Herkowitz, H.N. Alternatives to autologous bone harvest in spine surgery. *Univ. Penn. Orthop. J.* 12: 77–88, 1999.

[16] Brown, W.E. and Chow, L.C. Dental restorative cement pastes. U.S. Patent (4,518,430), 1985.

[17] Chow, L.C. Calcium phosphate cements. *Monogr. Oral Sci.* 18: 148–163, 2001.

[18] Schmitz, J.P., Hollinger, J.O., and Milam, S.B. Reconstruction of bone using calcium phosphate bone cements: a critical review. *J. Oral Maxillofac. Surg.* 57: 1122–1126, 1999.

[19] Ooms, E.M., Wolke, J.G., van de Heuvel, M.T., Jeschke, B., and Jansen, J.A. Histological evaluation of the bone response to calcium phosphate cement implanted in cortical bone. *Biomaterials* 24: 989–1000, 2003.

[20] Comuzzi, L., Ooms, E., and Jansen, J.A. Injectable calcium phosphate cement as a filler for bone defects around oral implants: an experimental study in goats. *Clin. Oral Implants Res.* 13: 304–311, 2002.

[21] Ooms, E.M., Wolke, J.G., van der Waerden, J.P., and Jansen, J.A. Trabecular bone response to injectable calcium phosphate (Ca-P) cement. *J. Biomed. Mater. Res.* 61: 9–18, 2002.

[22] Driessens, F.C., Boltong, M.G., Bermudez, O., and Planell, J.A. Formulation and setting times of some calcium orthophosphate cements: a pilot study. *J. Mater. Sci. Mater. Med.* 4: 503–508, 1993.

[23] Ishikawa, K., Takagi, S., Chow, L.C., Ishikawa, Y., Eanes, E.D., and Asaoka, K. Behavior of a calcium phosphate cement in simulated blood plasma *in vitro*. *Dent. Mater.* 10: 26–32, 1994.

[24] Takagi, S., Chow, L.C., Markovic, M., Friedman, C.D., and Costantino, P.D. Morphological and phase characterizations of retrieved calcium phosphate cement implants. *J. Biomed. Mater. Res.* 58: 36–41, 2001.

[25] Ducheyne, P. and Qiu, Q. Bioactive ceramics: the effect of surface reactivity on bone formation and bone cell function. *Biomaterials* 20: 2287–2303, 1999.

[26] Davies, J.E. Mechanisms of endosseous integration. *Int. J. Prosthodont.* 11: 391–401, 1998.

[27] Rosengren, A., Pavlovic, E., Oscarsson, S., Krajewski, A., Ravaglioli, A., and Piancastelli, A. Plasma protein adsorption pattern on characterized ceramic biomaterials. *Biomaterials* 23: 1237–1247, 2002.

[28] Rosengren, A., Oscarsson, S., Mazzocchi, M., Krajewski, A., and Ravaglioli, A. Protein adsorption onto two bioactive glass-ceramics. *Biomaterials* 24: 147–155, 2003.

[29] Radin, S., Ducheyne, P., Berthold, P., and Decker, S. Effect of serum proteins and osteoblasts on the surface transformation of a calcium phosphate coating: a physicochemical and ultrastructural study. *J. Biomed. Mater. Res.* 39: 234–243, 1998.

[30] Ong, J.L., Chittur, K.K., and Lucas, L.C. Dissolution/reprecipitation and protein adsorption studies of calcium phosphate coatings by FT-IR/ATR techniques. *J. Biomed. Mater. Res.* 28: 1337–1346, 1994.

[31] Combes, C. and Rey, C. Adsorption of proteins and calcium phosphate materials bioactivity. *Biomaterials* 23: 2817–2823, 2002.

[32] Maniatopoulos, C., Sodek, J., and Melcher, A.H. Bone formation *in vitro* by stromal cells obtained from bone marrow of young adult rats. *Cell Tissue Res.* 254: 317–330, 1988.

[33] Hanada, K., Dennis, J.E., and Caplan, A.I. Stimulatory effects of basic fibroblast growth factor and bone morphogenetic protein-2 on osteogenic differentiation of rat bone marrow-derived mesenchymal stem cells. *J. Bone Miner. Res.* 12: 1606–1614, 1997.

[34] Okumura, M., Ohgushi, H., Dohi, Y., Katuda, T., Tamai, S., Koerten, H.K., and Tabata, S. Osteoblastic phenotype expression on the surface of hydroxyapatite ceramics. *J. Biomed. Mater. Res.* 37: 122–129, 1997.

[35] Boo, J.S., Yamada, Y., Okazaki, Y., Hibino, Y., Okada, K., Hata, K., Yoshikawa, T., Sugiura, Y., and Ueda, M. Tissue-engineered bone using mesenchymal stem cells and a biodegradable scaffold. *J. Craniofac. Surg.* 13: 231–239, 2002.

[36] Dong, J., Uemura, T., Shirasaki, Y., and Tateishi, T. Promotion of bone formation using highly pure porous beta-TCP combined with bone marrow-derived osteoprogenitor cells. *Biomaterials* 23: 4493–4502, 2002.

[37] Uemura, T., Dong, J., Wang, Y., Kojima, H., Saito, T., Iejima, D., Kikuchi, M., Tanaka, J., and Tateishi, T. Transplantation of cultured bone cells using combinations of scaffolds and culture techniques. *Biomaterials* 24: 2277–2286, 2003.

[38] Hartman, E.H., Ruhe, P.Q., Spauwen, P.H., and Jansen, J.A. Ectopic bone formation in rats: comparison of biphasic ceramic implants seeded with cultured rat bone marrow cells in a pedicled and a revascularised muscle flap. Unpublished work, 2004.

[39] Kruyt, M.C., de Bruijn, J.D., Wilson, C.E., Oner, F.C., van Blitterswijk, C.A., Verbout, A.J., and Dhert, W.J. Viable osteogenic cells are obligatory for tissue-engineered ectopic bone formation in goats. *Tissue Eng.* 9: 327–336, 2003.

[40] Kadiyala, S., Young, R.G., Thiede, M.A., and Bruder, S.P. Culture expanded canine mesenchymal stem cells possess osteochondrogenic potential *in vivo* and *in vitro*. *Cell Transplant* 6: 125–134, 1997.

[41] Kruyt, M.C., Dhert, W.J., Yuan, H., Wilson, C.E., van Blitterswijk, C.A., Verbout, A.J., and de Bruijn, J.D. Bone tissue engineering in a critical size defect compared to ectopic implantations in the goat. *J. Orthop. Res.* 22: 544–551, 2004.

[42] Schliephake, H., Knebel, J.W., Aufderheide, M., and Tauscher, M. Use of cultivated osteoprogenitor cells to increase bone formation in segmental mandibular defects: an experimental pilot study in sheep. *Int. J. Oral Maxillofac. Surg.* 30: 531–537, 2001.

[43] Bruder, S.P., Kraus, K.H., Goldberg, V.M., and Kadiyala, S. The effect of implants loaded with autologous mesenchymal stem cells on the healing of canine segmental bone defects. *J. Bone Joint Surg. Am.* 80: 985–996, 1998.

[44] Kon, E., Muraglia, A., Corsi, A., Bianco, P., Marcacci, M., Martin, I., Boyde, A., Ruspantini, I., Chistolini, P., Rocca, M., Giardino, R., Cancedda, R., and Quarto, R. Autologous bone marrow stromal cells loaded onto porous hydroxyapatite ceramic accelerate bone repair in critical-size defects of sheep long bones. *J. Biomed. Mater. Res.* 49: 328–337, 2000.

[45] Petite, H., Viateau, V., Bensaid, W., Meunier, A., De Pollak, C, Bourguignon, M., Oudina, K., Sedel, L., and Guillemin, G. Tissue-engineered bone regeneration. *Nat. Biotechnol.* 18: 959–963, 2000.

[46] Shang, Q., Wang, Z., Liu, W., Shi, Y., Cui, L., and Cao, Y. Tissue-engineered bone repair of sheep cranial defects with autologous bone marrow stromal cells. *J. Craniofac. Surg.* 12: 586–593, 2001.

[47] Nakashima, M. and Reddi, A.H. The application of bone morphogenetic proteins to dental tissue engineering. *Nat. Biotechnol.* 21: 1025–1032, 2003.

[48] Reddi, A.H. and Cunningham, N.S. Initiation and promotion of bone differentiation by bone morphogenetic proteins. *J. Bone Miner. Res.* 8: 499–502, 1993.

[49] Uludag, H., Gao, T., Porter, T.J., Friess, W., and Wozney, J.M. Delivery systems for BMPs: factors contributing to protein retention at an application site. *J. Bone Joint Surg. Am.* 83-A: 128–135, 2001.

[50] Cunningham, N.S., Paralkar, V., and Reddi, A.H. Osteogenin and recombinant bone morphogenetic protein 2B are chemotactic for human monocytes and stimulate transforming growth factor beta 1 mRNA expression. *Proc. Natl Acad. Sci. USA* 89: 11740–11744, 1992.

[51] Li, R.H. and Wozney, J.M. Delivering on the promise of bone morphogenetic proteins. *Trends Biotechnol.* 19: 255–265, 2001.

[52] Louis-Ugbo, J., Kim, H.S., Boden, S.D., Mayr, M.T., Li, R.C., Seeherman, H., D'Augusta, D., Blake, C., Jiao, A., and Peckham, S. Retention of 125I-labeled recombinant human bone morphogenetic protein-2 by biphasic calcium phosphate or a composite sponge in a rabbit posterolateral spine arthrodesis model. *J. Orthop. Res.* 20: 1050–1059, 2002.

[53] Uludag, H., D'Augusta, D., Golden, J., Li, J., Timony, G., Riedel, R., and Wozney, J.M. Implantation of recombinant human bone morphogenetic proteins with biomaterial carriers: a correlation between protein pharmacokinetics and osteoinduction in the rat ectopic model. *J. Biomed. Mater. Res.* 50: 227–238, 2000.

[54] Uludag, H., D'Augusta, D., Palmer, R., Timony, G., and Wozney, J. Characterization of rhBMP-2 pharmacokinetics implanted with biomaterial carriers in the rat ectopic model. *J. Biomed. Mater. Res.* 46: 193–202, 1999.

[55] Li, R.H., Bouxsein, M.L., Blake, C.A., D'Augusta, D., Kim, H., Li, X.J., Wozney, J.M., and Seeherman, H.J. rhBMP-2 injected in a calcium phosphate paste (alpha-BSM) accelerates healing in the rabbit ulnar osteotomy model. *J. Orthop. Res.* 21: 997–1004, 2003.

[56] Herr, G., Hartwig, C.H., Boll, C., and Kusswetter, W. Ectopic bone formation by composites of BMP and metal implants in rats. *Acta Orthop. Scand.* 67: 606–610, 1996.

[57] Kirsch, T., Sebald, W., and Dreyer, M.K. Crystal structure of the BMP-2-BRIA ectodomain complex. *Nat. Struct. Biol.* 7: 492–496, 2000.

[58] Nickel, J., Dreyer, M.K., Kirsch, T., and Sebald, W. The crystal structure of the BMP-2 : BMPR-IA complex and the generation of BMP-2 antagonists. *J. Bone Joint Surg. Am.* 83-A: 7–14, 2001.

[59] Santos, E.M., Radin, S., Shenker, B.J., Shapiro, I.M., and Ducheyne, P. Si–Ca-P xerogels and bone morphogenetic protein act synergistically on rat stromal marrow cell differentiation *in vitro*. *J. Biomed. Mater. Res.* 41: 87–94, 1998.

[60] Blom, E.J., Klein-Nulend, J., Yin, L., van Waas, M.A., and Burger, E.H. Transforming growth factor-beta1 incorporated in calcium phosphate cement stimulates osteotransductivity in rat calvarial bone defects. *Clin. Oral Implants Res.* 12: 609–616, 2001.

[61] Lee, D.D., Tofighi, A., Aiolova, M., Chakravarthy, P., Catalano, A., Majahad, A., and Knaack, D. Alpha-BSM: a biomimetic bone substitute and drug delivery vehicle. *Clin. Orthop.* 396–405, 1999.

[62] Li, R.H., D'Augusta, D., Blake, C., Bouxsein, M., Wozney, J.M., Li, J., Stevens, M., Kim, H., and Seeherman, H. rhBMP-2 delivery and efficacy in an injectable calcium phosphate based matrix. *The 28th International Symposium on Controlled Release of Bioactive Materials*, San Diego, 2001.

[63] Isobe, M., Yamazaki, Y., Oida, S., Ishihara, K., Nakabayashi, N., and Amagasa, T. Bone morphogenetic protein encapsulated with a biodegradable and biocompatible polymer. *J. Biomed. Mater. Res.* 32: 433–438, 1996.

[64] Lu, L., Yaszemski, M.J., and Mikos, A.G. TGF-beta1 release from biodegradable polymer microparticles: its effects on marrow stromal osteoblast function. *J. Bone Joint Surg. Am.* 83-A (Suppl 1): 82–91, 2001.

[65] Miyamoto, S., Takaoka, K., Okada, T., Yoshikawa, H., Hashimoto, J., Suzuki, S., and Ono, K. Polylactic acid–polyethylene glycol block copolymer. A new biodegradable synthetic carrier for bone morphogenetic protein. *Clin. Orthop.* 333–343, 1993.

[66] Oldham, J.B., Lu, L., Zhu, X., Porter, B.D., Hefferan, T.E., Larson, D.R., Currier, B.L., Mikos, A.G., and Yaszemski, M.J. Biological activity of rhBMP-2 released from PLGA microspheres. *J. Biomech. Eng.* 122: 289–292, 2000.

[67] Pean, J.M., Venier-Julienne, M.C., Boury, F., Menei, P., Denizot, B., and Benoit, J.P. NGF release from poly(D,L-lactide-co-glycolide) microspheres. Effect of some formulation parameters on encapsulated NGF stability. *J. Control. Release.* 56: 175–187, 1998.

[68] Peter, S.J., Lu, L., Kim, D.J., Stamatas, G.N., Miller, M.J., Yaszemski, M.J., and Mikos, A.G. Effects of transforming growth factor beta1 released from biodegradable polymer microparticles on marrow stromal osteoblasts cultured on poly(propylene fumarate) substrates. *J. Biomed. Mater. Res.* 50: 452–462, 2000.

[69] Murata, M., Inoue, M., Arisue, M., Kuboki, Y., and Nagai, N. Carrier-dependency of cellular differentiation induced by bone morphogenetic protein in ectopic sites. *Int. J. Oral Maxillofac. Surg.* 27: 391–396, 1998.

[70] Ripamonti, U., Ma, S., and Reddi, A.H. The critical role of geometry of porous hydroxyapatite delivery system in induction of bone by osteogenin, a bone morphogenetic protein. *Matrix* 12: 202–212, 1992.

[71] Sumner, D.R., Turner, T.M., Urban, R.M., Turek, T., Seeherman, H., and Wozney, J.M. Locally delivered rhBMP-2 enhances bone ingrowth and gap healing in a canine model. *J. Orthop. Res.* 22: 58–65, 2004.

[72] Kuboki, Y., Saito, T., Murata, M., Takita, H., Mizuno, M., Inoue, M., Nagai, N., and Poole, A.R. Two distinctive BMP-carriers induce zonal chondrogenesis and membranous ossification, respectively; geometrical factors of matrices for cell-differentiation. *Connect. Tissue. Res.* 32: 219–226, 1995.

[73] Kroese-Deutman, H.C., Ruhe, P.Q., Spauwen P.H., and Jansen J.A. Bone inductive properties of rhBMP-2 loaded porous calcium phosphate cement implants inserted at an ectopic site in rabbits. *Biomaterials* 26: 1131–1138, 2005.

[74] Kuboki, Y., Takita, H., Kobayashi, D., Tsuruga, E., Inoue, M., Murata, M., Nagai, N., Dohi, Y., and Ohgushi, H. BMP-induced osteogenesis on the surface of hydroxyapatite with geometrically feasible and nonfeasible structures: topology of osteogenesis. *J. Biomed. Mater. Res.* 39: 190–199, 1998.

[75] Ohura, K., Hamanishi, C., Tanaka, S., and Matsuda, N. Healing of segmental bone defects in rats induced by a beta-TCP-MCPM cement combined with rhBMP-2. *J. Biomed. Mater. Res.* 44: 168–175, 1999.

[76] Glass, D.A., Mellonig, J.T., and Towle, H.J. Histologic evaluation of bone inductive proteins complexed with coralline hydroxylapatite in an extraskeletal site of the rat. *J. Periodontol.* 60: 121–126, 1989.

[77] Herr, G., Wahl, D., and Kusswetter, W. Osteogenic activity of bone morphogenetic protein and hydroxyapatite composite implants. *Ann. Chir. Gynaecol. Suppl.* 207: 99–107, 1993.

[78] Clarke, S.A., Brooks, R.A., Lee, P.T., and Rushton, N. Bone growth into a ceramic-filled defect around an implant. The response to transforming growth factor beta1. *J. Bone Joint Surg. Br.* 86: 126–134, 2004.

[79] Ono, I., Ohura, T., Murata, M., Yamaguchi, H., Ohnuma, Y., and Kuboki, Y. A study on bone induction in hydroxyapatite combined with bone morphogenetic protein. *Plast. Reconstr. Surg.* 90: 870–879, 1992.

[80] Ono, I., Inoue, M., and Kuboki, Y. Promotion of the osteogenetic activity of recombinant human bone morphogenetic protein by prostaglandin E1. *Bone* 19: 581–588, 1996.

[81] Ono, I., Gunji, H., Kaneko, F., Saito, T., and Kuboki, Y. Efficacy of hydroxyapatite ceramic as a carrier for recombinant human bone morphogenetic protein. *J. Craniofac. Surg.* 6: 238–244, 1995.

[82] Ono, I., Gunji, H., Suda, K., Kaneko, F., Murata, M., Saito, T., and Kuboki, Y. Bone induction of hydroxyapatite combined with bone morphogenetic protein and covered with periosteum. *Plast. Reconstr. Surg.* 95: 1265–1272, 1995.

[83] Ono, I., Tateshita, T., and Kuboki, Y. Prostaglandin E1 and recombinant bone morphogenetic protein effect on strength of hydroxyapatite implants. *J. Biomed. Mater. Res.* 45: 337–344, 1999.

[84] Koempel, J.A., Patt, B.S., O'Grady, K., Wozney, J., and Toriumi, D.M. The effect of recombinant human bone morphogenetic protein-2 on the integration of porous hydroxyapatite implants with bone. *J. Biomed. Mater. Res.* 41: 359–363, 1998.

[85] Ruhe, P.Q., Kroese-Deutman, H.C., Wolke, J.G., Spauwen, P.H., and Jansen, J.A. Bone inductive properties of rhBMP-2 loaded porous calcium phosphate cement implants in cranial defects in rabbits. *Biomaterials* 25: 2123–2132, 2004.

[86] Terheyden, H., Warnke, P., Dunsche, A., Jepsen, S., Brenner, W., Palmie, S., Toth, C., and Rueger, D.R. Mandibular reconstruction with prefabricated vascularized bone grafts using recombinant human osteogenic protein-1: an experimental study in miniature pigs. Part II: transplantation. *Int. J. Oral Maxillofac. Surg.* 30: 469–478, 2001.

[87] Terheyden, H., Knak, C., Jepsen, S., Palmie, S., and Rueger, D.R. Mandibular reconstruction with a prefabricated vascularized bone graft using recombinant human osteogenic protein-1: an experimental study in miniature pigs. Part I: prefabrication. *Int. J. Oral Maxillofac. Surg.* 30: 373–379, 2001.

[88] Sumner, D.R., Turner, T.M., Purchio, A.F., Gombotz, W.R., Urban, R.M., and Galante, J.O. Enhancement of bone ingrowth by transforming growth factor-beta. *J. Bone Joint Surg. Am.* 77: 1135–1147, 1995.

[89] Meraw, S.J., Reeve, C.M., Lohse, C.M., and Sioussat, T.M. Treatment of peri-implant defects with combination growth factor cement. *J. Periodontol.* 71: 8–13, 2000.

[90] Ripamonti, U., Ramoshebi, L.N., Matsaba, T., Tasker, J., Crooks, J., and Teare, J. Bone induction by BMPs/OPs and related family members in primates. *J. Bone Joint Surg. Am.* 83-A (Suppl 1): 116–127, 2001.

[91] Ripamonti, U., Crooks, J., and Rueger, D.C. Induction of bone formation by recombinant human osteogenic protein-1 and sintered porous hydroxyapatite in adult primates. *Plast. Reconstr. Surg.* 107: 977–988, 2001.

[92] Boden, S.D., Martin, G.J., Jr., Morone, M.A., Ugbo, J.L., and Moskovitz, P.A. Posterolateral lumbar intertransverse process spine arthrodesis with recombinant human bone morphogenetic protein 2/hydroxyapatite-tricalcium phosphate after laminectomy in the nonhuman primate. *Spine* 24: 1179–1185, 1999.

[93] Lane, J.M., Tomin, E., and Bostrom, M.P. Biosynthetic bone grafting. *Clin. Orthop.* 107–117, 1999.

[94] Laffargue, P., Fialdes, P., Frayssinet, P., Rtaimate, M., Hildebrand, H.F., and Marchandise, X. Adsorption and release of insulin-like growth factor-I on porous tricalcium phosphate implant. *J. Biomed. Mater. Res.* 49: 415–421, 2000.

[95] Damien, E., Hing, K., Saeed, S., and Revell, P.A. A preliminary study on the enhancement of the osteointegration of a novel synthetic hydroxyapatite scaffold *in vivo*. *J. Biomed. Mater. Res.* 66A: 241–246, 2003.

[96] Ono, I., Tateshita, T., Takita, H., and Kuboki, Y. Promotion of the osteogenetic activity of recombinant human bone morphogenetic protein by basic fibroblast growth factor. *J. Craniofac. Surg.* 7: 418–425, 1996.

10

Biomimetic Materials

Andrés J. García
Georgia Institute of Technology

Over the last decade, considerable advances in the engineering of biomaterials that elicit specific cellular responses have been attained by exploiting biomolecular recognition. These biomimetic engineering approaches focus on integrating recognition and structural motifs from biological macromolecules with synthetic and natural substrates to generate bio-inspired, biofunctional materials. These strategies represent a paradigm shift in biomaterials development from conventional approaches which deal with purely synthetic or natural materials, and provide promising schemes for the development of novel bioactive substrates for enhanced tissue replacement and regeneration. Because of the central roles that extracellular matrices (ECMs) play in tissue morphogenesis, homeostasis, and repair, these natural scaffolds provide several attractive characteristics worthy of "copying" or mimicking to convey functionality for molecular control of cell function, tissue structure, and regeneration. Four ECM "themes" have been targeted (i) motifs to promote cell adhesion, (ii) growth factor binding sites that control presentation and delivery, (iii) protease-sensitive sequences for controlled degradation, and (iv) structural motifs to convey mechanical properties.

10.1 Extracellular Matrices: Nature's Engineered Scaffolds

ECMs comprise a complex, insoluble, three-dimensional mixture of secreted macromolecules, including collagens and noncollagenous proteins, such as elastin and fibronectin, glycosaminoglycans, and proteoglycans that are present between cells. In addition, provisional fibrin-based networks constitute specialized matrices for wound healing and tissue repair. ECMs function to provide structure and order in the extracellular space and regulate multiple functions associated with the establishment, maintenance, and remodeling of differentiated tissues [1]. Matrix components, such as fibronectin and laminin, mediate adhesive interactions that support anchorage, migration, and tissue organization and activate signaling pathways directing cell survival, proliferation, and differentiation. ECM components also interact with growth and differentiation factors, chemotropic agents, and other soluble factors that regulate cell cycle progression and differentiation to control their availability and activity. By immobilizing and ordering these ligands, ECMs control the spatial and temporal profiles of these signals and generate gradients necessary for vectorial responses. Moreover, structural elements within ECMs, namely collagens, elastin, and proteoglycans, contribute to the mechanical integrity, rigidity, and viscoelasticity of skin, cartilage, vasculature, and other tissues. Finally, the composition and structure of ECMs are dynamically modulated by the cells within them, reflecting the highly regulated and bidirectional communication between cells and ECMs.

10.2   Bioadhesive Materials

Following the identification of adhesion motifs in ECM components, such as the arginine–glycine–aspartic acid (RGD) tri-peptide for fibronectin [2] and tyrosine–isoleucine–glycine–serine–arginine (YIGSR) oligopeptide for laminin [3], numerous groups have tethered short bioadhesive peptides onto synthetic or natural substrates and three-dimensional scaffolds to produce biofunctional materials that bind adhesion receptors and promote adhesion and migration in various cell types (as reviewed in References 4 to 6) (Figure 10.1). Nonfouling supports, such as poly(ethylene glycol), poly(acrylamide), and alginate, are often used to reduce nonspecific protein adsorption and background adhesion. The density of tethered peptide is an important design parameter as cell adhesion, focal adhesion assembly, spreading and migration [7–10], neurite extension and neural differenitiation [11,12], smooth muscle cell activities [13], and osteoblast and myoblast differentiation [14,15] exhibit peptide density-dependent effects. More importantly, tethering of bioadhesive ligands onto biomaterial surfaces and scaffolds enhances in vivo responses, such as bone formation and integration [16–18], nerve regeneration [19,20], and corneal tissue repair [21].

These results indicate that functionalization of biomaterials with short adhesive oligopeptides significantly enhances cellular activities. In addition to conveying biospecificity while avoiding unwanted interactions with other regions of the native ligand, short bioadhesive peptides allow facile incorporation into synthetic backbones and enhanced stability of the tethered motif. These strategies, however, are limited by (i) low activity of oligopeptides compared to native ligand due to the absence of modulatory domains, (ii) limited specificity for adhesion receptors and cell types, and (iii) inability to bind certain receptors due to conformational differences compared to the native ligand. Conformationally constrained (e.g., cyclic) RGD [22], oligopeptides mixtures [23,24], and recombinant protein fragments spanning the binding domains of native ligands [25]

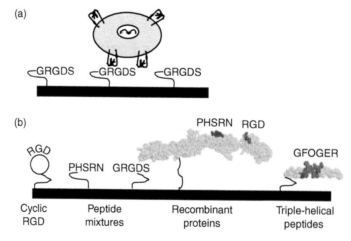

FIGURE 10.1   Biomimetic materials supporting cell adhesion. (a) Basic strategy focusing on RGD tethering to promote binding of cell adhesion receptors. (b) Biomolecular approaches to improve ligand activity and specificity.

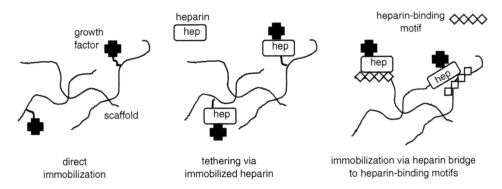

FIGURE 10.2   Biomimetic strategies for controlled growth factor interactions.

have been explored to improve ligand activity. Similarly, substrates presenting RGD in combination with other peptides, such as heparan sulfate-binding lyrine–arginine–serine–arginine (KRSR) [26] and Phenylalanine–histidine–arginine–arginine–isoleucine–lysine–alanine (FHRRIKA) [27], display improved cell adhesion and selectivity over materials modified with RGD alone. Finally, self-assembling peptides reconstituting the triple helical structure of type I collagen have been used to target collagen integrin receptors and promote enhanced osteoblastic differentiation and mineralization on biomaterial supports [28,29]. The improved activity and selectivity of these "second generation" biomolecular materials are expected to reconstitute additional features of natural ECMs and promote enhanced cellular activities.

10.3   Materials Engineered to Interact with Growth Factors

Natural ECMs interact with soluble growth and differentiation factors to control their activity by regulating their presentation, delivery kinetics, and stability [1]. Three general strategies have been pursued to convey growth factor activity to synthetic materials (Figure 10.2). Covalent immobilization of growth factors onto biomaterial supports, either via conventional peptide chemistry or enzymatic cross-linking, results in enhanced signaling activity [30,31]. Furthermore, enzyme-regulated release of tethered growth factors by incorporating protease-sensitive sequences in the tether has been applied to modulate release kinetics [32,33] and shown to improve blood vessel growth [34]. Mimicking the natural affinity of ECM glycosaminoglycans for heparin-binding growth factors, heparin has been covalently immobilized onto scaffolds to control the presentation and release kinetics of heparin-binding growth factors, such as basic fibroblast growth factor [35,36] and transforming growth factor-β [37]. Similarly, engineering of basic heparin-binding motifs, such as phenylalanine–alanine–lysine–leucine–alanine–alanine–arginine–lysine–tyrosine–arginine–lysine–alanine (FAKLAAR-LYRKA) and tyrosine–lysine–lysine–isoleucine–lysine–lysine–leucine (YKKIIKKL), into scaffolds and sequestering of heparin-binding growth factors via a heparin bridge results in enhanced delivery kinetics and nerve regeneration [38,39].

10.4   Protease-Degradable Materials

Cell-mediated local degradation of ECMs is critical to tissue development, maintenance, and remodeling [40]. To mimic this behavior, proteolytically sensitive substrate motifs specific for natural ECM enzymes, including matrix metalloproteinases and plasmin, have been incorporated into biomaterial scaffolds [41] (Figure 10.3). Consistent with migration behaviors in collagen and fibrin ECMs [40], cell migration and neurite extension through engineered networks require protease activity specific for the incorporated protease substrate motifs [42,43]. Structure–function analyses of cell invasion into synthetic hydrogels have shown that the extent of invasion depends on protease substrate activity, adhesion ligand concentration,

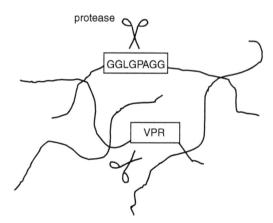

FIGURE 10.3   Enzyme-sensitive materials incorporating protease substrate motifs.

and network cross-linking density [44]. Furthermore, poly(ethylene glycol)-based hydrogels containing RGD oligopeptides and substrates for matrix metalloproteinases engineered to deliver BMP-2 to bone defects exhibit enhanced healing compared to standard treatments [44]. Bone regeneration in this model is dependent on the proteolytic sensitivity of the matrices and their architecture.

10.5   Artificial Proteins as Building Elements for Matrices

Artificial analogues of ECM proteins incorporating structural motifs to reconstitute secondary structures (e.g., coiled coil, α-helix) and convey controlled mechanical properties have been engineered via both synthetic routes and recombinant DNA technology [45]. These artificial proteins provide opportunities to generate novel hybrid macromolecules with additional or new functionalities and enhanced cost-efficiency, while overcoming limitations associated with natural ECMs, such as a restricted range in mechanical properties, processability, batch-to-batch variability, and the potential of pathogen transmission. For example, artificial proteins consisting of terminal leucine zipper domains flanking a central polyelectrolyte segment produce coiled-coil domains that render the material into a reversible, self-assembling hydrogel [46]. Polymers containing the glycine–valine–glycine–valine–proline (GVGVP) repeat from elastin have been designed to mimic the mechanical behavior of elastin [47]. These materials can be formulated as fibers and networks to create artificial ECMs [48,49]. Moreover, combination of elastin and fibronectin motifs has resulted in novel materials with biological and mechanical properties similar to those from natural ECMs [50]. Engineering of biofunctionalized and synthetic glycosaminoglycan polymers represents another promising avenue to artificial ECMs [51–53].

10.6   Conclusions and Outlook

Biomimetic materials incorporating bioactive motifs from natural ECMs have emerged as promising, novel biofunctional materials for tissue maintenance, repair, and regeneration. Although considerable progress has been attained in mimicking particular characteristics of ECMs, next-generation, bio-inspired materials must incorporate multiple characteristics from biological matrices to recapitulate the robust activities associated with these natural scaffolds. Furthermore, advances in materials engineering should provide routes for integrating multiple ligands, ligand gradients, nanoscale clustering, and dynamic, environment-responsive interfacial, and bulk properties. Finally, these biomimetic materials must be designed to support the activities of multiple cell types to successfully mimic desired tissue behavior and function.

References

[1] Reichardt, L.F., Introduction: extracellular matrix molecules, in *Guidebook to the Extracellular Matrix, Anchor, and Adhesion Proteins*, 2nd ed., Oxford University Press, New York, 1999, p. 335.

[2] Ruoslahti, E. and Pierschbacher, M.D., New perspectives in cell adhesion: RGD and integrins, *Science*, 238, 491, 1987.

[3] Graf, J. et al., A pentapeptide from the laminin B1 chain mediates cell adhesion and binds the 67,000 laminin receptor, *Biochemistry*, 26, 6896, 1987.

[4] Hubbell, J.A., Bioactive biomaterials, *Curr. Opin. Biotechnol.*, 10, 123, 1999.

[5] Shakesheff, K., Cannizzaro, S., and Langer, R., Creating biomimetic micro-environments with synthetic polymer–peptide hybrid molecules, *J. Biomater. Sci. Polym. Ed.*, 9, 507, 1998.

[6] Hubbell, J.A., Materials as morphogenetic guides in tissue engineering, *Curr. Opin. Biotechnol.*, 14, 551, 2003.

[7] Massia, S.P. and Hubbell, J.A., An RGD spacing of 440 nm is sufficient for integrin alpha v beta 3-mediated fibroblast spreading and 140 nm for focal contact and stress fiber formation, *J. Cell Biol.*, 114, 1089, 1991.

[8] Maheshwari, G. et al., Cell adhesion and motility depend on nanoscale RGD clustering, *J. Cell Sci.*, 113, 1677, 2000.

[9] Shin, H., Jo, S., and Mikos, A.G., Modulation of marrow stromal osteoblast adhesion on biomimetic oligo[poly(ethylene glycol) fumarate] hydrogels modified with Arg-Gly-Asp peptides and a poly(ethyleneglycol) spacer, *J. Biomed. Mater. Res.*, 61, 169, 2002.

[10] Sagnella, S.M. et al., Human microvascular endothelial cell growth and migration on biomimetic surfactant polymers, *Biomaterials*, 25, 1249, 2004.

[11] Schense, J.C. and Hubbell, J.A., Three-dimensional migration of neurites is mediated by adhesion site density and affinity, *J. Biol. Chem.*, 275, 6813, 2000.

[12] Silva, G.A. et al., Selective differentiation of neural progenitor cells by high-epitope density nanofibers, *Science*, 303, 1352, 2004.

[13] Mann, B.K. and West, J.L., Cell adhesion peptides alter smooth muscle cell adhesion, proliferation, migration, and matrix protein synthesis on modified surfaces and in polymer scaffolds, *J. Biomed. Mater. Res.*, 60, 86, 2002.

[14] Rezania, A. and Healy, K.E., The effect of peptide surface density on mineralization of a matrix deposited by osteogenic cells, *J. Biomed. Mater. Res.*, 52, 595, 2000.

[15] Rowley, J.A. and Mooney, D.J., Alginate type and RGD density control myoblast phenotype, *J. Biomed. Mater. Res.*, 60, 217, 2002.

[16] Ferris, D.M. et al., RGD-coated titanium implants stimulate increased bone formation *in vivo*, *Biomaterials*, 20, 2323, 1999.

[17] Eid, K. et al., Effect of RGD coating on osteocompatibility of PLGA-polymer disks in a rat tibial wound, *J. Biomed. Mater. Res.*, 57, 224, 2001.

[18] Alsberg, E. et al., Engineering growing tissues, *Proc. Natl Acad. Sci. USA*, 99, 12025, 2002.

[19] Schense, J.C. et al., Enzymatic incorporation of bioactive peptides into fibrin matrices enhances neurite extension, *Nat. Biotechnol.*, 18, 415, 2000.

[20] Yu, X. and Bellamkonda, R.V., Tissue-engineered scaffolds are effective alternatives to autografts for bridging peripheral nerve gaps, *Tissue Eng.*, 9, 421, 2003.

[21] Li, F. et al., Cellular and nerve regeneration within a biosynthetic extracellular matrix for corneal transplantation, *Proc. Natl Acad. Sci. USA*, 100, 15346, 2003.

[22] Humphries, J.D. et al., Molecular basis of ligand recognition by integrin alpha5beta 1. II. Specificity of arg-gly-Asp binding is determined by Trp157 of the alpha subunit, *J. Biol. Chem.*, 275, 20337, 2000.

[23] Kao, W.J. et al., Fibronectin modulates macrophage adhesion and FBGC formation: the role of RGD, PHSRN, and PRRARV domains, *J. Biomed. Mater. Res.*, 55, 79, 2001.

[24] Dillow, A.K. et al., Adhesion of alpha5beta1 receptors to biomimetic substrates constructed from peptide amphiphiles, *Biomaterials*, 22, 1493, 2001.

[25] Cutler, S.M. and García, A.J., Engineering cell adhesive surfaces that direct integrin alpha5beta1 binding using a recombinant fragment of fibronectin, *Biomaterials*, 24, 1759, 2003.

[26] Dee, K.C., Andersen, T.T., and Bizios, R., Design and function of novel osteoblast-adhesive peptides for chemical modification of biomaterials, *J. Biomed. Mater. Res.*, 40, 371, 1998.

[27] Rezania, A. and Healy, K.E., Biomimetic peptide surfaces that regulate adhesion, spreading, cytoskeletal organization, and mineralization of the matrix deposited by osteoblast-like cells, *Biotechnol. Prog.*, 15, 19, 1999.

[28] Reyes, C.D. and García, A.J., Engineering integrin-specific surfaces with a triple-helical collagen-mimetic peptide, *J. Biomed. Mater. Res.* 65A, 511, 2003.

[29] Reyes, C.D. and García, A.J., $\alpha2\beta1$ integrin-specific collagen-mimetic surfaces that support osteoblastic differentiation, *J. Biomed. Mater. Res.*, 69A, 591, 2004.

[30] Kuhl, P.R. and Griffith-Cima, L.G., Tethered epidermal growth factor as a paradigm for growth factor-induced stimulation from the solid phase, *Nat. Med.*, 2, 1022, 1996.

[31] Zisch, A.H. et al., Covalently conjugated VEGF — fibrin matrices for endothelialization, *J. Control. Release*, 72, 101, 2001.

[32] Kopecek, J., Controlled biodegradability of polymers — a key to drug delivery systems, *Biomaterials*, 5, 19, 1984.

[33] Sakiyama-Elbert, S.E., Panitch, A., and Hubbell, J.A., Development of growth factor fusion proteins for cell-triggered drug delivery, *FASEB J.*, 15, 1300, 2001.

[34] Ehrbar, M. et al., Cell-demanded liberation of VEGF121 from fibrin implants induces local and controlled blood vessel growth, *Circ. Res.*, 94, 1124, 2004.

[35] Edelman, E.R. et al., Controlled and modulated release of basic fibroblast growth factor, *Biomaterials*, 12, 619, 1991.

[36] Wissink, M.J. et al., Binding and release of basic fibroblast growth factor from heparinized collagen matrices, *Biomaterials*, 22, 2291, 2001.

[37] Schroeder-Tefft, J.A., Bentz, H., and Estridge, T.D., Collagen and heparin matrices for growth factor delivery, *J. Control. Release*, 49, 291, 1997.

[38] Sakiyama-Elbert, S.E. and Hubbell, J.A., Development of fibrin derivatives for controlled release of heparin-binding growth factors, *J. Control. Release*, 65, 389, 2000.

[39] Sakiyama-Elbert, S.E. and Hubbell, J.A., Controlled release of nerve growth factor from a heparin-containing fibrin-based cell ingrowth matrix, *J. Control. Release*, 69, 149, 2000.

[40] Chang, C. and Werb, Z., The many faces of metalloproteases: cell growth, invasion, angiogenesis and metastasis, *Trends Cell Biol.*, 11, S37, 2001.

[41] West, J.L. and Hubbell, J.A., Polymeric biomaterials with degradation sites for proteases involved in cell migration, *Macromolecules*, 32, 241, 1999.

[42] Gobin, A.S. and West, J.L., Cell migration through defined, synthetic ECM analogs, *FASEB J.*, 16, 751, 2002.

[43] Halstenberg, S. et al., Biologically engineered protein-graft-poly(ethylene glycol) hydrogels: a cell adhesive and plasmin-degradable biosynthetic material for tissue repair, *Biomacromolecules*, 3, 710, 2002.

[44] Lutolf, M.P. et al., Synthetic matrix metalloproteinase-sensitive hydrogels for the conduction of tissue regeneration: engineering cell-invasion characteristics, *Proc. Natl Acad. Sci USA*, 100, 5413, 2003.

[45] van Hest, J.C. and Tirrell, D.A., Protein-based materials, toward a new level of structural control, *Chem. Commun. (Camb.)*, 1897, 2001.

[46] Petka, W.A. et al., Reversible hydrogels from self-assembling artificial proteins, *Science*, 281, 389, 1998.

[47] Lee, J., Macosko, C.W., and Urry, D.W., Elastomeric polypentapeptides cross-linked into matrixes and fibers, *Biomacromolecules*, 2, 170, 2001.

[48] Nagapudi, K. et al., Photomediated solid-state cross-linking of an elastin-mimetic recombinant protein polymer, *Macromolecules*, 35, 1730, 2002.

[49] Trabbic-Carlson, K., Setton, L.A., and Chilkoti, A., Swelling and mechanical behaviors of chemically cross-linked hydrogels of elastin-like polypeptides, *Biomacromolecules*, 4, 572, 2003.

[50] Welsh, E.R. and Tirrell, D.A., Engineering the extracellular matrix: a novel approach to polymeric biomaterials. I. Control of the physical properties of artificial protein matrices designed to support adhesion of vascular endothelial cells, *Biomacromolecules*, 1, 23, 2000.

[51] Baskaran, S. et al., Glycosaminoglycan-mimetic biomaterials. 3. Glycopolymers prepared from alkene-derivatized mono- and disaccharide-based glycomonomers, *Bioconjug. Chem.*, 13, 1309, 2002.

[52] Dong, C.M. et al., Synthesis and characterization of glycopolymer-polypeptide triblock copolymers, *Biomacromolecules*, 5, 224, 2004.

[53] Shu, X.Z. et al., Attachment and spreading of fibroblasts on an RGD peptide-modified injectable hyaluronan hydrogel, *J. Biomed. Mater. Res.*, 68A, 365, 2004.

11

Nanocomposite Scaffolds for Tissue Engineering

Amit S. Mistry
Xinfeng Shi
Antonios G. Mikos
Rice University

11.1 Introduction

In recent years, a great deal of attention has been directed toward nanotechnology and the potential benefits that this growing field may bring to a wide variety of engineering applications. One of the many applications of nanotechnology toward the biomedical sciences is the advancement of biomaterials designed for tissue engineering, especially those intended for biological tissues with complex properties. Nanoscience will be particularly useful in tissue engineering since the interactions between cells and biomaterials occur in the nanoscale and the components of biological tissues are nanomaterials themselves.

Bone tissue, for example, is a nanocomposite composed of rigid hydroxyapatite (HA) nanocrystals (60%) precipitated onto collagen fibers (30%) (Figure 11.1) [1]. Hydroxyapatite, which occurs as small plates that are tens of nanometers in length and width and 2–3 nm in depth, impart compressive strength to bone. Collagen fibrils (1.5–3.5 nm in diameter) form triple helices and bundle into fibers (50–70 nm diameter) responsible for the unique tensile properties of composite bone tissue [2]. The unique and complex mechanical properties of bone tissue arise from the interaction of these two components in the nanoscale [3].

Similarly, nanomaterials possessing superior properties compared to conventional materials are capable of imparting some of their properties onto macroscopic materials to form nanocomposites. In this manner, biomaterials may gain enhanced properties for medical applications with the addition of nanomaterials. Biodegradable polymers, for example, are generally too weak for load-bearing tissue applications. However, the incorporation of nanofillers into the polymer matrix can greatly improve the polymer's

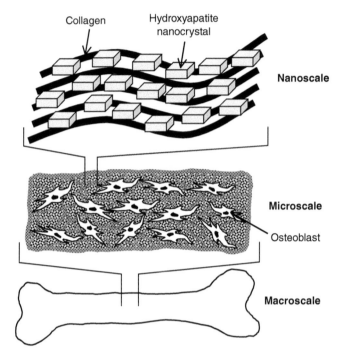

FIGURE 11.1 Nanocomposite structure of bone. The interaction between collagen fibers and HA nanocrystals in the nanoscale gives rise to the complex mechanical properties of bone tissue observed in the macroscale. Cells operating on this collagen/hydroxyapatite nanocomposite continually remodel bone on the microscale.

properties. More specifically, nanofillers have been shown to improve a composite material's flexural modulus [4], tensile strength, stiffness, toughness [5–8], fatigue resistance [9], wear resistance [10], thermal stability [11,12], and gas permeability properties [13]. Nanophase ceramic materials have been shown to also improve bone cell functions [14] and can impart osteoconductivity and improved biocompatibility to synthetic polymers [15,16]. Alternatively, HA nanocrystals can improve the osteoconductivity and biocompatibility of natural polymers, such as collagen, by mimicking the natural composition of bone [17–19]. The present chapter highlights current efforts toward nanocomposite scaffolds for tissue engineering applications.

11.2 Nanocomposite Materials

11.2.1 Nanomaterials Overview

Nanomaterials, as described here, are defined as materials with at least one of three dimensions <100 nm. Spherical nanoparticles, such as alumoxanes or silica nanoparticles, are nano-sized in all three dimensions. Nanotubes (carbon nanotubes), rods, or needles (HA) have two nanometer-sized dimensions. Nanosheets, such as layered silicates, have only one dimension in the nanoscale. Each of these nanomaterials offers mechanical reinforcement or osteoconductivity by dispersing into a matrix and chemically interacting with the macroscopic material. However, these particles are typically hydrophilic while the macroscopic material into which they are dispersed is usually hydrophobic. Thus, nanomaterial dispersion and promotion of interactions between the nanofillers and the macroscopic material are the two primary challenges for nanocomposite development.

Table 11.1 describes some of the many different nanomaterials currently being investigated for biomedical scaffolds. Each section of this chapter discusses the synthesis of a nanomaterial as well as the fabrication

TABLE 11.1 Nanomaterials for Biomedical Composites

Nanomaterial	Chemical formula	Composite materials for biomedical applications	References
Carboxylate Alumoxane	$[Al(O)_x(OH)_y(O_2CR)_z]_n$	PPF	Horch et al. [4]
Montmorillonite	$M_x(Al_{4-x}Mg_x)Si_8O_{20}(OH)_4$	PLLA	Lee et al. [11]
		Polyurethanes	Xu et al. [26]
Hydroxyapatite	$Ca_{10}(PO_4)_6(OH)_2$	PMMA	Fang et al. [39]
		PEG, PMMA, PBMA, PHEMA, PEG/PBT	Liu et al. [40–42]
		Chitosan	Hu et al. [43]
		PPF	Lewandrowskiet al. [44]
		Collagen	Zhang et al. [45]
Silica	SiO_2	PMMA, PCL	Rhee et al. [16, 46–47]
			Yoo et al. [48]
Alumina	Al_2O_3	Ce-TZP	Nawa et al. [8,49]
			Tanaka et al. [10]
			Uchida et al. [50]
Carbon nanotube	C	PMMA	Cooper et al. [65]
		PLA	Supronowicz et al. [67]

of nanocomposites from this material. A brief description of relevant issues and results of physical and biological testing is also included for each material.

11.2.2 Functionalized Alumoxane Nanocomposites

Carboxylate-alumoxanes are alumina-based nanoparticles developed as inorganic ceramic fillers for a variety of engineering applications. Alumoxanes are prepared directly from boehmite mineral in a "top-down" synthesis involving acid hydrolysis [20]. Nanoparticle size is controlled by conditions during synthesis and particles are easily functionalized based on the type of carboxylic acid used during synthesis [21]. Certain functional groups can be added to the hydrophilic alumoxane nanoparticles to aid in dispersion and covalent interaction with the composite medium. Vogelson et al. [6] modified alumoxanes with lysine and p-hydroxybenzoic acid to reinforce organic epoxy resins and yield sizable increases in thermal stability and tensile strength over blank resin [6].

In our laboratory, we have studied the effects of various surface-modified alumoxane nanoparticles on the mechanical properties of a biodegradable polymer for load-bearing bone tissue applications [4]. Alumoxane nanoparticles with three different surface modifications were tested — "activated" alumoxanes possessing two reactive double bonds available for interaction with the cross-link network of the polymer; "surfactant" alumoxanes modified with long fatty acid chains to aid in dispersion within the hydrophobic polymer; and "hybrid" alumoxanes modified with a surfactant chain and a reactive double bond within the same substituent (Figure 11.2). These nanoparticles were incorporated into a biodegradable poly(propylene fumarate)-based (PPF) system and the nanocomposites were tested for flexural and compressive mechanical properties.

Unmodified boehmite particle composites showed no significant improvement in mechanical properties compared to polymer resin alone and demonstrated a significant decrease in flexural fracture strength with increased loading. This is explained by the formation of large aggregates within the hydrophobic polymer, which promote crack formation (Figure 11.3a). The activated alumoxane nanocomposites were expected to covalently interact with the PPF matrix, but instead tended to aggregate into micron-sized clusters, which decreased flexural fracture strength with increased loading. Surfactant alumoxane nanocomposites demonstrated significant improvements in flexural modulus over blank polymer resin due to the fine dispersion of nanoparticles within the polymer matrix as determined by scanning electron microscopy (SEM). The hybrid alumoxane nanocomposites performed the best out of all materials by

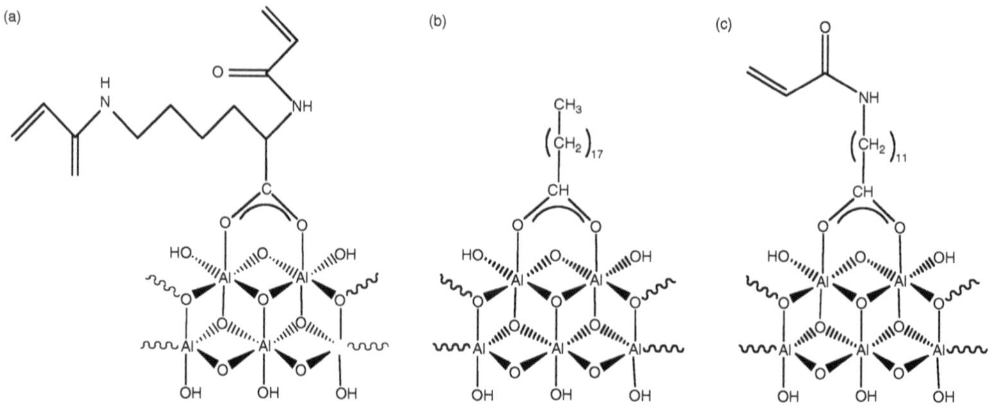

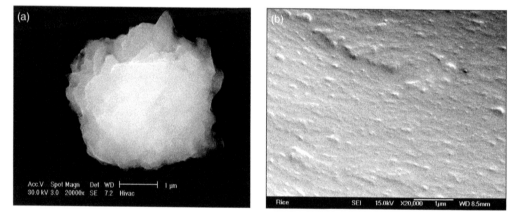

FIGURE 11.2 Chemical structures of modified alumoxanes: (a) diacryloyl lysine–alumoxane (activated), (b) stearic acid–alumoxane (surfactant), and (c) acryloyl undecanoic amino acid–alumoxane (hybrid).

FIGURE 11.3 SEM images of fracture planes of nanocomposite samples after mechanical testing: (a) unmodified boehmite crystals in polymer, bar is 1 μm; (b) hybrid alumoxane nanocomposite, bar is 1 μm.

improving the flexural modulus of PPF at loading concentrations between 0.5 and 5 wt.%. At a 1 wt.% loading, the flexural modulus reached 5410 ± 460 MPa, a factor of 3.5 greater than polymer alone (Figure 11.4a). Additionally, hybrid nanocomposites caused no significant loss of flexural or compressive fracture strength up to 5 wt. % loading (Figure 11.4b). SEM images revealed that the surfactant chain within the functional group of hybrid alumoxanes aided in dispersion within the polymer (Figure 11.3b) while the significant increase in mechanical properties may be explained by covalent interaction between PPF polymer and alumoxane nanoparticles. Thus, surface modification of alumoxane nanoparticles significantly increased the flexural modulus of polymer nanocomposites without a detrimental effect on fracture strength.

11.2.3 Polymer-Layered Silicate Nanocomposites

Layered silicates, derived from smectite clays, are commonly used as fillers for polymeric materials for many different applications as their chemistries have been extensively studied [22]. Polymer-layered silicate nanocomposites have shown improvements in mechanical, thermal, optical, physicochemical, and barrier properties as well as fire resistance compared to pure polymers or conventional composites (composed of micron-sized particles) [5,12].

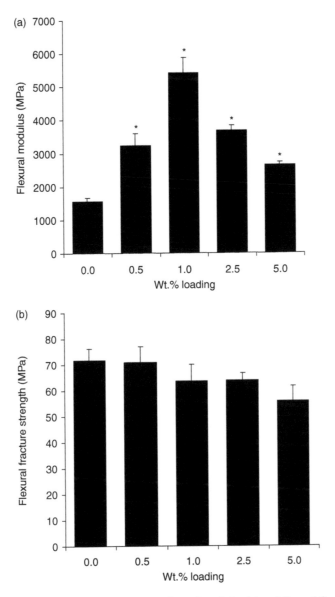

FIGURE 11.4 Flexural testing of hybrid alumoxanes. Flexural modulus (a) and flexural fracture strength (b) of hybrid alumoxane nanocomposites as a function of nanoparticle loading weight percentage. Error bars represent mean $\pm$ standard deviation for $n = 5$. The symbol "*" indicates a statistically significant difference compared to the pure polymer resin ($p < .05$).

These materials, unlike the other nanophase materials described in this chapter, are nano-sized in only one dimension and thereby act as nanoplatelets that sandwich polymer chains in composites. Montmorillonite (MMT) is a well-characterized layered silicate that can be made hydrophobic through either ionic exchange or modification with organic surfactant molecules to aid in dispersion [5,23]. Polymer-layered silicates may be synthesized by exfoliation adsorption, *in situ* intercalative polymerization, and melt intercalation to yield three general types of polymer/clay nanocomposites. Intercalated structures are characterized as alternating polymer and silicate layers in an ordered pattern with a periodic space between layers of a few nanometers [13]. Exfoliated or delaminated structure occurs when silicate layers are uniformly distributed throughout the polymer matrix. In some cases, the polymer does not intercalate the layers of silicate, resulting in a phase-separated structure containing micron-sized clusters of multiple silicate layers [5]. Typically, the greatest improvement of properties is observed in exfoliated structures based on the degree of layered-silicate dispersion within the polymer [13,24,25], though intercalated systems show more significant improvements in certain properties such as fracture behavior [23].

For biomedical applications, layered silicates have been incorporated into biodegradable lactide-based polymers to improve mechanical properties for hard-tissue applications. Lee et al. [11] incorporated MMT nanoplatelets into poly(L-lactic acid) (PLLA) with the aim of improving the intrinsic stiffness of porous polymer scaffolds. Exfoliated composites were prepared by the exfoliation-adsorption process and thermal and tensile mechanical properties were examined. The authors observed decreased glass transition temperatures with the addition of MMT to PLLA along with a larger amorphous region, which may have a positive effect on the biodegradation behavior of the composite. The tensile modulus of PLLA loaded with 5.79 wt.% MMT increased approximately 40% compared to pristine PLLA while maintaining more than 90% porosity.

Biomedical polyurethanes have also been modified with organically modified layered silicates (OLS) to improve mechanical properties and reduce gas permeability. Xu et al. [26] demonstrated an increase in tensile modulus with increased OLS concentration without the loss of strength and ductility that is typical for filler systems. Additionally, they observed a fivefold decrease in water vapor permeability, which is a major advantage for blood-contacting biomedical devices.

11.2.4 Hydroxyapatite Nanocomposites

As the nanostructure of bone was revealed, many researchers started to synthesize nanoscale HA and investigate its properties. Among various methods to prepare HA, the wet chemical method is most commonly used because it is well developed and easily adjusted for mass production [27]. Briefly, solutions of either calcium hydroxide and orthophosphoric acid or calcium salts and phosphate salts are mixed in a Ca/P ratio of $5:3$. Under these conditions, HA will precipitate from the solution [28]. Researchers can finely tune this simple method to make HA nanocrystals in various shapes such as spheres, rods, needles, and plates [28–33].

Due to its low flexural strength and toughness, commercial HA powders (of micron-sized particles) are usually limited to use as non-load-bearing implants or bioactive coatings on stronger materials, such as titanium alloys, to promote bone ingrowth [34]. Nanocrystalline HA with an average grain size of 100 nm possesses superior bending strength (182 MPa) and compressive strength (879 MPa) compared to conventional HA (in the range of 38–113 MPa in bending and 120–800 MPa in compression) [35]. Improved osteoblast adhesion, proliferation, and mineralization were also observed on the surface of nanoscale HA [14,36–38].

It is very attractive to introduce nanoscale HA as a filler for widely used biopolymers to improve bioactive and mechanical properties. As mentioned previously, however, the major challenge is achieving uniform dispersion of hydrophilic HA nanoparticles throughout hydrophobic polymer matrixes. Surfactants, such as lecithin, can be used to prevent aggregation of HA nanocrystals and homogeneously distribute them into poly(methylmethacrylate) (PMMA) polymer [39]. Liu et al. [40,41] successfully modified the surface of HA nanocrystals with poly(ethylene glycol) (PEG), PMMA, poly(butyl methacrylate) (PBMA), and poly(hydroxyethyl methacrylate) (PHEMA) to produce chemical bonding between the filler and matrix [40,41]. In another study, tensile strength and modulus of composites were significantly enhanced by the improved interface of HA nanocrystals and PEG/poly(butylenes terephthalate) (PEG/PBT) block copolymer [42].

Natural polymers have also been investigated for nanocomposites. In fact, a biodegradable chitosan/HA nanocomposite made by *in situ* hybridization exhibited higher bending strength and modulus than PMMA [43]. Moreover, nanoscale HA fillers can reduce water absorption, thus retaining material mechanical properties under moisture conditions for the potential application of internal fixation of bone fractures [43].

In an *in vivo* study, the biocompatibility and osteointegration of biodegradable PPF polymer grafts were improved when nanoscale HA was employed as opposed to micron-sized particles [44]. In another study, HA nanocrystals were reported to grow on the surface of self-assembled collagen triple helices along the longitudinal axes of their fibrils [45]. This designed hierarchical structure is very close to the actual nanostructure of bone. Thus, HA nanocomposites provide significant advantages for successful orthopaedic and dental applications in that they closely mimic natural bones and improve the bioactivity of many materials.

11.2.5 Other Ceramic Nanocomposites

In an effort to enhance the osteoconductivity of PMMA bone cement, Rhee and Choi [16,46] incorporated silica nanoparticles into the polymer. This composite was synthesized by sol–gel processing with the goal of improving binding at the bone-implant interface. The authors observed high mechanical properties in addition to crystalline apatite formation on implants in simulated body fluid.

While PMMA is useful as a bone cement, it is not ideal for tissue engineering applications as it is nondegradable. Researchers incorporated the same type of silica nanoparticles into the biodegradable poly(ε-caprolactone) (PCL) and also observed apatite formation and favorable mechanical properties [47,48].

Another noteworthy effort at nanocomposite fabrication applied ceramic nanoparticles to a ceramic material to enhance osteoconductivity and mechanical performance. Nawa et al. [49] developed a ceria-stabilized tetragonal zirconia polycrystal (Ce-TZP) ceramic and incorporated alumina (Al_2O_3) nanocrystals into it via wet chemistry methods for load-bearing bone applications. Further studies of this material investigated its ability to induce apatite formation [50], *in vivo* biocompatibility, and resistance to wear [10] with favorable results.

11.2.6 Carbon Nanotube Nanocomposites

Carbon nanotubes are among the strongest materials known because of their almost defect-free graphite architecture with the sp^2 type of carbon–carbon covalent bond, which is one of the strongest chemical bonds in nature [51,52]. There are two types of carbon nanotubes: multiwalled carbon nanotubes (MWNTs), first discovered in 1991 [53], and single-walled carbon nanotubes (SWNTs), first reported in 1993 [54,55]. Depending on the quality of nanotubes, elastic moduli can be as high as 1 TPa for SWNTs and 0.3 to 1 TPa for MWNTs, while strength ranges from 50 to 500 GPa for SWNTs and 10 to 60 GPa for MWNTs [56]. Owing to their very small diameters (ranging from 0.42 nm to dozens of nanometers) and lengths of more than several micrometers, the aspect ratio (length-to-diameter ratio) of carbon nanotubes can be more than 1000, while those of conventional carbon fibers are only about 100 [57]. Therefore, carbon nanotubes could become the best reinforcing fiber for composite materials.

Both types of carbon nanotubes are synthesized by three methods involving gas phase processing, namely, arc-discharge, laser ablation, and chemical vapor deposition (CVD) [58]. Subsequent purification procedures are required to remove impurities, such as catalyst particles, amorphous carbon, and nontubular fullerenes, from the nanotubes [59].

One of the greatest challenges for developing carbon nanotube nanocomposites is separating nanotube bundles, which aggregate into ropes due to strong inter-tube van der Waals attractions. This makes it quite difficult to obtain a uniform dispersion of individual nanotubes into a matrix material. Another significant challenge is effectively transferring load from a matrix to nanotubes, which have atomically smooth surfaces. The main dispersion methods include mechanical procedures, sonication of nanotubes in solvents and surfactants, and surface functionalization [60]. Among them, surface functionalization seems superior in that functional groups on the surfaces can not only isolate individual nanotubes from each other and therefore achieve uniform distribution throughout a matrix, but can also provide possible sites for covalent bonding between nanotubes and matrix to facilitate load transfer. Dyke and Tour [61] added various functional moieties to the surfaces of SWNTs by diazonium-based reactions and were able to separate bundles into individual nanotubes. Mitchell et al. [62] showed that such functionalized SWNTs demonstrated much better dispersion in polymer than nonfunctionalized ones.

Many researchers have reported significant improvements of mechanical properties in thermoplastics and epoxy resins by the addition of MWNTs or SWNTs [63,64]. Cooper et al. [65] found that impact strengths of both types of carbon nanotubes in PMMA were significantly improved compared to pure polymer. Carbon nanotubes provide another opportunity for creating dense ceramic composites with enhanced mechanical properties by absorbing energy through their highly flexible elastic behavior during deformation [66]. With the addition of 5 or 10% well-dispersed MWNTs, both the strength and fracture toughness of alumina were greatly increased [56]. In addition to their exceptional mechanical properties, carbon nanotubes also possess superior electric properties [58]. In an *in vitro* study, current-conducting composites of polylactic acid (PLA) and multiwalled carbon nanotubes were effectively used as substrates to expose osteoblasts to electrical stimulation, which promotes cellular functions for new bone formation [67]. Though carbon nanotubes are a relatively new material for biomedical applications, they show great potential for engineering biomaterials for hard tissue scaffolds.

11.3   Conclusions

As is evident by the described studies, a great deal of progress has been made toward improving bio-materials for tissue engineering through nanotechnology. Nanocomposite scaffolds have demonstrated enhanced mechanical properties and improved osteoconductivity of polymers as well as other materials. The challenge remains to design a nanocomposite scaffold with mechanical properties suitable for hard, cortical bone regeneration therapies. Future studies of nanocomposites should focus on answering an important question: How do these novel materials perform *in vivo*? The *in vivo* biocompatibility and osteoconductivity must be well characterized before the high potential of nanocomposite scaffolds for tissue engineering can be achieved.

Acknowledgments

This work was supported by the National Institutes of Health (R01 AR48756) (AGM) and the Nanoscale Science and Engineering Initiative of the National Science Foundation (EEC-0118001). ASM acknowledges the support of the NIH Biotechnology Training Grant (5 T32 GMO 08362).

References

[1] Athanasiou, K.A. et al., Fundamentals of biomechanics in tissue engineering of bone, *Tissue Eng.*, 6, 361, 2000.

[2] Rho, J.Y., Kuhn-Spearing, L., and Zioupos, P., Mechanical properties and the hierarchical structure of bone, *Med. Eng. Phys.*, 20, 92, 1998.

[3] Taton, T.A., Nanotechnology. Boning up on biology, *Nature*, 412, 491, 2001.

[4] Horch, R.A. et al., Nanoreinforcement of poly(propylene fumarate)-based networks with surface modified alumoxane nanoparticles for bone tissue engineering, *Biomacromolecules*, 5, 1990, 2004.

[5] Alexandre, M. and Dubois, P., Polymer-layered silicate nanocomposites: preparation, properties and uses of a new class of materials, *Mat. Sci. Eng. R*, 28, 1, 2000.

[6] Vogelson, C.T. et al., Inorganic–organic hybrid and composite resin materials using carboxylate-alumoxanes as functionalized cross-linking agents, *Chem. Mater.*, 12, 795, 2000.

[7] Kumar, S. et al., Synthesis, structure, and properties of PBO/SWNT composites, *Macromolecules*, 35, 9039, 2002.

[8] Nawa, M. et al., Tough and strong Ce-TZP/alumina nanocomposites doped with titania, *Ceram. Int.*, 24, 497, 1998.

[9] Bellare, A. et al., Improving the fatigue properties of poly(methyl methacrylate) orthopaedic cement containing radiopacifier nanoparticles, *Mater. Sci. Forum*, 426-4, 3133, 2003.

[10] Tanaka, K. et al., Ce-TZP/Al_2O_3 nanocomposite as a bearing material in total joint replacement, *J. Biomed. Mater. Res.*, 63, 262, 2002.

[11] Lee, J.H. et al., Thermal and mechanical characteristics of poly(L-lactic acid) nanocomposite scaffold, *Biomaterials*, 24, 2773, 2003.

[12] Torre, L. et al., Processing and characterization of epoxy-anhydride-based intercalated nanocomposites, *J. Appl. Polym. Sci.*, 90, 2532, 2003.

[13] Krook, M. et al., Barrier and mechanical properties of montmorillonite/polyesterarnide nanocomposites, *Polym. Eng. Sci.*, 42, 1238, 2002.

[14] Webster, T.J. et al., Enhanced functions of osteoblasts on nanophase ceramics, *Biomaterials*, 21, 1803, 2000.

[15] Liu, Q., de Wijn, J.R., and van Blitterswijk, C.A., Nano-apatite/polymer composites: mechanical and physicochemical characteristics, *Biomaterials*, 18, 1263, 1997.

[16] Rhee, S.H. and Choi, J.Y., Preparation of a bioactive poly(methyl methacrylate)/silica nanocomposite, *J. Am. Ceram. Soc.*, 85, 1318, 2002.

[17] Kikuchi, M. et al., Porous body preparation of hydroxyapatite/coliagen nanocomposites for bone tissue regeneration, *Key Eng. Mater.*, 254-2, 561, 2004.

[18] Du, C. et al., Three-dimensional nano-HAp/collagen matrix loading with osteogenic cells in organ culture, *J. Biomed. Mater. Res.*, 44, 407, 1999.

[19] Cui, F.Z. et al., Biodegradation of a nano-hydroxyapatite/collagen composite by peritoneal monocyte-macrophages, *Cells Mater.*, 6, 31, 1996.

[20] Callender, R.L. et al., Aqueous synthesis of water soluble alumoxanes: environmentally benign precursors to alumina and aluminum-based ceramics, *Chem. Mater.*, 9, 2418, 1997.

[21] Vogelson, C.T. and Barron, A.R., Particle size control and dependence on solution pH of carboxylate-alumoxane nanoparticles, *J. Non-Cryst. Solids*, 290, 216, 2001.

[22] Theng, B.K.G., *Formation and Properties of Clay–Polymer Complexes*, Elsevier Scientific Pub. Co., New York, 1979.

[23] Zerda, A.S. and Lesser, A.J., Intercalated clay nanocomposites: morphology, mechanics, and fracture behavior, *J. Polym. Sci. Polym. Phys.*, 39, 1137, 2001.

[24] Burnside, S.D. and Giannelis, E.P., Synthesis and properties of new poly(dimethylsiloxane) nanocomposites, *Chem. Mater.*, 7, 1597, 1995.

[25] Akelah, A. et al., Organophilic rubber — montmorillonite nanocomposites, *Mater. Lett.*, 22, 97, 1995.

[26] Xu, R. et al., Low permeability biomedical polyurethane nanocomposites, *J. Biomed. Mater. Res.*, 64, 114, 2003.

[27] Aoki, H., *Science and Medical Applications of Hydroxyapatite*, Ishiyaku EuroAmerica, St. Louis, MO, 1991.

[28] Pang, Y.X. and Bao, X., Influence of temperature, ripening time and calcination on the morphology and crystallinity of hydroxyapatite nanoparticles, *J. Eur. Ceram. Soc.*, 23, 1697, 2003.

[29] Li, Y. et al., Shape change and phase transition of needle-like non-stochiometric apatite crystals, *J. Mater. Sci.: Mater. Med.*, 5, 263, 1994.

[30] Li, Y. et al., Preparation and chracterization of nano-grade osteoapatite-like rod crystals, *J. Mater. Sci.: Mater. Med.*, 5, 252, 1994.

[31] Hsieh, M.F. et al., Organic–inorganic hybrids of collagen or biodegradable polymers with hydroxyapatite, *Key Eng. Mater.*, 254-2, 473, 2004.

[32] Liou, S.C. et al., Structural characterization of nano-sized calcium deficient apatite powders, *Biomaterials*, 25, 189, 2004.

[33] Wang, Y.F. et al., Preparation and characterization of nano hydroxyapatite sol, *T Nonferr. Metal. Soc.*, 14, 29, 2004.

[34] Catledge, S.A. et al., Nanostructured ceramics for biomedical implants, *J. Nanosci. Nanotechnol.*, 2, 293, 2002.

[35] Ahn, E. et al., Properties of nanostructured hydroxyapatite-based bioceramics, *Transactions of the Sixth World Biomaterials Congress*, p. 643, 2000.

[36] Webster, T.J. et al., Specific proteins mediate enhanced osteoblast adhesion on nanophase ceramics, *J. Biomed. Mater. Res.*, 51, 475, 2000.

[37] Webster, T.J., Siegel, R.W., and Bizios, R., Osteoblast adhesion on nanophase ceramics, *Biomaterials*, 20, 1221, 1999.

[38] Webster, T.J., Siegel, R.W., and Bizios, R., Nanoceramic surface roughness enhances osteoblast and osteoclast functions for improved orthopaedic/dental implant efficacy, *Scr. Mater.*, 44, 1639, 2001.

[39] Fang, L.R. et al., Preparation of nano-sized hydroxyapatite in chloroform medium, *J. Inorg. Mater.*, 18, 801, 2003.

[40] Liu, Q. et al., Surface modification of nano-apatite by grafting organic polymer, *Biomaterials*, 19, 1067, 1998.

[41] Liu, Q., de Wijn, J.R., and van Blitterswijk, C.A., Covalent bonding of PMMA, PBMA, and poly(HEMA) to hydroxyapatite particles, *J. Biomed. Mater. Res.*, 40, 257, 1998.

[42] Liu, Q., de Wijn, J.R., and van Blitterswijk, C.A., Composite biomaterials with chemical bonding between hydroxyapatite filler particles and PEG/PBT copolymer matrix, *J. Biomed. Mater. Res.*, 40, 490, 1998.

[43] Hu, Q.L. et al., Preparation and characterization of biodegradable chitosan/hydroxyapatite nano-composite rods via *in situ* hybridization: a potential material as internal fixation of bone fracture, *Biomaterials*, 25, 779, 2004.

[44] Lewandrowski, K.U. et al., Enhanced bioactivity of a poly(propylene fumarate) bone graft substitute by augmentation with nano-hydroxyapatite, *Bio-Med. Mater. Eng.*, 13, 115, 2003.

[45] Zhang, W., Liao, S.S., and Cui, F.Z., Hierarchical self-assembly of nano-fibrils in mineralized collagen, *Chem. Mater.*, 15, 3221, 2003.

[46] Rhee, S.H. and Choi, J.Y., Synthesis of a bioactive poly(methyl methacrylate)/silica hybrid, *Bioceramics* 14, 218-2, 433, 2002.

[47] Rhee, S.H., Effect of calcium salt content in the poly(epsilon-caprolactone)/silica nanocomposite on the nucleation and growth behavior of apatite layer, *J. Biomed. Mater. Res.*, 67A, 1131, 2003.

[48] Yoo, J.J. and Rhee, S.H., Evaluations of bioactivity and mechanical properties of poly (epsilon-caprolactone)/silica nanocomposite following heat treatment, *J. Biomed. Mater. Res.*, 68A, 401, 2004.

[49] Nawa, M. et al., The effect of TiO_2 addition on strengthening and toughening in intragranular type of 12Ce-TZP/Al_2O_3 nanocomposites, *J. Eur. Ceram. Soc.*, 18, 209, 1998.

[50] Uchida, M. et al., Apatite-forming ability of a zirconia/alumina nano-composite induced by chemical treatment, *J. Biomed. Mater. Res.*, 60, 277, 2002.

[51] Jin, Z. et al., Dynamic mechanical behavior of melt-processed multi-walled carbon nan-otube/poly(methyl methacrylate) composites, *Chem. Phys. Lett.*, 337, 43, 2001.

[52] Bernholc, J. et al., Mechanical and electrical properties of nanotubes, *Annu. Rev. Mater. Res.*, 32, 347, 2002.

[53] Iijima, S., Helical microtubules of graphitic carbon, *Nature*, 354, 56, 1991.

[54] Iijima, S. and Ichihashi, T., Single-shell carbon nanotubes of 1-nm diameter, *Nature*, 363, 603, 1993.

[55] Bethune, D.S. et al., Cobalt-catalyzed growth of carbon nanotubes with single-atomic-layerwalls, *Nature*, 363, 605, 1993.

[56] Ajayan, P.M., Schadler, L.S., and Braun, P.V., *Nanocomposite Science and Technology,* Wiley-VCH Verlag, New York, 2003.

[57] Calvert, P., Nanotube composites — a recipe for strength, *Nature*, 399, 210, 1999.

[58] Terrones, M., Science and technology of the twenty-first century: synthesis, properties and applications of carbon nanotubes, *Annu. Rev. Mater. Res.*, 33, 419, 2003.

[59] Haddon, R.C. et al., Purification and separation of carbon nanotubes, *Mrs Bull.*, 29, 252, 2004.

[60] Hilding, J. et al., Dispersion of carbon nanotubes in liquids, *J. Disper. Sci. Technol.*, 24, 1, 2003.

[61] Dyke, C.A. and Tour, J.M., Unbundled and highly functionalized carbon nanotubes from aqueous reactions, *NanoLetters*, 3, 1215, 2003.

[62] Mitchell, C.A. et al., Dispersion of functionalized carbon nanotubes in polystyrene, *Macromolecules*, 35, 8825, 2002.

[63] Qian, D. et al., Load transfer and deformation mechanisms in carbon nanotube–polystyrene composites, *Appl. Phys. Lett.*, 76, 2868, 2000.

[64] Zhu, J. et al., Improving the dispersion and integration of single-walled carbon nanotubes in epoxy composites through functionalization, *NanoLetters*, 3, 1107, 2003.

[65] Cooper, C.A. et al., Distribution and alignment of carbon nanotubes and nanofibrils in a polymer matrix, *Comp. Sci. Technol.*, 62, 1105, 2002.

[66] Yakobson, B.I. and Avouris, P., Mechanical properties of carbon nanotubes, *Carbon Nanotubes*, 80, 287, 2001.

[67] Supronowicz, P.R. et al., Novel current-conducting composite substrates for exposing osteoblasts to alternating current stimulation, *J. Biomed. Mater. Res.*, 59, 499, 2002.

12

Roles of Thermodynamic State and Molecular Mobility in Biopreservation

Alptekin Aksan

University of Minnesota
Center for Engineering in
Medicine/Surgical Services
Harvard Medical School
Massachusetts General Hospital
Shriners Hospital for Children

Mehmet Toner

Center for Engineering in
Medicine/Surgical Services
Harvard Medical School
Massachusetts General Hospital
Shriners Hospital for Children

In a very broad sense, preservation can be defined as the process of reversibly arresting the biochemical reactions and therefore the metabolism of an organism (in a state of suspended animation [1]) in order to sustain function after a "prolonged" exposure to otherwise lethal conditions. The lethal conditions are created by the inadequacy of the surrounding medium in supplying nutrients and removing by-products, exposure to draught, or the extremes of temperature that would disturb the biochemical processes vital to the organism.

The rates of biochemical reactions are dependent on the proximity and mobility of the reactants. Mobility is determined by the mutual interactions of the solvent with the solutes. The state of water (the solvent) determines the mobility of the solutes and in return, the solutes change the structural organization of nearby water molecules through hydrophilic and hydrophobic interactions. In the cytoplasm, the thermodynamic state of the medium (and therefore the molecular mobility) determines the rate of metabolic activity.

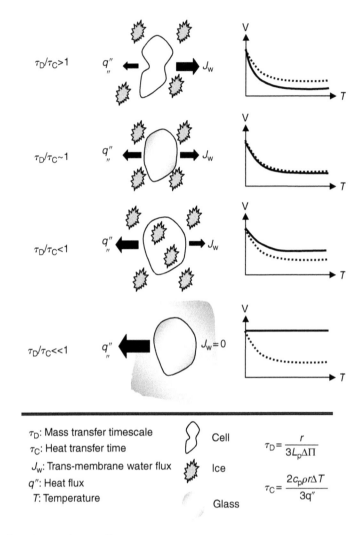

$\tau_D/\tau_C > 1$ q'' J_w

$\tau_D/\tau_C \sim 1$ q'' J_w

$\tau_D/\tau_C < 1$ q'' J_w

$\tau_D/\tau_C \ll 1$ q'' $J_w = 0$

τ_D: Mass transfer timescale   Cell $\tau_D = \dfrac{r}{3L_p\Delta\Pi}$

τ_C: Heat transfer time

J_w: Trans-membrane water flux   Ice

q'': Heat flux $\tau_C = \dfrac{2c_p\rho r\Delta T}{3q''}$

T: Temperature

Glass

FIGURE 12.1 Effect of timescales on cell response.

In this chapter, the mechanisms enabling preservation of biological systems will be examined from the perspective of "molecular mobility" exploring the effects of the timescales for cooling, freezing, crystallization, vitrification, structural relaxation, and diffusion. Following example underlines the importance of timescales in preservation.

The timescales of biochemical reactions and the preservation conditions applied to the organism play crucial roles in determining the success of preservation. For example, the ratio of the timescale of water diffusion, τ_D, across the cell membrane ($\tau_D = r/3L_p\Delta\Pi$, where r, L_p, and $\Delta\Pi$ are the cell radius, membrane permeability, and osmotic pressure differential, respectively) to the timescale of cooling the cell experiences, τ_C ($\tau_C = (2c_p\rho r\Delta T)/(3q'')$), where c_p, ρ, q'', and ΔT are the specific heat, mass density, heat flux, and temperature differential, respectively) determines the fate of a cell during freezing such that (Figure 12.1):

- $\tau_D/\tau_C > 1$ causes excessive dehydration of the cell.
- $\tau_D/\tau_C \sim 1$ establishes an intra/extracellular equilibrium such that the intracellular water transported across the membrane balances the extracellular osmotic increase induced by freezing (the solute-concentration effect [2]) minimizing the amount of intracellular free water.

- $\tau_D/\tau_C < 1$ results in rapid cooling (faster than the cell can reach equilibrium with its surroundings) inducing Intracellular Ice Formation (IIF) known to be lethal to most cells (see Figure 12.2 Toner [3], for the correlation between IIF and post-thaw viability of mammalian cells).
- $\tau_D/\tau_C \ll 1$ theoretically, yields to ultrafast cooling without ice crystallization (if as an additional constraint $\tau_\alpha/\tau_C \ll 1$ where, τ_α is the timescale of structural relaxations) enabling vitrification of the extracellular medium, and more importantly the cytosol.

12.1   Water–Solute Interactions and Intracellular Transport

Water is the most abundant substance in and around an organism, yet it is the least understood in terms of its role in biological function and preservation. Water has unique physical and chemical properties [4] (for a complete review, see Franks [5], for an extensive collection of the properties and the anomalies of water, see the excellent electronic source by Chaplin [6]). Hydrogen bonds ($E_a = 4$ to 7 kJ/mol [7]) with bond energies similar to the local thermal fluctuations are continuously formed and broken between neighboring water molecules organizing them into flickering clusters of minimum free energy. These loosely bonded hydrogen clusters have very short life spans ($\tau_W = 10^{-11}$ to 10^{-12} sec) and are quickly destroyed just to form new ones in a never-ending cycle. This behavior establishes the basis of molecular mobility of water such that even in pure liquid form, a single water molecule is not independent in its motion but, at any instant of time, moves in coordination with a cluster of molecules. It is therefore widely believed that for water a cluster (rather than an individual water molecule) is the elementary structural unit and the interactions of clusters are responsible for its unique chemical and physical properties [8].

There is a continuous tug-of-war between the hydrogen bonds trying to stabilize the network of water molecules and the temperature dependent random motions breaking these bonds. With decreasing temperature, the magnitude of local thermal fluctuations decrease, increasing the lifetime of the hydrogen

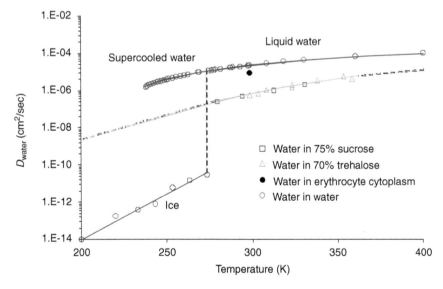

FIGURE 12.2 Self-diffusivity of water. Data of water diffusivity in 70% trehalose solution: NMR by Ekdawi-Sever et al. [112], NMR by Rampp et al. [42] and DMS by Conrad and de Pablo [41]; water diffusivity in 75% sucrose solution: Ekdawi-Sever et al. [112]; water diffusivity in the supercooled region: DMS by Paschek and Geiger [113] and NMR by Price et al. [114]; water diffusivity in ice: Onsager and Runnels [115] and Petrenko and Whitworth [116]; water diffusivity in liquid phase by Mills [117], NMR by Harris and Newitt [118]; water diffusivity in 75% sucrose: NMR by Moran et al. [119].

bonds among water molecules (i.e., the number of available neighboring hydrogen bonding sites per water molecule at any given time decreases). Water mobility (and its self-diffusion coefficient, D_w, as shown in Figure 12.2) therefore decreases [9,10] while the water clusters they participate in get more densely packed and grow [7]. Water mobility is not only a function of temperature but also the thermodynamic state. For example, D_w of liquid water decreases only by an O(2) over a range of 150 K whereas it drops by an O(6) upon freezing at 0°C (Figure 12.2). In the frozen state, each water molecule makes hydrogen bonds with only four neighboring molecules in a three-dimensional tetrahedron-like configuration. The degree of tetrahedricity (perfectness of the tetrahedral configuration) increases with decreasing temperature [10]. The strong interations between water molecules also cause an unexpected decrease in D_w when the density is decreased by increasing hydrostatic pressure. In water, density decrease lowers the hydrogen bonding possibility, therefore reduces mobility. In other liquids however, mobility is increased due to the increase in the free volume.

Any surface (hydrophilic or hydrophobic) or solute (charged or uncharged) disrupts the bonding patterns of the water molecules in its near vicinity causing local polarization and altering the life cycles of the surrounding water clusters [6,11]. This results in variations in water mobility, which can be detected by methods such as Nuclear Magnetic Resonance (NMR) and Fourier Transform Infrared Spectroscopy (FTIR). Close to a hydrophilic surface exerting a higher attraction force, water mobility decreases (the water molecules make stronger bonds with the surface and they are less available to join in a cluster). This causes depression of the freezing temperature and is the origin of the "unfreezable water" concept frequently used by the cryobiologists. Similarly, in close proximity to a hydrophobic surface or a solute, in this case entirely due to geometrical factors limiting hydrogen bonding possibility (that the water molecules can not make bonds with the hydrophobic surface), in the direction perpendicular to the surface, water mobility and therefore diffusion decreases. Parallel to the hydrophobic surface however, water diffusivity is not different from that of free water [12]. The coexistence of hydrophobic and hydrophilic surfaces on most proteins therefore creates large spatial gradients of water mobility, which may be closely related to protein function (e.g., the alternating regions of high and low water mobility within the hydration shells of actin filaments are thought to be contributing to the movement of myosin along these filaments [13]). Ions also affect nearby water molecules and alter their mobility [14]. For example, structure-breaking solutes such as urea [15] and large ions such as I^- and Cs^+ [14] increase the mobility of the water molecules in their immediate vicinity. Small ions such as Mg^{++} and F^-, on the other hand, have the opposite effect on their hydration layer. Interactions with nearby surfaces and solutes change the lifetime and the stability of each vicinal water cluster and change their physical properties (e.g., low mobility vicinal water has lower mass density, lower freezing point, and higher specific heat than bulk water).

The interaction of water with solids and surfaces is mutual. Water is not only a solvent but is also a react-ant itself. It is a substance functioning in cooperation with the solutes [16] altering their charge, conforma-tion, and reactivity. The range of water–solute interactions (the distance a water molecule should be from a surface or a solute to be fully isolated from its effects) is one of the most controversial topics in the literature, however it is widely accepted that vicinal water layers do not extend beyond 1 to 10 water molecules.

12.1.1 Intracellular Water and Molecular Mobility

In isotonic conditions, approximately 70% of the cell's volume is water. However, it would be wrong to think that the intracellular solutes and macromolecules bathe in a dilute solution. It has long been known that most, if not all, of the intracellular water exhibits physical properties unlike those in the bulk [17] (see the D_w in erythrocytes in Figure 12.2). This is attributed to the presence of high concentrations of proteins (200 to 300 g/l) [18], ions, amino acids, fatty acids, sugars, and other small solutes in the cytoplasm enmeshed in a network of cytoskeletal macromolecules (actin filaments, microtubules, and intermediate filaments). In individual organelles (such as mitochondria) the protein concentration may be even higher [19]. Within the cytoplasm, at any given time, water molecules are either a part of a tight cluster (bulk water) or in the close vicinity (vicinal water) of a surface (cell or organelle membrane) or a solute (a macromolecule, ion, or amino acid). There is not a consensus in the literature on the relative populations of vicinal and bulk water within the cytosol. The estimates vary in a range of 0 to 100% of the total intracellular water (for details, see Clegg [17] and the references therein). Similarly, the names given to the various subpopulations of water molecules in the close proximity of surfaces/solutes also vary from one source to another (hydration, bound, vicinal, essential, structural, ordered, unfreezable, osmotically inactive, etc.).

Overall cytosolic mobility is directly related to the metabolism and function of a cell [20,21]. However, the mobility of water in the cytosol is not spatially homogeneous [22,23] as evidenced by the presence of compartmentalization inside the cytoplasm (regions of solute aggregation and variable water mobility) using Fluorescence Recovery After Photobleaching (FRAP) [24] and Raman Scattering Microscopy (RSM) [25]. It is postulated that the intracellular mobility gradients determine the active and resting states of cells [26,27] and are altered in response to osmotic stress [28] and in the presence of carcinogens [26].

As opposed to dilute solutions, where the chemical reactions are transition-state-limited [29], most of the biochemical reactions in crowded environments are diffusion-limited. However, the diffusion mechanism in the cytoplasm is different from that in a dilute solution and is altered by the increased frequency of close-range interactions such as binding of and collisions between solutes and surfaces. In order to determine the hydrodynamic properties of the cytosol (translational, rotational diffusion coefficients and viscosity) various techniques have been utilized (NMR, FRAP, Electron Spin Resonance (ESR), etc. See Table 12.1 for details). The values reported in the literature lie in a very broad range (e.g., cytosolic viscosity values vary from 0.5 to 5 times that of water) and contradict each other (see reviews by Luby-Phelps et al. [24] and Arrio-Dupont et al. [30] for cytosolic diffusivity measurements using different methods and tracer molecules). The main reason for the discrepancy among the reported values is believed to be originating from the differences in the methodologies applied (such as the measurement of the translational diffusivity of a very large number of tracer molecules over a large volume [∼1/10 to 1/20 of the volume of an attached cell] with FRAP or the shortcoming of NMR in distinguishing the signals from the intermolecular and intramolecular bonds and the requirement for relatively long acquisition times [31]), the characteristics of the tracer used (e.g., its size [24]), and inability of most of these methods to distinguish among different molecular interactions (free diffusion, binding, or collision) in this crowded environment [32]. The differences observed between the cytoplasmic viscosity values measured by rotational vs. translational diffusion of tracers indicate that physical interactions (such as binding and

TABLE 12.1 Most Common Methods for Measurement of Molecular Mobility

Method	Quantity measured	Range/limitations
Nuclear magnetic resonance (NMR)	Relaxation times T_1, T_2 of proton (^{1}H) and carbon (^{13}C) nuclei of water–carbohydrate samples	Cannot distinguish between the intermolecular and intramolecular bond signals. Measurement times are higher than the measured relaxation times
Dielectric spectroscopy	Complex dielectric permittivity	Water dipole moment relaxations in the kHz–GHz range
Differential scanning calorimetry (DSC)	Specific heat change $\Delta C_p\|_{T=T_g}$	May be used in the 100–1500 K range. Measures the glass transition temperature of the bulk sample
Fluorescence recovery after photobleaching (FRAP)	Translational diffusivity of the tracer molecule	Measures mobility of very large number of molecules in a large area ($\sim$1 μm^3). Measurements in a glass are not feasible due to photobleaching
Electron spin resonance (ESR)	Spin relaxation of molecular probes (such as tempol)	Rotational mobility range [110]: $t = 10^{-12}$–10^{-8} sec (continuous-wave EPR), $t = 10^{-7}$–10^3 sec (saturation tranfser EPR). Probe properties change with hydration level [111]
Fourier Transform Infrared Spectroscopy (FTIR) Circular dichroism	Molecular bond vibration	Strong absorption of IR light by water
Quasielastic neutron scattering (QNS)		Measurement time $\sim$10^{-12} sec, Measurement distance $\sim$1A [17]

collisions) present a higher obstacle to diffusion when compared to fluid phase viscosity (see e.g., Figure 1 in Mastro and Keith [33]). Crowding and solute concentration affect larger macromolecules more than the small solutes and ions, and it is therefore not feasible to assign a single parameter for mobility. Even though the viscosity of the cytosol is not significantly higher than water, some large macromolecules do not diffuse at all in the timescale of hours [34]. This would limit the reaction rates of some of the intracellular biochemical processes, if they depended on diffusion only. Nature overcomes this problem by crowding certain reactants in small regions (compartmentalization) of the cytoplasm [35], which also explains the spatial heterogeneity of water mobility observed intracellularly [22,23].

12.1.2 Transmembrane Water Transport Effects

The cell membrane shows very low resistance to water transport. However, it is the biggest obstacle to the transport of solutes. Membrane permeability to solutes depends on the size, charge, and the hydrogen bonding characteristics of the solute (for a review of membrane transport phenomena, see McGrath [36]). Transport across the cell membrane in response to osmotic gradients is at the cornerstone of biopreservation studies since it is directly related to administration of preservation agents and to the amount and mobility of the intracellular water. Water is transported into the cell by three different methods (a) diffusive transport across the membrane, ($L_p \sim$ 2–50 $\times$ 10^4 cm/sec), (b) facilitated transport through membrane channels ($L_p \sim$ 200 $\times$ 10^4 cm/sec), and (c) cotransport through glucose transporters and ion channels ($L_p \sim$ 4 $\times$ 10^4 cm/sec) [37]. Methods for quantifying membrane transport are reviewed by Verkman [38].

Both desiccation and freezing (as well as their complementary processes; rehydration and thawing) induce very high osmotic gradients across the cell membrane. Cells are capable of responding to mild osmotic gradients by adjusting their volume, mainly by water transport. Applying an osmotic gradient almost all of the free water (called the osmotically active water) in a cell can be removed temporarily without any permanent damage. The water of hydration (participating in the osmotically inactive volume) on the other hand, is tightly associated with the solutes and surfaces and upon removal causes polarization of surfaces, aggregation and denaturation of the macromolecules [20,21].

12.2 Molecular Mobility in Preservation

In a dilute, nonreacting, binary solution diffusivities of the solvent and the solute depend on their relative molecular sizes [39] as well as their concentrations and temperature. For this system, Stokes–Einstein relationship correlates the hydrodynamical properties of the solution as,

$$D_{\text{translational}} = \frac{kT}{n\pi r\eta}, \tag{12.1}$$

where, D, k, T, r, and η are the diffusivity, Boltzmann's constant, absolute temperature, hydrodynamic radius of the diffusing particle (van der Waals radius), and the viscosity, respectively. The constant n, takes the value of 6 for a "stick (hydrophilic) boundary" condition and the value of 4, for a "slip (hydrophobic) boundary" condition. With increased solute concentration, diffusion becomes more restricted and different interactions such as collisions with other solutes and binding between molecules start to dominate and deviations from the Stokes–Einstein relationship is observed.

For a supersaturated solution, crystallization is the energetically most favorable path. However, if the concentration increases very rapidly (or the temperature drops very fast) a meta-stable "glassy" form can be reached. For a glass-forming system, the transition from a dilute to a concentrated solution diffusion mechanism is determined by the concentration corresponding to the crossover temperature, T_c, predicted by the Mode Coupling Theory [40]. At the crossover temperature there is a transition from liquid-like to solid-like dynamics. Note that $T_c \sim (1.14–1.6)T_g$ for most glass-forming solutions, where T_g is the glass transition temperature. Diffusion in very high concentration solutions (close to glass transition

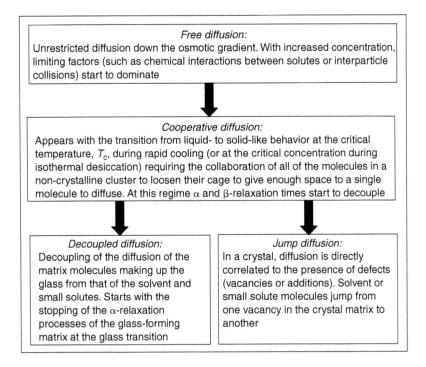

FIGURE 12.3 Mechanisms of diffusion.

temperature) is governed by the frequency of jumping between the cages surrounding the tagged molecule (either the solvent or a small solute) and is comparable to the time the molecule spends entrapped in the cage rattling (β-relaxation) [41]. This is similar to the mechanism of diffusion in crystalline systems, where the diffusing molecule jumps between the crystal defects (vacancies). Frequency of jumping is inversely related to the structural relaxation (α-relaxation) time, τ_α of the matrix. Temperature dependence of τ_α distinguishes between the "fragile" and "strong" glasses, where the variation in τ_α with temperature is steeper in the former case. In Figure 12.3 changes in the mechanism of diffusion with the thermodynamic state of the system is summarized.

In a concentrated and crowded environment such as in the cytosol, the motion of a small solute can be divided into two main components (Figure 12.4a) (1) the translational diffusive motion (governed by the α-relaxation timescale of the system), which results in a net displacement of the molecule down its osmotic gradient and (2) the random motion, which does not result in a net displacement. The random motion is governed by the physical and chemical interactions with the solvent and the surrounding solutes and is characterized by the β-relaxation timescale of the system, which includes rotation and Brownian motion. When the solvent is frozen, as a function of the storage temperature and the perfectness of the crystal structure formed, α-relaxation timescale increases. Depending on the relative magnitudes of the solvent and the solute molecules (and the size of the pores formed) β-relaxation may still continue (Figure 12.4b). Note the unfrozen bound water molecules in close proximity to the protein surface with lower mobility. If the system is desiccated (to a point where some of the water molecules in the hydration layer is also removed), both α and β-relaxations of the system may be stopped completely, however, due to removal of the hydration layer, the protein may denature and its active site may not be available for the binding of the ligand (Figure 12.4c). If denaturation of the protein is irreversible, even after rehydration (when molecular mobility is restored) the ligand can still not bind to the protein. Carbohydrates may be administered in order to prevent the denaturation of the protein while water is removed from the system lowering the mobility within the medium forming a glass (Figure 12.4d,e).

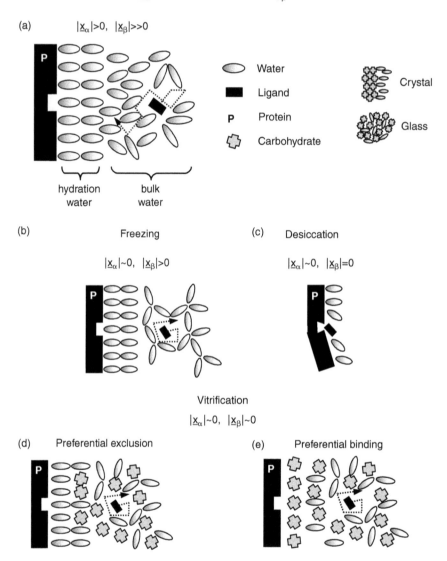

FIGURE 12.4   Molecular mobility in biopreservation.

For high solute concentrations in the absence of crystallization, Vogel–Tammann–Fulcher (VTF) equation predicts the changes in the timescales of molecular motion as:

$$\tau = \tau_0 \, e^{-(BT_0)/(T-T_0)}, \tag{12.2}$$

where τ is the timescale of molecular motion, T_0 is the Kauzmann temperature corresponding to the zero mobility state, τ_0 is the timescale of motion at the Kauzmann temperature (usually taken to be in the order of 10^{17} sec), and B is a constant related to the energy of activation of the relaxation process. The values of B, for different carbohydrate solutions can be found in Rampp et al. [42].

12.2.1 Molecular Mobility in Supercooling and Phase Change

At temperatures below the freezing temperature (0°C, 1 atm) water may exist as a supercooled liquid or ice. The theoretical limit for the presence of free water in the liquid form is −40°C, where homogeneous crystallization is initiated. For freezing to occur at any given temperature, certain number of water clusters should form at the same time and reach a critical size (known as the formation of a nucleation embryo). With decreasing temperature, the critical number of water molecules required to form a nucleation embryo for the initiation of freezing decreases (from approximately 16,000 at −10°C to 120 at −40°C [5]) and at −40°C, it becomes statistically impossible for free water to remain in the liquid phase. In biological systems, due to the presence of small hydrophobic solutes with low surface energy (such as ice nucleating proteins in certain plants and bacteria that survive freeze injury), ice nucleation in the supercooled state is initiated well before the theoretical limit is reached. This is believed to help protect against the freeze-induced damage by minimizing compartmentalization and creating a more uniform ice structure.

With decreasing temperature, the diffusivity of liquid water decreases (approximately O(2) 370 to 240 K, see Figure 12.2) due to change in the mechanism of diffusion from unrestricted to cooperative (Figure 12.3). Upon freezing, the drop in water diffusivity becomes even more significant (approximately O(6) as shown in Figure 12.2). The reduction in water mobility with supercooling and liquid-to-solid phase change in addition to the decrease in most chemical reaction rates at low temperatures, makes cryopreservation feasible.

12.2.2 Cryopreservation

Certain organisms are known to synthesize carbohydrates upon exposure to cold and desiccation (such as trehalose synthesis by *Escherichia coli* [43], yeast [44], and nematodes [45]), which is crucial for their survival [46]. It was discovered (by accident) that glycerol also protects against freeze injury. These findings have fueled researchers to explore ways to use these chemical agents (cryoprotectants) for the preservation of biological organisms, which are normally not freeze or desiccation resistant. Over the years, this has led to the discovery of other cryoprotectants such as dimethylsulfoxide (DMSO) and ethylene glycol.

Cryoprotectants traditionally are divided into two main groups as membrane permeable and impermeable. Most effective and widely used cryoprotectants, DMSO [47], ethylene glycol, and glycerol are highly membrane permeable whereas most of the carbohydrates (trehalose, hydroxyethyl starch, dextran, etc.), proteins, and polymers are normally not. Exposure to membrane impermeable (or low permeability when compared to that of water) cryoprotectants creates an osmotic gradient across the membrane, to which a cell responds by shrinking. If a membrane permeable cryoprotectant is present on the other hand, after initial shrinkage, with prolonged exposure and penetration of the chemical, the cell recovers to its original volume. Similarly after thawing, to remove intracellular cryoprotectants, the cells are exposed to hypotonic solutions. This results in swelling of the cell followed by return to its isotonic volume. It is widely accepted that a significant part of freeze damage is related to the uncontrolled swelling response during thawing, that the membrane stretches beyond its mechanical limit and ruptures. The volume response of the cell to cryoprotectants creates changes in the cytoplasmic molecular mobility due to the changes in (a) the amount of cytoplasmic free water present at any time, (b) the intracellular solute concentration, and (c) the changes in the electrical potential gradients due to proximity of macromolecular surfaces. Additionally, during freezing, depending on the freezing-rate-dependent solute concentration (as presented previously in the first part of this chapter) volume of the cell changes responding to osmotic gradients (Figure 12.1). Briefly, damage to cells during cryopreservation is attributed to various factors directly or indirectly correlated to the presence of intra/extracellular ice (such as solute concentration, membrane potential change, mechanical damage by ice crystals, steep electrical potential, and osmotic gradients, etc.), however the exact mechanism of freeze injury is not known.

Cryopreservation is a process, which inherently disrupts intra/extracellular continuum and introduces heterogeneity within the cytoplasm. During freezing of a complex solution, there always is a mutual interaction between the two phases present simultaneously: the frozen liquid phase, which rejects solids and the supercooled liquid phase, which cannot freeze due to the increased concentration of the solutes rejected by the ice. Presence of solutes depresses the freezing temperature of the water, as a function of their mole fraction. The tug-of-war between these two phases introduces compartmentalization yielding to very high osmotic and electrical gradients within the cytosol. This may explain the low survival rates recorded at relatively high subzero temperatures.

To this date numerous post-thaw viability and function experiments have been performed with virtually every kind of cell using cryoprotectants at different concentrations, freezing/thawing rates, and storage conditions. Interested readers are directed to reviews by Mazur [48] and McGrath [36]. Interestingly, the mechanism of cryoprotection offered by the most widely used chemicals, for example DMSO and glycerol, is not known [49,50] beyond the colligative action (that they replace the water molecules in the cytosol) and their "strong interaction potential with water" at low temperatures. The colligative action of cryopreservatives in the absence of water can be divided into the space-filling effect, which prevents structural collapse and the osmotic effect, which presents the cryoprotectant as an alternative solvent reducing solute concentration. DMSO has a very high hydrogen bonding affinity toward water creating a nonideal mixture behavior (e.g., even though the freezing point of water and DMSO are 0 and 18.6°C respectively, the freezing point of a 1 : 3 molar ratio DMSO–water solution is −70°C).

Given that, DMSO is toxic at high temperatures [51] and has been shown to cause gene activation (for a review, see Ashwood-Smith [52]) and have mutagenic potential [53] at high concentrations, it is surprising that it works so well for preservation at low temperatures in spite of its biological inadequacy: an indication that the preservation phenomena is directly related to the thermodynamical properties of most cryoprotectants (freezing point depression), their effects on the structure of water (such as eliminating ice crystallization), and to the degree of molecular mobility reduction they offer (e.g., DMSO increases cytoplasmic viscosity [50]).

A mechanism offered to explain the protective potential of membrane permeable cryoprotectants and certain solutes introduced artificially into the cell (such as carbohydrates) is that they have the ability to retard ice formation by replacing water and imposing an ordered water structure resistant to crystallization. Raman Scattering, NMR, dynamic molecular modeling, and Quasi Elastic Neutron Scattering data indicate that especially trehalose, even at low concentrations breaks the structure of adjacent water molecules and slows down their mobility [54,55] and therefore makes them more resistant to crystallization at low temperatures [56]. Similarly, DMSO can alter the structure and the rotational and translational mobilities of water (e.g., 10% DMSO reduces the self-diffusion coefficient of water by half [50]) as a function of its concentration and environmental temperature [57]. Analysis of osmotic stress injury to frozen cells [58] in the presence of DMSO, glycerol, and ethylene glycol [59] also point to different reasons to explain the protective mechanism offered by these cryoprotectants beyond their colligative action.

Even though the chemical reaction rates are expected to slow at low temperatures following Arrhenius kinetics, due to freeze concentration, which decreases the distance between reactants and changes the pH of the medium in addition to the possible catalytic effects of ice crystals [60] (and given that proton mobility is higher in ice than in water due to the organized crystal structure), some enzyme-catalyzed chemical reaction rates do increase by orders of magnitude in the frozen state as compared to supercooled state [61]. Combined with the detrimental effects of IIF, the factors stated above formed the scientific reasoning behind development of an alternative preservation method: preservation of cells in an undercooled state [62]. It is known that with the addition of each mole of solute to one liter of water suppresses the freezing temperature by approximately 2°C by increasing the entropy of the liquid phase. It was shown that if structural stability at subzero temperatures could be ensured so that the probability of intra/extracellular ice formation is at a minimum, short-term storage could be feasible for certain cells. This can be achieved for example, by adding ice nucleation retarding solutes, applying high pressures, or using smaller water volumes with minimum peripheral contact (such as utilization of small water droplets suspended in oil).

The intracellular ice formation (IIF) has been thought to be the main source of viability loss for preserved biological systems. It is postulated that the ice crystals in the cytosol mechanically disrupt the cytoskeleton and the macromolecules in addition to causing very high localized solute concentrations due to solute rejection from the frozen phase. It is also suggested that IIF can not form by itself and is usually induced by Extracellular Ice Formation (EIF) breaching the cell membrane through membrane pores, enlarging them and causing the membrane to rupture.

It was therefore suggested that if IIF and EIF can be eliminated altogether (in spite of the disadvantages of having to load very high concentrations of cryoprotectants into the cells and keeping them at liquid nitrogen temperatures) then the preservation of cells, organs, and even whole human bodies would be possible in a vitrified state [63]. With high cryoprotectant concentrations (as high as 9 M [63]) and low temperature storage, ice formation within the cells can be completely eliminated and also an intracellular amorphous (or glassy?) state may be reached. This line of work however, is not supported by detailed theoretical, experimental, and numerical analysis of molecular motion confirming the presence of intracellular glass. Given that the glass transition temperatures for DMSO, glycerol, and ethylene glycol (at 100% w/w) are -122.2, -187, and $-115°C$, respectively [64] when compared to that of pure water ($-135°C$) even if their primary mechanism of action in preservation at high concentration is by vitrification, then their only role is relaxing the high cooling rate (10^6 K/sec) requirement for the vitrification of water. Actually, for a 5 M solution of DMSO in water, the critical cooling rate for vitrification (the cooling rate at which crystal formation is minimum) is approximately 10 K/sec. This is achievable, however the critical warming rate is still unattainable (10^8 K/sec) [65].

12.2.3 Vitrification

The discovery of the presence of a vitrified state in the cytoplasms of desiccated plant seeds [66], Artemia cysts, and fungal spores started extensive research for desiccation preservation of mammalian cells and tissues. For plants and certain animals, it was shown that the transition to the glassy state was enabled through the accumulation of certain carbohydrates (glucose, sucrose, raffinose, and stachyose in plants, and trehalose in microorganisms and animals making up about 15 to 25% of their dry weight). Having the same chemical groups (–OH) as water, carbohydrates are hypothesized to replace the hydrogen bonds formed between water molecules and the macromolecules in the cytoplasm and the lipid membranes stabilizing their conformations in the absence of water [5]. This has formed the basis of the "water replacement hypothesis" [67] that was offered as an explanation for the protective capacity of certain carbohydrates against freezing and desiccation damage. Another theory proposed is based on the "preferential exclusion" of carbohydrates from macromolecule surfaces [68,69]. This theory predicts that when water is scarce in the cytoplasm (as would be the case during desiccation or freezing), water molecules in the hydration shell of the macromolecules remain undisturbed while the void left by the removed free water is filled by the carbohydrates eliminating structural collapse. It is not known to this date which one of these hypotheses explains the protection mechanism responsible for reducing macromolecule and membrane denaturation during desiccation.

Additionally, a recently introduced hypothesis based on molecular mobility measurements suggests that the main protective effect of carbohydrates is by breaking the structure of the surrounding water molecules (especially at low water concentrations) and creating more structured water clusters, therefore enabling encapsulation of biological macromolecules in a rigid matrix [70]. This hypothesis is in agreement with the additional claim made by the first two hypotheses that upon removal of the free water, a cytosolic glass with reduced molecular mobility is formed, virtually stopping all biochemical processes. In addition to carbohydrates, certain solutes abundant in the cytosol (such as proteins, amino acids, ions, and salts) are also thought to be participating in the formation of the cytoplasmic glass [71] (for a recent review, see Buitink et al. [72]). It should be noted that the glass transition temperature of an ideal mixture is determined by the glass transition temperatures of its constituents. Since it has a very low glass transition temperature ($T_g = -135°C$), water significantly reduces the glass transition temperature of carbohydrates as a function of its concentration.

Chemical reaction rates have been shown to decrease in the presence of high concentrations of carbohydrates, facilitating protein dynamics studies [73,74], slowing down oxygen diffusion therefore increasing the stability of fluorescent molecules [73], reducing degradation of enzymes [75,76] and proteins [74,77], and enabling lyophilization of bacteria [78] and viruses [79] supporting the cytosolic vitrification hypothesis by experimental evidence. However, contradicting reports also exist, where researchers did not find any difference between the state diagrams (glass transition temperature as a function of water content) of desiccation tolerant and intolerant plants [80]. This has been offered as the basis of why cytosolic vitrification alone can not be responsible for preservation in the desiccated state.

It was shown that for the carbohydrates to protect against desiccation damage, they should be present on both sides of the membrane [81]. Almost all of the glass-forming carbohydrates and polymers are membrane impermeable and should be introduced artificially with the exception of glucose. Currently applied reversible membrane breaching methods are microinjection [82], osmotic shock [83], electroporation [84,85], thermal shock [86], acoustical exposure [87,88], endocytosis [89], and transport through switchable membrane pores [90,91]. For details of these methods, see Acker et al. [81]. An alternative method utilized is the genetic modification of mammalian cells to express genes for coding carbohydrates intracellularly [92]. It is still perceived as a challenge to upload mammalian cells with high concentrations (0.2 to 0.3 M) of carbohydrates required for preservation.

A glass is a metastable liquid with very low molecular mobility of its main structural ingredient, which forms an amorphous matrix rather than an energetically more favorable crystal. Whether achieved by rapid cooling, desiccation, addition of cryoprotectants or carbohydrates, amorphous to glassy phase transition is characterized by a sharp discontinuity in the temperature–density curve. As an indication of very compact packing in the molecular arrangement in the glassy phase, density is a very weak function of temperature. At the glass transition temperature, specific heat also decreases significantly since a higher percentage of the thermal energy transferred to the system causes a temperature change without being dissipated by the exceedingly confined molecular motions of the matrix molecules. Even though vitrification is characterized by a very significant change in viscosity ($O(7–9)$), and the stopping of α-relaxation processes, depending on the relative sizes of the matrix molecules and the solvent remaining in the system, molecular motion does not completely stop [42]. During drying of thin films or small droplets of high molecular weight polymer and carbohydrate solutions for example, evaporation continues even though early on a glassy film is formed on the surface. The decoupling of matrix mobility from that of the solvent at glass transition may present a challenge that the diffusion of the solvent and the small solutes can not be prevented during storage of biological materials. The glass transition temperature increases with increasing molecular weight. Therefore, with increasing molecular weight of the matrix, the mobility of the solvent at matrix glass transition also increases [93,94]. This has led some researchers [95] to conclude that a glassy matrix (actin embedded in a glass formed by dextran, a very high molecular weight carbohydrate) can not protect against desiccation damage. Apparently, in this particular case even though the matrix was glassy, decoupling of matrix and solvent (and small solute) mobilities were very significant resulting in denaturation of actin.

It would be wrong to conclude however, that the diffusion of small solutes and the solvent in a glassy matrix would not be affected by the presence of the glass. With increasing solute concentration, the mobility of the solvent is increasingly limited and therefore the solvent switches from the free, unrestricted diffusion to jump-diffusion (Figure 12.3) between the cages formed by the glassy matrix [96]. There is evidence that the transition in the diffusion mode of water coincides with the point, where the carbohydrates have been shown to form a three-dimensional hydrogen bonded network [97]. The diffusion constant shows an Arrhenius type dependence on temperature above and below the glass transition temperature (with different activation energies) indicating decoupling of the solvent and the carbohydrate matrix mobility [98–100]. For example, during desiccation diffusivity of water in the Artemia cysts decreases with decreasing water content, however below a critical water content corresponding to 0.15 $g_{water}/g_{drymatter}$ [101], it starts to increase.

As opposed to diffusion in a glassy matrix, where translational mobility is dictated by cooperative diffusion, in cryopreservation, diffusion is dependent on the presence of defects in the crystalline structure (and the presence of grain boundaries). In a perfect crystal (without any defects or grain boundaries), mobility of the solvent and small solutes are exceedingly hindered (Figure 12.2).

For glass-forming liquids, with increasing cooling rate, the probability of reaching the glassy state increases (when compared to crystallization). However, since a glass is intrinsically a liquid with a higher energy than its crystal, given enough time (depending on the characteristics of the carbohydrate and the storage conditions this could be weeks, years, or even centuries) it will crystallize.

In the absence of chemical and electrical interactions, the size of a molecule determines its mobility in a glassy matrix. In dextran solutions, for example with increasing molecular weight of the dextran, water mobility at glass transition increases. For successful storage therefore, the size of the molecules to be preserved should be larger than that of the molecules making up the glassy matrix and the free volume of the matrix they form. Rigidity of the glassy matrix may also contribute to its storage potential by reducing solvent mobility. For example, it was shown by Elastic Incoherent Neutron Scattering measurements that the glassy matrix formed by trehalose (which is known to be superior in terms of its protective potential) is more rigid than the glasses formed by maltose and sucrose [102]. If the preserved molecules are polar, their mobility may further be reduced since they can form strong hydrogen bonds with the glassy matrix. Due to the facts presented above, we may conclude that a high glass transition temperature alone is not sufficient in determining the storage potential offered by a particular glass former. Even though dextran has a higher glass transition temperature than trehalose and sucrose, it has been repeatedly shown to have inferior protective capacity against desiccation and freezing damage. One reason for its low protective capacity may be its large size making it less flexible and therefore its hydration sites less accessible. However, there is reason to believe that the main factor is the high pore size of the glassy matrix it forms presenting lower resistance to the diffusion of solutes and solvents. With the technological advances making it feasible to measure molecular mobility in the glassy state, this hypothesis can be put to test.

12.2.4   Vitrification by Ultrafast Cooling

If the kinetic energy can be reduced faster than the rotational and translational diffusion timescales at any temperature, then vitrification without crystallization can be achieved for an aqueous glass former at any water content. If it can be cooled fast enough (10^6 to 10^{7}°C/sec), even water can be vitrified at a temperature of -135°C. For biological materials, these cooling rates can not be achieved using conventional techniques such as liquid nitrogen immersion. However, methods utilizing cooling with simultaneous heating are promising [103]. Ultrafast cooling, if achieved, does not expose cells to osmotic and dehydration stresses and therefore also eliminates compartmentalization (Figure 12.1). However, if storage at noncryogenic temperatures is desired, methods to reduce mobility at high temperatures should be developed.

12.2.5   Vitrification by Desiccation

During isothermal drying, the glass transition temperature of the solution increases. Evaporation of the solvent causes a decrease in the free volume increasing the viscosity and the structural relaxation times. Mode Coupling Theory [40] predicts a crossover temperature located at approximately 40 to 50 K above the glass transition temperature of a glass-forming solution, where there is a transition from liquid-like to solid-like molecular dynamics. The structural relaxation time, τ_α, is related to the collective motion of a group of molecules to loosen up the cage they form around a single molecule and enable it to make a translational diffusive motion. The short timescale relaxation time, τ_β, however involves the temperature dependent vibrational motion of a single molecule, which also involves rotation. τ_α reaches a value of 10^2 sec ($\tau_\alpha = 10^{-7}$ sec at the crossover temperature) very close to glass transition (for an extensive review, see Novikov, 2003 [104]). The O(9) decrease in the structural relaxation time with proximity to glassy state (each evaporating water molecule making it more difficult for the next one in the solution to evaporate) is thought to make vitrification improbable in the experimental timescales considered. However, vitrification of carbohydrate solutions can be accelerated using a cocktail of carbohydrates [105] and salts [120].

12.2.6 Lyophilization

Cryopreservation (either in the frozen or the vitrified state) requires very low temperatures for storage and transportation and therefore is not economical for many purposes. An alternative to cryopreservation is lyophilization, where the reduction of molecular mobility reached at low temperatures by cryopreservation can be achieved at relatively high temperatures by the removal of water (the universal plasticizer), thus increasing the glass transition temperature [106]. The product to be stored is initially frozen in the presence of cryoprotectants and carbohydrates. Then, by the application of vacuum, frozen water is sublimated at progressively increasing temperatures. Removal of water increases the glass transition temperature of the product and establishes stability at ambient temperatures. In the pharmaceutical and food industries, lyophilization is widely used for the preservation of proteins, enzymes, bacteria, and foodstuff.

The main advantage of lyophilization over desiccation is in minimizing the exposure to osmotic stresses. During diffusive or convective drying, by the removal of water, the solutes concentrate and the product to be preserved is exposed to increasing osmotic stresses over long periods of time. In lyophilization, the product is first frozen and then the water is removed, minimizing osmotic stress buildup. However, the removal of water from the frozen structure leaves pores within the protective matrix and the product. These pores in time may collapse and the structural integrity of the product may be lost. In order to prevent this during lyophilization, the primary and secondary drying temperatures are kept above the collapse temperature of the matrix. Certain high molecular weight carbohydrates (such as dextran) are also incorporated to increase the structural stability of the matrix.

12.3 Storage

During storage the preserved organism does not need a steady supply of nutrients (and removal of the by-products). On the down side, it does not have the metabolic activity to repair the accumulating damage. In the preserved state, the harmful environmental factors are, to a degree, physically isolated from the organism by the surrounding matrix (the frozen liquid crystal structure in cryopreservation or an amorphous matrix of carbohydrates during desiccation) reducing the amount of accumulated damage.

Regardless of the processing method, the final thermodynamic state of the product and the molecular mobility at that state determine the storage stability as a function of storage parameters (primarily humidity, temperature, and pressure). If at the storage condition, the product is not at equilibrium (internally or externally with the environment), the chemical processes will continue even at slow rates toward the minimum energy state. For example in vitrification, since the glass formed is inherently at a higher energy state it will crystallize in time. Crystallization is accompanied by compartmentalization [105] and heterogeneous devitrification, resulting in an increase in mobility. It also creates high mechanical stresses that can damage the product. The structural mobility of the glass determines its crystallization lifetime, while the mobilities of the solvents and small solutes within the glass determine the degradation of the stored product. One way to eliminate crystallization is to use cocktails of carbohydrates and salts. For example, raffinose and stachyose are very effective in inhibiting sucrose crystallization [107]. The presence of small solutes and proteins in the cytoplasm may also help reduce crystallization of sugars [108]. In cryopreservation, the grain boundaries between the frozen sections and the interfaces between the ice crystals and the solutes rejected during freezing have higher free energy and therefore mobility. Molecular mobility in these regions determine the stability of the stored product.

Vitrified products can be stored at higher temperatures, however their storage conditions still need to be regulated carefully since molecular mobility increases significantly with increasing water content. For a trehalose glass the glass transition temperature decreases by 50 K by a very slight change in the water content (0.05 $g_{water}/g_{drymatter}$). If the water activity in the environment is higher than that in glass, the product absorbs water and the molecular mobility within increases. On the opposite end, if the environmental water activity is lower, the stored product may desiccate further, losing its hydration water. For most lyophilized bacteria, it has been shown that irrespective of the preservation agent used, for successful revival, the water content within the product should not drop below a certain limit.

The other important factors that need to be considered for successful storage in vitrified state are oxygen and light. Free oxygen radicals have very high diffusivity and are extremely reactant. Even at very low mobility environments, they may jump-start enzymatic reactions. Some carbohydrates, such as trehalose are known to prevent oxygen radical damage [109] as well.

12.4 Summary

The "molecular mobility" hypothesis analyzed here suggests that, if all of the reactants in an organism can be reversibly immobilized in their native configuration, the biochemical processes can be stopped and preservation can be achieved (Figure 12.4). The condition set forth by the hypothesis is sufficient to halt the diffusion-limited biochemical reactions, which dominate in a crowded, confined environment such as in the cytoplasm [29]. However, not all of the chemical reactions in an organism is governed by diffusion. In order to respond to sudden changes in its environment (under certain conditions, the metabolic rate has been known to increase up to 35 times), the mammalian cell has devised strategies to accelerate certain reactions without being limited by diffusion timescales, for example by accumulating reactants in aggregates (compartmentalization), reducing the degree-of-freedom of the reactants (such as the case with membrane proteins) and advection (bulk mixing due to cell motion). The effects of the currently applied and proposed biopreservation methods on this kind of chemical reactions remain to be researched.

The important points that are highlighted in this chapter are:

(1) Glass transition is a collaborative phenomena involving the solvent and the solute. With increasing difference in the molecular sizes of the solvent and the solute mobility of the solute molecules making up the glassy matrix and that of the solvent (or other small solutes suspended in the solvent) decouple strongly at glass transition. Mobility of the solvent may be significantly higher than that of the matrix, even below the glass transition temperature of the system, enabling certain chemical reactions to proceed. High molecular weight carbohydrates increase the glass transition temperature of the system, however they do not necessarily decrease the solvent (or small solute) mobility in the same degree.

(2) If noncryogenic temperature storage is desired, the mobility within and surrounding the preserved product should be lowered. This may be achieved by loading high amounts of protectants (cryoprotectant solvents such as glycerol, DMSO, etc. or carbohydrates, proteins, etc.) into the product to displace water or change the properties of the intracellular water. It is unlikely that any organism can be successfully preserved by vitrification at conditions requiring complete removal of all of its water. The "essential water" in the product upon removal causes irreversible damage. This quantity is not easily measurable. The water affinities of the macromolecules and membranes in the preserved product and that of the added protectant are different and also change with the decrease in the availability of water. Decreasing the water content of the overall system comprised of the product to be preserved and the surrounding protectant matrix (frozen, amorphous, or glassy) does not necessarily mean that the water contents of the matrix and the product are equally reduced.

(3) The importance of diffusion-limited reactions for the metabolic activity of the product should be established. Reduction of mobility within the medium is most likely to reduce the diffusion of any size molecule, however there is increasingly more evidence collected showing that the majority of the biochemical reactions do not depend on diffusion alone. For preservation to be successful, mobility of all molecules in all size scales should be stopped without irreversibly disrupting their native configurations.

(4) The state of intracellular water is directly related to the metabolism and functioning of an organism. All of the chemical agents known to have protective capacity against freezing and desiccation damage modify the structure of the intracellular water. More research is required to establish the role of water in organisms in order to develop successful preservation methods and to increase the efficiency of the currently applied ones.

Acknowledgments

The authors thank Dr. James Clegg, Dr. John Bischof, and Dr. Xiang-Hong Liu for careful reading of the manuscript and their constructive feedback. This work was supported by the National Institutes of Health grants.

References

[1] Crowe, J.H. and Cooper, A.F., Cryptobiosis, *Sci. Am.*, 225, 30, 1971.

[2] Mazur, P., Leibo, S., and Chu, E.H.Y., A two-factor hypothesis of freezing injury, *Exp. Cell Res.*, 71, 345, 1972.

[3] Toner, M., Nucleation of ice crystals inside biological cells, in *Advances in Low-Temperature Biology*, Steponkus, P.L., Ed., JAI Press Ltd, Greenwich CN, 1993.

[4] Sastry, S., Order and oddities, *Nature*, 409, 300, 2001.

[5] Franks, F., *Water: A Matrix of Life*, Royal Society of Chemistry, Cambridge, UK, 2000.

[6] Chaplin, M., *Www.Lsbu.Ac.Uk/water/index.Html*, 2004.

[7] Szasz, A. et al., Water states in living systems: I. Structural aspects, *Physiol. Chem. Phys. Med. NMR*, 26, 299, 1994.

[8] Franks, F., *Water: A Comprehensive Treatise*, Plenum Press, New York, 1979.

[9] Sciortino, F., Geiger, A., and Stanley, H.E., Network defects and molecular mobility in liquid water, *J. Chem. Phys.*, 96, 3857, 1992.

[10] Geiger, A. et al., Mechanisms of the molecular mobility of water, *J. Mol. Liq.*, 106, 161, 2003.

[11] Granick, S., Motions and relaxations of confined liquids, *Science*, 253, 1374, 1991.

[12] Jensen, M.O., Mouritsen, O.G., and Peters, G.H., The hydrophobic effect: molecular dynamics simulations of water confined between extended hydrophobic and hydrophilic surfaces, *J. Chem. Phys.*, 120, 9729, 2004.

[13] Kabir, S.R. et al., Hyper-mobile water is induced around actin filaments, *Biophys. J.*, 85, 3154, 2003.

[14] Hribar, B. et al., How ions affect the structure of water, *J. Am. Chem. Soc.*, 124, 12302, 2002.

[15] Tovchigrechko, A., Rodnikova, M., and Barthel, J., Comparative study of urea and tetramethylurea in water by molecular dynamics simulations, *J. Mol. Liq.*, 79, 187, 1999.

[16] Watterson, J.G., A role of water in cell structure, *Biochem. J.*, 248, 615, 1987.

[17] Clegg, J.S., Intracellular water and the cytomatrix: some methods of study and current views, *J. Cell Biol.*, 99, 167S, 1984.

[18] Zimmerman, S.B. and Trach, S.O., Estimation of macromolecule concentrations and excluded volume effects for the cytoplasm of *Escherichia coli*, *J. Mol. Biol.*, 222, 599, 1991.

[19] Scalettar, B.A., Abney, J.R., and Hackenbrock, C.R., Dynamics, structure and function are coupled in the mitochondrial matrix, *Proc. Natl Acad. Sci. USA*, 88, 8057, 1991.

[20] Cayley, S. and Record, M.T., Roles of cytoplasmic osmolytes, water and crowding in the response of *Escherichia coli* to osmotic stress: biophysical basis of osmoprotection by glycine betaine, *Biochemistry*, 42, 12595, 2003.

[21] Clegg, J.S., Properties and metabolism of the aqueous cytoplasm and its boundaries, *Am. J. Physiol.*, 246, R133, 1984.

[22] Sehy, J.V., Ackerman, J.J.H., and Neil, J.J., Apparent diffusion of water, ions, and small molecules in the xenopus oocyte is consistent with brownian displacement, *Magn. Reson. Med.*, 48, 42, 2002.

[23] Ovádi, J. and Saks, V., On the origin of intracellular compartmentation and organized metabolic systems, *Mol. Cell. Biochem.*, 256/257, 5, 2004.

[24] Luby-Phelps, K., Taylor, D.L., and Lanni, F., Probing the structure of cytoplasm, *J. Cell Biol.*, 102, 2015, 1986.

[25] Potma, E.O. et al., Real-time visualization of intracellular hydrodynamics in single living cells, *Proc. Natl Acad. Sci. USA*, 98, 1577, 2001.

[26] Wiggins, P.M., High and low density water and resting, active and transformed cells, *Cell Biol. Int.*, 20, 429, 1996.

[27] Beall, P.T., Hazlewood, C.F., and Rao, P.N., Nuclear magnetic resonance patterns of intracellular water as a function of hela cell cycle, *Science*, 192, 904, 1976.

[28] Cameron, I.L., et al., A mechanistic view of the non-ideal osmotic and motional behavior of intracellular water, *Cell Biol. Int.*, 21, 99, 1997.

[29] Ellis, R.J., Macromolecular crowding: an important but neglected aspect of the intracellular environment, *Curr. Opin. Struct. Biol.*, 11, 114, 2001.

[30] Arrio-Dupont, M. et al., Translational diffusion of globular proteins in the cytoplasm of cultured muscle cells, *Biophys. J.*, 78, 901, 2000.

[31] Garcia-Martin, M.L., Ballesteros, P., and Cerdan, S., The metabolism of water in cells and tissues as detected by NMR methods, *Prog. Nucl. Magn. Reson. Spectrosc.*, 39, 41, 2001.

[32] Kao, H.P., Abney, J.R., and Verkman, A.S., Determinant of the translational mobility of a small solute in cell cytoplasm, *J. Cell Biol.*, 120, 175, 1993.

[33] Mastro, A.M. and Keith, A.D., Diffusion in aqueous compartment, *J. Cell Biol.*, 99, 180s, 1984.

[34] Welch, G.R. and Clegg, J.S., *The Organization of Cell Metabolism*, Welch, G.R. and Clegg, J.S., Eds., Plenum Press, New York, 1986.

[35] Minton, A.P., The influence of macromolecular crowding and macromolecular confinement on biochemical reactions in physiological media, *J. Biol. Chem.*, 276, 10577, 2001.

[36] McGrath, J.J., Membrane transport properties, in *Low Temperature Biotechnology*, McGrath, J.J. and Diller, K.R., Eds., ASME, New York, NY, 1988.

[37] Fettiplace, R. and Haydon, D.A., Water permeability of lipid membranes, *Physiol. Rev.*, 60, 510, 1980.

[38] Verkman, A.S., Water permeability measurement in living cells and complex tissues, *J. Membr. Biol.*, 173, 73, 2000.

[39] Bhattacharyya, S. and Bagchi, B., Anomalous diffusion of small particles in dense liquids, *J. Chem. Phys.*, 106, 1757, 1997.

[40] Gotze, W., Recent tests of the mode coupling theory for glassy dynamics, *J. Phys.: Condens. Matter*, 11, A1, 1999.

[41] Conrad, P.B. and de Pablo, J.J., Computer simulation of cryoprotectant disaccharide α,α-trehalose in aqueous solution, *J. Phys. Chem. A*, 103, 4049, 1999.

[42] Rampp, M., Buttersack, C., and Lüdemann, H.-D., C, t-Dependence of the viscosity and the self-diffusion coefficients in some aqueous carbohydrate solutions, *Carbohydr. Res.*, 328, 561, 2000.

[43] Kandror, O., DeLeon, A., and Goldberg, A.L., Trehalose synthesis is induced upon exposure of *Escherichia coli* to cold and is essential for viability at low temperatures., *Proc. Natl Acad. Sci. USA*, 99, 9727, 2002.

[44] Attfield, P.V., Trehalose accumulates in *Saccharomyces cerevisiae* during exposure to agents that induce heat shock response, *FEBS Lett.*, 225, 259, 1987.

[45] Madin, K.A.C. and Crowe, J.H., Anhydrobiosis in nematodes: carbohydrate and lipid metabolism during dehydration, *J. Exp. Zool.*, 193, 335, 1975.

[46] Crowe, J.H., Hoekstra, F.A., and Crowe, L.M., Anhydrobiosis, *Ann. Rev. Physiol.*, 54, 579, 1992.

[47] Arakawa, T. et al., The basis for toxicity of certain cryoprotectants — a hypothesis, *Cryobiology*, 27, 401, 1990.

[48] Mazur, P., Principles of cryobiology, in *Life in the Frozen State*, Fuller, B.J., Lane, N., and Benson, E.E., Eds., CRC Press, Boca Raton, FL, 2004.

[49] Murthy, S.S.N., Phase behavior of the supercooled aqueous solutions of dimethyl sulfoxide, ethylene glycol, and methanol as seen by dielectric spectroscopy, *J. Phys. Chem. B*, 101, 6043, 1997.

[50] Yu, Z.W. and Quinn, P.J., The modulation of membrane structure and stability by dimethyl sulphoxide (review), *Mol. Membr. Biol.*, 15, 59, 1998.

[51] Fahy, G.M., The relevance of cryoprotectant toxicity to cryobiology, *Cryobiology*, 23, 1, 1986.

[52] Ashwood-Smith, M.J., Genetic damage is not produced by normal cryopreservation procedures involving either glycerol or dimethyl sulfoxide: a cautionary note, however, on possible effects of dimethyl sulfoxide, *Cryobiology*, 22, 427, 1985.

[53] Preisler, H.D. and Lyman, G., Differentiation of erythroleukemia cells *in vitro* — properties of chemical inducers, *Cell Differ.*, 4, 179, 1975.

[54] Magazu, S. et al., Diffusive dynamics in trehalose aqueous solutions by qens, *Physica B*, 276, 475, 2000.

[55] Bordat, P. et al., Comparative study of trehalose, sucrose and maltose in water solutions by molecular modelling, *Europhys. Lett.*, 65, 41, 2004.

[56] Branca, C. et al., A,α-trehalose-water solutions. 3. Vibrational dynamics studies by inelastic light scattering, *J. Phys. Chem. B*, 103, 1347, 1999.

[57] Packer, K.J. and Tomlinson, D.J., Nuclear spin relaxation and self-diffusion in the binary system, dimethylsulfoxide (DMSO) + water, *Trans. Faraday Soc.*, 67, 1302, 1971.

[58] Takashashi, T., Hammett, M.F., and Cho, M.S., Multifaceted freezing injury in human polymorphonuclear cells at high subfreezing temperatures, *Cryobiology*, 22, 215, 1985.

[59] Takashashi, T. et al., Effect of cryoprotectants on the viability and function of unfrozen human polymorphonuclear cells, *Cryobiology*, 22, 336, 1985.

[60] Takenaka, N. et al., Acceleration mechanism of chemical reaction by freezing: the reaction of nitrous acid with dissolved oxygen, *J. Phys. Chem.*, 100, 13874, 1996.

[61] Hatley, R.H.M., Franks, F., and Mathias, S.F., The stabilization of labile biochemicals by undercooling, *Process Biochem.*, 22, 171, 1987.

[62] Mathias, S.F., Franks, F., and Hatley, R.H.M., Preservation of viable cells in the undercooled state, *Cryobiology*, 22, 537, 1985.

[63] Fahy, G.M. et al., Vitrification as an approach to cryopreservation, *Cryobiology*, 21, 407, 1984.

[64] Murthy, S.S.N., Some insight into the physical basis of the cryoprotective action of dimethyl sulfoxide and ethylene glycol, *Cryobiology*, 36, 84, 1998.

[65] Baudot, A., Lawrence, L., and Boutron, P., Glass-forming tendency in the system water–dimethyl sulfoxide, *Cryobiology*, 40, 151, 2000.

[66] Williams, R.J. and Leopold, A.C., The glassy state in corn embryos, *Plant Physiol.*, 89, 977, 1989.

[67] Webb, S.J., *Bound Water in Biological Integrity*, Charles C. Thomas Publisher, Springfield, IL, 1965.

[68] Xie, G. and Timasheff, S.N., Mechanism of the stabilization of ribonuclease a by sorbitol: preferential hydration is greater for the denatured than for the native protein, *Protein Sci.*, 6, 211, 1997.

[69] Arakawa, T. and Timasheff, S.N., Stabilization of protein structure by sugars, *Biochemistry*, 21, 6536, 1982.

[70] Migliardo, F., Magazu, S., and Migliardo, M., INS investigation on disaccharide/H_2O mixtures, *J. Mol. Liq.*, 110, 11, 2004.

[71] Sun, W.Q. and Leopold, A.C., Cytoplasmic vitrification and survival of anhydrobiotic organisms, *Compar. Biochem. Physiol.*, 117A, 327, 1997.

[72] Buitink, J. and Leprince, O., Glass formation in plant anhydrobiotes: survival in the dry state, *Cryobiology*, 48, 215, 2004.

[73] Mei, E. et al., Motions of single molecules and proteins in trehalose glass, *J. Am. Chem. Soc.*, 125, 2730, 2003.

[74] Sastry, G.M. and Agmon, N., Trehalose prevents myoglobin collapse and preserves its internal mobility, *Biochemistry*, 36, 7097, 1997.

[75] DePaz, R.A. et al., Effects of drying methods and additives on the structure, function and storage stability of subtilisin: role of protein conformation and molecular mobility, *Enzyme Microb. Technol.*, 21, 765, 2002.

[76] Miller, D.P., Anderson, R.E., and de Pablo, J.J., Stabilization of lactate dehydrogenase following freeze-thawing and vacuum-drying in the presence of trehalose and borate, *Pharma. Res.*, 15, 1215, 1998.

[77] Allison, S.D. et al., Hydrogen bonding between sugar and protein is responsible for inhibition of dehydration-induced protein folding, *Arch. Biochem. Biophys.*, 223, 289, 1999.

[78] Conrad, P.B. et al., Stabilization and preservation of *Lactobacillus acidophilus* in saccharide matrices, *Cryobiology*, 41, 17, 2000.

[79] Bieganski, R.M. et al., Stabilization of active recombinant retroviruses in an amorphous dry state with trehalose, *Biotechnol. Prog.*, 14, 615, 1998.

[80] Sun, W.Q., Irving, T.C., and Leopold, A.C., The role of sugar, vitrification and membrane phase transition in seed desiccation tolerance, *Physiol. Plant*, 90, 621, 1994.

[81] Acker, J.P. et al., Engineering desiccation tolerance in mammalian cells: tools and techniques, in *Life in the Frozen State*, Fuller, B.J., Lane, L., and Benson, E.E., Eds., CRC Press, Boca Raton, FL, 2004.

[82] Eroglu, A., Toner, M., and Toth, T.L., Beneficial effect of microinjected trehalose on the cryosurvival of human oocytes, *Fertil. Steril.*, 77, 152, 2002.

[83] Wolkers, W.F. et al., Temperature dependence of fluid phase endocytosis coincides with membrane properties of pig platelets, *Biochem. Biophys. Acta — Biomembr.*, 1612, 154, 2003.

[84] Mussauer, H., Sukhorukov, V.L., and Zimmermann, U., Trehalose improves survival of electrotransfected mammalian cells, *Cytometry*, 45, 2001.

[85] Shirakashi, R. et al., Intracellular delivery of trehalose into mammalian cells by electropermeabilization, *J. Membr. Biol.*, 189, 45, 2002.

[86] Puhlev, I. et al., Desiccation tolerance in human cells, *Cryobiology*, 42, 207, 2001.

[87] Lee, S. and Doukas, A.G., Laser-generated stress waves and their effects on the cell membrane, *IEEE J. Select. Top. Quant. Electron.*, 5, 997, 1999.

[88] Kodama, T., Hamblin, M.R., and Doukas, A.G., Cytoplasmic molecular delivery with shock waves: importance of impulse, *Biophys. J.*, 79, 1821, 2000.

[89] Oliver, A.E. et al., Loading human mesenchymal stem cells with trehalose by fluid-phase endocytosis, *Cell Preserv. Technol.*, 2, 35, 2004.

[90] Russo, M.J., Bayley, H., and Toner, M., Reversible permeabilization of plasme membranes with an engineered switchable pore, *Nat. Biotechnol.*, 15, 278, 1997.

[91] Eroglu, A. et al., Intracellular trehalose improves the survival of cryopreserved mammalian cells, *Nat. Biotechnol.*, 18, 163, 2000.

[92] Guo, N. et al., Trehalose expression confers desiccation tolerance on human cells, *Nat. Biotechnol.*, 18, 168, 2000.

[93] van den Dries, I.J. et al., Effects of water content and molecular weight on spin probe and water mobility in malto-oligamer glasses, *J. Phys. Chem. B*, 1004, 10126, 2000.

[94] Cicerone, M.T., Blackburn, F.R., and Ediger, M.D., How do molecules move near tg — molecular rotation of 6 probes in o-terphenyl across 14 decades in time, *J. Chem. Phys.*, 102, 471, 1995.

[95] Allison, S.D. et al., Effects of drying methods and additives on structure and function of actin: mechanisms of dehydration-induced damage and its inhibition, *Arch. Biochem. Biophys.*, 358, 171, 1998.

[96] Magazu, S., Maisano, G., and Majolino, D., Diffusive properties of alpha,alpha-trehalose-water solutions, *Prog. Theoret. Phys. Suppl.*, 195, 1997.

[97] Molinero, V., Çagin, T., and Goddard III, W.A., Mechanisms of nonexponential relaxation in supercooled glucose solutions: the role of water facilitation, *J. Phys. Chem. A*, 108, 3699, 2004.

[98] Tromp, R.H., Parker, R., and Ring, S.G., Water diffusion in glasses of carbohydrates, *Carbohydr. Res.*, 303, 199, 1997.

[99] Parker, R. and Ring, S.G., Diffusion in maltose–water mixtures at temperatures close to the glass-transition, *Carbohydr. Res.*, 273, 147, 1995.

[100] Aldous, B.J., Franks, F., and Greer, A.L., Diffusion of water within an amorphous carbohydrate, *J. Mater. Sci.*, 32, 301, 1997.

[101] Clegg, J.S. et al., Cellular response to extreme water loss: the water-replacement hypothesis, *Cryobiology*, 19, 306, 1982.

[102] Migliardo, F., Magazu, S., and Mondelli, C., Elastic incoherent neutron scattering studies on glass forming hydrogen-bonded systems, *J. Mol. Liq.*, 110, 7, 2004.

[103] Fowler, A.J. and Toner, M., Prevention of hemolysis in rapidly frozen erythrocytes by using a laser pulse, *Ann. NY Acad. Sci.*, 858, 245, 1998.

[104] Novikov, V.N. and Sololov, A.P., Universality of the dynamic crossover in glass-forming liquids: a "magic" relaxation time, *Phys. Rev. E*, 67, 031507, 2003.

[105] Aksan, A. and Toner, M., Isothermal desiccation and vitrification kinetics of trehalose–dextran solutions, *Langmuir*, 20, 5521, 2004.

[106] Rey, L. and May, J.C., *Freeze-Drying/Lyophilization of Pharmaceutical and Biological Products*, Rey, L. and May, J.C., Eds., New York, Marcel Dekker, Drugs and the pharmaceutical sciences, 1999.

[107] Smythe, B.M., Sucrose crystal growth, *Aust. J. Chem.*, 20, 1097, 1967.

[108] Sarciaux, J.M. et al., Influence of bovine somatotropin (BST) concentration on the physical/chemical stability of freeze-dried sucrose/BST formulations, *Pharmac. Res.*, 12, S, 1995.

[109] Benaroudj, N., Lee, D.H., and Goldberg, A.L., Trehalose accumulation during cellular stress protects cells and cellular proteins from damage by oxygen radicals, *J. Biol. Chem.*, 276, 24261, 2001.

[110] Buitink, J., Hemminga, M.A., and Hoekstra, F.A., Characterization of molecular mobility in seed tissues: an electron paramagnetic spin probe study, *Biophys. J.*, 76, 3315, 1999.

[111] Le Meste, M. and Voilley, A., Influence of hydration on rotational diffusivity of solutes in model systems, *J. Phys. Chem.*, 92, 1612, 1988.

[112] Ekdawi-Sever, N. et al., Diffusion of sucrose and α,α-trehalose in aqueous solutions, *J. Phys. Chem. A*, 107, 936, 2003.

[113] Paschek, D. and Geiger, A., Simulation study on the diffusive motion in deeply supercooled water, *J. Phys. Chem. B*, 103, 4139, 1999.

[114] Price, W.S., Ide, H., and Arata, Y., Self-diffusion of supercooled water to 238 K using PGSE NMR diffusion measurements, *J. Phys. Chem. A*, 103, 448, 1999.

[115] Onsager, L. and Runnels, L.K., Diffusion and relaxation phenomena in ice, *J. Chem. Phys.*, 50, 1089, 1969.

[116] Petrenko, V.F. and Whitworth, R.W., *Physics of Ice*, Oxford University Press Inc., New York, NY, 1999.

[117] Mills, R., Self-diffusion in normal and heavy water in the range 1–45°C, *J. Phys. Chem.*, 77, 685, 1973.

[118] Harris, K.R. and Newitt, P.J., Self-diffusion of water at low temperatures and high pressure, *J. Chem. Eng. Data*, 42, 346, 1997.

[119] Moran, G.R. and Jeffrey, K.R., A study of the molecular motion in glucose/water mixtures using deuterium nuclear magnetic resonance, *J. Chem. Phys.*, 110, 3472, 1999.

[120] Miller, D.P., dePablo, J.J., and Corti, H.R., Viscosity and glass transition temperature of aqueous mixtures of trehalose with borax and sodium chloride, *J. Phys. Chem. B*, 103, 10243–10249, 1999.

13

Drug Delivery

C. Becker
A. Göpferich
University of Regensburg

13.1 Introduction

13.1.1 Significance

The science of tissue engineering takes an integrated approach to replacing nonfunctional or missing tissue by gathering input from many different scientific disciplines [1]. An integral part of the emergence of a mature tissue from cells both *in vitro* and *in vivo* involves guiding cells through their development. Apart from the choice of materials used for cell and tissue culture, the use of pharmacologically active substances such as cytokines and growth factors has emerged as an important tool in tissue engineering research and development in recent years [2–4]. By taking advantage of these substances at the right time and in the right context, cells can be induced to proliferate, differentiate, or to migrate in a controlled way. The positive impact of growth factors, such as vascular endothelial growth factor (VEGF) and basic fibroblast growth factor (bFGF), which both stimulate the vascularization of tissue *in vivo*, and members of the bone morphogenic protein family, used to enhance bone formation, illustrate how useful drug-delivery strategies have been for tissue engineering applications in recent years [5,6].

Although many drugs can simply be added to cell- and tissue culture media or administered *in vivo* in the form of an injectable solution, we can take better advantage of many compounds, when they are administered in a controlled way. Protein and peptide drugs, for example, have short half-lives [7] and must, therefore, be administered continuously, which can be very cumbersome for large volumes of solution. In addition, undesired side effects may arise when substances intended for local delivery to a specific tissue are not spatially contained. It is obvious that for these and for many other reasons it has been deemed necessary to administer drugs according to regimes that cannot solely be achieved through the administration of solutions. The field of drug-delivery research has emerged over the last 30 years to overcome such limitations. In recent years, tissue engineering has profited tremendously from integrating drug-delivery technology into more traditional design schemes [8].

It is the goal of this chapter to elucidate the significance of drug delivery within the field of tissue engineering. We first review a number of definitions and basic drug-delivery technologies with a special focus on drug release mechanisms. We then briefly describe the fundamentals of tissue engineering within the realm of drug delivery. We then have a look at the specifics of drugs that are attractive for tissue engineering applications. Thereafter, we address classical drug-carrier systems that have already been used to deliver drugs to cells and tissues and concomitantly give an outlook on systems that hold great promise for similar applications in the future. We then give an overview over the area of using cell carriers for drug delivery. Finally, we review the specifics of some growth factors that have special requirements with respect to stability and pharmacokinetics. At the end, we provide a brief glance at the challenges that have still to be addressed, in an attempt to give an outlook on the further developments that are currently under way. More information can be found in a number of excellent reviews on related topics [9–12].

13.1.2 Goals of Drug Delivery

Defining the term "drug delivery" is not an easy task, because a plethora of definitions can be found in the contemporary literature. They range from brief statements such as "method and route used to provide medication" [13] to more detailed ones, such as "systems of administering drugs through controlled delivery so that an optimum amount reaches the target site. Drug-delivery systems encompass the carrier, route, and target." [14]. Some of them already imply that there is a close relation to tissue engineering research: "delivering agents to specific cells, tissues, or organs" [15].

While all of the definitions make it very clear that drug delivery is intended to control the kinetics of drug release, it is not at all obvious, why one should do so. The reasons are manifold and some of them date back more than 100 years to when Paul Ehrlich (*1854–†1915) coined the term of the "magic bullet" [16]. Since then it has been a well-accepted fact by the scientific community that hitting a biological target with high precision has significant advantages. While Ehrlich intended to avoid side effects during an antibacterial therapy by targeted drug delivery, we nowadays have more incentive than "only" avoiding collateral damage to control the delivery of drug in a biological environment. Thus, the pharmacokinetics, that is, the kinetics of drug resorption, distribution, and elimination processes, may force us to finely dose a drug over a defined time interval to account for its rapid clearance from an application site. Another motivation involves the short tissue half-lives of therapeutic agents, especially of protein and peptide drugs, which are frequently inactivated by proteases and peptidases [17,18] and need, therefore, to be substituted continuously by "fresh" compound from the drug-delivery device's reservoir. These are just a few of the many reasons that may illustrate the need for drug delivery. It is easy to imagine that in the post genomic era, when the number of drug candidates dramatically increases with continuing progress in proteomics [19], even more reasons may arise, leading us to use controlled drug-delivery devices for tissue engineering applications.

Drug delivery is often linked to the concepts of "sustained release" or "controlled release," meaning that a pharmaceutical compound is released over a certain period of time or in a somehow predetermined way, respectively. While "sustained" suggests only that the drug is released over an extended period of time, "controlled" suggests that the kinetics are well defined, but does not necessarily imply that the release runs over long time periods.

13.2 Mechanisms of Drug Delivery

The means that we can exploit to release a drug in a controlled way and which are an indispensable part of a drug-delivery device are manifold and different in nature. We can roughly classify them into three major mechanisms: diffusion, erosion, and swelling [20–23]. Besides these three major instruments that we have in our hands, a plethora of additional strategies have been used [24–27]. Among them are electric fields [28–30] that allow for the transport of charged molecules, osmosis whereby molecules are "squeezed" out of a reservoir due to a build up in osmotic pressure [31–33], convection by which the drug is moved along some stream of gas or fluid [34,35], or even the mechanical stimulation of a drug carrier [36,37]. This list of examples is not nearly complete, but it is beyond the scope of this chapter to describe all of them. As excellent reviews on the topic can be found in the contemporary literature [38–42], we only further describe the most important mechanisms with respect to applications in tissue engineering.

13.2.1 Diffusion

Diffusion is one of the most important transport mechanisms in drug-delivery applications. It is a thermodynamically driven process with well understood kinetics [43,44]. Diffusion is caused by Brownian motion, which is a random walk of particles that are typically micrometer-sized or smaller. Macroscopically, we observe that the particles appear to move along a concentration gradient. Fick's second law of diffusion describes the net transport of particles in time and space:

$$\frac{\partial C(x, y, z, t)}{\partial t} = \frac{\partial^2 C}{\partial x^2} + \frac{\partial^2 C}{\partial y^2} + \frac{\partial^2 C}{\partial z^2}$$

For this partial differential equation, numerous algebraic and numerical solutions have been derived, two of which are depicted in Figure 13.1. The solutions depend strongly on the initial conditions (such as the drug concentration inside a drug-delivery device) as well as the boundary conditions (such as the geometry of a drug-delivery device) of the diffusion problem. An advantage of diffusion controlled drug-delivery systems is certainly that we can usually predict the kinetics of drug release very well. One of its major disadvantages is that the flux (the mass transport per unit area and time) out of a device is usually, according to above outlined nature of the process, concentration dependent. Therefore, as the release progresses, the process can lead to a break down of concentration gradients and, as a result, the

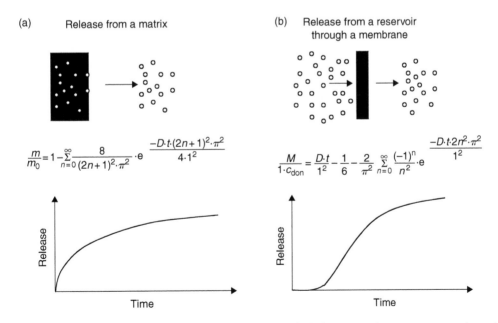

FIGURE 13.1 Drug release profiles resulting from diffusion controlled release. Equations represent exemplery solutions of Fick's second law of diffusion in one dimension describing release profiles (a) from a matrix type device, (b) via diffusion through a membrane.

release rates may decline. This is usually the case whenever we release a substance from a monolithic system (Figure 13.1a).

A method that allows the system to maintain a fairly constant flux over an extended period of time utilizes release through a membrane. In that case, an almost constant release over time is possible, especially when the drug reservoir contains large amounts of substance (Figure 13.1b). This is the more the case when the drug reservoir hosting the substance is oversaturated, so that a constant drug concentration can be maintained over an extended period of time.

13.2.2 Erosion

The materials and design of a drug carrier may be chosen to either encourage or hinder release via erosion. A carrier that erodes over time either precludes the necessity of postapplication removal or does not persist at the application site, which is an additional advantage for applications in tissue engineering.

Erosion is usually defined as the sum of all processes leading to a mass loss from a drug-delivery device [45]. In some cases, the material can simply dissolve in an aqueous environment, while in other cases, like with lipophilic degradable polymers, the material degrades into water soluble oligomers that allow for the initiation of matrix erosion. Polymers that are insoluble in water have been classified according to their erosion mechanism into surface-eroding and bulk-eroding materials [46,47]. While in the case of the first class, erosion phenomena are confined to the material surface [48], bulk-eroding materials tend to degrade over extended periods of time until a critical degree of degradation is reached, after which the whole material bulk is undergoing erosion (Figure 13.2) [49].

It is obvious that surface-eroding polymers allow for the control of drug release via the erosion process, which usually proceeds at constant velocity. A classification of polymers as surface and bulk eroding can be made based mainly on the nature of the bond between the monomers. As a rule of thumb, one can assume that the faster a bond is cleaved (usually by hydrolysis) the more the material is likely to undergo surface erosion. In recent years, models have been developed that allow for the calculation of a dimensionless "erosion number" that predicts the erosion mechanism of a polymer device [50].

13.2.3 Swelling

The swelling phenomena of polymers has widely been used to control the release of drugs from a device [46,51]. The major mechanism thereby is an increase of polymer chain mobility due to the uptake of water, which lowers the glass transition temperature of the polymer [52,53]. Drugs that are trapped inside a polymer matrix profit from the resulting increase in flexibility by an enhanced diffusivity and consequently an enhanced release rate [54]. In many cases, one can observe the formation of swelling fronts that separate a glassy polymer matrix core from the swollen polymer surface. It is obvious that diffusion plays a major overall role, as both the process of water uptake and drug release are diffusion controlled. Excellent reviews are available that provide more detail [55–57].

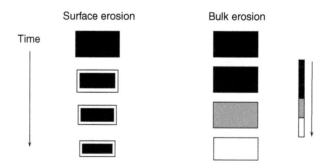

FIGURE 13.2 Schematic illustration of surface erosion and bulk erosion. (Reprinted from Gopferich, A. and Tessmar, J., *Adv. Drug Deliv. Rev.*, 54, 911, 2002.)

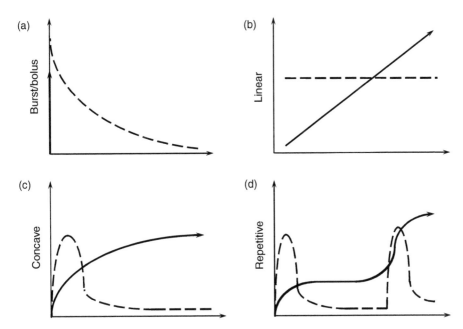

FIGURE 13.3 Schematic illustration of release profiles (closed lines) and resulting tissue concentrations (broken line). (a) Pulsatile release, (b) constant (zero order) release, (c) concave release kinetics, (d) repeated pulsatile release kinetics.

13.2.4 Competing Mechanisms and Overall Kinetics

In many drug-delivery systems, the mechanisms outlined above compete with one another. Release from a degradable polymer, for example, is usually a complex mixture of all three pathways of drug-release control. However, from a design point of view, it is highly desirable that one of the mechanisms dominates the overall release kinetics, as otherwise unfavorable release profiles may result.

A plethora of release profiles are available to choose from, ranging from the continuous delivery of a drug over an extended period of time to pulsatile release kinetics, by which an incorporated compound is set free according to a preprogrammed pattern [58]. Excellent overviews on the multitude of release kinetics can be found in the literature [59]. Figure 13.3 illustrates schematically typical drug-release profiles and the resulting tissue concentrations that one may expect, assuming that the substance is cleared from the application site according to first order kinetics, that is, in a concentration dependent way.

The kinetics that one should choose depends on the biological needs of the specific application. In tissue engineering, this may for example be defined by a desired cell phenotype or by the pharmacokinetics of the drug when applied to a tissue or administered *in vivo*. Typically, more continuous release profiles will lead to more constant tissue levels *in vivo*, while the pulsatile application of a drug will result in concentration peaks that rapidly fall.

13.3 Protein Drug Properties

Prior to a detailed look at the various drug-delivery strategies, it seems necessary to have a glance at the properties of active compounds that are currently in use for tissue engineering applications. Most of these drugs are growth factors and, therefore, peptides or proteins. A brief list of compounds that have been formulated to drug-delivery systems for the use in tissue engineering applications is given in Table 13.1.

TABLE 13.1 Growth Factor and Delivery Systems That Have Been Developed

Growth factor	Carrier/delivery system	Regenerated tissue	Ref.
BMP	PLA	Long bone	[60]
	Porous HA	Skull bone	[61]
rhBMP-2	Porous PLA	Spinal bone	[62]
		Skull bone	[63]
	Collagen sponge	Skull bone	[64]
	Gelatin	Skull bone	[65]
	PLA-coating	Long bone	[66]
	Porous HA	Skull bone	[67,68]
	PLA–PEG copolymer	Long bone	[69]
	Si–Ca–P xerogels		[70]
rhBMP-7/ OP-1	Collagen	Spinal bone	[71,72]
		Long bone	[73]
	HA	Spinal bone	[74]
aFGF	PVA	Angiogenesis	[75]
	Alginate	Angiogenesis	[76]
bFGF	Alginate/heparin	Angiogenesis	[77]
	Agarose/heparin	Angiogenesis	[78]
	Gelatin	Skull bone	[79]
	Fibrin gel	Long bone	[80]
	Chitosan	Dermis	[81]
	Collagen	Cartilage	[82]
EGF	Gelatin	Dermis	[83]
TGF-β1	Alginate	Cartilage	[84]
	Poloxamer; PEO gel	Dermis	[85]
	PLGA	Skull bone	[86]
	Collagen	Dermis	[87]
PDGF	Fibrin	Ligament	[88]
	CMC gel	Dermis	[89]
VEGF	Alginate	Angiogenesis	[90]
	Collagen	Angiogenesis	[91]
	PLGA scaffold	Angiogenesis	[92]
	Fibrin mesh	Angiogenesis	[93]
VEGF/PDGF	PLGA scaffold	Angiogenesis	[94]
NGF	PLGA	Nerve	[95]
	Collagen minipellet	Nerve	[96]

BMP, bone morphogenic protein; rhBMP; recombinant human bone morphogenic protein; OP-1, osteogenic protein 1; aFGF, acidic fibroblast growth factor; bFGF, basic fibroblast growth factor; EGF, endothelial growth factor; TGF-β1, transforming growth factor beta 1; PDGF, platelet derived growth factor; VEGF, vascular endothelial growth factor; NGF, nerve growth factor; PLA, poly lactic acid; PLGA, poly(lactic-co-glycolic) acid copolymer; PEG, poly(ethylene glycol); HA, hydroxyapatite; PVA, poly(vinyl alcohol).

In the case of these proteins we have to deal with highly efficient substances with often substantially limited stability, such as in physiological fluids, due to proteolytic enzyme activity [97]. The choice of an appropriate delivery system for these substances is, therefore, not only a matter of the intended therapeutic use or the pharmacokinetics but also of the physicochemical properties of the drug. Special attention must be paid to the chemical stability and physical integrity of a protein drug even during the formulation process. Examples for chemical and physical instabilities are deamidation, redox reactions, hydrolysis, and aggregation. These and other instability mechanisms have been extensively reviewed by numerous authors [98–100]. In addition environmental effects like pH, organic acids, metal ions, detergents, and temperature can induce a denaturation of proteins [101].

Another source of inactivation is the biological environment that we expose a protein to. Even if a growth factor of choice can be stabilized during formulation and inside a delivery system the biological environment can pose another severe threat. Half-lives of growth factors are typically in the range of minutes due to degradation by peptidases. Intravenous injections are indicative of this problem [102]. Platelet-derived growth factor (PDGF), bFGF, and VEGF, for example, have plasma half-lives of 2 [103], 3 [104], and 50 [105] minutes respectively. Besides rapid degradation, unfavorable distribution [104] or elimination by glomerular filtration [97] are problematic after intravenous administration. Direct systemic administration of growth factors, therefore, requires substantial drug doses that may lead to side effects [2,106], or if direct injection into the target site is possible, repeated administration to achieve biological efficacy is required. Prolonging the half-life of growth factors is, therefore, an important research goal. A promising approach to increase half-life is to conjugate proteins with poly(ethyleneglycol) (PEG) chains. Pegylation seems to cause a significant reduction of protein clearance from plasma [107–109]. The pegylation of growth hormone-releasing hormone (GRF) analogues, for example, led to an enhanced *in vivo* stability with acceptable biological activity [110]. In addition pegylation seems to reduce the velocity of enzymatic degradation [111]. Detailed reviews on protein and peptide pegylation can be found in the literature [112–115].

The incorporation of proteins into a controlled release drug-delivery device can protect it from inactivation, but the formulation process is sometimes another source of protein inactivation. One class of problem that occurs frequently is related to the interaction of proteins with interfaces. Considering the small quantities that have to be handled when growth factors are formulated to applicable systems, physical phenomena like protein adsorption to solid–liquid or liquid–liquid interfaces and subsequent denaturation can lead to massive protein loss in some instances [116–118]. Approaches to minimize protein adsorption to glass or plastic include competitive "coating" with bovine serum albumin [119,120], the use of PEG and glycerol solutions or the application of surfactants [121,122]. Enhanced denaturation and aggregation of proteins at interfaces have been observed under certain conditions like the application of mechanical forces [123]. It has, for example, been shown that simple vortex mixing of a 0.5 mg/ml porcine growth hormone solution for 1 min can lead to a loss of 70% of the protein due to aggregation [124,125]. Unfortunately the problem of adsorption on solid surfaces is not the only one that can occur during formulation. When microparticles are prepared by using emulsion techniques [126], for example, there is frequently the need for the use of organic solvent and vigorous mixing. The liquid–liquid interfaces and the high shear forces are two factors that may lead to denaturation and hence activity loss [125,127,128]. Moreover during the degradation of polymers a very harsh microclimate can be created [129] in which even chemical reactions such as an acylation of proteins and peptides can occur [130,131]. To stabilize proteins in polymers numerous approaches have been tried [132–135], many of them being very specific for an individual drug-delivery system.

The unfavorable interactions of growth factors with synthetic materials made many groups have a closer look at how they are created and stored in their natural biological environment. Transcription of genes encoding for growth factors create unstable mRNA [136–138], which leads to short half-life and only limited growth factor synthesis. Growth factors are released by cells either for instant signalling or are stored in the extracellular matrix (ECM). Some growth factors such as transforming growth factor beta (TGF-β) are secreted as biologically inactive latent precursors [139]. The release of growth factors from the ECM is controlled by matrix degradation. This natural environment provides a dynamic and responsive growth factor delivery system which can react according to specific cellular requirements and can, therefore, also serve as a cell guidance system [2,139,140]. This led various research groups to the development and use of ECM [141] derived materials for the design of controlled delivery systems for growth factors. Especially gel like systems can simulate the interaction of growth factors and the extracellular matrix [142,143] and can, therefore, provide similar stabilizing effects like the natural ECM. Interestingly, it was found that the bioactivity of vascular endothelial growth factor (VEGF) delivered from alginate microspheres was higher than VEGF alone [144]. It was speculated that that effect was due to stabilization of the factor in the system over alginate interaction. A detailed review about on the interactions between growth factors and the ECM can be found in the literature [145].

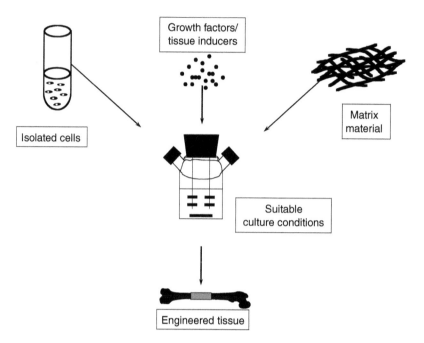

FIGURE 13.4   Schematic illustration of components of tissue engineering.

13.4   Drug Delivery in Tissue Engineering

It is certainly beyond the scope of this chapter to provide the reader with a complete overview of tissue engineering; however, it is necessary that we review several aspects that may be crucial (or beneficial) for a better understanding of drug-delivery applications in this field.

In the early days of tissue engineering, pioneers like Langer and Vacanti stressed its interdisciplinary character in their definitions [1]. At that time four fundamental components of tissue engineering were identified (Figure 13.4):

1. Cells that will form the desired type of tissue
2. A matrix that allows for cell proliferation and differentiation
3. Suitable cell culture conditions
4. Finally the use of cytokines and other drugs

While each of these components alone may be sufficient for an application, it is obvious that for *in vivo* as well as *in vitro* approaches it is important to guide cell- and tissue development into the right direction. Cytokines and growth factors have therefore emerged in recent years as precious substances for tissue engineering applications. That a controlled delivery of these and other pharmacologically active substances is very beneficial for tissue engineering applications is well accepted in the field.

The strategies to deliver a drug to cells or tissues are diverse. However, we can classify them into four major categories, which follow immediately from the blue print of tissue engineering as depicted in Figure 13.4.

A first strategy, which is all too obvious, is to use classical established technology for the delivery of substances either in cell culture media or near a tissue site *in vivo* (Figure 13.5a).

Doing so has the advantage that we can use systems that have been developed and described thoroughly in the contemporary literature [38,146–148]. Among these systems, we find essentially everything that has been developed for the parenteral application of drugs, such as implants, microspheres, and injectable gels, just to name a few. Despite their advantages, they may also cause some problems. They must always be administered in addition to tissue engineering specific components, such as cells and cell carriers.

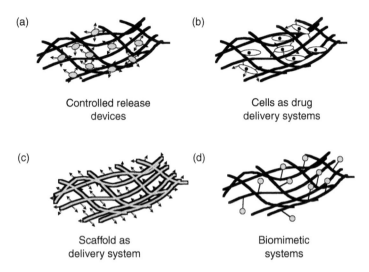

(a)

Controlled release
devices

(b)

Cells as drug
delivery systems

(c)

Scaffold as
delivery system

(d)

Biomimetic
systems

FIGURE 13.5 Drug-delivery strategies for tissue engineering applications: (a) Use of "classical" controlled delivery devices such as microspheres or monolithic systems incorporated into a cell carrier, (b) genetically altered cells serving for protein and peptide drug expression, (c) the cell carrier itself serving as a drug-delivery system, (d) covalent attachment of drugs to the polymer surface.

Moreover, they may not unequivocally allow for a homogeneous distribution in the tissue and may thereby cause undesirable drug concentration gradients.

A second approach is to deliver drugs via cells (Figure 13.5b) [149]. It is possible to genetically alter a portion of the cells or all of them to produce a certain protein or peptide drug. For this approach, numerous viral and nonviral gene delivery approaches are available [150–154].

A third approach is to use the scaffolds intended for cell attachment and proliferation as a drug-delivery system (Figure 13.5c). Many of these materials provide tremendous opportunities to control the release of drugs. Especially biodegradable polymers and hydrogel materials have been used extensively as porous scaffolds to allow for erosion controlled drug or DNA release.

Finally, we can implement the pharmacologically active substances that we want to use for signaling to cells into the design of materials that are used for cell proliferation and differentiation. In this scheme, drugs and cytokines are tethered to the surface of a material [155] (Figure 13.5d). The term "biomimetic" has been coined in recent years to describe materials that have been modified in this way [156,157]. There are numerous publications that illustrate the principle behind and advantages of this strategy [158–162].

In the following pages, we focus on aspects of classical drug-delivery systems and on the use of scaffolds as release systems, as the other two delivery strategies are covered in the sections on gene delivery and biomimetic materials of the handbook. We try to give a brief overview of the vast number of devices that have already been used for the delivery of drugs in tissue engineering applications and some of the more recent trends, such as the use of scaffolds concomitantly as a cell carrier and for the delivery of a drug.

13.4.1 The Use of Classical Drug-Delivery Systems

When designing a drug-delivery device for an application in tissue engineering, there are numerous aspects to be considered beyond the pharmacological aspects that are usually linked to the nature of the drug to be administered. Of utmost significance is the pharmacokinetics of the drug. The clearance rate from the site of application, in particular, should be determined as this, together with the drug efficacy and the intended time of application, dictates the dosing regimen. Once the required dosage is known, the corresponding release kinetics may be determined. The type of drug-delivery system to be used may then be chosen, taking the drug loading and release into account. Thereafter, further fundamental questions, such as, which material provides a maximum of drug stability, biocompatibility, and functionality with respect to a desired controlled-release pattern or with respect to biodegradability, can be addressed. Once these questions are answered, more individual aspects such as shaping the material to defect sites or to a specific type of application such as via injection may be considered.

Unfortunately, the information needed to address all of the problems listed above is often unavailable. In particular, there is usually little known about newly formulated compounds besides the pharmacological effect of the substance. At that stage of development, one goal of drug delivery is frequently to provide release systems that are capable of stabilizing and releasing the drug over an as long a period of time as possible. Controlled release technology can then help to find out if long-term applications are beneficial and what the reaction of a tissue to such an exposure is. In the case of proteins, this is usually not a trivial task. In the following section, we try to illustrate the options that we currently have to deliver drugs for tissue engineering applications. The reader should be aware that this is just a brief glance at the field that hosts a plethora of systems and technologies. To allow for better orientation, we classify the field roughly into:

- Monolithic systems of macroscopic geometries
- Particulate systems comprising microparticles and colloids with a size range from 1 to 1000 μm and from 1 to 1000 nm respectively
- Gel-like systems

13.4.1.1 Monolithic Systems

The foundations for the controlled release of bioactive substances especially of protein and peptide drugs from monolithic systems were laid in the 1970s when Folkman and Langer described the use of polymeric membrane systems made of poly(ethylene-co-vinyl acetate) (EVAc) for the release of bovine serum albumin, which served as a model compound. These delivery systems were among the first that achieved a controlled-release pattern for protein drugs [163]. Although these systems allowed for an excellent control of drug release [164,165], one of their major limitations was that they were not degradable. Since the 1980s a number of degradable polymer materials have been synthesized with the goal of enhancing the *in vivo* performance of devices made thereof. A plethora of polymer classes, such as poly(a-hydroxyesters) [42], polyanhydrides [166], and poly(orthoesters) [167], have been developed for this purpose, to mention just a few. More materials are described in the literature [46,168,169]. Materials that undergo biodegradation are typically significantly more complex than nondegradable materials, as they add several new variables to the already intricate system. In poly(lactic acid) (PLA) and poly(lactic-co-glycolic acid) (PLGA) matrices, for example, the degradation of the material can cause the pH of the surrounding region to drop below 2 [170], the osmotic pressure inside aqueous pores to increase far above physiological values [171], and chemical reactions between degradation products and peptides have been reported as well [172].

The first delivery systems that were made of these materials were monolithic matrices because they are easy to manufacture by techniques such as solvent casting, extrusion, or injection molding and usually provide excellent control over drug release. Furthermore, these matrices are able to carry large doses of drug and the release can be tailored to a required direction by changing the matrix material, drug loading, or use of co-dispersants [164]. The release of drug in nondegradable systems is primarily controlled by diffusion processes. The degradable systems, in contrast, offer the possibility of erosion-controlled release. Thus, there are more variables that may be adjusted in degradable systems, leading to more control over the release rate. In PLA/PLGA, for example, the degradation rate is influenced by factors such as crystallinity, copolymer composition [173,174], and molecular weight [175].

The materials that have been used for the manufacture of monolithic matrix-type delivery systems for tissue engineering have mainly been made of cross-linked hydrogels. Lipophilic degradable materials, such as PLA or PLGA, would also be suitable, however, over long time periods they are sometimes detrimental to the stability of a protein. Thus, gelatin has been used extensively for the manufacture of matrices by cross-linking gels either with glutaraldehyde or water-soluble carbodiimide [176]. Cross-linking can also be used with other hydrogel materials, such as alginate [177] or collagen. The latter has been used extensively for the manufacture of drug-delivery systems. Fujioka et al. extensively reviewed collagen-based systems that were developed for the delivery of cytokines and growth factors [178].

13.4.1.1.1 Particulate Systems

A major disadvantage of monolithic systems is the problem of application. A surgical procedure is usually needed for implantation. In the last 20 years, micro- and nanoparticulate systems have been developed as an alternative to facilitate the application of substances in a minimally invasive way. An excellent overview of the use of biodegradable microspheres has been given in recent reviews [179,180]. An advantage of the use of microparticles in tissue engineering is that they can be evenly distributed either in a tissue by injection or in a cell suspension, which guarantees that drug gradients, which often result from the application of monolithic systems, are avoided. An additional advantage for tissue engineering is the use of microencapsulation technology for the encapsulation of cells [181,182]. In addition to microparticles, nanoparticles and liposomes have found their way into tissue engineering applications [183,184], but we will focus on microparticulate systems in the next section because the have been applied more frequently.

Particulate systems have been used intensively in the past to deliver growth factors with the intention to regenerate tissue. Thus, Haller et al. reviewed efforts to deliver nerve growth factors [185]. The authors conclude that polymeric NGF release systems are useful for supporting cellular therapies for the treatment of neurodegenerative diseases. They found that microparticles made of EVAc and PLGA are suited for the release of nerve growth factor (NGF) [186]. Lipophilic degradable polymers have been especially popular for the delivery of protein drugs [179] and a multitude of further applications can be found in the contemporary literature, in many cases leading to very successful drug-delivery applications [179,180,187–189]. Despite the advantages that lipophilic degradable polymers such as PLA and PLGA and others offer, however, they are problematic materials for the reasons outlined above. De Weert et al. [190] summarized the problems with a special emphasis on the stability of protein drugs in an excellent review, in which they identify problems related to protein loading, microsphere formation, and drying. Despite the numerous countermeasures used to stabilize protein in these polymers, mounting problems led to tremendous research efforts to develop alternative systems.

Hydrogels, in particular, have emerged as a class of material that hold great promise for the release of sensitive protein and peptide drugs from microsphere systems. Among the first systems that were used are alginate beads, which can be manufactured by dropping or spraying alginate solutions into solutions containing divalent cations, such as calcium [191]. The ease of manufacture allows for a simple procedure to load growth factors. Gu et al. [192] could show, for example, that VEGF can be released at a fairly constant rate over a period of two weeks. Collagen and gelatin emerged as promising materials, as they can be considered biocompatible and biodegradable. Usually these materials are first processed in a gel state to microparticles, are then cross-linked to stabilize their geometry and finally loaded with proteins by immersion into a protein solution [193]. Tabata et al. [194] used this technology to deliver bFGF for de novo adipogenesis following subcutaneous microparticle injection into mice.

A problem that can exist with hydrogel-based microparticles is that they frequently require cross-linking procedures that can be detrimental to the stability of protein drugs. Therefore, alternative materials, such as lipids, have been intensively studied in the past to encapsulate protein and peptide drugs as well [195]. Like lipophilic degradable polymer microspheres, these substances offer the advantage that they can release drugs over an extended period of time [196]. Cortesi et al. [197] describe the manufacture of lipid microspheres from triglycerides and monoglycerides using emulsion and melt dispersion techniques. Reithmeier et al. achieved the manufacture of lipid microspheres that are able to release a tripeptide over a period of a few days [198] and insulin over a period of several days [199].

13.4.1.1.2 Gel-Like Systems

Like particulate systems, the ease of application of gel-like systems provides a significant benefit. Many systems have already been developed and have become precious materials for tissue engineering applications [200]. Gels offer the advantage of a fairly good compatibility with sensitive protein and peptide drugs. A disadvantage of these systems can be that they allow for drug release only over a short period of time.

Numerous systems with outstanding properties have been developed in recent years. In many cases, hydrogels have served as drug-delivery carriers for the delivery of growth factors. Cross-linked gelatin, for example, has served as a carrier for BMP-2 [201] and TGF-β1 [202], while cross-linked amylopectin was used to release bFGF [203]. Zisch et al. [204] reviewed the use of hydrogels for the release of angiogenic factors in more detail. A closer look at the literature reveals that more and more sophisticated systems have been developed in recent years. Kim et al., for example, has described a block copolymer that consists of poly(ethylene oxide), poly(L-lactic acid), and a urethane unit that exhibited a sol–gel transition at 37°C and was able to release dextrans that served as high molecular weight model compounds over a period of several days [205]. Similar principles can be used in tissue engineering applications to lift cells or tissues off of culture surfaces by decreasing the temperature below the lower critical solution temperature of the system at which the solidified gel to which the tissue is attached becomes liquid, leading to the detachment of cells from the culture plate [206]. Zisch et al. [207] describe the synthesis of a hydrogel network that consists essentially of branched PEG molecules that are linked by peptides; this material was then used as a substrate for matrix metalloproteinases. VEGF, which was covalently attached to this network, is liberated when tissue grows into the hydrogel, supporting tissue vascularization. Other systems take advantage of the binding of growth factors to heparin, which allows for the development of delivery systems in which heparin regulates the release of the protein by slowly releasing it from the complex [208].

13.4.2 The Delivery of Drugs via Cell Carriers

While the examples given above depend on the use of controlled-delivery devices in addition to a cell carrier, the carrier itself can serve as a delivery system. This can happen in two ways: we can either use a drug-delivery system, such as microspheres, and process it together with another material into a scaffold, or we can incorporate the drug directly into it. The use of multicomponent systems has the advantage that we can rely on off-the-shelf technology for the stabilization of a drug and for the control of its release. Doing so has, however, the disadvantage of being more complicated than loading a matrix directly with a drug.

13.4.2.1 Cell Carriers Loaded with Drug-Delivery systems

Holland et al. proposed the use of gelatin microspheres that have previously been developed for the delivery of proteins, such as bFGF [143,209], for the controlled release of transforming growth factor-β1 (TGF-β1) from oligo(poly(ethylene glycol) fumarate) (OPF) [210,211]. The authors took advantage of the fact that the polymer can be dissolved in water and suspend the microspheres therein. Upon chemical cross-linking, the particles were trapped inside the hydrogel and released the factor over a period of several weeks *in vitro*. The system can thus be used as an injectable scaffold material that also delivers a growth factor. Concomitantly, while gelatin degrades over time, this property allows cells to migrate into the biomaterial. Alternatively, such materials might be loaded with cells prior to cross-linking. Hollinger et al. [212] report a technique that allowed them to incorporate proteins into PLGA microparticles and to incorporate them into PLGA scaffolds.

13.4.2.2 Cell Carriers Loaded with Drugs

While the opportunities to incorporate a complete release system into a cell carrier are limited by the nature of the complexity of the endeavor, there are many options to incorporate a drug directly. Figure 13.6 gives an overview of the possibilities that we can choose from.

The choice of drug-loading method depends on the circumstances, such as the stability of the drug, the dose that we need, and the desired release kinetics, to mention just a few. In this section we will try to give a few examples that shed some light on the advantages and disadvantages of the individual approaches.

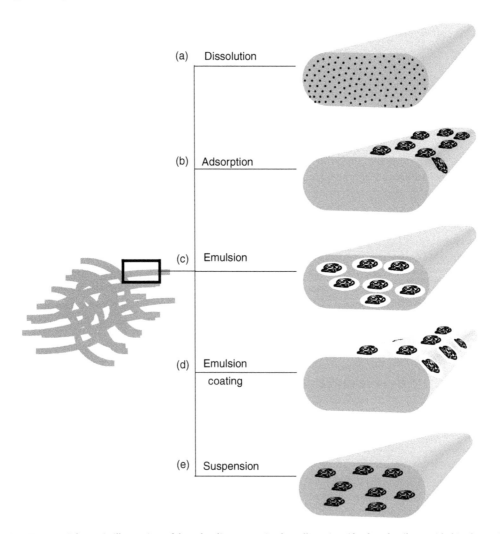

FIGURE 13.6 Schematic illustration of drug-loading strategies for cell carriers (further details provided in the text). (a) Dissolution of drug in the matrix material, (b) adsorption of drug to the matrix surface, (c) incorporation via emulsion droplets, (d) coating with drug–matrix emulsions, (e) suspension of drug inside the matrix.

13.4.2.2.1 Dissolution of Drug

In the simplest case, we can simply dissolve the drug in the bulk material prior to processing. This requires, however, that we find a solvent that dissolves both the matrix material and the drug. Kim et al. [213] describe the incorporation of ascorbate-2-phophate (AsAP) and dexamethasone into poly(D,L-lactic acid) (PLGA). Both drugs have been used to enhance the differentiation of mesenchymal stem cells to an osteoblastic phenotype. While dexamethasone dissolved in a PLGA, AsAP formed a suspension. The whole dispersion was then processed via solvent casting particulate leaching techniques. Both substances were released over a period of several weeks and enhanced the mineralization of the cells, which can be regarded as a differentiation marker.

Whenever we use hydrogels we have a fair chance that we will be able to dissolve them together with the drug in an aqueous solution. This is especially attractive when we intend to deliver a protein drug, which usually has very good stability in an aqueous environment. Lee et al. [141] used cross-linked alginate gels to control the release of VEGF. They could show that the mechanical stimulation of matrices lead to a significant increase in the release velocity.

13.4.2.2.2 Adsorption

Another simple strategy to "load" a scaffold with a drug is by simply adsorbing it to the material surface or incorporating it into the material bulk. The release of the drug is then controlled by the desorption kinetics, which are controlled by the drug/material interactions as well as transport mechanisms, such as diffusion. While polymers offer ample alternative options for the incorporation of drugs into a tissue engineering scaffold, inorganic materials, in particular, can typically only be loaded by adsorption. In one case, Ziegler et al. [214] adsorbed bone morphogenic protein 4 (BMP-4) and basic fibroblast growth factor (FGF) to a variety of materials, such as porous tricalcium phosphate ceramic and a neutralized glass ceramic (GB9N) as well as a composite of the latter with PLGA. They found that the adsorption behavior depends on the nature of both the material and the protein and that a phase of fast initial release during the first hours is followed by a phase of slow release from the investigated carriers.

Similar strategies can be used for loading polymer hydrogels after cross-linking as described earlier. These systems are usually capable of taking up drugs from solution by diffusion into the cross-linked gel. This has successfully been demonstrated by Tabata [215], who loaded cross-linked gelatin microspheres with bFGF. He could show that the release of growth factor takes place over a span of days and that the released drug is still bioactive. Ueda et al. [216] later transferred this strategy to the manufacture of porous collagen sponges. It could be shown in a cranial defect model that the growth factor was released in active form and that bone repair was enhanced significantly. It is interesting to note that the release of the proteins from such systems is not only a matter of diffusion, but also dependent on the electrostatic interactions between the charge protein carrier and the protein drug.

13.4.2.2.3 Emulsion Techniques

Whang et al. [217] describe a method that allows for the incorporation of a drug into lipophilic biodegradable polymer scaffolds made of poly(D,L-lactic-co-glycolic acid) (PLGA). This technique, which was originally developed for the incorporation of hydrophilic drugs into lipophilic microspheres, relies on the formation of a W/O emulsion. While the drug is dissolved in the aqueous phase, the polymer is dissolved in organic solvents such as methylene chloride. Porous polymer scaffolds were manufactured from the liquid formulation by lyophilization. The technology was successfully applied to the controlled release of rhBMP2 [218]. Another approach developed by Jang and Shea [219] uses a multiple emulsion technique to manufacture PLA microspheres loaded with DNA. The particles that were made via a W/O/W technique [220] were fused to a porous scaffold by mixing with sodium chloride crystals compressed into discs by incubation at 37°C and 95% relative humidity. After drying in a pressurized CO_2 atmosphere, the particles were leached in water. The scaffolds released the DNA over a period of more than 3 weeks.

An approach related to the use of emulsions as a raw material for scaffolding is the use of emulsions for coating. Sohier et al. [221] used a W/O emulsion of poly(ethylene glycol) terephthalate/poly(butylene terephthalate) multiblock copolymer loaded with lysozyme as a model protein to coat porous scaffolds made of the same material. They could release the protein over several days and were able to show that the activity of lysozyme was maintained. While the use of emulsions has the advantage of creating an even distribution of drugs inside the material, in some cases, sensitive drugs may not be stable in such an emulsion system [222–224].

13.4.2.2.4 Suspension of the Drug and Physical Mixtures

Drug suspensions and mixtures can be especially useful for the manufacture of a cell carrier. While the matrix material is dissolved in a solvent, the drug obviously needs to be insoluble in this mixture. This is especially attractive for the incorporation of proteins into lipophilic polymers, because proteins exhibit an extraordinary stability against organic solvents when they are in the solid state [225].

Physical mixtures have also been of interest to scientists working in tissue engineering. In this case, the matrix material and the drug are mixed in the solid state and subsequently processed into porous cell carriers. This process has been used by Shea et al. [226] to load PLGA scaffolds with plasmid DNA. For their studies, they used PLGA granules and suspended them in an aqueous solution of plasmid DNA at pH 7.4 in 10 M HEPES buffer containing mannitol, which was used as a lyophilization constituent.

After freeze drying, the mixture was compressed with NaCl particles into discs, the polymer particles fused using pressurized CO_2 and the porogen finally removed by leaching in water. The porous scaffolds allowed for the delivery of DNA over several weeks and showed excellent transfection efficiency.

13.5   Outlook

The field of tissue engineering has profited significantly from controlled release research and technology. Numerous growth factors and cytokines have been successfully applied via controlled drug-delivery devices in recent years. Despite this progress, there are numerous challenges ahead of us. First of all, we have to develop release systems that are on the one hand very flexible with respect to controlled drug release and that are on the other hand able to stabilize sensitive drugs, such as proteins and peptides.

Furthermore, more research will be needed on delivery systems that target intracellular compartments, cells, and tissues. This would allow for the delivery of DNA as well as drugs more specifically and safely.

We must also take better advantage of progress in medicinal chemistry, which will provide us with low molecular weight ligands that have a high affinity for well-defined targets. So far our efforts in tissue engineering research have profited only minimally from the opportunities that low molecular weight drugs offer.

Currently, release systems suffer from an inability to adequately mimic biology, in that typically only one drug is delivered. Guided by an increasingly better understanding of cell biology, drug-delivery systems that release several drugs according to a well-defined time pattern hold great promise.

Continuing progress will depend significantly on the availability of new materials for the development of drug-delivery systems. Biomimetic materials or materials that respond to biological stimuli are good examples. We have to overcome the problem that the rate-limiting step for the bench to beside process is not solely determined by our scientific progress, but also by tedious regulatory hurdles that new materials have to overcome.

References

[1] Langer, R. and Vacanti, J.P., Tissue engineering, *Science*, 260, 920, 1993.
[2] Chen, R.R. and Mooney, D.J., Polymeric growth factor delivery strategies for tissue engineering, *Pharm. Res.*, 20, 1103, 2003.
[3] Tabata, Y., Tissue regeneration based on growth factor release, *Tissue Eng.*, 9, S5, 2003.
[4] Boontheekul, T. and Mooney, D.J., Protein-based signaling systems in tissue engineering, *Curr. Opin. Biotechnol.*, 14, 559, 2003.
[5] Nomi, M. et al., Principals of neovascularization for tissue engineering, *Mol. Aspects Med.*, 23, 463.
[6] Nakashima, M. and Reddi, A.H., The application of bone morphogenetic proteins to dental tissue engineering, *Nat. Biotechnol.*, 21, 1025, 2003.
[7] Baldwin, S.P. and Mark Saltzman, W., Materials for protein delivery in tissue engineering, *Adv. Drug Deliv. Rev.*, 33, 71, 1998.
[8] Rose, F.R.A.J., Hou, Q., and Oreffo, R.O.C., Delivery systems for bone growth factors — the new players in skeletal regeneration, *J. Pharm. Pharmacol.*, 56, 415, 2004.
[9] Langer, R. and Peppas, N.A., Advances in biomaterials, drug delivery, and bionanotechnology, *AIChE J.*, 49, 2990, 2003.
[10] Crommelin, D.J.A. et al., Shifting paradigms: biopharmaceuticals versus low molecular weight drugs, *Int. J. Pharm.*, 266, 3, 2003.
[11] Gittens, S.A. and Uludag, H., Growth factor delivery for bone tissue engineering, *J. Drug Target.*, 9, 407, 2001.
[12] LaVan, D.A., McGuire, T., and Langer, R., Small-scale systems for *in vivo* drug delivery, *Nat. Biotechnol.*, 21, 1184, 2003.
[13] Definition drug delivery, *www.dictionarybarn.com*, 2004.

[14] Definition drug delivery systems, *www.dictionarybarn.com*, 2004.

[15] You, H.B. and Sung, W.K., Drug delivery, in *Frontiers in Tissue Engineering*, Patrick, C.W., Mikos, A.G., and McIntire, L.V., Eds., Pergamon, 261–277, 1998.

[16] Schwartz, R.S., Perspective: Paul Ehrlich's magic bullets, *N. Engl. J. Med.*, 350, 1079, 2004.

[17] Li, W. et al., Lysyl oxidase oxidizes basic fibroblast growth factor and inactivates its mitogenic potential, *J. Cell. Biochem.*, 88, 152, 2002.

[18] Klagsbrun, M. et al., Multiple forms of basic fibroblast growth factor: amino-terminal cleavages by tumor cell- and brain cell-derived acid proteinases, *Proc. Natl Acad. Sci. USA*, 84, 1839, 1987.

[19] Cunningham, M.J., Genomics and proteomics. The new millennium of drug discovery and development, *J. Pharmacol. Toxicol. Meth.*, 44, 291, 2001.

[20] Siepmann, J. and Gopferich, A., Mathematical modeling of bioerodible, polymeric drug delivery systems, *Adv. Drug Deliv. Rev.*, 48, 229, 2001.

[21] Gopferich, A., Mechanisms of polymer degradation and erosion, *Biomaterials*, 17, 103, 1996.

[22] Gopferich, A., Erosion of composite polymer matrices, *Biomaterials*, 18, 397, 1997.

[23] Gopferich, A. and Tessmar, J., Polyanhydride degradation and erosion, *Adv. Drug Deliv. Rev.*, 54, 911, 2002.

[24] Wright, S. and Huang, L., Antibody-directed liposomes as drug-delivery vehicles, *Adv. Drug Deliv. Rev.*, 3, 343, 1989.

[25] Wissing, S.A., Kayser, O., and Muller, R.H., Solid lipid nanoparticles for parenteral drug delivery, *Adv. Drug Deliv. Rev.*, 56, 1257, 2004.

[26] Mitragotri, S. and Kost, J., Low-frequency sonophoresis: a review, *Adv. Drug Deliv. Rev.*, 56, 589, 2004.

[27] Kalia, Y.N. et al., Iontophoretic drug delivery, *Adv. Drug Deliv. Rev.*, 56, 619, 2004.

[28] Sawahata, K. et al., Electrically controlled drug delivery system using polyelectrolyte gels, *J. Control. Release*, 14, 253, 1990.

[29] Murdan, S., Electro-responsive drug delivery from hydrogels, *J. Control. Release*, 92, 1, 2003.

[30] Gangarosa, S. and Hill, J.M., Modern iontophoresis for local drug delivery, *Int. J. Pharm.*, 123, 159, 1995.

[31] Verma, R.K., Krishna, D.M., and Garg, S., Formulation aspects in the development of osmotically controlled oral drug delivery systems, *J. Control. Release*, 79, 7, 2002.

[32] Santus, G. and Baker, R.W., Osmotic drug delivery: a review of the patent literature, *J. Control. Release*, 35, 1, 1995.

[33] Thombre, A.G., Zentner, G.M., and Himmelstein, K.J., Mechanism of water transport in controlled porosity osmotic devices, *J. Membr. Sci.*, 40, 279, 1989.

[34] Hamilton, J.F. et al., Heparin coinfusion during convection-enhanced delivery (CED) increases the distribution of the glial-derived neurotrophic factor (GDNF) ligand family in rat striatum and enhances the pharmacological activity of neurturin, *Exp. Neurol.*, 168, 155, 2001.

[35] Bankiewicz, K.S. et al., Convection-enhanced delivery of AAV vector in Parkinsonian monkeys; *in vivo* detection of gene expression and restoration of dopaminergic function using pro-drug approach, *Exp. Neurol.*, 164, 2, 2000.

[36] Qiu, Y. and Park, K., Environment-sensitive hydrogels for drug delivery, *Adv. Drug Deliv. Rev.*, 53, 321, 2001.

[37] Lee, K.Y., Peters, M.C., and Mooney, D.J., Controlled drug delivery from polymers by mechanical signals, *Adv. Mater. (Weinheim, Germany)*, 13, 837, 2001.

[38] Sah, H. and Chien, Y.W., Rate control in drug delivery and targeting: fundamentals and applications to implantable systems, in *Drug Delivery and Targeting*, Hillary, A.M., Llyod, A.W., and Swarbrick, J., Eds., London, Taylor & Francis Ltd., 83–115, 2001.

[39] Kalia, Y.N. et al., Iontophoretic drug delivery, *Adv. Drug Deliv. Rev.*, 56, 619, 2004.

[40] Huang, X. and Brazel, C.S., On the importance and mechanisms of burst release in matrix-controlled drug delivery systems, *J. Control. Release*, 73, 121, 2001.

[41] Kamath, K.R. and Park, K., Biodegradable hydrogels in drug delivery, *Adv. Drug Deliv. Rev.*, 11, 59, 1993.

[42] West, J.L., Drug delivery: pulsed polymers, *Nat. Mater.*, 2, 709, 2003.

[43] Cussler, E.L., *Diffusion, Mass Transfer in Fluid Systems*, p. 525, 1984.

[44] Gehrke, S.H. and Cussler, E.L., Mass transfer in pH-sensitive hydrogels, *Chem. Eng. Sci.*, 44, 559, 1989.

[45] Zhang, M. et al., Simulation of drug release from biodegradable polymeric microspheres with bulk and surface erosions, *J. Pharm. Sci.*, 92, 2040, 2003.

[46] Gombotz, W.R. and Pettit, D.K., Biodegradable polymers for protein and peptide drug delivery, *Bioconjug. Chem.*, 6, 332, 1995.

[47] Kumar, N. et al., Biodegradation of polyanhydrides, in *Biopolymers*, Matsamura, S. and Steinbuechel, A., Eds., Wiley-VCH Verlag GmbH, Weinheim, 2003, chap 9, 2003.

[48] Heller, J., Controlled drug release from poly(ortho esters) — a surface eroding polymer, *J. Control. Release*, 2, 167, 1985.

[49] Goepferich, A., Polymer bulk erosion, *Macromolecules*, 30, 2598, 1997.

[50] von Burkersroda, F., Schedl, L., and Gopferich, A., Why degradable polymers undergo surface erosion or bulk erosion, *Biomaterials*, 23, 4221, 2002.

[51] Rao, K.V.R. and Devi, K.P., Swelling controlled-release systems: recent developments and applications, *Int. J. Pharm.*, 48, 1, 1988.

[52] Kranz, H. et al., Physicomechanical properties of biodegradable poly(D,L-lactide) and poly(D,L-lactide-co-glycolide) films in the dry and wet states, *J. Pharm. Sci.*, 89, 1558, 2000.

[53] Bouillot, P., Petit, A., and Dellacherie, E., Protein encapsulation in biodegradable amphiphilic microspheres. I. Polymer synthesis and characterization and microsphere elaboration, *J. Appl. Polym. Sci.*, 68, 1695, 1998.

[54] Lee, W.F. and Chen, Y.J., Studies on preparation and swelling properties of the N-isopropylacrylamide/chitosan semi-IPN and IPN hydrogels, *J. Appl. Polym. Sci.*, 82, 2487, 2001.

[55] Colombo, P. et al., Swellable matrixes for controlled drug delivery: gel-layer behavior, mechanisms and optimal performance, *Pharm. Sci. Technol. Today*, 3, 198, 2000.

[56] Peppas, N. and Khare, A.R., Preparation, structure and diffusional behavior of hydrogels in controlled release, *Adv. Drug Deliv. Rev.*, 11, 1, 1993.

[57] Yoshida, T. et al., Newly designed hydrogel with both sensitive thermoresponse and biodegradability, *J. Polym. Sci., Part A: Polym. Chem.*, 41, 779, 2003.

[58] Vogelhuber, W. et al., Programmable biodegradable implants, *J. Control. Release*, 73, 75, 2001.

[59] Fan, L.S.S., *Controlled Release: A Quantitative Treatment*, Springer-Verlag, New York, 1989.

[60] Heckman, J.D. et al., The use of bone morphogenetic protein in the treatment of non-union in a canine model, *J. Bone Joint Surg. Am.*, 73, 750, 1991.

[61] Ono, I. et al., A study on bone induction in hydroxyapatite combined with bone morphogenetic protein, *Plast. Reconstr. Surg.*, 90, 870, 1992.

[62] Muschler, G.F. et al., Evaluation of human bone morphogenetic protein 2 in a canine spinal fusion model, *Clin. Orthop. Relat. Res.*, 308, 229, 1994.

[63] Kenley, R. et al., Osseous regeneration in the rat calvarium using novel delivery systems for recombinant human bone morphogenetic protein-2, *J. Biomed. Mater. Res.*, 28, 1139, 1994.

[64] Zellin, G. and Linde, A., Importance of delivery systems for growth-stimulatory factors in combination with osteopromotive membranes. An experimental study using rhBMP-2 in rat mandibular defects, *J. Biomed. Mater. Res.*, 35, 181, 1997.

[65] Hong, L. et al., Comparison of bone regeneration in a rabbit skull defect by recombinant human BMP-2 incorporated in biodegradable hydrogel and in solution, *J. Biomater. Sci., Polym. Ed.*, 9, 1001, 1998.

[66] Miyamoto, S. et al., Evaluation of polylactic acid homopolymers as carriers for bone morphogenetic protein, *Clin. Orthop. Relat. Res.*, 278, 274, 1992.

[67] Ripamonti, U., Yeates, L., and van den Heever, B., Initiation of heterotopic osteogenesis in primates after chromatographic adsorption of osteogenin, a bone morphogenetic protein, onto porous hydroxyapatite, *Biochem. Biophys. Res. Commun.*, 193, 509, 1993.

[68] Ripamonti, U. et al., Osteogenin, a bone morphogenetic protein, adsorbed on porous hydroxyapatite substrata, induces rapid bone differentiation in calvarial defects of adult primates, *Plast. Reconstr. Surg.*, 90, 382.

[69] Miyamoto, S. et al., Polylactic acid–polyethylene glycol block copolymer. A new biodegradable synthetic carrier for bone morphogenetic protein, *Clin. Orthop. Relat. Res.*, 294, 333, 1993.

[70] Santos, E.M. et al., Si–Ca–P xerogels and bone morphogenetic protein act synergistically on rat stromal marrow cell differentiation *in vitro*, *J. Biomed. Mater. Res.*, 41, 87, 1998.

[71] Cook, S.D. et al., *In vivo* evaluation of recombinant human osteogenic protein (rhOP-1) implants as a bone graft substitute for spinal fusions, *Spine*, 19, 1655, 1994.

[72] Grauer, J.N. et al., 2000 Young investigator research award winner. Evaluation of OP-1 as a graft substitute for intertransverse process lumbar fusion, *Spine*, 26, 127, 2001.

[73] Cook, S.D. et al., Effect of recombinant human osteogenic protein-1 on healing of segmental defects in non-human primates, *J. Bone Joint Surg. Am.*, 77, 734, 1995.

[74] Blattert, T.R. et al., Successful transpedicular lumbar interbody fusion by means of a composite of osteogenic protein-1 (rhBMP-7) and hydroxyapatite carrier: a comparison with autograft and hydroxyapatite in the sheep spine, *Spine*, 27, 2697, 2002.

[75] Fajardo, L.F. et al., The disc angiogenesis system, *Lab. Invest.*, 58, 718, 1988.

[76] Downs, E.C. et al., Calcium alginate beads as a slow-release system for delivering angiogenic molecules *in vivo* and *in vitro*, *J. Cell. Physiol.*, 152, 422, 1992.

[77] Laham, R.J. et al., Local perivascular delivery of basic fibroblast growth factor in patients undergoing coronary bypass surgery: results of a phase I randomized, double-blind, placebo-controlled trial, *Circulation*, 100, 1865, 1999.

[78] Edelman, E.R. et al., Controlled and modulated release of basic fibroblast growth factor, *Biomaterials*, 12, 619, 1991.

[79] Tabata, Y. et al., Bone regeneration by basic fibroblast growth factor complexed with biodegradable hydrogels, *Biomaterials*, 19, 807, 1998.

[80] Kawaguchi, H. et al., Stimulation of fracture repair by recombinant human basic fibroblast growth factor in normal and streptozotocin-diabetic rats, *Endocrinology*, 135, 774, 1994.

[81] Mizuno, K. et al., Effect of chitosan film containing basic fibroblast growth factor on wound healing in genetically diabetic mice, *J. Biomed. Mater. Res.*, 64A, 177, 2003.

[82] Fujisato, T. et al., Effect of basic fibroblast growth factor on cartilage regeneration in chondrocyte-seeded collagen sponge scaffold, *Biomaterials*, 17, 155, 1996.

[83] Ulubayram, K. et al., EGF containing gelatin-based wound dressings, *Biomaterials*, 22, 1345, 2001.

[84] Mierisch, C.M. et al., Transforming growth factor-beta in calcium alginate beads for the treatment of articular cartilage defects in the rabbit, *Arthroscopy: Journal of Arthroscopic & Related Surgery: Official Publication of the Arthroscopy Association of North America and the International Arthroscopy Association*, 18, 892, 2002.

[85] Puolakkainen, P.A. et al., The enhancement in wound healing by transforming growth factor-b1 (TGF-b1) depends on the topical delivery system, *J. Surg. Res.*, 58, 321, 1995.

[86] Gombotz, W.R. et al., Controlled release of TGF-b1 from a biodegradable matrix for bone regeneration, *J. Biomater. Sci., Polym. Ed.*, 5, 49, 1993.

[87] Mustoe, T.A. et al., Accelerated healing of incisional wounds in rats induced by transforming growth factor-b, *Science (Washington, DC, United States)*, 237, 1333, 1987.

[88] Hildebrand, K.A. et al., The effects of platelet-derived growth factor-BB on healing of the rabbit medial collateral ligament. An *in vivo* study, *Am. J. Sports Med.*, 26, 549, 1998.

[89] Nagai, M.K. and Embil, J.M., Becaplermin: recombinant platelet derived growth factor, a new treatment for healing diabetic foot ulcers, *Expert Opin. Biol. Ther.*, 2, 211, 2002.

[90] Lee, K.Y., Peters, M.C., and Mooney, D.J., Comparison of vascular endothelial growth factor and basic fibroblast growth factor on angiogenesis in SCID mice, *J. Control. Release*, 87, 49, 2003.

[91] Tabata, Y. et al., Controlled release of vascular endothelial growth factor by use of collagen hydrogels, *J. Biomater. Sci., Polym. Ed.*, 11, 915, 2000.

[92] Peters, M.C., Polverini, P.J., and Mooney, D.J., Engineering vascular networks in porous polymer matrices, *J. Biomed. Mater. Res.*, 60, 668, 2002.

[93] Kipshidze, N., Chawla, P., and Keelan, M.H., Fibrin meshwork as a carrier for delivery of angiogenic growth factors in patients with ischemic limb, *Mayo Clin. Proc.*, 74, 847, 1999.

[94] Richardson, T.P. et al., Polymeric system for dual growth factor delivery, *Nat. Biotechnol.*, 19, 1029, 2001.

[95] Camarata, P.J. et al., Sustained release of nerve growth factor from biodegradable polymer microspheres, *Neurosurgery*, 30, 313, 1992.

[96] Yamamoto, S. et al., Protective effect of NGF atelocollagen mini-pellet on the hippocampal delayed neuronal death in gerbils, *Neurosci. Lett.*, 141, 161, 1992.

[97] Robinson, S.N. and Talmadge, J.E., Sustained release of growth factors, *In Vivo*, 16, 535, 2002.

[98] Wang, W., Instability, stabilization, and formulation of liquid protein pharmaceuticals, *Int. J. Pharm.*, 185, 129, 1999.

[99] Manning, M.C., Patel, K., and Borchardt, R.T., Stability of protein pharmaceuticals, *Pharm. Res.*, 6, 903, 1989.

[100] Schwendeman, S.P. et al., Stability of proteins and their delivery from biodegradable polymer microspheres, in *Microparticulate Systems for Delivery of Proteins and Vaccines*, New York, Dekker, 1–49, 1996 1996.

[101] Tanford, C., Protein denaturation, *Adv. Protein Chem.*, 23, 121, 1968.

[102] Kubiak, T.M. et al., Position 2 and position 2/Ala15-substituted analogs of bovine growth hormone-releasing factor (bGRF) with enhanced metabolic stability and improved *in vivo* bioactivity, *J. Med. Chem.*, 36, 888, 1993.

[103] Bowen-Pope, D.F. et al., Platelet-derived growth factor *in vivo*: levels, activity, and rate of clearance, *Blood*, 64, 458, 1984.

[104] Edelman, E.R., Nugent, M.A., and Karnovsky, M.J., Perivascular and intravenous administration of basic fibroblast growth factor: vascular and solid organ deposition, *Proc. Natl Acad. Sci. USA*, 90, 1513, 1993.

[105] Lazarous, D.F. et al., Comparative effects of basic fibroblast growth factor and vascular endothelial growth factor on coronary collateral development and the arterial response to injury, *Circulation*, 94, 1074, 1996.

[106] Yancopoulos, G.D. et al., Vascular-specific growth factors and blood vessel formation, *Nature (London)*, 407, 242, 2000.

[107] Yang, B.B. et al., Polyethylene glycol modification of filgrastim results in decreased renal clearance of the protein in rats, *J. Pharm. Sci.*, 93, 1367, 2004.

[108] Eliason, J.F., Pegylated cytokines. Potential application in immunotherapy of cancer, *BioDrugs*, 15, 705, 2001.

[109] Edwards, C.K., III, PEGylated recombinant human soluble tumor necrosis factor receptor type I (r-Hu-sTNF-RI): novel high affinity TNF receptor designed for chronic inflammatory diseases, *Ann. Rheum. Dis.*, 58, I73, 1999.

[110] Archimbaud, E., Clinical trials of pegylated recombinant human megakaryocyte growth and development factor (PEG-rHuMGDF), *Haematol. Blood Transfus.*, 39, 343, 1998.

[111] Molineux, G., PEGylation: engineering improved pharmaceuticals for enhanced therapy, *Cancer Treat. Rev.*, 28, 13, 2002.

[112] Sato, H., Enzymatic procedure for site-specific pegylation of proteins, *Adv. Drug Deliv. Rev.*, 54, 487, 2002.

[113] Roberts, M.J., Bentley, M.D., and Harris, J.M., Chemistry for peptide and protein PEGylation, *Adv. Drug Deliv. Rev.*, 54, 459, 2002.

[114] Zalipsky, S., Chemistry of polyethylene glycol conjugates with biologically active molecules, *Adv. Drug Deliv. Rev.*, 16, 157, 1995.

[115] Veronese, F.M. and Harris, J.M., Introduction and overview of peptide and protein pegylation, *Adv. Drug Deliv. Rev.*, 54, 453, 2002.

[116] Zhdanov, V.P. and Kasemo, B., Monte Carlo simulations of the kinetics of protein adsorption, *Surf. Rev. Lett.*, 5, 615, 1998.

[117] Norde, W., Adsorption of proteins from solution at the solid–liquid interface, *Adv. Colloid Interface Sci.*, 25, 267, 1986.

[118] Malmsten, M., Protein adsorption at the solid–liquid interface, in *Protein Architecture*, Lvov, Y. and Moehwald, H., Eds., New York, Marcel Dekker, 1–23, 2000.

[119] Felgner, P.L. and Wilson, J.E., Hexokinase binding to polypropylene test tubes. Artifactual activity losses from protein binding to disposable plastics, *Anal. Biochem.*, 74, 631, 1976.

[120] Beyerman, H.C. et al., On the instability of secretin, *Life Sci.*, 29, 885, 1981.

[121] Kramer, K.J. et al., Purification and characterization of the carrier protein for juvenile hormone from the hemolymph of the tobacco hornworm *Manduca sexta* Johannson (Lepidoptera: Sphingidae), *J. Biol. Chem.*, 251, 4979, 1976.

[122] Suelter, C.H. and DeLuca, M., How to prevent losses of protein by adsorption to glass and plastic, *Anal. Biochem.*, 135, 112, 1983.

[123] Katakam, M. and Banga, A.K., Use of poloxamer polymers to stabilize recombinant human growth hormone against various processing stresses, *Pharm. Dev. Technol.*, 2, 143, 1997.

[124] Charman, S.A., Mason, K.L., and Charman, W.N., Techniques for assesing the effects of pharmaceutical excipients on the aggregation of porcine growth hormone, *Pharm. Res.*, 10, 954, 1993.

[125] Oliva, A. et al., Effect of high shear rate on stability of proteins: kinetic study, *J. Pharm. Biomed. Anal.*, 33, 145, 2003.

[126] Spenlehauer, G., Spenlehauer-Bonthonneau, F., and Thies, C., Biodegradable microparticles for delivery of polypeptides and proteins, *Prog. Clin. Biol. Res.*, 292, 283, 1989.

[127] van Erp, S.H., Kamenskaya, E.O., and Khmelnitsky, Y.L., The effect of water content and nature of organic solvent on enzyme activity in low-water media. A quantitative description, *Eur. J. Biochem./FEBS*, 202, 379, 1991.

[128] Khmelnitsky, Y.L. et al., Denaturation capacity: a new quantitative criterion for selection of organic solvents as reaction media in biocatalysis, *Eur. J. Biochem./FEBS*, 198, 31, 1991.

[129] Brunner, A., Mader, K., and Gopferich, A., The chemical microenvironment inside biodegradable microspheres during erosion, *Proceedings of the International Symposium on Controlled Release of Bioactive Materials*, Vol. 25, p. 154, 1998.

[130] Lucke, A. et al., The effect of poly(ethylene glycol)–poly(D,L-lactic acid) diblock copolymers on peptide acylation, *J. Control. Release*, 80, 157, 2002.

[131] Lucke, A., Kiermaier, J., and Gopferich, A., Peptide acylation by poly(.alpha.-hydroxy esters), *Pharm. Res.*, 19, 175, 2002.

[132] Schwendeman, S.P., Recent advances in the stabilization of proteins encapsulated in injectable PLGA delivery systems, *Crit. Rev. Ther. Drug Carrier Syst.*, 19, 73, 2002.

[133] Zhu, G., Mallery, S.R., and Schwendeman, S.P., Stabilization of proteins encapsulated in injectable poly(lactide-co-glycolide), *Nat. Biotechnol.*, 18, 52, 2000.

[134] Lam, X.M., Duenas, E.T., and Cleland, J.L., Encapsulation and stabilization of nerve growth factor into poly(lactic-co-glycolic) acid microspheres, *J. Pharm. Sci.*, 90, 1356, 2001.

[135] Pean, J.M. et al., Why does PEG 400 co-encapsulation improve NGF stability and release from PLGA biodegradable microspheres? *Pharm. Res.*, 16, 1294, 1999.

[136] Tang, B., Wang, M., and Wise, C., Nerve growth factor mRNA stability is controlled by a cis-acting instability determinant in the 3′-untranslated region, *Mol. Brain Res.*, 46, 118, 1997.

[137] Shaw, G. and Kamen, R., A conserved AU sequence from the 3′ untranslated region of GM-CSF mRNA mediates selective mRNA degradation, *Cell*, 46, 659, 1986.

[138] Schiavi, S.C., Belasco, J.G., and Greenberg, M.E., Regulation of proto-oncogene mRNA stability, *Biochim. Biophys. Acta*, 1114, 95, 1992.

[139] Nimni, M.E., Polypeptide growth factors: targeted delivery systems, *Biomaterials*, 18, 1201, 1997.

[140] Davis, G.E. and Camarillo, C.W., Regulation of endothelial cell morphogenesis by integrins, mechanical forces, and matrix guidance pathways, *Exp. Cell Res.*, 216, 113, 1995.

[141] Lee, K.Y. et al., Controlled growth factor release from synthetic extracellular matrices, *Nature*, 408, 998, 2000.

[142] Wallace, D.G. and Rosenblatt, J., Collagen gel systems for sustained delivery and tissue engineering, *Adv. Drug Deliv. Rev.*, 55, 1631, 2003.

[143] Drury, J.L. and Mooney, D.J., Hydrogels for tissue engineering: scaffold design variables and applications, *Biomaterials*, 24, 4337, 2003.

[144] Peters, M.C. et al., Release from alginate enhances the biological activity of vascular endothelial growth factor, *J. Biomater. Sci., Polym. Ed.*, 9, 1267, 1998.

[145] Taipale, J. and Keski-Oja, J., Growth factors in the extracellular matrix, *FASEB J.*, 11, 51, 1997.

[146] Langer, R. and Tirrell, D.A., Designing materials for biology and medicine, *Nature (London, United Kingdom)*, 428, 487, 2004.

[147] Majeti, N.V. and Ravi Kumar, M.N.V., Nano and microparticles as controlled drug delivery devices, *J. Pharm. Pharm. Sci. [online computer file]*, 3, 234, 2000.

[148] Lopez, V.C. and Snowden, M.J., The role of colloidal microgels in drug delivery, *Drug Deliv. Syst. Sci.*, 3, 19, 2003.

[149] Saltzman, W.M., Delivering tissue regeneration, *Nat. Biotechnol.*, 17, 534, 1999.

[150] Xu, R. et al., Diabetes gene therapy: potential and challenges, *Curr. Gene Ther.*, 3, 65, 2003.

[151] Merdan, T., Kopecek, J., and Kissel, T., Prospects for cationic polymers in gene and oligonucleotide therapy against cancer, *Adv. Drug Deliv. Rev.*, 54, 715, 2002.

[152] Lollo, C.P., Banaszczyk, M.G., and Chiou, H.C., Obstacles and advances in non-viral gene delivery, *Curr. Opin. Mol. Ther.*, 2, 136, 2000.

[153] Johnson-Saliba, M. and Jans, D.A., Gene therapy: optimising DNA delivery to the nucleus, *Curr. Drug Targets*, 2, 371, 2001.

[154] Bout, A. and Crucell, L., Gene therapy, in *Pharmaceutical Biotechnology*, 2nd ed., Commelin, D.J.A. and Sindelar, R.D., Eds., London, Taylor & Francis Ltd., 175–192, 2002.

[155] Kuhl, P.R. and Griffith-Cima, L.G., Tethered epidermal growth factor as a paradigm for growth factor-induced stimulation from the solid phase, *Nat. Med.*, 2, 1022, 1996.

[156] Blunk, T., Goepferich, A., and Tessmar, J., Special issue biomimetic polymers, *Biomaterials*, 24, 4335, 2003.

[157] Aldersey-Williams, H., Towards biomimetic architecture, *Nat. Mater.*, 3, 277, 2004.

[158] Ito, Y., Tissue engineering by immobilized growth factors, *Mater. Sci. Eng. C: Biomimetic Mater., Sensors Syst.*, C6, 267, 1998.

[159] Shakesheff, K.M., Cannizzaro, S.M., and Langer, R., Creating biomimetic micro-environments with synthetic polymer–peptide hybrid molecules, *J. Biomater. Sci., Polym. Ed.*, 9, 507, 1998.

[160] Thayumanavan, S. et al., Towards dendrimers as biomimetic macromolecules, *C. R. Chimie*, 6, 767, 2003.

[161] Hoffman, A.S. et al., Design of "smart" polymers that can direct intracellular drug delivery, *Polym. Adv. Technol.*, 13, 992, 2002.

[162] Peppas, N.A. and Huang, Y., Polymers and gels as molecular recognition agents, *Pharm. Res.*, 19, 578, 2002.

[163] Langer, R. and Folkman, J., Polymers for the sustained release of proteins and other macromolecules, *Nature*, 263, 797.

[164] Saltzman, W.M. and Langer, R., Transport rates of proteins in porous materials with known microgeometry, *Biophys. J.*, 55, 163, 1989.

[165] Bawa, R. et al., An explanation for the controlled release of macromolecules from polymers, *J. Control. Release*, 1, 259, 1985.

[166] Leong, K.W. et al., Polyanhydrides for controlled release of bioactive agents, *Biomaterials*, 7, 364, 1986.

[167] Heller, J., Controlled drug release from poly(ortho esters) — a surface eroding polymer, *J. Control. Release*, 2, 167, 1985.

[168] Goepferich, A., Mechanisms of polymer degradation and erosion, *Biomaterials*, 17, 103, 1996.

[169] Goepferich, A., Polymer degradation and erosion. Mechanisms and applications, *Eur. J. Pharm. Biopharm.*, 42, 1, 1996.

[170] Fu, K. et al., Visual evidence of acidic environment within degrading poly(lactic-co-glycolic acid) (PLGA) microspheres, *Pharm. Res.*, 17, 100, 2000.

[171] Brunner, A., Mader, K., and Gopferich, A., pH and osmotic pressure inside biodegradable microspheres during erosion, *Pharm. Res.*, 16, 847, 1999.

[172] Lucke, A., Kiermaier, J., and Gopferich, A., Peptide acylation by poly(a-hydroxy esters), *Pharm. Res.*, 19, 175, 2002.

[173] Athanasiou, K.A. et al., Orthopaedic applications for PLA–PGA biodegradable polymers, *Arthroscopy: The Journal of Arthroscopic and Related Surgery: Official Publication of the Arthroscopy Association of North America and the International Arthroscopy Association*, 14, 726, 1998.

[174] Biggs, D.L. et al., *In vitro* and *in vivo* evaluation of the effects of PLA microparticle crystallinity on cellular response, *J. Control. Release: Official Journal of the Controlled Release Society*, 92, 147, 2003.

[175] Sandor, M. et al., Effect of protein molecular weight on release from micron-sized PLGA microspheres, *J. Control. Release*, 76, 297, 2001.

[176] Tabata, Y. and Ikada, Y., Protein release from gelatin matrixes, *Adv. Drug Deliv. Rev.*, 31, 287, 1998.

[177] Gombotz, W.R. and Wee, S., Protein release from alginate matrixes, *Adv. Drug Deliv. Rev.*, 31, 267, 1998.

[178] Fujioka, K. et al., Protein release from collagen matrixes, *Adv. Drug Deliv. Rev.*, 31, 247, 1998.

[179] Sinha, V.R. and Trehan, A., Biodegradable microspheres for protein delivery, *J. Control. Release*, 90, 261, 2003.

[180] Sinha, V.R. et al., Poly-e-caprolactone microspheres and nanospheres: an overview, *Int. J. Pharm.*, 278, 1, 2004.

[181] Orive, G. et al., Cell microencapsulation technology for biomedical purposes: novel insights and challenges, *Trends Pharmacol. Sci.*, 24, 207, 2003.

[182] Orive, G., Hernandez, R.M., Gascón, A.R., and Pedraz, J.L., Challenges in cell encapsulation, in *Cell Immobilization Biotechnology. Part II. Applications*, Nedović, N. and Willaert, R., Eds., "Focus on Biotechnology" series, Dordrecht, The Netherlands, Kluwer, 9999, 1991.

[183] Giannoni, P. and Hunziker, E.B., Release kinetics of transforming growth factor beta1 from fibrin clots, *Biotechnol. Bioeng.*, 83, 121, 2003.

[184] Collier, J.H. and Messersmith, P.B., Phospholipid strategies in biomineralization and biomaterials research, *Ann. Rev. Mater. Res.*, 31, 237, 2001.

[185] Haller, M.F. and Saltzman, W.M., Nerve growth factor delivery systems, *J. Control. Release*, 53, 1, 1998.

[186] Saltzman, W.M. et al., Intracranial delivery of recombinant nerve growth factor: release kinetics and protein distribution for three delivery systems, *Pharm. Res.*, 16, 232, 1999.

[187] Lee, J.E. et al., Effects of the controlled-released TGF-b1 from chitosan microspheres on chondrocytes cultured in a collagen/chitosan/glycosaminoglycan scaffold, *Biomaterials*, 25, 4163, 2004.

[188] Cleland, J.L. et al., Recombinant human growth hormone poly(lactic-glycolic acid) (PLGA) microspheres provide a long lasting effect, *J. Control. Release*, 49, 193, 1997.

[189] Cleland, J.L., Protein delivery from biodegradable microspheres, *Pharm. Biotechnol.*, 10, 1, 1997.

[190] Van de Weert, M., Hennink, W.E., and Jiskoot, W., Protein instability in poly(lactic-co-glycolic acid) microparticles, *Pharm. Res.*, 17, 1159, 2000.

[191] Tonnesen, H.H. and Karlsen, J., Alginate in drug delivery systems, *Drug Dev. Ind. Pharm.*, 28, 621, 2002.

[192] Gu, F., Amsden, B., and Neufeld, R., Sustained delivery of vascular endothelial growth factor with alginate beads, *J. Control. Release*, 96, 463, 2004.

[193] Mladenovska, K. et al., BSA-loaded gelatin microspheres: preparation and drug release rate in the presence of collagenase, *Acta Pharm. (Zagreb, Croatia)*, 52, 91, 2002.

[194] Tabata, Y. et al., *De novo* formation of adipose tissue by controlled release of basic fibroblast growth factor, *Tissue Eng.*, 6, 279, 2000.

[195] Bummer, P.M., Physical chemical considerations of lipid-based oral drug delivery — solid lipid nanoparticles, *Crit. Rev. Ther. Drug Carrier Syst.*, 21, 1, 2004.

[196] Maschke, A. et al., Lipids: an alternative material for protein and peptide release, *ACS Symp. Ser.*, 879, 176, 2004.

[197] Cortesi, R. et al., Production of lipospheres as carriers for bioactive compounds, *Biomaterials*, 23, 2283, 2002.

[198] Reithmeier, H., Herrmann, J., and Gopferich, A., Lipid microparticles as parenteral controlled release device for peptides, *J. Control. Release*, 73, 339, 2001.

[199] Reithmeier, H., Gopferich, A., and Herrmann, J., Preparation and characterization of lipid micro-particles containing thymocartin, an immunomodulating peptide, *Proceedings of the International Symposium on Controlled Release of Bioactive Materials*, Vol. 26, p. 681, 1999.

[200] Lee, K.Y. and Mooney, D.J., Hydrogels for tissue engineering, *Chem. Rev.*, 101, 1869, 2001.

[201] Yamamoto, M., Takahashi, Y., and Tabata, Y., Controlled release by biodegradable hydrogels enhances the ectopic bone formation of bone morphogenetic protein, *Biomaterials*, 24, 4375, 2003.

[202] Yamamoto, M. et al., Bone regeneration by transforming growth factor b1 released from a biodegradable hydrogel, *J. Control. Release*, 64, 133, 2000.

[203] Tabata, Y., Matsui, Y., and Ikada, Y., Growth factor release from amylopectin hydrogel based on copper coordination, *J. Control. Release*, 56, 135, 1998.

[204] Zisch, A.H., Lutolf, M.P., and Hubbell, J.A., Biopolymeric delivery matrices for angiogenic growth factors, *Cardiovasc. Pathol.*, 12, 295, 2003.

[205] Jeong, B. et al., Biodegradable block copolymers as injectable drug-delivery systems, *Nature*, 388, 860, 1997.

[206] Ebara, M. et al., Temperature-responsive cell culture surfaces enable "On-Off" affinity control between cell integrins and RGDS ligands, *Biomacromolecules*, 5, 505, 2004.

[207] Zisch, A.H. et al., Cell-demanded release of VEGF from synthetic, biointeractive cell-ingrowth matrices for vascularized tissue growth, *FASEB J.*, 17, 2260, 2003.

[208] Sakiyama-Elbert, S.E. and Hubbell, J.A., Controlled release of nerve growth factor from a heparin-containing fibrin-based cell ingrowth matrix, *J. Control. Release*, 69, 149, 2000.

[209] Kimura, Y. et al., Adipose tissue engineering based on human preadipocytes combined with gelatin microspheres containing basic fibroblast growth factor, *Biomaterials*, 24, 2513, 2003.

[210] Holland, T.A., Tabata, Y., and Mikos, A.G., *In vitro* release of transforming growth factor-b1 from gelatin microparticles encapsulated in biodegradable, injectable oligo(poly(ethylene glycol) fumarate) hydrogels, *J. Control. Release*, 91, 299, 2003.

[211] Holland, T.A. et al., Transforming growth factor-b1 release from oligo(poly(ethylene glycol) fumarate) hydrogels in conditions that model the cartilage wound healing environment, *J. Control. Release*, 94, 101, 2004.

[212] Hu, Y., Hollinger, J.O., and Marra, K.G., Controlled release from coated polymer microparticles embedded in tissue-engineered scaffolds, *J. Drug Target.*, 9, 431, 2001.

[213] Kim, H., Kim, H.W., and Suh, H., Sustained release of ascorbate-2-phosphate and dexa-methasone from porous PLGA scaffolds for bone tissue engineering using mesenchymal stem cells, *Biomaterials*, 24, 4671, 2003.

[214] Ziegler, J. et al., Adsorption and release properties of growth factors from biodegradable implants, *J. Biomed. Mater. Res.*, 59, 422, 2002.

[215] Tabata, Y. and Ikada, Y., Vascularization effect of basic fibroblast growth factor released from gelatin hydrogels with different biodegradabilities, *Biomaterials*, 20, 2169, 1999.

[216] Ueda, H. et al., Use of collagen sponge incorporating transforming growth factor-beta1 to promote bone repair in skull defects in rabbits, *Biomaterials*, 23, 1003, 2002.

[217] Whang, K., Goldstick, T.K., and Healy, K.E., A biodegradable polymer scaffold for delivery of osteotropic factors, *Biomaterials*, 21, 2545, 2000.

[218] Whang, K. et al., Ectopic bone formation via rhBMP-2 delivery from porous bioabsorbable polymer scaffolds, *J. Biomed. Mater. Res.*, 42, 491, 1998.

[219] Jang, J.H. and Shea, L.D., Controllable delivery of non-viral DNA from porous scaffolds, *J. Control. Release*, 86, 157, 2003.

[220] Brunner A., Gopferich, A., *Characterization of Polyanhydride Microspheres*, p. 169, 1996.

[221] Sohier, J. et al., A novel method to obtain protein release from porous polymer scaffolds: emulsion coating, *J. Control. Release*, 87, 57, 2003.

[222] Li, X. et al., Influence of process parameters on the protein stability encapsulated in poly-DL-lactide-poly(ethylene glycol) microspheres, *J. Control. Release*, 68, 41, 2000.

[223] Crotts, G. and Park, T.G., Protein delivery from poly(lactic-co-glycolic acid) biodegradable microspheres: release kinetics and stability issues, *J. Microencapsul.*, 15, 699, 1998.

[224] Jiang, G. et al., Assessment of protein release kinetics, stability and protein polymer interaction of lysozyme encapsulated poly(D,L-lactide-co-glycolide) microspheres, *J. Control. Release*, 79, 137, 2002.

[225] Sirotkin, V.A. et al., Calorimetric and Fourier transform infrared spectroscopic study of solid proteins immersed in low water organic solvents, *Biochim. Biophys. Acta*, 1547, 359, 2001.

[226] Shea, L.D. et al., DNA delivery from polymer matrices for tissue engineering, *Nat. Biotechnol.*, 17, 551, 1999.

14

Gene Therapy

J.M. Munson
W.T. Godbey
Tulane University

14.1 Introduction

Gene therapy is the delivery of genetic material into cells for the purpose of altering cellular function. This seemingly straightforward definition encompasses a variety of situations that can at times seem unrelated. The delivered genetic material can be composed of deoxyribonucleic acid (DNA) or RNA, or even involve proteins in some cases. The alteration in cellular function can be an increase or decrease in the amount of a native protein that is produced, or the production of a protein that is foreign. The delivery of the genetic material can occur directly, as is the case with microinjection, or involve carriers that interact with cell membranes or membrane-bound proteins as a part of cellular entry. Polynucleotides can be single or double stranded, and can code for a message, or not (as is the case for antisense gene delivery). Even the location of cells at the time of gene delivery is not restricted. Cells can be part of a living organism, can exist as a culture on a plate, or can be removed from an organism, transfected, and replaced into the same or a different organism at a later time.

Gene therapy came into being after the development of recombinant DNA technology and initially moved forward using viruses to target sites *in vitro* [1]. With the understanding of retroviruses and their possible uses as vectors for delivery, gene therapy progressed to applications involving mammalian organisms during the 1980s [1]. Since then, new techniques for gene delivery have been developed that utilize both viral and nonviral carriers. These techniques have been successful enough that proposed disease treatments have evolved to the clinical trial stage with encouraging success. Although many cellular processing mechanisms remain unclear and the search for the ideal gene delivery vector remains ongoing, the tools and methods for gene therapy have provided a strong base from which to build.

14.2 Nucleotides for Delivery

14.2.1 DNA (deoxyribonucleic acid)

DNA is the building block of life and as such it remains a staple in the delivery of genetic material to the cell. There are many strategies as to how the DNA will interact with the existing genome and what is the best way to produce the desired effect.

14.2.1.1 Plasmids

A plasmid is a circular piece of DNA. Extrachromosomal plasmids can be replicated or transcribed independently of the rest of the DNA in the nucleus. This attribute makes plasmids an ideal tool for gene therapy. Plasmids can include a selectable marker for identification of transfectants/transductants, a multiple cloning site for ease of inserting/deleting additional nucleotide strands, the exon(s) of interest, and regulatory/binding elements such as promoters and enhancers. A common promoter used in gene therapy is the cytomegalovirus immediate early promoter (CMV_{ie}), used because of its strength in transcription initiation in nearly every mammalian cell. However, several other promoters have been investigated, such as the glucose-related protein promoter [2] and many cell-specific promoters. Enhancers include the simian virus 40 (SV_{40}) enhancer as well as long terminal repeats (LTRs) [2]. The exons that are delivered may code for necessary cell-specific proteins, engineered protein polymers [3], or what has become a common objective in cancer cell research: suicide genes [4].

14.2.1.2 Nucleotide Decoys

DNA and RNA decoys are small oligonucleotide fragments that mimic the start sequences of potentially harmful genes. By doing so, the decoys can effectively trap the transcriptional and translational machinery, thereby causing a sharp decrease in mRNA or protein production. There has been success with this approach in human immunodeficiency virus (HIV) research. DNA and RNA decoys have been used to halt the function of REV proteins, which are responsible for making late transcripts of RNA and exporting them to the cytoplasm [5], and the TAT protein, which binds and activates the natural promoter for HIV-1 [6].

14.2.2 RNA

RNA also holds a valuable place in gene therapy because it can be used to alter cellular function. Two examples of RNA use in gene therapy are antisense RNA and small interfering RNA (siRNA). Antisense RNA provides functional alterations in cells by acting on host gene products post-transcriptionally. Antisense RNA functions by binding to cellular mRNA transcripts, which prevents ribosomal binding and therefore translation. Antisense RNAs have been investigated for their use in applications such as beta-globin gene inhibition as a possible treatment for sickle cell anemia [7], bcr/abl interference in myeloid leukemia research [8], and CXCR4 disruption for HIV-1 gene therapy [9], among others. The use of siRNA can interfere directly with genomic DNA transcription. First described by Elbashir et al. [10] in 2001, siRNAs have very high specificity that can be used to target a single-nucleotide polymorphism (SNP) on a single mutant allele [11,12]. The result would be to silence the mutant allele while permitting continued transcription of the wild type gene. This technology has been used *in vivo* to successfully target spinocerebellar ataxia type 3 and frontotemporal dementia [13].

14.3 Gene Delivery

Simply administering naked genetic material in the vicinity of cell exteriors is usually not sufficient to bring about the desired cellular response at adequate levels. Carriers have been employed to aid in the transfer of genetic material from the cell exterior to the cytoplasm or nucleus. There is a wide variety of carriers available, each with links back to one of the main branches of natural science: biology, chemistry,

TABLE 14.1 Examples of Virus Families and Members Used for Gene Delivery

Family	Example	References
Adenoviridae	Adenovirus	[14]
Baculoviridae	Baculovirus	[15]
Herpesviridae	Pseudorabies virus	[16]
	Simplex virus	[17]
Parvoviridae	Adeno-associated	[18]
	Parvovirus	[19]
Poxviridae	Vaccinia	[20]
	Yaba monkey tumor virus	[21]
Retroviridae	Human immunodeficiency virus 1 and 2	[22,23]
	Lentivirus	[24]
	Murine leukemia virus	[25]
	Retrovirus	[26]

and physics. It is possible for hybrid systems that combine two or more basic technologies to exist, but the individual classifications will be discussed separately for clarity.

14.3.1 Biological Delivery Methods

Because they are a product of millions of years of biologic evolution, viruses are included here as the biological forms of gene delivery in spite of the fact that viruses are not technically alive (they do not respire). Viruses are the most efficient vectors for large-scale gene delivery. Some also offer permanent expression of delivered genes through the integration of genetic material into the transduced organism's genome. Table 14.1 lists some of the major viral families and specific members that have been used in gene therapy. Many of the references in the table cite review articles of the specific virus classification for further reading.

Currently, herpes simplex virus (HSV) is being used to transduce cells in the central nervous system. By altering the genome of the virus, researchers have been able to decrease cytotoxicity and increase transfection efficiency [27]. HSV is also desirable because it is relatively large to allow packaging of larger plasmids, does not integrate its DNA into the host genome which avoids the deleterious effects that are possible with random integration, and is easily deliverable into the brain due to its ability to actively travel towards the nuclei of the neurons through the axons in the peripheral nervous system. HSV has also been applied to the transduction of skeletal muscle. Recent reports have described successful HSV-mediated gene transfer into skeletal muscle cells *in vitro* and *in vivo* and outlined possible applications for the delivery of large genes such as that coding for dystrophin [28].

In the liver, the virus AcMNPV (of the family baculoviridae) has yielded promising transfection efficiencies [29]. The YABA-like virus, which exhibits very similar attributes to the vaccinia virus but does not produce immune responses in the host, has successfully targeted and treated ovarian cancer in mice [30].

Adeno-associated virus (AAV) requires coinfection with a helper virus (typically adenovirus) to be productive. AAV offers an advantage to the gene therapist in that it has been reported to provide permanent expression of delivered genes. *In utero* experiments in mice and nonhuman primates indicate that recombinant AAVs can successfully alter the genome of the host organism to produce stable transductants [31]. Adeno-associated viruses are commonly used and successful in targeting the central nervous system [32]. Although repeated viral transduction is associated with inflammatory and immunological responses in the host, it has been reported that these responses can be reduced by carefully balancing recombinant AAV dosing with prudent timing [33].

Retroviruses offer good transduction efficiencies *in vivo*. Hallmarks of retroviral transduction are that RNA is delivered into cells as opposed to DNA, cDNA is made using viral reverse transcriptase, and this cDNA is introduced into the host genome via the protein integrase to produce stable transfectants. Immunodeficiency viruses are commonly investigated retroviruses, and include HIV and feline immunodeficiency virus (FIV). Investigations into retroviral genome alterations, such as self-inactivation via promoter deletion, are being conducted in an attempt to render HIV-1 a safe vector for gene transfer [34]. Use of FIV has shown efficient transduction of nondividing cells without many of the inherent risks associated with delivering recombinant HIV to humans [35]. Lentivirus, which belongs to the same family as HIV and FIV, has been developed to produce concurrent regulated multiple gene delivery upon transduction [36].

14.3.2   Chemical Delivery Methods

Many chemical methods of transfection have been engineered and utilized in the past 20 years. The use of synthetic gene delivery vehicles, such as polymers or cationic lipids, offers several advantages. Nonviral gene delivery is not restricted to plasmid DNA sizes based upon the ability to fit into a viral head of finite size, as is evidenced by successful delivery of a 2.3 Mb artificial human chromosome using poly(ethylenimine)(PEI) [37], and a 60 Mb artificial chromosome into Chinese hamster ovary cells using liposomes [38]. Although nonviral gene delivery usually results in transient gene expression, an advantage of this characteristic is that there is little or no risk of random integration into host genomes. Random integration poses a problem in that it can knock out a vital gene such as a housekeeping gene or a tumor suppressor gene. An additional advantage of nonviral gene delivery is the reduced threat of immune response *in vivo* versus repeated viral injections.

Several cationic polymers exist that exhibit strong transfection qualities. Many of these polymers are based on polypeptide chains containing multiple lysine, histidine, or arginine residues. Historically, poly(L-lysine) has been the most commonly used peptide transfection agent. However, several arginine-based combinations of peptides have been formed into polyplexes of varying sizes and charge ratios which may yield better transfection results than the previously used polypeptide chains composed of a single amino acid [39]. Histidine–lysine polyplexes have also produced good transfection results that positively correlated with the amount of histidine in the complex [40]. There are also additions that can be made to these complexes in order to increase transfection efficiency and lower cytotoxic effects. These include ligands such as nerve growth factor, and hydrophilic polymers such as polyethylene glycol (PEG), which has been shown to increase transfection efficiency and lower cytotoxicity of poly(L-lysine) both *in vitro* and *in vivo* [41].

PEI is a highly cationic polymer that is available in both linear and branched forms. Each form of the polymer has a repeating $[-CH_2-CH_2-N<]$ structure, with each nitrogen taking the form of a primary (end group), secondary (middle of a straight chain), or tertiary (branch point) amine (Figure 14.1). The groups attached to secondary and tertiary amines are typically additional $[-CH_2-CH_2-N<]$, although terminal amines can be modified through the attachment of moieties for targeting or degradation purposes. Because of the large number of amines present in PEI, many of which carry a positive charge at physiological pH, DNA and RNA easily bind with the polymer via electrostatic interactions to form stable complexes. The transfection efficiency of branched PEI correlates with its molecular weight, with weight-average molecular weights of approximately 25,000 Da working best [42,43]. PEGylation of PEI polymers can, as was also the case for poly(L-lysine), increase transfection efficiency; this has been indicated in delivery of complexes to the central nervous system *in vivo* [44]. PEI has also been successfully used in *in vitro* transfections of rat endothelia and chicken embryonic neurons as well as *in vivo* in mice [45]. Conjugates of PEI are also being developed, such as silica microspheres coated with the polymer–DNA complexes. These conjugates have been used successfully for *in vitro* transfection, and offer the potential for relatively simple covalent surface modifications [46].

Dendrimers are highly branched polymers built around a single atom or molecule. Poly(amidoamine) (PAMAM) dendrimers are often used for gene delivery because of their high nitrogen content which aids in DNA binding and condensation. With a central atom of nitrogen, these polymers are built by the addition of amine-containing molecules, such as $[CH_2-CH_2-NH_2]$ (Figure 14.1). The result is a maximally branched molecule of known molecular weight; a sort of controlled version of PEI. PAMAM dendrimers, sometimes referred to as starburst dendrimers, have shown good transfection efficiencies in

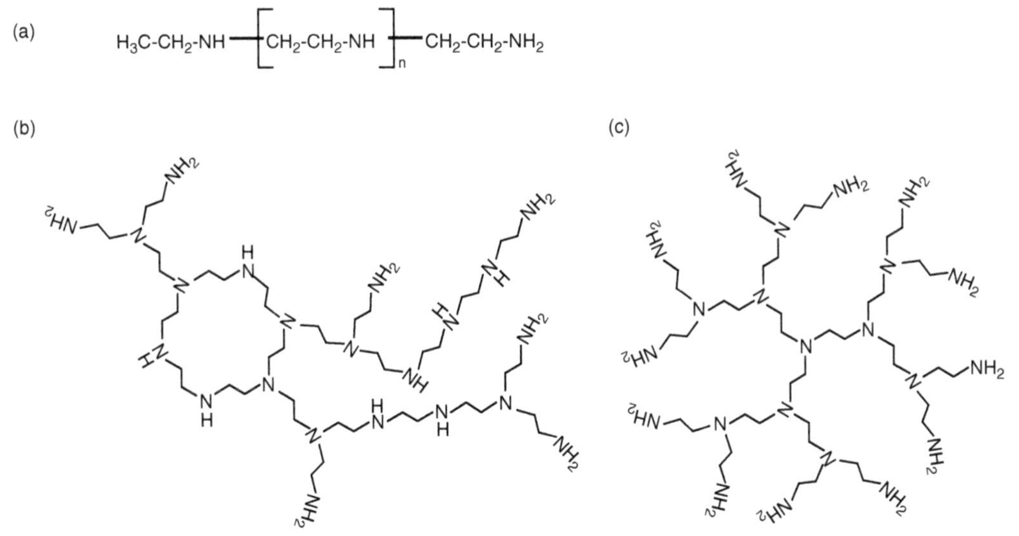

(a)

$$H_3C\text{-}CH_2\text{-}NH\left[CH_2\text{-}CH_2\text{-}NH\right]_n CH_2\text{-}CH_2\text{-}NH_2$$

(b)   (c)

FIGURE 14.1 Three polymers based upon the [–CH2–CH2–N<] basic unit. Amines are shown in the uncharged state for clarity. (a) *Linear PEI;* (b) *An example of a branched PEI.* The polymerization permits 2° amines and one cyclization via a back biting reaction. The polymer is somewhat irregular; (c) Third *generation PAMAM dendrimer.* All amines are tertiary amines (except for chain termini).

mammalian cells with little or no cytotoxicity *in vitro* [47]. A current application of this dendrimer is its coupling with an Epstein–Barr virus-based plasmid for suicide gene therapy in cancer cells [48].

Chitosan is another polymer that has been well characterized for use in transfection [49]. This natural nontoxic polysaccharide lends DNAase-resistance to its cargo while condensing the DNA to form stronger complexes [50]. The efficiency of chitosan is thought to rely upon its ability to swell and burst endolysosomes, which allows the delivered DNA to continue its path to the nucleus [51].

Alginate, a polycationic polysaccharide that can be used in the form of a hydrogel, has been used alone for transfection [52], as well as in combination with poly(L-lysine) complexes to form an oligonucleotide "sponge" [53]. Through slow degradation, these gels were shown to release embedded oligonucleotides over time. Possible applications of these gels to deliver antisense RNA are currently being pursued [53].

Microgels are similar to the alginate complex just mentioned. A thermosensitive microgel, composed of poly[ethylene glycol-(D, L-lactic acid-co-glycol acid)-ethylene glycol] (PEG–PLGA–PEG) triblock copolymers, has been created to have the property of being a liquid at room temperature but solidifying at 37°C. Once solidified, the gel slowly degrades over the course of 30 days, releasing its component oligonucleotides to the surrounding cells. Possible applications of this gel system to wound healing are being pursued [54].

Lipids, particularly cationic lipids, have been used for gene delivery in a process termed lipofection. Several lipofection agents have been used for successful transfection of cells both *in vitro* and *in vivo*. Because of the aqueous environment inside cells and tissues, the hydrophobic tails of the lipids will coalesce to form hollow micelles, the interiors of which can contain oligonucleotides for cellular delivery. The combination of more than one hydrophobic entity can often yield higher transfection efficiencies than using a single type of lipid. In primary cortical neurons, a combination of *N*-[1-(2,3-dioleoyloxy)propyl]-*N,N,N*-trimethylammonium methylsulfate and cholesterol (DOTAP : Chol) shows high transfection efficiency [55]. Using intravenous administration of *N*-[1-(2,3-dioleoyloxy)propyl]-*N,N,N*-trimethylammonium chloride–Tween 80 (DOTMA-Tween 80) complexes with high DOTMA : DNA ratios, transient transfection to several organs has also been accomplished [56]. The use of helper lipids such as cholesterol or dioleoyl phosphatidylethanolamine (DOPE) can increase transfection efficiency significantly through endosomal membrane destabilization [57].

14.3.3 Physical Delivery Methods

A highly efficient way of transferring naked DNA into a select cell or group of cells is via physical transfection methods. Microinjection by a skilled operator offers nearly perfect delivery efficiency on a cell-by-cell basis. This technique can be used to transfer oligonucleotides to cell cytoplasms or nuclei [58], or entire nuclei into enucleated eggs, as is the case with cloning [59,60]. However, the use of microinjection to transfect hundreds or thousands of somatic cells *in vivo* would be an infeasible venture, both in terms of getting to and visualizing the cells of interest, and the amount of time and effort required for the procedure. There exist alternative physical delivery methods that can be used to deliver genetic material directly into a larger number of cells, albeit in a slightly less precise fashion. These methods include electroporation and the gene gun.

Since 1982, electroporation has been a promising approach for both *in vitro* and *in vivo* gene delivery [61]. Electroporation causes disruption of the plasma membrane and formation of membrane-associated DNA aggregates, which enter the cytoplasm through transient pores [62]. These complexes have been shown to enter the cytoplasm 30 min after administration of current, and proceed to peri-nuclear locations by 24 h postadministration [62]. Electroporation has shown promising results in the administration of complexes to skeletal muscles, indicating a possible role for the technique in future *in vivo* muscular applications [63]. Electroporation can be combined with chemical transfection methods, including dendrimers, to increase the overall efficiency of transfection [64]. A similar approach, termed nucleofection, utilizes an electrical current in conjunction with specific solutions to deliver DNA not only into the cell but also into the cell nucleus [65].

The gene gun provides a ballistic means of transporting genetic material into cells. A typical application of the gene gun would entail attaching DNA to gold particles that are on the order of 1 μm in diameter. The coated gold is then loaded onto a cartridge or onto a disc and propelled down the barrel of the gun via pressurized helium at a force on the order of 200 to 300 psi. The end of the gun barrel is too small for the disk to exit, so its motion is halted abruptly before exit. The coated gold colloids detach from the disc and continue their path, passing through or becoming embedded within cells. When passing through cell membranes, cytoskeletons, and other structures, some of the DNA can become dislodged and remain within the cells for processing. This method provides good transfection of tissue surface layers and offers an advantage over microinjection in that many cells in a specified area are quickly transfected. However, cell damage and depth of gold penetration are two limiting factors of this method. The gene gun has been used successfully *in vivo* to transfect murine tumors with cytokines, successfully restricting cell growth [66]. Direct gene gun transfer of gold-adhered DNA particles has also been used in successful transfections of beating hearts with genes encoding the green fluorescent protein [67].

14.4 Intracellular Pathways

The mechanisms governing the transport of nonviral gene delivery complexes into the cytoplasm and on into the nucleus vary between different vectors. Endocytosis is the primary means for cellular import of nonviral gene carriers, which makes complex modification through the addition of specific ligands beneficial. Endocytosed complexes enter the cytoplasm in endosomes, which mature from early to late endosomes before meeting up with lysosomes to form endolysosomes. As this vesicle maturation occurs, membrane bound H^+/ATPases create an acidic environment within the compartments. It is this acidic environment which allows lysosomal enzymes (acid hydrolases) to remain active and potentially degrade the endocytosed material. The endolysosome therefore presents a major obstacle to nonvirally mediated gene therapy, as evidenced by increases in transfection efficiency in the presence of the lysosomotrophic agent chloroquine [68–70]. Research into lysosomal disruption has produced a peptide that can degrade the membrane of the late endosome, based on its cholesterol content. This method has reportedly increased the transfection efficiency of parenchymal liver by 30-fold [71].

PEI has been hypothesized to serve as a proton acceptor within endolysosomes, serving to buffer the pH of the endolysosome while additional H^+ is pumped in. According to this hypothesis, Cl^- ions will leak into the vesicles to alleviate the electric gradient, thereby creating an osmotic gradient which will be offset by an influx of water molecules into the endolysosomes. The result will be swelling and bursting of the endolysosomes [45]. PEI has been shown to enter cells via endocytosis, and the endosomes have been shown to swell within a period of under 6 h [72]. However, there exists a question of the degree of lysosomal involvement during PEI-mediated transfection. PEI/DNA-containing endosomes have been imaged while passing through lysosomal beds without lysosomal interaction [73]. In addition, PEI has been reported to utilize microtubules of the cytoskeleton as a direct pathway to the nuclear envelope, thereby eluding lysosomes [74]. It is perhaps the large excess of positive charges in PEI/DNA complexes that help direct the complexes in a seemingly atypical fashion during cellular processing. More research is required to elucidate the complete cellular mechanism of PEI-mediated transfection.

If the delivered genetic material is DNA, and the desired outcome of the transfection is expression of the delivered gene, then the next major hurdle to successful transfection is getting the delivered DNA into the cell nucleus for transcription. Many studies have monitored the effect of cell cycle stage upon transfection efficiency. Some viral vectors show strong transduction during all phases of the cell cycle because of their strong nuclear-importation machinery [75]. However, polyplex and lipoplex carriers show optimal transfection if they are administered during the G1 or S phases of the cell cycle [75]. It is known that the nuclear envelope is dismantled during late mitosis, hence perhaps these findings can be explained by the length of time required for transfection complexes to proceed from endocytosis to the outer nuclear membrane being roughly equal to the amount of time needed for the cell to progress from G1 or S phase to late mitosis. Other work indicates the barrier to transfection presented by the nuclear envelope through results that reveal very low transfection efficiencies in nondividing cells when cationic lipid-mediated vectors were used [76].

Aside from entering the nucleus by virtue of location during mitotic nuclear membrane disruption, genes and gene delivery complexes can be delivered to the nucleus via import. Possible mechanisms for nuclear import include interactions between the complexes and other proteins bound for nuclear import and interactions between the complexes and nuclear import receptors themselves. Proteins in the cytoplasm that are bound for nuclear import typically have a conserved peptide sequence that is recognized by the cell as a nuclear localization signal (NLS). Such signals can be manufactured in the laboratory for inclusion as an integral part of the gene delivery complexes [77,78]. Many factors govern the success of NLS-mediated nuclear import: the type of NLS, the manner in which the NLS is incorporated, DNA form, and the proper definition of the complexes [79].

14.5 Cell and Tissue Targeting

Each of the physical delivery methods listed in Section 14.3.3 is an excellent way to deliver genes to a specific region. Certainly, microinjection is a very straightforward way to introduce genes into a single cell of interest. However, physical delivery methods can be technically difficult, insubstantial in the number of cells that are transfected, or impossible for delivery to remote areas of the body. To provide feasible alternatives when such challenges exist, molecular targeting methods have been developed, yielding encouraging results.

The use of cell-specific ligands or receptors to target extracellular attributes is a good method for gene delivery to specific cell types or classes. Neuronal cells have been targeted for transduction via a neuron-specific viral envelope in conjunction with specific glycoproteins, leaving nonneuronal cells unaffected [80]. Other examples of ligands that have been used include LOX-1 to potentially target endothelial cells [81], the human hepatitis B virus preS1 peptide to target hepatocytes [82], plus transferrin [83], folate [84], anti-CD34 [85], and galactose [86], among others. The list of potential and realized target ligands is very large, which suggests the utility of this targeting technique.

Using cell-specific transcription regulation sequences (binding elements such as promoters or enhancers) is another way to target cells without modification of the gene delivery vehicle, which allows for cell targeting with known delivery and release kinetics without having to repeat the same work for modified vectors. This type of targeting, known as promoter or expression-targeting, works on the principle that many cell types could take up the delivered genes, but only the cells of interest will transcribe them because the transcriptional machinery that binds to the utilized promoters is present only in the target cells. This type of targeting has been used to target megakaryocytes via the promoter for integrin αIIβ (for a possible treatment for platelet disorders) [87]. Targeting liver cells has been achieved by using the promoter for liver-type pyruvate kinase and SV40 enhancer [88]. Expression-targeting has also been used to target an entire class of cells (Cox-2 overexpressing cells), which could be an important technique in the treatment of carcinomas or inflammation [89]. The use of two specific binding elements, a promoter/enhancer combination, has been employed to deliver the suicide gene for thymidine kinase to prostate tumors [90]. Expression-targeted gene delivery could prove to be a valuable tool in clinical gene therapy, alone or in conjunction with ligand-targeted systems.

14.6   Applications

14.6.1   *In Vitro*

In 1977, a chemically created somatostatin gene was fused to an *Escherichia coli* β-galactosidase gene and introduced into a bacterium resulting in expression of the gene [91]. The alteration of both cells and cellular functions has expanded greatly since then to produce many new experimental applications. *In vitro* modeling is valuable in learning more about living systems without the involvement of animals, while potentially leading to *in vivo* applications. For example, liposomal delivery of the gene for adenomatous polyposis coli (APC), delivered to produce a potential anticancer agent, has produced excellent results in diminishing a human duodenal cancer in the presence of bile *in vitro* [92].

The proliferation of cells *in vitro* not only serves as a model for *in vivo* studies but also serves as a valuable component of most *ex vivo* applications. However, it is often the case that cell behavior differs between *in vitro* and *in vivo* environments. Hematopoietic stem cells show this characteristic, displaying substantial proliferation *in vivo* but dividing at a more conservative rate *in vitro*. Building on the retroviral overexpression of the homeobox B4 gene, a 40-fold increase in *in vitro* growth has been observed in mouse bone marrow cells [93]. However, the difference between cell behavior inside and outside the living organism along with the involvement of a wider array of cell types and cellular proteins continue to be significant barriers to scientific research, and are reasons why *in vitro* investigations are not enough to produce treatments ready for the clinic. *In vitro* research is still important because it allows greater control over experimental conditions for molecular studies and reduces the need for research animals.

In vitro gene therapy also has applications in the creation of regulation vehicles for the complexes needed *in vivo*. The delivery of L-dopa and dopamine to the nervous system by genetically altered cells can be regulated by embedding the cells in hydrogels, created and currently being tested *in vitro*, and implanting these systems into the affected area [94]. Another regulation device has been created from freeze-dried PEI/DNA complexes along with sucrose and PLGA to create a porous medium filled with encapsulated transfection complexes [95]. *In vitro*, the use of this PLGA sponge has caused expression of a reporter gene in adherent fibroblasts for 15 days.

Besides delivery regulation mechanisms there are *in vitro* transfection applications that treat cells as microfactories to create pharmaceutical products. An example of using gene delivery to utilize cells as manufacturing plants is the injection of engineered plasmids into silkworm eggs for the production of type III procollagen [96]. This application is a useful way of producing protein-based polymers with high sequence specificity and low polydispersity. This technology could be used to create scaffolds for cell-seeding in tissue engineering applications.

14.6.2   *Ex Vivo*

Ex vivo gene therapy involves the methods of *in vitro* cell therapy coupled with the reintroduction of the altered cells into an organism. Likely candidates for this approach are bone marrow progenitor cells of the adult (mesenchymal and hematopoetic) embryonic and adult stem cells. Delivery of the transfecting (or transducing) gene can be accomplished via any of the methods mentioned previously in this chapter. There has been much success with this application of gene therapy, which has progressed to the clinical trial stage. Transplantation of nerve growth factor (NGF)-producing grafts of neural progenitor cells into the septum and nucleus basalis of the rat cerebrum has been shown to diminish the atrophy of the brain associated with aging [97]. Several more *ex vivo* investigations are outlined in Section 14.7 of this chapter.

14.6.3 *In Vivo*

To transport gene delivery complexes into living organisms, there exist several strategies that do not entail surgery. The first method that might come to mind is injection via syringe. Injections can be very simple, entailing introduction of complexes into a vein [98], into the peritoneum [99], or directly into a tissue such as muscle [98] or tumor [100]. Even the blood–brain barrier can be circumvented via direct injection [101] or injection into the jugular vein using the brain-specific promoter GFAP [102]. Additional approaches for gene delivery to the central nervous system include injection into the cranial nerves for retrograde transport of PEI/DNA complexes into the brainstem [103], and intraveneous administration of lipoplexes conjugated with transferrin to effectively cross the blood–brain barrier via its transferrin receptors [104]. Permanent transfection via nonrandom integration has even been achieved via high-pressure tail vein injection, in which plasmids coding for bacteriophage ϕC31 integrase were co-injected with plasmids containing attB and a gene for human factor IX to yield site-specific genomic integration of the factor IX gene in rat hepatocytes at native pseudo attP sites [105]. This concept could be applied to achieve permanent transfection in many nonviral applications, which lend themselves to IV administration.

Another method of complex administration *in vivo* is through inhalation. Inhalation of aerosols is an intuitive choice for gene delivery to the respiratory system. In mice, aerosol administration of a liposome–DNA complex has yielded transfection results that persisted for 21 days [106]. More recently, aerosol delivery of adeno-associated viral vectors through the nasal cavity produced positive transfection that was apparent in the bloodstream in addition to that in the lungs [107]. Aerosol delivery is a logical method of delivering transfection complexes to the lungs, while intravenous delivery might be better suited for organs such as the spleen or liver, as demonstrated by results from a recent study that compared the two delivery methods for PEI/DNA transfection complexes [108].

Mucosal administration of gene delivery complexes by oral or rectal routes is another option for *in vivo* gene therapy. Mice that were fed chitosan-coated plasmid particles showed expression of the delivered gene in the intestinal epithelium [109]. Another delivery method involves particle-mediated gene delivery directly into the oral mucosa [110]. In the cited study, several marker and cancer-targeting genes were delivered and shown to be expressed in the oral cavities of canines. Transrectal complex administration (enema) has also been used to deliver genes to canines, where a recombinant adenovirus vector carrying a marker gene (β-galactosidase) was used to demonstrate successful transduction in the colon [111].

14.7 Clinical Applications

To date, over 900 reported gene therapy clinical trials are underway worldwide; approximately 2% of these are phase III investigations [112]. Well over half of the current trials utilize viruses as the gene delivery vehicle. Physical and chemical (especially liposomal) delivery methods are also represented. Over half of the current trials are cancer investigations [112]. Following is an overview of some of the more visible trials.

Severe combined immunodeficiency (SCID) was the first clinical application of gene therapy to go into human trials, commencing in 1990. The treatment involved removing bone marrow stem cells, altering them by exposure to healthy T-cells in culture, and then reintroducing the differentiated marrow-derived cells to the host organism [113]. Another malady involving immunocompromise is the HIV infection, a retrovirally caused disease that affects T-helper cells and can eventually lead to acquired immunodeficiency syndrome (AIDS). Phase I, II, and III clinical trials are currently underway for several applications of HIV-related gene therapy. *Ex vivo* manipulation of T lymphocytes with ribozymes, followed by reintroduction of the cells into the host, shows promise in that the T-cells have a greater survival time and are clinically safe for the subject [114]. A separate study conducted on twins was completed in 2002, and showed the safety and efficacy of *ex vivo* lymphocyte manipulation for expression of an anti-HIV gene (revTD) [115].

Clinical investigations for tissue- or organ-related maladies are also underway. Ischemic heart disease is the result of poor circulatory perfusion of cardiac muscle, commonly from coronary artery blockage. The delivery of vascular endothelial growth factor (VEGF), first described as a tumor-produced vascular permeability factor, has been shown to stimulate angiogenesis in several tissues including cardiac muscle [116]. Phase I studies have indicated that direct injection of naked [117] or adenovirally delivered [118] plasmids encoding VEGF into cardiac tissue yields both safe and effective transfection/transduction.

Cystic fibrosis is another tissue/organ malady. The disease is the result of a mutation in a gene encoding a cellular cAMP-mediated chloride channel, known as the cystic fibrosis transmembrane conductance regulator (CFTR). With inadequate chloride transport capabilities, affected cells will develop an osmotic gradient that is relieved by internalization of water from the extracellular environment. In cells that are bathed in mucus, such as lung airway epithelia, the result is a loss of water from the coating mucus, raising mucus viscosity and lowering the body's ability to transport it via ciliary movement. Bacterial colonization is often the result of such stasis. Clinical trials for the treatment of cystic fibrosis have been conducted using different target administrations. A trial of liposome-mediated transfection for cystic fibrosis patients showed the efficacy and safety of delivery of the CFTR gene to nasal epithelium cells [119]. Aerosolized adenoviral particles carrying CFTR cDNA have also been used clinically for transduction in the lungs [120].

Neurologic disorders have also been the target of clinical gene therapy trials. A phase I study of Alzheimer's disease involved the removal of primary fibroblasts from subjects, effectively transducing the harvested cells with nerve growth factor and reintroducing the cells into the brain [121] (also reviewed in Reference 122). The use of nerve growth factor to target the cholinergic neurons was employed with the aim of preventing neural degradation and elevating levels of acetylcholine transferase.

Gene therapy applications for cancer treatment are very well represented in the group of clinical trial investigations. Phase I and II trials directed at malignant gliomas are using a modified herpes virus to target glioma cells in order to cause an immune response to diminish the number of cancerous cells [123]. The referenced study is currently focused on drug dose amount in subjects with recurrent glioblastoma. A phase II study of ovarian cancer commenced in the year 2000 to determine the safety and efficacy of genetically altered herpes simplex thymidine kinase-producing cells (HSV-TK) delivered into cancer-affected ovaries [124]. The amount of cancer research that is underway is too large to allow a proper presentation of the area in sufficient detail here.

14.8 Summary

Gene therapy is a very diverse and rapidly growing field. Researchers in many areas, including biology, chemistry, physics, engineering, and medicine can find new applications for their creations and discoveries in gene therapeutics. Since molecular biology and genetic recombination laid the foundation for controlled genetic alteration, the field of gene therapy has expanded dramatically. The ability to alter cellular function through the introduction of exogenous genetic material has drawn considerable hope that this technology can be used to discover new disease treatments as well as explicate some of the mysteries of life at the level of basic science. As laboratory and clinical trials proceed and knowledge of the area becomes more rational and objective within the general public, gene therapy should prove to deliver beneficial and lasting advances from the world of biomolecular science.

References

[1] Wolff, J. and Lederberg, J., An early history of gene transfer and therapy, *Hum. Gene Ther.*, 5, 469, 1994.

[2] Little E., et al., The glucose-regulated proteins (GRP78 and GRP94): functions, gene regulation, and applications, *Crit. Rev. Eukaryot. Gene Expr.*, 4, 1–18, 1994.

[3] Haider, M., Megeed, Z., and Ghandehari, H., Genetically engineered polymers: status and prospects for controlled release, *J. Control. Release*, 95, 1, 2004.

[4] Sun, X., Wu, Z., and Hu, J., Suicide gene therapy of hepatocellular carcinoma and delivery procedure and route of therapeutic gene *in vivo, HPBD Int.*, 1, 373, 2003.

[5] Nakaya, T. et al., Decoy approach using RNA–DNA chimera oligonucleotides to inhibit the regulatory function of human immunodeficiency virus type 1 Rev protein, *Antimicrob. Agents Chemother.*, 41, 319, 1997.

[6] Browning, C. et al., Potent inhibition of human immunodeficiency virus type 1 (HIV-1) gene expression and virus production by an HIV-2 Tat activation-response RNA decoy, *J. Virol.*, 73, 5191, 1999.

[7] Pace, B.S., Qian, X., and Ofori-Acquah, S.F., Selective inhibition of beta-globin RNA transcripts by antisense RNA molecules, *Cell. Mol. Biol.*, 50, 43, 2004.

[8] Li, M.J. et al., Specific killing of Ph+ chronic myeloid leukemia cells by a lentiviral vector-delivered anti-bcr/abl small hairpin RNA, *Oligonucleotides*, 13, 401, 2003.

[9] Anderson, J. et al., Potent suppression of HIV type 1 infection by a short hairpin anti-CXCR4 siRNA, *AIDS Res. Hum. Retroviruses*, 19, 699, 2003.

[10] Elbashir, S.M., Lendeckel, W., and Tuschl, T., RNA interference is mediated by 21- and 22-nucleotide RNAs, *Genes Dev.*, 15, 188, 2001.

[11] Elbashir, S.M. et al., Duplexes of 21-nucleotide RNAs mediate RNA interference in cultured mammalian cells, *Nature*, 411, 494, 2001.

[12] Miller, V.M. et al., Allele-specific silencing of dominant disease genes, *Proc. Natl Acad. Sci. USA*, 100, 7195, 2003.

[13] Miller, V. et al., Allele-specific silencing of dominant disease genes, *Proc. Natl Acad. Sci. USA*, 100, 7195, 2003.

[14] Douglas, J.T., Adenovirus-mediated gene delivery: an overview, *Meth. Mol. Biol.*, 246, 3, 2004.

[15] Huser, A. and Hofmann, C., Baculovirus vectors: novel mammalian cell gene-delivery vehicles and their applications, *Am. J. Pharmacogenom.*, 3, 53, 2003.

[16] Boldogkoi, Z. and Nogradi, A., Gene and cancer therapy — pseudorabies virus: a novel research and therapeutic tool? *Curr. Gene Ther.*, 3, 155, 2003.

[17] Goins, W.F. et al., Delivery using herpes simplex virus: an overview, *Meth. Mol. Biol.*, 246, 257, 2004.

[18] Lu, Y., Recombinant adeno-associated virus as delivery vector for gene therapy — a review, *Stem Cells Dev.*, 13, 133 2004.

[19] Gafni, Y. et al., Gene therapy platform for bone regeneration using an exogenously regulated, AAV-2-based gene expression system, *Mol. Ther.*, 9, 587, 2004.

[20] Hodge, J.W. et al., Modified vaccinia virus ankara recombinants are as potent as vaccinia recombinants in diversified prime and boost vaccine regimens to elicit therapeutic antitumor responses, *Cancer Res.*, 63, 7942, 2003.

[21] Hu, Y. et al., Yaba-like disease virus: an alternative replicating poxvirus vector for cancer gene therapy. *J. Virol.*, 75, 10300, 2001.

[22] Cockrell, A.S. and Kafri, T., HIV-1 vectors: fulfillment of expectations, further advancements, and still a way to go, *Curr. HIV Res.*, 1, 419, 2003.

[23] Cheng, L. et al., Human immunodeficiency virus type 2 (HIV-2) vector-mediated *in vivo* gene transfer into adult rabbit retina, *Curr. Eye Res.*, 24, 196, 2002.

[24] Kafri, T., Gene delivery by lentivirus vectors an overview, *Meth. Mol. Biol.*, 246, 367, 2004.

[25] Wolkowicz, R., Nolan, G.P., and Curran, M.A., Lentiviral vectors for the delivery of DNA into mammalian cells, *Meth. Mol. Biol.*, 246, 391, 2004.

[26] Rainov, N.G. and Ren, H., Clinical trials with retrovirus mediated gene therapy — what have we learned? *J. Neuro-oncol.*, 65, 227, 2003.

[27] Lilley, C. et al., Multiple immediate-early gene-deficient herpes simplex virus vectors allowing efficient gene delivery to neurons in culture and widespread gene delivery to the central nervous system *in vivo, J. Virol.*, 75, 4343, 2001.

[28] Cao, B. and Huard, J., Gene transfer to skeletal muscle using herpes simplex virus-based vectors, *Meth. Mol. Biol.*, 246, 301, 2004.

[29] Billelo, J. et al., Transient disruption of intercellular junctions enables baculovirus entry into nondividing hepatocytes, *J. Virol.*, 75, 9857, 2001.

[30] Hu, Y. et al., Yaba-like disease virus: an alternative replicating poxvirus vector for cancer gene therapy, *J. Virol.*, 75, 10500, 2001.

[31] Garrett, D. et al., In utero recombinant adeno-associated virus gene transfer in mice, rats, and primates, *BMC Biotech.*, 3, 16, 2003.

[32] Ahmed, B. et al., Efficient delivery of Cre-recombinase to neurons *in vivo* and stable transduction of neurons using adeno-associated and lentiviral vectors, *BMC Neurosci.*, 5, 4, 2004.

[33] Mastakov, M. et al., Immunological aspects of recombinant adeno-associated virus delivery to the mammalian brain, *J. Virol.*, 76, 8446, 2002.

[34] Zufferey, R., Self-inactivating lentivirus vector for safe and efficient *in vivo* gene delivery, *J. Virol.*, 72, 9873, 1998.

[35] Curran, M. et al., Efficient transduction of nondividing cells by optimized feline immunodeficiency virus vectors, *Mol. Ther.*, 1, 31, 2000.

[36] Reiser, J. et al., Development of multigene and regulated lentivirus vectors, *J. Virol.*, 74, 10589, 2000.

[37] Marschall, P., Malik, N., and Larin, Z., Transfer of YACs up to 2.3 Mb intact into human cells with polyethylenimine, *Gene Ther.*, 6, 1634, 1999.

[38] Vanderbyl, S., MacDonald, N., and de Jong, G., A flow cytometry technique for measuring chromosome-mediated gene transfer, *Cytometry*, 44, 100, 2001.

[39] Rossenberg, S. van. et al., Stable polyplexes based on arginine-containing oligopeptides for *in vivo* gene delivery, *Nature*, 11, 457, 2004.

[40] Chen, Q. et al., Branched co-polymers of histidine and lysine are efficient carriers of plasmids, *Nucleic Acids Res.*, 29, 1334, 2001.

[41] Toncheva V. et al., Novel vectors for gene delivery formed by self-assembly of DNA with poly(-lysine) grafted with hydrophilic polymers, *Biochim. Biophys. Acta*, 1380, 354, 1998.

[42] Godbey, W., Wu, K., and Mikos A., Size matters: molecular weight affects the efficiency of poly(ethylenimine) as a gene delivery vehicle, *J. Biomed. Mater. Res.*, 45, 268, 1999.

[43] Fischer, D. et al., A novel non-viral vector for DNA delivery based on low molecular weight, branched polyethylenimine: effect of molecular weight on transfection efficiency and cytotoxicity, *Pharm. Res.*, 16, 1273, 1999.

[44] Tang, G. et al., Polyethylene glycol modified polyethylenimine for improved CNS gene transfer: effects of PEGylation extent, *Biomaterials*, 24, 2351, 2004.

[45] Boussif, O. et al., A versatile vector for gene and oligonucleotide transfer into cells in culture and *in vivo*: polyethylenimine, *Proc. Natl Acad. Sci. USA*, 92, 7297, 1995.

[46] Manuel, W., Zheng, J., and Hornsby, P., Transfection by polyethyleneimine-coated microspheres, *J. Drug Target.*, 9, 15, 2001.

[47] Kuskowska-Latallo, J. et al., Efficient transfer of genetic material into mammalian cells using Starburst polyamidoamine dendrimers, *Proc. Natl Acad. Sci. USA*, 93, 4897, 1996.

[48] Maruyama-Tabata, H. et al., Effective suicide gene therapy *in vivo* by EBV-based plasmid vector coupled with polyamidoamine dendrimer, *Gene Ther.*, 7, 53, 2000.

[49] Mao, H.Q. et al., Chitosan-DNA nanoparticles as gene carriers: synthesis, characterization and transfection efficiency, *J. Control. Release*, 70, 399, 2001.

[50] Mansouri, S. et al., Chitosan-DNA nanoparticles as non-viral vectors in gene therapy: strategies to improve transfection efficacy, *Eur. J. Pharm. Biopharm.*, 57, 1, 2004.

[51] Ishii, T., Okahata, Y., and Sato, T., Mechanism of cell transfection with plasmid/chitosan complexes, *Biochim. Biophys. Acta*, 1514, 51, 2001.

[52] Padmanabhan, K. and Smith, T.A., Preliminary investigation of modified alginates as a matrix for gene transfection in a HeLa cell model, *Pharm. Dev. Technol.*, 7, 97, 2002.

[53] Gonzalez Ferreiro M. et al., Characterization of alginate/poly-L-lysine particles as antisense oligonucleotide carriers, *Int. J. Pharm.*, 239, 47 2002.

[54] Li, Z. et al., Controlled gene delivery system based on thermosensitive biodegradable hydrogel, *Pharm. Res.*, 20, 884, 2003.

[55] Girao Da Cruz, M., Simoes, S., and Pedroso De Lima, M., Improving lipoplex-mediated gene transfer into C6 glioma cells and primary neurons, *Exp. Neurol.*, 187, 65, 2004.

[56] Liu, F., Huang, L., and Liu, D., Factors controlling the efficiency of cationic lipid-mediated transfection *in vivo* via intravenous administration, *Gene Ther.*, 4, 517, 1997.

[57] Farhood, H., Serbina, N., and Huang, L., The role of dioleoyl phosphatidylethanolamine in cationic liposome mediated gene transfer, *Biochim. Biophys. Acta*, 1235, 289, 1995.

[58] King, R., Gene delivery to mammalian cells by microinjection, *Meth. Mol. Biol.*, 245, 167, 2004.

[59] Wilmut, I. et al., Viable offspring derived from fetal and adult mammalian cells, *Nature*, 385, 810, 1997.

[60] Hosaka, K. et al., Cloned mice derived from somatic cell nuclei, *Hum. Cell*, 13, 237, 2000.

[61] Neumann, E. et al., Gene transfer into mouse myeloma cells by electroporation in high electric fields, *EMBO J.*, 1, 841, 1982.

[62] Golzio, M., Teissié, J., and Rols, M., Direct visualization at the single-cell level of electrically mediated gene delivery, *Proc. Natl Acad. Sci. USA*, 99, 1292, 2002.

[63] Mir, L. et al., High-efficiency gene transfer into skeletal muscle mediated by electric pulses, *Proc. Natl Acad. Sci USA*, 96, 4262, 1999.

[64] Wang, Y. et al., Combination of electroporation and DNA/dendrimer complexes enhances gene transfer into murine cardiac transplants, *Am. J. Transplant.*, 1, 334, 2001.

[65] Schakowski, F. et al., Novel non-viral method for transfection of primary leukemia cells and cell lines, *Genet. Vaccines Ther.*, 2, 1, 2004.

[66] Sun, W. et al., *In vivo* cytokine gene transfer by gene gun reduces tumor growth in mice, *Proc. Natl Acad. Sci. USA*, 92, 2889, 1995.

[67] Matsuno Y. et al., Nonviral gene gun mediated transfer into the beating heart, *ASAIO J.*, 49, 641, 2003.

[68] Jeon, E., Kim, H.D., and Kim, J.S., Pluronic-grafted poly-(L)-lysine as a new synthetic gene carrier, *J. Biomed. Mater. Res.*, 66A, 854, 2003.

[69] Joubert, D. et al., A note on poly-L-lysine-mediated gene transfer in HeLa cells, *Drug Deliv.*, 10, 209, 2003.

[70] Tan, P.H. et al., Transferrin receptor-mediated gene transfer to the corneal endothelium, *Transplantation*, 71, 552, 2001.

[71] Van Rossenberg S.M. et al., Targeted lysosome disruptive elements for improvement of parenchymal liver cell-specific gene delivery, *J. Biol. Chem.*, 277, 45803, 2002.

[72] Godbey, W., Wu, K., and Mikos, A., Tracking the intracellular path of poly(ethylenimine)/DNA complexes for gene delivery, *Proc. Natl Acad. Sci. USA*, 96, 5177, 1999.

[73] Godbey, W.T. et al., Poly(ethylenimine)-mediated transfection: a new paradigm for gene delivery, *J. Biomed. Mater. Res.*, 51, 321, 2000.

[74] Suh, J., Wirtz, D., and Hanes, J., Efficient active transport of gene nanocarriers to the cell nucleus, *Proc. Natl Acad. Sci. USA*, 100, 3878, 2003.

[75] Brunner, S. et al., Cell cycle dependence of gene transfer by lipoplex, polyplex and recombinant adenovirus, *Gene Ther.*, 7, 7401, 2000.

[76] Escriou, V. et al., Critical assessment of the nuclear import of plasmid during cationic lipid-mediated gene transfer, *J. Gene Med.*, 3, 179, 2001.

[77] Zanta, M. et al., Gene delivery: a single nuclear localization signal peptide is sufficient to carry DNA to the cell nucleus, *Proc. Natl Acad. Sci. USA*, 96, 91, 1999.

[78] Aronsohn, A.I. and Hughes, J.A., Nuclear localization signal peptides enhance cationic liposome-mediated gene therapy, *J. Drug Target.*, 5, 163, 1998.

[79] Bremner, K. et al., Factors influencing the ability of nuclear localization sequence peptides to enhance nonviral gene delivery, *Bioconjug. Chem.*, 15, 152, 2004.

[80] Parveen, Z. et al., Cell-type-specific gene delivery into neuronal cells *in vitro* and *in vivo*, *Virology*, 314, 74, 2003.

[81] White, S.J. et al., Identification of peptides that target the endothelial cell-specific LOX-1 receptor, *Hypertension*, 37, 449, 2001.

[82] Argnani, R. et al., Specific targeted binding of herpes simplex virus type 1 to hepatocytes via the human hepatitis B virus preS1 peptide, *Gene Ther.*, 2004.

[83] Voinea, M., Binding and uptake of transferrin-bound liposomes targeted to transferrin receptors of endothelial cells, *Vascul. Pharmacol.*, 39, 13, 2002.

[84] Zhao, X.B. and Lee, R.J., Tumor-selective targeted delivery of genes and antisense oligodeoxyribo-nucleotides via the folate receptor, *Adv. Drug Deliv. Rev.*, 56, 1193, 2004.

[85] Yang, Q. et al., Development of novel cell surface CD34-targeted recombinant adenoassociated virus vectors for gene therapy, *Hum. Gene Ther.*, 9, 1929, 1998.

[86] Kunath, K. et al., Galactose–PEI–DNA complexes for targeted gene delivery: degree of substitution affects complex size and transfection efficiency, *J. Control. Release*, 88, 159, 2003.

[87] Wilcox, D.A. et al., Integrin alphaIIb promoter-targeted expression of gene products in mega-karyocytes derived from retrovirus-transduced human hematopoietic cells, *Proc. Natl Acad. Sci. USA*, 96, 9654, 1999.

[88] Park, C. et al., Targeting of therapeutic gene expression to the liver by using liver-type pyruvate kinase proximal promoter and the SV40 viral enhancer active in multiple cell types, *Biochem. Biophys. Res. Commun.*, 314, 131, 2004.

[89] Godbey, W.T. and Atala, A., Directed apoptosis in Cox-2-overexpressing cancer cells through expression-targeted gene delivery, *Gene Ther.*, 10, 1519, 2003.

[90] Ikegami, S. et al., Treatment efficiency of a suicide gene therapy using prostate-specific membrane antigen promoter/enhancer in a castrated mouse model of prostate cancer, *Cancer Sci.*, 95, 367, 2004.

[91] Itakura, K. et al., Expression in *Escherichia coli* of a chemically synthesized gene for the hormone somatostatin, *Science*, 198, 1056, 1977.

[92] Lee, J. et al., *In vitro* model for liposome-mediated adenomatous polyposis coli gene transfer in a duodenal model, *Dis. Col. Rec.*, 47, 219, 2004.

[93] Krosl, J., *In vitro* expansion of hematopoietic stem cells by recombinant TAT-HOXB4 protein, *Nat. Med.*, 9, 1428, 2003.

[94] Li, R.H. et al., Dose control with cell lines used for encapsulated cell therapy, *Tissue Eng.*, 5, 453, 1999.

[95] Huang, Y.C. et al., Fabrication and *in vitro* testing of polymeric delivery system for condensed DNA, *J. Biomed. Mat. Res.*, 67A, 1384, 2003.

[96] Tomita, M. et al., Transgenic silkworms produce recombinant human type III procollagen in cocoons, *Nat. Biotechnol.*, 21, 52, 2003.

[97] Martínez-Serrano, A. et al., *Ex vivo* nerve growth factor gene transfer to the basal forebrain in presymptomatic middle-aged rats prevents the development of cholinergic neuron atrophy and cognitive impairment during aging, *Proc. Natl Acad. Sci. USA*, 95, 1858, 1998.

[98] Magin-Lachmann, C. et al., *In vitro* and *in vivo* delivery of intact BAC DNA — comparison of different methods, *J. Gene Med.*, 6, 195, 2004.

[99] Brown, C.B., Use of the peritoneal cavity for therapeutic delivery, *Perit. Dial. Int.*, 19, 512, 1999.

[100] Shariat, S.F., Adenovirus-mediated transfer of inducible caspases: a novel "death switch" gene therapeutic approach to prostate cancer, *Cancer Res.*, 61, 2562, 2001.

[101] Xia, C.F. et al., Kallikrein gene transfer protects against ischemic stroke by promoting glial cell migration and inhibiting apoptosis, *Hypertension*, 43, 452, 2004.

[102] Shi, N. et al., Brain-specific expression of an exogenous gene after i.v. administration, *Proc. Natl Acad. Sci. USA*, 98, 12754, 2001.

[103] Wang, S. et al., Transgene expression in the brain stem effected by intramuscular injection of polyethylenimine/DNA complexes, *Mol. Ther.*, 3, 658, 2001.

[104] Shi, N. and Pardridge, W., Noninvasive gene targeting to the brain, *Proc. Natl Acad. Sci. USA*, 97, 7567, 2000.

[105] Olivares, E.C. et al., Site-specific genomic integration produces therapeutic factor IX levels in mice, *Nat. Biotechnol.*, 20, 1124, 2002.

[106] Stribling, R. et al., Aerosol gene delivery *in vivo*, *Proc. Natl Acad. Sci. USA*, 89, 11277, 1992.

[107] Auricchio, A. et al., Noninvasive gene transfer to the lung for systemic delivery of therapeutic proteins, *J. Clin. Invest.*, 110, 499, 2002.

[108] Koshkina, N.V. et al., Biodistribution and pharmacokinetics of aerosol and intravenously administered DNA-polyethyleneimine complexes: optimization of pulmonary delivery and retention, *Mol. Ther.*, 8, 249, 2003.

[109] Chen, J. et al., Transfection of mEpo gene to intestinal epothelium *in vivo* mediated by oral delivery of chitosan-DNA nanoparticles, *World J. Gastroenterol.*, 10, 112, 2004.

[110] Keller, E.T. et al., *In vivo* particle-mediated cytokine gene transfer into canine oral mucosa and epidermis, *Cancer Gene Ther.*, 3, 186, 1996.

[111] Weld, K.J. et al., Transrectal gene therapy of the prostate in the canine model, *Cancer Gene Ther.*, 9, 189, 2002.

[112] *Journal of Gene Medicine* Website, http://www.wiley.co.uk/genmed/clinical, Accessed 10/18/2005.

[113] Cavazzana-Calvo, M. et al., Gene therapy of human severe combined immunodeficiency (SCID)-X1 disease, *Science*, 288, 627, 2000.

[114] Wong-Staal, F., Poeschla, E.M., and Looney, D.J., A controlled, phase 1 clinical trial to evaluate the safety and effects in HIV-1 infected humans of autologous lymphocytes transduced with a ribozyme that cleaves HIV-1 RNA, *Hum. Gene Ther.*, 9, 2407, 1998.

[115] Twins Study of Gene Therapy for HIV Infection, National Human Genome Research Institute (NHGRI).

[116] Senger, D.R. et al., Tumor cells secrete a vascular permeability factor that promotes accumulation of ascites fluid, *Science*, 219, 983, 1983.

[117] Losordo, D. et al., Gene therapy for myocardial angiogenesis: initial clinical results with direct myocardial injection of phVEGF165 as sole therapy for myocardial ischemia, *Circulation*, 98, 2800, 1998

[118] Rosengart, T.K. et al., Six-month assessment of a phase I trial of angiogenic gene therapy for the treatment of coronary artery disease using direct intramyocardial administration of an adenovirus vector expressing the VEGF121 cDNA, *Ann. Surg.*, 230, 466, 1999.

[119] Sorscher, E., Study chair, Phase I Study of Liposome-Mediated Gene Transfer in Patients with Cystic Fibrosis, University of Alabama.

[120] Moss, R.B. et al., Repeated adeno-associated virus serotype 2 aerosol-mediated cystic fibrosis transmembrane regulator gene transfer to the lungs of patients with cystic fibrosis: a multicenter, double-blind, placebo-controlled trial, *Chest*, 125, 509, 2004.

[121] Tuszynski, M., Prin. invest., Gene therapy for Alzheimer's Disease Clinical Trial, University of California, San Diego.

[122] Tuszynski, M.H. and Blesch, A., Nerve growth factor: from animal models of cholinergic neuronal degeneration to gene therapy in Alzheimer's disease, *Prog. Brain Res.*, 146, 441, 2004.

[123] de Haan, Hans, Prin. invest., Gene Therapy in Treating Patients with Recurrent Malignant Glioma, University of Alabama at Birmingham Comprehensive Cancer Center, Birmingham, AL.

[124] Link, C., Prin. invest., Gene Therapy in Treating Women With Refractory or Relapsed Ovarian Epithelial Cancer, Fallopian Tube Cancer, or Peritoneal Cancer, Human Gene Therapy Research Institute, Des Moines, Iowa.

15

Tissue Engineering Bioreactors

Jose F. Alvarez-Barreto
Vassilios I. Sikavitsas
University of Oklahoma

15.1 Introduction

The main goal of tissue engineering is the creation of artificial tissue having the ability to repair or simply replace lost or damaged tissue. Common tissue engineering strategies involve the extraction of cells from a small piece of tissue and their *in vitro* culture for later implantation using a carrier that allows the formation of new tissue (Figure 15.1) [1]. Most approaches in this field are based on common bioactive factors, consisting of cells (generally stem cells, or progenitor cells), a scaffolding material, and growth and differentiation factors [2]. The *in vitro* creation of an efficient construct can be accelerated by applying certain stimuli that can elicit specific responses to the cells. Stimulation can be done in two major ways: chemically and electro/mechanically.

Chemical stimulation is carried out by using growth and differentiation factors specific for different responses. Growth factors play a major role in cell division, matrix synthesis, and tissue differentiation [3]. Examples of these proteins are: bone morphogenetic proteins (BMPs), which have been demonstrated to induce the differentiation of mesenchymal stem cells into an osteoblastic lineage (BMP-2 and BMP-7), and the vascular endothelial growth factor that greatly enhances angiogenesis [4,5]. The need for *in vitro* mechanical stimulation in tissue engineering is drawn from the fact that most tissues function under specific biomechanical environments *in vivo*. These environments play a key role in tissue remodeling and regeneration. The stresses can be translated into different kinds of forces that range from load bearing to hydrodynamic forces due to fluid flow [6]. Thus, the mechanochemical microenvironment that progenitor

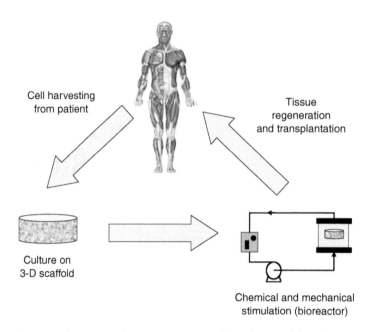

FIGURE 15.1 Basic steps in the process of tissue engineering: cells are harvested from the patient and proliferated in a three-dimensional environment, followed by the application of mechanical and chemical stimuli in a bioreactor prior to implantation.

cells grow into is expected to control the fate of these cells while undergoing differentiation toward different lineages.

A bioreactor is generally defined as a device capable of creating the proper environment for the creation of a certain biological product [21]. Therefore, a bioreactor is described as a simulator, a device in which biological as well as biochemical processes can be carried out. In tissue engineering, bioreactors are used to impart certain forces that imitate different electromagnetic and mechanical stimuli occurring in the body. However, these devices are not limited to the sole application of electromechanical stimuli; they must meet other requirements in order to create grafts that, when implanted, will lead to the regeneration of damaged organs. A bioreactor must efficiently transport nutrients and oxygen to the construct, maintaining an appropriate concentration in solution. In most tissue engineering applications, a scaffold is seeded with cells and supports the formation of extracellular matrix (ECM). Consequently, the bioreactor has to induce a homogeneous cell distribution throughout these structures in order to generate a uniformly distributed ECM. Tissue engineering bioreactors can be used for cell seeding and long-term cultures.

This chapter describes some of the most popular bioreactor designs available for the engineering of different tissues. Hydrodynamic conditions and transport phenomena considerations are addressed, as well as applications to functional tissues and unique devices for specific applications. Design considerations, challenges, and new directions are also presented.

15.2   Most Common Bioreactors in Tissue Engineering

The choice of a bioreactor to cultivate three-dimensional constructs depends upon the tissue to be engineered and its functional biomechanical environment. Emulation of physiological conditions is the main objective when developing these kinds of systems, and this issue has been addressed in different ways. The incorporation of convective forces has become a common characteristic among most bioreactors. In this section, we describe some of the most common bioreactors found in the engineering of several functional tissues such as bone, cartilage, and cardiovascular applications, among others.

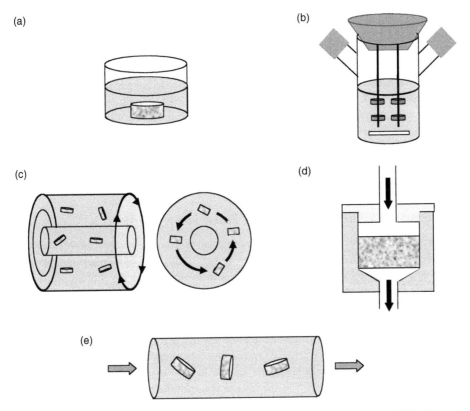

FIGURE 15.2 Common bioreactors used in tissue engineering. (a) Static culture, (b) spinner flask, (c) rotating wall, (d) perfusion system, and (e) perfused column.

15.2.1 Spinner Flask

The spinner flask (Figure 15.2b) represents one of the simplest bioreactor models. It was first designed with the idea to use convection in order to maintain a well-mixed system. The scaffolds are threaded into needles connected to the cover of the flask, and submerged in the culture medium. Convection is generated through the usage of a magnetic stir bar or a shaft that continuously mixes the media surrounding the scaffolds, providing a practically homogenous distribution of oxygen and nutrients [7,8]. The fluid dynamic environment at the external surface of the scaffolds is turbulent and characterized by the existence of eddies that may enhance the transport of nutrients into the porosity, and locally expose cells residing at the exterior of the construct to relatively high shear forces. The magnitude of the shear stresses can vary significantly between different locations; therefore, not all the cells are exposed to the same shear stresses. The presence of convective forces external to the scaffolds may not suppress concentration gradients appearing deep inside large three-dimensional constructs, where diffusion is the controlling mechanism of nutrient transport [9,25].

15.2.2 Rotating-Wall Vessels

Initially designed by NASA as a microgravity environment for cell culture, the rotating-wall bioreactor (Figure 15.2c) is now widely used in the formation of engineered bone, cartilage, and other tissues [7,8,10–12]. This device consists of two concentric cylinders whose annular space contains the cell culture medium [13]. The inner cylinder is static and permeable to allow gas exchange for oxygen supply. The outer cylinder, on the other hand, is impermeable and horizontally rotates at a speed that causes centrifugal forces that can balance, if tuned properly, the gravitational forces; thus, generating a pseudo microgravity environment [7,13,14]. Unlike the spinner flask, in the rotating-wall vessel the fluid flow is mostly laminar and the range of shear forces experienced by the cells at the outer surface is relatively narrow, with the existence of a stagnation zone at the upstream edge. As reported by Williams et al. [15], shear stresses decrease in the direction of flow and no significant variations from scaffold to scaffold are observed. Medium can be recirculated between the annular space and an external gas membrane. A modification of the original design, called rotating-wall perfused-vessel bioreactor, includes the rotation of the inner cylinder. In this model, media is perfused from the vessel's end cap to the pores of the inner cylinder [14].

The conditions of operation of a rotating-wall vessel must be carefully controlled. Large rotation speeds of the outer wall will affect mass transport since most of the inlet fluid bypasses the vessel volume, whereas an increase on the differential rotation enhances the radial and axial distribution of the fluid at low mean shear stresses [14].

15.2.3 Perfusion Chambers and Flow Perfusion Systems

Flow perfusion bioreactors provide continuous flow through chambers were the scaffolds are located. The perfusion column (Figure 15.2e) was one of the first designs of this kind of bioreactors. Culture medium is continuously recirculated through the chamber, thus improving the transport of nutrients and oxygen to the constructs [16,17,22]. Nevertheless, the flow of medium in these chambers is distributed between the inner network of the construct and its surroundings, minimizing convective flow through the scaffold [18]. To ensure that the flow of medium occurs exclusively through the porosity of the material, new designs of flow perfusion bioreactors include the confining of the construct in chambers (Figure 15.2d). In this way, a more controllable flow is achieved and nutrient transport limitations are virtually eliminated. Internal flow can also expose the cells inside the scaffold to fluid shear forces that have been known to be stimulatory for some cell types such as osteoblasts and endothelial cells [19,20]. A standard design of this kind of reactors does not exist, but all of them are based on the same principle. A more detailed description of a perfusion system is given later in this chapter.

15.3 Cell Seeding in Bioreactors

The first step to culturing cells in a three-dimensional environment is the seeding of scaffolds [21]. Along with the characteristics of the material, this process plays a crucial role in the development of efficient constructs for tissue engineering. Seeding of scaffolds determines the initial number of cells in the construct, as well as their spatial distribution throughout the matrix. Consequently, proliferation, migration, and the specific phenotypic expression of the engineered tissue will be affected by the utilized seeding technique [22]. In the case of tissues that require a fibrous or porous material, static seeding has been the most widely used method of cell seeding (Figure 15.2a). Burg et al. [26] compared different seeding techniques using rat aortic cells in polyglycolide fibrous meshes. Static seeding produced the poorest cellular distribution. In addition to preventing a homogeneous spatial distribution of the cells, static seeding also produces a low yield [23,26]. Holy et al. [23] reported a 25% efficiency of attachment after seeding 0.5 to 10×10^6 cells on porous PLGA 75/25 scaffolds. A low yield diminishes the development of specific functions related to cell–cell interactions and increases the required amount of cells; therefore, the usage of new seeding techniques becomes imperative.

In order to address these issues, researchers have incorporated convection into the process of cell seeding, suppressing some of the mass transfer limitations encountered in the static procedure. Spinner flask bioreactors (Figure 15.2b) have been implemented to create convection and, thereby, hydrodynamic forces that could help increase mass transport. Poly(glycolic acid) (PGA) scaffolds were threaded onto needles and chondrocytes suspensions with a total number of cells between 2×10^6 and 10×10^6 were used. A yield of 60% was obtained after 2 h of seeding. A more uniform distribution of the cells in the scaffold was seen (compared to the static seeding); nonetheless, the concentration of cells in the outer layer of the construct was 60 to 70% higher than that in the bulk [24]. This behavior may be due to the poor convection to the interior of the scaffold, making migration the only way for cells to reach the interior of the scaffold.

It has been reported that, in the spinner flask, the shear forces at the external surface of the scaffold are highly nonuniform. Such variability may influence the homogeneity of the seeded cells even when considering only the external surface area [25]. To avoid such problem, Mauney et al. rotated the scaffold every 2 h for the first 6 h of seeding so that each face of the construct could be exposed to the flow field, reaching a homogenous distribution of the cells and a higher efficiency than that of the static methodology. However, despite the high efficiency of seeding achieved with the spinner flask bioreactor, a more homogeneous distribution of the cells throughout the construct volume is still desired.

One way to guarantee mass transfer to the interior of the scaffold and a better distribution of cells is by applying perfusion [26,27]. In this technique, the construct is press fitted into a chamber, and the cell suspension is flowed through it (Figure 15.2c). Li et al. used a depth filtration system to seed poly (ethylene terephthalate) matrices at a rate of 1 ml/min. The cell suspension was recycled to increase the yield. Cell density increased along with the inoculation cell number, with an efficiency of about 65%, while with the static seeding, the yield stayed constant and lower than that achieved with the perfusion [22]. Similarly, Kim et al. seeded hepatocytes on polymeric matrices using a flow perfusion system. A suspension of rat hepatocytes at a density of 5×10^6 cells/ml was pumped through decellularized bone matrices at a flow rate of 1.5 ml/min for 4 h. A total of approximately 4.4×10^6 cells were attached to the matrix, which was considered successful. Furthermore, scanning electron microscopy and histology confirmed a uniform distribution of hepatocytes throughout the scaffold.

Wendt et al. [27] monitored seeding efficiency and uniformity of static, spinner flask, and perfusion systems. Using the same inoculation concentrations, there was not statistical significance among the efficiencies of the static and perfused techniques, both producing a larger yield than the spinner flask. Uniformity, however, was optimized by the perfusion apparatus, while the static and the spinner flask generated cell-scaffold constructs with low spatial uniformity [27].

15.4 Bioreactor Applications in Functional Tissues

After being seeded with the specific type of cells needed for the application, scaffolds must be subjected to longer periods of culture under the desired physical stimuli. The appropriate stimulation relies on the mechanical, biological, and ultrastructural characteristics of the native tissue. Bioreactors for engineered vascular grafts must mimic the natural fluid dynamic conditions of blood flow, including relatively high flow rates and pulsatile flow [42]. Bone tissue has also been shown to be stimulated by fluid flow [19]. Engineered ligaments and tendons need mechanical loading to emulate the conditions that they normally experience [87]. In this section we highlight some of the most important bioreactor applications in the engineering of different tissues; different designs and their efficiency in the development of inductive constructs are discussed.

15.4.1 Tissues of the Cardiovascular System

Several types of bioreactors have been used for the regeneration of tissues from the cardiovascular system. Spinner flasks and rotating-wall vessels have been employed in the regeneration of heart tissue. Cardiac myocytes were cultured on PGA scaffolds in a spinner flask, rotating-wall vessels and perfusion systems [28,29]. After 14 days of culture, the medium in the rotating-wall vessels showed higher levels of oxygen. The cell number in the spinner flask and rotating-wall reactor was larger than that of the static culture. Likewise, the uniformity of extracellular matrix throughout the scaffold was more homogeneous in the rotating-wall vessels. The engineered constructs had similar structural characteristics to that of the native tissue, and the cultured cells secreted proteins related to mature cardiac myocytes (myosin, troponin-T, tropomyosin, etc.). However, the cell density was always lower in the interior of the constructs [29].

15.4.1.1 Vascular Grafts

The emulation of physiological conditions in tissue-engineered vascular grafts is extremely important. Factors affecting the successful generation of a vascular graft include the selection of the appropriate scaffolding material, the cell type (smooth muscle cells, endothelial cells, and fibroblasts), and the culturing conditions [30,35]. Evaluation of mechanical properties is necessary to determine the quality of the engineered construct. Mechanical strength is directly related to the matrix structure, especially to the alignment of smooth muscle cells and collagen fibers [30,31].

Blood experiences rapid pulsating flow in the body with a velocity of about 33 cm/sec in the aorta and 0.33 mm/sec in capillaries [32]. Moreover, the conditions of flow will determine the differentiation path and properties of endothelial cells [6,33]. The extracellular matrix organization and mechanical properties of the graft, such as burst pressure and suture retention, are determinant factors of the graft quality [34]. Different systems with pulsatile flow have been designed in order to achieve these goals [35–38].

Static culture of endothelial cells seeded on different synthetic or natural polymers resulted in grafts with poor mechanical properties and morphological characteristics different from that of the native tissue [39]. In an attempt to mimic the natural environment of blood vessels, fluid flow has been employed during *in vitro* culture of smooth muscle cells seeded on PGA scaffolds. Incorporation of pulsation improved the mechanical properties of the constructs and their histological appearance resembled that of native arteries. It is important to point out that plain pulsation generated grafts with inferior mechanical properties [40]. The implementation of biomimetic bioreactors that utilize pulsatile flow in the physiological range appears to stimulate the formation of an organized matrix similar to that of the native tissue [41,42].

An example of a pulsatile-flow bioreactor for vascular grafts is shown in Figure 15.3. This design, used in the Laboratory of Cardiovascular Tissue Engineering at the University of Oklahoma, resembles the shell and tube concept of some heat exchangers. Flow circuits are separated into shell-side and tube-side to monitor (and control) system pressure and flow rates independently. Control valves downstream

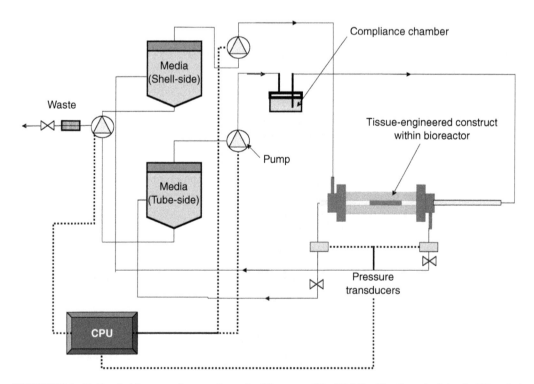

FIGURE 15.3 Shell–tube bioreactor for vascular grafts. (Courtesy of Dr. P. McFetridge from the School of Chemical Engineering and Materials Science, University of Oklahoma Bioengineering.)

of the bioreactor can be adjusted automatically in response to pressure variation. Typically, vascular bioreactors require different cell types seeded onto each side of the construct; as such, the dual-circuit-process flow allows different media types (endothelial and smooth muscle cell) to be run independently. Media is recirculated until nutrients, antibiotics, and pH require correction to remain within physiological parameters.

In another model designed by Thompson et al. [35], pressure is created by cyclical air inflow and the amount of pressure introduced in the system is controlled by a check valve. The air comes into contact with the culture medium in the so-called driving shaft. Tubing with medium is connected to the unit where the constructs are placed; this unit is called the manifold. Constructs can be accommodated in series or parallel, with a capacity of six scaffolds. Real-time flow and pressure can be monitored using an in-line flow meter and pressure transducer [35]. Hoerstrup et al. [43] designed a similar system in which the pulsatile flow is generated by the inflation/deflation of an elastic membrane.

Pulsatile flow, however, is not the only stimulation that has been used in the formation of tissue-engineered blood vessels. Selitkar et al. [44] designed a bioreactor to apply cyclic strain to cell-seeded scaffolds. After seeding aortic rat SMCs and culturing them for 8 days, circumferential orientation and a homogeneous distribution was observed in the scaffold. Collagen fibers were also produced in an organized fashion, forming bound assemblies. Mechanical properties were improved compared to unstrained constructs, achieving an ultimate stress of 58 kPa and modulus of 142 kPa, which are much greater than those achieved without any stimulation (16 and 68 kPa, respectively) [44].

Different stimuli have also been combined in order to enhance the formation of the extracellular matrix and mechanical properties. Peng et al. [45] used both cyclic stretching and pulsatile flow with perfusion to culture vascular cells. Endothelial cells were grown in distensible silastic tubes and subjected to pulse pressures and shear stresses comparable to the physiological values (up to 150 mmHg and 15 dyn/cm^2). Cells aligned in the direction of higher pulsation, almost parallel to the flow. Actin fibers showed a peripheral longitudinal orientation at greater pulsatilities.

15.4.1.2 Heart Valves

The environment in which heart valves perform is mechanically complex. Heart valves must operate at a frequency of approximately 1Hz, undergoing shear and bending stresses caused by blood flow [46,47]. Hoerstup et al. designed a pulsatile-flow system for the engineering of heart valves. This bioreactor basically consists of two chambers with a silicone diaphragm between them. The lower chamber maintains air, while the upper one is the culture medium chamber. Air is pumped, using a ventilator, into the lower chamber to displace the diaphragm and create the pulsation. Pulsatile flows from 50 to 2000 ml/min and systemic pressures from 10 to 240 mmHg can be achieved using this system [38]. Dumont et al. [37] designed a similar bioreactor for the formation of engineered aortic heart valves. This apparatus consists of a left ventricle (LV) made out of a silicone and an after-load with a compliance chamber and a resistance. The stroke and frequency of the machine is controlled by a piston that compresses and decompresses the LV. In this design, air is also incorporated in a compliance chamber, being pumped under conditions that resemble the standard systolic and diastolic pressures. The tissue-engineered aortic heart valve is placed between the left ventricle and the compliance chamber [37].

Schenke-Layland et al. [48] cultured endothelial cells and myofibroblasts on decellularized porcine heart valves under static and pulsatile conditions. Increased cellularity, collagen and elastin content, and mechanical strength were observed when the heart valves were cultured under pulsatile flow. The level of pulsation is also a critical parameter in the formation of functional heart valves, as demonstrated by Mol et al. [49].

15.4.2 Bone

Bone is a hard connective tissue that provides mechanical support to the human body and is a frame for locomotion. Bone grafts have been generated under a wide variety of culturing conditions, including static and dynamic systems. Among the most popular dynamic systems are spinner flasks, rotating-wall vessels, and perfusion systems [20,24,50,51]. As for every tissue, before deciding upon the kind of bioreactor to be used, considerations concerning the carrier matrix, cells (osteoblasts, mesenchymal stem cells, etc.), and growth factors must be taken into account. In the case of bone, the matrix to be used must be osteoconductive, provide mechanical support, deliver cells and allow their attachment, growth, migration, and osteoblastic differentiation [52]. Synthetic and natural polymers have been implemented. Among the synthetic polymers poly-α-hydroxy esters, poly(ε-caprolactone), poly(propylene fumarate), poly(sebacic acid), and their copolymers have been widely used. Materials such as ceramics and titanium have also been used for bone replacement [53,54]. Cell number and calcium deposition are good markers to evaluate the evolution of bone matrix. Furthermore, alkaline phosphatase activity (ALP) is used to assess early differentiation activity of osteoblastic cells. Production of extracellular matrix proteins such as osteocalcin, osteopontin, and bone sialoprotein is also taken into consideration [55].

As mentioned before, static culture was one of the first attempts to produce bone matrix. Ishaug et al. [56] seeded marrow stromal cells on top of poly (DL-lactic-co-glycolic acid) (PLGA) foams of different pore sizes at different densities. Cell proliferation was supported by the scaffold, and high level of ALP activity and calcium matrix deposition were observed. It was found that the depth of mineralized tissue increased over time, but the maximum penetration was only around 240 μm, resulting in a nonhomogeneous cell and matrix distribution [56].

Improvement in the development of bone matrix *in vitro* has been achieved with the addition of convection in the *in vitro* culture stage, which ultimately translates in a better transport of nutrients and gases. After statically seeding 1×10^6 marrow stromal cells on 75 : 25 PLGA scaffolds, Sikavitsas et al. [7] cultured these constructs under three different conditions: statically, in a spinner flask and in a rotating-wall vessel. The culture was carried out for 21 days, and samples were analyzed at 7, 14, and 21 days. Scaffolds cultured in the spinner flask bioreactor showed the largest number of cells at all time points, followed by the static culture. At the end of the culture period, constructs in the spinner flask presented higher calcium contents than those encountered in the static and rotating-wall vessel [7].

Shea et al. [57] also utilized a spinner flask to culture poly(lactic acid) foams seeded with MC3T3-E1 preosteoblasts and evaluated their differentiation. Cells were seeded statically and cultured for 12 weeks. Proliferation was observed over time; however, their distribution throughout the scaffold lacked homogeneity. Cells were densely located only at a thin layer of 200 μm near the scaffold's surface. The density dramatically decreased deeper into the construct. The same behavior was seen for the formation of extracellular matrix and calcium deposition.

It has been shown that mechanical stimulation augments the production of alkaline phosphatase, osteoblast proliferation, and mineral deposition in osteoblastic cells seeded on different scaffolding materials [58]. Osteoblastic cells have been shown to be responsive to shear stress induced by fluid flow. The stimulatory effect of shear stresses has shown to induce an increase in the release of important regulatory factors such as nitric oxide and prostaglandin E2 [59–61]. Interestingly, osteoblasts have been found to be more responsive to fluid shear forces than mechanical strain [62]. A question arises then, what is the physiological relevance of the stimulatory effect of fluid flow on bone cells? It has been hypothesized that mechanical strains on bone tissue cause fluid flow in the lacunar–canalicular porosity of bone [63–65]. Consequently, the incorporation of fluid flow through the porous network is desired in order to stimulate a faster and more efficient formation of bone matrix. This goal has been reached with the implementation of flow perfusion bioreactors [20,66,67].

Bancroft et al. [8] developed a perfusion system (Figure 15.4) where medium is pumped through the scaffold, thereby maintaining mechanical stimulation and transport of nutrients through the pores. The scaffolds are tightly fit into cassettes in order to ensure fluid flow exclusively through the porous network. Constructs are later placed in flow chambers that are capped and secured with o-rings to restrict the flow around them (Figure 15.4a). The medium is pumped from a flask to the top of the chamber and sent to another reservoir from the bottom. This direction of flow helps avoiding the entrance of air bubbles into the flow chamber. Both flasks are connected so that the medium is in continuous recirculation. The main body of the reactor consists of a total of six chambers and is made out of Plexiglas to allow the visualization and monitoring of the flow inside the chambers. Each chamber corresponds to an independent circuit

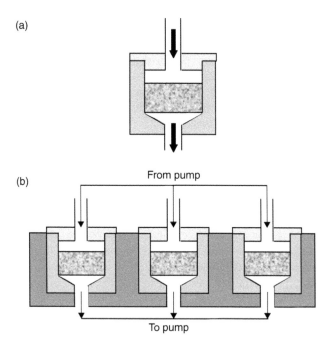

FIGURE 15.4 Schematics of a flow perfusion bioreactor. (a) Close up of the perfusion chamber where the scaffold is press fitted. (b) Lateral view of the main body of the bioreactor.

using one of the heads of a peristaltic pump that produces flow rates from 0.1 to 10 ml/min (Figure 15.4b). The tubing permits the exchange of carbon dioxide and oxygen with the atmosphere in the incubator. A complete change of medium can be done due to the two-reservoir set up [18].

To study how the shear rate affects the growth of bone matrix *in vitro*, Bancroft et al. [19] varied the flow rate when culturing titanium fiber meshes seeded with rat marrow stromal cells for 16 days. Controls have been cultured under static conditions, and the flow rates used in the perfusion culture were 0.3, 1.0, and 3.0 ml/min. It was found that the deposition of calcium was greatly increased in the flow perfusion culture as compared with the static conditions. It was also observed that increased medium flow improves the distribution of extracellular matrix throughout the construct volume [19]. The increased calcium deposition could have been due to the increased shear forces or increased chemotransport in the porosity of the scaffolds when higher flow rates were employed. To isolate the effects of shear forces from the mass transport effects, the shear forces were changed by varying the viscosity of the culture medium under constant flow rate [68]. An increase in viscosity, which translates into greater shear forces, was found to enhance the deposition of mineral matrix and the ECM distribution throughout the construct, demonstrating the importance of fluid-flow induced shear forces on the creation of bone tissue-engineered grafts.

Meinel et al. [67] cultured human mesenchymal stem cells on silk scaffolds for 5 weeks in a flow perfusion chamber (at 0.2 ml/min) and a spinner flask. Scaffolds cultured under flow perfusion showed a more homogenous distribution of the mineralized matrix throughout the construct although those cultured in the spinner flask produced a greater amount of deposited calcium.

Other kinds of stimulation include mechanical strain and electrical current. A bioreactor was developed that allowed the continuous exposure of cells to continuous cyclic stretching [69]. Primary osteoblast-like cells were seeded on silicone rubbers and subjected to 1000 microstrains at 1 Hz either continuously or in periods of 60 min. Cellularity and calcium deposition were enhanced under the presence of mechanical strain and the intermittent procedure was the most efficient of them [70]. Another bioreactor that can expose cells to mechanical load has been developed by Shimko et al. [71]. Unlike the previous study, mechanical loading had a negative impact in the mineral deposition.

Electrical current has been shown to stimulate bone regeneration and enhance its healing [72–75]. Bioreactors have been designed to electrically stimulate cells seeded on conductive surfaces [75]. After culturing calvarial osteoblasts under an electric field with 10 μA and 10 Hz for 6 h/day, the cell number was higher by 46% and the amount of calcium deposited was raised by 307% after 21 days, compared to nonstimulated cells. Upregulation of mRNA expression for collagen type I was also observed.

As demonstrated by several studies, the usage of flow perfusion greatly enhances the formation of bone matrix. By doing this, mechanotransduction mechanisms related to the differentiation of osteoblastic cells are potentially activated. Great challenges are still encountered however. What are the actual mechanisms that transduce the external shear forces and influence the cellular behavior? Currently, only estimates of the shear rates at the interior of three-dimensional scaffolds have been provided, and detailed mathematical modeling needs to be conducted. This will allow the determination of the actual values of the shear rate inside the constructs and their spatial distribution.

Using larger scaffolds can represent another problem; is it possible to achieve a completely homogeneous distribution of matrix in larger constructs? What would be the necessary culturing conditions to achieve this? How long must the cells be cultured to produce an osteoinductive enough matrix? And how long and under what conditions must the cells be precultured prior to the application of mechanical forces to avoid their detachment?

15.4.3 Cartilage

Cartilage is a tissue with limited capabilities of regeneration when damaged. Between the two general types of cartilage, articular cartilage is continuously exposed to mechanical forces and needs to dissipate loads under physiological conditions. On the other hand, in other parts of the body the elastic properties of the cartilage are more important. Cartilaginous tissue has been engineered using spinner flasks, rotating-wall bioreactors, perfusion systems, and compression bioreactors that better mimic the physiological environment of the cartilage tissue. The most widely used cell types for the regeneration of cartilage are mesenchymal stem cells or primary chondrocytes. Regarding the scaffold, different materials have been used; synthetic and natural hydrogels are among the most popular choices, including collagen, poly α-hydroxy esters, and their copolymers. Fibrinogen-based and glycosaminoglycan (GAG)-based matrices are also being employed [76,77]. Some of the markers used to evaluate the formation of cartilaginous matrix are cellularity, production of collagen type II, and production of GAGs [8]. It has to be pointed out that the synthesis of native cartilage dramatically varies in different locations of the body.

Vunjak-Novakovic et al. [10] compared the performance of different bioreactors (static, spinner flask, and rotating wall) in the formation of cartilaginous matrix. Highly porous PGA scaffolds were seeded with chondrocytes and cultured statically, in a spinner flask, and a rotating-wall vessel for 6 weeks. At the end of the culture period, GAG and collagen type II levels were five times greater than the values obtained in early culture. The spinner flask induced a larger production of GAGs and collagen type II; however, an even greater difference was observed in the rotating bioreactor compared to the static culture. Histomorphometric studies revealed the formation of tissue in the periphery of the constructs cultured statically, while those in the spinner flask had an outer fibrous capsule in spite of the increment in mass transport. Scaffolds cultured in the rotating-wall bioreactor, on the other hand, showed a better distribution of the matrix, but a gradient in concentration of GAGs was observed in all the constructs.

Gooch et al. [78] studied the effect of shear stress on the formation of cartilage matrix in a spinner flask and compared it with a static culture. Chondrocytes were dynamically seeded for 3 days on fibrous PGA matrices with a fiber diameter of 13 μm using a spinner flask bioreactor. Cultivation was carried out for 42 days at different mixing intensities. Production of GAGs was greater in the spinner flask, and increased along with the mixing intensity. Collagen production showed a similar behavior although no significant difference was observed among the different mixing intensities.

The effect of shear stress has also been conducted in a rotating-wall vessel [11]. Variation of the rotation speed of the mobile wall when culturing chondrocytes in poly(lactic acid) (PLA) foams for 28 days did not cause a change in cell density among scaffolds cultured at different rotation rates although they were larger than those found in static conditions. However, the deposition of matrix GAGs decreased at higher shear rates, whereas the collagen production showed a contrary behavior. The production of collagen increased along with the shear rate and time of cultivation, showing a more dramatic dependence on the rotation speed.

It is important to point out that the shear stress induced in a rotating-wall bioreactor occurs only at the surface of the scaffold; therefore, perfused cartridges have been used to produce stress stimulation at the interior of the construct. Pazzano et al. [79] cultured calf chondrocytes seeded on PGA matrices for 4 weeks under static and flow perfusion conditions. Not only did the perfusion system increase the amount of GAGs by 180% compared with the static conditions, but it also created an organized, homogeneous matrix. After the 4 weeks of culture, chondrocytes were aligned in the direction of flow, creating a structure similar to that of some regions of native articular cartilage [80].

Native articular cartilage withstands up to 20 MPa due to compression *in vivo*; therefore, bioreactors using this kind of stimulus have been implemented obtaining promising results [6,81–84]. Mizuno et al. [85] used a culture system that consisted of a perfusion column and a pressurizer. Medium is peristaltically pumped through the perfusion column were the scaffolds are kept. Downstream from the column there is a back-pressure regulator. Cyclic pressure was controlled by using a needle valve and a computer-controlled spring. Chondrocytes were statically seeded in porous collagen matrices for 3 days. Posteriorly, the sponges were introduced in the perfusion chamber, and flow was started at the rate of 0.33 ml/min. The culture conditions were pure perfusion, cyclic compression (0.015 Hz) 2.8 MPa, and constant compression at 2.8 MPa of hydrostatic fluid pressure. Production of GAGs and collagen type II were enhanced by the application of pressure load even though there were no significant differences in cellularity. Hydrostatic pressure produced about 3.1 times more sulfated GAGs than the controls, whereas the cyclic pressure produced 2.7 times more sulfated GAGs than the controls. In addition, a native-like matrix was observed, containing lacunae that entrapped the chondrocytes and a uniform spatial distribution [85]. Carver et al. [86] had found similar results when comparing the influences of cyclic pressure, flow perfusion, and culture in spinner flasks, in the formation of cartilaginous matrix. They found that the combination of flow perfusion and intermittent pressurization seemed to accelerate the matrix formation. In agreement with these findings, Hung et al. [84] reported the production of cartilage-like tissue after culturing chondrocytes seeded on agarose gels for 8 weeks under physiologic deformational loading.

The variability in the synthesis and the mechanical properties of cartilage in different locations of the body and zones of the same location introduces a great challenge in the regeneration of cartilage tissue. Is the mechanical environment the controlling factor in the formation of zone and location-specific ECM? Or in different locations of the body, are chondrocytes preprogrammed with different genetic information to generate a specific type of extracellular matrix? How do chondrocytes in the growth plate cartilage recognize a differentiation pattern that leads to bone formation? The elucidation of these questions will provide valuable information to cartilage tissue engineers.

15.4.4 Anterior Cruciate Ligament and Tendons

Ligaments and tendons operate in the presence of various types and levels of mechanical strains that are expected to influence the behavior of cells residing in these tissues. Mesenchymal stem cells (MSCs) have been shown to preferentially differentiate into ligament fibroblasts in the presence of mechanical forces when cultured in bioreactors that generated mechanical strains that resemble the native ligament stretching conditions [87].

The most widely used material as scaffolding for the regeneration of anterior cruciate ligament (ACL) and tendon is collagen; other materials such as polymeric fibrous matrices have been utilized [88]. However, the mechanical properties of collagen bundles are considerably inferior to those of the native tissue. This problem can be partially overcome by preferentially orienting the collagen fibers of the matrix through the seeding of cells and the application of mechanical strain [89].

The principle of a bioreactor that can mimic the biomechanical conditions of ligament and tendons is shown in Figure 15.5. The construct is attached to a moving platform that produces mechanical strain

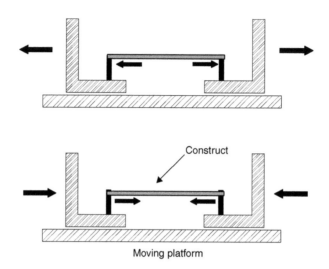

FIGURE 15.5   Principle of cyclic strain bioreactor for ligament and tendon engineering.

through stretching. Langelier et al. [89] designed a cyclic traction device that can produce these mechanical strains on cell-seeded constructs. This machine consists of an actuating unit that controls the cyclic motion frequency (0.01 to 1 cycle/min) and amplitude (0 to 2 cm) for collagenous matrices with a length between 2 and 10 cm. A transmission unit connects the actuating unit to the testing unit, and its purpose is to transmit motion to the testing unit through a rotating shaft. The testing compartment unit, shown in Figure 15.4, is where the matrix is located. The two walls at both extremes are made out of stainless steel, and the material chosen for the other surfaces was acrylic to allow visualization of the construct. The top of the compartment is a removable acrylic plate screwed to the walls and is removed when the scaffold needs to be manipulated. Generally, the collagen gel is formed over two bone anchors for adequate mechanical attachment to the traction machine [89]. The construct can be subjected to cyclic stretching at a frequency of up to 1 Hz and amplitude from 0 to 30 mm for any extended period of time. The system is controlled via special software, and the experimental conditions of stretching and amplitude can be easily changed [90].

MSCs seeded on collagen gels were cultured under translational and rotational strain concurrently. After 21 days of culture, a ligament-like morphology was observed in the constructs, with elongated, well-organized collagen types I and III in the longitudinal direction. Unlike the constructs cultured statically, those under mechanical strain presented an ECM that contained tenascin-C, which is a marker of ligament ECM [87]. A similar behavior was observed by Awad et al. [91] when MSC-seeded collagen scaffolds were cultured under a continuously strained environment; formation of organized collagen bundles was also observed.

Cyclic stretching has been also implemented in tendon tissue engineering and has demonstrated a significant effect on the alignment of collagen bundles [92]. Avian tendon internal fibroblasts were incorporated into type-I collagen matrices at a rate of 2×10^5 cells per scaffold. Mechanical loading was applied 1 h per day with 1% elongation and 1 Hz for up to 11 days. Loaded constructs presented aligned cells that spread throughout the matrix, with elongated nuclei and cytoplasmic extensions. Moreover, the modulus of elasticity of the mechanically loaded constructs was 2.9 times greater than those of the nonstimulated controls [92].

One of the challenges in tendon and ligament tissue engineering is the selection of cell source for their repair. Although MSCs have been shown to have the ability to differentiate into tendon and ligament cells in the presence of mechanical forces, the use of specific growth factor combinations that can initiate the differentiation toward these phenotypes in regular cell cultures would minimize the time of tissue regeneration in bioreactors. In addition to that, the specific cellular mechanisms that transduce the mechanical signals to these cells and control their behavior are not yet clearly understood.

15.4.5 Other Tissues

Apart from the previously mentioned applications where bioreactors have been widely utilized, bioreactors have been used for the culture of cell types involved in the regeneration of other tissues, including hepatocytes and neuronal cells. The high metabolic requirements of hepatic cells make the use of perfusion systems for their culture ideal. Powers et al. [93] cultured hepatocytes in the presence of continuous flow perfusion and the cells remained viable for up to 2 weeks. The use of electromagnetic fields in neuronal tissue regeneration has also provided elegant ways for the alignment of neuronal cells in the guided regrowth of axons [94].

15.5 Design Considerations

When designing a bioreactor, the first factor that must be taken into account is the kind of stimulation to be emulated. Therefore, the apparatus must be able to cause similar effects to those found physiologically and at matching conditions, in addition to ensuring proper transportation of nutrients. A temperature of 37°C and a humid atmosphere with 5% CO_2 are required [95]. Other important factors are:

1. Easy handling and assembling, lowering risks of contamination
2. Sterilizability
3. Fit into an incubator and operate at the given conditions
4. Allow change of culture medium aseptically
5. Facilitate monitoring of tissue formation
6. Automatically operated in order to ensure reproducibility of operation conditions
7. Ability to produce several constructs at the same time
8. Ease of scaling up

15.6 Challenges in Bioreactor Technologies

Implementation of bioreactors in the process of tissue engineering has been driven by the need to mimic the physiological conditions in which cells operate at different locations of the human body. Most bioreactors in reality are biomimetic environments that induce morphological arrangements similar to those found in native tissues. To a certain degree, the new technologies have been successful in achieving the initial goals: creation of native tissue-like matrices; nevertheless, many obstacles are yet to be overcome in order to produce efficient and practical constructs that can be used for the regeneration of damaged or lost organs.

The transport of nutrients and oxygen still remains an issue in the processing of many tissues. Some cells have a high metabolic demand; such is the case of hepatic tissue, thus requiring an efficient transportation of these components to the place of growth. As discussed in this chapter, part of this problem has been overcome by the utilization of perfusion systems. However, scaling up the size of the construct may pose new challenges. When increasing the dimensions of the scaffolding, other interrogatives arise. Would it still be possible to obtain a homogeneous distribution of cells and ECM throughout the scaffold? What is the flow rate necessary to guarantee the delivery of oxygen and nutrients to the interior of the construct?

Cell source is a factor that has not been studied in detail. It is not clear, for the engineering of some tissues, whether cells extracted from different locations could have different responses under identical stimulation. In the case of bone the question arises, as to whether cells extracted from an area that is not load bearing respond in similar fashion to those harvested from a load-bearing site like the femur? Similar questions are unanswered for cartilage and ligaments.

Bone, cartilage, vascular grafts, and ligaments, among others, require potent mechanical properties that have not been matched by the engineered constructs to this day. Establishing effective mechanical properties depends not only on the biomaterial used but also in the bioreactor design and time of culture.

Mechanical stimulations have proved to be the path to generate mechanically efficient implants, but not all biomaterials can be exposed to mechanical strains without generating some degree of structural damage.

Standard criteria for the evaluation of the quality of the final construct are necessary from a mechanical, chemical, biological, and morphological point of view. Large variations on the conditions of operation and culture in different bioreactor designs to engineer a given tissue prevent the comparison of different studies and thereby the drawing of definite conclusions regarding the optimization of these conditions. In addition, histological evaluation of the regenerated tissue (or *in vitro* generated graft) should always be complemented by appropriate mechanical tests.

Time also represents a significant concern when engineering tissue grafts. Given the fact that the engineered constructs will ultimately be transplanted into diseased patients, the time of production becomes a critical factor. The ultimate goal is to produce, in a short period of time, a construct that can induce the formation of new tissue *in vivo*. It is obvious that the definition of a "short period of time" depends upon the tissue being engineered and the urgency for a replacement. Some tissues like bone or cartilage could offer some flexibility in the time of production depending on the type of injury, but in life-threatening situations, such as failing heart valves, it is important to rapidly produce a construct. Consequently, tissue engineering strategies should always attempt to minimize the time of production.

The answer to shortening the time of generation of engineered tissues may lie in the combination of mechanical and chemical signals. Both principles have been studied separately; certain growth factors such as bone morphogenetic proteins, vascular endothelial growth factor, and insulin like growth factors, among others, have been shown to greatly help the formation of tissues *in vitro* and *in vivo*. Therefore, it is expected that the combination of growth factors and mechanical stimulation in a bioreactor could generate mature matrices at a faster pace. It is important as well to determine what the intracellular mechanisms related to the signaling pathways are. A better understanding of the biological mechanisms involved in the mechanosensing processes will give a new perspective in the design of new experiments and improvement of those already existent. Moreover, new directions could be taken based on these mechanisms.

Acknowledgment

We would like to acknowledge the Seed Grant Program of the Bioengineering Center at the University of Oklahoma.

References

[1] Bonassar, J. and Vacanti, C. Tissue engineering: the first decade and beyond. *J. Cell Biochem. Suppl.* 30/31, 297, 1998.

[2] Boden, S.D. Bioactive factors for bone tissue engineering. *Clin. Orthop.*, 367S, S84, 1999.

[3] Lieberman, J.R., Daluiski, A., and Einhorn, T.A. The role of growth factors in the repair of bone. *J. Bone Joint Surg.*, 84-A, 1032, 2002.

[4] Wang, E.A. et al. Recombinant human bone morphogenetic protein induces bone formation. *Proc. Natl Acad. Sci. USA*, 87, 2220, 1990.

[5] Trivedi, N. et al. Improved vascularization of planar diffusion devices following continuous infusion of vascular endothelial growth factor (VEGF). *Cell Transplant*, 8, 175, 1999.

[6] Stoltz, J.F. et al. Influence of mechanical forces on cells and tissues. *Biotechnology*, 37, 3, 2000.

[7] Sikavitsas, V.I., Bancroft, G.N., and Mikos, A.G. Formation of three-dimensional cell/polymer constructs for bone tissue engineering in a spinner flask bioreactor and a rotating wall Bessel bioreactor. *J. Biomed. Mater. Res.*, 62, 136, 2002.

[8] Darling, E.M. and Athanasiou, K.A. Articular cartilage bioreactors and bioprocesses. *Tissue Eng.*, 9, 9, 2003.

[9] Goldstein, A.S. et al. Effect of convection on osteoblastic cell growth and function in biodegradable polymer foam scaffolds. *Biomaterials*, 22, 1279, 2001.

[10] Vunjak-Novakovic, G., Obradovic, B., and Free, L.E. Bioreactor studies of native and tissue engineered cartilage. *Biorheology*, 39, 259, 2002.

[11] Saini, S. and Wick, T.M. Concentric cylinder bioreactor for production of tissue engineered cartilage: effect of cell density and hydrodynamic loading on construct development. *Biotechnol. Prog.*, 19, 510, 2003.

[12] Sutherland, F.W. et al. Advances in the mechanisms of cell delivery to cardiovascular scaffolds: comparison of two rotating cell culture systems. *ASAIO J.*, 48, 346, 2002.

[13] Vunjak-Novakovic, G. et al. Microgravity studies of cells and tissues. *Ann. N Y Acad. Sci.*, 974, 504, 2002.

[14] Begley, C.M. and Kleis, S.J. The fluid dynamic and shear environment in the NASA/JSC rotating wall perfused-vessel bioreactor. *Biotechnol. Bioeng.*, 70, 32, 2000.

[15] Williams, K.A. et al. Computational fluid dynamics modeling of steady-state momentum and mass transport in a bioreactor for cartilage tissue engineering. *Biotechnol. Prog.*, 18, 951, 2002.

[16] Mizuno, S., Allemann, F., and Glowacki, J. Effects of medium perfusion on matrix production by bovine chondrocytes in three-dimensional collagen sponges. *J. Biomed. Mater. Res.*, 56, 368, 2001.

[17] Navarro, F.A. et al. Perfusion of medium improves growth of human oral neomucosal tissue constructs. *Wound Repair Regen.*, 9, 507, 2001.

[18] Bancroft, G.N., Sikavitsas, V.I., and Mikos, A.G. Design of a flow perfusion bioreactor system for bone–tissue engineering applications. *Tissue Eng.*, 9, 549, 2003.

[19] Bancroft, G.N. et al. Fluid flow increases mineralized matrix deposition in 3D perfusion culture of marrow stromal osteoblasts in a dose-dependent manner. *Proc. Natl Acad. Sci. USA*, 99, 12600, 2002.

[20] Cartmell, S.H. et al. Effects of medium perfusion rate on cell seeded three-dimensional bone constructs *in vitro*. *Tissue Eng.*, 9, 1197, 2003.

[21] Martin, I., Wendt, D., and Herberer, M. The role of bioreactors in tissue engineering. *Trends Biotechnol.*, 22, 80–86, 2004.

[22] Li, Y. Effects of filtration seeding on cell density, spatial distribution, and proliferation in nonwoven fibrous matrices. *Biotechnol. Prog.*, 17, 935, 2001.

[23] Holy, C.E., Shoichet, M.S., and Davies, J.E. Engineering three-dimensional bone tissue *in vitro* using biodegradable scaffolds: investigating initial cell-seeding density and culture period. *J. Biomed. Mater. Res.*, 51, 376, 2000.

[24] Vunjak-Novakovic, G. et al. Dynamic cell seeding of polymer scaffolds for cartilage tissue engineering. *Biotechnol. Prog.*, 14, 193, 1998.

[25] Sucosky, P. et al. Fluid mechanics of a spinner-flask bioreactor. *Biotechnol. Bioeng.*, 85, 34, 2004.

[26] Burg, K.J.L. et al. Comparative study of seeding methods for three-dimensional polymeric scaffolds. *J. Biomed. Mater. Res.*, 51, 642, 2000.

[27] Wendt, D. et al. Oscilating perfusion of cell suspensions through three-dimensional scaffolds enhances cell seeding efficiency and uniformity. *Biotechnol. Bioeng.*, 84, 205, 2003.

[28] Sodian, R. et al. Tissue-engineering bioreactors: a new combined cell-seeding and perfusion system for vascular tissue engineering. *Tissue Eng.*, 8, 863, 2002.

[29] Carrier, R.L. et al. Cardiac tissue engineering: cell seeding, cultivation parameters, and tissue construct characterization. *Biotechnol. Bioeng.*, 64, 580, 1999.

[30] Tranquillo, R. The tissue-engineered small-diameter artery. *Ann. N Y Acad. Sci.*, 961, 251, 2002.

[31] L'Heureux, N. et al. A completely biological tissue-engineered human blood vessel. *FASEB J.*, 12, 47, 1998.

[32] Guyton, A. and Hall, J. *Textbook of Medical Physiology*, 10th ed., Saunders Company, 2000, p. 145.

[33] Braddock, M. et al. Fluid shear stress modulation of gene expression in endothelial cells. *New Physiol. Sci.*, 13, 241, 1998.

[34] Nerem, R. and Sliktar, D. Vascular tissue engineering. *Annu. Rev. Biomed. Eng.*, 3, 225, 2001.

[35] Thompson, C.A. et al. A novel pulsatile, laminar flow bioreactor for the development of tissue-engineered vascular structures. *Tissue Eng.*, 8, 1083, 2002.

[36] Sodian, R. et al. New pulsatile bioreactor for fabrication of tissue-engineered patches. *J. Biomed. Mater. Res. (Appl. Biomater.)*, 58, 401, 2001.

[37] Dumont, K. et al. Design of new pulsatile bioreactor for tissue engineered aortic heart valve formation. *Artif. Organs*, 26, 710, 2002.

[38] Hoerstrup, S.P. et al. New pulsatile bioreactor for *in vitro* formation of tissue engineered heart valves. *Tissue Eng.*, 6, 75, 2000.

[39] Weinberg, C.B. and Bell, E. A blood vessel model constructed from collagen and cultured vascular cells. *Science*, 23, 397, 1986.

[40] Niklason, L.E. et al. Functional arteries grown *in vitro*. *Science*, 284, 489, 1999.

[41] Lian, X. and Howard, P.G. Blood vessels, in *Principles of Tissue Engineering*, Lanza, R., Langer, R., and Vacanti, J. (Eds.), 2nd ed., Academic Press, New York, 2000, chap. 32.

[42] Niklason L.E. and Langer, R.S. Advances in tissue engineering of blood vessels and other tissues. *Transp. Immunol.*, 5, 303, 1997.

[43] Hoerstrup, S.P. et al. Tissue engineering of small caliber vascular grafts. *Eur. J. Cardiothorac. Surg.*, 20, 164, 2001.

[44] Selitkar, D. et al. Dynamic mechanical conditioning of collagen–gel blood vessel constructs induces remodeling *in vitro*. *Ann. Biomed. Eng.*, 28, 351, 2000.

[45] Peng, X. et al. *In vitro* system to study realistic pulsatile flow and stretch signaling in cultured vascular cells. *Am. J. Physiol. Cell Physiol.*, 279, C797, 2000.

[46] Barron, V. et al. Bioreactors for cardiovascular cell and tissue growth: a review. *Ann. Biomed. Eng.*, 31, 1017, 2003.

[47] Mann, B.K. and West, J.L. Tissue engineering in the cardiovascular system: progress toward a tissue engineered heart. *Anat. Rec.*, 263, 367, 2001.

[48] Schenke-Layland, K. Complete dynamic repopulation of decellularized heart valves by application of defined physical signals — an *in vitro* study. *Cardiovasc. Res.*, 60, 497, 2003.

[49] Mol, A. The relevance of large strains in functional tissue engineering of heart valves. *Thorac. Cardiovasc. Surg.*, 51, 78, 2003.

[50] Goldstain, A.S. et al. Effect of osteoblastic culture conditions on the structure of poly(DL-Lactic-co-Glycolic acid) foam scaffolds. *Tissue Eng.*, 5, 421, 1999.

[51] Botchwey, E.A. et al. Bone tissue engineering in a rotating bioreactor using a microcarrier matrix system. *J. Biomed. Mater. Res.*, 55, 242, 2001.

[52] Bancroft, G.N. and Mikos, A.G. Bone tissue engineering by cell transplantation, in *Polymer Based Systems on Tissue Engineering, Replacement and Regeneration*, Reis, R.L. and Cohn, D. (Eds.), Kluwer Academic Publishers, Boston, 2002, p. 251.

[53] Nam, Y.S., Yoon, J.J., and Park, T.G. A novel fabrication method of macroporous biodegradable polymer scaffolds using gas foaming salt as a porogen additive. *J. Biomed. Mater. Res. (Appl. Biomater.)*, 53, 1, 2000.

[54] Hollinger, J.O. and Battistone, G.C. Biodegradable bone repair materials. Synthetic polymers and ceramics. *Clin. Orthop. Relat. Res.*, 207, 290, 1986.

[55] Nefussi, J.R. et al. Sequential expression of bone matrix proteins during rat calvaria osteoblast differentiation and bone nobule formation *in vitro*. *J. Histochem. Cytochem.*, 45, 493, 1997.

[56] Ishaug, S.L. et al. Bone formation by three-dimensional stromal osteoblast culture in biodegradable polymer scaffolds. *J. Biomed. Mater. Res.*, 36, 17, 1997.

[57] Shea, L. et al. Engineered bone development from pre-osteoblast cell line on three-dimensional scaffold. *Tissue Eng.*, 6, 605, 2000.

[58] Pavlin D. et al. Mechanical loading stimulates differentiation of periodontal osteoblasts in a mouse osteoinduction model: effect on type I collagen and alkaline phosphatase genes. *Calcif. Tissue Int.*, 67, 163, 2002.

[59] Sikavitsas, V.I., Temenoff, J.S., and Mikos, A.G. Biomaterials and bone mechanotransduction. *Biomaterials*, 22, 2581, 2001.

[60] Reich, K.M. and Frangos, J.A. Effect of flow on prostaglandin E2 inositol trisphosphate levels in osteoblasts. *Am. J. Physiol.*, 261, C429, 1991.

[61] Klein-Nulend, J. et al. Nitric oxide response to shear stress by human bone cell cultures is endothelial nitric oxide synthase dependent. *Biochem. Biophys. Res. Commun.*, 250, 108, 1998.

[62] Owan, I. et al. Mechanotransduction in bone: osteoblasts are more responsive to fluid forces than mechanical strain. *Am. J. Physiol.*, 273, C810, 1997.

[63] Knothe-Tate, M.L., Knothe, U., and Niederer, P. Experimental elucidation of mechanical load-induced fluid flow and its potential role in bone metabolism and functional adaptation. *Am. J. Med. Sci.*, 316, 189, 1998.

[64] Burger, E.H. and Klein-Nulend, J. Mechanotransduction in bone: role of the lacuno-canalicular network. *FASEB J.*, 13S, S101, 1997.

[65] Cowin, S.C. Bone poroelasticity. *J. Biomech.*, 32, 217, 1999.

[66] Van den Dolder, J. et al. Flow perfusion culture of marrow stromal osteoblasts in titanium fiber mesh. *J. Biomed. Mater. Res.*, 64A, 235, 2003.

[67] Meinel, L. et al. Bone tissue engineering using human mesenchymal stem cells: effects of scaffold material and medium flow. *Ann. Biomed. Eng.*, 32, 112, 2004.

[68] Sikavitsas, V.I. et al. Mineralized matrix deposition by marrow stromal osteoblasts in 3D perfusion culture increases with increasing fluid shear forces. *Proc. Natl Acad. Sci. USA*, 100, 14683, 2003.

[69] Neidlinger-Wilke, C., Wilke, H.J., and Claes, L. Cyclic stretching of human osteoblasts affects proliferation and metabolism: a new experimental method and its application. *J. Orthop. Res.*, 12, 70, 1994.

[70] Winter, L.C. et al. Intermittent versus continuous stretching effects on osteoblast-like cells *in vitro*. *J. Biomed. Mater. Res.*, 67A, 1269, 2003.

[71] Shimko, D. et al. A device for long term, *in vitro* loading of three-dimensional natural and engineered tissues. *Ann. Biomed. Eng.*, 31, 1347, 2003.

[72] Yonernori, K. Early effects of electrical stimulation on osteogenesis. *Bone*, 19, 173, 1996.

[73] Brighton, C.T. et al. Signal transduction in electrically stimulated bone cells. *J. Bone Joint Surg. Am.*, 83-A, 1514, 2003.

[74] Bassett, C.A., Pawluk, R.J., and Pilla, A.A. Augmentation of bone repair by inductively coupled electromagnetic fields. *Science,* 184, 575–577, 1984.

[75] Supronowicz, P.R. et al. Novel current-conducting composite substrates for exposing osteoblasts to alternating current stimulation. *J. Biomed. Mater. Res.*, 59, 499, 2002.

[76] Temenoff, J.S. and Mikos, A.G. Review: tissue engineering for regeneration of articular cartilage. *Biomaterials*, 21, 431, 2000.

[77] Hendrickson, D.A. et al. Chondrocyte-fibrin matrix transplants for resurfacing extensive articular cartilage defects. *J. Orthop. Res.*, 12, 485, 1994.

[78] Gooch, K.J. et al. Effects of mixing intensity on tissue-engineered cartilage. *Biotechnol. Bioeng.*, 72, 402, 2001.

[79] Pazzano, D. et al. Comparison of chondrogenesis in static and perfused bioreactor culture. *Biotechnol. Prog.*, 16, 893, 2000.

[80] Kuettner, K.E. Biochemistry of articular cartilage in health and disease. *Clin. Biochem.*, 25, 155, 1992.

[81] Shieh, A.C. and Athanasiou, K.A. Principles of cell mechanics for cartilage tissue engineering. *Ann. Biomed. Eng.*, 31, 1, 2003.

[82] Buschmann, M.D. et al. Stimulation of aggrecan synthesis in cartilage explants by cyclic loading is localized to regions of high interstitial fluid flow. *Arch. Biochem. Biophys.*, 366, 1, 1999.

[83] Guilak, F. et al. The effects of matrix compression on PG metabolism in articular cartilage explants. *Osteoarthr. Cartil.*, 2, 91, 1994.

[84] Hung, C. et al. A paradigm for functional tissue engineering of articular cartilage via applied physiologic deformational loading. *Ann. Biomed. Eng.*, 32, 35, 2004.

[85] Mizuno, S. et al. Hydrostatic fluid pressure enhances matrix synthesis and accumulation by bovine chondrocytes in three-dimensional culture. *J. Cell Physiol.*, 39, 319, 2002.

[86] Carver, S.E. and Heath, C.A. Influence of intermittent pressure, fluid flow, and mixing on the regenerative properties of articular chondrocytes. *Biotechnol. Bioeng.*, 65, 274, 1999.

[87] Altman, G.H. et al. Cell differentiation by mechanical stress. *FASEB J.*, 16, 270, 2002.

[88] Goh, J.C. et al. Tissue-engineering approach to the repair and regeneration of tendons and ligaments. *Tissue Eng.*, 9, S31, 2003.

[89] Langelier, E. et al. Cyclic traction machine for long-term culture of fibroblast-populated collagen gels. *Ann. Biomed. Eng.*, 27, 67, 1999.

[90] Goulet, F. et al. Tendons and ligaments, in *Principles of Tissue Engineering*, Lanza, R., Langer, R., and Vacanti, J. (Eds.), 2nd ed., Academic Press, New York, 2000, chap. 50.

[91] Awad, H.A. et al. *In vitro* characterization of mesenchymal stem cell-seeded collagen scaffolds for tendon repair: effects of initial seeding density on contraction kinetics. *J. Biomed. Mater. Res.*, 51, 233, 2000.

[92] Garvin et al. Novel system for engineering bioartificial tendons and application of mechanical load. *Tissue Eng.*, 9, 130–132, 2003.

[93] Powers, M.J. et al. A microfabricated array bioreactor for perfused 3D liver cultures. *Biotechnol. Bioeng.*, 78, 257, 2002.

[94] Eguchi, Y., Ogiue-Ikeda, M., and Ueno, S. Control of orientation of rat Schwann cells using an 8-T static magnetic field. *Neurosci. Lett.*, 351, 130, 2003.

[95] Miller, W.M. Bioreactor design considerations for cell therapies and tissue engineering. In *WTEC Workshop on Tissue Engineering Research in the United States, Proceedings*, MD: International Technology Research Institute, Loyola College, Baltimore, 2000, p. 5.

16

Animal Models for Evaluation of Tissue-Engineered Orthopedic Implants

Lichun Lu
Esmaiel Jabbari
Michael J. Moore
Michael J. Yaszemski
Mayo Clinic College of Medicine

16.1 Introduction

Animal models are an indispensable tool in tissue engineering research as they provide important information that may lead to eventual development of clinically useful treatment of diseases. Current tissue engineering strategies often involve three main components: cells, scaffolds, and bioactive factors for the repair or regeneration of a specific tissue type. Research in animal models thus bridges the gap between *in vitro* studies (such as scaffold degradation, cell–scaffold interactions, and scaffold toxicity) and human clinical trials. Animal models have been and will continue to be used to develop an understanding of each of the primary components separately, in combination, and ultimately in pathologic orthopedic conditions.

As an example, the appropriate sequence of studies for the development of a new biodegradable bone regeneration material might be (1) biomaterial synthesis and structural characterization; (2) scaffold fabrication and measurements of mechanical and degradation properties *in vitro*; (3) biocompatibility and bone formation assessed by cell culture *in vitro*; (4) biocompatibility and degradation of the material *in vivo*, typically by subcutaneous implantation in a lower-level animal model such as a rat; (5) if no significant toxic effect is observed, the material is then evaluated for its intended application such as an appropriate bone defect model in a higher-level animal such as a rabbit; (6) if the material functions well in small animals, then a well-controlled study in large animals such as primates may be considered before human clinical trials.

In this chapter, we first discuss the general considerations when selecting an animal model and the most commonly used animals in orthopedic research. Then, specific animal models for testing the biocompatibility and biodegradation of tissue-engineered constructs are described. Well-established animal models for the evaluation of osteogeneic or chondrogenic potential of these constructs are introduced. Finally, the experimental design and evaluation methods involved in the animal studies are reviewed.

16.2   Animal Model Selection

Studies using animal models may be deemed necessary if there are no other *in vitro* alternatives, the knowledge gained can be applied for the benefit of humans or animals, and the procedure does not cause extreme pain or disability to the animals. In an ideal animal model, the anatomy and physiology should be suitable for the specific study design, the pathogenesis and disease progression should parallel that of humans, and there should be a similar histopathologic response to that seen in humans [1]. The preclinical animal models in which the tissue-engineered constructs are tested should mimic the clinical situation as closely as possible.

The use of animals in tissue engineering research has become a scientific as well as an ethical issue. Generally, the use of invertebrates is preferred over vertebrates and the use of rats, rabbits, goats, and sheep are preferred over dogs and cats due to their pet status. All animal protocols should be approved by the researcher's Institutional Animal Care and Use Committee to ensure that the experiments are appropriately designed, the number of animals is justified, and the animal procedures comply with the Animal Welfare Act.

From a practical perspective, animals of a particular species, strain, type, age, or weight should be easily available for the entire experimental study [1]. It is preferable that the local animal research facility has the capacity to house these animals. The animals should also be easy to handle in terms of transportation, housing, peri-operative care, specimen handling, and disposal. The susceptibility of animals to disease is also an important consideration especially for long-term survival studies. Finally, the costs of animal purchasing, transportation, time for quarantine, housing, surgical supply, and special equipment should be carefully calculated before the project begins. Generally, larger animals impose more housing and handling difficulties than small animals such as rats and rabbits, and are more expensive.

16.3   Commonly Used Animals

Although a lower level vertebrate, the rat is among the most popular animal subjects in tissue engineering orthopedic research and often the first to consider for a new study due to its low cost and easy care and handling. Rats have a mean healthy life span of 21 to 24 months. After bone elongation ceases by the age of 6 to 9 months, considerable useful lifespan remains for experimental evaluation [1]. Rats have been used extensively in studies of biocompatibility [2], fracture [3], and bone defect repair [4]. The mouse, also a small rodent, has gained popularity in tissue engineering research because its genome can be easily manipulated and investigated. Both regular and immunocompromised nude mice have been widely used for osteogenesis [5] and chondrogenesis [6] in subcutaneous tissue.

Rabbits are another commonly used animal in tissue engineering research. They are a relatively higher vertebrate, with an appropriate size for surgical operation and analysis. Rabbits are suitable for studies of bone defect repair [7], articular cartilage repair [8], ligament reconstruction [9], and spine fusion [10]. The dog, a higher-level vertebrate, has also played a dominant role in orthopedic research and has been extensively used in studies of bone defect repair [11], cartilage repair [12], ligament reconstruction [9], and meniscus repair [13].

Goats have become increasingly popular in tissue engineering research. They have been used for studies of biocompatibility [14], bone repair [15], cartilage repair [16], meniscus repair [17], ligament reconstruction [9], and spine fusion [18]. Sheep are large animals similar to goats, but less literature is available for their application in tissue engineering research. They have been used in studies of bone defect repair [19], cartilage defect repair [20], meniscus repair [13], and ligament reconstruction [9].

Pigs have been used in studies such as osteonecrosis of the femoral head [21] and fractures of cartilage and bone [22]. A miniature pig articular cartilage defect model has also been established [23]. Horses have been used mainly for the studies of cartilage and joint conditions due to their rich cartilage tissue [24]. Primates would be ideal for tissue engineering research because they are the closest to humans; however, due to lack of availability and high cost, they are only chosen when absolutely necessary as a step before human clinical trials. They have been used in studies of bone repair [25], cartilage repair [26], and spinal conditions [27].

Other animals in orthopedic research, though less frequently used, include hamsters for implant infection, chickens for scoliosis and tendon repair, turkeys for bone remodeling, emus for osteonecrosis of the femoral head, cats for osteoarticular transplantation, and guinea pigs for osteoarthritis [1].

16.4 Specific Animal Models

16.4.1 Biocompatibility

The biocompatibility study, as the first *in vivo* step of biomaterial evaluation, is typically performed by subcutaneous or intramuscular implantation in rats or rabbits. The rat is more economical and offers about a 15 month observation period, while the rabbit is used for longer-term evaluations of up to 6 months. Discs of 10 mm in diameter and 1 mm in thickness are commonly inserted in the dorsal subcutaneous tissue or back muscles (up to six implants per rat and eight to ten per rabbit) [28]. Alternatively, porous discs, biomaterials with seeded or encapsulated cells, and biomaterials contained in a stainless steel wire mesh cage have been studied [29,30].

The factors that determine the biocompatibility of a biomaterial include its chemistry, structure, morphology, and degradation. Biomaterial implantation often induces an inflammatory response through the activation of macrophages. The degree of fibrosis and vascularization of the tissue reaction dictate the nature of the response [29]. For instance, a mature fibrous capsule was observed after 12 weeks surrounding the implanted poly(propylene fumarate)/beta-tricalcium phosphate (PPF/beta-TCP) scaffolds in rats. Photocrosslinked PPF scaffolds also elicited a mild inflammatory response in rabbits [31]. The soft tissue response to oligo(poly(ethylene glycol)fumarate) (OPF) hydrogels was affected by the block length of poly(ethylene glycol) in a rabbit model [32]. In a granular tissue reaction, fibrovascular tissue ingrowth into porous poly(L-lactic acid) (PLLA) scaffolds was found to depend on the pore size [33].

A wide range of other biodegradable materials have also been assessed for their biocompatibility including commonly used poly(lactic acid) (PLA), poly(glycolic acid) (PGA), and their copolymers [34], as well as natural materials such as crosslinked gelatin [35] and hyaluronic acid [36].

16.4.2 Biodegradation

Biodegradable materials should be fully characterized for their degradation properties *in vitro*, but these studies cannot substitute for *in vivo* evaluation of degradation. Often, the rate of material degradation is significantly different *in vivo* compared to the *in vitro* condition. For example, poly(DL-lactic-co-glycolic acid) (PLGA) foams [37] and injectable PPF/beta-TCP composites [2] were both found to degrade more rapidly *in vivo* than *in vitro*. Differences in degradation rate *in vivo* vs. *in vitro* may be due to the extent of perfusion, pH levels, enzymatic or inflammatory mechanisms, or mechanical loading. Moreover, the apparent biocompatibility of a material may be confounded by greater toxicity of degradation products or material fragments.

Initial *in vivo* biodegradation studies may be undertaken in models similar or identical to those addressed in Section 16.4.1. Typically, discs, rods, foams, or other constructs are implanted in mice or rabbits and evaluated at multiple time intervals for changes in dry mass, molecular weight, mechanical properties, dimensions, and geometry. Standardization of sample geometry and dimensions is desirable to allow for comparison among research groups [28]. Samples may be implanted in various soft tissue locations, including the subdermal or subcutaneous space, intramuscular regions, or in the mesentery or intraperitoneal cavity.

Tests under loading conditions similar to the target tissue are ideal to include effects of mechanical loading on material degradation. If the material in question will have contact with bone, soft tissue evaluation must be followed or replaced by evaluation in bony tissue. Bone defect models are numerous [28], and generally involve drilled or otherwise excised bone segments. For example, self-reinforced PLLA and poly(DL-lactic acid) (PDLLA) [38] have been evaluated by implantation in mandibular bone or a femoral cavity of Sprague-Dawley rats. Because bone naturally remodels, it exhibits great healing capacity, and for this reason, critical sized defects are often necessary for long-term evaluation. These are defects that will predictably not heal spontaneously.

The choice of animal model to evaluate biodegradation *in vivo* also depends on the duration of the biodegradation study. For 1 to 6-month implantation studies, mice, rats, or rabbits can be used. Subdermal or subcutaneous implantation for up to 90 days in Sprague-Dawley rats has been used to evaluate biodegradation of poly(carbonate urethane), poly(ether urethane) [39], poly(ester amides) [40], poly(ether carbonates) [30], and poly(tetrahydrofuran) [41]. However, for long-term implantation studies, larger animals such as goats, sheep, or dogs should be used. As an example, sheep have been used as a model to evaluate biodegradation of self-reinforced poly(L-lactide) screws for up to 3 years *in vivo* [42,43].

16.4.3 Osteogenesis

In bone tissue engineering applications, after promising *in vitro* cell culture results are revealed, it is often a first step to use a heterotopic animal model for testing *in vivo* osteogenesis of the constructs. The major heterotopic models include the subcutaneous model, intramuscular model, intraperitoneal model, and mesentery model. Tissue-engineered constructs, either alone or placed in diffusion chambers can be implanted in mice, rats, rabbits, dogs, pigs, goats, or primates [44]. The rat subcutaneous model is the most frequently used. For example, bone formation was observed by marrow stromal osteoblast transplantation in a porous ceramic [45]. Ectopic bone formation was also demonstrated by transplantation of PLGA foams to the rat mesentery [46].

Although the heterotopic models are useful to provide some information helping to bridge the gap of *in vitro* and *in vivo* studies, the results obtained may not be the same once the constructs are implanted in actual bone defects. Therefore, further studies must be performed in bone defect models. Among them, the rabbit calvarial defect model is a very popular and reproducible bone defect model [44]. The calvarial bone is a plate which allows creation of a uniform defect that enables convenient radiographic and histological analysis. The calvarial bone has an appropriate size for easier surgical procedures and specimen handling. Because of the support by the dura and the overlying skin, no fixation is needed. This model has recently been used to evaluate the osteogenicity of a composite scaffold of PPF and PLGA [47], and growth-factor-coated PPF scaffolds [48].

Common long bone defect models include the rabbit radial model (most popular), rat femoral model, and the dog radial model. A composite bone scaffold consisting of nano-hydroxyapatite, collagen, and PLA was found to integrate the rabbit radial defect after 12 weeks [49]. The effect of bone morphogenetic protein was evaluated in the rat femoral defect model [4]. It should be noted that the rat model requires either internal or external fixation. In the dog model, ulnae fractures may occur, resulting in unexpected loss.

Due to the regenerative capability of bone defects, it is typical in tissue engineering research to consider critical sized defects. The critical size of the defect (CSD), defined as the smallest size that does not heal by itself if left untreated over a certain period of time, is 15 mm in diameter for adult New Zealand white rabbit calvarial defect model. The rat calvarial model is also popular with a CSD of 8 mm in diameter. A bone biomimetic device consisting of a porous biodegradable scaffold of poly(DL-lactide) and type I collagen, human osteoblast precursor cells, and rhBMP-2 was shown to promote bone regeneration in this model [50]. The CSD of long bones is at least two times the bone diameter (e.g., 12 to 15 mm for rabbit radial model and 5 to 10 mm for rat femoral model).

16.4.4 Chondrogenesis

Similar to bone tissue engineering, the first step to test the chondrogenic potential of tissue-engineered cartilage constructs *in vivo* is to use heterotopic models by subcutaneous, intramuscular, or intraperitoneal implantation in nude mice, syngeneic mice, rats, or rabbits. Among them, implantation of the constructs in dorsal subcutaneous pouches of nude mice is the most popular. For example, constructs containing genetically engineered chondrocytes were implanted into nude mice to demonstrate transgene expression and synthesis of glycosaminoglycan [51].

Commonly used cartilage defect models include partial-thickness (chondral) and full-thickness (osteo-chondral) defects in rabbits and dogs [52]. The rabbit distal femoral defect model is a well-established, reproducible model. Creating a partial-thickness defect can be challenging because the cartilage thickness in the rabbit femoral condyles is only 0.25 to 0.75 mm (compared to 2.2 to 2.4 mm in humans) [53]. The dog distal femur defect model, by contrast, offers a larger defect size with which to work. Besides species, the age of the animal is also an important consideration since the potential for cartilage repair as well as the response to various treatments varies with different animals [53]. Skeletal maturity of the commonly used New Zealand white rabbits typically occurs between 4 and 6 months.

Tissue-engineered articular cartilage is generally anchored to a defect by press fitting, suture, fibrin glue, or by using a periosteal flap. The local mechanical environment is well known to influence chondrogenesis and tissue healing. In the rabbit knee model, a defect created on the distal femoral surface is considered weight bearing, while a defect in the intercondylar groove is partial weight bearing [52]. The type of postoperative treatment should also be considered; the use of continuous passive motion can enhance cartilage healing [54] whereas joint immobilization may lead to decreased articular cartilage regeneration [55].

Many naturally derived and synthetic polymers are currently used as scaffolds for regeneration of articular cartilage [56]. They function not only as a vehicle for the delivery and retention of chondrogenic cells, but also as a substrate for cell attachment and matrix production. Natural polymeric materials such as collagen, hyaluronic acid, alginate, and fibrin glue have been extensively studied, as well as many synthetic polymers including PLA, PGA, and poly(ethylene oxide) (PEO) [53]. More recently, a new synthetic hydrogel based on oligopolyethylene glycol fumarate (OPF) was developed for cartilage tissue engineering [57].

16.5 Experimental Studies

16.5.1 Experimental Design

The experimental design includes the experimental groups to be used, sample size, sampling error, and control groups. The number of animals depends on the intrinsic variability among the animals, the consistency of the surgical procedure, the accuracy of the evaluation methods, and the choice of statistical method for data analysis. The number of animals needed in a study can often be determined from the results of a preliminary pilot study. For example, six to eight animals can be used in a preliminary study with histomorphometry or mechanical testing as the evaluation methods [58,59]. From this preliminary study, the standard deviation of the mean, coefficient of variation, and mean difference among the groups are determined. The sample size is affected by the desired power, the acceptable significance level, and the expected effect size. To reduce the interanimal variance, the animals should be of the same strain, sex, age, and weight. Some of the experimental animal models such as Sprague-Dawley rats and New Zealand white rabbits have identical genetic traits, but dogs and cats are relatively heterogeneous [60]. Therefore larger numbers of animals are often required for studies employing dogs or cats. The number of animals also depends on the evaluation methods.

Sampling error reflects the inherent uncertainty of results about a population based on information gained from sample data, which is a subset of the population. Two sampling procedures used in biomedical research are random and stratified sampling [61,62]. In simple random sampling, each element of a population has an equal probability of being included. In stratified sampling, the elements are divided into nonoverlapping blocks, and specimens are chosen from each block by simple random sampling. Stratified sampling provides greater control over the distribution of specimens.

Controls in an experimental study include normal controls, treatment controls, and time controls. Unilateral, bilateral, unicortical, and bicortical bone models are being used in bone tissue engineering studies. Because of more efficient comparison between control and experimental groups, bilateral models are extensively used for evaluating biocompatibility and function of foreign materials in bones [63,64]. When major procedures are performed involving a joint, a unilateral model should be designed to allow the animal to heal over time. The bilateral cortical model is a design that doubles the number of specimens for testing if it is appropriate from animal use perspective [65,66].

16.5.2   Evaluation Methods

The evaluation method should be valid and capable of measuring the parameter of interest. The development of new evaluation methods requires assessment of its accuracy. Sources of error in an evaluation method are inadequate surgical procedures or specimen preparation, systemic error of testing systems, and data collection error. Clinical evaluation, necroscopy, morphological or structural analysis, biochemical evaluation, mechanical testing, and the use of specialized devices are used in orthopedic research for evaluation. The most commonly used methods in orthopedic animal research are clinical observation, radiography, histological evaluation, and mechanical testing [60].

One method of biocompatibility evaluation is the implantation of disk-shaped samples of the material into the right and left epididymal fat pads or the right and left rear haunch subcutaneous tissue [67]. The following are removed at autopsy: the implant with its surrounding capsule, skin and subcutaneous tissue, the axillary and popliteal lymph nodes, heart, kidneys, lungs, liver, spleen, brain, and aorta, as well as any macroscopically unusual findings in the stomach, bowels, and mesenteric lymph nodes [68,69]. For evaluation of bone growth in explanted skeletal specimens, the total cross-sectional area of newly formed trabecular bone is determined on sequential longitudinal sections [70–72]. Confocal microscopy is utilized for cell visualization and morphology [73]. Alkaline phosphatase activity and calcium content are measured to assess the extent of mineralization in newly formed bone [74,75]. Western analyses of the newly synthesized collagen types I, II, and IX from the explanted samples are done using monoclonal antibodies and chemiluminescent substrate [76,77].

The explanted bone samples should be tested intact to establish reference mechanical properties [78]. Loads should be applied in a manner that minimizes focal loading of the specimen, so that accurate, representative material properties are measured. Preconditioning the specimens by cyclic loading, over several complete load-unload cycles, to a peak compressive force of approximately one average body weight (70 kg) is reasonable to do. This technique tends to smooth out variability in the data, and increases the likelihood of obtaining reproducible data.

References

[1] An, Y.H. and Friedman, R.J., Animal selections in orthopaedic research, in *Animal Models in Orthopaedic Research*, An, Y.H. and Friedman, R.J. (Eds.), CRC Press, Boca Raton, FL, 1999, p. 39.
[2] Peter, S.J. et al., *In vivo* degradation of a poly(propylene fumarate)/beta-tricalcium phosphate injectable composite scaffold, *J. Biomed. Mater. Res.* 41, 1, 1998.
[3] An, Y. et al., Production of a standard closed fracture in the rat tibia, *J. Orthop. Trauma* 8, 111, 1994.

[4] Yasko, A.W. et al., The healing of segmental bone defects, induced by recombinant human bone morphogenetic protein (rhBMP-2). A radiographic, histological, and biomechanical study in rats, *J. Bone Joint Surg. Am.* 74, 659, 1992.

[5] Miyazawa, K., Kawai, T., and Urist, M.R., Bone morphogenetic protein-induced heterotopic bone in osteopetrosis, *Clin. Orthop. Relat. Res.* 324, 259, 1996.

[6] Paige, K.T. et al., *De novo* cartilage generation using calcium alginate-chondrocyte constructs, *Plast. Reconstr. Surg.* 97, 168, 1996.

[7] Roy, T.D. et al., Performance of degradable composite bone repair products made via three-dimensional fabrication techniques, *J. Biomed. Mater. Res.* 66A, 283, 2003.

[8] Wakitani, S. et al., Repair of large full-thickness articular cartilage defects with allograft articular chondrocytes embedded in a collagen gel, *Tissue Eng.* 4, 429, 1998.

[9] Carpenter, J.E. and Hankenson, K.D., Animal models of tendon and ligament injuries for tissue engineering applications, *Biomaterials* 25, 1715, 2004.

[10] Boden, S.D., Biology of lumbar spine fusion and use of bone graft substitutes: present, future, and next generation, *Tissue Eng.* 6, 383, 2000.

[11] Johnson, K.D. et al., Evaluation of ground cortical autograft as a bone graft material in a new canine bilateral segmental long bone defect model, *J. Orthop. Trauma* 10, 28, 1996.

[12] Lee, C.R. et al., Effects of a cultured autologous chondrocyte-seeded type II collagen scaffold on the healing of a chondral defect in a canine model, *J. Orthop. Res.* 21, 272, 2003.

[13] Buma, P. et al., Tissue engineering of the meniscus, *Biomaterials* 25, 1523, 2004.

[14] Mendes, S.C. et al., Biocompatibility testing of novel starch-based materials with potential application in orthopaedic surgery: a preliminary study, *Biomaterials* 22, 2057, 2001.

[15] Kruyt, M.C. et al., Optimization of bone–tissue engineering in goats, *J. Biomed. Mater. Res.* 69B, 113, 2004.

[16] Butnariu-Ephrat, M. et al., Resurfacing of goat articular cartilage by chondrocytes derived from bone marrow, *Clin. Orthop. Relat. Res.* 330, 234, 1996.

[17] Port, J. et al., Meniscal repair supplemented with exogenous fibrin clot and autogenous cultured marrow cells in the goat model, *Am. J. Sports Med.* 24, 547, 1996.

[18] Kruyt, M.C. et al., Bone tissue engineering and spinal fusion: the potential of hybrid constructs by combining osteoprogenitor cells and scaffolds, *Biomaterials* 25, 1463, 2004.

[19] Viljanen, V.V. et al., Xenogeneic mouse (*Alces alces*) bone morphogenetic protein (mBMP)-induced repair of critical-size skull defects in sheep, *Int. J. Oral Maxillofac. Surg.* 25, 217, 1996.

[20] Homminga, G.N. et al., Repair of sheep articular cartilage defects with a rabbit costal perichondrial graft, *Acta Orthop. Scand.* 62, 415, 1991.

[21] Seiler, J.G., III et al., Posttraumatic osteonecrosis in a swine model. Correlation of blood cell flux, MRI and histology, *Acta Orthop. Scand.* 67, 249, 1996.

[22] Tomatsu, T. et al., Experimentally produced fractures of articular cartilage and bone. The effects of shear forces on the pig knee, *J. Bone Joint Surg. Br.* 74, 457, 1992.

[23] Hunziker, E.B., Driesang, I.M., and Morris, E.A., Chondrogenesis in cartilage repair is induced by members of the transforming growth factor-beta superfamily, *Clin. Orthop. Relat. Res.* 391S, S171, 2001.

[24] Litzke, L.E. et al., Repair of extensive articular cartilage defects in horses by autologous chondrocyte transplantation, *Ann. Biomed. Eng.* 32, 57, 2004.

[25] Cancian, D.C. et al., Utilization of autogenous bone, bioactive glasses, and calcium phosphate cement in surgical mandibular bone defects in *Cebus apella* monkeys, *Int. J. Oral Maxillofac. Implants* 19, 73, 2004.

[26] Buckwalter, J.A. et al., Osteochondral repair of primate knee femoral and patellar articular surfaces: implications for preventing post-traumatic osteoarthritis, *Iowa Orthop. J.* 23, 66, 2003.

[27] Boden, S.D., Grob, D., and Damien, C., Ne-osteo bone growth factor for posterolateral lumbar spine fusion: results from a nonhuman primate study and a prospective human clinical pilot study, *Spine* 29, 504, 2004.

[28] An, Y.H. and Friedman, R.J., Animal models for testing bioabsorbable materials, in *Animal Models in Orthopaedic Research*, An, Y.H. and Friedman, R.J. (Eds.), CRC Press, Boca Raton, FL, 1999, p. 219.

[29] Babensee, J.E. et al., Host response to tissue engineered devices, *Adv. Drug Deliv. Rev.* 33, 111, 1998.

[30] Dadsetan, M. et al., *In vivo* biocompatibility and biodegradation of poly(ethylene carbonate), *J. Control. Release* 93, 259, 2003.

[31] Fisher, J.P. et al., Soft and hard tissue response to photocrosslinked poly(propylene fumarate) scaffolds in a rabbit model, *J. Biomed. Mater. Res.* 59, 547, 2002.

[32] Shin, H. et al., *In vivo* bone and soft tissue response to injectable, biodegradable oligo(poly(ethylene glycol) fumarate) hydrogels, *Biomaterials* 24, 3201, 2003.

[33] Wake, M.C., Patrick, C.W., Jr., and Mikos, A.G., Pore morphology effects on the fibrovascular tissue growth in porous polymer substrates, *Cell Transplant* 3, 339, 1994.

[34] Behravesh, E. et al., Synthetic biodegradable polymers for orthopaedic applications, *Clin. Orthop. Relat. Res.* 367S, S118, 1999.

[35] Hong, S.R. et al., Biocompatibility and biodegradation of cross-linked gelatin/hyaluronic acid sponge in rat subcutaneous tissue, *J. Biomater. Sci. Polym. Ed.* 15, 201, 2004.

[36] Baier Leach, J. et al., Photocrosslinked hyaluronic acid hydrogels: natural, biodegradable tissue engineering scaffolds, *Biotechnol. Bioeng.* 82, 578, 2003.

[37] Lu, L. et al., *In vitro* and *in vivo* degradation of porous poly(DL-lactic-co-glycolic acid) foams, *Biomaterials* 21, 1837, 2000.

[38] Majola, A. et al., Absorption, biocompatibility, and fixation properties of polylactic acid in bone tissue: an experimental study in rats, *Clin. Orthop. Relat. Res.* 268, 260, 1991.

[39] Christenson, E.M. et al., Poly(carbonate urethane) and poly(ether urethane) biodegradation: *in vivo* studies, *J. Biomed. Mater. Res.* 69A, 407, 2004.

[40] Tsitlanadze, G. et al., Biodegradation of amino-acid-based poly(ester amide)s: *in vitro* weight loss and preliminary *in vivo* studies, *J. Biomater. Sci. Polym. Ed.* 15, 1, 2004.

[41] Pol, B.J. et al., *In vivo* testing of crosslinked polyethers. I. Tissue reactions and biodegradation, *J. Biomed. Mater. Res.* 32, 307, 1996.

[42] Jukkala-Partio, K. et al., Biodegradation and strength retention of poly-L-lactide screws *in vivo*. An experimental long-term study in sheep, *Ann. Chir. Gynaecol.* 90, 219, 2001.

[43] Suuronen, R. et al., A 5-year *in vitro* and *in vivo* study of the biodegradation of polylactide plates, *J. Oral Maxillofac. Surg.* 56, 604, 1998.

[44] An, Y.H. and Friedman, R.J., Animal models of bone defect repair, in *Animal Models in Orthopaedic Research*, An, Y.H. and Friedman, R.J. (Eds.), CRC Press, Boca Raton, FL, 1999, p. 241.

[45] Gao, J. et al., Tissue-engineered fabrication of an osteochondral composite graft using rat bone marrow-derived mesenchymal stem cells, *Tissue Eng.* 7, 363, 2001.

[46] Ishaug-Riley, S.L. et al., Ectopic bone formation by marrow stromal osteoblast transplantation using poly(DL-lactic-co-glycolic acid) foams implanted into the rat mesentery, *J. Biomed. Mater. Res.* 36, 1, 1997.

[47] Dean, D. et al., Poly(propylene fumarate) and poly(DL-lactic-co-glycolic acid) as scaffold materials for solid and foam-coated composite tissue-engineered constructs for cranial reconstruction, *Tissue Eng.* 9, 495, 2003.

[48] Vehof, J.W. et al., Bone formation in transforming growth factor beta-1-coated porous poly(propylene fumarate) scaffolds, *J. Biomed. Mater. Res.* 60, 241, 2002.

[49] Liao, S.S. et al., Hierarchically biomimetic bone scaffold materials: nano-HA/collagen/PLA composite, *J. Biomed. Mater. Res.* 69B, 158, 2004.

[50] Winn, S.R. et al., Tissue-engineered bone biomimetic to regenerate calvarial critical-sized defects in athymic rats, *J. Biomed. Mater. Res.* 45, 414, 1999.

[51] Madry, H. et al., Gene transfer of a human insulin-like growth factor I cDNA enhances tissue engineering of cartilage, *Hum. Gene Ther.* 13, 1621, 2002.

[52] An, Y.H. and Friedman, R.J., Animal models of articular cartilage defect, in *Animal Models in Orthopaedic Research*, An, Y.H. and Friedman, R.J. (Eds.), CRC Press, Boca Raton, FL, 1999, p. 309.

[53] Reinholz, G.G. et al., Animal models for cartilage reconstruction, *Biomaterials* 25, 1511, 2004.

[54] O'Driscoll, S.W. and Salter, R.B., The repair of major osteochondral defects in joint surfaces by neo-chondrogenesis with autogenous osteoperiosteal grafts stimulated by continuous passive motion. An experimental investigation in the rabbit, *Clin. Orthop. Relat. Res.* 208, 131, 1986.

[55] Vanwanseele, B., Lucchinetti, E., and Stussi, E., The effects of immobilization on the characteristics of articular cartilage: current concepts and future directions, *Osteoarthr. Cartil.* 10, 408, 2002.

[56] Temenoff, J.S. and Mikos, A.G., Injectable biodegradable materials for orthopedic tissue engineering, *Biomaterials* 21, 2405, 2000.

[57] Holland, T.A. et al., Transforming growth factor-beta1 release from oligo(poly(ethylene glycol) fumarate) hydrogels in conditions that model the cartilage wound healing environment, *J. Control Release* 94, 101, 2004.

[58] Munro, B.H., Jacobson, B.J., and Braitman, L.E., Introduction to inferential statistics and hypothesis testing, in *Statistical Methods for Health Care Research*, 2nd ed., Munro, B.H. and Page, E.B. (Eds.), Lippincott, Philadelphia, 1993.

[59] Matthews, D.E. and Farewell, V.T., *Using and Understanding Medical Statistics*, 3rd ed, Basel Karger, London, 1996.

[60] An, Y.H. and Bell, T.D., Experimental design, evaluation methods, data analysis, publication, and research ethics, in *Animal Models in Orthopaedic Research*, An, Y.H. and Friedman, R.J. (Eds.), CRC Press, Boca Raton, FL, 1999, p. 15.

[61] Manly, B.J., *The Design and Analysis of Research Studies*, Cambridge University Press, Cambridge, 1992.

[62] Forthofer, R.N. and Lee, E.S., *Introduction to Biostatistics. A Guide to Design, Analysis, and Discovery*, Academic Press, San Diego, CA, 1995.

[63] Laberge, M. and Powers, D.L., Scientific basis for bilateral animal models in orthopaedics, *J. Invest. Surg.* 4, 109, 1991.

[64] An, Y.H. et al., Fixation of osteotomies using bioabsorbable screws in the canine femur, *Clin. Orthop. Relat. Res.* 355, 300, 1998.

[65] Thomas, K.A. and Cook, S.D., An evaluation of variables influencing implant fixation by direct bone apposition, *J. Biomed. Mater. Res.* 19, 875, 1985.

[66] Thomas, K.A. et al., The effect of surface treatments on the interface mechanics of LTI pyrolytic carbon implants, *J. Biomed. Mater. Res.* 19, 145, 1985.

[67] Kidd, K.R. et al., A comparative evaluation of the tissue responses associated with polymeric implants in the rat and mouse, *J. Biomed. Mater. Res.* 59, 682, 2002.

[68] Nary Filho, H. et al., Comparative study of tissue response to polyglecaprone 25, polyglactin 910 and polytetrafluorethylene structure materials in rats, *Braz. Dent. J.* 13, 86, 2002.

[69] Eltze, E. et al., Influence of local complications on capsule formation around model implants in a rat model, *J. Biomed. Mater. Res.* 64A, 12, 2003.

[70] Lewandrowski, K.U. et al., Effect of a poly(propylene fumarate) foaming cement on the healing of bone defects, *Tissue Eng.* 5, 305, 1999.

[71] Schantz, J.T. et al., Induction of ectopic bone formation by using human periosteal cells in combination with a novel scaffold technology, *Cell Transplant* 11, 125, 2002.

[72] Yasin, M. and Tighe, B.J., Polymers for biodegradable medical devices. VIII. Hydroxybutyrate-hydroxyvalerate copolymers: physical and degradative properties of blends with polycaprolactone, *Biomaterials* 13, 9, 1992.

[73] Behravesh, E. and Mikos, A.G., Three-dimensional culture of differentiating marrow stromal osteo-blasts in biomimetic poly(propylene fumarate-co-ethylene glycol)-based macroporous hydrogels, *J. Biomed. Mater. Res.* 66A, 698, 2003.

[74] Temenoff, J.S. et al., *In vitro* osteogenic differentiation of marrow stromal cells encapsulated in biodegradable hydrogels, *J. Biomed. Mater. Res.* 70A, 235, 2004.

[75] Temenoff, J.S. et al., Thermally cross-linked oligo(poly(ethylene glycol) fumarate) hydrogels support osteogenic differentiation of encapsulated marrow stromal cells *in vitro*, *Biomacromolecules* 5, 5, 2004.

[76] Adkisson, H.D. et al., *In vitro* generation of scaffold independent neocartilage, *Clin. Orthop. Relat. Res.* 391S, S280, 2001.

[77] Gibson, G. et al., Type X collagen is colocalized with a proteoglycan epitope to form distinct morphological structures in bovine growth cartilage, *Bone* 19, 307, 1996.

[78] Elder, S. et al., Biomechanical evaluation of calcium phosphate cement-augmented fixation of unstable intertrochanteric fractures, *J. Orthop. Trauma* 14, 386, 2000.

17

The Regulation of Engineered Tissues: Emerging Approaches

Kiki B. Hellman
The Hellman Group, LLC

David Smith
Teregenics, LLC

17.1 Introduction

A critical step in translating tissue engineering research into product applications for the clinic and marketplace is understanding the strategies developed by government agencies for providing appropriate regulatory oversight. Since the eventual goal is the establishment of a global industry with the ability of companies to market products across national boundaries, a harmonized international regulatory approach would be ideal. However, while groups actively work toward that end, and, recognizing that market acceptance can be influenced by local cultural, ethical, and legal concerns, it is essential to recognize and appreciate the approaches of the regulatory agencies of those countries where engineered tissue research is already moving into product development and clinical application. While the science in the field is now worldwide in scope [1], we will limit our discussion to the regulatory approaches of the United States, with attention to emerging trends in Europe and Japan.

The U.S. Food and Drug Administration (FDA) has recognized that an important segment of the products it regulates arises from applications of new technology such as tissue engineering, and that, the product applications pose unique and complex questions. As a result, the agency has devoted considerable resources, from the 1990s on, to the regulatory considerations of what have been termed human cellular and tissue-based products (HCT/Ps).

In February 1997, FDA proposed a comprehensive approach (the "Proposed Approach") for regulation of these products, which was tier-based with the level of product review proportionate to the degree of risk [2]. Tissues recovered and processed by methods which did not change tissue function or characteristics (i.e., tissues which have not been more than "minimally manipulated") did not require premarket application review and approval. These "banked" human tissues, such as cornea, skin, umbilical cord blood stem cells, cartilage, and bone would be regulated under Section 361 of the Public Health Service (PHS) Act (42 USC §264), under which the agency is empowered to take action to prevent the spread of infectious disease. All other products would be regulated as medical products for which clinical testing and premarket applications would be required, with the expectation that most would be classified as either biological drugs ("biologics") (Section 351 of the PHS Act) [3] or medical devices ("devices") [Food, Drug and Cosmetic (FD&C) Act, 1976 Amendments] [4]. Specific jurisdiction over biologics and devices is generally assigned within the FDA to the Center for Biologics Evaluation and Research (CBER) or the Center for Devices and Radiological Health (CDRH), respectively, two of the agency's internal medical product review centers.

However, as these products typically consist of more than one component, such as biomolecules, cells, or tissues combined with a biomaterial, they are considered "Combination Products" and regulated under the jurisdiction of CBER, CDRH, or the FDA's third medical product review center, the Center for Drug Evaluation and Research (CDER). A determination of the product's primary mode of action dictates the jurisdictional authority for the product and the primary reviewing center; other centers act as consultants in the review process.

The FDA followed its issue of the Proposed Approach with a series of proposed rules to begin the process of translating its thinking about the regulation of HCT/Ps into law. In May 1998, the agency published two proposed rules to implement aspects of the proposed approach which would (1) create a unified system for registering establishments that manufacture HCT/Ps and for the listing of their products [5]. The following year it released a proposed rule to require most cell and tissue donors to be tested for relevant communicable diseases [6]. A proposed rule to require compliance with current good tissue practice by manufacturers of Human Cellular and Tissue-Based Products and to provide for inspection and enforcement was issued in 2001 [7]. Over the last few years, FDA has converted these proposed rules into final rules and new regulations having the force of law (Table 17.1). The last of these final rules, regarding donor eligibility and current good tissue pracices, became effective as of May 25, 2005.

Because of the continuing concern of communicable disease transmission, the FDA is now in the process of establishing a comprehensive new system for HCT/Ps formerly considered as "banked" human tissue. New regulations would require manufacturers of HCT/Ps to follow current good tissue practices (cGTP), such as proper product handling, processing, storage and labeling, as well as recordkeeping and establishment of a quality program [7]. In addition, manufacturers would be subject to inspection and enforcement activity by the agency.

17.2 FDA Regulation

Broad authority to control the distribution and sale of medical products in the United States has been granted to the FDA under the federal Food, Drug, and Cosmetic Act (FD&C Act) and the Public Health Service Act (PHS Act). The FD&C Act contains numerous provisions regarding the development and distribution of medical products classifiable as "drugs" or "devices." The PHS Act contains just two sections of particular importance to FDA regulation of medical products, especially those derived through tissue engineering: §351 prohibits the distribution of unlicensed "biological products" and establishes criteria and procedures the FDA shall observe in issuing such licenses; and §361 empowers the FDA to prevent the spread of communicable diseases.

TABLE 17.1   Key FDA Documents Concerning Regulation of Human Tissue and Cell Therapies[a]

---

1.   FDA Notice: Application of Current Statutory Authorities to Human Somatic Cell Therapy Products and Gene Therapy Products (58 FR 53248; October 14, 1993)

2.   FDA Notice of Interim Rule: Human Tissue Intended for Transplantation (58 FR 65514; December 14, 1993)

3.   FDA Notice of Public Hearing: Products Comprising Living Autologous Cells Manipulated *ex vivo* and Intended for Implantation for Structural Repair or Reconstruction (60 FR 36808; July 18, 1995)

4.   FDA Final Rule: Elimination of Establishment License Application for Specified Biotechnology and Specified Synthetic Biological Products (61 FR 24227; May 14, 1996)

5.   FDA Notice: Availability of Guidance on Applications for Products Comprising Living Autologous Cells … (etc.) (61 FR 26523; May 28, 1996)

6.   FDA Guidance on Applications for Products Comprised of Living Autologous Cells Manipulated of *ex vivo* and Intended for Structural Repair or Reconstruction (61 FR 24227; May 14, 1996)

7.   FDA Proposed Approach to Regulation of Cellular and Tissue-Based Products (February 28, 1997)

8.   FDA Notification of Proposed Regulatory Approach Regarding Cellular and Tissue-Based Products (62 FR 9721; March 4, 1997)

9.   FDA Final Rule: Human Tissue Intended for Transplantation (62 FR 40429; July 29, 1997)

10.   FDA Notice: Availability of Guidance on Screening and Testing of Donors of Human Tissue Intended for Transplantation (62 FR 40536; July 29, 1997)

11.   FDA Guidance to Industry: Screening and Testing of Donors of Human Tissue Intended for Transplantation (July 29, 1997)

12.   FDA Guidance for Industry: Guidance for Human Somatic Cell Therapy and Gene Therapy (March, 1998)

13.   FDA Proposed Rule: Establishment Registration and Listing for Manufacturers of Human Cellular and Tissue-Based Products (63 FR 26744; May 14, 1998)

14.   FDA Proposed Rule: Eligibility Determination for Donors of Human Cellular and Tissue-Based Products (64 FR 52696; September 30, 1999)

15.   FDA Proposed Rule: Current Good Tissue Practice for Manufacturers of Human Cellular and Tissue-Based Products; Inspection and Enforcement (66 FR 1508; January 8, 2001)

16.   FDA Final Rule: Human Cells, Tissues, and Cellular and Tissue-Based Products; Establishment Registration and Listing (66 FR 5447; January 19, 2001)

17.   FDA Final Rule: Human Cells, Tissues, and Cellular and Tissue-Based Products; Establishment Registration and Listing; Delay of Effective Date (68 FR 2689; January 21, 2003)

18.   FDA Interim Final Rule: Human Cells, Tissues, and Cellular and Tissue-Based Products; Establishment Registration and Listing (69 FR 3823; January 27, 2004)

19.   FDA Final Rule: Eligibility Determination for Donors of Human Cellular and Tissue-Based Products (69 FR 29786; May 25, 2004)

20.   FDA Final Rule: Current Good Tissue Practice for Human Cell, Tissue and Cellular Tissue-Based Product Establishments: Inspection and Enforcement (69 FR 68612; November 24, 2004)

---

[a] With the exception of Document #1, each document listed here can be obtained through the FDA Web site (www.fda.gov/cber). While provisions of the FD&C and PHS Acts and the Final Rules, codified as part of the Code of Federal Regulations (CFR), promulgated thereunder by the FDA, have the force of law and are binding on the agency, FDA guidance documents are not. Nevertheless, guidances are clearly helpful in anticipating the agency's response to particular marketing approval and other regulatory issues.

Exercising its authority under these statutes, the FDA has adopted a set of regulations that control virtually every aspect of the development and marketing of a medical product according to the potential risk of harm the product may pose to patients or the public health. Thus, the FDA regulates the introduction, manufacture, advertising, labeling, packaging, marketing and distribution of, and record-keeping for, such products.

As a rule, the FDA requires a sponsor of a new medical product to submit a formal application for approval to market the product after the completion of preclinical studies and phased clinical trials that demonstrate to the agency's satisfaction that the product is safe and effective. The form and review of that request to initiate human trials and the subsequent marketing application vary according to the classification of the product with reference to categories established in the statutes granting regulatory authority to the FDA.

17.2.1 Classification of Medical Products

Under current federal law, every medical product is classifiable as a drug, device, biological product (a "biologic"), or "combination product" (i.e., a combination device/drug, device/biologic, etc.). The classification of the product determines the particular processes of review and approval the FDA may employ in determining the safety and efficacy of the product for human use.

Under the FD&C Act (at §201(g)(1)), a "*drug*" is broadly defined as

> *[an article] intended for use in the diagnosis, cure, mitigation, treatment, or prevention of disease ... [or] ... intended to affect the structure or any function of the body.*

The FD&C Act (at §201(h)) defines a "*device*" largely by what it is not (i.e., neither a drug nor a biologic):

> *an instrument, apparatus, implement, machine, contrivance, implant, in vitro reagent, or other similar related article ... intended for use in the diagnosis of disease or other conditions, or in the cure, mitigation, treatment or prevention of disease ... or intended to affect the structure or any function of the body ... and which does not achieve any of its primary intended purposes through chemical action within or on the body ... and which is not dependent upon being metabolized for the achievement of any of its primary intended purposes.[Emphasis added.]*

Finally, the PHS Act (at §351(a)) defines a "*biologic*" as

> *any virus, therapeutic serum, toxin, antitoxin, vaccine, blood, blood component or derivative, allergenic product, or analogous product"* applicable to the prevention, treatment or cure of diseases or injuries.

Not surprisingly, the advance of medical technology has generated products not readily classifiable as drugs, devices, or biologics as those terms have been defined by federal statute. To provide for the expanding varieties of products expressing features of more than one of those classifications, the FDA has been authorized to recognize "combination products." A combination product is classified, assigned to a particular center, and regulated as a drug, device, or biologic according to its "primary mode of action," as determined by the FDA. Disputes over the classification of a combination product between a sponsor and the FDA or between centers are submitted to the FDA's ombudsman in the Office of the Combination Products for resolution. In fact, the FDA's current approach to the regulation of engineered tissue products began with the ombudsman's consideration of the classification of the Carticel™ autologous cartilage repair service developed by Genzyme Tissue Repair in 1995.

The Office of the Combination Products was established under the aegis of the Medical Device User Fee and Modernization Act of 2002. This office has responsibilities over the regulatory life cycle of combination products, and (1) assigns the FDA Center for primary review jurisdiction of the product; (2) works with FDA Centers to develop regulatory guidance and clarify regulation of these products; and (3) serves as a focal point for combination product-related issues for internal and external stakeholders.

With respect to engineered tissue products, the consequences inuring to the device and biologic classifications deserve particular attention. First, and most importantly, a medical product cannot be a device if its therapeutic or diagnostic benefit is obtained through metabolization, a limitation in the statutory definition of a device that might appear to exclude any product incorporating and depending on the function of any living human tissues. Nevertheless, allogeneic skin products such as Organogenesis's Apligraf have been classified and granted market approval as devices. They were considered interactive wound and burn dressings and, hence, classified as devices, since CDRH has regulating jurisdiction over wound and burn dressings. As engineered tissue products become less "structural" and more "functional" that is, more interactive with the host or require metabolism by the host, in nature, a "device" classification may become more difficult to square with the current statutory definition.

17.2.2 Special Product Designations

The FD&C Act recognizes that demand for all new medical products is not equally large or robust, such that the cost of obtaining marketing approval for a given product may be prohibitive in view of the relatively small size of the population it will benefit. To reduce the likelihood that a financial cost–benefit analysis applied to rarer diseases will leave them untreated, the FDA is authorized to grant special considerations and exceptions to reduce the economic burden upon developers of products under such conditions. Thus, the FDA may be petitioned to grant a "humanitarian device exemption" for certain devices (FD&C Act, §520(m)) or to recognize certain drugs or biologics as "orphan drugs" (FD&C Act, §525, et. seq.). However, the significance or value of these designations — especially for sponsors of tissue products — varies considerably according to the classification of the product in question.

Humanitarian use devices are those intended to treat a disease or condition that affects fewer than 4000 people in the United States. The FDA is authorized to exempt a sponsor from the obligation to demonstrate the effectiveness of such a device to obtain marketing approval; however, the sponsor is precluded from selling the product for more than the cost to develop and produce it.

Orphan drugs are those intended to treat a disease or condition affecting fewer than 200,000 persons in the United States, or for which there is little likelihood that the cost of developing and distributing it in the United States will be recovered from sales of the drug in the United States. The orphan drug designation was established through an amendment of the FD&C Act by the 1982 Orphan Drug Act (ODA) prior to the creation of the humanitarian device exemption. In contrast to the humanitarian use device designation, the orphan drug designation could be important to sponsors of certain engineered tissue products classifiable as biologics, illustrating the larger implications of the classification process. An orphan drug is defined to include biologics specifically licensed under §351 of the PHS Act, a distinction which may be relevant under the FDA's proposed plan for regulating engineered tissue products (see below). The FDA is empowered, under certain conditions, to grant marketing exclusivity for an orphan drug in the United States for a period of seven years from the date the drug is approved for clinical use; this exclusivity is stronger than and far less expensive to maintain than that provided by a patent. Additional benefits of the orphan drug designation include: certain tax credits for clinical research expenses; cash grant support for clinical trials; and waiver of the expensive prescription drug filing fee. A petition for orphan drug designation must be filed before any application for marketing approval.

17.2.3 Human Cellular and Tissue-Based Products

While a broad range of potential therapeutic applications for engineered tissues is being developed, they fall into two general categories, those providing either a structural/mechanical or a metabolic function. Structural/mechanical applications include approaches for skin, musculoskeletal, cardiovascular, and central nervous systems, among them, while metabolic applications include therapies for conditions such as diabetes and liver disease.

The FDA defines an HCT/P as a "product containing or consisting of human cells or tissue that is intended for implantation, transplantation, infusion, or transfer into a human recipient" and, in addition to those indicated previously, includes cadaveric ligaments, dura mater, heart valves, manipulated autologous chondrocytes, and spermatozoa [7]. Since determining the regulatory process for certain HCT/Ps may be complicated, the agency will rely on the Tissue Reference Group (TRG) composed of representatives from CBER, CDRH, and Office of Combination Products in the Office of the Commissioner to provide a single reference point and make recommendations to the center directors regarding product jurisdiction of specific tissue products.

Much of the regulatory framework for engineered tissues has been promulgated by the FDA through formal, binding, rule-making procedures. Previously the FDA had issued a number of documents which, while not binding upon the Agency, provide the public with a formal expression of its evolving thinking regarding the future regulation of human cellular or tissue-based products (see Table 17.1). Of these documents, by far the most important has been the Proposed Approach to Regulation of Cellular and Tissue-Based Products ("Proposed Approach") that the FDA issued on February 28, 1997.

The Proposed Approach outlined a plan of regulatory oversight, which may include a premarket approval requirement for such tissue products based upon a matrix ranking the products, classified by certain characteristics, within identified areas of regulatory concern. These tissue products would be classified according to the relationship between the donor and the recipient of the biological material used to produce the tissue product; the degree of *ex vivo* manipulation of the cells comprising the tissue product; and whether the tissue product is intended for a homologous use, for metabolic or structural purposes, or is to be combined with a device, drug, or another biologic.

Since issuing the Proposed Approach more than seven years ago, the FDA has been working to formalize its regulation of human tissue and cell therapies through a rule-making process to amend the U.S. Code of Federal Regulations ("CFR"). This process was completed as of May 25, 2005, when the last of the regulations implementing the proposed approach went into effect (see Table 17.1).

In introducing the February 1997 "Proposed Approach," the FDA identified five areas of regulatory concern raised by the development of new medical products derived from the manipulation of human biological materials: communication of infectious disease; processing and handling; clinical safety and efficacy; indicated uses and promotional claims; and monitoring and education (Table 17.2).

The FDA has proposed that autologous tissue that is banked, processed, or stored should be tested for infection agents and it will require companies to keep appropriate records to assure that patient tissues are not mismatched or commingled. The agency proposes that allogeneic tissue be tested for infection agents, that donors be screened, and that appropriate records be kept. Periodic submissions to the agency showing compliance with the testing or record-keeping requirements will not be necessary; the FDA assumes that a company's observation of these requirements will be assured through the accreditation they can be expected to maintain with professional tissue banking or processing societies.

The extent of the FDA's regulatory intervention in the areas of processing and handling and clinical safety and efficacy according to the characteristics of the particular tissue product in question. To the extent that a tissue product undergoes more than minimal manipulation in processing, is intended for a non-homologous use, is combined with nontissue components, or is intended to achieve a metabolic outcome, the agency will require a greater demonstration of safety and efficacy through appropriate clinical trials.

"Manipulation," in the agency's regulatory salience, is a measure of the extent to which the biological characteristics of a tissue have been changed *ex vivo*. The FDA has stated it presently considers cell selection or separation, or the cutting, grinding, or freezing of tissue, to constitute minimal manipulation. Cell expansion and encapsulation are examples of more than minimal manipulation.

To the extent that the tissue product only undergoes minimal manipulation, is intended for a homologous application to achieve a structural outcome (or reproductive or metabolic outcome, as between family members related by blood), and does not combine with nontissue components, the FDA will expect "good tissue practices" to be observed but will not impose any reporting duties or, consistent with its authority under §361 of the PHS Act, any product licensing or premarket approval requirements. Any other tissue product requires submission of appropriate chemistry, manufacturing, and controls information and BLA approval for any tissue product that does not incorporate nontissue components. Tissue products that are combinations of tissue and devices or tissue and drugs may be regulated according to established premarket approval (PMA or PDP) or new drug application (NDA) schemes.

Regulations now in effect compel the registration of sponsors and other persons engaged in production and distribution of such products, to screen donors of tissues used to produce HCT/Ps, and to observe Good Tissue Practices in their manufacturing or processing of such products.

17.2.4 Marketing Review and Approval Pathways

As discussed above, the particular program(s) of regulatory review applicable to a medical product are predetermined according to its FDA classification. Thus, the FD&C Act requires a sponsor to submit a device Pre-Market Application (PMA) or Product Development Protocol (PDP) to market a device, or

TABLE 17.2 First HCT/Ps Approved in the United States

HCT/P (Sponsor)	Approval	Approved indicated use	Composition; mode of action	Pivotal clinical studies
Dermagraft (Advanced Tissue Science, Inc.)	September 28, 2001 (approved as a Device)	Treatment of full-thickness diabetic foot ulcers	Cryopreserved human fibroblast-derived dermal substitute; it is composed of neonatal foreskin fibroblasts, extracellular matrix, and a bioabsorbable scaffold. Fibroblasts proliferate to fill the interstices of this scaffold and secrete human dermal collagen, matrix proteins, growth factors, and cytokines, to create a three-dimensional human dermal substitute containing metabolically active, living cells	595 Patients (2 Studies: 281/142 Control; 314/151 Control)
Cultured composite skin (Ortec, Inc.)	February 2, 2001 (approved as a Device under HDE Designation)	Adjunct to standard autograft procedures for covering wounds and donor sites created after the surgical release of hand contractions	Collagen matrix in which allogeneic human skin cells (i.e., epidermal keratinocytes and dermal fibroblasts) are cultured in two distinct layers. Designed to enhance wound healing in part by release of cytokines	145 Patients (63 control)
Apligraf (Organogenesis, Inc.)	May 22, 1998 (approved as a Device)	Standard therapeutic compression for the treatment of noninfected partial and full-thickness skin ulcers	Viable, bilayered, skin construct, which contains Type I bovine collagen, extracted and purified from bovine tendons and viable allogeneic human fibroblast and keratinocyte cells isolated from human infant foreskin	297 Patients (136 control)
Carticel (Genzyme Corporation)	August 22, 1997 (approved as a Biologic)	Repair of clinically significant, symptomatic, cartilaginous defects of the femoral condyle (medial, lateral, or trochlear) caused by acute or repetitive trauma	Used in conjunction with debridement, placement of a periosteal flap, and rehabilitation. The independent contributions of the autologous cultured chondrocytes and other components of the therapy to outcome are unknown	Retrospective analysis of Swedish Study (153 Patients) and US Registry data (241 Patients)

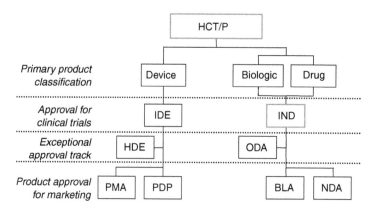

FIGURE 17.1   Product approval pathways.

a new drug application (NDA) to market a drug. The PHS Act provides that marketing approval for a biologic shall be obtained through the submission of a Biologics License Application (BLA). Certain drugs or biologics may qualify for special designation as orphan drugs under the Orphan Drug Act.

In addition, the FDA requires that sponsors of regulated products must first obtain preliminary approval for the clinical trials on humans that will support a subsequent application for full marketing approval. Clinical trials in support of a PMA application or as part of a PDP for a device may be conducted only after the FDA has issued an investigational device exemption (IDE); clinical trials in support of an application for marketing approval of a drug or biologic cannot be initiated until the FDA has approved an investigational new drug (IND) application (Figure 17.1).

17.2.4.1   Devices

The FDA has divided devices into three classes to identify the level of regulatory control applicable to them. The highest category, Class III, includes those devices for which premarket approval is or will be required to determine the safety and effectiveness of the device (21 CFR, §860.3(c); 21 U.S.C., §360c(a)(1)(C)). Absent a written statement of reasons to the contrary, the FDA classifies any "implant" or "life-supporting or life-sustaining device" as Class III (21 CFR, §860.93; 21 U.S.C., §360c(c)(2)(C)).

There are two primary pathways by which the FDA permits a medical device to be marketed: premarket clearance by means of a 510(k) notification, or premarket approval by means of a PMA ("Premarket Approval Application") or PDP ("Product Development Protocol") submission.

A sponsor may seek clearance for a device by filing a 510(k) premarket notification with the FDA, which demonstrates that the device is "substantially equivalent" to a device that has been legally marketed or was marketed before May 28, 1976, the enactment date of the Medical Device amendments to the FD&C act. The sponsor may not place the device into commercial distribution in the United States until the FDA issues a substantial equivalence determination notice. This notice may be issued within 90 days of submission but usually takes longer. The FDA, however, may determine that the proposed device is not substantially equivalent, or require further information such as additional test data or clinical data, or require a sponsor to modify its product labeling, before it will make a finding of substantial equivalence.

If a sponsor cannot establish to the FDA's satisfaction that a new device is substantially equivalent to a legally marketed device, it will have to seek approval to market the device through the PMA or PDP process. This process involves preclinical studies and clinical trials to demonstrate that the device is safe and effective.

FDA regulations (21 CFR, §860.7(d)) provide that, based on "valid scientific evidence," a device shall be found to be "safe"

> *… when it can be determined … that the probable benefits to health from use of the device for its intended uses and conditions of use … outweigh any probable risks[,]*

and that a device shall be found to be "effective"

> *... when it can be determined ... that in a significant portion of the target population, the use of the device for its intended uses and conditions of use ... will provide clinically significant results.*

Testing in humans to obtain clinical data demonstrating these qualities in support of a PMA or pursuant to a PDP must be conducted pursuant to an investigational device exemption (IDE). The IDE is the functional equivalent of the IND that governs clinical trials of drugs and biologics. As with other medical products, clinical testing is typically conducted in multiple phases, with the earliest phases primarily intended to demonstrate safety and later phases addressing both safety and effectiveness considerations. The sponsor of the device must also demonstrate compliance with applicable current good manufacturing practices (cGMPs, now also known as Quality System Regulations) before the FDA may approve the product for marketing by granting the PMA or accepting the completion of the PDP.

17.2.4.2 Biologics

Until recently, permission to market a biologic required two applications: one to obtain a product license application (PLA) for the biologic itself and another for approval of the facility where the biologic would be prepared, that is, an establishment license application. The 1997 FDA Modernization Act amended the PHS Act by eliminating the separate product and establishment license applications in favor of a single biologics license application (BLA), which, like the PMA or PDP for devices, includes an evaluation of compliance with appropriate quality controls and current cGMP as part of the assessment of the safety and efficacy of the product in question.

§351 of the PHS Act directs the FDA to approve a BLA on the basis of a determination that the biologic in question is "safe, pure, and potent." Those terms are defined in FDA regulations promulgated to give effect to that statutory authority:

> *safety means the relative freedom from harmful effect to persons affected, directly or indirectly, by a product when prudently administered, taking into consideration the character of the product in relation to the condition of the recipient at the time[;]*

> *purity means relative freedom from extraneous matter in the finished product, whether or not harmful to the recipient or deleterious to the product ... [and] includes but is not limited to relative freedom from residual moisture or other volatile substances and pyrogenic substances[;]*

> *potency is interpreted to mean the specific ability or capacity of the product, as indicated by appropriate laboratory tests or by adequately controlled clinical data obtained through the administration of the product in the manner intended, to effect a given result.*

Testing in humans to obtain clinical data demonstrating these qualities in support of a BLA must be conducted pursuant to an investigational new drug application (IND). The IND is the functional equivalent of the IDE that governs clinical trials of devices. As with other medical products, clinical testing is typically conducted in multiple phases, with the earliest phases primarily intended to demonstrate safety, and later phases intended to address both safety and efficacy considerations.

The emphasis given to process by the earlier requirement of a separate approval of the manufacturing facility illuminates the dual nature of the regulatory authority created under the PHS Act and ultimately exercised by the FDA. Besides assuring that only safe, pure, and potent biologics are marketed in the United States, the FDA is also charged with a general duty to prevent the introduction, transmission, or spread of communicable disease (PHS Act, §361(a)). While the BLA is an amalgam of product and process quality criteria, a particular emphasis upon the authority to eliminate sources of dangerous infection reappears in the context of the FDA's proposed regulatory triage for engineered tissues.

17.3   Regulation of Pharmaceutical/Medical Human Tissue Products in Europe

Regulation of medical products incorporating viable human tissue products among or within the member states of the European Union (EU) is marked by inconsistency but is presently the subject of substantial discussion and debate. As part of the overall coordination of national laws and governmental activities within the EU, the regulation of the marketing of certain medical products by national authorities is being consolidated within designated EU agencies, especially the European Medicines Evaluation Agency (EMEA). Within the scope of what medical products are considered pharmaceutical and regulated, there are two broad subcategories, medicinal products and medical devices.

The EMEA was established in 1993 by the European Economic Community (EEC, now EU) Council Regulation No. 2309/93 to implement procedures to give effect to a single market for "medicinal products" among the member states. In conjunction with three directives adopted concurrently (Council Directives 93/39EEC, 93/40EEC, and 93/41EEC), the regulation authorized EMEA to manage a "centralized procedure" for an EEC authorization to market medicinal products for either human or veterinary use. The directives also established a "mutual recognition procedure" for marketing authorization of medicinal products based upon the principle of mutual recognition of authorizations granted by national regulatory bodies. These procedures came into effect on January 1, 1995, with a three-year transition period until December 31, 1997. As of January 1, 1998, the independent authorization procedures of the member states are strictly limited to the initial phase of mutual recognition (i.e., granting marketing authorization by the "reference Member State") and to medicinal products that are not marketed in more than one member state. Consequently, sponsors seeking marketing authorization for medicinal products throughout the EU are obliged to seek such approval through the centralized procedure administered by EMEA.

The concept of a "medicinal product" in EEC legislation substantially predated the organization of EMEA. Council Directive 65/65EEC of January 26, 1965, defined the term medicinal product to include

> *any substance or combination of substances presented for treating or preventing disease in human beings or animals.*

[and]

> *any substance or combination of substances which may be administered to human beings or animals with a view to making a medical diagnosis or to restoring, correcting or modifying physiological functions in human beings or in animals ... .*

A "substance" is further defined to include "[a]ny matter irrespective of origin which may be human ... animal ... vegetable ... [or] chemical" (Directive 65/65/EEC, Article 1). However, the directive also makes clear that its regulation of medicinal products (and, through amendments to the directive recognizing the authority of EMEA, the "centralized procedure") does not apply to products "intended for research and development trials" (Directive 65/65/EEC, Article 2).

Sponsors of medical products derived through tissue engineering have reported substantial inconsistency among the regulatory bodies of EU member states regarding the classification of such products for purposes of determining the applicability of national or EU marketing authorization requirements. A determination that engineered tissue products are "medicinal products" subject to the centralized procedure for authorization administered by EMEA will substantially clarify and rationalize the process by which such products may be marketed throughout the European Community.

The EMEA has in place a Biotechnology Working Party that has considered, among other things, safety issues in the delivery of human somatic cell therapies and a definition of a "cell therapy medicinal product" (CPMP/BWP/41450/98 draft). This definition would consider engineered human tissues to be "medicinal products" within the meaning of Directive 65/65/EEC, provided the engineered tissue was the product of both the following:

a. ... *an industrial manufacturing process carried out in dedicated facilities. The process encompasses expansion or more than minimal manipulation designed to alter the biological or physiological characteristics of the resulting cells, and*
b. ... *further to such manipulation, the resulting cell product is definable in terms of qualitative and quantitative composition including biological activity.*

Points to Consider on Human Somatic Cell Therapy, CPMP/BWP/41450/98 draft, page 3/9.

The Biotechnology Sector of EMEA is likely to have primary responsibility for considering the authorization of engineered tissue products in the event they are classifiable as "medicinal products" [8].

Human tissue and cellular products may not be presently definable as "medicinal products" subject to regulation, to the extent that they are the result of modest manipulation of autologous tissues in the course of treating a fairly small patient population. Under these circumstances, the regulation of such cellular products is more likely to remain with the competent authorities of the Member States (with substantial variability in the classification and resulting regulation of such products, as outlined in Table 8.2 of book titled "The Biomedical Engineering Handbook: Tissue Engineering and Artificial Organs, 3rd Edition). Nevertheless, an EMEA decision to accept an engineered tissue product as a "medicinal product" could occur in response to a petition from a sponsor of such a product. To be successful, such a petition should probably stress the "industrial" nature of the fabrication process and the extent of manipulation of the human biological material to develop the engineered tissue product. Assuming an engineered tissue product could be established to be a "medicinal product," there does not appear to be any EU rule that could limit the ability of EMEA to grant market authorization according to the type or source of tissue from which the product had been derived.

EMEA is aligned with Enterprise DG (formerly DG III; the department of the European Commission primarily responsible for establishing and implementing rules promoting the Single Market for products). A unit of Enterprise DG oversees application of EU directives regulating marketing authorization of medical devices. Providing for engineered tissue products could require some reconsideration of the specific areas of responsibility of the units or agencies involved in regulating medical products.

17.4 Regulation of Pharmaceutical/Medical Human Tissue Products in Japan

It appeared at the time of the World Technology Evaluation Center (WTEC) panel's visit to Japan that the Government of Japan was only beginning to focus on codifying regulation of engineered human tissue products within its scheme of regulating other medical products [1]. The WTEC panel was unable within the scope of this study to provide an analysis of Japan's medical product approval process as potentially applied to engineered human tissue products. However, presented here is an outline of Japan's process and agencies responsible for regulation of medical products generally.

The Pharmaceutical and Medical Safety Bureau (PMSB) has primary responsibility within the Japanese Ministry of Health, Labour and Welfare for administering the requirements established for the safety and efficacy of medical products under Japan's Pharmaceutical Affairs Law. This legislation was substantially amended in 1996 (with the reforms made effective in April 1997) to provide for the present medical product review and approval system.

Applications for approval of new drugs and medical devices are referred by PMSB to the Central Pharmaceutical Affairs Council (CPAC) to obtain its recommendation. The CPAC, in turn, is advised by the Pharmaceutical and Medical Devices Evaluation Center (PMDEC), an expert body organized in July 1997 to evaluate the quality, efficacy, and safety of medical products administered to humans. Specific authority within PMSB to approve recommendations received from CPAC regarding the discrete aspects of the clinical testing, licensing, and use of new medical products is distributed among relevant divisions, such as the Evaluation and Licensing Division (premarketing and supplemental application approvals) and the Safety Division (adverse reaction measures). A regulatable medical product in Japan is classified as either a medical device or a pharmaceutical.

Advice concerning the design and conduct of clinical trials, as well as the adequacy of applications for approval of pharmaceuticals, is provided to PMDEC and to the product sponsor by the Drug Organization, a quasi-governmental agency established in 1979 as a fund to support patients experiencing adverse drug reactions. It is not clear whether the Drug Organization serves a similar function with respect to medical devices, or if there exists an equivalent medical device organization. However, applications for approval of "copy-cat" devices are referred to the Japan Association for the Advancement of Medical Equipment for a determination of the equivalence of the new device to devices already approved for clinical use. For a more detailed description of Japan's general medical product approval process, see, for example, Hirayama [9] and Yamada [10].

17.5   Other Considerations Relevant to Engineered Tissues

17.5.1   FDA Regulation and Product Liability

Protection from product liability lawsuits, in the form of an immunity from such litigation, may come from satisfying the federal regulations which govern the design and manufacture of, as well as the warnings to be supplied with, medical products.

By virtue of the Supremacy Clause of the Constitution, the federal government is permitted to regulate certain affairs free of state interference. State civil litigation is a form of regulation, so it is a form of interference. If Congress elects to exclusively regulate certain conduct, then litigation under state law regarding the same conduct is prohibited, as it may produce inconsistent or conflicting standards regulating that conduct.

The public policy arguments in favor of federal preemption with respect to the regulation of medical products are readily discernible: while both state and federal regulation would have the enhancement of public health and safety as their goal, the establishment of nationwide labeling and design criteria for medical products will promote uniformity and regularity in the interpretation of applicable regulations and will ensure enforcement of these regulations is conducted in the public interest, rather than through isolated lawsuits that may produce inconsistent results. In addition, the natural preeminence of a federal administration administering such regulations will simplify and improve communication between the regulators and the medical product sponsors. Federal preemption, then, is not a shield for bad medical products; rather, it protects a process of reasoned, scientific inquiry.

Congressional and FDA silence on the question of preemption in statutes and regulations governing drugs and biologics going back to the earliest federal Food and Drug Acts has meant that sponsors of these products have been left to argue only "implied" or "conflict" preemption with largely inconsistent — though often disappointing — results.

The situation for sponsors of devices would seem to be decidedly different. When Congress passed the 1976 Medical Device Amendments to the FD&C Act, it explicitly authorized preemption of any state requirement as to the safety or effectiveness of a medical device in favor of FDA regulations governing the same product. Because of the high degree of scrutiny to which Class III devices are subjected through the FDA's premarket approval (PMA) process, courts have shown a general willingness to recognize that personal injury litigation regarding these devices has been preempted. However, the FDA may also permit the marketing of devices which have been shown to be "substantially equivalent" to devices either previously approved or marketed before the enactment of the Medical Device Amendments in 1976. This "510(k)" clearance involves significantly less regulatory review of the device than would have been given to a PMA application.

In *Medtronic v. Lohr*, though, the U.S. Supreme Court concluded that the apparent preemption language of the Medical Device Amendments does not actually preclude product liability actions against sellers of "510(k)" devices. The court reasoned that the 510(k) premarket clearance process did not involve a comprehensive, in-depth analysis of the safety, efficacy, and labeling of a medical product prior to clinical use. While a number of justices suggested they might not reach the same conclusion regarding "PMA" devices, the *Lohr* decision has weakened — but certainly has not obliterated — a major defense of those engineered tissue products which may be classified and regulated as devices. Indeed, to the extent the further viability of the preemption defense is predicated upon the degree to which the FDA has given specific attention to and approval of the design of the device in question, premarket approval by means of the PDP process would seem to offer the greatest chance of immunity from product liability litigation.

17.5.2 Ownership of Human Tissues

While critical to the general advance of medical research, access to human tissues for research or product development is highly sensitive to public disclosure of practices where tissues are taken or used without consent or under circumstances suggesting a commercial market in body parts. The absence of comprehensive federal or state legislation governing "research" tissues deprives the biomedical community of clear, consistent guidelines to follow in acquiring and using tissues, while simultaneously representing a legislative vacuum that may be filled with substantial adverse unintended consequences if done suddenly in response to some public outcry. Absent effective coordination, the initiatives of individual federal agencies to establish policies for research involving human tissues or subjects may impose conflicting requirements or expectations.

Significant advances in medical research over the past several years has contributed substantially to the commercial utility of human biological materials. Consequently, the source of such materials used in the creation of engineered tissue products may become important for reasons beyond — and certainly removed from — the possible transfer of adventitious agents or the management of immunological responses. Simply put, the use of allogeneic materials raises issues of ownership, donation, and consent not to be found with respect to autologous tissues.

The common law of the United States recognizes a severely restricted property interest in human bodies or organs. In a broad sense, a "property interest" in something may be thought of as a "bundle of rights" to possess, to use, to profit from, to dispose of, and to deal in that thing. Courts have granted next of kin nothing more than a "quasiproperty" right — or right of sepulcher — in a decedent's body for the purposes of burial or other lawful disposition. In place of an exegesis of the religious or cultural prohibitions against recognizing a property interest in a dead body, it is clear that the limited right which has been fashioned by the courts has been intended to offer nothing more than that some interested person may ensure the remains are disposed of with dignity.

Organ transplantation has certainly created the conditions for a market for human body parts. In response, Congress and state legislatures have enacted statutes prohibiting the sale of any human organ. The National Organ Transplant Act was passed to regulate the availability of organs for transplantation through voluntary donation exclusively by explicitly prohibiting organ purchases. The same prohibition has been passed into law by the 15 states, to date, which have adopted the Uniform Anatomical Gift Act (1987) [11]. Other state statutes have imposed criminal penalties for the purchase of organs or tissue from either living or cadaveric providers.

These federal and state statutes effectively banning purchases of human organs were enacted in the mid-1980s in immediate response to the prospect of a widespread trade in these body parts to supply the growing demand for transplant material. The vision of a vendor selling livers and kidneys — or worse, a patient harvesting one of his or her own organs for money clearly hovered over the debate leading to the passage of this legislation. But that vision imagined people selfdismantling for cash; it did not really allow for a trade in renewable body parts, especially cells.

17.6 Conclusion

No part of the process of bringing new biomedical products from the laboratory to the patient occurs in isolation from or independent of all of the other aspects of organizing and maintaining that technology development effort: including intellectual property protection and financing market and end user acceptance, and public perception just to mention a few. While FDA premarket approval is the most obvious form of external control over the introduction of new medical products in the United States, it is not the only one; healthcare reimbursement under U.S. centers for Medicare and Medicaid Services (CMS) regulations and private insurer practices is critical, and the evidence necessary to secure reimbursement should be considered in planning clinical trials for FDA approval. The approach to FDA regulatory oversight itself requires careful analysis of product classification (including special designation) options. The novelty, variety, and potential complexity of forms of tissue engineering compel strategic analysis of external controls over the commercial development of human cellular and tissue-based products.

The FDA has adopted a cooperative approach across appropriate FDA Centers and the Office of Commissioner in developing its regulatory strategies for engineered tissue products. These strategies have attempted to simplify and facilitate the administrative process for evaluating products and for resolving product regulatory jurisdiction questions. However, it is apparent that, just as the science continues to evolve, so too must the regulatory approaches. As new and different research directions are pursued, such as the apparent shift toward the use of stem cell technology [12] new and different products will be developed, with the expectation of unique product-specific issues. The FDA, given its legislative authority and regulatory responsibility will certainly continue to build on current initiatives, apply lessons learned from previous products as applicable, and look to the best scientific minds and methods to achieve innovative, flexible, and appropriate resolutions that are transparent to the research and industrial community as well as the public. The challenge for the tissue engineering community is to maintain awareness of the regulatory landscape in the United States and abroad and to be an active voice for articulating issues in order to maintain a productive dialogue with the regulatory agencies and consumers so that engineered tissue products find their proper place in clinical medicine.

References

[1] Hellman, K.B. and Heineken, F.G. Introductory letter. In *World Technology Evaluation Center (WTEC) Panel Report on Tissue Engineering Research*, McIntyre, J.L. et al. (Eds.), Academic Press, San Diego, CA, 2003.

[2] Reinventing the Regulation of Human Tissue: A Proposed Approach to Regulation of Cellular and Tissue-based Products (the "Proposed Approach") (62 FR 9721) March 4, 1997.

[3] PHS Act. Effective 1999. Public Health Service Act, 42 U.S. Code Section 201. See also FDA website: http://www.fda.gov/opacom/laws/phsvcact/phsvcact.htm

[4] U.S. Government. *Food, Drug, and Cosmetic Act* (as amended by the FDA Modernization Act of 1997). See also http://www.fda.gov/opacom/laws/fdcact/fdctoc.htm

[5] Establishment Registration and Listing for Manufacturers of Human Cellular and Tissue-based Products (63 FR 26744), May 14, 1998.

[6] Suitability Determination for Donors of Human Cellular and Tissue-based Products (64 FR 52696), September 30, 1999.

[7] Current Good Tissue Practice for Manufacturers of Human Cellular and Tissue-based Products: Inspection and Enforcement.

[8] EMEA Biotechnology Working Party, *Points to Consider on Human Somatic Cell Therapy*, CPMP/BWP/41450/98 draft.

[9] Hirayama, Y. Changing the review process: the view of the Japanese Ministry of Health and Welfare. *Drug Inform. J.* 1998; 32:111–117.

[10] Yamada, M. The approval system for biological products in Japan. *Drug Inform. J.* 1997; 3:1385–1393.

[11] *Uniform Anatomical Gift Act of 1987*, Section 10. Reproduced at 8A Uniform Laws Annotated 58 (Master Edition, 1993).

[12] Lysaght, M. *Tissue Engineering: Great Expectations; Presentation at Engineering Tissue Growth Conference and Exposition*, Pittsburgh, PA, March 17–20, 2003.

III

Tissue Engineering Applications

18

Bioengineering of Human Skin Substitutes

Dorothy M. Supp
Shriners Hospital for Children
Cincinnati Burns Hospital

Steven T. Boyce
University of Cincinnati

18.1 Introduction

The field of tissue engineering has grown in response to the many medical needs for tissue replacement. For the skin, two distinctly different types of wounds have stimulated the development of various tissue-engineered skin substitutes. On one end of the spectrum are burn wounds, which represent an acute injury to the skin. Over 1 million burn injuries occur annually, resulting in over 50,000 acute hospital admissions and more than 5,000 deaths [1]. Partial-thickness wounds have the capacity to heal without the need for tissue replacement from stem cells present in skin appendages. However, full-thickness burn wounds require grafting of skin to replace the destroyed tissue. The grafting of split-thickness autologous skin has been the prevailing standard for permanent closure of excised full-thickness burn wounds. This can be accomplished in patients with relatively small wounds, but in patients with massive burns involving a large total body surface area (TBSA), permanent wound closure is problematic because of the lack of donor sites for skin autografting. Delayed wound coverage increases the likelihood of infection and sepsis, major causes of burn mortality [2].

On the other end of the spectrum are chronic wounds, which tend to involve a relatively small area of skin, but represent a major medical need because they affect a large population of patients. The most common chronic wounds include pressure ulcers and leg ulcers, which are estimated to affect over 2 million people in the United States alone [3]. This figure may be expected to rise as the average age of the population increases. In some patients, wound closure can be enhanced with topical agents, such as growth factors to stimulate healing and antimicrobial agents to minimize infection. However, a large percentage of patients require grafting for permanent closure of chronic wounds. Autograft may not be a feasible option in these patients due to underlying deficiencies in wound healing, which compromise healing of donor sites.

The need for timely wound closure in these diverse clinical settings has led to the development of skin substitutes as alternatives to split-thickness or full-thickness autograft. Although none of the skin substitutes currently available can replace all of the structures and functions of native skin, they can be used to provide wound coverage and facilitate healing of both acute and chronic wounds. Further, the technologies that have yielded skin substitutes for grafting have also been used to produce materials for *in vitro* irritancy and toxicology studies.

18.2   Objectives of Skin Substitutes

Normal human skin performs a wide variety of protective, perceptive, and regulatory functions that help the body maintain homeostasis. The outermost layer of the skin, the epidermis, is comprised mainly of keratinocytes, which provide the barrier function of the skin. Other epidermal cells and structures include adnexal cells that comprise the glands, hair, and nails; melanocytes, which contribute pigment; Langerhans cells of the immune system; and sensory structures of nerves. The epidermis contains only very small amounts of extracellular matrix, predominantly carbohydrate polymers and stratum corneum lipids that form a barrier to permeability of aqueous fluids [4–6]. In contrast, the underlying dermis consists mainly of extracellular matrix molecules, including collagens, elastin, reticulin, and polysaccharides, that give mechanical strength to the skin. Dermal cells include fibroblasts, which synthesize collagen and other matrix components; vascular components, such as endothelial cells and smooth muscle cells; nerve cells; and mast cells of the immune system.

To be useful in a clinical setting, the primary goal for any skin substitute is restoration of skin barrier, normally a function of the epidermis, which helps to minimize protein and fluid loss and prevent infection [7]. As outlined in Table 18.1, there are many products that meet this need, but most provide only temporary wound coverage or partial skin replacement. Currently, limitations of bioengineered skin substitutes, compared with native skin grafts, include: reduced rates of engraftment, increased microbial contamination, mechanical fragility, increased time to healing, increased requirement for regrafting, and very high cost [59–63]. These complications may increase, rather than decrease, the risks to patients and delay recovery. Therefore, the use of skin substitutes may be suitable as an adjunctive therapy in cases without other alternatives, such as very large burns or chronic wounds that have failed to respond to conventional treatments.

Ideally, skin replacements should promote permanent engraftment without the need for regrafting; allow rapid healing to replace both dermal and epidermal layers; be ready to use when needed; achieve acceptable functional and cosmetic outcome; and be free from risk of disease transmission and immunological reaction [2]. These many goals have not been satisfied by any skin substitute, but ongoing research in biomedical and genetic engineering may someday make an "off-the-shelf" skin replacement a reality.

18.3   Composition of Skin Substitutes

Skin substitutes currently available vary in complexity, ranging from temporary synthetic wound dressings to permanent skin replacements, either with or without incorporation of cultured skin cells (Table 18.1). Some of the acellular materials have components that incorporate into the wound bed and may become populated with dermal cells from the host. These dermal substitutes replace the dermal component of the

TABLE 18.1   Bioengineered Skin Substitutes

Skin substitute	Dermal component	Epidermal component	Indications for use	Refs.
AlloDerm® (LifeCell Corp.)	Allogeneic acellular human dermis	None	Burn wounds; repair of soft tissue defects	[8–11]
Apligraf® (Novartis)	Collagen gel with allogeneic fibroblasts	Allogeneic keratinocytes	Burn wounds; chronic foot ulcers; venous leg ulcers; epidermolysis bullosa	[12–20]
Biobrane® (Bertek Pharmaceuticals)	Nylon mesh coated with collagen	Semipermeable silicone membrane	Partial-thickness burn wounds and donor sites	[21–24]
Cultured Skin Substitutes (Univ. Cincinnati/Shriners Hospitals	Collagen-based polymer with autologous fibroblasts	Autologous keratinocytes	Burn wounds; congenital nevi; chronic wounds (using allogeneic cells)	[25–29]
Dermagraft® (Smith and Nephew)	Polyglactin mesh scaffold with allogeneic fibroblasts	None	Chronic/diabetic foot ulcers	[30–32]
Epicel® (Genzyme Biosurgery)	None	Autologous keratinocyte sheet	Burn wounds; congenital nevi	[33–36]
Epiderm™(MatTek Corporation)	Collagen coated cell culture insert	Allogeneic keratinocytes	*In vitro* assessment for irritancy and toxicology testing	[37–40]
EpidermFT™ (MatTek Corporation)	Allogeneic fibroblasts on cell culture insert	Allogeneic keratinocytes	*In vitro* testing model for assessment of dermal–epidermal interactions	
Epidex™ (Modex Therapeutics)	None	Autologous keratinocytes with silicone membrane support	Chronic leg ulcers	[41,42]
Integra® (Integra Life Sciences)	Collagen-chondroitin-6-sulfate matrix	Silicone polymer coating	Burn wounds	[43–49]
Melanoderm™(MatTek Corporation)	Collagen coated cell culture insert	Allogeneic keratinocytes and melanocytes	*In vitro* testing model for assessment of pigmentation	
OrCel® (OrTec International)	Collagen sponge with allogeneic fibroblasts	Allogeneic keratinocytes	Donor sites in burn patients; epidermolysis bullosa	[50,51]
SkinEthic® (SkinEthic Laboratories)	Cell culture insert	Allogeneic keratinocytes (without or with melanocytes)	*In vitro* testing model	[52–54]
SureDerm (Hans Biomed)	Allogeneic acellular human dermis	None	Burn wounds; repair of soft tissue defects	
TranCell (CellTran Limited)	None	Autologous keratinoctyes on acrylic acid polymer	Chronic diabetic foot ulcers	[55]
TransCyte® (Smith and Nephew)	Nylon mesh with dermal matrix secreted by allogeneic fibroblasts (cells nonviable at grafting)	None	Burn wounds	[56–58]

skin, but may require grafting of autologous epidermis in a second surgical procedure. Skin substitutes that contain allogeneic cells have greater similarity to native skin at grafting than acellular materials, but these are considered temporary skin substitutes [64]. The allogeneic cells are eventually replaced by host-derived cells [7]. Allogeneic skin substitutes containing cells can secrete extracellular matrix and growth factors that improve healing, and can provide wound coverage until autograft is available. For small wounds, such as chronic wounds, allogeneic skin substitutes may stimulate healing from the wound bed and margins, without further grafting. These characteristics, coupled with essentially immediate availability, are advantages of the allogeneic skin substitutes. However, for permanent closure of large wounds grafting of autologous skin or engineered skin containing autologous cells is required. A disadvantage of bioengineered skin substitutes containing autologous cells is the increased time required for cell culture and graft preparation [65]. However, they can act as permanent skin replacements once engraftment is achieved. Therefore, the increased time required for preparation may be offset by a reduction in the number of surgical procedures required for permanent wound closure [66].

Skin replacements can be used as adjunctive therapies in patients who are also receiving conventional skin grafts to facilitate wound closure [25–27,67,68]. By increasing availability of skin grafts, skin substitutes can provide several advantages over conventional therapy, including reduced donor site area required to close wounds permanently, decreased number of surgical procedures and hospitalization time, and reduced scarring [66,69].

18.3.1 Acellular Skin Substitutes

Biobrane, an acellular biocomposite dressing, has been used for over two decades for treatment of partial-thickness burn wounds [21,70]. It has been shown to control pain and reduce frequency of dressing changes, decrease the length of hospitalization, and improve healing times [21,23,70]. Biobrane is composed of a collagen peptide-coated nylon mesh material attached to a semipermeable silicone membrane. It is applied to clean, debrided wounds with the mesh side down; the mesh adheres to the wound, and the material is removed after the underlying tissue has healed. Integra is also a synthetic acellular skin replacement, but one that combines both temporary and permanent components. It consists of a cross-linked collagen–glycosaminoglycan membrane as a dermal matrix, and a silastic coating that provides a synthetic epidermal structure for barrier replacement [43,44]. It has been widely used for coverage of excised burn wounds. The artificial dermis does not contain cells at the time of grafting, but within a few weeks it becomes populated with cells from the wound bed and is vascularized. Thus, the dermal component incorporates into the wound, generating a neodermis; the temporary silastic coating can then be removed and replaced with a thin sheet of split-thickness skin [45,46].

Another example of a bilayer temporary skin substitute is TransCyte, which has been used for coverage of partial-thickness wounds during re-epithelialization, or for full-thickness wounds prior to autograft placement. Transcyte is comprised of a synthetic semipermeable epidermal layer, and a dermal nylon mesh layer seeded with allogeneic human fibroblasts [56–58]. The fibroblasts are allowed to proliferate within the synthetic dermal construct, where they secrete extracellular matrix components and growth factors that can facilitate wound healing. Prior to grafting, the material is frozen, without cryoprotection of the cells. Thus, at the time of grafting, Transcyte does not contain any viable cells.

AlloDerm is an acellular dermal matrix derived from donated human skin [8]. The skin is processed to remove cells, circumventing tissue rejection but preserving much of the dermis's three-dimensional dermal structure. Vascular channels with basement membrane are present, even though the cells have been destroyed, facilitating graft vascularization [10]. It has been most useful for soft tissue replacement, such as abdominal wall reconstruction [10]. For treatment of burn wounds, it has been used in conjunction with split-thickness autograft to yield results similar to split-thickness grafts of greater thickness [8].

18.3.2 Allogeneic Cellular Skin Substitutes

Other temporary wound covers, such as Dermagraft, Apligraf, and OrCel, contain viable cells at the time of grafting. Because the cells are allogeneic, typically isolated from donated human neonatal foreskin, these products are considered temporary skin substitutes rather than permanent skin replacements. Dermagraft is composed of a polymer mesh scaffold seeded with allogeneic fibroblasts [30–32]. Apligraf and OrCel both contain allogeneic keratinocytes and fibroblasts, coupled with a biopolymer consisting of either a bovine type I collagen gel or collagen sponge, respectively [12,13,51]. Grafts that contain allogeneic cells provide wound coverage and supply growth factors that facilitate wound repair, but the allogeneic cells do not persist on the patient after healing is complete. It is generally believed that allogeneic cells are replaced within 1 to 6 weeks after grafting by the patient's own cells.

Cultured skin models containing allogeneic cells have been developed for *in vitro* testing purposes [37,40,52–54]. SkinEthic Laboratories has developed a Reconstituted Human Epidermis model, comprised of stratified epidermal keratinocytes supplied on cell culture inserts, for *in vitro* test applications [52–54]. In addition, several tissue-specific models of Reconstructed Human Epithelium have been prepared. These include models of oral, vaginal, corneal, and alveolar epithelium. Similar models designed for *in vitro* toxicology testing include EpiDerm, a differentiated model of human epidermis, and Melanoderm, a stratified coculture of human melanocytes and keratinocytes [37]. EpiDermFT (EpiDerm "Full Thickness") is comprised of neonatal foreskin-derived fibroblasts and keratinocytes grown on cell culture inserts. These skin models closely parallel human skin at the ultrastructural level, and can be reproducibly manufactured, offering attractive alternatives to *in vivo* animal testing for irritancy, toxicology, and gene expression studies.

18.3.3 Autologous Cellular Skin Substitutes

Autologous keratinocytes and fibroblasts have generally been derived from biopsies of the patient's uninjured skin [71–73]. Recently, keratinocytes have been isolated from the outer root sheath of plucked hair follicles [41,42]. The cells can be expanded in culture and used to populate skin substitutes *in vitro*. Grafted as epithelial sheets or as composites of biopolymers and cells, these are theoretically permanent skin substitutes once engraftment is achieved [25–27,33,41].

Relatively few commercial skin substitutes contain autologous cells. Epicel was the first commercial product comprised of cultured autologous cells for wound transplantation. It is indicated for use in burn patients with very large deep partial-thickness or full-thickness wounds and in congenital nevi patients [33,36,74]. Also referred to as cultured epithelial autograft (CEA), Epicel consists of a sheet of autologous keratinocytes that are grown in culture and transplanted to patients with a petrolatum gauze backing. The keratinocytes are isolated from a full-thickness biopsy of the patient's uninjured skin, and grafts are generally ready within 3 weeks time [74]. Epicel can be grafted on top of vascularized allogeneic dermis or directly onto debrided wounds; however, engraftment is higher when used in conjunction with allodermis [34,74]. A major problem encountered with Epicel has been epidermal blistering. Small blisters may resolve spontaneously, but larger blisters result in graft failure [34]. This phenomenon has been attributed to absence of dermal–epidermal junction components at grafting. Despite this limitation, the availability of Epicel has provided a much needed treatment option for patients with massive burns where donor skin for autograft is extremely limited [34].

A similar product, EpiDex, is a cultured epidermal sheet graft that is indicated for the treatment of small chronic wounds. The unique aspect of EpiDex is that the keratinocytes are not cultured from skin biopsies, but from scalp hair follicles [42]. For preparation of grafts, hair is plucked from the patient and the outer root sheath cells are placed in explant culture [42]. The keratinocytes that are cultured in this fashion are highly proliferative, apparently regardless of donor age [41]. After approximately 5 weeks of cell expansion, a secondary culture is initiated for preparation of epithelial sheet grafts. These are transplanted to wounds as discs measuring approximately 1 cm diameter, attached to a silicone backing to facilitate handling. Early clinical results have been favorable, though the wound areas treated with EpiDex are relatively small [41].

A limitation of these autologous skin replacements is that they contain only keratinocytes, and thus they supply only an epidermal layer. For large full-thickness wounds, such as burns, replacement of both dermal and epidermal layers is beneficial, to reduce scarring and improve the functional and cosmetic outcome. This can be accomplished by combining dermal substitutes and epidermal skin replacements, but this generally requires multiple surgical procedures. A composite cultured skin substitute (CSS) that contains both autologous keratinocytes and fibroblasts *in vitro* is currently in clinical trials [26,27,67,68,75]. CSS are comprised of bovine collagen-glycosaminoglycan sponges that are populated with autologous fibroblasts and keratinocytes (Figure 18.1) [76,77]. The most extensive experience with autologous CSS

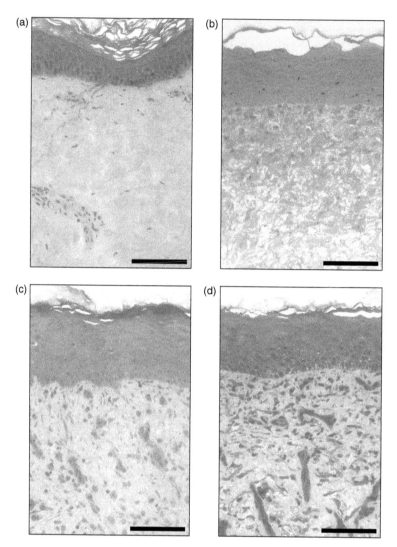

FIGURE 18.1 Histological comparison of native skin and cultured skin substitute prepared for treatment of pediatric burn patient. (a) Native human skin. (b) Cultured skin substitute *in vitro*, shown after 6 days of incubation. Graft was transplanted to patient after 10 days of *in vitro* incubation. (c),(d) Healed autograft (c), and healed cultured skin substitute (d), biopsied 3 weeks after grafting to excised full-thickness burn wounds. Scale bars = 0.1 mm.

has been in the treatment of patients with burns affecting greater than 50% TBSA. In these patients, donor sites for autografting are extremely limited and adjunctive treatments for wound coverage are required. For preparation of CSS, primary cultures of keratinocytes and fibroblasts are isolated using standard techniques from a small split-thickness skin biopsy that is usually taken during a patient's first autografting procedure [26,27,71,72,78,79]. Selective *in vitro* culture of keratinocytes and fibroblasts stimulates exponential increases in cell numbers, resulting in very large populations of cells in only 2 to 3 weeks of culture [78]. Grafting to patients can generally be performed within 2 weeks of inoculation of CSS, which corresponds to 4 to 5 weeks after the initial patient biopsy. Because the CSS contains both dermal and epidermal layers and develops a functional basement membrane *in vitro*, grafting of CSS can replace both skin layers in a single surgical procedure (Figure 18.1) [80]. Culture of CSS at the air–liquid interface promotes development of a stratified epidermal layer with functional barrier properties *in vitro*, providing protection of the wound immediately after grafting [68,80,81]. CSS have been used with favorable results as an adjunctive treatment for the healing of large burn wounds, and have been shown to significantly reduce the requirement for autograft and shorten the number of surgical procedures needed for definitive wound closure [27,67,68]. CSS have also been used to a limited extent for treatment of chronic wounds and congenital giant nevi [28,29].

18.4 Clinical Considerations

Multiple clinical factors can determine whether treatment of wounds with engineered skin substitutes will result in skin repair. Modifications of care protocols for wounds must be used to compensate for the anatomic and physiologic deficiencies in engineered skin. Currently available skin substitutes are avascular, tend to heal more slowly than skin autograft, and may be mechanically fragile. Factors that affect the clinical outcome with bioengineered skin replacements include, but are not limited to: composition at the time of grafting; wound bed preparation; control of microbial contamination; dressings and nursing care; and survival of transplanted cells during vascularization of grafts.

Attachment of cultured epithelium to a dermal substitute *in vitro* is advantageous because both epidermal and dermal components can be applied in a single procedure, similar to skin autograft. Culture conditions can be optimized to promote deposition of basement membrane proteins at the dermal–epidermal junction prior to grafting, thereby eliminating the problem of blistering that is frequently observed after grafting of epithelial sheets [33,80]. Alternatively, dermal and epidermal components of skin substitutes may be applied in two stages: first, application of a dermal substitute followed by vascularization; and second, grafting of an autologous epidermal substitute [45,82,83]. This two-step approach increases the density of blood vessels and extracellular matrix in the wound bed, and has been reported to improve engraftment of cultured keratinocyte sheets. However, it requires two surgical procedures to achieve permanent wound closure.

The lack of a vascular plexus is a major limitation of all skin replacements currently available. Split-thickness skin contains a vascular plexus and adheres to debrided wounds by coagulum. Inosculation of vessels in the graft to vessels in the wound occurs within 2 to 3 days [84]. In the absence of microbial contamination or mechanical damage, autograft skin is generally engrafted and reperfused within 1 week after transplantation. In contrast, current clinical models of engineered skin substitutes are avascular, requiring reperfusion from *de novo* angiogenesis. The time required for perfusion is proportional to the thickness of the dermal component of the skin substitute, and is longer than perfusion of split-thickness skin. Vascularization can be accelerated by secretion of angiogenic factors from engineered skin containing keratinocytes and fibroblasts, but growth factors alone cannot compensate for the lack of a vascular plexus prior to grafting [85,86]. The additional time required for vascularization may contribute to epithelial loss from microbial destruction and nutrient deprivation.

Due to the delayed vascularization of skin replacements, control of contamination is critical for engraftment. Topical antimicrobials are more effective for control of wound contamination than parenteral agents [87]. Topical treatments must provide effective coverage of a broad spectrum of microorganisms, but must have low cytotoxicity to allow healing to proceed. It is also important to avoid overlap of topical agents with parenteral drugs used for treatment of sepsis, which could facilitate development of resistant organisms. Several studies have identified individual agents, and mixtures of multiple agents, which are effective against common wound organisms but are not inhibitory to proliferation of keratinocytes and fibroblasts [88–90].

An additional limitation of engineered skin substitutes is mechanical fragility, which contributes to graft failure due to shear and maceration. For delicate grafts, a backing material can facilitate handling and attachment to the wound. For example, CEA are routinely attached to petrolatum-impregnated gauze for surgical application [91]. However, this material may not be compatible with wet dressings used to manage infection. CSS may be handled and stapled to wounds with a backing of N-Terface™, a relatively strong, nonadherent, highly porous material [25,26]. Porous dressings do not interfere with the delivery of topical solutions and permit drainage of wound exudate.

An important practical obstacle to the routine clinical use of skin substitutes, as with any engineered tissue, remains the high cost of their preparation and care. Estimates for the cost of keratinocyte sheets range from $1,000 to $13,000 for each percent of absolute TBSA [63,92]. If a dermal substitute is also included, the cost can be expected to approximately double [8,45]. Therefore, expense can become a limiting factor for treatment of very large wounds, such as those seen in severely burned patients. Unfortunately, these are the patients most in need of skin substitutes. Although the use of skin substitutes can theoretically reduce the number of surgeries required to heal large burns, which should decrease the total time of hospitalization, there are currently no studies that clearly demonstrate a decrease in costs by use of skin substitutes of any kind. Engineered skin grafts remain an important adjunct to conventional skin grafting, particularly in the treatment of burns, but cannot be used as a primary modality of wound closure except in the most extreme cases [61].

18.5 Assessment

The outcome after treatment of wounds with bioengineered skin substitutes must be measured to determine whether the benefits justify any risks associated with the therapy. Qualitative outcome, which relies heavily on the trained eye of the clinician, can be assessed through clinical evaluation integrating multiple properties of the wound. For example, the Vancouver Scale is used for assessment of burn scar by trained clinicians, and provides an ordinal score for properties of skin including pigmentation, vascularity, pliability, and scar height [93]. Such scales assign quantitative values to qualitative measurements and can provide a relative comparison for evaluation, but they are inherently subjective and dependent on the examiner.

Objectivity may be increased by use of noninvasive instruments to measure biophysical properties of skin, including vascular perfusion, epidermal barrier, pliability, color, and surface pH. Quantitative assessment of skin substitutes can highlight deficiencies compared to normal skin or split-thickness autograft, or can be used to assess the advantage of skin substitutes to the patient without interfering with recovery. Although no single biophysical property is definitive, multiple measurements can provide a general assessment for evaluation of outcome. For example, measurements of surface electrical capacitance (SEC) can be used to define the degree of skin barrier development [94]. SEC is measured using a dermal phase meter, an instrument that is easily used in a clinical setting, with minimal pain or discomfort for the patient [95]. Pigmentation of grafted wounds treated with engineered skin substitutes can be measured using a chromameter. Multiple parameters of skin function must be measured to quantify overall benefit from treatment with skin substitutes.

18.6 Regulatory Issues

In the United States, it is the responsibility of the Food and Drug Administration (FDA) to protect the public from health risks associated with new medical therapies. FDA approval requires that new therapies be safe and effective, and that the probable benefits to health outweigh the probable risks of the therapy or of the untreated disease or condition [96]. Safety considerations for engineered skin substitutes must take several factors into account, including media composition, tissue acquisition, graft fabrication and storage, and sterility testing of the final product [96]. For example, cell culture media must be of the highest purity and free from toxic contaminants. Cells derived from allogeneic donors must test negative for transmissible pathogens. Autologous cells must be handled carefully as well. Because autologous tissues are not routinely screened for pathogens, universal precautions to protect laboratory personnel must be practiced. Xenogeneic components, such as bovine collagen, must not only be free from pathogens that can cross species boundaries, but must also be nonimmunogenic. If xenogeneic cells, such as irradiated 3T3 mouse fibroblasts, are used to facilitate initiation of keratinocytes cultures, compliance with safety standards for xenogeneic transplantation must be assured [97]. Although these common cells are generally thought to be free from risk of disease transmission, unknown risks may exist, and hence patients are excluded from future donation of blood or body parts [36,98].

Bioengineered skin substitutes may be regulated as either devices or biologics, depending on their composition and "primary mode of action." Skin substitutes consisting of autologous cells only, or an acellular human tissue matrix, may not require collection of effectiveness data for regulatory approval. Living autologous cell populations intended for structural repair are considered to be inherently efficacious [99]. However, if no effectiveness data are collected, no claims of effectiveness can be made. Skin substitutes that combine cells with biopolymers are currently considered class III (significant risk) devices that require demonstration of effectiveness in addition to safety [96].

18.7 Future Directions

Despite encouraging clinical results with bioengineered skin for the adjunctive treatment of burns, chronic wounds, and other skin deficiencies, skin substitutes containing just two cell types are limited by anatomic and physiologic deficiencies compared to split-thickness skin autograft. Several areas of preclinical investigation suggest that skin substitutes can be further engineered to increase homology to native human skin. These include the incorporation of additional cell types to improve functional and cosmetic outcome, and the use of genetically modified skin cells to enhance performance after grafting.

18.7.1 Pigmentation

Normal skin pigmentation results from the appropriate epidermal distribution and function of melanocytes. The most critical function of melanocytes is protection from ultraviolet irradiation, but they have psychological importance as well, as a patient's body image and personal identity can impact recovery from massive skin injury [100–102]. Pigmentation of cultured skin may result from transplantation of "passenger" melanocytes, which may persist in selective cultures of epidermal keratinocytes [67,79]. Melanocytes can survive under conditions used for keratinocyte culture, though they proliferate at slower rates and are depleted upon serial passage or cryopreservation [103–105]. In CSS grafted to excised burns, pigmented areas resulting from passenger melanocytes have been observed as individual foci within two months after transplantation [79]. By 1 to 2 years after healing, the foci increase in area, occasionally fusing together to form larger pigmented regions. Uniform pigmentation was demonstrated in preclinical studies with CSS deliberately populated with selectively cultured human melanocytes [106–108]. Future studies will be needed to address regulation of the level of pigmentation in uniformly pigmented cultured skin.

18.7.2 *In Vitro* Angiogenesis

The absence of a vascular plexus in bioengineered skin necessitates vascularization to occur *de novo*, rather than through inosculation of the graft with the wound, increasing the time of nutrient deprivation and susceptibility to microbial contamination after grafting. This limitation can be indirectly addressed in a clinical setting by irrigating the graft with solutions of nutrients and antimicrobial agents for several days after transplantation [27,67,109]. A direct approach would be to initiate angiogenesis in the skin substitutes *in vitro*, prior to grafting. This would permit vascularization to occur through both inosculation of existing vessels and also neovascularization, as occurs for grafted split-thickness skin [84].

Initiation of angiogenesis *in vitro* requires the addition of endothelial cells to the engineered skin. Endothelial cells may organize into vascular structures in culture with the aid of biomaterial supports and coculture with accessory cells. For example, engineered blood vessels have been constructed *in vitro* using mixed cultures of fibroblasts, human umbilical vein endothelial cells (HUVEC), and vascular smooth muscle cells in a collagen matrix [110]. In preclinical studies, transplantation of engineered blood vessels constructed by culture of HUVEC in three-dimensional collagen/fibronectin gels has been reported [111]. More recently, a composite cultured skin containing HUVEC and keratinocytes in a human dermal matrix was reported, which displayed evidence of perfusion after grafting to mice [112]. These studies illustrate the feasibility of grafting synthetic vessels in cultured skin, but overexpression of Bcl-2 through retroviral modification was required to promote survival of the transplanted endothelial cells [111]. In addition, another potential limitation of these studies that will impede their clinical application is the reliance on nondermal or nonautologous endothelial cells. Ideally, multiple cell types (keratinocytes, fibroblasts, and human dermal microvascular endothelial cells, or HDMEC) could be derived from a single autologous skin sample. Transplantation of HDMEC in a composite skin substitute containing isogenic keratinocytes and fibroblasts was demonstrated in an athymic mouse model, though perfusion was not observed [113]. The transplantation of HDMEC in a clinically relevant cultured skin model showed the feasibility of preparing autologous cultured skin containing HDMEC, but future studies must demonstrate inosculation of vessels in the graft with vessels in the wound bed to yield improved performance after grafting.

18.7.3 Cutaneous Gene Therapy

Keratinocytes and fibroblasts are amenable to genetic modification *in vitro* by a variety of methods. Genetically modified cells can be used to populate engineered skin substitutes. This is termed "*ex vivo*" gene therapy because cells are removed from the body and genetically modified in culture before being transplanted back to the recipient. There has been a great deal of interest in the use of genetically modified skin substitutes for the treatment of cutaneous diseases. For example, preclinical studies suggest that *ex vivo* gene therapy can be useful for treatment of lamellar ichthyosis, a condition characterized by defective epidermal barrier, and the blistering skin disease junctional epidermolysis bullosa (JEB) [114–116]. Theoretically, genetically modified keratinocytes can be transplanted for secretion of circulating factors to treat systemic diseases. Keratinocytes have been genetically modified to secrete human growth hormone and clotting factor IX, but therapeutic protein levels have been difficult to obtain after grafting [117–121].

Another application of cutaneous gene therapy is the regulation of wound healing. Hypothetically, genetic modification may be used to overcome anatomic limitations or to enhance their biological activity. For example, keratinocytes modified to overexpress the mesenchymal cell mitogen Platelet Derived Growth Factor-A (PGDF-A), seeded on an acellular dermal matrix, showed increased cellularity, vascularization, and collagen deposition after grafting to mice, suggesting improved function due to PDGF-A overexpression [122]. In other studies, keratinocytes were genetically modified by retroviral transduction to overexpress the angiogenic cytokine Vascular Endothelial Growth Factor (VEGF) [85,86]. After transplantation to athymic mice, skin substitutes containing fibroblasts and VEGF-modified keratinocytes showed enhanced and accelerated vascularization, decreased contraction, and increased engraftment compared to control grafts containing unmodified cells [85,86]. Thus, genetic modification of keratinocytes can hypothetically be used to overcome the lack of a vascular plexus in engineered skin grafts.

A particularly promising avenue of research involves the genetic modification of cells in cultured skin grafts to reduce or eliminate immune rejection. Preclinical studies have shown that reduced expression of major histocompatibility complex (MHC) class I and II antigens can prolong engraftment of skin grafts in mouse allograft models [123]. In one study, fetal skin was used because it exhibited substantially reduced MHC class I and II expression compared with neonatal skin. In another study, keratinocytes were genetically modified to overexpress indoleamine 2,3-deoxygenase (IDO), a tryptophan-catalyzing enzyme that functions to prevent fetal rejection during pregnancy [124]. Increased IDO expression in keratinocytes led to a down-regulation of MHC class I expression. These studies suggest a possible mechanism for preparation of allogeneic skin substitutes that would escape immune rejection, and may someday lead to universal donor cultured skin grafts.

18.8 Conclusions

Technological advances in the fabrication of biomaterials and the culture of skin cells have permitted the production of bioengineered skin substitutes. These have provided improved therapeutic options for patients suffering from acute or chronic wounds, and offer the promise of new treatments for inherited cutaneous diseases. Continued research will be needed to identify more efficient methods to utilize precious autologous tissue, which will provide greater amounts of skin substitutes for grafting as well as shorten the time required for their preparation. Increasing the complexity of skin substitutes, from acellular biopolymers to composite materials with multiple cell types, will result in continued improvements in anatomy and physiology, working toward greater homology to native human skin. These improvements will lead to enhanced performance of engineered skin grafts, greater clinical efficacy, and reduction of morbidity and mortality for patients with wounds or cutaneous disease.

References

[1] Brigham, P. and McLoughlin, E., Burn incidence and medical care use in the United States: estimates, trends, and data sources, *J. Burn Care Rehabil.*, 17, 95, 1996.

[2] Berthod, F. and Damour, O., *In vitro* reconstructed skin models for wound coverage in deep burns, *Br. J. Dermatol.*, 136, 809, 1997.

[3] Phillips, T.J., Chronic cutaneous ulcers: etiology and epidemiology, *J. Invest. Dermatol.*, 102, 38S, 1994.

[4] Sorrell, J.M., Caterson, B., Caplan, A.I., Davis, B., and Schafer, I.A., Human keratinocytes contain carbohydrates that are recognized by keratan sulfate-specific monoclonal antibodies, *J. Invest. Dermatol.*, 95, 347, 1990.

[5] Schurer, N.Y. and Elias, P.M., The biochemistry and role of epidermal lipid synthesis, *Adv. Lipid Res.*, 24, 27, 1991.

[6] Elias, P.M., Stratum corneum architecture, metabolic activity and interactivity with subjacent cell layers, *Exp. Dermatol.*, 5, 191, 1996.

[7] Gallico, G.G.I., Biologic skin substitutes, *Clin. Plast. Surg.*, 17, 519, 1990.

[8] Wainwright, D., Madden, M., Luterman, A., Hunt, J., Monafo, W., Heimbach, D., Kagan, R., Sittig, K., Dimick, A., and Herndon, D., Clinical evaluation of an acellular allograft dermal matrix in full-thickness burns, *J. Burn Care Rehabil.*, 17, 124, 1996.

[9] Sheridan, R., Choucair, R., Donelan, M., Lydon, M., Petras, L., and Tompkins, R., Acellular allodermis in burns surgery: 1-year results of a pilot trial, *J. Burn Care Rehabil.*, 19, 528, 1998.

[10] Menon, N.G., Rodrigues, E.D., Byrnes, C.K., Girotto, J.A., Goldberg, N.H., and Silverman, R.P., Revascularization of human acellular dermis in full-thickness abdominal wall reconstruction in the rabbit model, *Ann. Plast. Surg.*, 50, 523, 2003.

[11] Lorenz, R.R., Dean, R.L., Hurley, D.B., Chuang, J., and Citardi, M.J., Endoscopic reconstruction of anterior and middle cranial fossa defects using acellular dermal allograft, *Laryngoscope*, 113, 496, 2003.

[12] Falanga, V., Margolis, D.J., Alvarez, O., Auletta, M., Maggiacomo, F., Altman, M., Jensen, J., Sabolinski, M., and Hardin-Young, J., Rapid healing of venous ulcers and lack of clinical rejection with an allogeneic cultured human skin equivalent, *Arch. Dermatol.*, 134, 293, 1998.

[13] Eaglstein, W.H., Iriondo, M., and Laszlo, K., A composite skin substitute (Graftskin) for surgical wounds: a clinical experience, *Dermatol. Surg.*, 21, 839, 1995.

[14] Sams, H.H., Chen, J., and King, L.E., Graftskin treatment of difficult to heal diabetic foot ulcers: one center's experience, *Dermatol. Surg.*, 28, 698, 2002.

[15] Curran, M.P. and Plosker, G.L., Bilayered bioengineered skin substitute (Apligraf): a review of its use in the treatment of venous leg ulcers and diabetic foot ulcers, *BioDrugs*, 16, 439, 2002.

[16] Phillips, T.J., Manzoor, J., Rojas, A., Isaacs, C., Carson, P., Sabolinski, M., Young, J., and Falanga, V., The longevity of a bilayered skin substitute after application to venous ulcers, *Arch. Dermatol.*, 138, 1079, 2002.

[17] Falabella, A.F., Schachner, L.A., Valencia, I.C., and Eaglstein, W.H., The use of tissue-engineered skin (Apligraf) to treat a newborn with epidermolysis bullosa, *Arch. Dermatol.*, 135, 1219, 1999.

[18] Falabella, A.F., Valencia, I.C., Eaglstein, W.H., and Schachner, L.A., Tissue-engineered skin (Apligraf) in the healing of patients with epidermolysis bullosa wounds, *Arch. Dermatol.*, 136, 1225, 2000.

[19] Ozerdem, O.R., Wolfe, S.A., and Marshall, D., Use of skin substitutes in pediatric patients, *J. Craniofac. Surg.*, 14, 517, 2003.

[20] Fivenson, D.P., Scherschun, L., and Cohen, L.V., Apligraf in the treatment of severe mitten deformity associated with recessive dystrophic epidermolysis bullosa, *Plast. Reconstr. Surg.*, 112, 584, 2003.

[21] Tavis, M.N., Thornton, N.W., Bartlett, R.H., Roth, J.C., and Woodroof, E.A., A new composite skin prosthesis, *Burns*, 7, 123, 1980.

[22] Purdue, G.F., Hunt, J.L., Gillespie, R.W., Hansbrough, J.F., Dominic, W.J., Robson, M.C., Smith, D.J., MacMillan, B.G., Waymack, J.P., Heradon, D.N. et al., Biosynthetic skin substitute versus frozen human cadaver allograft for temporary coverage of excised burn wounds, *J. Trauma*, 27, 155, 1987.

[23] Lal, S., Barrow, R.E., Wolf, S.E., Chinkes, D.L., Hart, D.W., Heggers, J.P., and Herndon, D.N., Biobrane improves wound healing in burned children without increased risk of infection, *Shock*, 14, 314, 2000.

[24] Arevalo, J.M. and Lorente, J.A., Skin coverage with Biobrane biomaterial for the treatment of patients with toxic epidermal necrolysis, *J. Burn Care Rehabil.*, 20, 406, 1999.

[25] Hansbrough, J.F., Boyce, S.T., Cooper, M.L., and Foreman, T.J., Burn wound closure with cultured autologous keratinocytes and fibroblasts attached to a collagen–glycosaminoglycan substrate, *JAMA*, 262, 2125, 1989.

[26] Boyce, S.T., Greenhalgh, D.G., Kagan, R.J., Housinger, T., Sorrell, J.M., Childress, C.P., Rieman, M., and Warden, G.D., Skin anatomy and antigen expression after burn wound closure with composite grafts of cultured skin cells and biopolymers, *Plast. Reconstr. Surg.*, 91, 632, 1993.

[27] Boyce, S.T., Goretsky, M.J., Greenhalgh, D.G., Kagan, R.J., Rieman, M.T., and Warden, G.D., Comparative assessment of cultured skin substitutes and native skin autograft for treatment of full-thickness burns, *Ann. Surg.*, 222, 743, 1995.

[28] Boyce, S.T., Glatter, R., and Kitzmiller, W.J., Treatment of chronic wounds with cultured cells and biopolymers: a pilot study, *Wounds*, 7, 24, 1995.

[29] Passaretti, D., Billmire, D., Kagan, R., Corcoran, J., and Boyce, S., Autologous cultured skin substitutes conserve donor site autograft in elective treatment of congenital giant melanocyte nevus, *Plast. Reconstr. Surg.* 114, 1523, 2004.

[30] Cooper, M.L., Hansbrough, J.F., Spielvogel, R.L., Cohen, R., Bartel, R.L., and Naughton, G., *In vivo* optimization of a living dermal substitute employing cultured human fibroblasts on a biodegradable polyglycolic or polyglactin mesh, *Biomaterials*, 12, 243, 1991.

[31] Hansbrough, J.F., Dore, C., and Hansbrough, W.B., Clinical trials of a living dermal tissue replacement placed beneath meshed, split-thickness skin grafts on excised wounds, *J. Burn Care Rehabil.*, 13, 519, 1992.

[32] Hanft, J.R. and Surprenant, M.S., Healing of chronic foot ulcers in diabetic patients treated with a human fibroblast-derived dermis, *J. Foot Ankle Surg.*, 41, 291, 2002.

[33] Carsin, H., Ainaud, P., Le Bever, H., Rives, J., Lakhel, A., Stephanazzi, J., Lambert, F., and Perrot, J., Cultured epithelial autografts in extensive burn coverage of severely traumatized patients: a five year single-center experience with 30 patients, *Burns*, 26, 379, 2000.

[34] Gobet, R., Raghunath, M., Altermatt, S., Meuli-Simmen, C., Benathan, M., Dietl, A., and Meuli, M., Efficacy of cultured epithelial autografts in pediatric burns and reconstructive surgery, *Surgery*, 121, 654, 1997.

[35] Compton, C.C., Current concepts in pediatric burn care: the biology of cultured epithelial autografts: an eight-year study in pediatric burn patients, *Eur. J. Pediatr. Surg.*, 2, 216, 1992.

[36] Gallico III, G.G., O'Connor, N.E., Compton, C.C., Remensynder, J.P., Kehinde, O., and Green, H., Cultured epithelial autografts for giant congenital nevi, *Plast. Reconstr. Surg.*, 84, 1, 1989.

[37] Koria, P., Brazeau, D., Kirkwood, K., Hayden, P., Klausner, M., and Andreadis, S.T., Gene expression profile of tissue engineered skin subjected to acute barrier disruption, *J. Invest. Dermatol.*, 121, 368, 2003.

[38] Oren, A., Ganz, T., Liu, L., and Meerloo, T., In human epidermis, beta-defensin 2 is packaged in lamellar bodies, *Exp. Mol. Pathol.*, 74, 180, 2003.

[39] Liu, L., Roberts, A.A., and Ganz, T., By IL-1 signaling, monocyte-derived cells dramatically enhance the epidermal antimicrobial response to lipopolysaccharide, *J. Immunol.*, 170, 575, 2003.

[40] Liebsch, M., Traue, D., Barrabas, C., Spielmann, H., Uphill, P., Wilkins, S., McPherson, J.P., Wiemann, C., Kaufmann, T., Remmele, M., and Holzhutter, H.-G., The ECVAM prevalidation study on the use of EpiDerm for skin corrosivity testing, *ATLA*, 28, 371, 2000.

[41] Limat, A., Mauri, D., and Hunziker, T., Successful treatment of chronic leg ulcers with epidermal equivalents generated from cultured autologous outer root sheath cells, *J. Invest. Dermatol.*, 107, 128, 1996.

[42] Yang, J.S., Lavker, R.M., and Sun, T.T., Upper human hair follicle contains a subpopulation of keratinocytes with superior *in vitro* proliferative potential, *J. Invest. Dermatol.*, 101, 652, 2003.

[43] Yannas, I.V. and Burke, J.F., Design of an artificial skin. I. Basic design principles, *J. Biomed. Mater. Res.*, 14, 65, 1980.

[44] Yannas, I.V., Burke, J.F., Gordon, P.L., Huang, C., and Rubenstein, R.H., Design of an artificial skin. II. Control of chemical composition, *J. Biomed. Mater. Res.*, 14, 107, 1980.

[45] Heimbach, D., Luterman, A., Burke, J.F., Cram, A., Herndon, D., Hunt, J., Jordon, M., McManus, W., Solem, L., Warden, G., and Zawacki, B., Artificial dermis for major burns; a multi-center randomized clinical trial, *Ann. Surg.*, 208, 313, 1988.

[46] Sheridan, R.L., Hegarty, M., Tompkins, R.G., and Burke, J.F., Artificial skin in massive burns — results to ten years, *Eur. J. Plast. Surg.*, 17, 91, 1994.

[47] Heimbach, D.M., Warden, G.D., Luterman, A., Jordan, M.H., Ozobia, N., Ryan, C.M., Voigt, D.W., Hickerson, W.L., Saffle, J.R., DeClement, F.A., Sheridan, R.L., and Dimick, A.R., Multicenter postapproval clinical trial of Integra dermal regeneration template for burn treatment, *J. Burn Care Rehabil.*, 24, 42, 2003.

[48] Wisser, D. and Steffes, J., Skin replacement with a collagen based dermal substitute, autologous keratinocytes and fibroblasts in burn trauma, *Burns*, 29, 375, 2003.

[49] Kopp, J., Magnus, N.E., Rubben, A., Merk, H.F., and Pallua, N., Radical resection of giant congenital melanocyte nevus and reconstruction with meek-graft covered integra dermal template, *Dermatol. Surg.*, 29, 653, 2003.

[50] Stephens, R., Wilson, K., and Silverstein, P., A premature infant with skin injury successfully treated with bilayered cell matrix, *Ostomy Wound Manage.*, 48, 34, 2002.

[51] Still, J., Glat, P., Silverstein, P., Griswold, J., and Mozingo, D., The use of a collagen spong/living cell composite material to treat donor sites in burn patients, *Burns*, 29, 837, 2003.

[52] Doucet, O., Robert, C., and Zastrow, L., Use of a serum-free reconstituted epidermis as a skin pharmacological model, *Toxicol. In Vitro*, 10, 305, 1996.

[53] De Fraissinette, A.D.B., Picarles, V., Chibout, S.D., Kolopp, M., Medina, J., Burtin, P., Ebelin, M.-E., Osborne, S., Mayer, F.K., Spake, A., Rosdy, M., De Wever, B., Ettline, R.A., and Cordier, A., Predictivity of an *in vitro* model for acute and chronic skin irritation (SkinEthic) applied to the testing of topical vehicles, *Cell Biol. Toxicol.*, 15, 121, 1999.

[54] Medina, J., Elsaesser, C., Picarles, V., Grenet, O., Kolopp, M., Chibout, S.D., and De Fraissinette, A.D.B., Assessment of the phototoxic potential of compounds and finished topical products using a human reconstructed epidermis, *In Vitro Mol. Toxicol.*, 14, 157, 2001.

[55] Higham, M.C., Dawson, R., Szabo, M., Short, R., Haddow, D.B., and MacNeil, S., Development of a stable chemically defined surface for the culture of human keratinocytes under serum-free conditions for clinical use, *Tissue Eng.*, 9, 919, 2003.

[56] Hansbrough, J.F., Mozingo, D.W., Kealey, G.P., Davis, M., Gidner, A., and Gentzkow, G.D., Clinical trials of a biosynthetic temporary skin replacement, Dermagraft-transitional covering, compared

with cryopreserved human cadaver skin for temporary coverage of excised burn wounds, *J. Burn Care Rehabil.*, 18, 43, 1997.

[57] Purdue, G.F., Hunt, J.L., Still Jr, J.M., Law, E.J., Herndon, D.N., Goldfarb, I.W., Schiller, W.R., Hansbrough, J.F., Hickerson, W.L., Himel, H.N., Kealey, G.P., Twomey, J., Missavage, A.E., Solem, L.D., Davis, M., Totoritis, M., and Gentzkow, G.D., A multicenter clinical trial of a biosynthetic skin replacement, Dermagraft-TC, compared with cryopreserved human cadaver skin for temporary coverage of excised burn wounds, *J. Burn Care Rehabil.*, 18, 52, 1997.

[58] Noordenbos, J., Dore, C., and Hansbrough, J.F., Safety and efficacy of TransCyte for the treatment of partial-thickness burns, *J. Burn Care Rehabil.*, 20, 275, 1999.

[59] Odessey, R., Addendum: multicenter experience with cultured epithelial autografts for treatment of burns, *J. Burn Care Rehabil.*, 13, 174, 1992.

[60] Pittelkow, M.R. and Scott, R.E., New techniques for the *in vitro* culture of human skin keratinocytes and perspectives on their use for grafting of patients with extensive burns, *Mayo Clin. Proc.*, 61, 771, 1986.

[61] Desai, M.H., Mlakar, J.M., McCauley, R.L., Abdullah, K.M., Rutan, R.L., Waymack, J.P., Robson, M.C., and Herndon, D.N., Lack of long term durability of cultured keratinocyte burn wound coverage: a case report, *J. Burn Care Rehabil.*, 12, 540, 1991.

[62] Williamson, J., Snelling, C., Clugston, P., Mac Donald, I., and Germann, E., Cultured epithelial autograft: five years of clinical experience with twenty-eight patients, *J. Trauma*, 39, 309, 1995.

[63] Rue, L.W., Cioffi, W.G., McManus, W.F., and Pruitt, B.A., Wound closure and outcome in extensively burned patients treated with cultured autologous keratinocytes, *J. Trauma*, 34, 662, 1993.

[64] Arons, M.S., Management of giant congenital nevi, *Plast. Reconstr. Surg.*, 110, 352, 2002.

[65] Boyce, S.T., Cultured skin substitutes: a review, *Tissue Eng.*, 2, 255, 1996.

[66] Boyce, S.T. and Warden, G.D., Principles and practices for treatment of cutaneous wounds with cultured skin substitutes, *Am. J. Surg.*, 183, 445, 2002.

[67] Boyce, S.T., Kagan, R.J., Meyer, N.A., Yakuboff, K.P., and Warden, G.D., *The 1999 Clinical Research Award*, Cultured skin substitutes combined with Integra to replace native skin autograft and allograft for closure of full-thickness burns, *J. Burn Care Rehabil.*, 20, 453, 1999.

[68] Boyce, S.T., Kagan, R.J., Yakuboff, K.P., Meyer, N.A., Rieman, M.T., Greenhalgh, D.G., and Warden, G.D., Cultured skin substitutes reduce donor skin harvesting for closure of excised, full-thickness burns, *Ann. Surg.*, 235, 269, 2002.

[69] Lukish, J.R., Eichelberger, M.R., Newman, K.D., Pao, M., Nobuhara, K., Keating, M., Golonka, N., Pratsch, G., Misra, V., Valladares, E., Johnson, P., Gilbert, J.C., Powell, D.M., and Hartman, G.E., The use of a bioactive skin substitute decreases length of stay for pediatric burn patients, *J. Pediatr. Surg.*, 36, 1118, 2001.

[70] Klein, R.L., Rothmann, B.F., and Marshall, R., Biobrane — a useful adjunct in the therapy of outpatient burns, *J. Pediatr. Surg.*, 19, 846, 1984.

[71] Boyce, S.T. and Ham, R.G., Cultivation, frozen storage, and clonal growth of normal human epidermal keratinocytes in serum-free media, *J. Tissue Cult. Meth.*, 9, 83, 1985.

[72] Boyce, S.T., Methods for serum-free culture of keratinocytes and transplantation of collagen-GAG based composite grafts, in *Methods in Tissue Engineering*, Morgan, J.R. and Yarmush, M., Eds., Humana Press, Inc., Totowa, NJ, 1998, p. 365.

[73] Boyce, S.T., Methods for the serum-free culture of keratinocytes and transplantation of collagen-GAG-based skin substitutes, in *Methods in Molecular Medicine, Vol. 18: Tissue Engineering Methods and Protocols*, Morgan, J.R. and Yarmush, M.L., Eds., Humana Press Inc., Totowa, NJ, 2001, p. 365.

[74] Munster, A.M., Cultured skin for massive burns: a prospective, controlled trial, *Ann. Surg.*, 224, 372, 1996.

[75] Boyce, S.T., Foreman, T.J., English, K.B., Stayner, N., Cooper, M.L., Sakabu, S., and Hansbrough, J.F., Skin wound closure in athymic mice with cultured human cells, biopolymers, and growth factors, *Surgery*, 110, 866, 1991.

[76] Boyce, S.T., Cultured skin for wound closure, in *Skin Substitute Production by Tissue Engineering: Clinical and Fundamental Applications*, Rouahbia, M., Ed., R.G. Landes, Austin, TX, 1997, p. 75.

[77] Boyce, S.T., Skin substitutes from cultured cells and collagen-GAG polymers, *Med. Biol. Eng. Comp.*, 36, 791, 1998.

[78] Boyce, S.T. and Ham, R.G., Calcium-regulated differentiation of normal human epidermal keratinocytes in chemically defined clonal culture and serum-free serial culture, *J. Invest. Dermatol.*, 81 (Suppl 1), 33s, 1983.

[79] Harriger, M.D., Warden, G.D., Greenhalgh, D.G., Kagan, R.J., and Boyce, S.T., Pigmentation and microanatomy of skin regenerated from composite grafts of cultured cells and biopolymers applied to full-thickness burn wounds, *Transplantation*, 59, 702, 1995.

[80] Boyce, S.T., Supp, A.P., Swope, V.B., and Warden, G.D., Vitamin C regulates keratinocyte viability, epidermal barrier, and basement membrane formation *in vitro*, and reduces wound contraction after grafting of cultured skin substitutes, *J. Invest. Dermatol.*, 118, 565, 2002.

[81] Boyce, S.T., Supp, A.P., Harriger, M.D., Pickens, W.L., Wickett, R.R., and Hoath, S.B., Surface electrical capacitance as a noninvasive index of epidermal barrier in cultured skin substitutes in athymic mice, *J. Invest. Dermatol.*, 107, 82, 1996.

[82] Cuono, C., Langdon, R., Birchall, N., Barttelbort, S., and McGuire, J., Composite autologous-allogeneic skin replacement: development and clinical application, *Plast. Reconstr. Surg.*, 80, 626, 1987.

[83] Burke, J.F., Yannas, I.V., Quinby, W.C., Bondoc, C.C., and Jung, W.K., Successful use of a physiologically acceptable skin in the treatment of extensive burn injury, *Ann. Surg.*, 194, 413, 1981.

[84] Young, D.M., Greulich, K.M., and Weier, H.G., Species-specific *in situ* hybridization with fluorochrome-labeled DNA probes to study vascularization of human skin grafts on athymic mice, *J. Burn Care Rehabil.*, 17, 305, 1996.

[85] Supp, D.M., Supp, A.P., Bell, S.M., and Boyce, S.T., Enhanced vascularization of cultured skin substitutes genetically modified to overexpress Vascular Endothelial Growth Factor, *J. Invest. Dermatol.*, 114, 5, 2000.

[86] Supp, D.M. and Boyce, S.T., Overexpression of vascular endothelial growth factor accelerates early vascularization and improves healing of genetically modified cultured skin substitutes, *J. Burn Care Rehabil.*, 23, 10, 2002.

[87] Monafo, W.W. and West, M.A., Current treatment recommendations for topical burn therapy, *Drugs*, 40, 364, 1990.

[88] Boyce, S.T. and Holder, I.A., Selection of topical antimicrobial agents for cultured skin for burns by combined assessment of cellular cytotoxicity and antimicrobial activity, *Plast. Reconstr. Surg.*, 92, 493, 1993.

[89] Boyce, S.T., Warden, G.D., and Holder, I.A., Cytotoxicity testing of topical antimicrobial agents on human keratinocytes and fibroblasts for cultured skin grafts, *J. Burn Care Rehabil.*, 16, 97, 1995.

[90] Boyce, S.T., Warden, G.D., and Holder, I.A., Non-cytotoxic combinations of topical antimicrobial agents for use with cultured skin, *Antimicrob. Agents Chemother.*, 39, 1324, 1995.

[91] Compton, C.C., Wound healing potential of cultured epithelium, *Wounds*, 5, 97, 1993.

[92] Munster, A.M., Weiner, S.H., and Spence, R.J., Cultured epidermis for coverage of burn wounds: a single center experience, *Ann. Surg.*, 211, 676, 1990.

[93] Sullivan, T., Smith, H., Kermode, J., Mclver, E., and Courtemanche, D.J., Rating the burn scar, *J. Burn Care Rehabil.*, 11, 256, 1990.

[94] Supp, A.P., Wickett, R.R., Swope, V.B., Harriger, M.D., Hoath, S.B., and Boyce, S.T., Incubation of cultured skin substitutes in reduced humidity promotes cornification *in vitro* and stable engraftment in athymic mice, *Wound Repair Regen.*, 7, 226, 1999.

[95] Goretsky, M.J., Supp, A.P., Greenhalgh, D.G., Warden, G.D., and Boyce, S.T., *The 1995 Young Investigator Award*: Surface electrical capacitance as an index of epidermal barrier properties of composite skin substitutes and skin autografts, *Wound Repair Regen.*, 3, 419, 1995.

[96] Boyce, S.T., Regulatory issues and standardization, in *Methods of Tissue Engineering*, Atala, A. and Lanza, R., Eds., Academic Press, San Diego, 2001, p. 3.

[97] US Food and Drug Administration, Guidance for industry: precautionary measures to reduce the possible risk of transmission of zoonoses by blood and blood products from xenotransplantation product recipients and their contacts, *Fed. Regist.*, 64, 73562, 1999.

[98] Rheinwald, J.G. and Green, H., Serial cultivation of strains of human epidermal keratinocytes: the formation of keratinizing colonies from single cells, *Cell*, 6, 331, 1975.

[99] US Food and Drug Administration, Guidance on applications for products comprised of living autologous cells manipulated *ex vivo* and intended for structural repair or reconstruction, *Fed. Regist.*, 61, 26523, 1996.

[100] Abdel-Malek, Z.A., Endocrine factors as effectors of integumental pigmentation, *Dermatol. Clin.*, 6, 175, 1988.

[101] Nordlund, J.J., Abdel-Malek, Z.A., Boissy, R.E., and Rheins, L.A., Pigment cell biology: an historical review, *J. Invest. Dermatol.*, 92, 53S, 1989.

[102] Fauerbach, J.A., Heinberg, L.J., Lawrence, J.W., Munster, A.M., Palombo, D.A., Richter, D., Spence, R.J., Stevens, S.S., Ware, L., and Muehlberger, T., Effect of early body image dissatisfaction on subsequent psychological and physical adjustment after disfiguring injury, *Psychosom. Med.*, 62, 576, 2000.

[103] Compton, C.C., Gill, J.M., Bradford, D.A., Regauer, S., Gallico G.G., and O'Connor, N.E., Skin regenerated from cultured epithelial autografts on full-thickness burn wounds from 6 days to 5 years after grafting, *Lab. Invest.*, 60, 600, 1989.

[104] DeLuca, M., Franzi, A., D'Anna, F., Zicca, A., Albanese, E., Bondanza, S., and Cancedda, R., Co-culture of human keratinocytes and melanocytes: differentiated melanocytes are physiologically organized in the basal layer of the cultured epithelium, *Eur. J. Cell Biol.*, 46, 176, 1988.

[105] Compton, C.C., Warland, G., and Kratz, G., Melanocytes in cultured epithelial autografts are depleted with serial subcultivation and cryopreservation: implications for clinical outcome, *J. Burn Care Rehabil.*, 19, 330, 1998.

[106] Boyce, S.T., Medrano, E.E., Abdel-Malek, Z.A., Supp, A.P., Dodick, J.M., Nordlund, J.J., and Warden, G.D., Pigmentation and inhibition of wound contraction by cultured skin substitutes with adult melanocytes after transplantation to athymic mice, *J. Invest. Dermatol.*, 100, 360, 1993.

[107] Swope, V.B., Supp, A.P., Cornelius, J.R., Babcock, G.F., and Boyce, S.T., Regulation of pigmentation in cultured skin substitutes by cytometric sorting of melanocytes and keratinocytes, *J. Invest. Dermatol.*, 109, 289, 1997.

[108] Swope, V.B., Supp, A.P., and Boyce, S.T., Regulation of cutaneous pigmentation by titration of human melanocytes in cultured skin substitutes grafted to athymic mice, *Wound Repair Regen.*, 10, 378, 2002.

[109] Boyce, S.T., Supp, A.P., Harriger, M.D., Greenhalgh, D.G., and Warden, G.D., Topical nutrients promote engraftment and inhibit wound contraction of cultured skin substitutes in athymic mice, *J. Invest. Dermatol.*, 104, 345, 1995.

[110] L'Heureux, N., Germian, L., Labbe, R., and Auger, F.A., *In vitro* construction of a human blood vessel from cultured vascular cells: a morphologic study, *J. Vasc. Surg.*, 14, 499, 1993.

[111] Schechner, J.S., Nath, A.K., Zheng, L., Kluger, M.S., Hughes, C.C.W., Sierra-Honigmann, M.R., Lorber, M.I., Tellides, G., Kashgarian, M., Bothwell, A.L.M., and Pober, J.S., *In vivo* formation of complex microvessels lined by human endothelial cells in an immunodeficient mouse, *Proc. Natl Acad. Sci. USA*, 97, 9191, 2000.

[112] Schechner, J.S., Crane, S.K., Wang, F., Szeglin, A.M., Tellides, G., Lorber, M.I., Bothwell, A.L., and Pober, J.S., Engraftment of a vascularized human skin equivalent, *FASEB J.*, 17, 2250, 2003.

[113] Supp, D.M., Wilson-Landy, K., and Boyce, S.T., Human dermal microvascular endothelial cells form vascular analogs in cultured skin substitutes after grafting to athymic mice, *FASEB J.*, 16, 797, 2002.

[114] Choate, K.A., Medalie, D.A., Morgan, J.R., and Khavari, P.A., Corrective gene transfer in the human skin disorder lamellar ichthyosis, *Nat. Med.*, 2, 1263, 1996.

[115] Vailly, J., Gagnouz-Palacios, L., Dell'Ambra, E., Romero, C., Pinola, M., Zambruno, G., De Luca, M., Ortonne, J.P., and Meneguzzi, G., Corrective gene transfer of keratinoyctes from patients with junctional epidermolysis bullosa restores assembly of hemidesmosomes in reconstructed epithelia, *Gene Ther.*, 5, 1322, 1998.

[116] Dellambra, E., Vailly, J., Pellegrini, G., Bondanza, S., Golisano, O., Macchia, C., Zambruno, G., Meneguzzi, G., and De Luca, M., Corrective transduction of human epidermal stem cells in laminin-5-dependent junctional epidermolysis bullosa, *Hum. Gene Ther.*, 9, 1359, 1998.

[117] Vogt, P.M., Thompson, S., Andree, C., Liu, P., Breuing, K., Hatzis, D., Brown, H., Mulligan, R.C., and Ericksson, E., Genetically modified keratinocytes transplanted to wounds reconstitute the epidermis, *Proc. Natl Acad. Sci. USA*, 91, 9307, 1994.

[118] Morgan, J.R., Barrandon, Y., Green, H., and Mulligan, R.C., Expression of an exogenous growth hormone gene in transplantable human epidermal cells, *Science*, 237, 1476, 1987.

[119] Gerrard, A.J., Hudson, D.L., Brownlee, G.G., and Watt, F.M., Towards gene therapy for haemophilia B using primary human keratinocytes, *Nat. Gen.*, 3, 180, 1993.

[120] Page, S.M. and Brownlee, G.G., An *ex vivo* keratinocyte model for gene therapy of hemophilia B, *J. Invest. Dermatol.*, 108, 139, 1997.

[121] White, S.J., Page, S.M., Margaritis, P., and Brownlee, G.G., Long-term expression of human clotting factor IX from retrovirally transduced primary human keratinocytes *in vivo*, *Hum. Gene Ther.*, 9, 1187, 2002.

[122] Eming, S.A., Medalie, D.A., Tompkins, R.G., Yarmush, M.L., and Morgan, J.R., Genetically modified human keratinocytes overexpressing PDGF-A enhance the performance of a composite skin graft, *Hum. Gene Ther.*, 9, 529, 1998.

[123] Erdag, G. and Morgan, J.R., Survival of fetal skin grafts is prolonged on the human peripheral blood lymphocyte reconstituted-severe combined immunodeficient mouse/skin allograft model, *Transplantation*, 73, 519, 2002.

[124] Li, Y., Tredget, E.E., and Ghahary, A., Cell surface expression of MHC class I antigen is suppressed in indoleamine 2,3-dioxygenase genetically modified keratinoycytes: implications in allogeneic skin substitute engraftment, *Hum. Immunol.*, 65, 114, 2004.

19

Nerve Regeneration: Tissue Engineering Strategies

Jennifer B. Recknor
Surya K. Mallapragada
Iowa State University

19.1 Introduction

The nervous system of the adult mammal is divided into two main components: the peripheral nervous system (PNS) and central nervous system (CNS). The PNS, consisting of cranial and spinal nerves and their associated ganglia, has the intrinsic ability for repair and regeneration. However, the adult mammalian central nervous system (CNS), consisting of the brain and spinal cord, is viewed as largely incapable of self-repair or regeneration of correct axonal and dendritic connections after injury [1–3]. When a peripheral nerve is injured and the nerve retracts, or tissue is lost, preventing an end-to-end repair, grafting is a commonly performed. Grafting methods, involving autografts and allografts, provide trophic factors and guidance for regenerating axons [4,5]. However, there are major limitations to grafting including multiple surgeries, lack of donor tissue, and immune rejection. The CNS is a more complex environment that proves to be not as amenable to healing as the PNS. Adult CNS injury is typically followed by neuronal degeneration, cell death, and the breakdown of synaptic connections. It is established that fish, amphibia, mammalian peripheral nerves as well as developing central nerves respond differently to injury than the adult mammalian CNS. In these systems, functional axons can regrow after they have been damaged [1]. However, currently, in the mammalian CNS, there is no treatment for the restoration of human nerve function due to the intricate series of events that must take place in order for regeneration to occur.

Growth across the PNS–CNS transition zone has not been successful in many regeneration attempts [6]. The limitations with current strategies for repair of the PNS and CNS can eventually be minimized using tissue-engineering applications to restore biological function by providing a permissive environment for regeneration. The purpose of this chapter is to review recent research applications involving guidance and molecular and cellular strategies for nerve repair that have demonstrated the use of neural tissue engineering to achieve therapeutic goals for nervous system regeneration.

19.2   Neural Regeneration

Injured axons in the PNS are capable of regenerating long distances and have the capacity to establish connections with their targets. In adult mammals, injury to the CNS does not usually result in regeneration. When nerve damage occurs, the portion of the axon isolated from the cell body is referred to as the distal segment. The portion connected to the cell body is the proximal segment. The response of the native environment surrounding these segments following injury is the key reason behind the inherent capacity of the PNS for regeneration, which is not characteristic of the CNS.

19.2.1   Peripheral Nervous System

Nerve injury results in a characteristic chain of degenerative and regenerative events. Complete transection of a nerve is the most severe PNS injury. At the site of nerve damage, the myelin and axons break up, triggering the migration of phagocytic cells, Schwann cells and macrophages, into the PNS to clean up debris. After an axon is transected, the proximal segment of the nerve swells and experiences retrograde degeneration near the wound site. After myelin is cleared and the Schwann cells and macrophages remove debris, the proximal end begins to sprout axons and growth cones emerge [7]. After transection, Wallerian degeneration occurs distal to the injury within hours. The axons and myelin degenerate completely but the endoneurium is left intact forming natural conduits [8]. The endoneurial channel directs the axon regrowth in the reparative phase, and axons of surviving neurons grow into the endoneurial tube. Axons of the proximal segment can regenerate back to previous synaptic sites as long as their cell bodies remain alive and the proximal axons have made successful contact with neurolemmocytes in the endoneurial channel. The proliferation of Schwann cells at this time retains the endoneurial tube structure as these cells form ordered columns. Neurotrophic factors are also produced by Schwann and macrophages, and growth-promoting substances are upregulated providing an optimal environment for axon growth. In humans, regeneration proceeds at approximately 2 mm/d in the smaller nerve and 5 mm/d in the larger nerve. Functional reinnervation only occurs when axons find the Schwann cell column and reach their distal target or effector organ. Successful axonal regeneration depends on the distance of the lesioned site from the cell body, the nature of the injury and axonal contact with neurolemmocytes in the distal segment. For additional reviews on peripheral nerve regeneration, refer to Fawcett and Keynes [9].

19.2.2   Central Nervous System

CNS axons do not regenerate due to a lack of support by the endogenous environment following injury. In the CNS of mammals, axonal regeneration is limited by inhibitory influences of the glial and extracellular environment [10]. The CNS environment is inhibitory in nature making axonal regrowth from adult neurons nearly impossible. Myelin-associated inhibitors of neurite growth, astrocytes, oligodendrocytes, oligodendrocyte precursors, and microglia migrate to the injury site making the environment nonpermissive for axonal growth. In the distal axonal segment, degeneration is slower in the CNS compared to the PNS, and inhibitory myelin and axonal debris are not cleared away as quickly. The axons that survive axotomy are surrounded by unfavorable glial reactions at the lesion site, known as the glial scar. Neurons cannot regenerate beyond the glial scar where axonal outgrowth is essentially stopped [11]. Proximal axons initially demonstrate a spontaneous attempt to regenerate, but the surrounding environment rapidly hinders growth [12]. Consequently, regenerating axons in the CNS cannot reach synaptic targets and reestablish their original connections. Mechanisms for removing or neutralizing the inhibitory components of cellular debris cannot be found in the CNS. However, severed mammalian CNS axons will regrow in more permissive environments [13–15]. They are able to recognize target areas and to re-establish functional synapses with target neurons [16,17]. There has been much recent evidence that suggests that the mature CNS is a less hostile environment for regeneration than was previously thought. If the axons can transverse the injury site, there is possibility of regrowth in the unscarred areas and of functional recovery [18,19]. The use of scaffolds [18,20–23], cell implantation and replacement therapies involving neural stem [20,24], Schwann [25,26] and olfactory ensheathing cells [27], as well as delivery of growth factors [28,29] have provided greater potential for production of new neurons and the repair of injured CNS regions. For additional reviews on central nervous system regeneration, refer to Fry [3] and Schwab [30].

19.3 Guidance Strategies for Regeneration

For successful regeneration, damaged axons must be prevented from dying, the sprouting axons from the proximal nerve stump must extend axons toward their targets, across the injury site, into the distal nerve stump and make synaptic connections to the correct target regions. Common repair techniques facilitating regeneration include grafting using natural materials and entubulization using nerve conduits or scaffolds. These methods connect proximal and distal nerve stumps using a synthetic or biologically derived conduit. Such conduits optimize regeneration by allowing for both physical and chemical guidance and reducing cellular invasion and scarring of the nerve. Entubulization minimizes unregulated axonal growth at the site of injury by providing a distinct environment, and allows for diffusion of trophic factors emitted from the distal stump to reach the proximal segment, which enhances physiological conditions for nerve regeneration. The transplantation of tissue engineered nerve conduits based on polymers and alternative methods to engineer an artificial environment to mimic natural physical and chemical stimulus promotes nerve regeneration and minimizes difficulties associated with grafting.

19.3.1 Autografts

A nerve autograft (or autologous tissue graft) is the transfer of nerve segments from an uninjured part of the body to another. This graft tissue provides endoneurial tubes that enable guided regeneration of axons. Autologous nerve grafts are currently used in the treatment of peripheral nerve injuries where the gap of a transected nerve is large [31]. This method for restoring tissue results in partial deinnervation of the donor site to repair the injury site. Tissue availability is a concern as well as the need for multiple surgeries and potential differences in size and shape leading to difficulties with scale-up.

The capacity of the CNS axons to regrow in a permissive environment has been demonstrated using autologous PNS tissue grafts to bridge the adult rat medulla oblongata and the spinal cord following injury to the CNS. Axons from neurons at the medulla and spinal cord levels grew along the bridge. However, the axons did not re-enter the host tissue [14]. So and Aguayo [32] also demonstrated that transected axons of retinal ganglion cells could regrow into autologous PNS segments grafted directly into the rat retina. Regenerating axons were able to recognize target areas and re-establish functional synapses with target neurons [16,17]. Such autologous PNS grafts provide essential components for the facilitation of growth and functional recovery that are not found in the native CNS.

19.3.2 Allografts and Acellular Nerve Matrices

Allografts involve the transfer of nerve tissue from donor to recipient. These nerve grafts make use of allogenic and xenogeneic tissues to replace lost function. Allografts eliminate the need for harvesting patient tissue. However, tissue rejection involving an undesirable immune response and lack of donor tissue are two major disadvantages of using allografts. Disease transmission is also a risk. Using allografts with immunosuppression has been considered. However, results obtained with allografts have not matched the performance observed with autografts [33].

Acellular nerve matrix allografts have also been observed as useful biomaterials for nerve regeneration [34]. These allografts preserve extracellular matrix (ECM) components and, therefore, mimic the ECM of peripheral nerves mechanically and physically. In this way, acellular nerve matrices aid in the reconstruction of the peripheral nerve. However, decellularization techniques, such as thermal decellularization, can damage the ECM structure and fail to extract all cellular components leading to inflammation upon implantation [35].

19.3.3 Entubulization Using Nerve Conduits

Current limitations with autografts and allografts have led to exploration of possible alternatives for the repair of nerve injury. Tissue-engineered nerve conduits based on polymers have been created for implantation mimicking the three-dimensional and biological environment that is necessary for enhanced regeneration. While bridging the gap between nerve segments, these conduits can preserve neurotropic and neurotrophic communication between the nerve stumps, repel external inhibitory molecules, and provide physical guidance for the regenerating axons similar to the grafts (Figure 19.1) [36]. The spatial cues provided by conduits also induce a change in tissue-architecture cabling cells within the microconduit [37]. Polymers are being extensively investigated to help facilitate nerve regeneration and provide physical and chemical stimulus to regenerating axons [38,39]. These materials vary in composition from entirely synthetic to naturally derived biomaterials. Synthetic conduits are fabricated from metals and ceramics, biodegradable (i.e., poly(esters), such as poly(lactic acid) [40–42], poly(lactic-co-glycolic acid) [42,43], and poly(caprolactone) [44,45], or polyhydroxybutarate [46]) and nonbiodegradable (i.e., methacrylate-based hydrogels [47], polystyrene [48], silicone [49,50], expanded poly(tetrafluoroethylene) or ePTFE (Gore-Tex®: W.L. Gore & Associates, Flagstaff, AZ) [51,52] or poly(tetrafluroethylene) (PTFE) [53]) synthetic polymers. These materials are especially advantageous because specific chemical and physical properties can be readily changed depending on the application for which they are used. Such properties include microgeometry, degradation rate, porosity, and mechanical strength. Biologically derived materials include proteins and polysaccharides (i.e., ECM-based proteins including fibronectin [54], laminin [55] and collagen [56,57], fibrin and fibrinogen [58–60], hyaluronic acid derivatives [61], and agarose [62]). Collagen has been the most widely used natural polymer [57]. These natural materials are biocompatible and enhance migration of support cells [33]. However, batch-to-batch variability needs to be considered when using biological materials such as these. Selecting the appropriate material for a particular application is an essential part of the scaffold design. There are also certain physical properties and that are most desirable for nerve conduits (Figure 19.2) [63]. General requirements for scaffold design are typically followed, which include being biocompatible, having a high surface area/volume ratio with sufficient mechanical integrity and having the ability to provide a suitable environment for axonal growth that can integrate with the surrounding neural environment. These biomaterial-based (synthetic or natural) conduits can also be environmentally enhanced with chemical stimulants, such as laminin and nerve growth factor (NGF), biological or cellular cues such as from neural stem cells as well as Schwann cells and astrocytes, the satellite cells of the peripheral and central nervous systems, and lastly, physical guidance cues.

19.4 Enhancing Neural Regeneration Using Entubulization Strategies

Providing a permissive environment for damaged neural tissues that have suffered trauma is essential for the regeneration of the injured nervous system. In order to generate an environment that supports regeneration, physical, chemical, and biological manipulations must be made. Guidance cues must be presented that aid in control of nerve outgrowth and navigate neuronal growth cones to distant targets

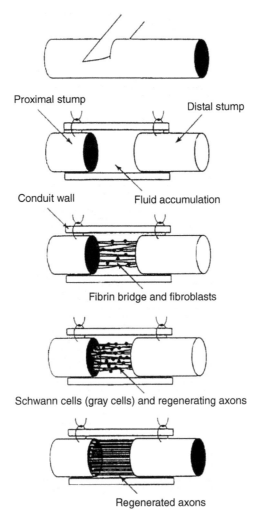

FIGURE 19.1   Silicone-chamber model showing the progression of events during peripheral-nerve rgeneration. After bridging the proximal and distal nerve stumps, the silicone tube becomes filled with serum and other extracellular fluids. A fibrin bridge containing a variety of cell types connects the two stumps. Schwann cells and axons processes migrate from the proximal end to the distal stump along the bridge. The axons continue to regenerate through the distal stump to their final contacts. (Reproduced from Heath, C. and Rutkowski, G.E. *Trends Biotechnol.*, 16, 163–168, 1998. With permission.)

in vivo. "Intelligent" nerve conduits having the appropriate combination of such cues will provide insight into the mechanisms behind axon growth and regeneration in nervous system. These scaffolds can enhance regeneration and help repair severed or injured neural tissue.

19.4.1   Physical Modifications

19.4.1.1   Microtexturing

Scaffolds are extremely useful for evaluating nerve regeneration processes for many reasons. Among these is that the properties of these conduits can be physically altered to optimize nerve regeneration. The microtexture of the surface of the lumen within the conduit affects the outgrowth of neurons and regulates regeneration. Smooth inner surfaces allow the formation of an organized, discrete nerve cable having many myelinated axons. In contrast, inside rough inner surfaces, nerve fascicles are dispersed throughout

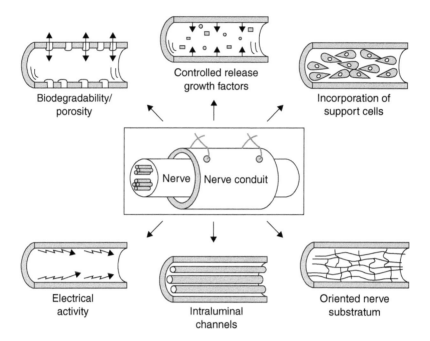

FIGURE 19.2 Properties of the ideal nerve conduit. The desired physical properties of a nerve conduit include (*clockwise from top left*): a biodegradable and porous channel wall; the ability to deliver bioactive factors, such as growth factors; the incorporation of support cells; an internal oriented matrix to support cell migration; intraluminal channels to mimic the structure of nerve fascicles; and electrical activity. (Reproduced from Hudson, T.W., Evans, G.R.D., and Schmidt, C.E. *Clin. Plast. Surg.*, 26, 617, 1999. With permission.)

the lumen and unorganized resulting in little regeneration. When comparing expanded microfibrillar poly(tetrafluoroethylene) (Gore–Tex®: W.L. Gore & Associates, Flagstaff, AZ) tubes having different internodal distances (1, 5, and 10 μm) to smooth-walled, impermeable PTFE tubes, it was discovered that rougher the texture of the surface, the greater the spread of nerve fascicles [64]. Furthermore, the molecular and cellular makeup of the regenerating tissue is affected the stability of the wall structure [41] and channel geometry [65].

19.4.1.2 Micropatterning

In addition to texture, scaffolds can exert control over other aspects of the neural environment. The structure of these scaffolds can be precisely defined for a particular application. Relying on knowledge of structure and function of cells and tissues in the nervous system, specific biomaterial "architecture" can be created and applied to the reconstruction of tissue function. The susceptibility of a cell to topographical structure is determined by the organization of the cytoskeleton, cell adhesion, and cell-to-cell interactions [66]. Cell growth can be controlled at the cellular level through the fabrication of microgrooves and other patterns on substrate surfaces [67]. The development of microfabrication and nanofabrication techniques involving photolithography and reactive ion etching has allowed precise control over patterned features using a variety of materials. Recent developments in manufacturing techniques have included the move from silicon-based fabrication to polymer-based biomaterial scaffolds and the creation of three-dimensional constructs based on success with two-dimensional fabrication methods. Isolating large numbers of individual cells and having control over their shape and distribution is extremely valuable in the analyzing functional changes in individual cells and their relationship with their environment in culture. Fabrication techniques producing substrates with various feature shapes and dimensions are used to study cell behavior and morphology *in vitro* before integrating similar techniques into a scaffold design. In the recent past, micropatterned biodegradable and nonbiodegradable substrates have been developed using microfabrication and transfer patterning techniques [48,68–70]. Microcontact printing techniques involving elastomeric polydimethylsiloxane (PDMS) stamps have been used to create adhesive islands for the control of cell shape, growth, and function [71,72]. Microfluidic patterning has also been used for developing topographical cues on substrates [73]. The effects of the microenvironment on cell behavior has been studied on different substrate materials, including Perspex [66,67,74], silicon wafers [75–77], quartz [78,79], and polymers such as polystyrene [48,80] and biodegradable polymers, including poly (lactide-*co*-glycolide) and poly(DL-lactide) [69,70].

Cell adhesion and proliferation has also been examined using various shapes and feature sizes with several neural cell types. Rectangular shapes [68,81], hexagonal [76], as well as circular features [82] have been successful in controlling cellular behavior. In experiments using lithographically patterned quartz, hippocampal neurites grew parallel to deep, wide grooves but perpendicular to those that were shallow and narrow. Neurites also grew faster in the favored direction of orientation and turned through large angles to align on grooves [81]. The shape and expression of a differentiated phenotype of retinal pigment epithelium (RPE) was controlled using octadecyltrichlorosilane (OTS)-modified glass micropatterned substrates [82]. Webb et al. [79] demonstrated that rat optic nerve astrocytes aligned on surface features as small as 100 nm depths with 260 nm pattern spacing on quartz discs. The oligodendrocyte lineage displayed a high degree of sensitivity to topography as well [79]. Schwann cell and neurite alignment has been demonstrated on micropatterned biodegradable substrates with evidence that groove depth affects the proportion of neurites aligned. Deeper grooves have a stronger effect on cellular behavior [68,70]. Furthermore, these micropatterned biodegradable polymer films were inserted inside poly(D,L-lactide) (PDLA) conduits. The micropatterned surfaces were pre-seeded with Schwann cells in order to provide guidance to axons at the cellular level. Over 95% alignment of the axons and Schwann cells was observed on the micropatterned surfaces with laminin selectively attached to the microgrooves [68]. Mechanisms of contact guidance as well as the intracellular distribution of cytoskeletal elements such as microtubules, microfilaments, intermediate filaments, and adhesive structures on cells as they respond to various geometric configurations, including pillars, columns, and spikes, has been analyzed on microfabricated substrates [73,83].

19.4.2 Biochemical Modifications: Creating an "Active" Nerve Conduit

Eliciting a desirable reaction from the host tissue after nerve injury has a profound influence on the regenerative capacity of the nervous tissue. It has been determined that the response of the host tissue is related to not only the mechanical and physical properties of the implanted biomaterial but the chemical properties also play a strong role in promoting a beneficial response from the native environment [38]. Manipulating the natural repair process of the nervous system by engineering a specific biochemical response from the matrix within the conduit or through delivery of growth and neurotrophic factors is an attractive strategy for enhanced nerve regeneration.

19.4.2.1 Chemical Patterning

Providing an adhesive substrate for cells and neurites to grow on is an important mechanism for guidance. To control cell adhesion, migration as well as tissue growth and repair, scaffolds for neural tissue engineering incorporate specific bioactive chemicals. Several studies have been performed using different methods for generating patterns of adhesive domains on various materials. These studies have largely consisted of two-dimensional substrates where adhesive areas are patterned adjacent to nonadhesive areas. However, chemical patterning of three-dimensional substrates has also been demonstrated. The techniques developed for these *in vitro* studies are then applied to create precise patterns of adhesive domains in conduits [84] to be used in *in vivo* experimentation. Patterning techniques manipulating surface chemistry and using photolithographic photomasks have aided in the reproducible creation of desired patterns of biological molecules on surfaces. Microstamping hexadecanethiol on gold in a self-assembled monolayer (SAM) has been used to create islands of various shapes that support the adsorption of many proteins [72]. Similar methods have been used to print poly-L-lysine, laminin, and bovine serum albumin directly on surfaces using texturized silicone stamps dipped in protein, dried, and transferred onto chemically modified substrates [85,86]. Microfabrication techniques have also been used in the photolithographic patterning of organosilanes to silicon wafers [75,76], quartz, and standard glass [82,87]. Several chemicals and proteins have been employed to create regions of two-dimensional patterns of adhesive domains: laminin [88]; nerve growth factor [89]; fibronectin; collagen; albumin; and laminin paired with other chemicals including laminin–denatured laminin, laminin–albumin, polylysine-conjugated laminin [90], and laminin–collagen [91]. Results from cell experimentation have suggested that biochemical patterning might play a stronger role in inducing cellular response (i.e., attachment, spreading, and alignment) than topography. Therefore, efforts have been made to combine multiple guidance cues by chemically patterning three-dimensional substrates [48,69,70]. For more information on the topographical and biochemical patterning of scaffolds see Flemming et al. [92]; Curtis and Wilkinson [93]; Craighead et al. [73]; and Bhatia and Chen [84].

19.4.2.2 Matrices Within Polymer Conduits

Neural tissue-engineering applications focus on mimicking the nerve and the supporting extracellular matrix (ECM) in order to repair or regenerate axons following damage or disease. Experimentation with ECM molecules [94] and the tailoring of matrices within conduits has led to support of axonal regrowth following injury. Evidence has been presented that the basement membrane protein laminin provides pathways of adhesiveness in both peripheral and central nervous system tissues [95]. This ECM protein is capable of initiating and supporting neurite extension on glass and biodegradable polymers of poly (lactide-*co*-glycolide) and poly(DL-lactide) [68,87] as well as glial cell outgrowth on polystyrene [48] (Figure 19.3). Agarose gels derivatized with laminin oligopeptides have enabled three-dimensional neurite outgrowth *in vitro* from cells containing receptors to the laminin peptides [96]. Regeneration of transected peripheral nerves was enhanced using agarose gels having specific laminin peptides inside the scaffold lumen [97,98]. These gel matrices provide support, create an environment that supports growth and incorporate materials that alter surface area. Collagen, fibronectin, and fibrin have also been used to enhance cell-substrate interaction [55–57,99–103]. Gels using magnetic fields have been shown to orient fibers of collagen within the gels. Compared to randomly oriented fibers, these gels promoted neurite extension both *in vitro* and *in vivo* [56]. Magnetically aligned fibrin gels (MAFGs) having different fibril diameters but similar alignment (Figure 19.4) resulted in drastic changes in the contact guidance response of neurites. In gels formed in 1.2 mM Ca^{2+} and having a smaller fibril diameter, there was no response from chick dorsal root ganglia. However, a strong response in gels formed in 12 and 30 mM Ca^{2+} with a larger fibril diameter enhanced neurite length twofold [103]. Inosine, a purine analog that promotes axonal extension, was loaded into PLGA conduits for controlled release during sciatic nerve regeneration. Inosine-loaded PLGA foams were fashioned into cylindrical nerve guidance channels using a novel low-pressure injection molding technique. After ten weeks, a higher percentage cross-sectional area composed of neural tissue was found in the inosine-loaded conduits compared with controls [104]. Furthermore, nerve conduits can be filled with specific bioactive molecules to elicit new axonal growth following injury. Such molecules including axon guidance and pathfinding molecules (netrins, semaphorins, ephrins, Slits) [105], cell adhesion molecules (CAMs) that promote neurite growth (NCAM, L1, *N*-cadherin, tenascin) [106,107], as well as proteins involved in synaptic differentiation (agrin, laminin beta 2, and ARIA) [108] have numerous applications for nerve regeneration *in vivo*.

Synthetic hydrogels have served as artificial matrices for neural tissue reconstruction, for the delivery of cells and for the promotion of axonal regeneration required for successful neurotransplantation. Cultured neurons were found to attach to hydrogel substrates prepared from poly (2-hydroxyethylmethacrylate) (PHEMA) but grow few nerve fibers unless fibronectin, collagen, or nerve growth factor was incorporated into the hydrogel. This provides a mechanism to provide controlled growth on hydrogel surfaces [109]. Hydrogels have been created with bioactive characteristics for neural cell adhesion and growth [110]. Arg–Gly–Asp (RGD) peptides were synthesized and chemically coupled to the bulk of poly (*N*-(2-hydroxypropyl) methacrylamide) (PHPMA) based polymer hydrogels. These RGD-grafted polymers implanted into the striata of rat brains promoted and supported the growth and spread of glial tissue

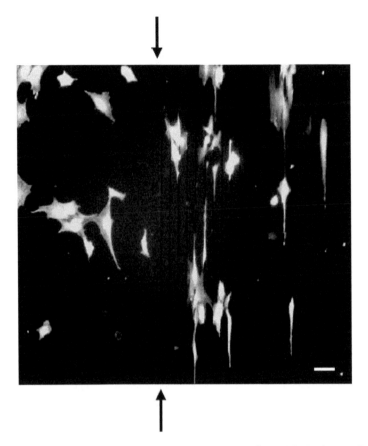

FIGURE 19.3 Astrocytes cultured on a laminin (LAM)-coated (0.01 mg/ml in EBSS) PS substrate. On the 10/20/3 μm LAM-PATTERN/LAM–NO PATTERN substrate, astrocytes are aligned in the direction of the groove on the patterned side (grooves at 90°; right of the arrows) while astrocytes were oriented randomly on the nonpatterned side (left of the arrows) of the substrate. Astrocytes were stained with CFDA SE in cell suspension prior to seeding. Images were taken from PS substrate fixed 24 h after seeding. Scale bar = 30 μm. (Reproduced from Recknor, J.B. et al. *Biomaterials*, 25, 2753–2767, 2004. With permission.)

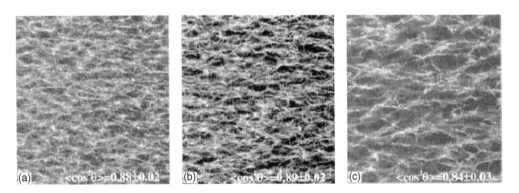

FIGURE 19.4 Confocal fluorescence images of aligned fibrin gels for varied Ca^{2+} concentration. Representative confocal images are shown along with the calculated fibrin fibril alignment parameter for magnetically aligned fibrin gels (MAFGs) formed for three different Ca^{2+} concentrations used in the fibrin-forming solution: (a) 1.2 mM, (b) 12 mM, and (c) 30 mM. Fibril diameter and inter-fibril spacing (porosity) is seen to increase with increasing Ca^{2+} concentration. (Reproduced from Dubey, N., Letourneau, P.C., and Tranquillo, R.T., *Biomaterials*, 22, 1065–1075, 2001. With permission.)

onto and into the hydrogels [111]. Cultured Schwann cells, neonatal astrocytes or cells dissociated from embryonic cerebral hemispheres were also dispersed within PHPMA hydrogel matrices and found to promote cellular ingrowth *in vivo* [112]. These polymer hydrogel matrices were found to have neuroinductive and neuroconductive properties and the potential to repair tissue defects in the central nervous system by promoting the formation of a tissue matrix and axonal growth by replacing lost tissue [47,113–115]. Furthermore, in the injured adult and developing rat spinal cord, these biocompatible porous hydrogels (NeuroGels) promoted axonal growth within the hydrogels, and supraspinal axons migrated into the reconstructed cord segment [116].

19.4.2.3 Neurotrophins

Neurotrophins are proteins in the nervous system that regulate neuronal survival and outgrowth, synaptic connectivity and neurotransmission. These growth-promoting factors are used to functionalize guidance conduits and create a desired response from the regenerating neural environment. Manipulation of polymer conduits for growth factor administration may be a useful treatment for neurodegenerative diseases, such as Alzheimer's disease or Parkinson's disease, which are characterized by the degeneration of neuronal cell populations. There is also much potential for overcoming severe tissue loss using growth factors released from nerve conduits in cases where there is a large gap as a result of axotomy. Various neurotrophic factors, including nerve growth factor (NGF), brain- derived neurotrophic factor (BDNF), neurotrophin-3 (NT-3), and neurotrophin-4/5 (NT-4/5) as well as other important growth factors and cytokines, including ciliary neurotrophic factor (CNTF), glial cell line-derived growth factor (GDNF), and acidic and basic fibroblast growth factor (aFGF, bFGF), have been "trapped" inside polymer conduits that control their release. For further information on these neurotrophic factors and their effects on neural regeneration, refer to Blesch et al. [117]; Jones et al. [119]; Terenghi [118]; Stichel and Muller [120].

Nerve growth factor (NGF) was one of the earliest neurotrophic factors identified and is one of the most thoroughly studied neurotrophins. In early experiments incorporating NGF into silicone chambers, the effects of NGF on nerve regeneration were positive but limited. This was attributed to the rapid decline in NGF concentrations in the conduit due to degradation in aqueous media and leakage from the conduit [121]. The method of delivery of these factors is a challenge and such limitations were overcome by providing controlled release of NGF. Controlled-release polymer delivery systems may be an important technology in enabling the prevention of neuronal degeneration, or even the stimulation of neuronal regeneration, by providing a sustained release of growth factors to promote the long-term survival of endogenous or transplanted cells [122]. Polymeric implants providing controlled release of NGF for one month were developed and found to improve neurite extension in cultured PC12 cells [123]. Continuous delivery of NGF has been shown to increase regeneration in both the PNS [124,125] and the CNS [126,127]. Furthermore, in an effort to readily provide for prolonged, site-specific delivery of NGF to the tissue, without adverse effects on the conduit, biodegradable polymer microspheres of poly (L-lactide) *co*-glycolide containing NGF were fabricated. Biologically active NGF was released from the microspheres, as assayed by neurite outgrowth in a dorsal root ganglion tissue culture system [128–130]. NGF co-encapsulated in PLGA microspheres along with ovalbumin was found to be bioactive for over 90 days [131]. Sustained release of NGF within nerve guide conduits has also been tested. NGF release from biodegradable poly(phosphoester) microspheres produced using a double emulsion technique exhibited a lower burst effect but similar protein entrapment levels and efficiencies when compared with those made of PLGA [132]. These NGF-loaded poly(phosphoester) microspheres were successfully implanted to bridge a 10 mm gap in a rat sciatic nerve model. Furthermore, the exogenous NGF had long-term morphological regeneration effects in the sciatic nerve [133].

Basic fibroblast growth factor (b-FGF) has been shown to enhance the *in vitro* survival and neurite extension of various types of neurons including dorsal root ganglia. One of the earliest studies involved controlled release of b-FGF and alpha-1 glycoprotein (α1-GP) from synthetic nerve guidance channels fabricated using the dip molding technique. After an initial burst in the first day, linear release was obtained from the conduits for a period of at least 2 weeks afterward. Only the tubes releasing b-FGF or b-FGF and alpha 1-GP displayed regenerated cables bridging both nerve stumps, which contained nerve fascicles with myelinated and unmyelinated axons [134]. Biodegradable polymer foams modified with a-FGF and used for controlled release and the provision of a permissive environment for spinal cord regeneration were formed using freeze-drying techniques [135]. Furthermore, evidence has been shown that a-FGF and b-FGF promote angiogenesis and may aid in the repair of damaged nerves [136].

Other neurotrophic factors such as GDNF, BDNF, and NT-3 have been released from synthetic guidance channels. In an effort to study facial nerve regeneration, the effects of the cytokine growth factor, GDNF, and NT-3 on nerve regeneration were assessed after rat facial nerve axotomy. Nerve cables regenerated in the presence of GDNF showed a large number of myelinated axons while no regenerated axons were observed in the absence of growth factors, demonstrating that GDNF, as previously described for the sciatic nerve, a mixed sensory and motor nerve, is also very efficient in promoting regeneration of the facial nerve, an essentially pure motor nerve [137]. Exogenous BDNF and NT-3 were delivered simultaneously into Schwann cell-grafted semipermeable guidance channels by an Alzet minipump to test the ability of these neurotrophins to promote axonal regeneration. This novel experiment elegantly demonstrated that BDNF and NT-3 infusion enhanced propriospinal axonal regeneration and enhanced axonal regeneration of specific distant populations of brain stem neurons into grafts in the adult rat spinal cord [138].

A pharmacotectonics concept, in which drug-delivery systems were arranged spatially in tissues to shape concentration fields for potent agents, has been presented. NGF-releasing implants placed within 1 to 2 mm of the treatment site enhanced the biological function of cellular targets, whereas identical implants placed approximately 3 mm from the target site of treatment produced no beneficial effect [139]. Due to certain limitations with controlled-delivery systems, alternatives such as the encapsulation of cells that secrete these factors are discussed in the next section.

19.4.3 Cellular Modifications

Due to certain limitations with the control of growth factor delivery systems, cells have been manipulated for the direct delivery of certain neurotrophic factors using a variety of therapeutic strategies. Genetic engineering has been used to modify cells for neurotrophic factor delivery [140]. (For a review of gene therapy, refer to Tresco et al. [141] and Tinsley and Eriksson [142]) Cells that produce specific neurotrophic and growth factors have been encapsulated using polymeric biomaterials. Other experiments have involved the direct seeding of cells known to secrete neurotrophic factors, such as Schwann cells, olfactory ensheathing cells (OECs) and neural stem cells (NSCs), into nerve conduits.

19.4.3.1 Cell Encapsulation

Implanting polymer-encapsulated cells for secreting growth and neurotrophic factors has been used for treatment for neurodegenerative disorders and to promote nerve regeneration. As an experimental therapy for Parkinsonian patients, enhanced benefit from neural transplantation can be provided through the combination of grafting with trophic factor treatment. This strategy ultimately results in improved survival and growth of grafted embryonic dopaminergic neurons. It has been demonstrated that the implantation of polymer-encapsulated cells genetically engineered to continuously secrete glial cell line-derived neurotrophic factor to the adult rat striatum improves dopaminergic graft survival and function. This shows that polymer encapsulation of cells can be used as an effective vehicle for long-term trophic factor supply [143]. A number of proteins have specific neuroprotective activities *in vitro*; however, the local delivery of these factors into the central nervous system over the long term at therapeutic levels has been difficult to achieve. Direct administration at the target site is a logical alternative, particularly in the central nervous system, but the limits of direct administration have not been defined clearly. For instance NGF must be delivered within several millimeters of the target to be effective in treating Alzheimer's disease [139]. Cells engineered to express the neuroprotective proteins, encapsulated in immunoisolation polymeric devices and implanted at the site of lesions have the potential to alter the progression of neurodegenerative disorders. The polymers used for encapsulation should allow transport of nutrients and oxygen to the cells, but also afford immunoprotection. Long-term cell viability *in vivo* in these constructs due to diffusional limitations has been the major drawback of this approach.

Ciliary neurotrophic factor (CNTF) decreases naturally occurring and axotomy-induced cell death and has been evaluated as a treatment for neurodegenerative disorders such as amyotrophic lateral sclerosis (ALS) and Huntington's disease [144]. Effective administration of this protein to motoneurons has been hampered by the exceedingly short half-life of CNTF, and the inability to deliver effective concentration into the central nervous system after systemic administration *in vivo*. BHK cells stably transfected with a plasmid construct containing the gene for human or mouse CNTF were encapsulated in polymer fibers and found to continuously release CNTF and slow down motoneuron degeneration following axotomy [145]. Implantation of polymer-encapsulated cells genetically engineered to continuously secrete GDNF to the adult rat striatum was found to improve dopaminergic graft survival and function [143]. Therefore cell encapsulation is a potentially important method in nerve regeneration, and can be used alone or in conjunction with other methods such as entubulization. For a review on the microencapsulation of neuroactive compounds and living cells producing these substances, see Maysinger and Morinville [146].

19.4.3.2 Cell Implantation

19.4.3.2.1 Schwann Cells

Schwann cells play an important role in supporting axonal regeneration after damage or disease. These cells clean debris from the injury site and secrete regulatory proteins that aid in neuronal survival and axonal growth. In the PNS and CNS, Schwann cells organize the regenerating environment through the myelination of axons, production of ECM, CAMs, and neurotrophins as well as aiding in the guidance of regenerating axons. The use of PNS grafts in the CNS by David and Aguayo [14] in the early 1980s distinguished Schwann cells as essential for CNS repair [14]. Conduits incorporating these cells have enhanced PNS regeneration [147]. CNS regeneration has also been induced after implantation of the semipermeable polyacrylonitrile/polyvinylchloride (PAN/PVC) and Matrigel conduits seeded with Schwann cells into the transected rat spinal cord [26,148]. Poly (α-hydroxy acids) with seeded Schwann cells or Schwann cell grafts were also found to be effective candidates for spinal cord regeneration [22,149]. However, limited regeneration has been demonstrated between the implant and the distal end of the host spinal cord [150] and myelination beyond the injury site has not been observed [151]. Recent research has shown that rolled Schwann cell monolayer grafts implanted into the transected rat sciatic nerve increased functional regeneration compared to acellular controls [152]. Research applicable to human applications has demonstrated that implantation of human Schwann cells in the nude (T cell deficient) rat spinal cord allowed axonal growth across the graft and re-entry into the spinal cord. For reviews of these and other Schwann cell therapies, see Jones et al. [119] and Bunge [150].

19.4.3.2.2 Olfactory Ensheathing Cells

Olfactory bulb ensheathing cells (OECs) are the primary glial cells found in both the PNS and CNS of the olfactory system. This system differs from other CNS tissues in that axons continue to grow throughout adulthood. These cells share characteristics with astrocytes of the CNS and Schwann cells from the PNS. Like astrocytes, these cells express glial fibrillary acidic protein (GFAP), yet they ensheath and myelinate axons and support axonal regrowth, which are features of Schwann cells. Most work in this area has involved OEC transplantations performed to promote remyelination of demyelinated rat spinal cord axons [153,154] and foster regeneration of damaged axons in the mature CNS [27,155–157]. Incorporating OECs into conduits has promoted axonal regeneration in both the PNS [158] and CNS [156] using Schwann cell-filled guidance channels. The results from these transplantations have demonstrated a distinct advantage of these cells over Schwann cells in creating a regenerative environment within the CNS.

19.4.3.2.3 Neural Stem Cells

Neural stem cells (NSCs) that have the potential to produce new neurons and glia are present in the mammalian CNS. These cells can remain in certain regions of the adult CNS after development even though neurogenesis no longer occurs in most areas after birth. Neural stem cells are described as generating neural tissue or being derived from the neural system, having capacity for self-renewal, and they are multipotent or possess the ability to adopt a variety of cellular fates. Neural progenitors with more limited capacities in terms of growth and differentiation have been known to proliferate throughout life in a variety of

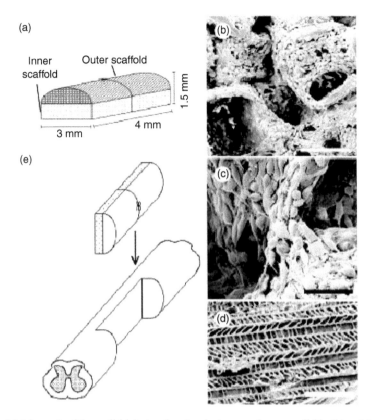

FIGURE 19.5   (a) Schematic of the scaffold design showing the inner and outer scaffolds. (b) and (c) Inner scaffolds seeded with NSCs. (Scale bars: 200 and 50 μm, respectively.) The outer section of the scaffold was created by means of a solid–liquid phase separation technique that produced long, axially oriented pores for axonal guidance as well as radial pores to allow fluid transport and inhibit the ingrowth of scar tissue. (d) Scale bar, 100 μm. (e) Schematic of surgical insertion of the implant into the spinal cord. (Reproduced from Teng, Y.D. et al., *Proc. Natl Acad. Sci. USA*, 99, 3024–3029, 2002. With permission.)

mammalian species, including humans [159,160]. NSCs were seeded into a dual scaffold structure made of biodegradable polymers to address the issues of spinal cord injury. Unique biodegradable polymer scaffolds were fabricated where the general design of the scaffold was derived from the structure of the spinal cord with an outer section that mimics the white matter with long axial pores to provide axonal guidance and an inner section seeded with neural stem cells for cell replacement and mimic the general character of the gray matter (Figure 19.5) [161]. The seeded scaffold improved functional recovery as compared with the lesion control or cells alone following spinal cord injury. Implantation of the scaffold-neural stem cells unit into an adult rat hemisection model of spinal cord injury promoted long-term improvement in function that was persistent up to one year in some animals, relative to a lesion-control group [18].

Human embryonic stem (hES) cells hold promise as an unlimited source of cells for transplantation therapies [162]. However, control of their proliferation and differentiation into complex, viable 3D tissues is challenging. Combining physical support with chemical cues created a supportive environment for the control of differentiation and organization of hES cells. Langer et al. developed biodegradable poly(lactic-*co*-glycolic acid)/poly(L-lactic acid) polymer scaffolds to promote hES cell growth and differentiation and formation of 3D structures. Complex structures with features of various committed embryonic tissues were generated *in vitro* using early differentiating hES cells and using the supportive three-dimensional environment to further induce differentiation. Growth factors such as retinoic acid, transforming growth factor β activin-A, or insulin-like growth factor directed hES cell differentiation and organization within the scaffold resulting in the formation of structures with characteristics of developing neural tissues, cartilage, or liver as well as the formation of a 3D vessel-like network. These constructs were transplanted into severe combined immunodeficient mice and continued to express specific human proteins in defined differentiated structures. This recent study presents a novel mechanism for creating viable human tissue structures for therapeutic applications [20]. For a review on stem cells in tissue engineering, refer to Bianco and Robey [163].

19.5 Conclusions

Engineering regeneration in the nervous system presents many challenges. Many strategies that may enhance regeneration in the PNS cannot be applied directly to the CNS due to the complexity of the environment. Novel tissue-engineering strategies and mechanisms, including the use of three dimensional polymer constructs with or without biological components (i.e., cells) and products fabricated for the induction of specific responses (i.e., regeneration), and the manipulation of biological cells *in vitro* (i.e., stem cells or cells for neuronal support), hold much promise for the enhancement of functional repair and replacement of tissue function. The successful use of polymeric nerve conduits in facilitating peripheral nerve regeneration has been demonstrated and polymers have shown great promise in addressing spinal cord injuries as well. This regeneration process, with various polymers, both degradable as well as nondegradable, has been enhanced further by promoting directed growth and by the addition of chemical cues such as ECM molecules, nerve growth factors and neurotrophins and other agents incorporated in the conduits to be released in a controlled fashion. Polymers have also played an important role in encapsulating cells and in the transplantation of neuronal support cells that release factors to promote nerve regeneration. Multidiscliplinary tissue-engineering approaches involving such biomaterials to mimic the native neural environment of the body have resulted in significant progress in gaining some understanding of the systems and signals involved in nerve regeneration. Incorporating multiple guidance cues and experimental strategies will allow further opportunities to elucidate the mechanisms behind nerve regeneration and specific considerations for the efficient repair of the nervous system.

Acknowledgments

The authors would like to acknowledge NSF (BES-9983735) and USDOD for funding.

References

[1] Horner, P.J. and Gage, F.H., Regenerating the damaged central nervous system, *Nature* 407, 963–970, 2000.

[2] Bjorklund, A. and Lindvall, O., Cell replacement therapies for central nervous system disorders, *Nat. Neurosci.* 3, 537–544, 2000.

[3] Fry, E.J., Central nervous system regeneration: mission impossible? *Clin. Exp. Pharmacol. Physiol.* 28, 253–258, 2001.

[4] Keeley, R.T., Atagi, T., Sabelman, E., Padilla, J., Kadlcik, S., Keeley, A., Nguyen, K., and Rosen, J., Peripheral nerve regeneration across 14-mm gaps: a comparison of autograft and entubulation repair methods in the rat, *J. Reconstruc. Microsurg.* 9, 349–358, 1993.

[5] Wang, G., Hirai, K., and Shimada, H., The role of laminin, a component of Schwann cell basal lamina, in rat sciatic nerve regeneration within antiserum-treated nerve grafts, *Brain Res.* 570, 116–125, 1992.

[6] Carlstedt, T., Nerve fibre regeneration across the peripheral-central transitional zone, *J. Anat.* 190, 51–56, 1997.

[7] Goodrum, J.F. and Bouldin, T.W., The cell biology of myelin degeneration and regeneration in the peripheral nervous system, *J. Neuropathol. Exp. Neurol.* 55, 943–953, 1996.

[8] Tortora, G.J., *Principles of Human Anatomy: Nervous System*, 6th ed. Harper Collins, New York, 1992, pp. 459–466.

[9] Fawcett, J.W. and Keynes, R.J., Peripheral nerve regeneration, *Annu. Rev. Neurosci.* 13, 43–60, 1990.

[10] Bahr, M. and Bonhoeffer, F., Perspectives on axonal regeneration in the mammalian CNS, *Trends Neurosci.* 17, 473–479, 1994.

[11] Silver, J. and Miller, J.H., Regeneration beyond the glial scar, *Nat. Rev. Neurosci.* 5, 146–156, 2004.

[12] Tatagiba, M., Brosamle, C., and Schwab, M.E., Regeneration of injured axons in the adult mammalian central nervous system, *Neurosurgery* 40, 541–546, 1997.

[13] Bray, G.M., Villegasperez, M.P., Vidalsanz, M., and Aguayo, A.J., The use of peripheral-nerve grafts to enhance neuronal survival, promote growth and permit terminal reconnections in the central-nervous-system of adult-rats, *J. Exp. Biol.* 132, 5–19, 1987.

[14] David, S. and Aguayo, A.J., Axonal elongation into peripheral nervous-system bridges after central nervous-system injury in adult-rats, *Science* 214, 931–933, 1981.

[15] Villegas-Perez, M.P., Vidal-Sanz, M., Bray, G.M., and Aguayo, A.J., Influences of peripheral nerve grafts on the survival and regrowth of axotomized retinal ganglion cells in adult rats, *J. Neurosci.* 8, 265–280, 1988.

[16] Aguayo, A.J., Vidalsanz, M., Villegasperez, M.P., and Bray, G.M., Growth and connectivity of axotomized retinal neurons in adult-rats with optic nerves substituted by Pns grafts linking the eye and the midbrain, *Anna. NY Acad. Sci.* 495, 1–9, 1987.

[17] Keirstead, S.A., Rasminsky, M., Fukuda, Y., Carter, D.A., Aguayo, A.J., and Vidalsanz, M., Electrophysiologic responses in hamster superior colliculus evoked by regenerating retinal axons, *Science* 246, 255–257, 1989.

[18] Teng, Y.D., Lavik, E.B., Qu, X., Park, K.I., Ourednik, J., Zurakowski, D., Langer, R., and Snyder, E.Y., Functional recovery following traumatic spinal cord injury mediated by a unique polymer scaffold seeded with neural stem cells, *Proc. Natl Acad. Sci. USA* 99, 3024–3029, 2002.

[19] Neumann, S. and Woolf, C.J., Regeneration of dorsal column fibers into and beyond the lesion site following adult spinal cord injury, *Neuron* 23, 83–91, 1999.

[20] Levenberg, S., Huang, N.F., Lavik, E., Rogers, A.B., Itskovitz-Eldor, J., and Langer, R., Differentiation of human embryonic stem cells on three-dimensional polymer scaffolds, *Proc. Natl Acad. Sci. USA* 100, 12741–12746, 2003.

[21] Vacanti, M.P., Leonard, J.L., Dore, B., Bonassar, L.J., Cao, Y., Stachelek, S.J., Vacanti, J.P., O'Connell, F., Yu, C.S., Farwell, A.P., and Vacanti, C.A., Tissue-engineered spinal cord, *Transplant Proc.* 33, 592–598, 2001.

[22] Gautier, S.E., Oudega, M., Fragoso, M., Chapon, P., Plant, G.W., Bunge, M.B., and Parel, J.M., Poly(alpha-hydroxyacids) for application in the spinal cord: resorbability and biocompatibility with adult rat Schwann cells and spinal cord, *J. Biomed. Mater. Res.* 42, 642–654, 1998.

[23] Holmes, T.C., de Lacalle, S., Su, X., Liu, G.S., Rich, A., and Zhang, S.G., Extensive neurite outgrowth and active synapse formation on self-assembling peptide scaffolds, *Proc. Natl Acad. Sci. USA* 97, 6728–6733, 2000.

[24] Park, K.I., Teng, Y.D., and Snyder, E.Y., The injured brain interacts reciprocally with neural stem cells supported by scaffolds to reconstitute lost tissue, *Nat. Biotechnol.* 20, 1111–1117, 2002.

[25] Xu, X.M., Zhang, S.X., Li, H.Y., Aebischer, P., and Bunge, M.B., Regrowth of axons into the distal spinal cord through a Schwann-cell-seeded mini-channel implanted into hemisected adult rat spinal cord, *Eur. J. Neurosci.* 11, 1723–1740, 1999.

[26] Xu, X.M., Chen, A., Guenard, V., Kleitman, N., and Bunge, M.B., Bridging Schwann cell transplants promote axonal regeneration from both the rostral and caudal stumps of transected adult rat spinal cord, *J. Neurocytol.* 26, 1–16, 1997.

[27] Li, Y., Field, P.M., and Raisman, G., Repair of adult rat corticospinal tract by transplants of olfactory ensheathing cells, *Science* 277, 2000–2002, 1997.

[28] Menei, P., Montero-Menei, C., Whittemore, S.R., Bunge, R.P., and Bunge, M.B., Schwann cells genetically modified to secrete human BDNF promote enhanced axonal regrowth across transected adult rat spinal cord, *Eur. J. Neurosci.* 10, 607–621, 1998.

[29] Teng, Y.D., Mocchetti, I., Taveira-DaSilva, A.M., Gillis, R.A., and Wrathall, J.R., Basic fibroblast growth factor increases long-term survival of spinal motor neurons and improves respiratory function after experimental spinal cord injury, *J. Neurosci.* 19, 7037–7047, 1999.

[30] Schwab, M.E., Increasing plasticity and functional recovery of the lesioned spinal cord, *Prog. Brain Res.* 137, 351–359, 2002.

[31] Lundborg, G., *Nerve Injury and Repair*, Longman Group, UK, New York, 1988.

[32] So, K.F. and Aguayo, A.J., Lengthy regrowth of cut axons from ganglion-cells after peripheral-nerve transplantation into the retina of adult-rats, *Brain Res.* 328, 349–354, 1985.

[33] Evans, G.R.D., Challenges to nerve regeneration, *Semin. Surg. Oncol.* 19, 312–318, 2000.

[34] Kim, B.-S. and Atala, A., Periperal nerve regeneration, in *Methods of Tissue Engineering*, Atala, A. and Lanza, R. (Eds.), Academic Press, New York, 2002, pp. 1135–1142.

[35] Frerichs, O., Fansa, H., Schicht, C., Wolf, G., Schneider, W., and Keilhoff, G., Reconstruction of peripheral nerves using acellular nerve grafts with implanted cultured Schwann cells, *Microsurgery*, 22, 311–315, 2002.

[36] Heath, C.A. and Rutkowski, G.E., The develpoment of bioartifical nerve grafts for peripheral-nerve regeneration, *Trends Biotechnol.* 16, 163–168, 1998.

[37] Pearson, R.G., Molino, Y., Williams, P.M., Tendler, S.J., Davies, M.C., Roberts, C.J., and Shakesheff, K.M., Spatial confinement of neurite regrowth from dorsal root ganglia within nonporous microconduits, *Tissue Eng.* 9, 201–208, 2003.

[38] Bellamkonda, R. and Aebischer, P., Review: tissue engineering in the nervous system, *Biotechnol. Bioeng.* 43, 543–556, 1994.

[39] Schmidt, C.E. and Leach, J.B., Neural tissue engineering: strategies for repair and regeneration, *Annu. Rev. Biomed. Eng.* 5, 293–347, 2003.

[40] Evans, G.R., Brandt, K., Katz, S., Chauvin, P., Otto, L., Bogle, M., Wang, B., Meszlenyi, R.K., Lu, L., Mikos, A.G., and Patrick, C. W., Jr., Bioactive poly(L-lactic acid) conduits seeded with Schwann cells for peripheral nerve regeneration, *Biomaterials* 23, 841–848, 2002.

[41] Evans, G.R.D., Brandt, K., Widmer, M.S., Lu, L., Meszlenyi, R.K., Gupta, P.K., Mikos, A.G., Hodges, J., Williams, J., and Gurlek, A., *In vivo* evaluation of poly(-lactic acid) porous conduits for peripheral nerve regeneration, *Biomaterials* 20, 1109–1115, 1999.

[42] Hadlock, T., Elisseeff, J., Langer, R., Vacanti, J., and Cheney, M., A tissue-engineered conduit for peripheral nerve repair, *Arch. Otolaryngol. Head Neck Surg.* 124, 1081–1086, 1998.

[43] Widmer, M.S., Gupta, P.K., Lu, L., Meszlenyi, R.K., Evans, G.R., Brandt, K., Savel, T., Gurlek, A., Patrick, C.W., Jr., and Mikos, A.G., Manufacture of porous biodegradable polymer conduits by an extrusion process for guided tissue regeneration, *Biomaterials* 19, 1945–1955, 1998.

[44] Nicoli Aldini, N., Fini, M., Rocca, M., Giavaresi, G., and Giardino, R., Guided regeneration with resorbable conduits in experimental peripheral nerve injuries, *Int. Orthop.* 24, 121–125, 2000.

[45] Nicoli Aldini, N., Perego, G., Cella, G.D., Maltarello, M.C., Fini, M., Rocca, M., and Giardino, R., Effectiveness of a bioabsorbable conduit in the repair of peripheral nerves, *Biomaterials* 17, 959–962, 1996.

[46] Hazari, A., Wiberg, M., Johansson-Ruden, G., Green, C., and Terenghi, G., A resorbable nerve conduit as an alternative to nerve autograft in nerve gap repair, *Br. J. Plast. Surg.* 52, 653–657, 1999.

[47] Woerly, S., Pinet, E., De Robertis, L., Bousmina, M., Laroche, G., Roitback, T., Vargova, L., and Sykova, E., Heterogeneous PHPMA hydrogels for tissue repair and axonal regeneration in the injured spinal cord, *J. Biomater. Sci. Polym. Ed.* 9, 681–711, 1998.

[48] Recknor, J.B., Recknor, J.C., Sakaguchi, D.S., and Mallapragada, S.K., Oriented astroglial cell growth on micropatterned polystyrene substrates, *Biomaterials* 25, 2753–2767, 2004.

[49] Danielsen, N., Pettmann, B., Vahlsing, H.L., Manthorpe, M., and Varon, S., Fibroblast growth factor effects on peripheral nerve regeneration in a silicone chamber model, *J. Neurosci. Res.* 20, 320–330, 1988.

[50] Lundborg, G., Dahlin, L.B., Danielsen, N., Gelberman, R.H., Longo, F.M., Powell, H.C., and Varon, S., Nerve regeneration in silicone chambers — influence of gap length and of distal stump components, *Exp. Neurol.* 76, 361–375, 1982.

[51] Pitta, M.C., Wolford, L.M., Mehra, P., and Hopkin, J., Use of gore-tex tubing as a conduit for inferior alveolar and lingual nerve repair: experience with 6 cases, *J. Oral Maxillofac. Surg.* 59, 493–496, 2001.

[52] Miloro, M., Halkias, L.E., Mallery, S., Travers, S., and Rashid, R.G., Low-level laser effect on neural regeneration in Gore-Tex tubes, *Oral Surg. Oral Med. Oral Pathol. Oral Radiol. Endodontics* 93, 27–34, 2002.

[53] Tong, Y.W. and Shoichet, M.S., Enhancing the interaction of central nervous system neurons with poly(tetrafluoroethylene-co-hexafluoropropylene) via a novel surface amine-functionalization reaction followed by peptide modification, *J. Biomater. Sci. Polym. Ed.* 9, 713–729, 1998.

[54] Ahmed, Z. and Brown, R.A., Adhesion, alignment, and migration of cultured Schwann cells on ultrathin fibronectin fibres, *Cell Motil. Cytoskel.* 42, 331–343, 1999.

[55] Toba, T., Nakamura, T., Lynn, A.K., Matsumoto, K., Fukuda, S., Yoshitani, M., Hori, Y., and Shimizu, Y., Evaluation of peripheral nerve regeneration across an 80-mm gap using a polyglycolic acid (PGA) — collagen nerve conduit filled with laminin-soaked collagen sponge in dogs, *Int. J. Artif. Organs* 25, 230–237, 2002.

[56] Dubey, N., Letourneau, P.C., and Tranquillo, R.T., Guided neurite elongation and Schwann cell invasion into magnetically aligned collagen in simulated peripheral nerve regeneration, *Exp. Neurol.* 158, 338–350, 1999.

[57] Tong, X.J., Hirai, K.I., Shimada, H., Mizutani, Y., Izumi, T., Toda, N., and Yu, P., Sciatic-nerve regeneration navigated by laminin–fibronectin double coated biodegradable collagen grafts in rats, *Brain Res.* 663, 155–162, 1994.

[58] Ahmed, Z., Underwood, S., and Brown, R.A., Low concentrations of fibrinogen increase cell migration speed on fibronectin/fibrinogen composite cables, *Cell Motil. Cytoskel.* 46, 6–16, 2000.

[59] Herbert, C.B., Nagaswami, C., Bittner, G.D., Hubbell, J.A., and Weisel, J.W., Effects of fibrin micromorphology on neurite growth from dorsal root ganglia cultured in three-dimensional fibrin gels, *J. Biomed. Mater. Res.* 40, 551–559, 1998.

[60] Schense, J.C., Bloch, J., Aebischer, P., and Hubbell, J.A., Enzymatic incorporation of bioactive peptides into fibrin matrices enhances neurite extension, *Nat. Biotechnol.* 18, 415–419, 2000.

[61] Seckel, B.R., Jones, D., Hekimian, K.J., Wang, K.K., Chakalis, D.P., and Costas, P.D., Hyaluronic acid through a new injectable nerve guide delivery system enhances peripheral nerve regeneration in the rat, *J. Neurosci. Res.* 40, 318–324, 1995.

[62] Balgude, A.P., Yu, X., Szymanski, A., and Bellamkonda, R.V., Agarose gel stiffness determines rate of DRG neurite extension in 3D cultures, *Biomaterials* 22, 1077–1084, 2001.

[63] Hudson, T.W., Evans, G.R.D., and Schmidt, C.E., Engineering strategies for peripheral nerve repair, *Clin. Plastic Surg.* 26, 617–628, 1999.

[64] Valentini, R.F., Nerve guidance channels, in *The Biomedical Engineering Handbook*, 2nd ed., Bronzino, J.D. Ed., CRC Press, Boca Raton, FL, 2000, pp. 135-1–135-12.

[65] Sundback, C., Hadlock, T., Cheney, M., and Vacanti, J., Manufacture of porous polymer nerve conduits by a novel low-pressure injection molding process, *Biomaterials* 24, 819–830, 2003.

[66] Clark, P., Connolly, P., Curtis, A.S.G., Dow, J.A.T., and Wilkinson, C.D.W., Topographical control of cell behaviour: II. Multiple grooved substrata, *Development* 108, 635–644, 1990.

[67] Clark, P., Connolly, P., Curtis, A.S.G., Dow, J.A.T., and Wilkinson, C.D.W., Topographical control of cell behaviour. I. Simple step cues, *Development* 99, 439–448, 1987.

[68] Miller, C., Jeftinija, S., and Mallapragada, S., Synergistic effects of physical and chemical guidance cues on neurite alignment and outgrowth on biodegradable polymer substrates, *Tissue Eng.* 8, 367–378, 2002.

[69] Miller, C.A., Jeftinija, S., and Mallapragada, S.K., Micropatterned Schwann cell-seeded biodegradable polymer substrates significantly enhance neurite alignment and outgrowth, *Tissue Eng.* 7, 705–715, 2001.

[70] Miller, C., Shanks, H., Witt, A., Rutkowski, G., and Mallapragada, S., Oriented Schwann cell growth on micropatterned biodegradable polymer substrates, *Biomaterials* 22, 1263–1269, 2001.

[71] Chen, C., Mrkisch, M., Huang, S., Whitesides, G., and Ingber, D., Geometric control of life and death, *Science* 276, 1425–1428, 1997.

[72] Singhvi, R., Kumar, A., Lopez, G., Stephanopoulous, G., Wang, D.I.C., Whitesides, G., and Ingber, D., Engineering cell shape and function, *Science* 264, 696–698, 1994.

[73] Craighead, H.G., Turner, S.W., Davis, R.C., James, C., Perez, A.M., St. John, P.M., Isaacson, M.S., Kam, L., Shain, W., Turner, J. N., and Banker, G., Chemical and topographical surface modification for control of central nervous system cell adhesion, *Biomed. Microdev.* 1, 49–64, 1998.

[74] Clark, P., Cell and neuron growth cone behavior on micropatterned surfaces, in *Nanofabrication and Biosystems: Integrating Materials, Science, Engineering, and Biology*, Hoch, H.C., Jelinski, L.W., and Craighead, H.G. (Eds.), Cambridge University Press, New York, 1996, pp. 356–366.

[75] Kleinfeld, D., Kahler, K., and Hockberger, P., Controlled outgrowth of dissociated neurons on patterned substrates, *J. Neurosci.* 8, 4098–4120, 1988.

[76] Kam, L., Shain, W., Turner, J.N., and Bizios, R., Correlation of astroglial cell function on micropatterned surfaces with specific geometric parameters, *Biomaterials* 20, 2343–2350, 1999.

[77] Brunette, D.M., Fibroblasts on micromachined substrata orient hierarchically to grooves of different dimensions, *Exp. Cell Res.* 164, 11–26, 1986.

[78] Kawana, A., Formation of a simple model brain on microfabricated electrode arrays, in *Nanofabrication and Biosystems: Integrating Materials, Science, Engineering, and Biology*, Hoch, H.C., Jelinski, L.W., and Craighead, H.G. Ed., Cambridge University Press, New York, 1996, pp. 258–275.

[79] Webb, A., Clark, P., Skepper, J., Compston, A., and Wood, A., Guidance of oligodendrocytes and their progenitors by substratum topography, *J. Cell Sci.* 108, 2747–2760, 1995.

[80] Ohara, P.T. and Buck, R.C., Contact guidance *in vitro*. A light, transmission, and scanning electron microscopic study, *Exp. Cell Res.* 121, 235–249, 1979.

[81] Rajnicek, A., Britland, S., and McCaig, C., Contact guidance of CNS neurites on grooved quartz: influence of groove dimensions, neuronal age and cell type, *J. Cell Sci.* 110, 2905–2913, 1997.

[82] Lu, L., Kam, L., Hasenbein, M., Nyalakonda, K., Bizios, R., Gopferich, A., Young, J.F., and Mikos, A.G., Retinal pigment epithelial cell function on substrates with chemically micropatterned surfaces, *Biomaterials* 20, 2351–2361, 1999.

[83] Turner, A.M.P., Dowell, N., Turner, S.W.P., Kam, L., Isaacson, M., Turner, J.N., Craighead, H.G., and Shain, W., Attachment of astroglial cells to microfabricated pillar arrays of different geometries, *J. Biomed. Mater. Res.* 51, 430–441, 2000.

[84] Bhatia, S.N. and Chen, C.S., Tissue engineering at the micro-scale, *Biomed. Microdev.* 2, 131–144, 1999.

[85] Wheeler, B.C., Corey, J.M., Brewer, G.J., and Branch, D.W., Microcontact printing for precise control of nerve cell growth in culture, *J. Biomech. Eng.-Trans. ASME* 121, 73–78, 1999.

[86] Bernard, A., Delamarche, E., Schmid, H., Michel, B., Bosshard, H.R., and Biebuyck, H., Printing patterns of proteins, *Langmuir* 14, 2225–2229, 1998.

[87] Clark, P., Britland, S., and Connolly, P., Growth cone guidance and neuron morphology on micropatterned laminin surfaces, *J. Cell Sci.* 105, 203–212, 1993.

[88] Tai, H.C. and Buettner, H.M., Neurite outgrowth and growth cone morphology on micropatterned surfaces, *Biotechnol. Prog.* 14, 364–370, 1998.

[89] Gundersen, R.W., Sensory neurite growth cone guidance by substrate adsorbed nerve growth factor, *J. Neurosci. Res.* 13, 199–212, 1985.

[90] Kam, L., Shain, W., Turner, J.N., and Bizios, R., Axonal outgrowth of hippocampal neur- ons on micro-scale networks of polylysine-conjugated laminin, *Biomaterials* 22, 1049–1054, 2001.

[91] Buettner, H.M., Microcontrol of neuronal outgrowth, in *Nanofabrication and Biosystems: Integrating Materials, Science, Engineering, and Biology*, Hoch, H.C., Jelinski, L.W., and Craighead, H.G. (Eds.), Cambridge University Press, New York, 1996, pp. 300–314.

[92] Flemming, R.G., Murphy, C.J., Abrams, G.A., Goodman, S.L., and Nealey, P.F., Effects of synthetic micro- and nano-structured surfaces on cell behavior, *Biomaterials* 20, 573–588, 1999.

[93] Curtis, A. and Wilkinson, C., Topographical control of cells, *Biomaterials* 18, 1573–1583, 1998.

[94] Grimpe, B. and Silver, J., The extracellular matrix in axon regeneration, *Prog. Brain Res.* 137 (Spinal Cord Trauma), 333–349, 2002.

[95] Letourneau, P.C., Madsen, A.M., Palm, S.L., and Furcht, L.T., Immunoreactivity for laminin in the developing ventral longitudinal pathway of the brain, *Develop. Biol.* 125, 135–144, 1988.

[96] Bellamkonda, R., Ranieri, J.P., and Aebischer, P., Laminin oligopeptide derivatized agarose gels allow 3-dimensional neurite extension *in-vitro, J. Neurosci. Res.* 41, 501–509, 1995.

[97] Yu, X.J. and Bellamkonda, R.V., Tissue-engineered scaffolds are effective alternatives to autografts for bridging peripheral nerve gaps, *Tissue Eng.* 9, 421–430, 2003.

[98] Yu, X.J., Dillon, G.P., and Bellamkonda, R.V., A laminin and nerve growth factor-laden three-dimensional scaffold for enhanced neurite extension, *Tissue Eng.* 5, 291–304, 1999.

[99] Itoh, S., Takakuda, K., Kawabata, S., Aso, Y., Kasai, K., Itoh, H., and Shinomiya, K., Evaluation of cross-linking procedures of collagen tubes used in peripheral nerve repair, *Biomaterials* 23, 4475–4481, 2002.

[100] Yoshii, S. and Oka, M., Peripheral nerve regeneration along collagen filaments, *Brain Res.* 888, 158–162, 2001.

[101] Chen, Y.S., Hsieh, C.L., Tsai, C.C., Chen, T.H., Cheng, W.C., Hu, C.L., and Yao, C.H., Peripheral nerve regeneration using silicone rubber chambers filled with collagen, laminin and fibronectin, *Biomaterials* 21, 1541–1547, 2000.

[102] Ceballos, D., Navarro, X., Dubey, N., Wendelschafer-Crabb, G., Kennedy, W.R., and Tranquillo, R.T., Magnetically aligned collagen gel filling a collagen nerve guide improves peripheral nerve regeneration, *Exp. Neurol.* 158, 290–300, 1999.

[103] Dubey, N., Letourneau, P.C., and Tranquillo, R.T., Neuronal contact guidance in magnetically aligned fibrin gels: effect of variation in gel mechano-structural properties, *Biomaterials* 22, 1065–1075, 2001.

[104] Hadlock, T., Sundback, C., Koka, R., Hunter, D., Cheney, M., and Vacanti, J., A novel, biodegradable polymer conduit delivers neurotrophins and promotes nerve regeneration, *Laryngoscope* 109, 1412–1416, 1999.

[105] Dickson, B.J., Molecular mechanisms of axon guidance, *Science* 298, 1959–1964, 2002.

[106] Webb, K., Budko, E., Neuberger, T.J., Chen, S., Schachner, M., and Tresco, P.A., Substrate-bound human recombinant L1 selectively promotes neuronal attachment and outgrowth in the presence of astrocytes and fibroblasts, *Biomaterials* 22, 1017–1028, 2001.

[107] Thanos, P.K., Okajima, S., and Terzis, J.K., Ultrastructure and cellular biology of nerve regeneration, *J. Reconstruct. Microsurg.* 14, 423–436, 1998.

[108] Ruegg, M.A., Agrin, laminin beta 2 (s-laminin) and ARIA: their role in neuromuscular development, *Curr. Opin. Neurobiol.* 6, 97–103, 1996.

[109] Carbonetto, S.T., Gruver, M.M., and Turner, D.C., Nerve fiber growth on defined hydrogel substrates, *Science* 216, 897–899, 1982.

[110] Woerly, S., Hydrogels for neural tissue reconstruction and transplantation, *Biomaterials* 14, 1056–1058, 1993.

[111] Woerly, S., Laroche, G., Marchand, R., Pato, J., Subr, V., and Ulbrich, K., Intracerebral implantation of hydrogel-coupled adhesion peptides: tissue reaction, *J. Neural Transplant Plast.* 5, 245–255, 1995.

[112] Woerly, S., Plant, G.W., and Harvey, A.R., Cultured rat neuronal and glial cells entrapped within hydrogel polymer matrices: a potential tool for neural tissue replacement, *Neurosci. Lett.* 205, 197–201, 1996.

[113] Woerly, S., Doan, V.D., Evans-Martin, F., Paramore, C.G., and Peduzzi, J.D., Spinal cord reconstruction using NeuroGel implants and functional recovery after chronic injury, *J. Neurosci. Res.* 66, 1187–1197, 2001.

[114] Woerly, S., Doan, V.D., Sosa, N., de Vellis, J., and Espinosa, A., Reconstruction of the transected cat spinal cord following NeuroGel implantation: axonal tracing, immunohistochemical and ultrastructural studies, *Int. J. Dev. Neurosci.* 19, 63–83, 2001.

[115] Woerly, S., Petrov, P., Sykova, E., Roitbak, T., Simonova, Z., and Harvey, A.R., Neural tissue formation within porous hydrogels implanted in brain and spinal cord lesions: ultrastructural, immunohistochemical, and diffusion studies, *Tissue Eng.* 5, 467–488, 1999.

[116] Woerly, S., Pinet, E., de Robertis, L., Van Diep, D., and Bousmina, M., Spinal cord repair with PHPMA hydrogel containing RGD peptides (NeuroGel™), *Biomaterials* 22, 1095–1111, 2001.

[117] Blesch, A., Lu, P., and Tuszynski, M.H., Neurotrophic factors, gene therapy, and neural stem cells for spinal cord repair, *Brain Res. Bull.* 57, 833–838, 2002.

[118] Terenghi, G., Peripheral nerve regeneration and neurotrophic factors, *J. Anatomy* 194, 1–14, 1999.

[119] Jones, L.L., Oudega, M., Bunge, M.B., and Tuszynski, M.H., Neurotrophic factors, cellular bridges and gene therapy for spinal cord injury, *J. Physiol. (Lond.)* 533, 83–89, 2001.

[120] Stichel, C.C. and Muller, H.W., Experimental strategies to promote axonal regeneration after traumatic central nervous system injury, *Prog. Neurobiol.* 56, 119–148, 1998.

[121] Hollowell, J.P., Villadiego, A., and Rich, K.M., Sciatic nerve regeneration across gaps within silicone chambers: long-term effects of NGF and consideration of axonal branching, *Exp. Neurol.* 110, 45–51, 1990.

[122] Haller, M.F. and Saltzman, W.M., Nerve growth factor delivery systems, *J. Control. Release* 53, 1–6, 1998.

[123] Powell, E.M., Sobarzo, M.R., and Saltzman, W.M., Controlled release of nerve growth factor from a polymeric implant, *Brain Res.* 515, 309–311, 1990.

[124] Fine, E.G., Decosterd, I., Papaloizos, M., Zurn, A.D., and Aebischer, P., GDNF and NGF released by synthetic guidance channels support sciatic nerve regeneration across a long gap, *Eur. J. Neurosci.* 15, 589–601, 2002.

[125] Bloch, J., Fine, E.G., Bouche, N., Zurn, A.D., and Aebischer, P., Nerve growth factor- and neurotrophin-3-releasing guidance channels promote regeneration of the transected rat dorsal root, *Exp. Neurol.* 172, 425–432, 2001.

[126] Ramer, M.S., Priestley, J.V., and McMahon, S.B., Functional regeneration of sensory axone into the adult spinal cord, *Nature* 403, 312–316, 2000.

[127] Nakahara, Y., Gage, F.H., and Tuszynski, M.H., Grafts of fibroblasts genetically modified to secrete NGF, BDNF, NT-3, or basic FGF elicit differential responses in the adult spinal cord, *Cell Transplant* 5, 191–204, 1996.

[128] Camarata, P.J., Suryanarayanan, R., Turner, D.A., Parker, R.G., and Ebner, T.J., Sustained release of nerve growth factor from biodegradable polymer microspheres, *Neurosurgery* 30, 313–319, 1992.

[129] Pean, J.M., Boury, F., Venier-Julienne, M.C., Menei, P., Proust, J.E., and Benoit, J.P., Why does PEG 400 co-encapsulation improve NGF stability and release from PLGA biodegradable microspheres? *Pharm. Res.* 16, 1294–1299, 1999.

[130] Pean, J.M., Venier-Julienne, M.C., Boury, F., Menei, P., Denizot, B., and Benoit, J.P., NGF release from poly(D,L-lactide-co-glycolide) microspheres. Effect of some formulation parameters on encapsulated NGF stability, *J. Control. Release* 56, 175–187, 1998.

[131] Cao, X. and Schoichet, M.S., Delivering neuroactive molecules from biodegradable microspheres for application in central nervous system disorders, *Biomaterials* 20, 329–339, 1999.

[132] Xu, X., Yu, H., Gao, S., Ma, H.Q., Leong, K.W., and Wang, S., Polyphosphoester microspheres for sustained release of biologically active nerve growth factor, *Biomaterials* 23, 3765–3772, 2002.

[133] Xu, X., Yee, W.C., Hwang, P.Y., Yu, H., Wan, A.C., Gao, S., Boon, K.L., Mao, H.Q., Leong, K.W., and Wang, S., Peripheral nerve regeneration with sustained release of poly(phosphoester) microencapsulated nerve growth factor within nerve guide conduits, *Biomaterials* 24, 2405–2412, 2003.

[134] Aebischer, P., Salessiotis, A.N., and Winn, S.R., Basic fibroblast growth factor released from synthetic guidance channels facilitates peripheral nerve regeneration across long nerve gaps, *J. Neurosci. Res.* 23, 282–289, 1989.

[135] Maquet, V., Martin, D., Scholtes, F., Franzen, R., Schoenen, J., Moonen, G., and Jer me, R., Poly(D,L-lactide) foams modified by poly(ethylene oxide)-block-poly(D,L-lactide) copolymers and a-FGF: *in vitro* and *in vivo* evaluation for spinal cord regeneration, *Biomaterials* 22, 1137–1146, 2001.

[136] Friesel, R.E. and Maciag, T., Molecular mechanisms of angiogenesis — fibroblast growth-factor signal-transduction, *FASEB J.* 9, 919–925, 1995.

[137] Barras, F.M., Pasche, P., Bouche, N., Aebischer, P., and Zurn, A.D., Glial cell line-derived neurotrophic factor released by synthetic guidance channels promotes facial nerve regeneration in the rat, *J. Neurosci. Res.* 70, 746–755, 2002.

[138] Xu, X.M., Guenard, V., Kleitman, N., Aebischer, P., and Bunge, M.B., Combination of Bdnf and Nt-3 promotes supraspinal axonal regeneration into Schwann-cell grafts in adult-rat thoracic spinal-cord, *Exp. Neurol.* 134, 261–272, 1995.

[139] Mahoney, M.J. and Saltzman, W.M., Millimeter-scale positioning of a nerve-growth-factor source and biological activity in the brain, *Proc. Natl Acad. Sci. USA.* 96, 4536–4539, 1999.

[140] Grill, R., Murai, K., Blesch, A., Gage, F.H., and Tuszynski, M.H., Cellular delivery of neurotrophin-3 promotes corticospinal axonal growth and partial functional recovery after spinal cord injury, *J. Neurosci.* 17, 5560–5572, 1997.

[141] Tresco, P.A., Biran, R., and Noble, M.D., Cellular transplants as sources for therapeutic agents, *Adv. Drug Deliv. Rev.* 42, 3–27, 2000.

[142] Tinsley, R. and Eriksson, P., Use of gene therapy in central nervous system repair, *Acta. Neurol. Scand.* 109, 1–8, 2004.

[143] Sautter, J., Tseng, J.L., Braguglia, D., Aebischer, P., Spenger, C., Seiler, R.W., Widmer, H.R., and Zurn, A.D., Implants of polymer-encapsulated genetically modified cells releasing glial cell line-derived neurotrophic factor improve survival, growth, and function of fetal dopaminergic grafts, *Exp. Neurol.* 149, 230–236, 1998.

[144] Emerich, D.F., Winn, S.R., Hantraye, P.M., Peschanski, M., Chen, E.Y., Chu, Y., McDermott, P., Baetge, E.E., and Kordower, J.H., Protective effect of encapsulated cells producing neurotrophic factor CNTF in a monkey model of Huntington's disease, *Nature* 386, 395–399, 1997.

[145] Tan, S.A., Deglon, N., Zurn, A.D., Baetge, E.E., Bamber, B., Kato, A.C., and Aebischer, P., Rescue of motoneurons from axotomy-induced cell death by polymer encapsulated cells genetically engineered to release CNTF, *Cell Transplant* 5, 577–587, 1996.

[146] Maysinger, D. and Morinville, A., Drug delivery to the nervous system, *Trends Biotechnol.* 15, 410–418, 1997.

[147] Guenard, V., Kleitman, N., Morrissey, T.K., Bunge, R.P., and Aebischer, P., Syngeneic Schwann-cells derived from adult nerves seeded in semipermeable guidance channels enhance peripheral-nerve regeneration, *J. Neurosci.* 12, 3310–3320, 1992.

[148] Xu, X.M., Guenard, V., Kleitman, N., and Bunge, M.B., Axonal regeneration into Schwann cell-seeded guidance channels grafted into transected adult-rat spinal-cord, *J. Comp. Neurol.* 351, 145–160, 1995.

[149] Oudega, M., Gautier, S.E., Chapon, P., Fragoso, M., Bates, M.L., Parel, J.M., and Bunge, M.B., Axonal regeneration into Schwann cell grafts within resorbable poly(alpha-hydroxyacid) guidance channels in the adult rat spinal cord, *Biomaterials* 22, 1125–1136, 2001.

[150] Bunge, M.B., Bridging areas of injury in the spinal cord, *Neuroscientist* 7, 325–339, 2001.

[151] Franklin, R.J.M. and Barnett, S.C., Do olfactory glia have advantages over Schwann cells for CNS repair? *J. Neurosci. Res.* 50, 665–672, 1997.

[152] Hadlock, T.A., Sundback, C.A., Hunter, D.A., Vacanti, J.P., and Cheney, M.L., A new artificial nerve graft containing rolled Schwann cell monolayers, *Microsurgery* 21, 96–101, 2001.

[153] Franklin, R.J.M., Gilson, J.M., Franceschini, I.A., and Barnett, S.C., Schwann cell-like myelination following transplantation of an olfactory bulb-ensheathing cell line into areas of demyelination in the adult CNS, *Glia* 17, 217–224, 1996.

[154] Imaizumi, T., Lankford, K.L., Waxman, S.G., Greer, C.A., and Kocsis, J.D., Transplanted olfactory ensheathing cells remyelinate and enhance axonal conduction in the demyelinated dorsal columns of the rat spinal cord, *J. Neurosci.* 18, 6176–6185, 1998.

[155] Ramon-Cueto, A. and Nietosampedro, M., Regeneration into the spinal-cord of transected dorsal-root axons is promoted by ensheathing glia transplants, *Exp. Neurol.* 127, 232–244, 1994.

[156] Ramon-Cueto, A., Plant, G.W., Avila, J., and Bunge, M.B., Long-distance axonal regeneration in the transected adult rat spinal cord is promoted by olfactory ensheathing glia transplants, *J. Neurosci.* 18, 3803–3815, 1998.

[157] Navarro, X., Valero, A., Gudino, G., Fores, J., Rodriguez, F.J., Verdu, E., Pascual, R., Cuadras, J., and Nieto-Sampedro, M., Ensheathing glia transplants promote dorsal root regeneration and spinal reflex restitution after multiple lumbar rhizotomy, *Ann. Neurol.* 45, 207–215, 1999.

[158] Verdu, E., Navarro, X., Gudino-Cabrera, G., Rodriguez, F.J., Ceballos, D., Valero, A., and Nieto-Sampedro, M., Olfactory bulb ensheathing cells enhance peripheral nerve regeneration, *Neuroreport* 10, 1097–1101, 1999.

[159] Temple, S., The development of neural stem cells, *Nature* 414, 112–117, 2001.

[160] Lowenstein, D.H. and Parent, J.M., Brain, heal thyself, *Science* 283, 1126–1127, 1999.

[161] Lavik, E., Teng, Y.D., Zurakowski, D., Qu, X., Snyder, E., and Langer, R., Functional recovery following spinal cord hemisection mediated by a unique polymer scaffold seeded with neural stem cells, *Materials Research Society Symposium Proceedings* 662 (Biomaterials for Drug Delivery and Tissue Engineering), OO1.2/1-OO1.2/5, 2001.

[162] Thomson, J.A., Itskovitz-Eldor, J., Shapiro, S.S., Waknitz, M.A., Swiergiel, J.J., Marshall, V.S., and Jones, J.M., Embryonic stem cell lines derived from human blastocysts, *Science* 282, 1145–1147, 1998.

[163] Bianco, P. and Robey, P.G., Stem cells in tissue engineering, *Nature* 414, 118–121, 2001.

20

Gene Therapy and Tissue Engineering Based on Muscle-Derived Stem Cells: Potential for Musculoskeletal Tissue Regeneration and Repair

Johnny Huard
Baohong Cao
Yong Li
Hairong Peng
University of Pittsburgh

20.1 Introduction

Recent discoveries in cell biology have significantly advanced our knowledge of disease mechanisms. Newly developed techniques, coupled with advances in cell biology and polymer chemistry, are enabling the evolvement of novel tissue engineering approaches for the treatment of various disorders. Surgeons, who have traditionally used the tools of excision and reconstruction to treat patients, may now act as surgical gardeners who create microenvironments conducive to tissue regeneration. Our research team has isolated a novel population of muscle-derived stem cells (MDSCs) that display enhanced regenerative capability due to their multipotency, self-renewal ability, and immune-privileged behavior. This paper presents evidence supporting the existence of MDSCs and reviews some current MDSC-based gene therapy and tissue engineering applications designed to improve the healing of various musculoskeletal tissues, including skeletal muscle, bone, and intra-articular tissues. This review updates readers on the foundational principles and current advances in muscle-based gene therapy and tissue engineering for the musculoskeletal system.

20.1.1 Tissue Engineering

"Tissue engineering" refers to the science of creating living tissue to replace, repair, or augment diseased tissue. The engineered tissue may be created *in vitro* and subsequently implanted into the patient or the tissue may be created entirely *in vivo*. Regardless of the technique, tissue engineering requires at least three components: a growth-inducing stimulus (induction), a scaffold to support tissue formation (conduction), and responsive cells (production). The mechanical stimulation that occurs after the appropriate tissue engineering constructs have been introduced also strongly influences the remodeling of the newly regenerated tissue.

20.1.2 Growth Factors and Gene Therapy

"Growth factors" are soluble proteins that promote cell division, maturation, and differentiation. Recombinant DNA techniques have increased our understanding of the function of these proteins, which influence various tissue regeneration and repair processes. These techniques also have enabled the production of large quantities of growth factors for use in experimental investigations. Currently, a great deal of research is focused on the identification of optimal growth factors for particular applications. After identifying the ideal growth factor, scientists must determine the optimal method for delivering the growth factor within the surgical microenvironment. Direct application of a particular recombinant protein is the most straightforward means of delivering growth factors *in situ*. Research conducted during the 36 years since the discovery of bone morphogenetic proteins (BMPs) has demonstrated that one of the most important factors for *in vivo* healing is the presence of high, sustained doses of growth factors. Most growth factor proteins have half-lives of minutes and are cleared rapidly by the bloodstream. Even when the protein is bound to a collagen scaffold, its half-life is limited to several days. Although microgram doses are sufficient for *in vitro* manipulation of cells, milligram doses typically are required for *in vivo* regional regeneration. Researchers thus have turned to regional gene therapy for the sustained delivery of a growth factor to a local site *in vivo*.

One common gene therapy approach involves the transfer of a gene encoding a desired protein into cells via viral or nonviral vectors. The transduced cells subsequently secrete the desired protein into the microenvironment. Gene therapy generally assumes one of two forms: *in vivo* or *ex vivo* (indirect). The *in vivo* approach, which involves the direct injection or implantation of a gene-carrying vector into a recipient or patient, is attractive due to its technical simplicity. However, when using *in vivo* gene-therapy techniques it is impossible to perform *in vitro* safety testing on the transduced cells. In the *ex vivo* approach, cells are isolated from a tissue biopsy, expanded, and then transfected or transduced *in vitro*. Before being introduced into the recipient or patient, the genetically altered cells can be tested *in vitro* to assess the efficiency of gene transfer or to identify any abnormal behavior.

Clinical application of a particular gene therapy or tissue engineering technique requires identification of the protein required for proper tissue function, the gene that expresses this protein, and a reproducible method by which to deliver the protein to the diseased or injured tissue in sufficient quantities with adequate persistence. During the last few years, muscle cells have emerged as promising vehicles for gene therapy and tissue engineering in the musculoskeletal system. Several attributes make muscle an ideal tissue for tissue engineering such as, (1) Obtaining a muscle biopsy or tissue harvest involves a simple surgical procedure that can be repeated without compromising patients' health; (2) Muscle-derived cells tolerate *ex vivo* manipulation well and it is possible to obtain a large number of muscle-derived cells rather quickly (within 1 to 2 weeks) via expansion [1,2]; (3) After reinjection *in vivo*, these cells naturally fuse to form permanent postmitotic myotubes that allow for long-term expression of the desired protein; (4) Muscle cells can be transduced easily by a variety of viral vectors [3–6]; and (5) Muscle tissue contains a population of muscle stem cells that display an enhanced regenerative capacity when used in tissue engineering applications [1,2,7,8].

Because of the aforementioned characteristics, muscle cells have been used extensively as vehicles in gene therapy protocols for various muscle-related diseases, including Duchenne muscular dystrophy (DMD) [9–12]. Researchers also have used these cells as vehicles in many nonmuscle-related gene-therapy applications, such as the transfer and expression of factor IX for hemophilia B [13]; systemic delivery of human growth hormone for growth retardation [14]; gene delivery of human adenosine deaminase for the adenosine deaminase deficiency syndrome [15]; gene transfer of human proinsulin for diabetes mellitus [16]; expression of tyrosine hydroxylase for Parkinson's disease [17]; expression of FasL to prevent immunorejection of pancreatic islet cell transplants [18]; and injection of muscle cells into the joint (i.e., the meniscus, synovium, and ligament) for the treatment for arthritis and improvement of intra-articular tissue healing (see the sections that follow in this chapter) [19–21].

20.1.3 Isolation of Muscle-Derived Stem Cells

Muscle is composed of a heterogeneous population of cell types. Skeletal muscle contains satellite cells, which are resting, mononucleated, myogenic precursor cells capable of fusing to form postmitotic, multinucleated myotubes and myofibers. Skeletal muscle also contains stem cells. These unique cells are identifiable based on their capacity for protracted renewal; production of daughter cells, which may proceed toward lineage commitment and terminal differentiation; multipotency; and the ability to exhibit such behaviors throughout life [7]. Accordingly, muscle-derived stem cells produce populations of daughter cells that can undergo terminal differentiation into muscle cells, some of which subsequently can differentiate into myofibers while others remain undifferentiated [22]. These stem cells do not express markers of mature muscle tissue, but are capable of prolonged production of progeny and are multipotent [7,8,23–25]. Of the various cells that constitute any given postmitotic muscle sample, approximately 1% are committed satellite cells; the percentage of muscle-derived stem cells is substantially lower — roughly 1% of the number of satellite cells [7,8].

Researchers have used a variety of techniques to isolate stem cells from skeletal muscle digest. These cells can differentiate into myogenic, hematopoietic, osteogenic, adipogenic, chondrogenic, neural, and endothelial lineages [7,26–33]. Importantly, these stem cells also appear to display distinct marker profiles that distinguish them from satellite cells. Expression profiles for satellite cells vary depending on the activated state and the degree of differentiation. In our laboratory, we have used the preplate technique to isolate a novel population of highly purified MDSCs [7,8]. These cells express both early myogenic markers (e.g., desmin, c-met, MNF, and Bcl-2) and stem cell markers (e.g., Sca-1, Flk-1, and CD34) [7,8]. Because these cells can differentiate into various lineages (mesodermal, endodermal, and ectodermal) *in vitro* and *in vivo* and display an enhanced ability to regenerate various tissues, they constitute ideal cellular vehicles for gene therapy and tissue engineering applications. During the last few years, we have investigated the use of this technology to improve the healing of different components of the musculoskeletal system, including skeletal muscle, bone, and various intra-articular structures.

20.2 Skeletal Muscle

20.2.1 Muscle Disease: Duchenne Muscular Dystrophy

Patients with DMD lack the dystrophin protein in the sarcolemma of the muscle fiber; its absence disrupts the integration of the cytoskeleton with the extracellular matrix, a process normally controlled by the dystrophin associated proteins (DAPs), and consequently renders muscle fibers more susceptible to damage after contraction. Intensive efforts have been made to develop an effective cell therapy for DMD [9–12]. Although myoblast transplantation can transiently deliver dystrophin and improve the strength of injected dystrophic muscle, this approach is hindered by immune rejection, poor cellular survival rates, and limited dissemination of the injected cells. Animal and human clinical trial results have demonstrated that myoblast transplantation, although feasible, is rather inefficient [10–12]. In animal experiments, immunodeficient animals and immunosuppressive regimens, preirradiation of the injected muscle, or myonecrotic agents have been used extensively to improve the efficiency of this technique [34–43]. Although these approaches have ameliorated the restoration of dystrophin in mdx mice, success remains quite limited. More importantly, these attempts to improve myoblast transplantation in mdx mice are not applicable to human DMD patients and are therefore clinically irrelevant.

Our research team has used a modified preplate technique to isolate a novel population of MDSCs that display an improved transplantation capacity when injected into skeletal muscle [8]. These cells can be expanded *in vitro* for more than 30 passages while preserving their phenotype (Sca-1[+], CD34[low/−], c-kit[−], CD45[−]) and their ability to differentiate into various lineages both *in vitro* and *in vivo*. The transplantation of these MDSCs, in contrast to transplantation of other myogenic cells such as satellite cells, improved the efficiency of muscle regeneration and dystrophin delivery to dystrophic muscle in mice. The number of dystrophin-positive myofibers observed 90 days after injection of these MDSCs was not significantly different than the high number observed 10 days after transplantation, despite the fact that the mdx mice used as recipients in this experiment were not immunosuppressed and the injected muscles were not preirradiated or injured with a myonecrotic agent. Examination at 30 and 90 days after cell transplantation revealed that the injection of these MDSCs resulted in ten times more dystrophin-positive myofibers in the injected muscle than did the transplantation of satellite cells, although the same number of cells was injected in both groups. The ability of these MDSCs to proliferate *in vivo* for an extended period of time — combined with their strong capacity for self-renewal, their multipotency, and their immune-privileged behavior — reveals, at least in part, a basis for the benefits associated with their use in cell transplantation [8].

Although the above results show that allogeneic transplantation of normal MDSCs can restore dystrophin within dystrophic animals, autologous transplantation of MDSCs genetically engineered to express dystrophin may be more suitable for DMD patients. Our research group currently is investigating such autologous transplantation of genetically engineered MDSCs to determine the feasibility of this technology.

20.2.2 Sports-Related Muscle Injuries

Muscle injuries, particularly pulls and strains, pose a challenging problem in traumatology and are among the most common and most frequently disabling injuries to afflict athletes [44,45]. Although injured muscles can heal, such healing occurs very slowly and often results in incomplete functional recovery [44,45]. The regeneration initiated in injured muscle shortly after injury is inefficient and hindered by fibrosis, that is, scar tissue formation [44,45]. The scar tissue that often replaces the damaged myofibers is a potential contributing factor in the tendency of strains to recur.

We have identified various growth factors with the ability to enhance muscle regeneration by increasing myoblast proliferation and differentiation. The delivery of these growth factors via genetically engineered muscle cells improves the regeneration of injured muscle, but fibrosis still impedes full recovery [46–48]. The overexpression of transforming growth factor (TGF)-β1 in various injured tissues has been identified as the major cause of fibrosis in animals and humans [49]. We have observed that TGF-β1 plays a central role in skeletal muscle fibrosis [50] and, more importantly, that the use of antifibrotic agents that inactivate this molecule (e.g., decorin, suramin, γIFN, and relaxin) can reduce muscle fibrosis and consequently improve muscle healing to near-complete recovery levels [51–53].

The overall goal of our research on sports-related muscle injuries is the development of biological approaches to improve muscle healing after common muscle injuries. Because the development of muscle fibrosis has greatly hindered our attempts to improve muscle healing by enhancing muscle regeneration, the prevention of such fibrosis has become the major emphasis of our ongoing research. Although numerous studies have demonstrated the involvement of TGF-β1 in fibrosis in various tissues, very few have examined the significance of fibrosis or the role played by TGF-β1 in the scarring process in injured skeletal muscle. As described above, our recent data show that TGF-β1 plays a central role in skeletal muscle fibrosis and that the use of antifibrotic agents (e.g., decorin, suramin, γIFN, and relaxin) to inactivate this molecule can reduce muscle fibrosis and consequently improve muscle healing to near-complete recovery levels. We believe that the ultimate approach to improving muscle healing after sports-related muscle injuries may involve the transplantation of muscle cells genetically engineered to express antifibrotic agents in order to block scar tissue formation while promoting muscle regeneration.

20.3   Bone Healing

20.3.1   The Problem of Bone Healing

Fracture healing constitutes a fundamental problem in orthopaedic surgery. Although the majority of fractures heal well, difficulty in the form of delayed healing or nonunions can be devastating. Fractures in anatomically compromised locations (e.g., the talar neck or scaphoid), with inadequate fixation or infection, resulting from high-energy-type injuries with soft tissue stripping or segmental bone loss, or occurring in poor bone stock (osteoporosis) often result in delayed healing. According to a recent review, 5 to 10% of the 5.6 million fractures that occur annually in the United States exhibit delayed or impaired healing [54]. Treatment options for the orthopaedic surgeon confronted with these complex fractures, particularly fractures involving segmental bone loss, include bone autograft, vascularized bone grafting, allograft supplemented with osteogenic proteins, bone transport, or amputation [55,56]. Unfortunately, patients generally must endure a lengthy recovery period, numerous procedures, potential donor site morbidity, and, often, a less-than-satisfactory end result [57]. Similar difficulties are associated with closure of the residual bony defects in craniofacial reconstruction and oncology or trauma surgery. Consequently, research directed toward improving fracture treatment and the development of optimal bone substitutes continues to garner intense interest.

20.3.2   The Use of Muscle-Derived Cells to Improve Bone Healing

We have demonstrated that a population of MDSCs isolated from mouse skeletal muscle produced alkaline phosphatase in a dose-dependent manner after stimulation with recombinant human BMP-2 and BMP-4 [1,2,7,58,59]. After being genetically engineered to express the osteogenic proteins BMP-2 and BMP-4 and injected into skeletal muscle, these mouse muscle-derived cells induced and participated in ectopic bone formation [1,2,7,58,59]. These MDSCs were also able to differentiate toward the osteogenic lineage and consequently improve bone healing in calvarial defects in both immunodeficient mice and immunocompetent mice [1,2,7,58,59].

Although angiogenesis is integral to normal bone development, it remains largely unknown whether bone regeneration can be further improved by the use of angiogenic factors, either alone or in combination with BMPs—the most promising growth factors for inducing bone formation. To address this question, we transduced MDSCs to express human BMP-4, vascular endothelial growth factor (VEGF), or both growth factors. VEGF significantly improved the efficacy of BMP-4-elicited bone formation and regeneration by enhancing angiogenesis, but VEGF alone did not improve bone regeneration. We have also characterized the mechanism by which VEGF accelerates the bone formation mediated by genetically engineered

MDSCs expressing BMP-4. In addition to improved vasculature, we observed increased mesenchymal cell infiltration and cartilage formation, decreased cell apoptosis at the injured site, earlier cartilage resorption, and a subsequent increase in mineralized bone formation within the regenerated bone that had been treated with the combination of BMP-4 and VEGF compared to that treated with BMP-4 alone [58]. Thus VEGF appears to enhance BMP-4-induced endochondral bone formation by influencing steps before and after cartilage formation. Intriguingly, we have also demonstrated that the beneficial effect of VEGF on the bone healing elicited by BMP-4 requires proper dosing, with high doses of VEGF leading to detrimental effects on bone healing [58]. These findings open a new avenue by which to improve bone repair based on the synergistic effects of osteogenic and angiogenic factors delivered via MDSCs.

20.3.3 Can We Isolate Human Muscle-Derived Stem Cells That Display Similar Osteogenic Behavior?

To translate this research to the clinical setting, our research group is pursuing the isolation of a similar population of MDSCs from human skeletal muscle. We have recently evaluated whether muscle-derived cells isolated from the skeletal muscle of humans (~50 years of age) via the preplate technique can be used in gene therapy and tissue engineering applications to support bone regeneration and repair. We have explored the use of primary human muscle-derived cells (isolated via the preplate technique) for *ex vivo* gene therapy to deliver BMP-2 and produce bone *in vivo*. Our findings indicate that certain cells isolated from a primary cell culture taken from human skeletal muscle responded to BMP-2 by differentiating into osteogenic cells, which suggests that these cells possessed osteocompetence [60]. Using *ex vivo* gene transfer, we were able to transduce these cells to secrete BMP-2 at levels sufficient to induce radiographically detectable ectopic bone formation within skeletal muscle [60]. Histological assessment of the injected, transduced, human muscle-derived cells suggests that the cells may have responded to the secreted BMP-2 in an autocrine fashion by becoming osteoblasts, thereby contributing to the induction of bone formation [60].

We also have used adenoviral and retroviral constructs to genetically engineer these freshly isolated human skeletal muscle cells to express human BMP-2. We implanted these cells into nonhealing bone defects (skull defects) in severe combined immunodeficient (SCID) mice and monitored the closure of the defect grossly and histologically. Mice implanted with BMP-2–producing human muscle-derived cells displayed full closure of the defect by 4 to 8 weeks after transplantation [61]. Remodeling of the newly formed bone was histologically evident during the 4 to 8 week period [61]. Analysis by fluorescent *in situ* hybridization (FISH) revealed a small fraction of the injected human muscle-derived cells within the newly formed bone in the location where osteocytes normally reside [61]. These results indicate that genetically engineered human muscle-derived cells enhance bone healing primarily by delivering BMP-2, although a small fraction of the cells seems to differentiate into osteogenic cells. We are investigating whether this small fraction of human muscle cells is the human counterpart of the mouse MDSCs that display osteogenic behavior.

20.3.4 Can We Regulate the Expression of Osteogenic Proteins and the Resultant Bone Formation?

Regulated therapeutic gene expression has become increasingly important to the success of various gene therapy applications — particularly stem cell-based ones — due to the potential long-term persistence of genetically engineered cells within the body. To develop a system by which to regulate the expression of osteogenic proteins, we first investigated the use of antagonist proteins to limit bone formation. Heterotopic ossification (HO) of muscles, tendons, and ligaments is a common problem encountered by orthopaedic surgeons. We evaluated the ability of noggin, a BMP antagonist, to inhibit HO in various mouse models. First, we developed a retroviral vector carrying the gene encoding human noggin and used this vector to transduce MDSCs. MDSCs transduced with BMP-4 were implanted into both hind limbs of mice along with 100,000, 500,000, or 1,000,000 noggin-expressing MDSCs (treated limb) or with an equivalent number of nontransduced MDSCs (control limb). The mice were sacrificed and radiographed 4 weeks after implantation to look for evidence of HO. In a second set of experiments, 80 mg of human demineralized bone matrix (DBM) was implanted into the hind limbs of SCID mice along with 100,000, 500,000 or 1,000,000 noggin-expressing MDSCs (treated limb) or nontransduced MDSCs (control limb). The mice were sacrificed and radiographed 8 weeks after cell implantation. In a final set of experiments, immunocompetent mice underwent bilateral Achilles tenotomy along with the implantation of 1,000,000 noggin-expressing MDSCs (treated limb) or nontransduced MDSCs (control limb). The mice were sacrificed and radiographed 10 weeks after surgery.

Our *in vitro* results showed that the MDSCs expressed noggin at a high level (280 ng/million cells/24 h) [62]. Four weeks after implantation, we found that the BMP-4-expressing MDSCs combined with the three varying doses of noggin-expressing MDSCs had formed 53, 74, and 99% less HO in a dose-dependent manner ($p < .05$) [62]. Similarly, DBM formed 91, 99, and 99% less HO when combined with the three varying doses of noggin-expressing MDSCs ($p < .05$) [62]. Additionally, all 11 animals that underwent Achilles tenotomy developed HO at the site of injury in the control limbs, whereas the limbs treated with the noggin-expressing MDSCs exhibited 84% less HO formation; 8 of the 11 animals exhibited no radiographic evidence of HO formation ($p < .05$) [62]. These findings suggest that the delivery of noggin mediated by MDSCs can inhibit HO in three different animal models. Gene therapy to deliver noggin may consequently serve as a powerful method to inhibit HO and perhaps regulate the amount of bone formation mediated by osteogenic proteins delivered by MDSCs [62].

We have also developed a retroviral vector that appears to be suitable for regulated gene therapy to improve bone healing. To date, one of the most intensively studied inducible gene expression systems comprises the tetracycline (tet)-controlled gene and its corresponding synthetic promoter [63]; researchers have used this system for either tet-on (gene expression activated by the presence of tet) or tet-off (gene expression activated in the absence of tet) applications. Scientists have developed retroviral vectors containing tet-on- or tet-off-controlled therapeutic genes with varying degrees of success [64]. We tried to optimize the most promising design to date — the self-inactivating (SI) tet-on retroviral vector — to develop a retroviral vector suitable for regulated gene therapy to improve bone healing. After extensive investigation, we identified and optimized a retroviral vector that confers a high level of inducible transgene expression (i.e., BMP-4 expression) to transduced, muscle-derived stem cells (MDSCs) [65]. This novel retroviral vector also enables us to regulate the amount of bone formation elicited by the transduced cells after transplantation [65]. We believe that the findings generated by this study could be translated to a wide variety of other gene-therapy applications that require stringent regulation of a transgene's therapeutic effect, such as the codelivery of inducible VEGF and its specific antagonist to achieve regulated angiogenesis.

20.3.5 Is the Bone Regenerated by BMP-Producing Muscle-Derived Cells Biomechanically Relevant?

We have evaluated the ability of muscle-derived cells transduced with retroBMP4 to heal a long bone defect both structurally and functionally. Primary muscle-derived rat cells were genetically engineered to express BMP-4 and implanted into 7-mm rat femoral defects; muscle-derived cells transduced with retroLacZ were used as the control. Bone healing was monitored via radiography, histology, and biomechanical testing. Our results indicate that 78.6% of the defects treated with muscle-derived cells expressing BMP-4 formed bridging callous by 6 weeks after surgery, and 12 of the 13 femora exhibited radiographically evident union 12 weeks after implantation. Histological analysis at the 12-week time point revealed that the medullary canal of the femur was restored and the cortex was remodeled between the proximal and distal ends of each BMP-4-treated defect. In contrast, nonunions were observed at all tested time points in the defects treated with muscle-derived cells expressing β-galactosidase (a marker gene). The maximum torque-to-failure in the treatment group (BMP-4–producing cells) indicated up to 73% strength of the contralateral intact femur, while the torsional stiffness and energy-to-failure were not significantly different between the treated and the intact limbs. This study demonstrates that retroBMP4-transduced muscle-derived cells can elicit both structural and functional healing of critical sized segmental defects in rat femora [66].

20.4 Articular Disorders

Various articular disorders, including arthritis, chondral and osteochondral defects, meniscal tears, and damaged ligaments, are potential candidates for novel treatment using gene therapy and tissue engineering applications.

20.4.1 Articular Cartilage

Articular cartilage defects and progressive osteoarthritis are among the most frequent and challenging conditions encountered in orthopaedics. In contrast to bone, articular cartilage heals poorly. Current repair techniques include cartilage debridement and resurfacing [67,68], subchondral drilling [69], arthroscopic abrasion, and microfracture. Common to these techniques is the disruption of subchondral bone to allow osteochondral progenitors from the marrow to gain access to and participate in repair of the cartilage defect. The repaired cartilage, however, is structurally inferior to native cartilage and contains significantly less proteoglycan [70]. Newer strategies for cartilage repair seek to provide an alternate source of cells to stimulate healing. Strategies have included the use of autologous chondrocyte transplantation [71] and transplantation of cartilage plugs [72]. Although adult or embryonic chondrocytes may be used for transplantation, the limiting factor is chondrocyte supply. Other researchers have tried to overcome this limitation by generating chondrocytes from mesenchymal stem cells [73]. These cells then are induced to become osteochondral progenitor cells in culture before reimplantation.

Partly because muscle-derived cells are easily accessible, we have explored their capacity for repairing articular defects. The ability of muscle-derived cells to promote cartilage formation/restoration and bone formation through an endochondral pathway (see earlier sections) further justifies the evaluation of their potential use for articular cartilage repair. For these reasons, we have investigated the utilization of transplanted allogeneic muscle-derived cells embedded in collagen gels for the repair of full thickness articular cartilage defects [74]. The results were compared to chondrocyte transplantation and control (nontreated) groups. Although grafted cells were found in the defects only up to 4 weeks after transplantation, histological assessment at 12 and 24 weeks after transplantation revealed that the tissue repaired with muscle-derived cells or chondrocytes displayed comparably higher levels of healing than observed in the control groups [74]. Notably, at 24 weeks the repaired tissues in the muscle-derived cell and chondrocyte groups were composed primarily of type II collagen, indicating the presence of hyaline articular cartilage [74]. These results demonstrated that allogeneic muscle-derived cells could be used as both gene-delivery vehicles and a cell source to promote full-thickness cartilage repair. The implantation of muscle-derived cells appeared to improve the healing of the defects with an efficiency equivalent to that of chondrocyte cartilage transplantation. Muscle-derived cell transplantation is a particularly attractive option given the fact that the muscle biopsy required to obtain muscle cells is less invasive than the arthroscopy required to obtain chondrocytes [74].

20.4.2 Meniscus

Allograft meniscal transplantation is one of the few available treatment options after meniscectomy. Despite acceptable early results, considerable controversy exists due to the subsequent poor graft regeneration, shrinkage, and biomechanical failure of transplanted menisci. Meniscal injuries are promising candidates for treatment via gene therapy and tissue engineering applications. The possible *in vitro* creation of a custom, replacement meniscus by using scaffolds, cells, and gene therapy for subsequent *in vivo* implantation is intriguing. Alternatively, gene therapy could be used to genetically engineer meniscal cells in order to promote the healing of certain injuries. Meniscal cells are amenable to gene transfer of both marker genes and various growth factor genes via either direct or *ex vivo* gene therapy, with gene expression persisting for up to 6 weeks [75]. Successful *ex vivo* gene transfer to the meniscus has been accomplished using either muscle-derived cells [20] or meniscal cells [75]. Identification of the growth factors that best promote meniscal healing, techniques to improve long-term gene expression, and the optimal scaffold required to create de novo menisci is the focus of ongoing research.

We have recently investigated the feasibility of using gene transfer in meniscal allografts in rabbits to eventually deliver specific growth factor genes that may improve the regeneration of meniscal allograft [76]. Four different viral vectors encoding marker genes, including LacZ, luciferase, and green fluorescence protein, were used to investigate viral transduction in 50 lapine menisci for 4 weeks *in vitro*. Subsequently, 16 unilateral meniscus replacements were performed with *ex vivo* retrovirally transduced meniscal allografts; the expression of the LacZ gene was examined histologically at 2, 4, 6, and 8 weeks after transplantation. Gene expression in the superficial cell layers of the menisci was detected for up to 4 weeks *in vitro*, but the level of gene transfer declined over time [76]. The retrovirally transduced cells displayed more persistent transgene expression and penetrated the menisci more deeply after implantation [76]. *In vivo*, declining numbers of β-galactosidase-positive cells were detected in the retrovirally transduced allografts for up to 8 weeks. Transduced cells consistently were found at the meniscosynovial junction of the transplants and in deeper layers of the menisci. There was no evidence of cellular immune response in the transduced transplants. This investigation demonstrates the promise of growth factor delivery in auto- and allografts. We will conduct additional experiments to assess the potential of vectors expressing therapeutic proteins (e.g., growth factors) to improve remodeling and healing of meniscal allografts [76].

20.4.3 Ligaments

Researchers also are investigating the use of gene-therapy techniques for treatment of injured ligaments. LacZ gene transfer to ligaments via either direct or *ex vivo* adenoviral or retroviral approaches has proven feasible in animal models. *Ex vivo* gene transfer to ligaments has been achieved using either ligament fibroblasts or skeletal muscle-derived cells [21]. Many growth factors, such as basic fibroblast growth factor (bFGF), platelet-derived growth factor (PDGF), VEGF, insulin-like growth factor (IGF)-1 and -2, TGF-β, and BMP-12, are believed to possibly play roles in ligament healing. Data suggest that PDGF stimulates cell division and migration, whereas TGF-β and the IGFs promote extracellular matrix synthesis [77]. Use of a viral-liposome conjugate vector for the direct gene transfer of PDGF-β into rat patellar ligaments was found to initially improve angiogenesis and subsequently enhance extracellular matrix synthesis [78].

The integration of tendon grafts typically used for anterior cruciate ligament replacement continues to be unsatisfactory and may be associated with postoperative anterior–posterior laxity. We have evaluated whether BMP-2 gene transfer can improve the integration of semitendinosus tendon grafts at the tendon–bone interface after anterior cruciate ligament reconstruction in rabbits [79]. Anterior cruciate ligament reconstruction using autologous double-bundle semitendinosus tendon grafts was performed in 46 adult rabbits. The semitendinosus tendon grafts were genetically engineered *in vitro* with adenovirus–luciferase, adenovirus–LacZ (AdLacZ), or adenovirus–BMP-2 (AdBMP-2); untreated grafts served as controls. The anterior cruciate ligament grafts were examined histologically 2, 4, 6, and 8 weeks after surgery. In an additional series of experiments, the structural properties of the femur–anterior cruciate ligament graft–tibia complexes were investigated in 10 untreated controls and 10 specimens with AdBMP-2-transduced semitendinosus tendon grafts. The animals were sacrificed 8 weeks after anterior cruciate ligament surgery, and the femur–anterior cruciate ligament graft–tibia complexes were tested under uniaxial tension. The stiffness (N/mm) and ultimate load at failure (N) were determined from load-elongation curves.

Our results indicate that genetically engineered semitendinosus tendon grafts expressed reporter genes and BMP-2 *in vitro* [79]. The AdLacZ-infected anterior cruciate ligament grafts showed two different histological patterns of transduction. Intra-articular, infected cells were mostly aligned along the surface and decreased in number during the 6 weeks after surgery. In the intra-tunnel portions of the anterior cruciate ligament grafts, the number of infected cells did not decrease during the observation period. Moreover, we observed a high number of transduced cells in the deeper layers of the tendons. In the control group, granulation-type tissue at the tendon–bone interface showed progressive reorganization into a dense connective tissue, and subsequent establishment of fibers resembling Sharpey's fibers.

In the specimens receiving AdBMP-2-infected anterior cruciate ligament grafts, a broad zone of newly formed matrix resembling chondro-osteoid was observed at the tendon–bone interface 4 weeks after surgery. This area had increased by 6 weeks after surgery, showing a transition from bone to mineralized and nonmineralized cartilage. In addition, in the BMP-2-treated specimens, the tendon–bone interface in the osseous tunnel was similar to that of a normal anterior cruciate ligament insertion. The stiffness (29.0 ± 7.1 vs. 16.7 ± 8.3 N/mm) and the ultimate load at failure (108.8 ± 50.8 vs. 45.0 ± 18.0 N) were significantly enhanced in the specimens with BMP-2-transduced anterior cruciate ligament grafts when compared with controls. This study demonstrated that BMP-2 gene transfer significantly improves the integration of semitendinosus tendon grafts in the tunnels after anterior cruciate ligament reconstruction in rabbits [79]. Before introducing gene transfer as a therapeutic method in orthopaedics, however, researchers must answer questions regarding safety and regulatory issues.

20.5   Summary

This paper summarizes the current knowledge and most recent achievements in gene therapy and tissue engineering for the musculoskeletal system; many of these advances are attributable to or strongly influenced by the research findings of our laboratory. Although this review focuses on a population of muscle-derived stem cells isolated by our research team, we are not excluding the possible use of alternative cell populations, including stem cells from various sources (e.g., hematopoietic cells, mesenchymal stem cells, or fat-derived cells) for future applications. Bringing tissue engineering technology to clinical fruition, however, will require multidisciplinary collaboration to identify and optimize the appropriate cell type, growth factor, and scaffold construct for each application. Future investigations should elaborate upon the utility of stem cells derived from various adult tissues in autologous, allogeneic, or perhaps xenogeneic settings. Details regarding these cells' proliferative potential, susceptibility to genetic engineering, and immunogenic potential remain to be characterized. Furthermore, researchers must learn to combine stem cells with growth factors in a viable spatiotemporal relationship in order to re-create the *in situ* microenvironment. Much work remains before the regeneration of tissue for healing the musculoskeletal system — a complicated endeavor with vast potential applicability — becomes a clinical reality.

Acknowledgments

The authors wish to thank Marcelle Pellerin, James Cummins, and Brian Gearhart (Children's Hospital of Pittsburgh, Pittsburgh, PA) for technical support and Ryan Sauder (University of Pittsburgh, Pittsburgh, PA) for editorial assistance with the manuscript. This work was supported by grants to Dr. Johnny Huard from the Muscular Dystrophy Association (USA), the National Institutes of Health (NIH P01 AR 45925-01, R01 AR 49684-01, R01 AR 47973-01, R01 DE 13420-01A2), the Orris C. Hirtzel and Beatrice Dewey Hirtzel Memorial Foundation and the William F. and Jean W. Donaldson Chair at Children's Hospital of Pittsburgh, and the Henry J. Mankin Endowed Chair at the University of Pittsburgh.

The authors also want to thank the clinical fellows Chan Y.S., Fukushima K., Shen H.C., Adachi N., and Martinek V. for their contribution to this work.

References

[1] Bosch, P. et al., Osteoprogenitor cells within skeletal muscle. *J. Orthop. Res.* 2000; 18: 933–944.

[2] Musgrave, D.S. et al., *Ex vivo* gene therapy to produce bone using different cell types. *Clin. Orthop.* 2000: 290–305.

[3] Booth, D.K. et al., Myoblast-mediated *ex vivo* gene transfer to mature muscle. *Tissue Eng.* 1997; 3: 125.

[4] Floyd, S.S. Jr. et al., *Ex vivo* gene transfer using adenovirus-mediated full-length dystrophin delivery to dystrophic muscles. *Gene Ther.* 1998; 5: 19–30.

[5] van Deutekom, J.C. et al., Implications of maturation for viral gene delivery to skeletal muscle. *Neuromuscul. Disord.* 1998; 8: 135–148.

[6] van Deutekom, J.C., Hoffman, E.P., and Huard, J., Muscle maturation: implications for gene therapy. *Mol. Med. Today* 1998; 4: 214–220.

[7] Lee, J.Y. et al., Clonal isolation of muscle-derived cells capable of enhancing muscle regeneration and bone healing. *J. Cell Biol.* 2000; 150: 1085–1100.

[8] Qu-Petersen, Z. et al., Identification of a novel population of muscle stem cells in mice: potential for muscle regeneration. *J. Cell Biol.* 2002; 157: 851–864.

[9] Huard, J. et al., Dystrophin expression in myotubes formed by the fusion of normal and dystrophic myoblasts. *Muscle Nerve* 1991; 14: 178–182.

[10] Karpati, G. et al., Dystrophin is expressed in mdx skeletal muscle fibers after normal myoblast implantation. *Am. J. Pathol.* 1989; 135: 27–32.

[11] Partridge, T.A., Invited review: myoblast transfer: a possible therapy for inherited myopathies? *Muscle Nerve* 1991; 14: 197–212.

[12] Tremblay, J.P. et al., Results of a triple blind clinical study of myoblast transplantations without immunosuppressive treatment in young boys with Duchenne muscular dystrophy. *Cell Transplant* 1993; 2: 99–112.

[13] Dai, Y., Roman, M., Naviaux, R.K., and Verma, I.M., Gene therapy via primary myoblasts: long-term expression of factor IX protein following transplantation *in vivo*. *Proc. Natl Acad. Sci. USA* 1992; 89: 10892–10895.

[14] Dhawan, J. et al., Systemic delivery of human growth hormone by injection of genetically engineered myoblasts. *Science* 1991; 254: 1509–1512.

[15] Lynch, C.M. et al., Long-term expression of human adenosine deaminase in vascular smooth muscle cells of rats: a model for gene therapy. *Proc. Natl Acad. Sci. USA* 1992; 89: 1138–1142.

[16] Simonson, G.D., Groskreutz, D.J., Gorman, C.M., and MacDonald, M.J., Synthesis and processing of genetically modified human proinsulin by rat myoblast primary cultures. *Hum. Gene Ther.* 1996; 7: 71–78.

[17] Jiao, S., Gurevich, V., and Wolff, J.A., Long-term correction of rat model of Parkinson's disease by gene therapy. *Nature* 1993; 362: 450–453.

[18] Lau, H.T., Yu, M., Fontana, A., Stoeckert, C.J. Jr., Prevention of islet allograft rejection with engineered myoblasts expressing FasL in mice. *Science* 1996; 273: 109–112.

[19] Day, C.S. et al., Myoblast-mediated gene transfer to the joint. *J. Orthop. Res.* 1997; 15: 894–903.

[20] Kasemkijwattana, C. et al., The use of growth factors, gene therapy and tissue engineering to improve meniscal healing. *Mater. Sci. Eng.* 2000; 13: 19–28.

[21] Menetrey, J. et al., Direct-, fibroblast- and myoblast-mediated gene transfer to the anterior cruciate ligament. *Tissue Eng.* 1999; 5: 435–442.

[22] Baroffio, A. et al., Identification of self-renewing myoblasts in the progeny of single human muscle satellite cells. *Differentiation* 1996; 60: 47–57.

[23] Miller, J.B., Schaefer, L., and Dominov, J.A., Seeking muscle stem cells. *Curr. Top. Dev. Biol.* 1999; 43: 191–219.

[24] Nicolas, J.F., Mathis, L., Bonnerot, C., and Saurin, W., Evidence in the mouse for self-renewing stem cells in the formation of a segmented longitudinal structure, the myotome. *Development* 1996; 122: 2933–2946.

[25] Slack, J.M.W., *From Egg to Embryo*. 2nd ed. New York: Cambridge University Press, 1991.

[26] Cao, B. et al., Muscle stem cells differentiate into haematopoietic lineages but retain myogenic potential. *Nat. Cell Biol.* 2003; 5: 640–646.

[27] Gussoni, E. et al., Dystrophin expression in the mdx mouse restored by stem cell transplantation. *Nature* 1999; 401: 390–394.

[28] Kawada, H. and Ogawa, M., Bone marrow origin of hematopoietic progenitors and stem cells in murine muscle. *Blood* 2001; 98: 2008–2013.

[29] McKinney-Freeman, S.L. et al., Muscle-derived hematopoietic stem cells are hematopoietic in origin. *Proc. Natl Acad. Sci. USA* 2002; 99: 1341–1346.

[30] Torrente, Y. et al., Intraarterial injection of muscle-derived CD34(+)Sca-1(+) stem cells restores dystrophin in mdx mice. *J. Cell Biol.* 2001; 152: 335–348.

[31] Williams, J.T. et al., Cells isolated from adult human skeletal muscle capable of differentiating into multiple mesodermal phenotypes. *Am. Surg.* 1999; 65: 22–26.

[32] Young, H.E. et al., Pluripotent mesenchymal stem cells reside within avian connective tissue matrices. *In Vitro Cell Dev. Biol. Anim.* 1993; 29A: 723–736.

[33] Young, H.E. et al., Human reserve pluripotent mesenchymal stem cells are present in the connective tissues of skeletal muscle and dermis derived from fetal, adult, and geriatric donors. *Anat. Rec.* 2001; 264: 51–62.

[34] Beauchamp, J.R., Morgan, J.E., Pagel, C.N., and Partridge, T.A., Dynamics of myoblast transplantation reveal a discrete minority of precursors with stem cell-like properties as the myogenic source. *J. Cell Biol.* 1999; 144: 1113–1122.

[35] Fan, Y., Maley, M., Beilharz, M., and Grounds, M., Rapid death of injected myoblasts in myoblast transfer therapy. *Muscle Nerve* 1996; 19: 853–860.

[36] Guerette, B. et al., Control of inflammatory damage by anti-LFA-1: increase success of myoblast transplantation. *Cell Transplant* 1997; 6: 101–107.

[37] Huard, J. et al., Gene transfer into skeletal muscles by isogenic myoblasts. *Hum. Gene Ther.* 1994; 5: 949–958.

[38] Huard, J. et al., High efficiency of muscle regeneration after human myoblast clone transplantation in SCID mice. *J. Clin. Invest.* 1994; 93: 586–599.

[39] Kinoshita, I. et al., Very efficient myoblast allotransplantation in mice under FK506 immunosuppression. *Muscle Nerve* 1994; 17: 1407–1415.

[40] Morgan, J.E., Pagel, C.N., Sherratt, T., and Partridge, T.A., Long-term persistence and migration of myogenic cells injected into pre-irradiated muscles of mdx mice. *J. Neurol. Sci.* 1993; 115: 191–200.

[41] Petersen, Z.Q. and Huard, J., The influence of muscle fiber type in myoblast-mediated gene transfer to skeletal muscles. *Cell Transplant* 2000; 9: 503–517.

[42] Qu, Z. et al., Development of approaches to improve cell survival in myoblast transfer therapy. *J. Cell Biol.* 1998; 142: 1257–1267.

[43] Qu, Z. and Huard, J., Matching host muscle and donor myoblasts for myosin heavy chain improves myoblast transfer therapy. *Gene Ther.* 2000; 7: 428–437.

[44] Huard, J., Li, Y., and Fu, F.H., Muscle injuries and repair: current trends in research. *J. Bone Joint Surg. Am.* 2002; 84-A: 822–832.

[45] Li, Y., Cummins, J., and Huard, J., Muscle injury and repair. *Curr. Opin. Orthop.* 2001; 12: 409–415.

[46] Kasemkijwattana, C. et al., Use of growth factors to improve muscle healing after strain injury. *Clin. Orthop.* 2000: 272–285.

[47] Kasemkijwattana, C., Menetrey, J., Day, C.S., and Huard, J., Biologic intervention in muscle healing and regeneration. *Sports Med. Arthrosc. Rev.* 1998; 6: 95–102.

[48] Menetrey, J. et al., Growth factors improve muscle healing *in vivo. J. Bone Joint Surg. Br.* 2000; 82: 131–137.

[49] Border, W.A. and Noble, N.A., Transforming growth factor beta in tissue fibrosis. *N. Engl. J. Med.* 1994; 331: 1286–1292.

[50] Li, Y. and Huard, J., Differentiation of muscle-derived cells into myofibroblasts in injured skeletal muscle. *Am. J. Pathol.* 2002; 161: 895–907.

[51] Chan, Y.S. et al., Antifibrotic effects of suramin in injured skeletal muscle after laceration. *J. Appl. Physiol.* 2003; 95: 771–780.

[52] Foster, W. et al., Gamma interferon as an antifibrosis agent in skeletal muscle. *J. Orthop. Res.* 2003; 21: 798–804.

[53] Fukushima, K. et al., The use of an antifibrosis agent to improve muscle recovery after laceration. *Am. J. Sports Med.* 2001; 29: 394–402.

[54] Einhorn, T.A., Enhancement of fracture-healing. *J. Bone Joint Surg. Am.* 1995; 77: 940–956.

[55] Fiebel, R.J. et al., Simultaneous free-tissue transfer and Ilizarov distraction osteosynthesis in lower extremity salvage: case report and review of the literature. *J. Trauma* 1994; 37: 322–327.

[56] Prokuski, L.J. and Marsh, J.L., Segmental bone deficiency after acute trauma. The role of bone transport. *Orthop. Clin. North Am.* 1994; 25: 753–763.

[57] De Oliveira, J.C., Bone grafts and chronic osteomyelitis. *J. Bone Joint Surg. Br.* 1971; 53: 672–683.

[58] Peng, H. et al., Synergistic enhancement of bone formation and healing by stem cell-expressed VEGF and bone morphogenetic protein-4. *J. Clin. Invest.* 2002; 110: 751–759.

[59] Wright, V. et al., BMP4-expressing muscle-derived stem cells differentiate into osteogenic lineage and improve bone healing in immunocompetent mice. *Mol. Ther.* 2002; 6: 169–178.

[60] Musgrave, D.S. et al., Human skeletal muscle cells in *ex vivo* gene therapy to deliver bone morphogenetic protein-2. *J. Bone Joint Surg. Br.* 2002; 84: 120–127.

[61] Lee, J.Y. et al., Enhancement of bone healing based on *ex vivo* gene therapy using human muscle-derived cells expressing bone morphogenetic protein 2. *Hum. Gene Ther.* 2002; 13: 1201–1211.

[62] Hannallah, D. et al., Retroviral delivery of Noggin inhibits the formation of heterotopic ossification induced by BMP-4, demineralized bone matrix, and trauma in an animal model. *J. Bone Joint Surg. Am.* 2004; 86-A: 80–91.

[63] Gossen, M. and Bujard, H., Tight control of gene expression in mammalian cells by tetracycline-responsive promoters. *Proc. Natl Acad. Sci. USA* 1992; 89: 5547–5551.

[64] Iida, A., Chen, S.T., Friedmann, T., and Yee, J.K., Inducible gene expression by retrovirus-mediated transfer of a modified tetracycline-regulated system. *J. Virol.* 1996; 70: 6054–6059.

[65] Peng, H. et al., Converse relationship between *in vitro* osteogenic differentiation and *in vivo* bone healing elicited by different populations of muscle-derived cells genetically engineered to express BMP4. *J. Bone Miner. Res.* 2004; 19: 630–641.

[66] Shen, H.C. et al., Structural and functional healing of critical-size segmental bone defects by transduced muscle-derived cells expressing BMP4. *J. Gene Med.* 2004; 6: 984–991

[67] Insall, J., The Pridie debridement operation for osteoarthritis of the knee. *Clin. Orthop.* 1974; 101: 61–67.

[68] Pridie, K., A method of resurfacing osteoarthritic knee joints. *J. Bone Joint Surg. Br.* 1959; 41: 618–619.

[69] Mitchell, N. and Shepard, N., The resurfacing of adult rabbit articular cartilage by multiple perforations through the subchondral bone. *J. Bone Joint Surg. Am.* 1976; 58: 230–233.

[70] Shapiro, F., Koide, S., and Glimcher, M.J., Cell origin and differentiation in the repair of full-thickness defects of articular cartilage. *J. Bone Joint Surg. Am.* 1993; 75: 532–553.

[71] Brittberg, M. et al., Treatment of deep cartilage defects in the knee with autologous chondrocyte transplantation. *N. Engl. J. Med.* 1994; 331: 889–895.

[72] Hangody, L. et al., Mosaicplasty for the treatment of articular cartilage defects: application in clinical practice. *Orthopedics* 1998; 21: 751–756.

[73] Wakitani, S. et al., Repair of rabbit articular surfaces with allograft chondrocytes embedded in collagen gel. *J. Bone Joint Surg. Br.* 1989; 71: 74–80.

[74] Adachi, N. et al., Muscle derived, cell based *ex vivo* gene therapy for treatment of full thickness articular cartilage defects. *J. Rheumatol.* 2002; 29: 1920–1930.

[75] Goto, H. et al., Transfer of lacZ marker gene to the meniscus. *J. Bone Joint Surg. Am.* 1999; 81: 918–925.

[76] Martinek, V. et al., Genetic engineering of meniscal allografts. *Tissue Eng.* 2002; 8: 107–117.

[77] Evans, C.H., Cytokines and the role they play in the healing of ligaments and tendons. *Sports Med.* 1999; 28: 71–76.

[78] Nakamura, N. et al., Early biological effect of *in vivo* gene transfer of platelet-derived growth factor (PDGF)-B into healing patellar ligament. *Gene Ther.* 1998; 5: 1165–1170.

[79] Martinek, V. et al., Enhancement of tendon–bone integration of anterior cruciate ligament grafts with bone morphogenetic protein-2 gene transfer: a histological and biomechanical study. *J. Bone Joint Surg. Am.* 2002; 84-A: 1123–1131.

21

Tissue Engineering Applications — Bone

Ayse B. Celil
Scott Guelcher
Jeffrey O. Hollinger
Carnegie Mellon University

Michael Miller
University of Texas

21.1 Biology of the Bone

Bone is a complex living tissue that provides internal support for all higher vertebrates. It develops by osteogenesis, the process of ossification, starting out as a highly specialized form of connective tissue. Two major players in the formation of bone are the bone cells called osteoblasts (bone-forming) and the osteoclasts (bone-resorbing). During the process of ossification, osteoblasts secrete type I collagen, in addition to many noncollagenous proteins such as osteocalcin, bone sialoprotein, and osteopontin. Osteoblast-secreted extracellular matrix may initially be amorphous and noncrystalline, but it gradually transforms into more crystalline forms [1]. Mineralization is a process of bone formation promoted by osteoblasts and is thought to be initiated by the matrix vesicles that bud from the plasma membrane [2] of osteoblasts to create an environment for the concentration of calcium and phosphate, allowing crystallization. Collagen serves as a template and may also initiate and propagate mineralization independent of the matrix vesicles [3,4]. Eventually, some osteoblasts are surrounded by the bone matrix that they help to form and are called osteocytes. Despite their location, osteocytes are not metabolically inactive; they dissolve and resorb some bone mineral though osteolysis [5]. Bone resorption is in fact the primary function of another bone cell, the osteoclast, which can also digest calcified cartilage and is then called the chondroclast. Formation by the osteoblasts and resorption by the osteoclasts maintains bone in constant renewal as a dynamic tissue.

Bone formation in the developing embryo occurs by two developmental processes: intramembranous and endochondral ossification. Intramembranous ossification is the direct differentiation of osteoblasts from mesenchymal cells. Several craniofacial bones and parts of the mandible and clavicle develop by intramembranous ossification [6]. Most other bones are formed by endochondral ossification. In this process of bone formation, mesenchymal cells condense and differentiate into cartilage. The cartilage anlage matures, undergoes hypertrophy, mineralizes, and subsequently becomes invaded by blood vessels as well as osteoprogenitor cells [7,8].

Woven and lamellar bone are the two types of bone observed at the microscopic level. Woven bone is a disoriented arrangement of collagen fibers [9] and is the first bone formed in the embryo. After birth, woven bone is gradually replaced by lamellar bone except in a few places (e.g., tooth sockets). Lamellar bone is highly organized and has collagen bundles oriented in the same direction [9]. Structural organization of woven and lamellar bone is in two categories:

1. Trabecular bone (spongy, cancellous)
2. Cortical bone (compact)

The inherent architecture of bone is influenced by the mechanical stresses. The structure–function relationship of bone was described in 1892 by Wolff's law [10], which states that mechanical stress is responsible for the architecture of the bone. Bone undergoes adaptive changes in response to the demands of mechanical stress from the environment [5]. It has been shown that bone formation is upregulated in response to increased load application [11,12] and that bone tissue is removed from the skeleton in response to reduced loading as in microgravity [13,14]. The mechanical signals to the osteoprogenitor cells of the bone are also important in determining the formation of the tissue that forms during development and healing [14]. Several noninvasive techniques are currently used to evaluate bone mechanical properties in the clinic such as quantitative computer tomography (QCT), magnetic resonance imaging (MRI), ultrasound, and dual-energy x-ray absorptiometry (DEXA) [15–17]. In a laboratory setting several bone marker genes are used to evaluate differentiation into the osteoblast phenotype and formation of bone. Among these marker genes are: alkaline phosphatase (alp), type I collagen (type I col), osteopontin (opn), osteocalcin (ocn) and bone sialoprotein (bsp). Alp and type I col are induced earlier in differentiation, whereas ocn and bsp are considered to be late markers. Techniques such as conventional and quantitative real-time PCR and northern blot allow the detection of these genes at the RNA level. For genome wide analysis microarray technology can be used for the analysis of a large sample population.

Alkaline phosphatase (Alp) is an enzyme that catalyzes the hydrolysis of phosphate esters at an alkaline pH. It has several different isoforms: tissue nonspecific, placental, and intestinal [18]. Three isoforms exist for the tissue nonspecific isoenzyme including bone, liver, and kidney. The skeletal isoform is a glycoprotein on the cell membrane of osteoblasts [19]. Alp is important in bone matrix mineralization and its activity is recognized as an indicator of osteoblast function.

Type I collagen (type I col) is the major organic component of bone matrix. Collagen has a basic structure of repeating primary amino acid sequence of -gly-X-Y. Osteoblasts synthesize type I collagen molecules to form fibrils, which give the characteristic cross-banding pattern. Collagen secretion by osteoblasts promotes their differentiation into a more mature phenotype [20].

Osteopontin (Opn) is a noncollagenous, acidic, sialic acid-rich phosphorylated glycoprotein; it binds to hydroxyapatite and is abundant in the mineral matrix of bones [21].

Osteocalcin (Ocn) is the most abundant noncollagenous protein in bone, comprising approximately 2% of total protein in the human body. It is important in bone metabolism and its recently revealed structure indicates a negatively charged protein surface that places calcium ions in positions complementary to those in hydroxyapatite. Ocn could potentially modulate the crystal morphology and growth of hydroxyapatite [22].

Bone sialoprotein (Bsp) is a noncollagenous protein that promotes RGD (ARG-GLY-ASP) dependent cell attachment via integrins. Bsp is thought to be involved in the nucleation of hydroxyapatite for mineralization of bone [23].

The restoration of skeletal tissue to its normal state and function, also known as fracture repair, is dependent on the careful arrangement of the action of several factors. A number of growth factors, cytokines, and their receptors are present around the fracture site to start the repair process. Whereas many of these components are expressed in the skeletal tissue at all times, several other molecules are released from the inflammatory cells at the site of injury. Fracture repair has been discussed in several reviews and some of the factors and signaling pathways involved in this process are illustrated in Figure 21.1 [24,25]. In Section 21.2 we will overview some of the signaling molecules in bone formation and repair.

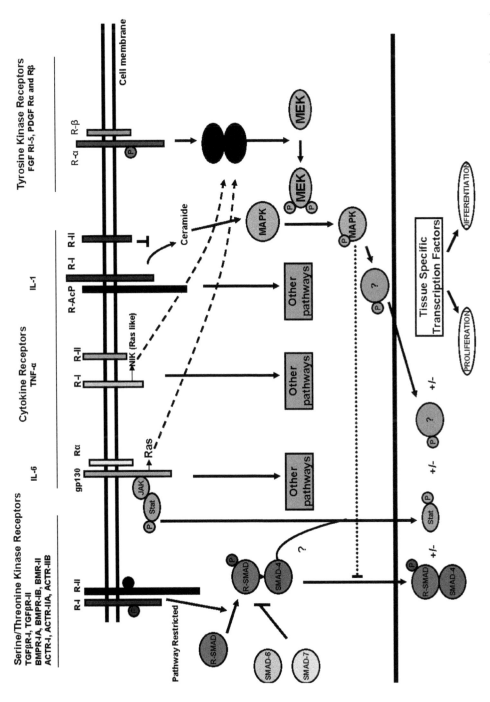

FIGURE 21.1   Major growth factors and cytokines involved in fracture repair. (Reproduced from *J. Bone Miner. Res.* 1999, 14:1808 with permission of the American Society for Bone and Mineral Research.)

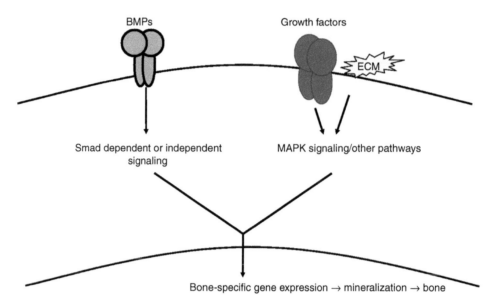

FIGURE 21.2   Growth factors, ECM proteins, BMPs exert their effects on cells to promote osteoblastic differentiation and bone formation.

21.2   Signaling Molecules for Bone

21.2.1   Growth Factors

Bone formation and development is the arrangement of the actions of a wide variety of signaling molecules (Figure 21.2). These signaling molecules include growth factors, hormones, vitamins, and cytokines. Hormones may be several categories including amino acid derivatives, peptides, steroids and fatty acid derivatives. Growth factors are peptide hormones, which induce cellular proliferation. Cytokines, which are also peptide-based can affect inflammatory and immune responses of the metabolism. The orchestration of these factors regulates mitogenesis, cell shape, movement, differentiation, and apoptosis.

Growth factor effects are concentration-dependent and are exerted through their receptors on the cell surface. A secreted growth factor may bind to matrix molecules, carrier molecules, or binding proteins to regulate its activity and stabilization [26]. Growth factors can associate with specific binding proteins that limit access to their receptors to control the bioavailability of the growth factor (e.g., IGF-I, IGF-II, TGFβ, BMPs) [27–29]. The conversion of a growth factor to a bioactive state requires an activation event. Using IGF-I as an example, greater than 99% are bound by IGFBPs (IGF binding proteins) in fluid and solid phase [26,30] and requires protease activation to release IGF-I. This mechanism of sequestration allows temporal and spatial regulation. For tissue engineering applications it is crucial to control the concentration and physical placement and sequestration of a growth factor. Several methods are available for controlling the physical placement of a growth factor including, but not limited to microfluidics, microencapsulation, entrapment and release from polymeric systems, and nonspecific adsorption to matrices [26,31]. Growth factors can also be immobilized to engineered matrices to localize delivery [32], but spatial patterning remains a challenge.

The design of a tissue engineering application requires an appreciation of the mechanism by which a factor elicits its signal and the downstream effect originating from this signal. In this section we will briefly overview some of the factors involved in bone formation and regulation.

21.2.1.1   Insulin-Like Growth Factors

The putative functions of insulin-like growth factors (IGF-I and IGF-II) include embryonic and natal growth, bone matrix mineralization, cartilage development and homeostasis [33,34]. IGFs can stimulate collagen production and prevent collagen degradation by reducing collagenase synthesis [35]. IGF-I is known to activate osteocalcin expression [36] inside the cell. It has been reported to be important for maintaining bone mass and promoting longitudinal bone growth [37].

The IGFs transduce their signals via two different receptors known as IGF-I and IGF-II receptors [38]. IGF-II mediates its signal through the type II IGF receptor although it can tissue-specifically activate the type I IGF receptor. IGF-I is a single chain peptide with a structure similar to proinsulin, but consisting of four domains (insulin contains three domains) [39]. During posttranslational modification of the molecule, one of the peptide domains is cleaved from the rest of the domains and the rest are joined to one another by forming disulfide bonds [39,40]. Growth hormone induces IGF-I synthesis [41]. The kidney, muscle, and bone also contribute to the circulating IGF-I levels [41].

IGF-II has 60% homology to IGF-I and acts independent of growth hormone [41,42]. It appears to be more important during fetal growth than in postnatal growth [43,44]. In a study where it was systemically administered to rats, IGF-II was found to be less potent than IGF-I in stimulating skeletal growth [45]. A study on chimeric mice carrying one inactivated IGF-II allele indicated that these mice were smaller than their wild type littermates [46] indicating the significance of IGF-II on the skeleton during early stages of growth. More studies have been conducted on IGF-I than IGF-II for tissue engineering applications. Studies in rats suggest that IGF-I increases intramembranous ossification [47], improves the effects of age-related osteopenia [48,49], and accelerates functional recovery from Achilles tendon injury [50]. IGF-I has also been used for spinal fusion application in sheep, giving a successful outcome when delivered via poly-(D, L-lactide) (PDLLA)-coated titanium [51]. In another study, IGF-I was delivered via polyacetate (PLLA) microspheres to metacarpal defects in calves of pigs and showed enhancement of bone formation [52].

21.2.1.2 Fibroblast Growth Factors

Fibroblast growth factors FGF-1 and -2, also known as acidic and basic FGF, respectively, belong to a family of growth factors with heparin binding domains. FGFs regulate mitogenesis, differentiation, protease production, receptor modulation, and cell maintenance [53]. FGF-2 (also called basic FGF or bFGF) is produced by the osteoblasts and stored in skeletal tissues [54]. FGF-1 has been associated with chondrocyte proliferation [55].

The FGF-1 and FGF-2 systemically and locally administered to ovariectomized rats increased new bone formation and bone density [56,57]. Systemic delivery of FGF-1 appeared to be effective in restoring the microarchitecture of bone and preventing bone loss associated with estrogen withdrawal [56]. In a rabbit ulcer model, FGF-1 delivery within a modified fibrin matrix stimulated angiogenic and fibroblastic responses in addition to an increased epithelialization rate [57]. Several studies indicated that scaffold mediated delivery rather than direct injection of FGF-2 was more effective in improving bone healing in rats [58], rabbits [59], and dogs [49,60]. Local infusion of recombinant FGF-2 increased bone ingrowth in a rabbit tibia model in the presence of polyethylene particles [61].

21.2.1.3 Vascular Endothelial Growth Factors

Vascular endothelial growth factors, of which there are six different isoforms [62], are vascular cytokines that promote angiogenesis, increased vascular permeability, and vasodilation [62]. Prosthetic vascular grafts that were coated with VEGF supported endothelial cell proliferation and migration [63]. Several studies indicated increased capillary density and vasodilator-induced blood flow in response to VEGF treatment [64–66]. Synergistic effects of VEGF and FGF-2 have also been demonstrated in the production of new blood vessels [67]. Macroporous scaffolds with poly(lactide-co-glycolide) which were designed to release VEGF increased the generation of mineralized tissue due to an increase in vascularization, but did not increase osteoid formation [68]. VEGF is a crucial factor for tissue engineering due to its role in angiogenesis; it is the main provider of nutrients and growth factors to the wound repair site.

21.2.1.4 Transforming Growth Factor-ß

Transforming growth factor-ß family consists of five members, bone morphogenetic proteins (BMP), growth and differentiation factors (GDF), activins, inhibins, and Mullerian substance [69]. Osteoblasts and chondrocytes express TGFß receptors [70,71]. Mainly found in bone, platelets, and cartilage TGFß triggers growth, differentiation, and extracellular matrix synthesis [72]. TGFß is thought to be a coupler between bone formation and resorption. Studies on the use of TGFß as a therapeutic reagent have been difficult to assess due to the use of different isoforms and superphysiological doses of TGFß.

21.2.1.5 Bone Morphogenetic Proteins

Bone morphogenetic proteins are members of the TGFß superfamily. These proteins are highly conserved with sequence homology across species. BMPs play a critical role in embryonic development and regulate a wide range of cellular activities including cell proliferation, differentiation, cell determination, and apoptosis. At the cellular level BMPs bind to their transmembrane receptors and initiate a cascade of phosphorylation events which transduces the signal to upregulate downstream genes. BMPs elicit their signals through phosphorylation of Smad molecules, which translocate to the nucleus when activated and regulate the transcription of specific target genes. Recently Hassel et al. [73] have demonstrated that a Smad independent pathway is activated if the BMP-2 signaling pathway is initiated via BMP-2 induced signaling receptor complexes instead of preformed receptor complexes (see Figure 21.3). Several studies on animals have demonstrated the osteoinductive (inducing bone formation) properties of BMPs in healing nonunions and enhancing spinal fusion. Recombinant human BMP-2 protein delivered to diaphyseal defects in dogs was able to achieve union (heal a fracture) [74]. A similar study in dogs also showed promising results using BMP-7 [75]. BMP-7 has been shown to enhance spinal fusion in rabbits [76] and sheep [77]. Recombinant human BMP-2 successfully healed critical-sized defects in sheep [78], rabbit [79], and rat [80].

Since their discovery by Dr. Marshall Urist, BMPs have been used in a number of human clinical trials as well. BMP-7 was effective in healing critical sized defects in the fibula [81] whereas BMP-2 promoted lumbar interbody fusion [82]. Transgenic BMP-2 produced by human mesenchymal stem cells was effective in bone regeneration [83]. Currently, only BMP-2 is available for clinical use. BMP-7 (rhOP-1) may be approved to treat long bone nonunions secondary to trauma. BMP-2 is approved for tibial nonunions and in a spinal fusion construct which consists of a spinal fusion cage and rhBMP-2 on a type I collagen scaffold [84].

21.2.1.6 Platelet Derived Growth Factors

Platelet derived growth factor (PDGF) is composed of two polypeptide chains that may exist as a homodimer (PDGF-AA, PDGF-BB) or heterodimer (PDGF-AB) [85]. Several reports have indicated that PDGF-AA and PDGF-BB can enhance wound repair [86], support angiogenesis [87–90], and stimulate cell proliferation in the fetal rat calvarial system and in cultures of osteoblast-like cells derived from adult human bone explants [91,92]. The PDGF A and B genes act as regulators of cell growth and have been shown to be chemotactic [88,89,91]. PDGF-BB has been reported to be the most potent PDGF isoform in skeletal and nonskeletal cells [93]. As a consequence of its role in cell growth, PDGF may exert its effect by increasing the number of collagen synthesizing cells, although it does not increase collagen synthesis on a cellular basis [93]. The role of PDGF in osteoblast differentiation may be to increase the number of cells that can progress into osteoblastic lineage and express the osteoblast phenotype [93]. PDGF expression at fracture sites in addition to its mitogenic effects indicates a role for PDGF in wound healing and fracture repair. Systemic administration of PDGF in an osteoporotic animal model demonstrated that it could stimulate bone formation and improve mechanical strength in long bones and vertebrae [94]. Howes and colleagues showed that subcutaneously implanted demineralized bone matrix augmented with PDGF could enhance bone healing in a rat model [95]. Locally administered recombinant human PDGF-BB (rhPDGF-BB) delivered with an injectable collagen gel to rabbit tibial osteotomies enhanced fracture repair and stimulated osteogenesis [96]. Currently, PDGF-BB is approved by FDA for soft tissue

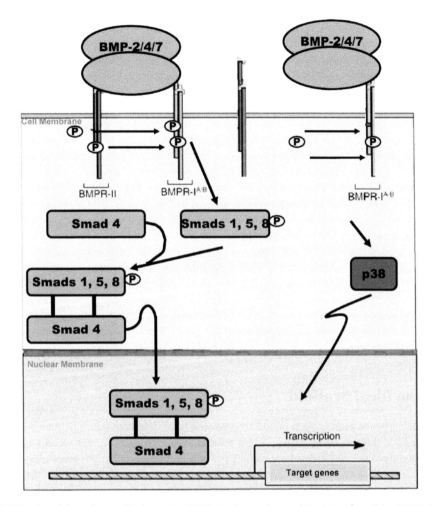

FIGURE 21.3   Smad dependent and independent BMP signaling pathways. (Reprinted from Schmitt J.M., Hwang K., Winn S.R., and Hollinger J.O. 1999. *J. Orthop. Res.* 17: 269–278. With permission from the Orthopedic Research Society.)

healing and remains as a compelling agent for tissue engineering applications especially in the treatment of osteoporotic fractures.

Despite the indispensable roles of these factors in osteoblast differentiation and bone formation, there are a number of concerns about their use in human patients. For instance, one concern with the injection of a growth factor like IGF-I is hypoglycemia, as reported in a study [97]. Another concern has been the use of superphysiological doses of these factors in order to trigger a response from the host. In the case of BMP-2, milligram doses (1.7 to 3.4 mg/dl) have to be used in patients due to diffusion from the wound site and instability *in vivo* [98,99]. Another reason for the rapid degradation of BMPs may be the presence of its natural inhibitors such as noggin and chordin at the fracture site [100,101]. Excessive bone formation remains a concern due to the risk of bony overgrowth leading to inadvertent fusion of adjacent levels or compression of the neural elements [102]. The potential side effects also need to be studied in longer time course experiments and trials for a fair assessment of the outcome.

21.2.2   Stem Cells and Gene Therapy

Mesenchymal stem cells (MSCs) are a population of self-renewing, undifferentiated cells. They can progress into a number of different cell fates, for example, adipocytic, osteogenic, chondrogenic, fibroblastic.

Stem cells can be harvested from fat, muscle, and bone marrow and can be genetically engineered to express bone signaling molecules [103–105]. Bruder et al. [106] implanted human bone marrow MSCs seeded on a ceramic carrier into the critical-sized defects of the femora of adult athymic rats and observed evidence for bone formation within 8 weeks. Parietal bone defects in adult sheep were repaired with MSCs added to a calcium alginate composite [107].

Gene therapy is the process by which genetic material is transferred into a cell's genome. A gene of interest can be delivered with the use of nonviral or viral carriers into targeted cell lines. Nonviral delivery methods include the use of naked plasmid DNA, liposomes, or gene gun. Examples of viral vectors include adenovirus, adeno-associated virus, lentivirus, herpes simplex virus, Moloney murine leukemia virus and retrovirus [108]. A method known as *ex vivo* gene therapy allows the viral infection of the cells to take place outside the body, thus increasing control over the system. A number of *ex vivo* studies have been performed to treat various bone defects in mouse, rabbit, and rat [109]. Human MSCs genetically engineered with an adenoviral construct expressing BMP-2 were able to form bone and cartilage and regenerate nonunion fractures in a mouse radius model [53]. An adenoviral construct carrying the BMP-2 gene was able to achieve spinal fusion in athymic rats [110]. However, the same research group also reported an immune response with adenoviral BMP-2 delivery into immunocompetent rats rather than athymic rats [111,112]. Furthermore, FDA has placed a hold on certain gene therapy applications due to safety concerns.

One method used to improve the efficacy of the delivery of signaling molecules is to use matrix scaffolds. In Section 21.3 we will overview the importance of biomaterials in tissue engineering and some current developments in this field.

21.3   The Ideal Scaffold

Autograft bone remains as the gold standard treatment of bone defects in the clinic due to its capability of providing cells, differentiative factors, and a reliable matrix required for fusion. Autograft bone is usually isolated from the iliac crest of the patient and can lead to a number of complications including chronic pain, infection, and fracture. One other option is to use allograft bone in the clinic; however, in this case immune response and disease transmission remain major concerns. The inadequacies of the current treatment methods have generated a need for alternative therapeutics for the treatment of bone defects.

Delivery of an osteoblast-specific gene or a signaling molecule remains a desirable option, but the therapeutic molecule requires an osteoconductive scaffold for its delivery. The ideal scaffold should encourage cell attachment, promote and support vascularization, and resist soft tissue forces [113]. The scaffold needs to be osteoinductive, osteoconductive, and biodegradable; this means that it needs to have the ability to induce bone formation at a nonbony site, the ability to provide a scaffold for new bone formation at the delivered site, and the ability to decompose without any toxic components to the cells and tissues, respectively.

Biomaterials for scaffold design can be classified under two major groups: acellular and cellular systems. Absorbable filter materials that can promote bone formation without a cellular component are called acellular systems. Cellular systems have cells embedded in the matrix to guide bone development [112].

Naturally derived matrices are derived from primary components of bone matrix and have natural affinity to growth factors as well as other osteoinductive factors; however they are difficult to sterilize and can trigger immune response from the host. Examples include hyaluronic acid, chitosan, and collagen matrices [69,113,114–117]. Inorganic materials include hydroxyapatite, porous coralline, calcium phosphate cements, and calcium sulfate. Their major advantage is the resemblance to bone structure, they can be resorbable or nonresorbable, but they are difficult to mold and are brittle [115,118–121]. Synthetic polymers are easy to manufacture and sterilize and they can be designed with controlled release parameters; however, they may degrade into toxic components and may be difficult to get recognized by the cells. Examples of synthetic polymers include poly (α-hydroxy acids), polypropylene fumarate, polyanhidrides, polyphosphazenes, polyethylene glycol, and poloxamers [115,122–128]. Ceramics such as tri-calcium phosphate and hydroxyapatite are biocompatible and display osteoconductivity [129–132]; however, resorption and porosity remain as concerns in these types of matrices. Oxidized cellulose and oxidized cellulose esters are also biocompatible polymers and have application in surgically implantable materials [133,134].

New biomaterial composites are being created to overcome the limitations of the different types of scaffolds mentioned above. Tsuchiya et al. [135] reported that they achieved osteogenesis with the design of a web-like structured biodegradable hybrid sheet composed of PLGA sheets containing collagen microsponges in their openings when seeded with bone marrow stem cells. These biodegradable hybrid sheets could be laminated or rolled into any shape. Photoencapsulation of hydrogels in different layers may help to mimic zonal organization of tissues which may be crucial in tissue engineering of the cartilage [136]. Recent fabrication techniques allow the synthesis of biomaterials that contain signal recognition ligands such as RGD domains to enable molecular and cellular responses.

Biocompatible and biodegradable polyurethane scaffolds have also been prepared for bone tissue engineering applications [137–144]. Polyurethanes are prepared by reacting diisocyanates with diols, whereas polyureas are the reaction product of diisocyanates with diamines (Figure 21.4). To be useful as resorbable scaffolds, conventional diisocyanates, such as methylene bis diphenylisocyanate (MDI) and toluene diisocyanate (TDI), which degrade to carcinogenic and mutagenic compounds [145], cannot be used. To avoid the toxicity problems associated with aromatic diisocyanates, aliphatic diisocyanates, such as lysine ethyl ester diisocyanate (LDI), and 1,4-diisocyanatobutane (BDI), have been reacted with polyether and polyester polyols to synthesize resorbable polyurethanes [139,146–148].

Diisocyanates react with water to form a disubstituted urea and carbon dioxide gas, which acts as blowing agent [149]. The water reaction is exploited commercially to manufacture flexible and rigid polyurethane foams. Zhang et al. [139,140] have prepared porous scaffolds by adding water to an isocyanate-terminated prepolymer (i.e., the low molecular weight reaction product of a diisocyanate with a polyol). By varying the concentration of water added, the pore size distribution was controlled to support the growth and proliferation of rabbit bone marrow stromal cells. Porous scaffolds for the knee-joint meniscus have also been prepared by the solvent casting/salt leaching technique [150,151].

Bioactive molecules can be incorporated into polyurethanes through the reaction of diisocyanate with primary amine and hydroxyl groups. Following this approach, Zhang et al. [141] have recently synthesized a bioactive polyurethane scaffold from lysine ethyl ester diisocyanate (LDI), glycerol, polyethylene glycol (PEG), water, and ascorbic acid [142]. As the polyurethane degraded, ascorbic acid was released to the extracellular matrix and stimulated both cell proliferation and type I col and Alp synthesis *in vitro*. Other degradation products included lysine and PEG, which are biocompatible. Polyurethanes are potentially useful biomaterials for preparing bioactive porous scaffolds from both high (e.g., proteins) and low (e.g., signal recognition ligands) molecular weight bioactive molecules.

In the previous sections, we overviewed some of the required components for bone tissue regeneration: signaling molecules, cells, and scaffolds. In Section 21.4, we will summarize applications for reconstructive medicine in a clinical setting. We will review some of the current techniques applied to augment bone deficits.

21.4   Clinical Reconstruction of Bone Defects

Physical deformities caused by missing or defective tissues affect people of all ages. Most often they are due to cancer, trauma, or congenital abnormalities. Each year over 500,000 reconstructive procedures to correct these deformities are performed by plastic surgeons in the United States [152]. There are two fundamental types of deformities. One is when all tissue elements are present but not according to normal anatomy. An example is a fracture that heals with bone segments in improper orientations (i.e., fracture malunion). The second type is when the tissues are significantly impaired or absent altogether.

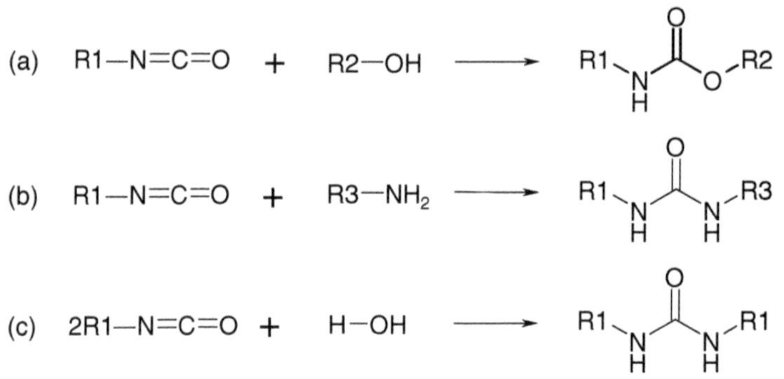

(a) $R1-N{=}C{=}O$ + $R2-OH$ ⟶ $R1\underset{H}{\overset{O}{\underset{N}{\|}}} \, O^{R2}$

(b) $R1-N{=}C{=}O$ + $R3-NH_2$ ⟶ $R1\underset{H}{\overset{O}{\underset{N}{\|}}} \underset{H}{\underset{N}{}}^{R3}$

(c) $2R1-N{=}C{=}O$ + $H-OH$ ⟶ $R1\underset{H}{\overset{O}{\underset{N}{\|}}} \underset{H}{\underset{N}{}}^{R1}$

FIGURE 21.4   Reactions of isocyanates with (a) alcohols to form urethane groups, (b) amines to form urea groups, and (c) water to form disubstituted ureas.

An example is damage caused by cancer radiotherapy (i.e., osteoradionecrosis) or massive trauma (e.g., a shotgun blast). When a deformity contains all essential elements, repair is possible by rearranging or augmenting the local tissues. On the other hand, when useful tissues are absent then new tissue must be supplied. Ideally, tissue replacements should be readily available, easily implanted, reliably incorporated into the surrounding normal tissues. This is the goal of tissue engineering for reconstructive surgery.

It is important to understand current clinical techniques of reconstructive surgery in order to develop practical new methods. Repairing every deformity involves similar steps of planning, tissue manipulation, and patient care. During the planning phase, the clinician will assess the defect to determine the exact form and amount of the deficient tissue. Conventional radiographs and computed tomography (CT) scanning are the most useful diagnostic tools for the osseous component. Magnetic resonance imaging (MRI) best demonstrates the soft tissue component. It is important to always consider both elements. A bone defect due to trauma may have the same anatomic appearance as one caused by cancer, but the best reconstructive method may be different in each case because of the health and stability of the surrounding soft tissues. After characterizing the defect, the next step is to select a source of replacement tissue. The clinician must balance the tissue requirements with the potential morbidity related to harvest. After considering all of these issues, the clinician and patient discuss them and agree upon a plan for surgery.

Repairing deformities by tissue replacement is a two-step process that first involves tissue transfer followed by tissue modification [153]. In the transfer step, tissue is harvested from an uninjured location (i.e., tissue donor site) and moved into the defect (i.e., recipient site). It is important to understand the principles that govern this manipulation in natural tissues because they also apply to engineered tissues. Living tissue may be transferred either as a surgical graft or a surgical flap. Grafts derive a blood supply from the tissues that surround them in the new location. Except for cartilage and thin pieces of skin, only small amounts of tissue can be transferred as grafts because survival of the cellular elements by simple diffusion of oxygen and nutrients is limited to volumes of less than 0.3 cm^3 [154,155]. Success depends on the potential for angiogenesis and specialized tissue formation that exists in the tissues surrounding the graft in the new location. During the initial 48 h after transfer, the graft must survive by diffusion alone [156]. Afterward, blood vessels arising from the tissues surrounding graft begin to align with and make connections to remnants of blood vessels found in the graft [157]. It takes up to 5 days for revascularization to occur, depending on the grafted tissue and condition of the tissue bed into which it is placed. This far exceeds the time during which most whole tissues can survive without a blood supply. Skeletal muscle tissue, for example, undergoes degeneration within 4 h after being deprived of blood supply. As new vessels penetrate the graft there is cell-mediated destruction of the degenerating muscle fibers. The basal laminae and some of the satellite cells appear to be the only elements of the muscle tissue persists [158]. Bone grafts are unique, however. They essentially are porous, calcified, degradable scaffolds that guide tissue formation (i.e., osteoconduction) and deliver a set of bioactive molecules to induce new bone formation (i.e., osteoinduction). Only a limited number of cellular elements survive and contribute to healing because of diffusion limitations [159,160]. The surviving cells are mostly located on the surfaces of the calcified matrix and appear to consist mainly of endosteal lining cells and marrow stromal cells [161,162]. Even though most cellular components do not survive transfer, bone grafts still contribute to bone healing because they supply the other essential components of tissue repair. Bone grafts must be completely replaced over time in order to achieve healing. This can require up to 2 years. They have limited utility for defects greater than 6 cm in length and are avoided when there is significant local tissue impairment due to significant bacterial contamination, poor vascularity, unstable soft tissues, or radiation injury in the area of the defect [163]. Tissue engineered bone created without a capillary system *ex vivo* in a bioreactor can be expected to perform clinically in a way analogous to a bone graft.

In contrast to grafts, surgical flaps are tissues transferred with a blood supply independent of the tissues surrounding the defect. They are called flaps because originally they were actual flaps of skin elevated with an attachment at the base and rotated into an adjacent area. Over time the definition broadened to include any unit of skin, muscle, fat, bone, or viscera (e.g., small intestine) that has a discreet vascular source permitting surgical isolation from the donor site and transfer to a distant recipient site. The most advanced method of transfer is by detaching the flap completely and reestablishing the blood supply by suturing the blood vessels of the flap to vessels adjacent to the defect. This is called microvascular surgery because it involves operating on blood vessels less than 5 mm in diameter using an operating microscope. Bone transferred in this way allows survival of all tissue elements and yields the most reliable healing. Bone flaps incorporate more rapidly than grafts and do not require resorption and replacement before achieving full strength [161]. They are the treatment of choice in circumstances with large defects, significant soft tissue deficits, or impaired local tissues due to severe trauma, infection, or exposure to radiation [163–166]. The ultimate goal of bone tissue engineering for reconstructive surgery is to fabricate surgical bone flaps.

After transfer, the tissue must then be modified to simulate the missing parts. Bones have a complex three-dimensional shape and must tolerate powerful deforming forces. The relative importance of shape and load bearing differs based on the anatomic site. The long bones of the extremities function primarily as load bearing members. Minor shape discrepancies are tolerable as long as there is no significant loss of strength. On the other hand, craniofacial bones have a limited load bearing function. Their shape is critical to support and protect the complex and delicate soft tissue structures of the head and neck. They also determine human facial appearance and play a major role in psychosocial health [168,169]. Therefore, bone replacements in the craniofacial skeleton must maintain a durable shape. In addition, craniofacial bones have thin soft tissue coverage, and they are located in close proximity to heavily bacteria-contaminated surfaces of oral and nasal cavities. The oral cavity is one of the most heavily contaminated areas of the body with numerous bacterial species present including aerobes, anaerobes, fungi, viruses, and protozoa [170]. The paranasal sinuses are normally not sterile, although the bacterial load appears much less [171]. It is impossible to perform tissue implantation surgery in this area without a high probability of bacterial contamination. Therefore, the tissue replacement must be compatible with the thin soft tissues to avoid erosion and have an intrinsic resistance to infection. These features of the craniofacial skeleton make fabricating replacements by tissue engineering particularly challenging.

After the surgery is complete, the patient requires special care to recover. The reconstructed areas must be protected from disruption and infection. Grafted tissues must be stably fixed to prevent shearing motion at the interface with the tissue bed, which slows the process of revascularization. Flaps must be closely monitored to rapidly detect thrombosis and occlusion of the blood vessels that supply the tissues. After complete healing and tissue incorporation, the final step is often a period of rehabilitation to ensure maximum restoration of function. Finally, additional surgery may be required to make revisions and improve minor deficiencies that are not able to be avoided during the primary reconstruction or which may have appeared later due to scar contracture, for example. The entire process can require many months to completely restore the patient (Figure 21.5a,b and Figure 21.6a–d).

(a)

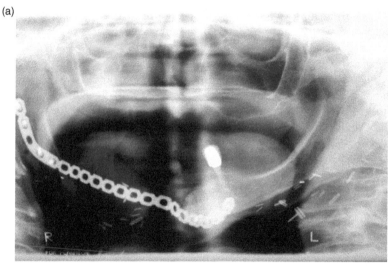

(b)

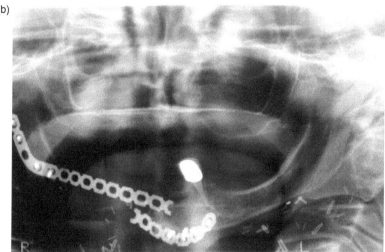

FIGURE 21.5   Panoramic radiographs of a patient after right mandible reconstruction with only a titanium reconstruction plate showing appearance after surgery (a) and appearance several months later after fracture of the plate
due to metal fatigue (b).

21.5   Conclusion

Tissue engineering may offer a new dimension of therapeutic care for patients. The design and development of the tissue engineered therapeutics must be guided by basic, fundamental biological pathways
targeting specific clinical applications. Targeted clinical performance standards for the tissue engineered
therapeutic must be achieved in the design, validated by the appropriate standardized characterization
protocols (e.g., ASTM, ISO 10993), preclinical and clinical phase I–III studies. To fulfill stringent clinical
requirements for a tissue engineered bone therapeutic, specific clinical targets must be identified. Is the
target a nonload bearing or load bearing bone? Is the target a fracture in the tibia of a healthy, young male?
Or is the clinical target a distal radius fracture in a postmenopausal osteoporotic?

In this chapter, we have briefly reviewed the pathways of bone formation and some of the key signaling
molecules cuing discrete outcome events. Tissue engineered products must integrate into their design
fundamental biological elements consistent with bone formation pathways. Many of the key temporal and

(a)

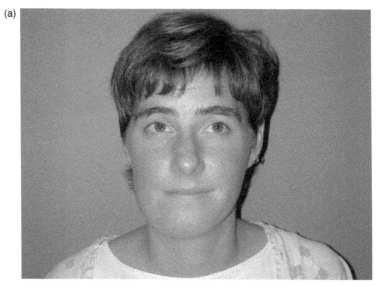

(b)

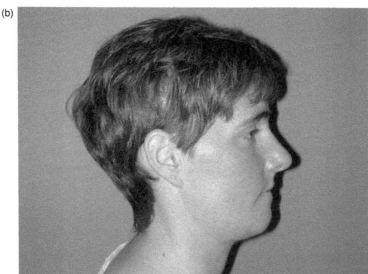

FIGURE 21.6 Appearance of a young woman treated with surgery for a tumor involving the right mandible. The entire mandible on the right was removed and replaced using a bone flap harvested from the fibula. Appearance before surgery (a),(b) and 1 year after surgery (c),(d). Although surgery resulted in a durable reconstruction, note the changes in appearance due to the difference in the shape of the fibula and native mandible.

spatial aspects of bone formation (developmental, homeostatic, healing) remain a mystery. Unless these key aspects are elucidated, rational, effective tissue engineered therapeutics will not develop.

A broad understanding by tissue engineers of the complex anatomical and physiological issues challenging surgeons and patients will provide tissue engineers with an important foundational head start for robust design options that may include uniquely engineered compositions of cells, genes, and biomimetic extracellular matrices (e.g., bioactive matrices). The biomimetic extracellular matrix was especially emphasized in this chapter to stress the importance of spatially and temporally directing the process of tissue regeneration. Matrix design is the pivotal tissue engineering challenge.

We provided some clinical examples where bone tissue engineering will benefit craniofacial reconstruction. The craniofacial surgeon faces different clinical challenges than the orthopedic surgeon. Tissue

(c)

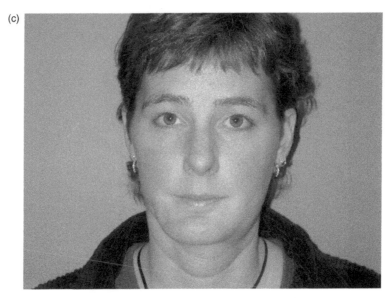

(d)

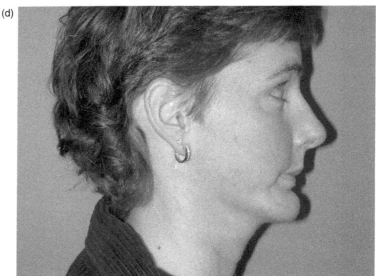

FIGURE 21.6 Continued.

engineering must emphasize a regional-anatomic and physiological approach to design and development. A single platform treatment for a tibial diaphyseal fracture in the healthy adult male may be inadequate for the patient with an avulsive bone wound of the oral–antral complex.

There are significant opportunities for the tissue engineer to improve the quality of patient care. The bedrock underscoring tissue engineering is basic biology, clinical performance standards, and patient application focus.

References

[1] Boskey A.L. 2003. Biomineralization: an overview. *Connect Tissue Res.* 44: 5.

[2] Barckhaus R.H. and Hohling H.J. 1978. Electron microprobe analysis of freeze dried and unstained mineralized epiphyseal cartilage. *Cell Tissue Res.* 18693: 541–549.

[3] Marks S.C. and Odgreen P.R. 2002. Structure and development of the skeleton. In J.P. Bilezikian, L.G. Raisz, and G.A. Rodan (Eds.), *Principles of Bone Biology*, Vol. 1, pp. 3–14, New York, Academic Press.

[4] Gay C.V., Donahue H.J., Siedlecki C.A., and Vogler E. 2004. Cellular elements of the skeleton: osteoblasts, osteocytes, osteoclasts, bone marrow stromal cells. In J.O. Hollinger, T.A. Einhorn, and B.A. Doll (Eds.), *Bone Tissue Engineering*, Chapter 3, Boca Raton, FL, CRC Press.

[5] Olsson S.E. and Ekman S. 2002. Morphology and physiology of the growth cartilage under normal and pathologic conditions. In G.E. Fackelman (Ed.), *Bone in Clinical Orthopedics*, p. 117, Stuttgart, Germany, AO Publishing.

[6] Scott, C.K. and Hightower J.A. 1991. The matrix of endochondral bone differs from the matrix of intramembranous bone. *Calcif. Tissue Int.* 49: 349–354.

[7] Sandberg M.M. 1991. Matrix in cartilage and bone development: current views on the function and regulation of the major organic components. *Ann. Med.* 23: 207–217.

[8] Tuan R.S. 1994. Developmental skeletogenesis. In C.T. Brighton, G. Friedlaender, and J.M. Lane (Eds.), *Bone Formation and Repair*, pp. 13, Rosemont, IL, AAOS.

[9] Doll, B.A. 2004. Developmental biology of the skeletal system. In J.O. Hollinger, T.A. Einhorn, and B.A. Doll (Eds.), *Bone Tissue Engineering*, Chapter 1, Boca Raton, FL, CRC Press.

[10] Wolff J. 1892. *The Law of Bone Remodelling*. Translated by Maquet P. and Furlong R. 1986. New York, NY, Springer-Verlag.

[11] Mikuni-Takagaki Y. 1999. Mechanical responses and signal transduction pathways in stretched osteocytes. *J. Bone Miner. Metab.* 17: 57–60.

[12] Dehority W., Halloran B.P., Bikle D.D., Curren T., Kostenuik P.J., Wronski T.J., Shen Y., Rabkin B., Bouraoui A., and Morey-Holton E. 1999. Bone and hormonal changes induced by skeletal unloading in the mature male rat. *Am. J. Physiol.* 276: E62–E69.

[13] Montufar-Solis D., Duke P.J., and Morey-Holton E. 2001. The Spacelab 3 stimulation: basis for a model of growth plate response in microgravity in the rat. *J. Gravit. Physiol.* 8: 67–76.

[14] Cullinane D.M. and Salisbury K.T. 2004. Biomechanics. In J.O. Hollinger, T.A. Einhorn, and B.A. Doll (Eds.), *Bone Tissue Engineering*, Chapter 10, Boca Raton, FL, CRC Press.

[15] Ferretti J.L., Capozza R.F., and Zanchetta J.R. 1996. Mechanical validation of a tomographic QCT index for noninvasive estimation of rat femur bending strength. *Bone* 18: 97–102.

[16] Beck T.J., Mourtada F.A., Ruff C.B., Scott W.W., and Kao G. 1998. Experimental testing of a DEXA-derived curved beam model of the proximal femur. *J. Orthop. Res.* 16: 394–398.

[17] Toyras J., Nieminen M.T., Kroger H., and Jurvelin J.S. 2002. Bone mineral density, ultrasound velocity, and broadband attention predict mechanical properties of trabecular bone differently. *Bone* 31: 503–507.

[18] Stigbrand T. 1984. *Present Status and Future Trends of Human Alkaline Phosphatases*, pp. 3–14, New York, NY, Alan R. Liss.

[19] Cole D.E. and Cohen M.M. 1990. Mutations affecting bone forming cells. In B.K. Hall (Ed.), *Bone: The Osteoblast and the Osteocyte*, pp. 442–452, New Jersey, The Telford Press.

[20] Risteli L. and Risteli J. 1993. Biochemical markers of bone metabolism. *Ann. Med.* 25: 385–393.

[21] Butler W.T. 1989. The nature and significance of osteopontin. *Connect Tissue Res.* 23: 123–136.

[22] Hoang Q.Q., Sicheri F., Howard A.J., and YANG D.S.C. 2003. Bone recognition mechanism of porcine osteocalcin from crystal structure. *Nature* 425: 977–980.

[23] Hunter G.K. and Goldberg H.A. 1993. Nucleation of hydroxyapatite by bone sialoprotein. *Proc. Natl Acad. Sci. USA* 90: 8562–8565.

[24] Barnes G.L., Kostenuik P.J., Gerstenfeld L.C., and Einhorn T.A. 1999. Growth factor regulation of fracture repair. *J. Bone Miner. Res.* 14: 1805–1815.

[25] Learnmoth I.D. 2004. The management of periprosthetic fractures around the femoral stem. *J. Bone Joint Surg. Br.* 86: 13–19.

[26] Sfeir C., Jadlowiec J.J., Koch H., and Campbell P.G. 2004. Signaling molecules for tissue engineering. In J.O. Hollinger, T.A. Einhorn, and B.A. Doll (Eds.), *Bone Tissue Engineering*, Chapter 5, Boca Raton, FL, CRC Press.

[27] Mohan S. and Baylink D.J. 2002. IGF-binding proteins are multifunctional and act via IGF-dependent and -independent mechanisms. *J. Endocrinol.* 175: 19–31.

[28] Nunes I., Gleizes P.E., Metz C.N., and Rifkin D.B. 1997. Latent transforming growth factor-beta binding protein domains involved in activation and transglutaminase-dependent cross-linking of latent transforming growth factor-beta. *J. Cell Biol.* 136: 1151–1163.

[29] Balemans W. and Hul W.V. 2002. Extracellular regulation of BMP signaling in vertebrates: a cocktail of modulators. *Dev. Biol.* 250: 231–250.

[30] Baxter, R.C. 2000. Insulin-like growth factor (IGF)-binding proteins: interactions with IGFs and intrinsic bioactivities. *Am. J. Physiol. Endocrinol. Metab.* 278: E967–E976.

[31] Saltzman W.M. and Olbreicht W.L. 2002. Building drug delivery into tissue engineering. *Nat. Rev.: Drug Dis.* 1: 177–186.

[32] Richardson T.P., Peters M.C., Ennett A.B., and Mooney D.J. 2001. Polymeric system for dual growth factor delivery. *Nat. Biotechnol.* 19: 1029–1034.

[33] Nixon A.J., Brower-Toland B.D., Bent S.J., Saxer R.A., Wilke M.J., Robbins P.D., and Evans C.H. 2000. Insulin-like growth factor-I gene therapy applications for cartilage repair. *Clin. Orthop.* 379: S201–1113.

[34] Zhang M., Xuan S., Bouxsein M.L., von Stechow D., Akeno N., and Fougere M.C. et al. 2002. Osteoblast-specific knockout of the IGF receptor gene reveals an essential role of IGF signaling in bone matrix mineralization. *J. Biol. Chem.* 277: 44005–44012.

[35] McCarthy T.L., Centrella M., and Canalis E. 1989. Regulatory effects of insulin-like growth factors I and II on bone collagen synthesis in rat calvarial cultures. *Endocrinology* 124: 301–309.

[36] Ogata N., Chikazu D., Kubota N., Terauchi Y., and Tobe K. et al. 2000. Insulin receptor substrate-1 in osteoblast is indispensable for maintaining bone turnover. *J. Clin. Invest.* 105: 935–943.

[37] Zhang M., Faugere M.C., Malluche H., Rosen C.J., Chernausek S.D., and Clemens T.L. 2003. Paracrine overexpression of IGFBP-4 in osteoblasts of transgenic mice decreases bone turnover and causes global growth retardation. *J. Bone Miner. Res.* 18: 836–843.

[38] Raile K., Hoflich A., Hessler U., Yang Y., Pfuender M., Blum W.F., Kolb H., Schwartz H.B., and Kiess W. 1994. Human osteosarcoma (U-2 OS) cells express both insulin-like growth factor-I (IFG-I) receptors and insulin-like growth factor-II/mannose-6-phosphate (IGF-II/M6P) receptors and synthesize IGF-II: autocrine growth stimulation by IGF-II via the IGF-I receptor. *J. Cell Physiol.* 159: 531–541.

[39] Rosenfeld R.G. and Roberts C.T. 1999. *The IGF System: Molecular Biology, Physiology and Clinical Applications*, p. 19, NJ, Homana Press.

[40] Chan S.J., Magamatsu S., Cao Q.-P., and Steiner D.F. 1992. Structure and evolution of insulin and insulin-like growth factors in chordates. *Prog. Brain Res.* 1992: 15–24.

[41] Mohan S. and Baylink D.J. 1996. The I.G.Fs as Potential therapy for metabolic bone diseases. In J.P. Bilezikian, L.G. Raisz, and G.A. Rodan (Eds.), *Principles of Bone Biology*, pp. 1100–1107, New York, Academic Press.

[42] Gray A., Tam A., Dull W., Hayflick T.J., Pintar J., Cavenee W.K., Koufos A., and Ulrich A. 1987. Tissue-specific and developmentally-regulated transcription of the IGF 2 gene. *DNA* 6: 283–295.

[43] Trippel, S.B. 1994. Biologic regulation of bone growth. In C.T. Brighton, G. Friedlaender, and J.M. Lane (Eds.), *Bone Formation and Repair*, p. 43, Rosemont, IL, AAOS.

[44] Underwood L.E. and Van Wyk J.J. 1992. Normal and aberrant growth. In J.D. Wilson and D.W. Foster (Eds.), *Williams Textbook of Endocrinology*, pp. 1079–1138, Philadelphia, PA, WB Saunders.

[45] Schoenle E., Zapf J., Hauri C. et al. 1985. Comparison of *in vivo* effects of insulin-like growth factors I and II and of growth hormone in hypophysectomized rats. *Acta Endocrinol.* 108: 167–174.

[46] DeChiara T.M., Efstratiadis A., and Robertson E.J. 1990. A growth factor deficiency phenotype in heterozygous mice carrying an insulin-like growth factor II gene disrupted by targeting. *Nature*. 345: 78–80.

[47] Thaler S.R., Dart A., and Tesluk H. 1993. The effect of insulin-like growth factor-I on critical sized calvarial defects in Sprague-Dawley rats. *Ann. Plast. Surg.* 31: 429–433.

[48] Tanaka H., Quarto R., Williams S., Barnes J., and Liang C.T. 1994. *In vivo* and *in vitro* effects of insulin-like growth factor-I (IGF-I) on femoral mRNA expression in old rats. *Bone* 15: 647–653.

[49] Jadlowiec J.J., Celil A.B., and Hollinger J.O. 2003. Bone tissue engineering: recent advances and promising therapeutic agents. *Exp. Opin. Biol. Ther.* 3: 409–423.

[50] Kurtz C.A., Loebig T.G., Anderson D.D., Demeo P.J., and Campbell P.G. 1999. Insulin-like growth factor I accelerates functional recovery from Achilles tendon injury in a rat model. *Am. J. Sports Med.* 27: 363–369.

[51] Kandziora F., Schmidmaier G., Schollmeier G. et al. 2002. IGF-I and TGF-beta application by a poly-(D,L-lactide)-coated cage promotes intervertebral bone matrix formation in the sheep cervical spine. *Spine* 27: 1710–1723.

[52] Illi O.E. and Feldmann C.P. 1998. Stimulation of fracture healing by local application of humoral factors integrated in biodegradable implants. *Eur. J. Pediatr. Surg.* 8: 251–255.

[53] Trippel S.B., Whelan M.C., Klagsbrun M. et al. 1992. Interaction of basic fibroblast growth factor with bovine growth plate chondrocytes. *J. Orthop. Res.* 10: 638–646.

[54] Canalis E., McCarthy T.L., and Centrella M. 1989. Effects of platelet-derived growth factor on bone formation *in vitro*. *J. Cell Physiol.* 140: 530–537.

[55] Jingushi S., Heydemann A., Kana S.R., Macey L.R., and Bolander M.E. 1990. Acidic fibroblast growth factor (aFGF) injection stimulates cartilage enlargement and inhibits cartilage gene expression in rat fracture healing. *J. Orthop. Res.* 8: 364–371.

[56] Dunstan C.R., Boyce R., Boyce B.F. et al. 1999. Systemic administration of acidic fibroblast growth factor (FGF-1) prevents bone loss and increases new bone formation in ovariectomized rats. *J. Bone Miner. Res.* 14: 953–959.

[57] Pandit A.S., Feldman D.S., Caulfield J. et al. 1998. Stimulation of angiogenesis by FGF-1 delivered through a modified fibrin scaffold. *Growth factors* 15: 113–123.

[58] Kimoto T., Hosokawa R., Kubo T., Maeda M., Sano A., and Akagawa Y. 1998. Continuous administration of basic fibroblast growth factor (FGF-2) accelerates bone induction on rat calvaria — an application of a new drug delivery system. *J. Dent. Res.* 77: 1965–1969.

[59] Inui K., Maeda M., Sano A. et al. 1998. Local application of basic fibroblast growth factor minipellet induces the healing of segmental bony defects in rabbits. *Calcif. Tissue Int.* 63: 490–495.

[60] Hosokawa R., Kikuzaki K., Kimoto T. et al. 2000. Controlled local application of basic fibroblast growth factor (FGF-2) accelerates the healing of GBR. An experimental study in beagle dogs. *Clin. Oral Implants Res.* 11: 345–353.

[61] Goodman S.B., Song Y., Yoo J.Y., Fox N., Trindale M.C., Kajiyama G., Ma T., Regula D., Brown J., and Smith R.l. 2003. Local infusion of FGF-2 enhances bone ingrowth in rabbit chambers in the presence of polyethylene particles. *J. Biomed. Res.* A: 454–461.

[62] Bates D.O., Lodwick D., and Williams B. 1999. Vascular endothelial growth factor and microvascular permeability. *Microcirculation* 6: 83–96.

[63] Stone D., Phaneuf M., Sivamurthy N. et al. 2002. A biologically active VEGF construct *in vitro*: implications for bioengineering-improved prosthetic vascular grafts. *J. Biomed. Mater. Res.* 59: 160–165.

[64] Bauters C., Asahara T., Zheng L.P. et al. 1995. Recovery of disturbed endothelium-dependent flow in the collateral-perfused rabbit ischemic hindlimb after administration of vascular endothelial growth factor. *Circulation* 91: 2802–2809.

[65] Takeshita S., Pu L.Q., Stein L.A. et al. 1994. Intramuscular administration of vascular endothelial growth factor induces dose-dependent collateral artery augmentation in a rabbit model of chronic limb ischemia. *Circulation* 90: II228–II234.

[66] Takeshita S., Tsurumi Y., Couffinahl T. et al. 1996. Gene transfer of naked DNA encoding for three isoforms of vascular endothelial growth factor stimulates collateral development *in vitro*. *Lab. Invest.* 75: 487–501.

[67] Pepper M.S., Ferrara N., Orci L. et al. 1992. Potent synergism between vascular endothelial growth factor and basic fibroblast growth factor in the induction of angiogenesis *in vitro*. *Biochem. Biophys. Res. Commun.* 189: 824–831.

[68] Murphy W.L., Simmons C.A., Kaigler D., and Mooney D.J. 2004. Bone regeneration via a mineral substrate and induced angiogenesis. *J. Dent. Res.* 83: 204–210.

[69] Massague J. and Wotton D. 2000. Transcriptional control by the TGF-beta/Smad signaling system. *EMBO J.* 19: 1745–1754.

[70] Bourque W.T., Gross M., and Hall B.K. 1993. Expression of four growth factors during fracture repair. *Int. J. Dev. Biol.* 37: 573–579.

[71] Robey P.G., Young M.F., Flanders K.C. et al. 1987. Osteoblasts synthesize and respond to transforming growth factor type beta (TGF-beta) *in vitro*. *J. Cell Biol.* 105: 457–463.

[72] Abe N., Lee Y.P., Sato M. et al. 2002. Enhancement of bone repair with a helper-dependent adenoviral transfer of bone morphogenetic protein-2. *Biochem. Biophys. Res. Commun.* 297: 523–527.

[73] Hassel S., Schmitt S., Hartung A. et al. 2003. Initiation of Smad-dependent and Smad independent signaling via distinct BMP-receptor complexes. *J. Bone Joint Surg. Am.* 85: 44–51.

[74] Sciadini M.F. and Johnson K.D. 2000. Evaluation of recombinant human bone morphogenetic protein-2 as a bone graft substitute in a canine segmental defect model. *J. Orthop. Res.* 18: 289–302.

[75] Salkeld S.L., Parton L.P., Barrack R.L., and Cook S.D. The effect of osteogenic protein-1 on the healing of segmental bone defects treated with autograft or allograft bone. *JBJS Am.* 2001: 83–1: 803–816.

[76] Grauer J.N., Patel T.C., Erulkar J.S., Troiano N.W., Panjabi M.M., and Friedlaender G.E. 2001. Young Investigator research award winner: evaluation of OP-1 as a graft substitute for intertransverse process lumbar fusion. *Spine* 26: 127–133.

[77] Magin M.N. and Delling G. 2001. Improved lumbar vertebral interbody fusion using rhOP-1: a comparison of autogenous bone graft, bovine hydroxyapatite and BMP-7 in sheep. *Spine* 26: 469–478.

[78] Gerhart T.N., Kirker-Head C.A., Kriz M.J. et al. 1993. Healing segmental defects in sheep using recombinant human bone morphogenetic protein. *Clin. Orthop.* 293: 317–326.

[79] Bostrom M., Lane J.M., Tomin E. et al. 1996. Use of bone morphogenetic protein-2 in the rabbit ulnar nonunion model. *Clin. Orthop.* 327: 272–282.

[80] Yasko A.W., Lane J.M., Fellinger E.J., Rosen V., Wozney J.M., and Wang E.A. 1992. The healing of segmental defects, induced by human bone morphogenetic protein (rhBMP-2). A radiographic, histological, and biomechanical study in rats. *J. Bone Joint Surg. Am.* 74: 659–670.

[81] Geesink R.G., Hoefnagels N.H., and Bulstra S.K. 1999. Osteogenic activity of OP-1 bone morphogenetic protein (BMP-7) in a human fibular defect. *J. Bone Joint Surg. Br.* 81: 710–718.

[82] Boden S.D., Zdeblick T.A., Sandhu H.S., and Heim S.E. 2000. The use of rhBMP-2 in interbody fusion cages: definitive evidence of osteoinduction in human — a preliminary report. *Spine* 25: 376–381.

[83] Turgeman G., Pittman D.D., Muller R. et al. 2001. Engineered human mesenchymal stem cells: a novel platform for skeletal cell mediated gene therapy. *J. Gene Med.* 3: 240–251.

[84] Issack P.S. and DiCesare P.E. 2003. Recent advances toward the clinical application of bone morphogenetic proteins in bone and cartilage repair. *Am. J. Orthop.* 32: 429–436.

[85] Westermark B. and Heldin C.H. 1993. Platelet-derived growth factor. *Acta Oncol.* 32: 101–105.

[86] Pierce G.F. and Musote T.A. 1995. Pharmacologic enhancement of wound healing. *Annu. Rev. Med.* 46: 467–481.

[87] Centrella M., McCarthy T., Kusik W., and Canalis E. 1992. Isoform specific regulation of platelet derived growth factor activity and binding in osteoblast-enriched cultures from fetal rat bone. *J. Clin. Invest.* 89: 1076–1084.

[88] Rodan S.B. and Rodan G.A. 1992. Fibroblast growth factor and platelet derived growth factor. In Gowan M. (Ed.), *Cytokines and Bone Metabolism*, pp. 116–140, Boca Raton, FL, CRC Press.

[89] Andrew J.J., Hoyland J., Freemont A., and Marsh D. 1995. Platelet derived growth factor expression in normally healing human fractures. *Bone* 16: 455–460.

[90] Gruber R., Varga F., Fishcer M., and Watzek G. 2002. Platelets stimulate proliferation of bone cells: involvement of platelet derived growth factors, microparticles and membranes. *Clin. Oral Implant. Res.* 3: 529–535.

[91] Lynch, S.E., Colvin R.B., and Antoinades H.N. 1989. Growth factors in wound healing: single and synergistic effects on partial thickness porcine skin wounds. *J. Clin. Invest.* 84: 640–646.

[92] Pierce G.F., Mustoe T.A., Altrock B.W., Deuel T.F., and Thomason A. 1991. Role of platelet derived growth factor in wound healing. *J. Cell Biochem.* 10: 131–138.

[93] Canalis E. and Rydziel S. 1996. Platelet-derived growth factor and the skeleton. In J.P. Bilezikian, L.G. Raisz, and G.A. Rodan (Eds.), *Principles of Bone Biology*, p. 621, New York, Academic Press.

[94] Mitlak B., Finkelman R., Hill E., Li J., Martin B., Smith T., D'Andrea M., Antoniades H., and Lynch S. 1996. The effect of systemically administered PDGF-BB on the rodent skeleton. *J. Bone Miner. Res.* 11: 238–247.

[95] Howes R., Bowness J., Grotendorst G., Martin G., and Reddi A. 1988. Platelet derived growth factor enhances demineralized bone matrix-induced cartilage and bone formation. *Calcif. Tissue Int.* 42: 34–38.

[96] Nash T., Howlett C., Martin C., Steele J., Johnson K., and Hicklin D. 1994. Effect of platelet derived growth factor on tibial osteotomies in rabbit. *Bone* 15: 203–208.

[97] Guler H.P., Zapf J., and Froesch E.R. 1987. Short-term metabolic effects of recombinant human insulin-like growth factor I in healthy adults. *N. Engl. J. Med.* 317: 137–140.

[98] Boden S.D., Zdeblick T.A., Sandhu H.S., and Heim S.E. 2002. Use of recombinant human bone morphogenetic protein-2 to achieve postrolateral lumbar spine fusion in humans: a prospective, randomized clinical pilot trial. *Spine* 27: 2661–2673.

[99] Boyne P.J., Marx R.E., Nevins M. et al. 1997. A feasibility study evaluating rhBMP-2/absorbable collagen sponge for maxillary sinus floor augmentation. *Int. J. Periodont. Restorat. Dent.* 17: 11–25.

[100] Abe E., Yamamoto M., Taguchi Y. et al. 2000. Essential requirement of BMPs-2/4 for both osteoblast and osteoclast formation in murine bone marrow cultures from adult mice: antagonism by noggin. *J. Bone Miner. Res.* 15: 663–673.

[101] Aspenberg P., Jeppson C., and Economides A.N. 2001. The bone morphogenetic proteins antagonist Noggin inhibits membranous ossification. *J. Bone Miner. Res.* 3: 15.

[102] Poynton A.R. and Lane J.M. 2002. Safety profile for the clinical use of bone morphogenetic proteins in the spine. *Spine* 27: S40–S48.

[103] Morizono K., De Ugarte D.A., Zhu M. et al. 2003. Multilineage cells from adipose tissue as gene delivery vehicles. *Hum. Gene Ther.* 14: 59–66.

[104] Jankowski R.J., Deasy B.M., and Huard J. 2002. Muscle-derived stem cells. *Gene Ther.* 9: 642–647.

[105] Lee J.Y., Peng H., Usas A. et al. 2002. Enhancement of bone healing based on *ex vivo* gene therapy using human muscle-derived cells expressing bone morphogenetic protein-2. *Hum. Gene Ther.* 13: 1201–1211.

[106] Bruder S.P., Kurth A.A., Shea M., Hayes W.C., Jaiswal N., and Kadiyala S. 1998. Bone regeneration by implantation of purified, culture-expanded human mesenchymal stem cells. *J. Orthop. Res.* 16: 155–162.

[107] Shang Q., Wang Z., Liu W., Shi Y., Cui L., and Cao Y. 2001. Tissue engineered bone repair of sheep cranial defects with autologous bone marrow stromal cells. *J. Craniofac. Surg.* 12: 586–593.

[108] Hannallah D., Peterson B., Lieberman J.R., Fu F.H., and Huard J. 2003. Gene therapy in orthopedic surgery. *Inst. Course Lect.* 52: 753–768.

[109] Gamradt S.C. and Lieberman J.R. 2004. Genetic modification of stem cells to enhance bone repair. *Ann. Biomed. Eng.* 32: 136–147.

[110] Alden T.D., Pittman D.D., Beres E.J. et al. 1999. Percutaneous spinal fusion using bone morphogenetic protein-2 gene therapy. *J. Neurosurg.* 90: 109–114.

[111] Alden T.D., Pittman D.D., Hankins G.R., Beres E.J. et al. 1999. *In vivo* endochondral bone formation using a bone morphogenetic protein-2 adenoviral vector. *Hum. Gene Ther.* 10: 2245–2253.

[112] Kakar S. and Einhorn T.A. 2004. Tissue engineering of bone. In J.O. Hollinger, T.A. Einhorn, and B.A. Doll (Eds.), *Bone Tissue Engineering*, Chapter 11, Boca Raton, FL, CRC Press.

[113] Hutmacher D.W. 2000. Scaffolds in tissue engineering bone and cartilage. *Biomaterials* 21: 2529–2543.

[114] Gerhart T.N., Kirker-Head C.A., Kriz M.J. et al. 1993. Healing segmental femoral defects in sheep using recombinant human bone morphogenetic protein. *Clin. Orthop.* 293: 317–326.

[115] Li R.H. and Wozney J.M. 2001. Delivering on the promise of bone morphogenetic proteins. *Trends Biotechnol.* 19: 255–265.

[116] Radomsky M.L., Aufdemorte T.B., Swain L.D. et al. 1999. Novel formulation of fibroblast growth factor-2 in a hyaluronan gel accelerates fracture healing in nonhuman primates. *J. Orthop. Res.* 17: 607–614.

[117] Sellers R.S., Zhang R., Glasson S.S. et al. 2000. Repair of articular cartilage defects one year after treatment with recombinant human bone morphogenetic protein-2 (rhBMP-2). *J. Bone Joint Surg. Am.* 82: 151–160.

[118] Boden S.D. 1999. Bioactive factors for bone tissue engineering. *Clin. Orthop.* 367: S84–S94.

[119] Wang S. 2000. Study on the mechanism of bone formation of bioactive materials BMP/Beta-TCP restoring bone defects by using quantitative analysis methods. Society for Biomaterials, *6th World Biomaterials Conference*, Hawaii, USA.

[120] Jin Q.M., Takita H., Kohgo T. et al. 2000. Effects of geometry of hydroxyapatite as a cell substratum in BMP-induced ectopic bone formation. *J. Biomed. Mater. Res.* 51: 491–499.

[121] Ijiri S., Nakamura T., Fujisawa Y. et al. 1997. Ectopic bone induction in porous apatite-woolastonite-containing glass ceramic combined with bone morphogenetic protein. *J. Biomed. Mater. Res.* 35: 421–432.

[122] Winn S.R., Uludag H., and Hollinger J.O. 1999. Carrier systems for bone morphogenetic proteins. *Clin. Orthop.* 3679: S95–S106.

[123] Whang K., Goldstick T.K., and Healy K.E. 2000. A biodegradable polymer scaffold for delivery of osteotropic factors. *Biomaterials* 21: 2545–2551.

[124] Lee S.C., Shea M., Battle M.A. et al. 1994. Healing of large segmental defects in rat femurs is aided by rhBMP-2 in PLGA matrix. *J. Biomed. Mater. Res.* 28: 1149–1156.

[125] Hollinger J.O. and Leong K. 1996. Poly(alpha-hydroxy acids): carriers for bone morphogenetic proteins. *Biomaterials* 17: 187–194.

[126] Lucas P.A., Laurencin C., Syftestad G.T. et al. 1990. Ectopic induction of cartilage and bone by water soluble proteins from bovine bone using a polyanhydride delivery vehicle. *J. Biomed. Mater. Res.* 24: 901–911.

[127] Clokie C.M. and Urist M.R. 2000. Bone morphogenetic protein excipients: comparative observations on poloxamer. *Plast. Reconstr. Surg.* 105: 628–637.

[128] Saito N., Okada T., Toba S. et al. 1999. New synthetic absorbable polymers as BMP carriers: plastic properties of poly-D,L-lactic acid-polyethylene glycol block copolymers. *J. Biomed. Mater. Res.* 47: 104–110.

[129] Kenley R., Marden L., Turek T. et al. 1994. Osseous regeneration in the rat calvarium using novel delivery systems for recombinant human bone morphogenetic protein-2 (rhBMP-2). *J. Biomed. Mater. Res.* 28: 1139–1147.

[130] Gao T., Lindholm T.S., Marttinen A. et al. 1996. Composite of bone morphogenetic protein (BMP) and Type I. V. collagen, coral-derived coral hydroxyapatite, and tricalcium phosphate ceramics. *Int. Orthop.* 20: 321–325.

[131] Higuchi T., Kinoshita A., Takahashi K. et al. 1999. Bone regeneration by recombinant human bone morphogenetic protein-2 in rat mandibular defects. An experimental model of defect filling. *J. Periodontol.* 70: 1026–1031.

[132] Ripamonti U. 1996. Osteoinduction in porous hydroxyapatite implanted in heterotropic sites of different animal models. *Biomaterials* 17: 31.

[133] Galgut P.N. 1994. A technique for treatment of extensive periodontal defects: a case study. *J. Oral Rehabil.* 21: 27–32

[134] Askar I., Gultan S.M., Erden E., and Yormuk E. 2003. Effects of polyglycolic acid bioabsorbable membrane and oxidized cellulose on the osteogenesis in bone defects: an experimental study. *Acta Chir. Plast.* 45: 131–138.

[135] Tsuchiya K., Mori T., Chen G., Ushida T., Tateishi T., Matsuno T., Sakamoto M., and Umezawa A. 2004. Custom shaping system for bone regeneration by seeding marrow stromal cells onto a web-like biodegradable hybrid sheet. *Cell Tissue Res.* 316: 141–153.

[136] Sharma B. and Elisseeff J.H. 2004. Engineering structurally organized cartilage and bone tissues. *Ann. Biomed. Eng.* 32: 148–159.

[137] Aurell C.J. and Flodin P. 2002. *New Linear Block Polymer.* July 11, 2002.

[138] Aurell C.J. and Flodin P. 2003. *A Method for Preparing an Open Porous Polymer Material and an Open Porous Polymer Material.* EP1353982.

[139] Zhang J.Y., Beckman E.J., Hu J., Yuang G.G., Agarwal S., and Hollinger J.O. 2002. Synthesis, biodegradability, and biocompatibility of lysine diisocyanate-glucose polymers. *Tissue Eng.* 8: 771–785.

[140] Zhang J.Y., Beckman E.J., Piesco N.J., and Agarwal S. 2002. A new peptide-based urethane polymer: synthesis, biodegradation, and potential to support cell growth *in vitro. Biomaterials* 21: 1247–1258.

[141] Zhang J.Y., Doll B.A., Beckman E.J., and Hollinger J.O. 2003. Three-dimensional biocompatible ascorbic acid-containing scaffold for bone tissue engineering. *Tissue Eng.* 9: 1143–1157.

[142] Doll B.A., Beckman E.J., and Hollinger J.O. 2003. A biodegradable polyurethane–ascorbic acid scaffold for bone tissue engineering. *J. Biomed. Mater. Res.* 67A: 389–400.

[143] De Groot J.H., Kuijper H.W., and Pennings A.J. 1997. A novel method for fabrication of biodegradable scaffolds with high compression moduli. *J. Mater. Sci.: Mater. Med.* 8: 707–712.

[144] De Groot J.H., de Vrijer R., Pennings A.J., Klompmaker J., Veth R. P.H., and Jansen H.W.B. 1996. Use of porous polyurethanes for meniscal reconstruction and meniscal prosthesis. *Biomaterials* 17: 163–174.

[145] Szycher M. Biostability of polyurethane elastomers: a critical review. 1988. *J. Biomater. Appl.* 3: 297–402.

[146] Spaans C.J., De Groot J.H., Van der Molen L.M., and Pennings A.J. 2001. New biodegradable polyurethane-ureas, polyurethane and polyurethane-amide for *in-vivo* tissue engineering: structure–properties relationships. *Polym. Mater. Sci. Eng.* 85: 61–62.

[147] Spaans C.J., De Groot J.H., Belgraver V.W., and Pennings A.J. 1988. A new biomedical polyurethane with a high modulus based on 1,4-butanediisocyanate and e-caprolactone. *J. Mater. Sci.: Mater. Med.* 9: 675–678.

[148] Skarja G.A. and Woodhouse K.A. 2000. Structure–property relationships of degradable polyurethane elastomers containing an amino acid-based chain extender. *J. Appl. Polym. Sci.* 75: 1522–1534.

[149] Oertel G. *Polyurethane Handbook.* 1994. 2nd ed., Berlin, Germany, Hanser Gardner Publications.

[150] Spaans C.J., Belgraver V.W., Rienstra O., De Groot J.H., Veth R.P.H., and Pennings A.J. 2000. Solvent-free fabrication of micro-porous polyurethane-amide and polyurethane-urea scaffolds for repair and replacement of the knee-joint meniscus. *Biomaterials* 21: 2453–2460.

[151] Spaans C.J., De Groot J.H., Dekens F.G., Veth R.P.H., and Pennings A.J. 1999. Development of new polyurethanes for repair and replacement of the knee joint meniscus. *Polym. Preprints* 40: 589–590.

[152] Surgeons, A.S.O.P., *Procedural Statistics.* 2004.

[153] Miller M.J. and Patrick C.W. 2003. Tissue engineering. *Clin. Plast. Surg.* 30: 91–103, vii.

[154] Folkman J. and Hochberg M.M. 1973. Self-regulation of growth in three dimensions. *J. Exp. Med.* 138: 745.

[155] Vacanti J.P. et al. 1988. Selective cell transplantation using bioabsorbable artificial polymers as matrices. *J. Pediatr. Surg.,* 23: 3.

[156] Smahel J. 1971. Biology of the stage of plasmatic imbibition. *Br. J. Plast. Surg.* 24: 140–143.

[157] Converse J.M. et al. 1975. Inosculation of vessels of skin graft and host bed: a fortuitous encounter. *Br. J. Plast. Surg.* 28: 274–282.

[158] Hansen-Smith F.M. and Carlson B.M. 1979. Cellular responses to free grafting of the extensor digitorum longus muscle of the rat. *J. Neurol. Sci.* 41: 149–173.

[159] Heslop B.F., Zeiss I.M., and Nisbet N.W. 1960. Studies on transference of bone: I. A comparison of autologous and homologous implants with reference to osteocyte survival, osteogenesis, and host reaction. *Br. J. Exp. Pathol.* 41: 269.

[160] DeLeu J. and Trueta J. 1965. Vascularization of bone grafts in the anterior chamber of the eye. *J. Bone Joint Surg.* 47B: 319.

[161] Gray J.C. and Elves M.W. 1979. Early osteogenesis in compact bone isografts: a quantitative study of contributions of the different graft cells. *Calcif. Tissue Int.* 29: 225–37.

[162] Dell P.C., Burchardt H., and Glowczewskie F.P. 1985. A roentgenographic, biomechanical, and histological evaluation of vascularized and non-vascularized segmental fibular canine autografts. *J. Bone Joint Surg. Am.* 67: 105–112.

[163] Finkemeier C.G. 2002. Bone-grafting and bone-graft substitutes. *J. Bone Joint Surg. Am.* 84-A: 454–464.

[164] Han C.S. et al. 1992. Vascularized bone transfer. *J. Bone Joint Surg. Am.* 74: 1441–1449.

[165] Pogrel M.A. et al. 1997. A comparison of vascularized and nonvascularized bone grafts for reconstruction of mandibular continuity defects. *J. Oral Maxillofac. Surg.* 55: 1200–1206.

[166] Chang D.W. et al. 2001. Management of advanced mandibular osteoradionecrosis with free flap reconstruction. *Head Neck* 23: 830–835.

[167] Ang E. et al., 2003. Reconstructive options in the treatment of osteoradionecrosis of the craniomaxillofacial skeleton. *Br. J. Plast. Surg.* 56: 92–99.

[168] Pertschuk M.J. and Whitaker L.A. 1985. Psychosocial adjustment and craniofacial malformations in childhood. *Plast. Reconstr. Surg.* 75: 177–184.

[169] Pruzinsky T. 1992. Social and psychological effects of major craniofacial deformity. *Cleft Palate-Craniofacial J.* 29: 578–584; discussion 570.

[170] Schuster G.S. 1999. Oral flora and pathogenic organisms. *Infect. Dis. Clin. North Am.* 13: 757–774, v.

[171] Jiang R.S. et al. 1999. Bacteriology of endoscopically normal maxillary sinuses. *J. Laryngol. Otol.* 113: 825–828.

22

Cartilage Tissue Engineering

Fan Yang
Jennifer H. Elisseeff
Johns Hopkins University

22.1 Introduction

This chapter provides a review of important developments in the field of cartilage tissue engineering, with an emphasis on articular cartilage. Clinical significance of cartilage repair is discussed and the necessity of using tissue engineering approach to solve this problem is presented. Structure, function, and biochemical components of cartilage are described. This chapter will compare and contrast the various options available for cartilage tissue engineering. Four major components of tissue engineering in relation to cartilage are discussed including cells, scaffolds, growth factors, and mechanical stimuli.

Cartilage is a highly specialized connective tissue consisting of dispersed chondrocytes embedded in a rich extracellular matrix [1]. The matrix, primarily composed of proteoglycans, collagen, and water, is directly responsible for the unique functional properties of cartilage and provides shape, resilience, and resistance against compression and shear [2,3]. Despite the relatively simple structure, cartilage has little capacity for repair and regeneration due to many factors such as lack of vascularity and progenitor cells, a sparse and highly differentiated cell population, and a slow matrix turnover [4–6].

There are three types of cartilage: hyaline cartilage, fibrocartilage, and elastic cartilage. Each type of cartilage has different functional properties and hence different biochemical contents. Hyaline cartilage is rich in Type II collagen and proteoglycans and is found mainly on the articular surface of joints where it serves as a shock absorber. Elastic cartilage consists of elastic fibers, not exclusively collagen. The elastic fibers give this type of cartilage the ability to be deformed and return to shape. Examples of elastic cartilage include external ear, epiglottis, and upper portion of larynx. Fibrocartilage is rich in Type I collagen and is found in tissues that are subject to tensile forces such as the intervertebral disk.

More than 1 million surgical procedures in the United States involve cartilage replacement. Conventional therapies include drilling and debriding, autologous cartilage transplantation, and implantation of artificial polymers or metal prostheses [7]. Drilling and debriding may provide immediate relief of symptoms but usually result in the production of a fibrocartilaginous tissue with poor long-term outcome. Transplantation can be an effective treatment for patients with small cartilage defects, but its application is limited due to insufficient donor tissue supply, donor site morbidity, and technical difficulties to shape host cartilage into desired, delicate three-dimensional shapes. Artificial prostheses can be a source of infection, unwanted protein deposition, and immune responses without being able to adapt to environmental stresses.

22.1.1 Prospects for Tissue Engineering

Given the clinical significance of cartilage injury and the poor outcomes of the conventional therapies outlined above, investigators have been motivated to seek other approaches for improving cartilage treatment. One possible solution is to create new cartilage using tissue engineering strategies. Tissue engineering is an interdisciplinary area that combines biology, material sciences, and clinical medicine to regenerate or restore tissue structure and function that is lost due to trauma, disease, or congenital abnormalities. Cartilage contains only one cell type and has no blood supply. This relatively simple cartilage structure has allowed significant progress in the area of cartilage tissue engineering [8,9].

22.2 Cells

Cells are the building blocks of tissues and the fundamental source of extracellular matrix that forms tissues. Identification and creation of a suitable cell population is the first critical step toward cartilage tissue engineering and often requires a large number of cells. Numerous cell types have been explored including chondrocytes [10,11], bone-marrow derived mesenchymal stem cells (MSCs) [12,13], stem cells isolated from adipose tissue [14], from muscle [15], and embryonic stem cells [16,17]. All these cell types have been shown to exhibit a chondrogenic potential under appropriate culture conditions. When choosing the optimal cell type, it is important to consider the cell proliferative capacity, phenotype stability, and immunogenicity.

22.2.1 Chondrocytes

Chondrocytes, the major cell type that is present in differentiated cartilage, is the most obvious cell option for cartilage tissue engineering. Thus far, autologous chondrocyte transplantation (ACT) is the only Food and Drug Administration (FDA) approved cartilage repair product in the United States. This procedure uses *in vitro* expanded autologous chondrocytes, harvested from a biopsy, combined with a periosteal membrane to repair a cartilage defect (Figure 22.1).

The ACT was first used in humans in 1987 and results of the first pilot study was published in 1994 [10]. Since then more than 950 patients have been treated with this technique with 2 to 10 years follow-up [11]. Histology of the repair tissue shows that this technique gives stable long-term results with a higher success rate in femoral condyle (84 to 90%) than in other locations (average 74%). Despite the encouraging results that ACT has shown, several problems remain with the approach. First, the age or disease state of the patient may limit the availability and function of desired chondrocytes. *In vitro* expansion is also limited due to replicative senescence that chondrocytes exhibit. More importantly, once removed from their extracellular environment and expanded in monolayer, chondrocytes would rapidly lose their differentiated phenotype, characterized by transforming into a fibroblastic morphology, decreasing expression of Type II Collagen and aggrecan with an increase in Type I Collagen expression [18,19]. Studies have shown that chondrocyte phenotype can be retained or reexpressed by suspending

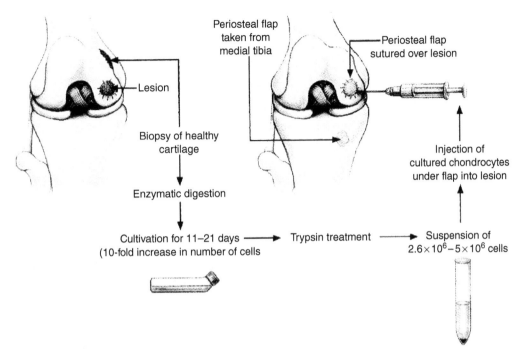

FIGURE 22.1 Flow diagram for ACT articular chondrocyte harvest and implantation. (From Brittberg, M., Lindahl, A., Nilsson, A., Ohlsson, C., Isaksson, O., and Peterson, L. *N. Engl. J. Med.* 1994; 331: 889–895. With permission.)

chondrocytes in a three-dimensional environment such as agarose gel [20], alginate beads [21], and collagen gel [22]. Optimal combinations of monolayer and three-dimensional cultures for cartilage tissue engineering are currently being investigated [23]. Challenges remain for patients who do not have healthy cartilage where autologous cell transplantation cannot be used and alternate cell sources need to be explored.

22.2.2 Mesenchymal Stem Cells

Mesenchymal stem cells are multipotent progenitor cells that have the capacity to differentiate into a variety of connective tissue cells including bone [24], cartilage [25], and adipose tissue [12] both *in vitro* and *in vivo*. MSCs are present in various tissue types during human development and are prevalent in adult bone marrow. As a readily available source, MSCs can be isolated, expanded in culture, and differentiated into chondrogenic cells under appropriated tissue conditions. Numerous studies have shown that transforming growth factor-beta (TGF-β) plays an important role in inducing undifferentiated MSCs into the chondrogenic pathway. In the presence of TGF-β, MSCs will gradually transform from a fibroblastic morphology to a mature chondrocyte morphology, accompanied by a production of cartilage-specific extracellular matrix proteins including Type II collagen, glycoaminoglycan, and proteoglycans. Unlike osteogenesis, which can be induced in monolayer culture, chondrogenesis of MSCs can only be achieved in micromass pellet culture [25] or high density suspension in a three-dimensional environment [26]. This close cell–cell contact requirement resembles the natural environment of embryonic cartilage development, which suggests that MSCs are undergoing a similar pathway. These results indicate that successful tissue-engineered cartilage repair based on MSCs can be achieved by recapitulating aspects of embryonic tissue formation. Studies so far demonstrated the great potential of using MSCs for cartilage regeneration as an alternate resource, which may eliminate the need harvest cartilage from the patient. More *in vivo* studies on cartilage regeneration using MSCs will occur as they reach clinical application.

22.2.3 Other Cell Sources

The discovery of human embryonic stem cells (hESCs) capable of indefinite replication and the ability to differentiate into all somatic cells of the body brings optimism that great progress will be made to cell transplantation therapies [27,28]. Theoretically, hESCs can be used to generate any desired cell or tissue for the treatment of degenerative diseases. So far, very few studies have been performed inducing hESCs into chondrogenic lineage [16,17]. Challenges with hESCs include potential of undesired growth, difficulty to achieve homogeneous differentiation *in vivo*, and related ethics issues. Other potential cell sources were found by Mizuno [29], who demonstrated that human dermal fibroblasts seeded onto Type I collagen sponges containing demineralizd bone matrix can be stimulated to produce a cartilaginous tissue expressing Type II collagen. Also, Lorenz [30] recently reported that human adipose tissue derived cells can also be induced into chondrogenic pathway *in vitro*.

22.3 Scaffolds

Although some approaches to cartilage tissue engineering use cells alone (i.e., ACT), most studies indicate that the scaffold is helpful for promoting desired cell phenotypes and regeneration of cartilage. Ideally, the scaffold serves as a structural support for cells at the early stage of tissue development and degrades into nontoxic products at a rate related to the formation of a functional extracellular matrix. Numerous scaffolds and matrices have been explored for cartilage tissue engineering including both natural and synthetic polymers. Natural polymers are usually porous, biocompatible, and can degrade without immunogenicity. Problems related to naturally derived scaffolds include batch-to-batch quality variance and weak structural properties. By contrast, synthetic polymers provide a more consistent quality and more control over the material properties such as degradation rate, mechanical properties, and surface biology. When choosing a scaffold, tissue engineers must decide which genre of materials, or combination, is more appropriate depending on the specific purpose and location of use.

22.3.1 Natural Polymers

As a major component of cartilage extracellular matrix, collagen was the first natural material explored as scaffold for cartilage tissue engineering. Collagen can be easily extracted from tissues such as tendon and can be readily processed into multiple forms including sponges or meshes. Collagen is liquid at 4°C and can be polymerized by raising the temperature. Further crosslinking after gelation introduces extended degradation times and greater mechanical strength [31]. Chondrocytes have been shown to maintain their morphology and phenotype when seeded in collagen gels [32–34]. Grande et al. [35] showed that collagen sponges stimulate collagen synthesis while bioabsorbable polymer such as polyglycolic acid (PGA) enhanced proteoglycan synthesis.

Hyaluronan comprises 1 to 10% of the cartilage glycosaminoglycan (GAG) and plays an important role in many biological processes such as cell attachment, hydration, and matrix synthesis. HYAFF® 11 (Fidia Advanced Biopolymers, Abano Terme, Italy) is a semisynthetic derivative of Hyaluronan by esterification of carboxyl groups of the glucuronic acid with benzyl alcohol. This processing greatly improves the originally poor ductility of hyaluronan, enabling HYAFF® 11 to be processed into multiple forms such as fibers, membranes, or microspheres. HYAFF and ACP, two hyaluronan-based biomaterials with different degradation profiles, were tested as chondrogenic delivery vehicles for rabbit MSCs and compared with a well-characterized porous calcium phosphate ceramic delivery vehicle [36]. Results showed that HYAFF 11 sponges incorporated twice as many cells and produced a 30% increase in the relative amount of bone and cartilage per unit area.

Agarose and alginate are examples of natural materials that are based on linear polysaccharide polymers that are extracted from seaweed [37]. These polymers are highly water-soluble due to a large number of hydroxyl groups. Agarose gels can be easily formed by pouring a warm liquid solution of the polymer into a mold, which gels as the solution cools to room temperature. Benya et al. [38] demonstrated that dedifferentiated rabbit articular chondrocytes can survive and redifferentiate during suspension culture in firm agarose gels. Recently, it has also been shown that dynamic pressure transmission through agarose gel was complete and immediate, which validates the use of agarose gels in studies employing dynamic pressurization in cartilage tissue engineering [39].

Alginate is a block copolymer of alternating residues of M (D-mannuronic acid) and G (L-guluronic acid). Alginate forms a gel in the presence of divalent ions such as calcium or magnesium. Alginate has been widely used as scaffold in cartilage tissue engineering because its solubility, gelation rate, and mechanical properties can be adjusted by the G/M ratio in the polymer backbone. Hauselmann showed that human chondrocytes cultured in alginate formed a similar matrix to native human articular cartilage [40]. When exposed to reduced oxygen tension (5%), chondrocytes encapsulated in an alginate gel has been shown to upregulate aggrecan and Type II collagen gene expression while downregulating Type I collagen expression [41].

22.3.2 Synthetic Polymers

No single existing polymer can meet the demands of all tissue-engineering applications, given the unique biochemical and mechanical requirements for different tissues or sites of implantation. Poly(lactic acid, PLA), PGA, and poly(lactic-co-glycolic acid, PLGA) are among the earliest synthetic polymers that were investigated for cartilage tissue engineering [7]. Originally designed for use as biodegradable surgical sutures, these polyesters can form highly porous lattice structures that allow cell encapsulation and proliferation with effective nutrient and waste transfer. A number of cell types, such as chondrocytes and bone-marrow derived MSCs have been seeded [42,43] in these polyester materials and cartilage matrix production have been achieved both *in vitro* and *in vivo*. Most studies involve implantation of cell-seeded polymer constructs into subcutaneous pockets in nude mice. Grande has shown neocartilage formation when implanted a chondrocyte-PGA construct into the rabbit knee defects [44]. PGA and PLA scaffolds are also biodegradable via nonspecific hydrolytic cleavage at the ester linkage site. Although polyester scaffolds have been widely used, they too have drawbacks such as poor cell attachment, difficulty in molding, and mild inflammatory reaction once implanted *in vivo*.

Injectable hydrogels are a genre of scaffolds that can be administered to fill defects of any shape and polymerize *in situ* with precise temporal and spatial control in a minimally invasive manner. Sims initially investigated the utilization of polyethylene oxide, a linear polyether-$(CH_2CH_2O)_n$ as encapsulating polymer scaffolds for delivering large numbers of isolated chondrocytes via injection [45]. New cartilage formation was observed as early as 6 weeks, with specimens exhibiting a white opalescence and histological characteristics consistent with those of hyaline cartilage.

Moreover, polyethylene oxide gels can be modified by adding acrylate end group that can potentially allow the transformation from a liquid to a solid with a defined structure in the presence of ultraviolet light. Elisseeff demonstrated that this transdermal photopolymerization technique can be used to implant the chondrocyte-seeded scaffolds, allowing simple exposure of the skin surface to light to induce gelation of scaffolds injected subcutaneously in nude mice [46,47] (see Figure 22.2). Williams et al. [26] has further shown the feasibility of applying photopolymerizing poly(ethylene glycol)-based hydrogels for chondrogenesis of adult goat MSCs, in the presence of growth factors. Pluronics is a commercially available copolymer of poly-(ethylene oxide) and poly-(propylene oxide) (Pluronics; BASF Corporation; Mount Olive, NJ). Pluronics forms a gel through physical crosslinks by hydrophobic group association at body temperature [48]. The main limitation of hydrogels for cartilage tissue engineering is the relatively weak mechanical properties when compared to other solid fibrous scaffolds. Mechanical properties of hydrogels may be improved by creating composite materials that incorporate a solid and hydrogel.

Ideally, polymers used in tissue engineering should be not only biocompatible, but also bioresponsive and biodegradable. The common feature of early polymers in biomedical applications was their biological inertness. Research in the past decade has been focusing more on the development of bioresponsive polymers that can respond to the local biological and physical microenvironment and elicit a controlled action and reaction. To enhance the interaction between the scaffold and the encapsulated cells, various

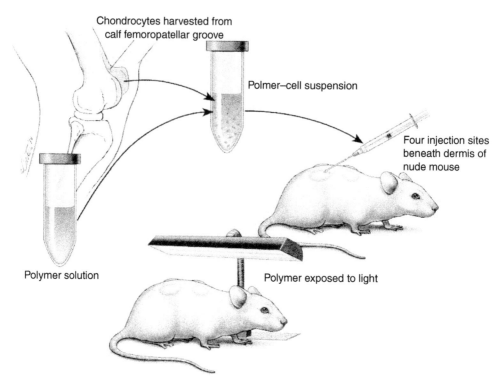

FIGURE 22.2 Schematic of procedure for cartilage tissue engineering by using transdermal photopolymerization depicting (1) isolation of bovine chondrocytes from the femoropatellar groove and combination with polymer (10 to 20% poly[ethylene oxide]-dimethacrylate) to form (2) a polymer/chondrocyte suspension. The polymer/chondrocyte suspension is subsequently (3) injected subcutaneously on the dorsal surface of a nude mouse, and (4) photopoly-merized by placement of the mouse under an ultraviolet A lamp for 3 min. (From Elisseeff, J., Anseth, K., Sims, D., McIntosch, W. et al. *Plast. Reconstr. Surg.* 1999; 104: 1014–1022. With permission.)

biological signals such as cell adhesion peptides (i.e., RGD — arginine–glycine–aspartic acid), and growth factors have been incorporated into scaffolds. Enhanced cell attachment and extracellular deposition was achieved using scaffolds with biological activity [49–52]. As tissue mass increases and more extracellular matrix is deposited, it is also very important for the scaffold to degrade in a similar rate to allow effective diffusion and space for the newly formed tissue. Also, biodegradable scaffolds can play an important role as delivery vehicles for sustained growth factor release [53].

22.4 Mechanical Stimuli

Cartilage, especially articular cartilage, is a tissue that is accustomed to mechanical stimulation. During joint loading, articular cartilage undergoes compression and a plowing motion, and chondrocytes imbed-ded in the cartilage thus experience direct compressive deformation, shear, and hydrostatic pressure [54]. To understand the mechanotransduction pathway, various bioreactors have been designed to mimic the conditions necessary for functional cartilage tissue engineering. Freed and Vunjak-Novakovic [55] have demonstrated that dynamic stimuli is critical for cartilage matrix production by chondrocytes seeded on a PGA scaffold. Under static conditions, cell growth rates on the PGA scaffolds decreased due to limited diffusion caused by increased cell mass and new matrix deposition. Therefore, to engineer cartilage tissue that is clinically useful in size, it is essential to improve the *in vitro* culture conditions [42,56,57]. Carver and Heath [58] examined the effect of intermittent pressure on chondrocytes seeded on PGA scaffolds using a semiperfusion bioreactor. It was found that intermittent pressure increased the GAG content to native levels but the compressive modulus of the construct, though increased, did not reach the native level. This suggested that although ECM synthesis has been increased, matrix organization may be incomplete. Therefore, a current challenge in engineering functional cartilage tissue using bioreactor approach includes not only upregulation of ECM synthesis, but also its retention in the matrix and reorganization to restore the native-level mechanical function. More work is still needed to determine the type of mechanical forces, amplitude, and duty cycle that would induce the chosen cell type to form a functional cartilage tissue.

22.5 Growth Factors

Research advances in the fields of embryonic development, molecular and cellular biology have identified a variety of proteins that can affect cell chondrogenic differentiation and phenotypic expression, including transforming growth factor TGF-βs, insulin-like growth factor (IGF-1), bone morphogenetic proteins (BMPs), fibroblast growth factors (FGFs), and epidermal growth factor (EGF) [59]. These molecules have a broad range of activities, including proliferation induction, increasing the synthesis and deposition of cartilage extracellular matrix by chondrocytes, and inducing chondrogenesis by MSCs. Transforming growth factor (TGF-β) superfamily, including TGF-βs 1–3 and insulin-like growth factor-1 (IGF-1), stimulate GAG synthesis while chondrocyte synthesis of collagen II is strongly stimulated by IGF-1. It has also been demonstrated that MSC cultures exposed to TGF-βs show increased cell density, nodule formation, GAG, and Type II collagen synthesis [26,60].

The majority growth factor studies have been *in vitro* and focused on simply adding a soluble growth factor to the medium [26]. However, for clinical application, a continuous and stable effect of growth factors is a challenge. This challenge may be overcome by several possible approaches. Growth factors can be tethered to the scaffold by chemical modification and being released as the scaffold degrades [51]. They can also be encapsulated in polymer microspheres to achieve a sustained release [61]. The major limitation of both the above methods is that growth factors are proteins that can easily denature and it is challenging to retain biological activities during relative harsh chemical synthesis or microsphere suspension process.

Advances in gene transfer technology provide another novel alternative to achieve sustained and localized presentation of growth factors or gene products [62]. A variety of cDNAs have been cloned to stimulate biological processes that could promote chondrogenesis. Direct injection of a vector or genetically modified cells into the defect is conceptually the simplest approach to gene transfer. Besides the growth factors mentioned above, this method can also be applied to deliver chondrogenesis-related transcriptional factors such as Sox-9, signal transduction molecules such as SMADS [63]. Since these molecules function intracellularly and cannot be delivered to cells in a soluble form, gene transfer is a unique method for translation of these technologies in a clinical setting.

22.6 Zonal Organization

Articular cartilage is comprised of three zones that exhibit significant differences in their metabolic rate, matrix synthesis, and response to mechanical stimuli [54]. As opposed to current tissue engineering strategies where chondrocytes from all zones are seeded homogeneously, isolation of chondrocyte subpopulations from specific cartilage zones and their organization into a scaffold to mimic the native organization may provide engineered cartilage with enhanced structure and functional properties [64]. The superficial zone has a relatively low glycoaminoglycan content and compressive modulus, which facilitates the conformation of superficial zone to the opposing interface and allows stress distribution under compressive conditions. Klein isolated chondrocytes from the superficial and middle zones of immature bovine cartilage and cultured in alginate for 1 to 2 weeks. Analysis indicated that chondrocyte subpopulations play an important role in determining the structure, composition, metabolism, and mechanical function of tissue-engineered cartilage constructs [65]. Sharma has demonstrated the feasibility of using multilayered photopolymerizing hydrogel to recreate the zonal organization of articular cartilage [66].

Chondrocytes from the upper, middle, and lower zones of bovine articular cartilage were isolated and photoencapsulated in a trilayered poly(ethylene oxide)-diacrylate, PEODA, hydrogel. Cell tracking and viability assay showed that cells remained in their respective polymerized layer and cell viability was comparable throughout all three layers. Biochemical analysis further demonstrated that distinct metabolic characteristics are maintained when the chondrocytes are cultured in the layers, suggesting that this multilayer culture system may provide a native tissue mimicking environment that promotes regulatory signal molecules exchange.

22.7 Conclusion

Advances in tissue-engineering research provided an alternative possible therapeutic approach to restore lost tissue and function. Many researchers have focused on cartilage tissue engineering with encouraging results that will continue to motivate research efforts to engineer cartilage tissue with biological and mechanical properties similar to native tissue. The zonal organization of cartilage can be recreated using layered scaffolds, allowing a better mimic of the native cartilage tissue. Further development of cartilage tissue engineering is also closely related to the advances in polymer science to synthesize bioactive and biodegradable polymer with suitable biocompatibility. More research is required to achieve a sustained delivery of biological signals such as growth factors and transcriptional factors to promote tissue development *in vivo*. Identifying and providing suitable mechanical stimuli may also help engineering tissue to produce cartilage-specific biochemical compositions in an organized manner and hence acquire the mechanical properties that would meet the physiological needs of clinical application.

References

[1] Moss, M.L. and Moss-Salentijn, L. Vertebrate cartilages. In: Hall, B.K. (Ed.), *Cartilage: Structure, Function and Biochemistry*. Academic Press, New York, 1983; pp. 1–24.

[2] Heinegard, D., Bayliss, M., and Lorenzo, P. Biochemistry and metabolism of normal and osteoarthritis cartilage. In: Brandt, K.D., Doherty, M., and Lohmander, L.S. (Eds.), *Osteoarthritis*. Oxford University Press, Oxford, 1998; pp. 74–84.

[3] Mow, V.C. and Setton, L.A. Mechanical properties of normal and osteoarthritic articular cartilage. In: Brandt, K.D., Doherty, M., and Lohmander, L.S. (Eds.), *Osteoarthritis*. Oxford University Press, Oxford, 1998; pp. 108–122.

[4] Mankins, H.J. Currents concepts review: the response of articular cartilage to mechanical injury. *J. Bone Joint Surg.* 1982; 64A: 460–466.

[5] Silver, F.H. and Glasgold, A.I. Cartilage wound healing: an overview. *Otolaryngol. Clin. North Am.* 1995; 28: 847–864.

[6] Hunziker, E.B. Articular cartilage repair: problems and prospectives. *Biorheology* 2000; 37: 163–164.

[7] Langer, R. and Vacanti, J.P. Tissue engineering, *Science* 1993; 260: 920–926.

[8] Vacanti, C.A. and Upton, J. Tissue-engineered morphogenesis of cartilage and bone by means of cell transplantion using synthetic biodegradable polymer matrices [review]. *Clin. Plas. Surg.* 1994; 21: 445–462.

[9] Vacanti, C.A. and Vacanti, J.P. Bone and cartilage reconstruction with tissue engineering approaches [review]. *Otolaryngol. Clin. North Am.* 1994; 27: 263–276.

[10] Brittberg, M., Lindahl, A., Nilsson, A., Ohlsson, C., Isaksson, O., and Peterson, L. Treatment of deep cartilage defects in the knee with autologous chondrocyte transplantation. *N. Engl. J. Med.* 1994 331: 889–895.

[11] Brittberg, M., Tallheden, T., Sjogren-Jansson, B., Lindahl, A., and Peterson, L. Autologous chondrocytes used for articular cartilage repair: an update. *Clin. Orthop.* 2001; (391 Suppl.): S337–S348.

[12] Pittenger, M.F. et al. Multilineage potential of adult human mesenchymal stem cells. *Science* 1999; 284: 143–147.

[13] Jiang Y. et al. Pluripotency of mesenchymal stem cells derived from adult marrow. *Nature* 2002; 418: 41–49.

[14] Wickham, M.Q., Erickson, G.R., Gimble, J.M., Vail, T.P., and Guilak, F. Multipotent stromal cells derived from the infrapatellar fat pad of the knee. *Clin. Orthop.* 2003; (412): 196–212.

[15] Deasy, B.M. and Huard, J. Gene therapy and tissue engineering based on muscle-derived stem cells. *Curr. Opin. Mol. Ther.* 2002; 4: 382–389.

[16] Kramer, J. et al. Embryonic stem cell-derived chondrogenic differentiation *in vitro*: Activation by BMP-2 and BMP-4. *Mech. Dev.* 2000; 92: 193–205.

[17] Nakayama, N. et al. Macroscopic cartilage formation with embryonic-stem-cell-derived meso-dermal progenitor cells. *J. Cell Sci.* 2003; 116: 2015–2028.

[18] Schnabel, M. et al. Dedifferentiation-associated changes in morphology and gene expression in primary human articular chondrocytes in cell culture. *Osteoarthr. Cartil.* 2002; 10: 62–70.

[19] Benya, P.D., Padilla, S.R., and Nimni, M.E. Independent regulation of collagen types by chondrocytes during the loss of differentiated function in culture. *Cell* 1978; 15: 1313–1321.

[20] Benya, P.D. and Shaffer, J.D. Dedifferentiated chondrocytes reexpress the differentiated collagen phenotype when cultured in agarose gels. *Cell* 1982; 30: 215–224.

[21] Bonaventure, J. et al. Reexpression of cartilage-specific genes by dedifferentiated human articular chondrocytes cultured in alginate beads. *Exp. Cell Res.* 1994; 212: 97–104.

[22] Gibson, G.J. et al. Effects of matrix molecules on chondrocyte expression: synthesis of a low molecular weight collagen species by cells cultured within collagen gels. *J. Cell Biol.* 1982; 93: 767–774.

[23] Kuriwaka, M. et al. Optimum combination of monolayer and three-dimensional cultures for cartilage-like tissue engineering. *Tissue Eng.* 2003; 9: 41–49.

[24] Jaiswal, N., Haynesworth, S.E., Caplan, A.I., and Bruder, S.P. Osteogenic differentiation of purified, culture-expanded human mesenchymal stem cells *in vitro*. *J. Cell Biochem.* 1997; 64: 295–312.

[25] Barry, F., Boynton, R.E., Liu, B., and Murphy, J.M. Chondrogenic differentiation of mesenchymal stem cells from bone marrow: differentiation-dependent gene expression of matrix components. *Exp. Cell Res.* 2001; 268: 189–200.

[26] Williams, C.G., Kim, T.K., Taboas, A., Malik, A., Manson, P., and Elisseeff, J. *In vitro* chondrogenesis of bone marrow-derived mesenchymal stem cells in a photopolymerizing hydrogel. *Tissue Eng.* 2003; 9: 679–688.

[27] Thomson, J.A., Itskovitz-Eldor, J., Shapiro, S.S. et al. Embryonic stem cell lines derived from human blastocysts. *Science* 1998; 282: 1145–1147.

[28] Amit, M., Carpenter, M.K., Inokuma, M.S. et al. Clonally derived human embryonic stem cell lines maintain pluripotency and proliferative potential for prolonged periods of culture. *Dev. Biol.* 2000; 227: 271–278.

[29] Mizuno, S. and Glowacki, J. Three-dimensional composite of demineralized bone powder and collagen for *in vitro* analysis of chondroinduction of human dermal fibroblasts. *Biomaterials* 1996; 17: 1819–1825.

[30] Zuk, P.A., Zhu, M., Mizuno, H., Huang, J., Futrell, J.W., Katz, A.J. et al. Multilineage cells from human adipose tissue: implications for cell-based therapies. *Tissue Eng.* 2001; 7: 211–228.

[31] Bell, E. Deterministic models for tissue engineering. *Mol. Eng.* 1995; 1: 28–34; Campoccia, D., Doherty, P., Radice, M., Brun, P., Abatangelo, G., and Williams, D.F. Semisynthetic resorbable matrices from hyaluronan esterification. *Biomaterials* 1998; 19: 2101–2127.

[32] Weiser, L., Bhargava, M., Attia, E., and Torzilli, P.A. Effect of serum and platelet-derived growth factor on chondrocyte growth in collagen gels. *Tissue Eng.* 1999; 5: 533–544.

[33] Ochi, M., Uchio, Y., Tobita, M., and Kuriwaka, M. Current concepts in tissue engineering techniques for repair of cartilage defect. *Artif. Organs* 2001; 25: 172–179.

[34] Yamamoto, T., Katoh, M., Fukushima,R., Kurushima, T., and Ochi, M. Effect of glycosaminoglycan production on hardness of cultured cartilage fabricated by the collagen-gel embedding method. *Tissue Eng.* 2002; 8: 119–129.

[35] Grande, D.A., Halberstadt, C., Naughton, G., Schwartz, R., and Manji, R. Evaluation of matrix scaffolds for tissue engineering of articular cartilage grafts. *J. Biomed. Mater. Res.* 1997; 34: 211–220.

[36] Solchaga, L.A., Dennis, J.E., Goldberg, V.M., and Caplan, A.I. Hyaluronic acid-based polymers as cell carriers for tissue-engineered repair of bone and cartilage. *J. Orthop. Res.* 1999; 17: 205–213.

[37] Paige, K.T., Cima, L.G., Yaremchuk, M.J., Vacanti, J.P., and Vacanti, C.A. Injectable Cartilage. *Plast. Reconstr. Surg.* 1995; 96: 168–178.

[38] Benya, P.D. and Shaffer, J.D. Dedifferentiated chondrocytes reexpress the differentiated collagen phenotype when cultured in agarose gels. *Cell* 1982; 30: 215–224.

[39] Saris, D.B., Mukherjee, N., Berglund, L.J., Schultz, F.M., An, K.N., and O'Driscoll, S.W. Dynamic pressure transmission through agarose gels. *Tissue Eng.* 2000; 6: 531–537.

[40] Hauselmann, H.J. et al Adult human chondrocytes cultured in alginate form a matrix similar to native human articular cartilage. *Am. J. Physiol.* 1996; 271: C742–C752.

[41] Murphy, C.L. and Sambanis, A. Effect of oxygen tension and alginate encapsulation on restoration of the differentiated phenotype of passaged chondrocytes. *Tissue Eng.* 2001; 7: 791–803.

[42] Agrawal, C.M. and Ray, B. Biodegradable polymeric scaffolds for musculoskeletal tissue engineering. *J. Biomed. Mater. Res.* 2001; 55: 141–150.

[43] Martin, I. et al. Selective differentiation of mammalian bone marrow stromal cells cultured on three-dimensional polymer foams. *J. Biomed. Mater. Res.* 2001; 55: 229–235.

[44] Grande, D.A., Pitman, M.I., Peterson, L., Menche, D., and Klein, M. The repair of experimentally produced defects in rabbit articular cartilage by autologous chondrocyte transplantation. *J. Orthop. Res.* 1989; 7: 208–218.

[45] Sims, C.D., Butler, P.E., Casanova, R., Lee, B.T., Randolph, M.A., Lee, W.P. et al. Injectable cartilage using poly-ethylene oxide polymer substrates. *Plast. Reconstr. Surg.* 1996; 98: 843–850.

[46] Elisseeff, J., Anseth, K., Sims, D., McIntosh, W., Randolph, M., Yaremchuk, M. et al. Transdermal photopolymerization of poly(ethylene oxide)-based injectable hydrogels for tissue-engineered cartilage. *Plast. Reconstr. Surg.* 1999; 104: 1014–1022.

[47] Elisseff, J., Anseth, K., Sims, D., McIntosh, W., Randolph, M., and Langer, R. Transdermal photopolymerization for minimally invasive implantation. *Proc. Natl Acad. Sci. USA* 1999; 96: 3104–3107.

[48] Cao, Y., Rodriguez, A., Vacanti, M., Ibarra, C., Arevalo, C., and Vacanti, C.A. Comparative study of the use of poly(glycolic acid), calcium alginate and pluronics in the engineering of autologous porcine cartilage. *J. Biomater. Sci. Polym. Ed.* 1998; 9: 475–487.

[49] Hern, D.L. and Hubbell, J.A. Incorporation of adhesion peptides into nonadhesive hydrogels useful for tissue resurfacing. *J. Biomed. Mater. Res.* 1998; 39: 266–276.

[50] Mann, B.K. et al. Modification of surfaces wit cell adhesion peptides alters extracellular matrix deposition. *Biomaterials* 1999; 20: 2281–2286.

[51] Mann, B.K., Schmedlen, R.H., and West, J.L. Tethered TGF-beta increases extracellular matrix production of vascular smooth muscle cells. *Biomaterials* 2001; 22: 439–444.

[52] Rowley, J.A. and Mooney, D.J. Alginate type and RGD density control myoblast phenotype. *J. Biomed. Mater. Res.* 2002; 60: 217–223.

[53] Holland, T.A., Tessmar, J.K., Tabata, Y., and Mikos, A.G. Transforming growth factor-beta1 release from oligo(poly(ethylene glycol) fumarate) hydrogels in conditions that model the cartilage wound healing environment. *J. Controll. Release* 2004; 94: 101–114.

[54] Mow, V.C. and Guo, X.E. Mechano-electrochemical properties of articular cartilage: their inhomogeneities and anisotropies. *Annu. Rev. Biomed. Eng.* 2002; 4: 175–209. Epub 2002, March 22.

[55] Freed, L.E., Hollander, A.P., Martin, I., Barry, J.R., Langer, R., and Vunjak-Novakovic, G. Chondrogenesis in a cell-polymer-bioreactor system. *Exp. Cell Res.* 1998; 240: 58–65.

[56] Vunjak-Novakovic, G., Martin, I., Obradovic, B., Treppo, S., Grodzinsky, A.J., Langer, R., and Freed, L.E. Bioreactor cultivation conditions modulate the composition and mechanical properties of tissue-engineering cartilage. *J. Orthop. Res.* 1999; 17: 130–138.

[57] Freed, L.E., Vunjak-Novakovic, G., and Langer, P. Cultivation of cell-polymer cartilage implants in bioreactors. *J. Cell Biochem.* 1993; 51: 257–264.

[58] Carver, S.E. and Heath, C.A. 1999. Influence of intermittent pressure, fluid flow, and mixing on the regenerative properties of articular chondrocytes. *Biotechnol. Bioeng.* 1999; 65: 274–281.

[59] Trippel, S.B. Growth factor actions on articular cartilage. *J. Rheumatol. Suppl.* 1995; 43: 129–132.

[60] Worster, A.A., Nixon, A.J., Brower-Toland, B.D., and Williams, J. Effect of transforming growth factor beta1 on chondrogenic differentiation of cultured equine mesenchymal stem cells. *Am. J. Vet. Res.* 2000; 61: 1003–1010.

[61] Elisseeff, J., McIntosh, W., Fu, K., Blunk, T., and Langer, R. Controlled Released IGF-I and TGF-beta on Bovine Chondrocytes Encapsulated in a Photopolymerizing Hydrogel. *J. Orthop. Res.* 2001; 19: 1098–1104 .

[62] Trippel, S.B., Ghivizzani, S.C., and Nixon, A.J. Gene-based approaches for the repair of articular cartilage. *Gene Ther.* 2004; 11, 351–359.

[63] Tsuchiya, H., Kitoh, H., Sugiura, F., and Ishiguro, N. Chondrogenesis enhanced by overexpression of sox9 gene in mouse bone marrow-derived mesenchymal stem cells. *Biochem. Biophys. Res. Commun.* 2003; 301: 338–343.

[64] Schinagl, R.M., Gurskis, D., Chen, A.C., and Sah, R.L. Depth-dependent confined compression modulus of full-thickness bovine articular cartilage. *J. Orthop. Res.* 1997; 15: 499–506.

[65] Klein, T.J., Schumacher, B.L., Schmidt T.A., and Sah, R.L. Tissue engineering of stratified articular cartilage from chondrocyte subpopulations. *Osteoarth. Cartil.* 2003; 11: 595–602.

[66] Sharma, B. and Elisseeff, J. Engineering structurally organized cartilage and bone tissues. *Ann. Biomed. Eng.* 2004; 32: 148–159.

23

Tissue Engineering of the Temporomandibular Joint

Mark E.K. Wong
Kyriacos A. Athanasiou
University of Texas Dental Branch
Rice University

Kyle D. Allen
University of Texas Dental Branch

23.1 Introduction

Tissue engineering of the temporomandibular joint (TMJ) is very much in its infancy. Similar to efforts directed at the replacement of other diseased joints, tissue engineering of the TMJ requires a detailed understanding of anatomy, normal joint activity and various pathologies. The body of knowledge available for the TMJ is relatively incomplete compared to other joints, which accounts for many of the problems encountered in previous reconstructive attempts. To date, very little is known about the manner in which the joint functions and the environment most conducive to its physiology. Even less is known about the impact of skeletal morphology and the presence or absence of a dentition. However, it is believed that both these parameters affect the type and range of motion as well as the forces created during joint function. The manner by which disease affects the TMJ has been derived largely from orthopedic and rheumatology studies of other diarthrodial joints. As various investigators attempt to apply basic tissue engineering techniques to produce replacement components for the TMJ, aggressive efforts are also underway to characterize its normal and abnormal behavior. These studies will provide an essential basis for the development of successful and lasting tissue engineering techniques for the reconstruction of the TMJ.

23.2 Gross Anatomy and Function of the TMJ

Movement of the lower jaw, or mandible, can occur through three planes of space as the result of *rotation* around a transcranial axis (vertical opening and closure), *translation* relative to the skull base (protrusion, retrusion, and side-to-side motion), or a combination of the two actions. These movements are possible, because of the presence of paired joints that separate the lower jaw from the skull base on either side of the cranium. The joints are named for the two bones that provide the articulating surfaces, namely the temporal bone (*temporo*) which houses the superior articulation in an area known as the glenoid fossa and articular eminence, and the mandible (*mandibular*), whose condylar processes provide the inferior portion of the articulation.

The TMJs are diarthrodial structures, which is to say that within the joint, a complete interpositional disc creates separate superior and inferior joint spaces. While this nomenclature suggests that the disc is isolated from its superior and inferior relationships by true "spaces," under normal functional conditions, the separations are extremely small and filled with synovial fluid. Conceptually, it would be more accurate to characterize these areas as *potential* spaces, with superior and inferior volumes approximating 1.0 and 0.5 ml, respectively [1]. This arrangement allows the fibrocartilaginous disc to fill the void between the condylar head and the glenoid fossa, promoting congruity between two dissimilarly shaped and sized structures [2] (Figure 23.1). Alterations in discal properties as a result of disease, improper position, or perforation are common precursors to degeneration of the articular surfaces of the joint, as mechanisms for the dissipation of normal physiological loads are lost. These degenerative conditions provide ample reasons for employing tissue engineering techniques to fabricate replacements for damaged discs [3] as well as condylar and fossa surfaces.

The disc sits astride the condylar head and together with its peripheral attachments and surrounding joint capsule, produces a closed space separating intra- and extra-articular environments. Three morphological zones have been described for the disc [4] and these are best appreciated in a sagittal view of the joint (Figure 23.1). The thickest region of the disc is the posterior band, followed by the anterior band. The intermediate zone represents the thinnest portion of the disc. The junctions between the zones are indistinct by gross examination and appear to blend with each other. In sagittal section, the disc has been described as a biconcave structure (Figure 23.1), but this is not a true depiction of its morphology. Instead, the disc resembles a cap attached along its entire peripheral margin to both the condylar neck and the skull base through a variety of different connective tissues. The superior attachments are less tenuous and allow the condylar head to slide forwards and from side to side in relation to the fossa during *translatory* movements of the joint. These attachments are composed of reflections of the superior surface of the disc, in association with the synovial membrane anteriorly and the superior laminar of the retrodiscal tissue (*bilaminar zone*) posteriorly. The synovial membrane in turn attaches to the skull base anterior to the articular eminence while the superior laminar of the retrodiscal tissue inserts into the petrous portion of the temporal bone. Medially and laterally, the superior surface of the disc turns inferiorly, blending with fibers of the capsule producing medial and lateral gutters. The inferior surface of the disc is more closely adapted to the condylar head, especially in its medial and lateral aspects. This relationship promotes *rotary* movements of the condyle in relation to the disc and glenoid fossa. Anteriorly, inferior reflections of the disc are interspersed with tendinous insertions of the superior head of the lateral pterygoid muscle, which attach to a concavity in the condylar head termed the *pterygoid fovea*. Posteriorly, the disc attaches to the inferior lamina of the retrodiscal tissue, which contains elastic fibers and blood vessels (Figure 23.2). As a result of these attachments, rotation occurs in the inferior space while translation of the joint occurs primarily involves the superior joint space. The different movements of the jaw rely upon the contraction of the lateral pterygoid muscle and its angle of attachment to the mandibular condyle (Figure 23.3).

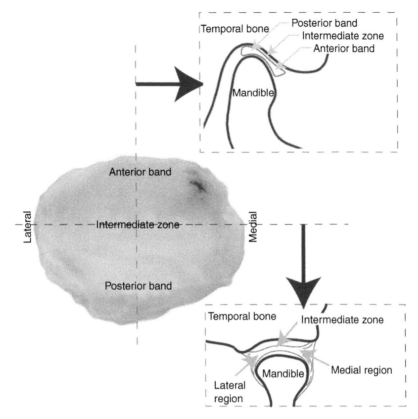

FIGURE 23.1 Superior, sagittal, and coronal views of the TMJ disc. The TMJ disc divides the joint into superior and inferior spaces. From a superior view, the disc displays an ellipsoidal shape, being thicker in the mediolateral dimension than anteroposteriorly. In a sagittal view of the disc, thicker anterior and posterior bands and a substantially thinner intermediate zone (central regions of the disc) can be appreciated. In coronal sections, the disc presents a similar biconcave shape, with thicker medial and lateral regions compared with the intermediate zone. However, regional variations in thickness are less obvious in the coronal plane.

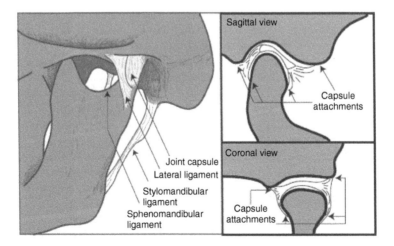

FIGURE 23.2 Ligament and capsule attachments in the TMJ. The TMJ capsule and various mandibular ligaments serve to guide and constrain mandibular movement. The capsule also separates intra- and extra-articular environments and produces a closed space containing synovial fluid, which serves both metabolic and lubricative functions. Various specialized neural elements within the capsule provide sensory input to guide joint position and movement.

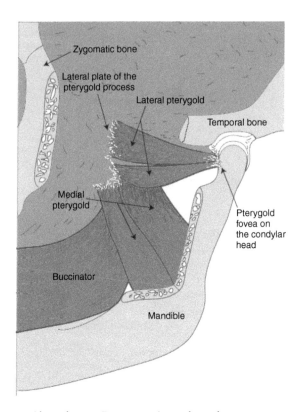

FIGURE 23.3 Lateral pterygoid attachment. Rotary opening and translatory movements of the mandible are the result of contraction of the superior and inferior bellies of the lateral pterygoid muscle. This muscle originates from the lateral surface of the lateral pterygoid plate and inserts into the pterygoid fovea, a depression in the anterior surface of the condylar head.

While the mouth is capable of opening widely, an interincisal opening of greater than 40 mm is the result of condylar rotation and translation. Most physiological opening movements are the result of smaller rotations of the condylar head. Under normal conditions, these movements produce complex compressive loads between the anterior surface of the condyle and the posterior slope of the articular eminence [5]. In the presence of a normally positioned disc, forces in this region are reduced and dissipated by the interposed intermediate zone of the disc along with the lubricating actions of synovial fluid. Only with excessive opening are shear forces generated at the interface of the superior surface of the disc and the glenoid fossa–eminence complex. The temporomandibular joints' tissues, particularly the TMJ disc, appear to have been designed to distribute both tensile and compressive forces during jaw movement.

The TMJ is enclosed by a fibrous capsule, which encloses the joint on its medial and lateral aspects. Laterally, the capsule is reinforced by the temporomandibular ligament. As described previously, attachments between the capsule and disc produce closed superior and inferior joint spaces (Figure 23.2). The capsule and ligament are highly innervated structures, containing a significant number of receptors which are capable of detecting stretch and pressure. Pacinian corpuscles, Ruffini nerve endings, Golgi tendon bodies, and free nerve endings have all been identified in the joint capsule [6]. These receptors provide important proprioceptive signals that guide the position of the mandible through neural relays with the muscles of mastication. While joint sensation is commonly ascribed to the auriculotemporal nerve, the role of other neurological mechanisms governing joint motion and mandibular position are not well understood. This lack of characterization contributes to the morbidity associated with reconstructive procedures.

The inner surfaces of the capsule are lined by the synovial membrane, a specialized structure containing cells and vessels responsible for both the phagocytosis of foreign particles and potentially pathological organisms and the production of synovial fluid. Synovial fluid serves two principal functions within a joint. As a lubricating fluid, synovial fluid is extremely effective exhibiting a coefficient of friction of approximately 0.00001. The lubricating action of synovial fluid is complex in itself and involves the interaction of multiple proteins [7]. Synovial fluid also acts a medium for the transmission of nutrients and the removal of metabolic waste from the cells present within the joint. As a dialysate of plasma, it contains all the components of plasma with the exception of large coagulation proteins. This composition suggests that synovial fluid is capable of performing most of the physiological functions of blood and is essential to the viability of joint structures. Tissue engineering products aimed at reconstruction of the TMJ or its components must be coupled with efforts to restore the functions of the synovium.

Potential movement of the mandible through three planes of space has already been described. Of interest is the considerably larger bulk of musculature responsible for closing movements as opposed to opening and excursive actions. Closure of the mandible is under the influence of the masseter, temporalis, and medial pterygoid muscles. These muscles form a robust sling around and above the mandible that generate a considerable amount of force. In contrast, opening movements are produced through the relatively small contractions of the superior head of the lateral pterygoid and (possibly) suprahyoid muscles (Figure 23.4). This muscular arrangement suggests that under normal conditions, opening movements constitute relatively passive actions through smaller ranges, while closure of the mandible invokes a power stroke to resist loads imposed by gravity and the presence of a bolus of food between the teeth. Consideration of the functional loads imposed upon the TMJ is critical in the design of tissue-engineered constructs [3]. Constructs must survive joint loading conditions, especially if protective neuromuscular controls are compromised by disease or surgical reconstructive procedures.

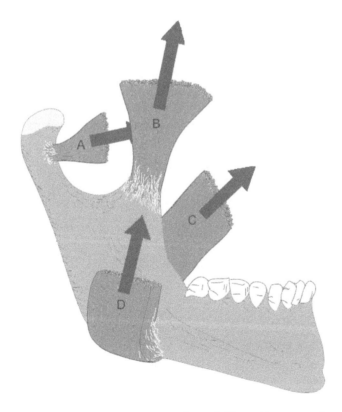

FIGURE 23.4 Muscles of mastication. In contrast to opening and side-to-side movements, mandibular closure is produced by contraction of three large muscles: the temporalis muscle (B), masseter muscle (C), and medial pterygoid muscle (D). These muscles, along with the lateral pterygoid (A), are collectively referred to as muscles of mastication since their principal activity is associated with eating.

23.3 Microscopic Anatomy of the TMJ Disc

Extensive histological studies of the TMJ utilizing light and scanning electron microscopy have been conducted in both human [8,9] and animal specimens [10,11]. Unlike most diarthrodial joints, the articular surfaces of the TMJ appear to be formed by nonvascularized and noninnervated fibrous connective tissue instead of hyaline cartilage, but these structures have not been completely characterized. The interpositional disc, on the other hand, is the subject of considerable interest in both human and animal models and these studies confirm that the disc is composed of fibrocartilage with an organic matrix composed primarily of collagen fibers enclosing a mixed fibroblastic and chondrocytic cell population [12–14]. In studies using porcine disc specimens, the ratio of fibroblasts to fibrochondrocytes is approximately 7 : 3 [14]. The presence of true chondrocytes, which are characteristic of hyaline cartilage, is debatable and indicative of the difficulty in phenotyping cells on the basis of their light microscopic appearance, especially when artifactual errors are taken into consideration. Using collagen typing, which provides strong evidence of the type of cell present based on its secretory product, a predominance of type I collagen in the TMJ disc [12,13,15] implies that the major cellular constituent is derived from a fibroblastic lineage. Trace amounts of type II collagen and a procollagen precursor have also been detected in human discs near the superior and inferior surfaces [16]. The cellular population of the disc appears to vary with age and also with the animal model studied. This is not surprising considering the different functional loads applied by various dietary patterns present in old and young ruminants and carnivores alike, and suggests that discal cells possess the ability to differentiate (or dedifferentiate) according to function. In general, higher cell numbers have been identified in the anterior and posterior bands relative to the intermediate zone [14]. The population of cells in the intermediate zone, subjected to greater compressive forces, appears to be more chondrocytic than the anterior and posterior bands [14]. In the anterior and posterior attachment regions, fibroblasts and fibrocytes predominate [14]. Smaller amounts of elastin fibers and other noncollagenous matrix proteins complete the organic structure of the disc [17].

The histological organization of the collagen fibers in the disc shows considerable regional variation between the intermediate zone and peripheral attachment regions [18]. In the central area of the intermediate zone, the collagen fibers are oriented primarily in an anterior–posterior direction. In areas where the disc attaches to the condylar neck and skull base, the fibers assume a ring-like structure (Figure 23.5). The distribution of the elastin fibers also appears to follow a circumferential pattern [17] with smaller

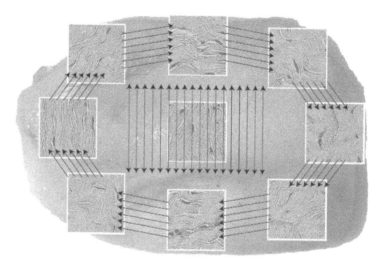

FIGURE 23.5 Collagen fiber alignment in the TMJ disc. The collagen fibers of the TMJ disc are arranged in a ring-like structure around the periphery of the disc. In the intermediate zone, the fibers are predominantly aligned in an anterior-posterior direction. In transition regions between the intermediate zone and the anterior and posterior bands, the collagen fibers curve medially, laterally, inferiorly, and superiorly.

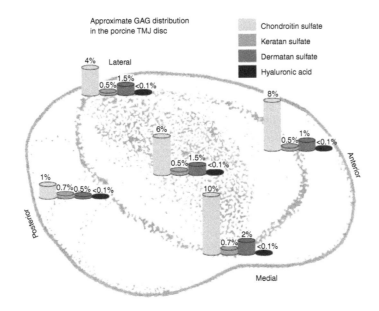

FIGURE 23.6 GAG content and distribution in the TMJ disc. Chondroitin sulfate is the predominant GAG in the TMJ disc, followed by dermatan sulfate. Keratan sulfate and hyaluronic acid GAGs are also found in the TMJ disc. Regionally, the medial portion of the disc appears to possess the largest concentration of GAGs followed by the anterior region. (Data from Detamore, M.S., et al. [2004] *Matrix Biol.* 24, 45–57. With permission.)

amounts of elastin present near the center of the disc and larger numbers of fibers in the posterior band [19]. By characterizing the disc according to the orientation of fibers, a different functional view of the disc is derived. Instead of dividing the disc into the intermediate zone and anterior and posterior bands, the disc can be regarded as possessing a central zone surrounded by peripheral attachment regions.

The collagen fibers also vary in appearance between the superior and inferior surfaces of the disc. Fibers within the inferior surface of the disc exhibit a more wavy appearance than those closer to the superior surface. This arrangement supports the need to dissipate larger compressive forces within the inferior joint space as opposed to tensile forces in the superior joint space.

At the molecular level, the presence, type, and distribution of glycosaminoglycans (GAGs) helps determine the response of articular tissue to function. The literature is not entirely in agreement regarding the total amount of GAGs present, the type and relative proportion of GAG, and the zonal distribution of each GAG [10,20–24]. The variable observations may be the result of differences in the animal model selected as well as different analytical techniques, which have included inhibition ELISA assays, ion exchange chromatography, electrophoresis, high-performance liquid chromatography, and colorimetric analysis [10,20–24].

Most studies suggest that GAGs represent between 0.5 and 10% of the dry weight of the TMJ disc [10,20–25]. The predominant GAG appears to be either chondroitin sulfate or dermatan sulfate, with small amounts of keratan sulfate and hyaluronic acid also detected [22,26]. Similar to the knee meniscus, water content of the disc is high, approximately 70% [27]. Antero-posteriorly, the highest concentration of chondroitin sulfate and keratan sulfate is found in the intermediate zone [10]. Medio-laterally, a higher concentration of GAGs is found in the medial regions [10]. The approximate values of GAG content in the TMJ disc are presented in Figure 23.6.

23.4 Biomechanical Behavior and Failure of the Disc

The TMJ disc under tension and compression exhibits significant anisotropy [28–31]. This behavior is believed to be directly related to the heterogeneous structure and organization of its cellular and molecular components. Furthermore, the disc appears to have much more tensile integrity than compressive; tensile moduli are 100- to 1000-fold larger than compressive moduli [28,29]. The bands of the disc also show a larger capacity to resist compressive loading than the central portions of the disc [28]. While the disc experiences both tensile and compressive loading, its role in the joint appears to be primarily compressive, yet its material properties under compressive loading are much smaller than other compressive tissues such as articular cartilage. The material properties of the TMJ disc support the hypothesis that the TMJ disc operates in a "trampoline-like" fashion. As the disc experiences increased levels of tensile deformation, the mesh-like collagen fiber network increases its resistance to deformations from the mandibular condyle.

The TMJ disc's material properties vary regionally [28–31]. The posterior and anterior bands appear to possess the largest resistance to instantaneous compressive loading, while the medial portions of the disc exhibits a significant relaxed resistance [28]. Under antero–posterior loading, the TMJ disc demonstrates considerable stiffness in all regions, however, the lateral portions of the disc appear to be less stiff than the central and medial portions [29]. Under medio-lateral tension, the disc behaves very differently. The anterior and posterior bands each display large resistance to tensile deformation, but the intermediate zone appears to be much weaker and consequently, more prone to deformation. From this account, we see strong correlations between a disc's tensile properties and its collagen fiber alignment [29]. Correlations between the disc's compressive properties and its matrix constituents are less defined. While GAG content is believed to increase the hydrostatic pressure within a matrix, and thus its compressive stiffness, the GAG content of the TMJ disc may not be large enough to account for significant variations throughout the tissue. Other factors, such as the specific type of GAGs or the size of collagen fibers present, account for the regional variations [28].

23.5 Common Conditions Affecting the TMJ

Within the context of clinical disease, four categories of pathology have been recognized to affect the TMJ to the extent that treatment is required. *Internal derangement of the TMJ* is a condition characterized by incoordinate movements of the disc relative to the condyle. It is believed to be the result of several inter-related pathological processes, including softening (*chondromalacia*) which produces either a deformed or malpositioned disc (most commonly in an anteromedial direction), disc *perforation* as a result of functional loads applied to diseased tissue, intra-articular reparative processes which limit joint and disc motion (*adhesions or scar bands*), and an alteration of synovial fluid properties which affect lubrication and metabolism (*synovitis*). When a disc is displaced from its normal position, a variety of responses occur depending on the extent and duration of the displacement. Minimally displaced discs that are capable of reduction often provoke adaptive responses within a joint, consisting of remodeling of the articular surfaces and discal tissue. However, if the disc is displaced significantly for a prolonged period, degeneration occurs as the result of unfavorable loading forces and the consequences of reduced mobility [32].

Degenerative joint disease represents a different state where catabolic loss of articular tissue exceeds anabolic attempts to restore joint structure. Various etiologies are capable of producing degeneration and these include excessive and repetitive loading patterns (*osteoarthritis*) and autoimmune conditions, where inflammatory antibodies against joint matrix proteins destroy the structural organization of the joint (*rheumatoid and psoriatic arthritis*). The etiology of osteoarthritis is unknown, but the interaction of several factors such as heredity, prolonged loading, trauma to a joint, immobility, obesity, joint instability, and advanced age appear to affect the normal mechanism by which cartilage cells replenish matrix macromolecules responsible for cartilage integrity [33]. The capacity for internal remodeling of both the discal and articular cartilage diminishes as a result of changes in matrix composition, biosynthetic activity of cartilage cells, and responsiveness of these cells to stimulation.

The cause of rheumatoid arthritis is also unknown, and while a strong genetic predisposition has been suggested by the presence of the HLA-DR4/DR1 epitope in over 90% of patients, the concordance of disease in monozygotic twins is between 15 and 20%, suggesting the significant influence of nongenetic factors. Macroscopic hallmarks of disease include synovial proliferation (*rheumatoid pannus*) and uncontrolled inflammation, reflected on a microscopic level by synovial cell hyperplasia and endothelial cell proliferation. The inflammatory process is also associated with activation of various cells of the reticuloendothelial system, including CD_4 T lymphocytes, B cells, phagocytes, and neutrophils. Fibroblast and osteoclast activation result in both cartilage and bone resorption and thickening of the periarticular tissue.

Both internal derangement of the TMJ and degenerative joint disease can produce a reduction in the range of motion of a TMJ. These conditions may also be associated with significant pain. The origins of pain remain controversial, often because of the difficulty in localizing the sources of pain. Activation of nerve endings by inflammatory mediators and stimulation of neural elements present within the capsule, TMJ ligament, synovial membrane, and retrodiscal tissue by distortion or compression have been implicated. In addition, compromised joint function is capable of altering the normal behavior of the mandibular musculature. This sequela of joint disease produces reactive muscular responses and the phenomenon has been characterized as *myofascial pain dysfunction* (MPD), a condition which includes myospasm and muscle fatigue. The combination of pain derived from multifactorial origins and reduction in joint mobility produces the principal signs and symptoms associated with *temporomandibular dysfunction* (TMD), a clinical syndrome associated with pain and an inability to achieve normal mandibular motion.

The third category of TMJ disease differs from the preceding conditions, because it reflects a disorder of bone metabolism, rather than cartilage. Here, the local intra- and extra-articular environment favors the formation of heterotopic bone. When the dystrophic bone bridges the joint space (temporal–condyle interphase), complete immobility of the joint (and mandible) occurs. This condition is referred to as *ankylosis* of the TMJ.

The last category of conditions affecting the TMJ consists of different neoplasms or cysts that affect the mandible. These diseases rarely affect the joint without involvement of the adjacent mandibular bone and as such, will not be considered in this discussion on tissue engineering of the TMJ, because the defect requiring reconstruction usually extends beyond the confines of the joint.

The successful application of tissue engineering strategies to reconstruct either individual joint components or the entire joint mandates a full appreciation of the pathological condition(s) responsible for the destruction of the joint in the primary instance. Ignoring the etiology of joint disease may result in a diminished capacity for successful implementation of an engineered construct and an inability of tissue engineering to fulfill its promise as a reconstructive modality.

23.6 Current Methods for TMJ Reconstruction

Reconstruction of the TMJ can involve the replacement of a disc, reconstruction of the superior articulation (*hemi-arthroplasty*), or replacement of the superior and inferior articulating surfaces (*total joint reconstruction*). A variety of autogenous and alloplastic materials have been used for joint reconstruction with varying and conflicting results. It is beyond the scope of this chapter to discuss comparative outcomes from these procedures. Analyzing the relative success and failure of each technique or material is complicated by an inability to fully appreciate the normal function of a particular patient's joint, which has lead to difficulty in gauging outcomes. Criteria used to evaluate treatment of patients with TMJ pathology usually include a comparison between the mandibular range of motion and pain levels before and after surgery, which are imprecise measurements fraught with subjective error.

Removal of a diseased or severely displaced disc (*discectomy or meniscectomy*) is a procedure employed when repositioning maneuvers of the malposed disc fail or are not possible because of advanced degeneration of the disc. Following removal, treatment alternatives include leaving the joint empty, replacement of the disc with an autogenous fat, fascia (e.g., fascia lata), cartilage (e.g., auricular cartilage), dermal graft, or rotating a fascia flap (e.g., temporalis fascia) into the defect. Whichever modality is selected, it appears that the formation of an interarticular scar band either through a *de novo* process or metaplastic transformation of grafts and flaps into fibrous tissue provides a common endpoint.

Experiences with different alloplastic materials for discal replacement have been characterized by a number of significant failures resulting in severe joint resorption, alteration of mandibular skeletal relationships, compromised motion, pain, and allegations of systemic immune compromise. These surgical disasters and the resultant lawsuits have unfortunately tainted all forms of TMJ surgery and discouraged many surgeons from seeking alternative methods to reconstruct the joint. Before the controversy surrounding the implantation of medical-grade silicone developed, silicone (*Silastic*) interpositional implants were available for disc replacement. As permanent replacements, these devices were capable of fragmentation, but when used as a temporary interpositional implant ("pull-out" technique), they were observed to provoke the formation of a dense fibrous tissue capsule, which served as an interarticular cushion. Their relatively successful use following discectomy might be attributed to this reaction.

One of the most litigious experiences associated with the alloplastic replacement of a disc occurred with the use of a Teflon–Proplast implant in the late 1980s and early 1990s. Produced by the Vitek® Corporation, fragmentation of the implant under functional and parafunctional loads was associated with an exuberant foreign body giant cell response and significant osteoclastic activity, resulting in the resorption of condylar and fossa surfaces and severe local inflammatory events. The lessons learned from this experience are essential, and include the significance of characterizing the loading patterns within a joint, and the importance of recognizing the effects of degradation products upon the local joint environment.

The TMJ hemi-arthroplasty was a procedure popularized by Christensen and Morgan in the 1960s, in which the superior articulation of the joint was replaced with an implant fabricated out of chrome–cobalt alloy. The Christensen implant reconstructed both the fossa and articular eminence while the Morgan implant covered the eminence only. Concerns over accelerated degeneration of the natural condyle articulating against a less-deformable surface eventually resulted in the replacement of the hemi-arthroplasty with total joint reconstructive procedures utilizing both prosthetic fossa and condylar components. Several systems are currently licensed by the Food and Drug Administration (FDA) (ca. 2004) for implantation into patients, though limitations have been imposed on surgeons wishing to use these devices and the selection of patient candidates. Stringent follow-up of patients treated with these implants form the basis of various clinical trials designed to test not only the ability of the procedure to improve a patient's condition, but also the integrity of the devices over time.

As an alternative to prosthetic devices, autogenous grafting techniques are available. Reconstruction of the TMJ with autogenous tissue involves the harvesting of a costochondral graft (fourth to sixth rib) and attaching the graft to the remaining mandibular ramus following removal of the diseased joint. The rib is usually placed on the lateral surface of the ramus and fixated with multiple transcortical screws or a bone plate (Figure 23.7). The rib may also be placed on the posterior border of the ramus so that its widest dimension occupies the fossa, but fixation with either bone plates or screws becomes a little more difficult. When the disc is removed along with the condyle, alternatives for discal reconstruction are similar to those following discectomy and range from leaving the interarticular space empty to using some form of autogenous soft tissue to fill the void.

In general, reconstruction following joint removal for neoplastic or cystic disease enjoys a higher rate of success than similar procedures performed in patients with temporomandibular dysfunction. Several reasons have been proposed for this dichotomy, including persistence of parafunctional loading of the articulation, excessive scarring from multiple operations, the physical and psychological consequences of chronic illness, and degenerative disease states which are directed against bone and cartilage.

The principal indications for considering an alloplastic total joint over an autogenous costochondral graft include the treatment of joints where heterotopic bone formation is a concern (e.g., ankylosis of the TMJ), joints affected by a resorptive inflammatory process (e.g., rheumatoid arthritis or joints affected by a foreign body response to previously implanted alloplasts), and multiply operated patients where the soft tissue bed is excessively scarred and relatively avascular. Under these adverse conditions, autogenous grafts may be compromised by further disease or conditions which do not favor graft survival. The ability to customize an alloplastic device is also useful for correcting a skeletal discrepancy that may occur in

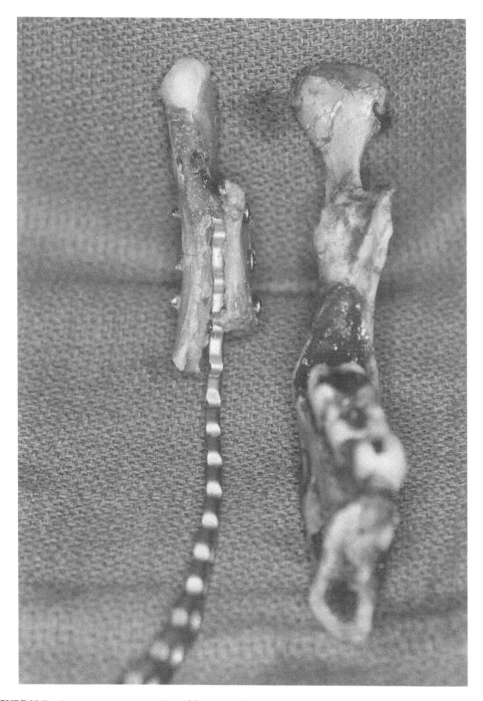

FIGURE 23.7 Autogenous reconstruction of the joint with a costochondral graft.

patients with severe degenerative joint disease, where retrusion and rotation of the mandible is the result of decreased posterior vertical support (Figure 23.8a,b).

Customized prosthetic devices involve complex surgical techniques. In order to accurately reproduce the skeletal bases to which the devices will be attached, two separate surgeries are ideally required. During the first procedure, the diseased joint (or failed implant) is removed and the area derided. A temporary alloplastic space maintaining implant is used to reduce the amount of soft tissue in-growth into the

(a)

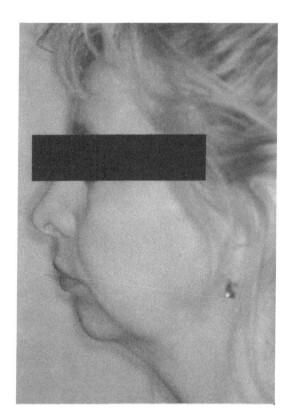

(b)

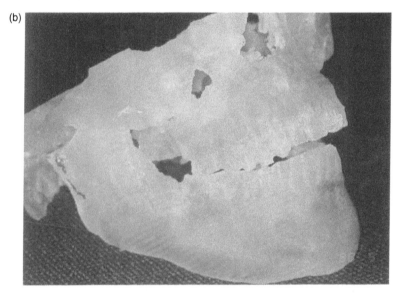

FIGURE 23.8 Mandibular retrusion secondary to bilateral condylar resorption. (a) Profile view of patient with severe, bilateral rheumatoid degeneration of the mandibular condyles. Retrusion of the mandible can be appreciated from the posterior position of the chin; (b) stereolithographic model of the same patient demonstrating the resulting malocclusion created by loss of condylar height. Correction of mandibular skeletal position with bilateral alloplastic total joint reconstruction. (c) Patient with bilateral degenerative rheumatoid disease of the mandibular condyles following surgical reconstruction of the joints with alloplastic total joint prosthesis. Note restoration of mandibular projection; and (d) Postero-anterior (PA) skull radiograph demonstrating bilateral alloplastic total joints.

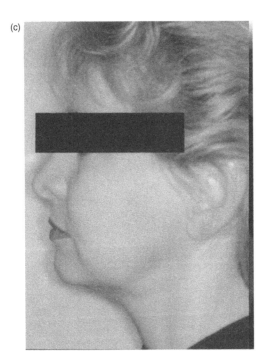

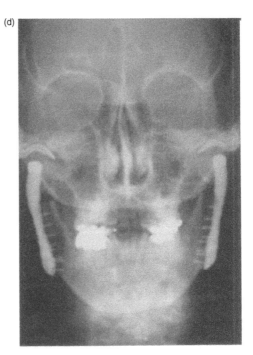

FIGURE 23.8   *Continued.*

resulting space. Following this surgery, a thin-cut Computerized Tomography (CT) scan is obtained using a protocol devised by companies specializing in the fabrication of stereolithographic models. The anatomically accurate model is sent to a joint fabrication company where CAD–CAM technology is used to produce a prototype of the final device. Each device is composed of a prosthetic fossa and eminence as well as a condylar head attached to a ramus component. The prototype is returned to the surgeon who confirms that the surgical defect has been correctly reconstructed. At this time, minor modifications to the skeletal defect may be proposed to better accommodate and fit the implant. Once the customized implant has been completed, the patient undergoes a second surgery during which time, the surgical sites are adjusted to match the defect created on the stereolithographic model before the prosthetic fossa and condyle are attached to their bony bases. The entire process is time-consuming and expensive, but justification for its use lies in the magnitude of the problem requiring correction (Figure 23.8c,d).

Single surgery reconstruction of a joint is also possible with custom devices. In this procedure, a stereolithographic model of the diseased joint is prepared. If an alloplastic device is already in place, digital subtraction technology is employed to artificially remove the prosthesis. Otherwise, the surgeon creates the anticipated surgical defect on the model and the device is fabricated. Minor adjustments to the skeletal remnants can be made at the time of implantation to promote a close adaptation of the device to the defect.

The Christensen total joint system has been available since 1965, though the current device, which employs a Vitallium fossa articulating against a chrome–cobalt condylar head, is significantly modified from the original design which used a metallic fossa matched with a condylar head composed of poly-methylmethacrylate (PMMA) (Figure 23.9a). Concerns over the development of a giant cell mediated foreign body response to particulate PMMA prompted this change. Both the fossa and condylar prosthesis are fixated to the temporal bone and mandibular ramus, respectively, with screws (Figure 23.9b). Both patient-specific implants, customized according to computerized tomographic data, as well as stock devices with different sizes and shapes for the fossa and condyle are available.

Another system currently available is offered by TMJ Concepts® (Figure 23.10). This system utilizes a chromium–cobalt–molybdenum condylar head attached to a ramus framework made out of titanium alloy

(a)

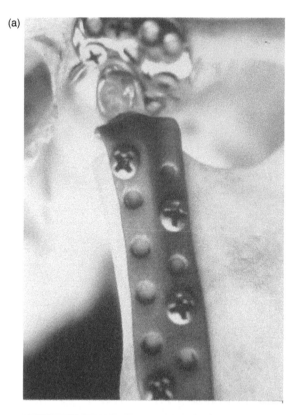

(b)

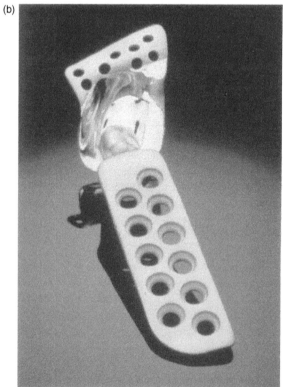

FIGURE 23.9   Christensen® total joint prosthesis with (a) acrylic condyle and (b) metal condyle.

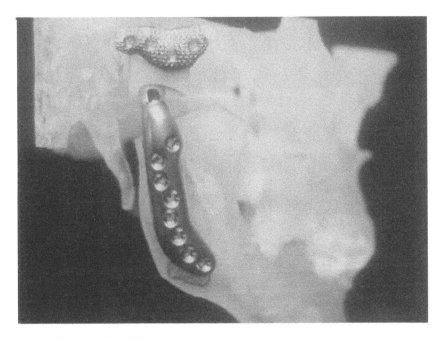

FIGURE 23.10   TMJ Concepts® total joint prosthesis.

(6AL–4V) and a fossa component composed of ultra-high molecular weight polyethylene (UHMWPE) with a nonalloy titanium mesh backing. The respective components are customized to the individual patient's anatomical defect and are produced with advanced CAD–CAM technology. After fabrication, the condyle and fossa are attached to their respective skeletal components with multiple screws (titanium alloy) placed in a nonlinear fashion to promote maximum stability. There are no stock components available for this system.

The third device is currently available only to surgeons who are formally registered to participate in an FDA-sanctioned clinical trial for the Lorenz-Biomet prosthetic total joint. The fossa consists of a UHMWPE articular surface mounted on a metallic base which is used to secure the prosthesis with screws to the lateral margins of the glenoid fossa. Since this is a stock device available in three sizes, the patient's anatomy requires preparation to conform with the prosthetic contours. This is achieved in part by removing most of the articular eminence and if indicated, filling the space between the fossa and device with orthopedic cement (e.g., Simplex P). A significant difference between the design of this fossa prosthesis and those used in the Christensen or TMJ Concepts systems is the thickness of the articular surface. The Lorenz-Biomet total joint system attempts to shift the center of rotation of the reconstructed joint inferiorly to simulate translatory movements by increasing the thickness of the UHMWPE to a minimum of 4 mm. The condylar portion of this device is composed of a cobalt–chromium–molybdenum alloy and is available in 45, 50, and 55 mm lengths to accommodate various ramus heights. Several design features have also been incorporated including augmenting the peripheral margin of the fossa prosthesis to reduce the likelihood of re-ankylosis and the in-setting of the condylar head so that it articulates against the middle of the fossa instead of the lateral aspect.

23.7   Tissue Engineering Strategies for TMJ Reconstruction

23.7.1   Tissue Engineering Motivation and Philosophy

The motivations for tissue engineering in the TMJ are abundant [3,34]. While surgical removal of diseased articular surfaces or a pathologically affected disc with or without reconstruction remains a viable treatment option, the resulting articulation is left without the benefit of physiologically adaptive tissues. Tissue

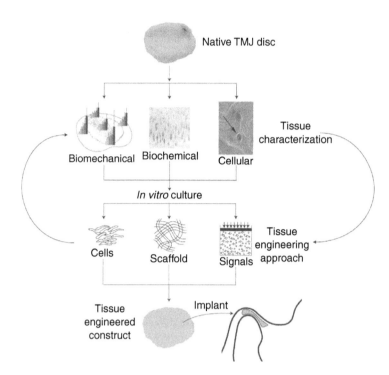

FIGURE 23.11 Developmental approach toward tissue engineering. The combination of cells, scaffolding, and signals provide the foundations for the creation of a tissue engineered construct. However, before designing such an implant, the material characteristics of the native tissue should be fully characterized and incorporated into the design.

engineering offers the potential to replace tissues that have been destroyed and removed [3]. Laboratory fabricated biological constructs would be capable of immediate structural reconstruction, reproducing normal joint physiology, and reducing the need for multiple surgical procedures. Current tissue engineering efforts in the TMJ are focused on the replacement of bony, cartilage, and fibrocartilage structures as separate components. There are currently no documented efforts to reconstruct the capsule or synovium, which places the field at a considerable distance from the engineering of a true total joint system.

The tissue engineering approach to articular reconstruction combines cells, a scaffold, and biological agents that control the formation and maintenance of tissue with properties comparable to native structures [3]. Several variations to this general approach include the use of different cell lineages [35], various scaffold materials and shapes [36], and the methods by which biological signals are applied [37]. In order for engineered constructs to succeed, characterization of the native tissue properties (cellular, biochemical, and mechanical) must first be conducted, such that tissue engineering efforts may be designed to create constructs which emulate the native tissues' functions [18]. Figure 23.11 summarizes the progression and philosophy associated with tissue engineering efforts. Each level of tissue characterization, as well as general efforts to engineer joint components, is the focus of active research. A comprehensive understanding of the challenges in these research areas will lead to the development of viable, implantable tissue engineered construct for TMJ reconstruction.

23.7.2 Cellular Issues

Cells are the factories responsible for the formation and maintenance of replacement tissues. Through the action of proteins responsible for creating or degrading the surrounding extracellular matrix, cells produce and remodel tissue structure to comply with functional requirements. Cells for TMJ tissue engineered constructs are derived from one of three sources: native cells isolated from healthy mature tissue, undifferentiated progenitor cells isolated from a variety of autogenous sites, and cells which infiltrate a noncellular, implanted scaffold from the surrounding tissue *in vivo*. Each cell source is associated with distinct advantages balanced against phenotypical limitations. The selection of an appropriate cell source is therefore based upon the requirements of a particular tissue engineering application.

Several issues govern the selection of cells for tissue engineering in the TMJ. For example, the presence of a healthy native cell source may be impossible due to joint pathologies. Also, these mature cells may have fewer propensities to produce ECM proteins. For these reasons, the promotion of an undifferentiated cellular phenotype has been suggested as a possible cell source for diseased joints. However, this promotion may be unnecessary if healthy, native cells which have retained their phenotype are available.

Further complicating the selection of the cell source is a tissue's phenotypic nature. In the case of the TMJ disc, a heterogeneous group of cells are found throughout the disc, thus native cell harvesting techniques may yield the best approximation of the disc's cellular constituency. Mature cells could have a reduced ability to produce extracellular matrix proteins, but these effects may be a result of other factors such as passaging effects. In addition to these issues are the multiple phenotypic origins which may assist in TMJ development. Cell populations in the TMJ have been shown to possess origins associated with mesenchymal and neural crest derivatives. These lineages may play very different and interrelated roles in TMJ development [38].

When a reconstruction aims to replace the entire TMJ and not the disc alone, multiple cell lines, scaffolds, and molecular factors have to be considered. In the most extensive form of joint reconstruction, it may be necessary to replace bone, cartilage, fibrocartilage, capsule, and synovium simultaneously. The concurrent regeneration of these different tissues in a single scaffold with multiple cell lines and specialized molecular signals may prove extremely difficult. An alternative approach would attempt to engineer components of the joint separately and to combine these elements after they have been formed, in a process similar to the use of current alloplastic devices.

To complete the reconstruction, an implantable construct may require the incorporation of a neural network capable of mediating joint neuromuscular responses as well as a vascular supply to support various metabolic requirements of the reconstructed joint. Current tissue engineering studies are not sufficiently developed to handle this level of complexity and ultimately, this effort might prove unnecessary if the capsule, synovium, and temporomandibular ligament are preserved and re-attached after the implantation of the construct. Currently, the focus of TMJ tissue engineering is the separate regeneration of the major tissue constituents of the joint, namely bone, cartilage, or fibrocartilage, but success in this area does not necessarily translate into a successful reconstruction of the total joint.

23.7.3 Scaffold Issues

The scaffold is used to support cells while they produce a three-dimensional biological matrix. The choice of material for the scaffold is based upon how the construct will be utilized. Important considerations include the characteristics of the tissue being regenerated, the mechanical environment to which the scaffold is subjected, interactions between cells and the scaffolding material, the biocompatibility of the scaffold and its degradation products, and the ability of a scaffold to convey various biological signals. The scaffold should also elicit a minimal immunogenic response to reduce rejection and facilitate the formation of biological tissue. Current TMJ tissue engineering efforts have focused on scaffolds composed of hydrogels and meshes fabricated from different natural and synthetic materials [39].

Hydrogels are scaffolds used in current tissue engineering applications for bone, cartilage, and fibrocartilage regeneration. Some of the advantages of hydrogels include controlled rates of degradation, the ability to be injected, photopolymerization capabilities, and the assembly of protein sequences conducive to cell attachment [40,41]. Specific hydrogels such as oligo(poly(ethyleneglycol)fumarate) (OPF) have shown promise in the regeneration of bone and might prove effective in the promotion of cartilage regeneration [35]. Alginate hydrogels have also been used in the regeneration of the TMJ disc using porcine TMJ disc fibrochondrocytes [39]. However, the cellular population of the construct was not stable and observed to decrease with time [39]. The importance of a scaffold to the production of bone and cartilage from cells has been well documented. When cells remain unattached, tissue formation is absent, while suspension in hydrogels can promote both bone and cartilage formation [42].

Meshes can be made from degradable or nondegradable synthetic materials. Natural porous meshes, such as coral hydroxyapatite have also been utilized as scaffolds [43]. Nonwoven meshes of polyglycolic acid (PGA) and polylactic acid (PLA), used alone or as copolymers, are popular in tissue engineering, because they have already received approval from the FDA for medical implantation into humans [39]. These materials can be formed into complex shapes by polymerizing or fashioning the material within anatomic molds. The rate of degradation of the copolymeric scaffolds can also be adjusted by changing the ratio of the copolymers. While complete hydrolysis of many polymeric meshes will occur, acidic degradation products may incite an inflammatory response which can adversely affect the formation of bone or cartilage. Nonwoven meshes have already demonstrated their ability to support TMJ disc fibrochondrocytes [39], osteoblast-like cells [44], and chondrocytes [45] *in vitro*. These studies demonstrate the promise of these materials for the reconstruction of an interpositional disc and articular cartilage.

23.7.4 Signals and Stimulation Issues

Molecular signals and nonbiological stimuli may be used to induce positive and negative gene expression, and thus biosynthetic responses, in cells. These modulators can take a variety of forms such as growth factor application, mechanical stimulation, the use of adhesion peptides, and the generation of oxygen gradients, electrical currents, and sonic stimulation. The effects of these agents can be direct or synergistic and may produce single or inter-related responses. By varying parameters, such as the bioactivity of growth factors and proteins, the size and frequency of mechanical and sonic stimulations, the length of stimulations, and dosage and saturation effects, the regeneration of tissue can be optimized.

Several studies have investigated the effect of growth factors, mechanical stimulation, and oxygen on tissue engineered constructs. Each stimulant, acting through the modulation of the local environment, is capable of producing tissue-like ECM proteins. However, the formation of an organized, three-dimensional tissue array with biochemical and biomechanical properties similar to native tissue will probably require a combination of signals. One method for creating a multisignal environment employs bioreactors to supply mechanical stimulation, such as hydrostatic pressure, in combination with a variety of other signals such as growth factors. Bioreactors capable of perfusion are also able to deliver molecular signals more efficiently by dispersing the biochemical agent throughout the tissue engineered construct.

23.7.5 Specific Issues Related to Joint Reconstruction

Since a tissue engineered construct is composed of biological tissue, consideration must be given to the original disease process or joint loading patterns that initially produced a dysfunctional joint. If parafunctional forces exist, strategies to decrease joint loading must be employed. These include the use of dental orthotic devices, which reduce forces generated by the muscles of mastication and provide protective responses through neuromuscular afferent signals from the periodontium surrounding the teeth.

Another challenge is the maintenance of construct viability following implantation. In order to survive, cells require nutrients diffusing from the surrounding vasculature and a method to remove metabolic waste. Establishing a vascular supply in a tissue engineered construct is critical, especially for vascular tissues like bones. Fortunately, fibrocartilage has even lower metabolic requirements, an advantage where TMJ reconstruction is concerned. A separate concern involves the creation and connection of the nervous system in a tissue engineered construct to guide the complex movements of the joint and provide proprioceptive input to the neuromuscular system. Meeting this requirement is important for the return of normal function to the reconstructed joint. It may be possible to utilize native tissue for these two functions, if efforts are made at the time of joint or disc removal to preserve the capsule, synovium, and vascular supply from the bony resection margins. When these structures are compromised, such as the case with multiply operated patients, the success of any reconstruction diminishes. Avoidance of multiple surgeries should therefore be part of the overall strategy for using tissue engineering to provide improved results compared to current modalities.

There are also concerns with the immunogenicity of tissue engineered constructs if nonautologous cells and a synthetic matrix are used for the construct. This is especially significant if xenogeneic sources are employed. Possible sources for TMJ cells include the contralateral joint, provided that it is not also diseased. Alternatively, fibroblasts and fibrochondrocytes, the lineages found in TMJ tissue, may be harvested from another fibrocartilaginous joint such as the sternoclavicular joint. It should be kept in mind that biological tissue is capable of varying degrees of adaptation according to the local environment. While this tenet has not been explored in tissue engineering reconstruction of the TMJ, it would not be unreasonable to expect that an engineered construct composed of cells from other diarthrodial joints or cartilaginous structures are capable of phenotypical transformation into cells more closely resembling those of the native TMJ. The successful use of such unrelated tissues as scar bands, fascia, auricular cartilage, costochondral cartilage, dermis, and fat support this contention.

The time required to create a viable tissue engineered construct for the reconstruction of a disc or total joint must also be considered. At this stage, an estimation of the period necessary for the population of an implant with sufficient cells of the desired lineage with properties suitable for handling, surgical implantation and attachment, is unknown. While TMJ reconstruction is generally an elective procedure, it would be unreasonable to consider tissue engineering techniques requiring more than several months as viable alternatives to current modalities.

Another important concern focuses on the ability to stabilize a construct within the joint. Even after the successful creation of replacement tissue, the implant must be anchored to existing structures. To facilitate this process, the implant must possess sufficient bulk, surface area, strength, and tenacity to attach it to the remaining skeleton. Current methods for repositioning a disc include the use of sutures threaded through holes made in the adjacent bone or attaching it to bone anchors such as the Mitek® device inserted into the condylar head. Single point attachment of structures which usually require circumferential fixation for positional anchorage are inherently unstable. Methods for providing multiple attachments of a disc replacement need to be developed for successful function to be restored. This is especially true when the zonal differences and functional properties of each zone are considered. Attachment of a condylar or fossa engineered implant is a little easier, provided that the construct possesses sufficient bulk and strength. Bone plates and screws are currently available for the fixation of skeletal structures and these could be used for a tissue engineered construct.

With the reconstruction of a joint, lubrication of the articulating surfaces must be considered. This function is normally performed by the synovial fluid, but the effects of disease or surgery may compromise the normal secretion and composition of the fluid available. Synthetic adjuvants, such as hyaluronic acid substitutes, may be used in conjunction with joint replacement procedures to protect the implant from unwanted forces, while the synovium heals.

The last issue concerning the successful application of tissue engineering techniques for the reconstruction of TMJ components concerns the unavoidable consequences of any surgical maneuver, specifically the development of intra-articular scars. Steroid injections can help reduce this natural consequence of surgery, but perhaps the most effective results are seen through the fastidious employment of postoperative physical therapy exercises. These maneuvers promote the formation of long vs. short, scars. They also encourage the necessary plasmatic diffusion of fluids responsible for joint nutrition and metabolism.

23.7.6 The Future of Tissue Engineering in the TMJ

The challenges associated with tissue engineering are immense, but the benefit of utilizing tailored biological constructs without the morbidity of an autogenous donor site, are obvious. Tissue engineered initiatives for the reconstruction of the TMJ and disc avoids the problems associated with the implantation of prosthetic devices or tissues dissimilar from natural joint components. Efforts toward tissue engineering in the TMJ are relatively new, however, emergent technologies and studies add promise to the goal of tissue engineering TMJ replacement tissues. Along the way, characteristics about TMJ tissues are revealed and our comprehension of this complex joint's function and structure increases. Solutions for several challenges facing biomedical engineers will be developed with each progressive study. Currently, TMJ tissue engineering is in its formative years. Only four original studies on tissue engineering of the TMJ disc had been published at the date of this article's submission. However, the future promises new knowledge and innovative technologies which may yet allow for the creation of viable, implantable tissue engineered constructs.

References

[1] Dolwick, M.F. (1983) The temporomandibular joint: normal and abnormal anatomy. In *Internal Derangements of the Temporomandibular Joint* (Helms, C.A. et al., eds.), pp. 1–14, Radiology Research and Education Foundation.

[2] Gillbe, G.V. (1975) The function of the disc of the temporomandibular joint. *J. Prosthet. Dent.* 33, 196–204.

[3] Detamore, M.S. and Athanasiou, K.A. (2003) Motivation, characterization, and strategy for tissue engineering the temporomandibular joint disc. *Tissue Eng.* 9, 1065–1087.

[4] Rees, L.A. (1954) The structure and function of the mandibular joint. *Br. Dent. J.* 96, 125–133.

[5] Gallo, L.M. et al. (2000) Stress-field translation in the healthy human temporomandibular joint. *J. Dent. Res.* 79, 1740–1746.

[6] Zimny, M.L. (1988) Mechanoreceptors in articular tissues. *Am. J. Anat.* 182, 16–32.

[7] Hills, B.A. and Monds, M.K. (1998) Enzymatic identification of the load-bearing boundary lubricant in the joint. *Br. J. Rheumatol.* 37, 137–142.

[8] Minarelli, A.M. et al. (1997) The structure of the human temporomandibular joint disc: a scanning electron microscopy study. *J Orofac. Pain* 11, 95–100.

[9] Minarelli, A.M. and Liberti, E.A. (1997) A microscopic survey of the human temporomandibular joint disc. *J. Oral Rehabil.* 24, 835–840.

[10] Detamore, M.S. et al. (2004) Quantitive analysis and comparative regional investigation of the extracellular matrix of the porcine temporomandibular joint disc. *Matrix Biol.* 24, 45–57.

[11] Berkovitz, B.K. and Pacy, J. (2000) Age changes in the cells of the intra-articular disc of the temporomandibular joints of rats and marmosets. *Arch. Oral Biol.* 45, 987–995.

[12] Mills, D.K. et al. (1994) Morphologic, microscopic, and immunohistochemical investigations into the function of the primate TMJ disc. *J. Orofac. Pain.* 8, 136–154.

[13] Landesberg, R. et al. (1996) Cellular, biochemical and molecular characterization of the bovine temporomandibular joint disc. *Arch. Oral Biol.* 41, 761–767.

[14] Detamore, M.S. et al. (2003) Cell type and distribution in the porcine temporomandibular joint disc. *J. Oral Maxillofac. Surg.* Accepted.

[15] Milam, S.B. et al. (1991) Characterization of the extracellular matrix of the primate temporo-mandibular joint. *J. Oral Maxillofac. Surg.* 49, 381–391.

[16] Kondoh, T. et al. (1998) Prevalence of morphological changes in the surfaces of the temporo-mandibular joint disc associated with internal derangement. *J. Oral Maxillofac Surg.* 56, 339–343; discussion 343–334.

[17] Keith, D.A. (1979) Elastin in the bovine mandibular joint. *Arch. Oral Biol.* 24, 211–215.

[18] Detamore, M.S. et al. (2003) Structure and function of the temporomandibular joint disc: implications for tissue engineering. *J. Oral Maxillofac. Surg.: Offic. J. Am. Assoc. Oral Maxillofac. Surg.* 61, 494–506.

[19] Gross, A. et al. (1999) Elastic fibers in the human temporo-mandibular joint disc. *Int. J. Oral Maxillofac. Surg.* 28, 464–468.

[20] Almarza, A.J. et al. (2005) Biochemical content and distribution in the porcine temporomandibular joint disc. *Br. J. Oral Maxillofac. Surg.*

[21] Nakano, T. and Scott, P.G. (1989) A quantitative chemical study of glycosaminoglycans in the articular disc of the bovine temporomandibular joint. *Arch. Oral Biol.* 34, 749–757.

[22] Nakano, T. and Scott, P.G. (1989) Proteoglycans of the articular disc of the bovine temporomandibular joint. I. High molecular weight chondroitin sulphate proteoglycan. *Matrix* 9, 277–283.

[23] Sindelar, B.J. et al. (2000) Effects of intraoral splint wear on proteoglycans in the temporomandibular joint disc. *Arch. Biochem. Biophys.* 379, 64–70.

[24] Axelsson, S. et al. (1992) Glycosaminoglycans in normal and osteoarthrotic human temporomandibular joint disks. *Acta Odontol. Scand.* 50, 113–119.

[25] Scott, P.G. et al. (1989) Proteoglycans of the articular disc of the bovine temporomandibular joint. II. Low molecular weight dermatan sulphate proteoglycan. *Matrix* 9, 284–292.

[26] Nakano, T. and Scott, P.G. (1996) Changes in the chemical composition of the bovine temporomandibular joint disc with age. *Arch. Oral Biol.* 41, 845–853.

[27] Gage, J.P. et al. (1995) Collagen type in dysfunctional temporomandibular joint disks. *J. Prosthet. Dent.* 74, 517–520.

[28] Allen, K.D. and Athanasiou, K.A. (2005) A surface-regional and freeze-thaw characterization of the porcine temporomandibular joint disc. *Ann. Biomed. Eng.* 33, 859–897.

[29] Detamore, M.S. et al. (2003) Tensile properties of the porcine temporomandibular joint disc. *J. Biomech. Eng.* 125, 558–565.

[30] del Pozo, R. et al. (2002) The regional difference of viscoelastic property of bovine temporomandibular joint disc in compressive stress-relaxation. *Med. Eng. Phys.* 24, 165–171.

[31] Kim, K.W. et al. (2003) Biomechanical characterization of the superior joint space of the porcine temporomandibular joint. *Ann. Biomed. Eng.* 31, 924–930.

[32] Berteretche, M.V. et al. (2001) Histologic changes associated with experimental partial anterior disc displacement in the rabbit temporomandibular joint. *J. Orofac. Pain* 15, 306–319.

[33] Hinton, R. et al. (2002) Osteoarthritis: diagnosis and Therapeutic Considerations. *Am. Family Phys.* 65, 841–847.

[34] Feinberg, S.E. et al. (2001) Image-based biomimetic approach to reconstruction of the temporomandibular joint. *Cells Tissues Organs* 169, 309–321.

[35] Temenoff, J.S. et al. (2004) Thermally cross-linked oligo(poly(ethylene glycol) fumarate) hydrogels support osteogenic differentiation of encapsulated marrow stromal cells *in vitro*. *Biomacromolecules* 5, 5–10.

[36] Fisher, J.P. et al. (2004) Effect of biomaterial properties on bone healing in a rabbit tooth extraction socket model. *J. Biomed. Mater. Res.* 68A, 428–438.

[37] Holland, T.A. et al. (2004) Transforming growth factor-beta1 release from oligo(poly(ethylene glycol) fumarate) hydrogels in conditions that model the cartilage wound healing environment. *J. Controll. Release: Offic. J. Control. Release Soc.* 94, 101–114.

[38] Chai, Y. et al. (2000) Fate of the mammalian cranial neural crest during tooth and mandibular morphogenesis. *Development* 127, 1671–1679.

[39] Almarza, A.J. and Athanasiou, K. (2004) Seeding techniques and scaffolding choice for the tissue engineering of the temporomandibular joint disc. *Tissue Eng.* 10, 1787–1795.

[40] Poshusta, A.K. and Anseth, K.S. (2001) Photopolymerized biomaterials for application in the temporomandibular joint. *Cells Tissues Organs* 169, 272–278.

[41] Behravesh, E. et al. (2003) Three-dimensional culture of differentiating marrow stromal osteoblasts in biomimetic poly(propylene fumarate-co-ethylene glycol)-based macroporous hydrogels. *J. Biomed. Mater. Res.* 66A, 698–706.

[42] Weng, Y. et al. (2001) Tissue-engineered composites of bone and cartilage for mandible condylar reconstruction. *J. Oral Maxillofac. Surg.: Offic. J. Am. Assoc. Oral Maxillofac. Surg.* 59, 185–190.

[43] Roux, F.X. et al. (1988) Madreporic coral: a new bone graft substitute for cranial surgery. *J. Neurosurg.* 69, 510–513.

[44] Lu, H.H. et al. (2003) Three-dimensional, bioactive, biodegradable, polymer-bioactive glass composite scaffolds with improved mechanical properties support collagen synthesis and mineralization of human osteoblast-like cells *in vitro. J. Biomed. Mater. Res.* 64A, 465–474.

[45] Ma, P.X. and Langer, R. (1999) Morphology and mechanical function of long-term *in vitro* engineered cartilage. *J. Biomed. Mater. Res.* 44, 217–221.

24

Engineering Smooth Muscle

Yu Ching Yung
University of Michigan

David J. Mooney
University of Michigan
Harvard University

24.1 Introduction

Tissue engineering has emerged over the past two decades to address the growing need for biological substitutes to restore or replace damaged tissues and organs [1]. Current approaches to organ repair rely primarily on transplantation of whole or partial organs and tissues from autogeneic, allogeneic, and less frequently, xenogeneic sources. The imbalance of need versus availability of organs poses as a significant and inherent limitation to this method [2]. Tissue engineering promises an alternative via rebuilding tissues or organs from targeted cell populations, often with the participation of matrices that guide tissue regeneration while providing specific instructions with signaling molecules.

Diseases related to the malfunction of cardiovascular, gastrointestinal and urinary tissues account for millions of deaths annually [1], and smooth muscle (SM) tissue has a critical role in the structure and function of a number of these tissues. In blood vessels of the cardiovascular system, the smooth muscle cell (SMC) component (medial layer) provides mechanical strength and elasticity. SM tissue plays a key role in the gastrointestinal system through its ability to drive the transport of solids and liquids, and malfunction of these tissues typically result in malnutrition. The SM in the intestinal tract, in contrast to SM tissue in blood vessels, is organized in two layers of opposing orientation, and provides for propulsive movements of food through peristalsis. The SM of the bladder also regulates the reservoir function of this tissue, as it controls the storage and release of urine. In all of these tissues, the SMCs contain a highly organized structure of actin and myosin filaments that allow the cells to efficiently modulate and respond to a mechanically dynamic environment and regulate the operation of the organ. A functional smooth muscle component will be critical to the success of engineered tissues intended to replace any of these tissues or organs.

24.2 Cell Source

A critical question in the engineering of smooth muscle tissues is the appropriate source of the SMCs that will comprise the tissue. The majority of research to date has utilized smooth muscle procured from the tissue of interest. However, the isolation of smooth muscle progenitors may allow for a less invasive and destructive approach. In addition, it may be possible to directly recruit SMCs to the site at which one wants a tissue to form.

24.2.1 Differentiated SMCs

The most direct approach to form smooth muscle tissues is to utilize SMCs obtained from the tissue that one desires to engineer. In this approach, smooth muscle containing tissue is typically explanted and dissociated into individual cells. The cells are then directly transplanted or expanded in culture, and subsequently transplanted (Figure 24.1a,b). Direct transplantation may be advantageous as it bypasses *in vitro* culturing, which can alter the contractile phenotype of SMCs. In contrast, culture of SMCs prior to transplantation may lead to phenotypic changes [3,4], but this approach allows one to greatly expand the cell population. This may allow a relatively small explant to ultimately yield sufficient cells to engineer a large tissue. The phenotypic changes noted in SMCs, as they revert to a synthetic phenotype [5] may be reversed or prevented through appropriate culture conditions (i.e., cyclic mechanical loading) [6].

Although smooth muscle tissues characteristically contain contractile apparatus and form the muscular components of visceral structures, there are differences between SM in various tissues [7]. These likely relate to the specific microenvironment and physiology of each tissue type. For this reason, SM biopsies are typically procured for the specific type of tissue being engineered. The artery is the most commonly excised tissue for vascular regeneration [8–11] primarily because it is the largest blood vessel, and hence contains the thickest medial layer. Current methods for bladder replacement require a biopsy to obtain

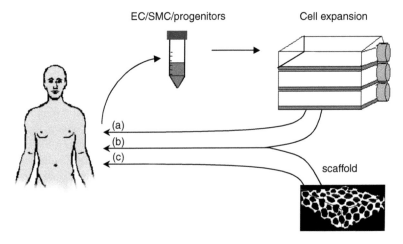

FIGURE 24.1   Scheme of approaches to engineer SM tissue. (a) The direct cell transplantation approach first involves the procurement of the appropriate cell types for the tissue of interest (e.g., EC, SM, or progenitor cells). The cells are typically expanded in culture, and induced to differentiate into SMCs before implantation if necessary. (b) Alternatively, the expanded cells may be seeded onto a scaffold, and subsequently implanted as a cell–matrix construct. (c) Third, acellular matrices may be implanted without seeded cells, and in this situation one relies on the recruitment of neighboring SMCs or progenitors to infiltrate scaffold and form the new SM tissue.

a small specimen of the donor or host bladder tissue, which is then used to expand separate cultures of urothelial and SMCs [12]. Ureter or renal pelvis cells can be similarly harvested. Regeneration of gastrointestinal organs, specifically the stomach and the intestine, has most commonly utilized organoid units which contain SM precursors [13,14].

Differentiated SMCs have shown tremendous utility for the successful regeneration of SM tissues. However the invasive nature of this cell procurement, the inherent limited proliferation capability of primary cells, and maintenance of smooth muscle phenotype are all limitations to this cell source.

24.2.2 Smooth Muscle Progenitor Cells

Smooth muscle progenitors may potentially be isolated using minimally invasive techniques, and subsequently induced to differentiate down a smooth muscle lineage. Cells isolated from bone marrow are termed bone marrow stromal cells (BMSCs), or mesenchymal stem cells (MSCs) depending on the mode of cell purification selection *in vitro*.

Bone marrow can be obtained easily from the medullary canals of long bones or the cancellous cavities [15], and the resultant BMSC can be readily expanded in culture. BMSCs have demonstrated the ability to differentiate into multiple mesenchymal cell lineages, and offer an alternative source of SMCs [7,16–23]. Recent studies have shown that BMSCs are inducible down a smooth muscle pathway, and this process is regulated by an interplay between stimulatory molecules [24,25], with TGF-β and PDGF as the main modulators [24]. Mechanical stimulation has also been shown to effect differentiation of bone marrow stromal cells [26]. However, the mechanism of this effect is still unclear. SM progenitors can also be derived from embryonic stem cells [27,28], circulating blood [29,30], bone marrow [31,32], and other tissues [33].

24.2.3 Recruitment of SMCs

The recruitment of SMCs or progenitors from a surrounding tissue to an engineered tissue provides an alternative to SM transplantation (Figure 24.1c). Signaling molecules such as PDGF and TGF-β have chemotactic effects on SMCs [34], and growth factors (Figure 24.2) released by endothelial cells (ECs) can also induce the migration of MSCs and their subsequent differentiation into SM like cells [35,36]. Similarly, myoblast recruitment can be modulated by a gradient of a chemotactic agent [37]. This recruitment approach greatly simplifies the process of SM tissue engineering, as it eliminates the isolation and expansion of cells *in vitro*. In addition, this approach could have utility in applications such as blood

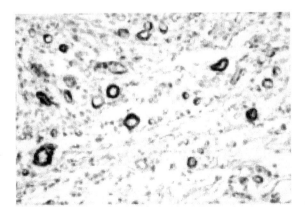

FIGURE 24.2 Sustained and localized delivery of PDGF can lead to the recruitment of SMCs to nascent blood vessels. Photomicrograph of tissue section stained using standard immunohistochemical techniques to detect (brown stain) cells containing α-SMA (SMCs) associating with blood vessels. (Photomicrograph was taken from Richardson, T.P. et al., *Nat. Biotechnol.*, 2001, **19**: 1029–1034 used with permission from the Nature Publishing Group.)

vessel repair where direct placement of smooth cells in the lumen could cause a thrombogenic effect. The use of signaling molecules to recruit circulating progenitor cells may provide a useful alternative in this situation.

24.3 Extracellular Signaling

The extracellular environment plays a pivotal role in SM tissue engineering by providing instructions to the SMCs. Both signaling molecules and mechanical stimuli similar to that found in the native extracellular environment of SM tissues may be used to induce specific SM responses.

Mitogenic factors that have shown to possess a significant effect on SMCs in culture include platelet derived growth factor (PDGF), TGF-β, angiopoietin, and to a lesser extent, heparin-binding epidermal growth factor (HB-EGF) and fibroblast growth factor-2 (FGF-2) [38,39]. PDGF has been shown to have potent effects on proliferation, migration, and matrix production by SMCs [27,28,34,35,37,40–42] and potentially has a significant role in cell transplantation where SMC proliferation is necessary. An increase in the synthesis of SM extracellular proteins is stimulated by TGF-β and these matrix components provide mechanical integrity to engineered SM tissues [34,36,43–48]. Particularly in strategies where SM recruitment is important, angiopoietin and HB–EGF have both shown to mediate EC and SMC interactions [39,49–51]. However, SM tissue engineering strategies that utilize growth factors must consider the mode of delivery. Polymeric encapsulation of growth factors is a common approach to deliver the molecules to the developing SM tissues in a controlled and sustained manner [52].

Mechanical signals modulate the phenotype of SMCs [40,53,54] and enhance the development and function of engineered tissues [55–57]. Cyclic mechanical strain of engineered SM tissues [53,56] result in an increase of matrix production and increased mechanical strength. These mechanical signals are transmitted intracellularly though various signaling pathways that initiate at the transmembrane receptors known as integrins [58], which link the extracellular matrix to the cytoskeleton [59,60]. Integrins activate mitogen activated protein kinase (MAPK) cascades via signaling through extracellular signal-regulated kinase (ERK) [60–63], either through focal adhesion kinases (FAKs) or through coactivation of growth factor receptors (GFRs). Cyclic mechanical strain increases levels of focal contact components [64], which may increase integrin clustering [61], and provide a potential mechanism for the role of mechanical signals in activation of FAKs. It is currently unclear whether these pathways act synergistically with growth factor stimulation, but optimal development and function of engineered SM tissues will likely require appropriate chemical and mechanical signals.

24.4 Synthetic Extracellular Matrix

Tissue engineering utilizes synthetic extracellular matrices (ECMs) to provide an infrastructure for the formation of tissues by providing a predefined space to localize tissue growth and the mechanical support necessary to facilitate this growth. Synthetic ECMs may also provide specific signals to the SMCs. Two general designs of synthetic ECMs for SM tissue regeneration are being pursued, one involving a biological approach where the matrix is assembled by the resident cells and the other utilizing predefined polymeric structures.

24.4.1 Self-Assembly

SMCs maintained for extended times in culture will synthesize, secrete, and assemble an ECM with sufficient mechanical integrity to allow a sheet of confluent SMCs to be manipulated and formed into a three-dimensional tissue [65]. This technique is attractive for tissue engineering because it eliminates the need for exogenous biomaterials, and thereby eradicates any potential inflammatory issues related to the material [66,67]. Self-assembly approaches have focused on engineering vascular grafts by individually culturing cellular sheets to model the defined layers of the blood vessel. A sheet of SMCs is used to form the medial layer, which is subsequently wrapped with a sheet composed of fibroblasts to form the adventitial layer, and finally seeded with endothelial cells to create the lumen. Initial studies on tissues formed utilizing this approach reported poor mechanical strength [68,69], which is indicative of a deficient medial and, or adventitial layers. A revised approach increased the mechanical strength of tissues formed with this approach [65]. However, a limitation to this approach is the extensive time required to form the cellular sheets.

24.4.2 Polymer Scaffolds: Synthetic and Naturally Derived

Most approaches to engineer SM tissues have utilized three-dimensional, biodegradable polymeric scaffolds. Polymeric scaffolds formed from exogenous biomaterials provide mechanical stability and can deliver signaling molecules or adhesion peptides to induce appropriate tissue development. These polymeric biomaterials are fabricated from either synthetic or naturally derived materials. Synthetic polymers typically used for engineering SM tissues include several forms of polyesters, elastomeric polymers, and hydrogels. The most common used naturally derived polymer used to engineer SMC is type I collagen.

24.4.2.1 Synthetic Polymers

The most prevalent synthetic polymers used to engineer smooth muscle tissues are the polyesters poly(glycolic acid) (PGA) (Figure 24.3a), poly(L-lactic acid) (PLLA), and poly (lactic-co-glycolic acid) (PLGA). Advantageous features of these polymers include their reproducible and readily altered mechanical properties and degradation rates [70,71]. These polymer scaffolds provide temporary mechanical support [72] sufficient to resist cellular contractile forces *in vitro* [73–76], and scaffolds exhibiting partial elastic properties under cyclic strain enabled induction of a more contractile, differentiated smooth muscle phenotype from attached SMCs [56]. In addition to structural stability, appropriate signals may be required to guide the development of smooth muscle tissues. Synthetic polymers can be modified to incorporate signals to alter cellular function, including cell adhesion molecules [77–79] and growth factors [80,81].

Hydrogel forming polymers have also been investigated for engineering SM tissues. Polyethylene glycol (PEG) hydrogels intrinsically resist protein adsorption and cell adhesion [82] and this characteristic

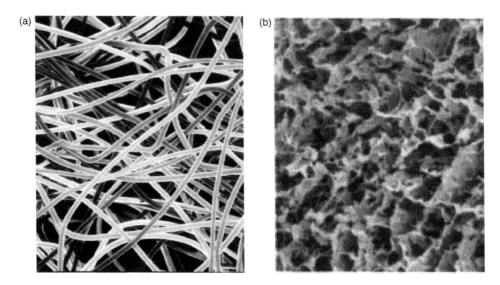

FIGURE 24.3 Scanning electron photomicrographs of typical polymeric scaffolds utilized for engineering SM tissue. (a) PGA fiber based scaffold, and (b) type I collagen scaffold. (Photomicrograph was taken from Kim, B.S. and D.J. Mooney, *J. Biomech. Eng.*, 2000, **122**: 210–215; Kim, B.S. and D.J. Mooney, *J. Biomed. Mater. Res.*, 1998, **41**: 322–332, and used with permission from John Wiley & Sons and ASME, respectively.)

offers advantages for studying the effects of specific bioactive ligands or peptides presented from the scaffold [83,84]. Studies utilizing surface modified PEGs have demonstrated that a number of cellular functions, including adhesion [83], migration [85], and matrix production [46] can be regulated by ligand presentation. In general, hydrogels are an appealing scaffold material because they are structurally similar to the highly hydrated ECM of many tissues [86]. However, the use of hydrogels is often constrained by their limited range of mechanical properties.

The elasticity provided by elastin in SM tissues has motivated the development of elastomeric scaffolds that can similarly provide this property to engineered SM. Elastomeric polymers can recover from extensive deformation [87–89] and are designed to resemble the incompressible nature of the ECM [90]. This property of biomaterials may be ideal to engineer functional SM tissues that require transduction of mechanical signals from the extracellular environment in order to elicit and activate key cellular functions [40,53,54]. This type of biomaterial resolves the limitations of lack of pliancy that limits many synthetic polymer scaffolds (i.e., poly[lactic acid] [PLA]).

24.4.2.2 Naturally Derived Biomaterials

Type I collagen (Figure 24.3b) has been frequently used to create polymer scaffolds for engineering SM tissues [56,73,91,92]. Naturally derived collagen is an attractive biologic material because collagen is the primary constituent of the ECM [58], and contains adhesion ligands that facilitate cell attachment. Although type I collagen does not require additional surface modification to promote tissue formation, glycosaminoglycans (GAGs) [93] and growth factors [45] can be incorporated to improve mechanical properties and to induce specific cellular functions. Type I collagen matrices used to engineer SM tissues have demonstrated partial elasticity and are capable of withstanding cyclic stain [56]. The high tensile strength of type I collagen can be attributed to its molecular structure, while the elasticity is conferred by the intermolecular cross-linking. The degradation of type I collagen scaffolds is dependent on the extent of cross-linking, pore structure and the apparent density, which are variables that can be readily altered to meet a desired target. Although type I collagen is typically extracted from xenogeneic sources, it is considered biocompatible and exhibits low immunogenic responses, likely due to the similarity of this molecule between species [94]. However, naturally derived materials may suffer from batch to batch variations.

Another collagen based biomaterial, small intestinal submucosa (SIS), has also been widely used in tissue engineering research [95–97]. This xenogeneic matrix is harvested from the submucosal layer of the intestine. SIS may provide functional growth factors [98]. that contribute to SM tissue formation. In addition, SIS matrices maintain elasticity and high strength [99]. SIS has typically been obtained from porcine sources, but isolation from rats [100] and canines [101] has also been attempted. SIS has been used to promote regeneration of several SM tissues, in the blood vessels [102,103] and in the bladder [99,101,104].

24.5   Engineered Smooth Muscle Tissues

A number of studies to date have utilized a combination of scaffolding technologies and cells to reconstruct the smooth muscle component of cardiovascular, gastrointestinal, and urinary tissues. The two primary tissue-engineering approaches used to regenerate tissues are cell transplantation and cell recruitment from surrounding tissue. Cell transplantation requires an initial step of procuring cells, often via biopsy from the host, followed by dissociation and expansion *in vitro*. The cells are then seeded onto a scaffold and implanted as a cell–matrix construct. Alternatively, an implanted acellular matrix may be implanted to promote the recruitment of neighboring SMCs and possibly other cell types of interest (e.g., ECs, urothelial cells). Work to date in engineering SM tissues is briefly summarized in this section.

A great deal of research has been performed with the goal of developing blood vessel substitutes, due to the large impact this advance would have on the millions of patients that annually suffer from diseases of blood vessels [105]. Strategies to engineer blood vessel must provide adequate mechanical properties,

(a) 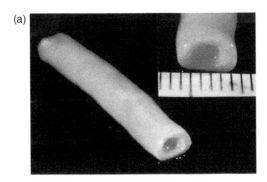 (b)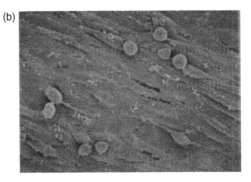

FIGURE 24.4 Engineered blood vessel substitutes. (a) Self-assembly approach to blood vessel formation relies on the ability of sheets of cells to form their own ECM, and multiple cells sheets can subsequently be combined to form tissues (b) Cyclic mechanical strain can play a prominent role in the development of engineered vascular tissues, as it can lead to alignment of cells, as shown in this photomicrograph, and also leads to an increase in tissue mechanical properties. (Photomicrograph was taken from Kim, B.S. et al., *Nat. Biotechnol.*, 1999, **17**: 979–983; L'Heureux, N. et al., *FASEB J.*, 1998, **12**: 47–56, and used with permission from FASEB and the Nature Publishing Group, respectively.)

to avoid catastrophic failure in this mechanically demanding site, and appropriate cellular components to form the complex vascular wall. An early approach to engineer the blood vessels involved the culture of different vascular cell populations in collagen gels to form three distinct layers, resembling the three layers of native blood vessel [68]. However, this model did not lead to tissues with adequate mechanical strength. A later approach exploited the ability of fibroblasts and SM cells to synthesize and secrete their own ECM and form self assembled sheets. These sheets were subsequently wrapped around a mandrel to form distinct layers of the native vessels [65] (Figure 24.4a). This method led to tissues with much greater mechanical strength, comparable to that of human vessels [69]. The increased mechanical strength of these tissues my be partially attributed to paracrine effects between ECs and SMCs [35,39,51] that contribute to the stability of nascent blood vessels by increasing matrix production. Also, implantation of a decellularized SIS with additional type I bovine collagen into a rabbit artery led to the formation of a blood vessel characterized by reasonable burst strength, cell and matrix organization [102].

Several groups have utilized externally applied mechanical stimulation to improve the mechanical integrity of engineered SM tissues (Figure 24.4b). Blood vessel substitutes formed from allogeneic vascular SMCs and ECs cultured on biodegradable PGA scaffolds were maintained under pulsatile stress, and this resulted in an increased matrix production [57]. These engineered constructs were subsequently implanted into swine for seven weeks and the explanted vessels exhibited adequate burst pressures and histology. Several studies document that one can improve the properties of constructs engineered using collagen through the use of mechanical stimulation [55,106]. The significance of mechanical stimulation was also demonstrated by studies where synthetic SMCs cultured with ECs on collagen gels were found to undergo a phenotypic reversion under contractile forces [5,107].

Currently, a common approach to replace or repair bladder tissue utilizes gastrointestinal segments, but this method can result in mucus production, stone formation, and other abnormalities that may be attributed to the different physiologic role of each tissue type. These complications have motivated investigation into new methods for bladder replacement utilizing tissue engineering techniques. One common acellular approach to engineer bladder tissue utilizes SIS membranes, as SIS membranes grafted during partial cystectomy of canines have displayed development of all three layers of the bladder (urothelium, SM, and serosa) [95]. Additionally, these regenerated tissues demonstrated contractile nerve regeneration. Acellular biomaterial extracted from rat bladder also resulted in a well integrated construct when implanted, but one that developed a compromised SM layer [108]. While these studies utilizing acellular scaffold implantation resulted in bladder regeneration, but function of the resultant tissues was not reported. In contrast to this approach, PGA–PLGA scaffolds seeded with autologous canine cells led to formation of a new tissue (Figure 24.5) that regained bladder function [12].

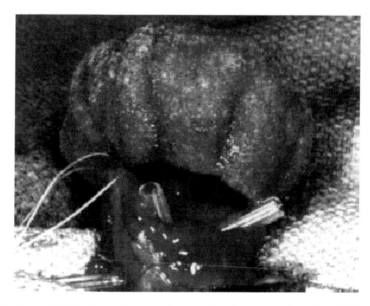

FIGURE 24.5 Engineered bladder formed from autologous cells cultured on polymeric scaffolds *in vitro*. The neo-bladders can be implanted to replace lost bladder tissue. (Photomicrograph was taken from Oberpenning, F. et al., *Nat. Biotechnol.*, 1999, **17**: 149–155, and used with permission from the Nature Publishing Group.)

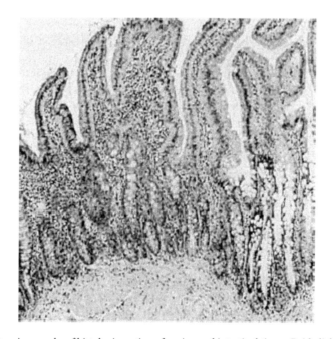

FIGURE 24.6 Photomicrographs of histologic section of engineered intestinal tissue. Epithelial organoid units were isolated, seeded onto polymeric scaffolds, and implanted. Ultimately, the transplanted cells differentiate and form new tissue. (Photomicrograph was taken from Vacanti, J.P., *J. Gastrointest. Surg.*, 2003, **7**: 831–835., and used with permission from Elsevier Inc.)

A number of studies suggest it may be feasible to engineer functional gastrointestinal tissues. Isolated crypt cells implanted on PGA tubes formed epithelial-lined tubular structures lacking in a SM component [109] (Figure 24.6). The transplantation of intestinal organoid units, in place of individual cells on PLGA scaffolds led to the development of a neomucosa layer [14] and SM layers [109]. However, the neomucosa may have been lacking in its ability to control nutrient. SIS patch implantation, without cells, into canines led to formation of both the epithelial and SM layer. However, a large percentage of animals died and a large number of inflammatory cells were found in explanted SIS patches. One study utilized transplantation of precursor cells derived from bone marrow in the place of differentiated SMCs [110], but the engineered tissues did not regenerate a functional muscle layer, potentially due to a lack of appropriate extracellular signals to induce the differentiation of the mesenchymal stem cells into SMCs.

24.6 Conclusion/Future Directions

Impressive progress has been made to date in SM engineering, and these tissues may have significant clinical impact and provide models to study basic biological processes. However our current understanding of the complex interplay of factors that modulate the SM function are far from comprehensive, and advances in this knowledge will likely translate to improved systems for engineering functional SM tissues. One important issue yet to be addressed is whether approaches that successfully regenerate one type of SM tissue will necessarily be successful for other organ systems. Physiologically, there are distinct differences between the SM in cardiovascular, gastrointestinal, and urinary tissues. Similar approaches have been utilized to date in most SM engineering approaches. However, inherent differences exist in the microenvironment of different SM tissues that may have a significant effect on the developing SM tissue. The substratum to which SMCs adhere also plays an important role in the presentation of paracrine signals and that may regulate SMC phenotype [35,37,111]. An ideal cell source is also currently lacking, but stem cells may fill this need. BMSCs contain a population of SM progenitors, but the difficulties in reproducibly isolating and regulating the differentiation of these cell populations pose as an obstacle to their use. Embryonic stem cells may provide a more homogeneous population of pluripotential cells, but these cells have not yet been demonstrated to form functional SM tissue. In addition, for engineered SM tissues to be truly functional, they must also be innervated and vascularized. Current research in the therapeutic angiogenesis field may provide methods to induce formation of vascular networks [81]. However, development of fully functional SM tissues will also require a means to transduce neural signals critical for blood vessel vasoactivity.

References

[1] Langer, R. and J.P. Vacanti, Tissue engineering. *Science*, 1993, **260**: 920–926.
[2] Gridelli, B. and G. Remuzzi, Strategies for making more organs available for transplantation. *N. Engl. J. Med.*, 2000, **343**: 404–410.
[3] Owens, G.K., Regulation of differentiation of vascular smooth muscle cells. *Physiol. Rev.*, 1995, **75**: 487–517.
[4] Campbell, J.H. and G.R. Campbell, *Vascular Smooth Muscle in Culture*. Boca Raton, FL: CRC Press, 2 vol, 1987.
[5] Kanda, K., H. Miwa, and T. Matsuda, Phenotypic reversion of smooth muscle cells in hybrid vascular prostheses. *Cell Transplant.*, 1995, **4**: 587–595.
[6] Nikolovski, J., B.S. Kim, and D.J. Mooney, Cyclic strain inhibits switching of smooth muscle cells to an osteoblast-like phenotype. *FASEB J.*, 2003, **17**: 455–457.
[7] Young, B. et al., *Wheater's Functional Histology: A Text and Colour Atlas*. Churchill Livingstone: Edinburgh; New York, 2000.
[8] Stegemann, J.P. and R.M. Nerem, Altered response of vascular smooth muscle cells to exogenous biochemical stimulation in two- and three-dimensional culture. *Exp. Cell Res.*, 2003, **283**: 146–155.
[9] McKee, J.A. et al., Human arteries engineered *in vitro*. *EMBO Rep.*, 2003, **4**: 633–638.
[10] Stock, U.A. et al., Tissue engineering of heart valves — current aspects. *Thorac. Cardiovasc. Surg.*, 2002, **50**: 184–193.
[11] Shinoka, T. et al., Creation of viable pulmonary artery autografts through tissue engineering. *J. Thorac. Cardiovasc. Surg.*, 1998, **115**: 536–545; discussion 545–546.

[12] Oberpenning, F. et al., *De novo* reconstitution of a functional mammalian urinary bladder by tissue engineering. *Nat. Biotechnol.*, 1999, **17**: 149–155.

[13] Maemura, T. et al., A tissue-engineered stomach as a replacement of the native stomach. *Transplantation*, 2003, **76**: 61–65.

[14] Kaihara, S. et al., Long-term follow-up of tissue-engineered intestine after anastomosis to native small bowel. *Transplantation*, 2000, **69**: 1927–1932.

[15] Junqueira, L.C.U. and J. Carneiro, Basic Histology: Text & Atlas. 10th ed. 2003, New York: Lange Medical Books McGraw-Hill Medical Pub. Division, p. viii, 515.

[16] Pittenger, M.F. et al., Multilineage potential of adult human mesenchymal stem cells. *Science*, 1999, **284**: 143–147.

[17] Bianco, P. and P.G. Robey, Stem cells in tissue engineering. *Nature*, 2001, **414**: 118–121.

[18] Bianco, P. et al., Bone marrow stromal stem cells: nature, biology, and potential applications. *Stem Cells*, 2001, **19**: 180–192.

[19] Krebsbach, P.H. et al., Bone marrow stromal cells: characterization and clinical application. *Crit. Rev. Oral Biol. Med.*, 1999, **10**: 165–181.

[20] Ferrari, G. et al., Muscle regeneration by bone marrow-derived myogenic progenitors. *Science*, 1998, **279**: 1528–1530.

[21] Ferrari, G. and F. Mavilio, Myogenic stem cells from the bone marrow: a therapeutic alternative for muscular dystrophy? *Neuromuscul. Disord.*, 2002, **12**(Suppl 1): S7–S10.

[22] Hirschi, K.K. and M.A. Goodell, Hematopoietic, vascular and cardiac fates of bone marrow-derived stem cells. *Gene Ther.*, 2002, **9**: 648–652.

[23] Kadner, A. et al., A new source for cardiovascular tissue engineering: human bone marrow stromal cells. *Eur. J. Cardiothorac. Surg.*, 2002, **21**: 1055–1060.

[24] Dennis, J.E. and P. Charbord, Origin and differentiation of human and murine stroma. *Stem Cells*, 2002, **20**: 205–214.

[25] Kinner, B., J.M. Zaleskas, and M. Spector, Regulation of smooth muscle actin expression and contraction in adult human mesenchymal stem cells. *Exp. Cell Res.*, 2002, **278**: 72–83.

[26] Altman, G.H. et al., Cell differentiation by mechanical stress. *FASEB J.*, 2002, **16**: 270–272.

[27] Yamashita, J. et al., Flk1-positive cells derived from embryonic stem cells serve as vascular progenitors. *Nature*, 2000, **408**: 92–96.

[28] Koike, N. et al., Tissue engineering: creation of long-lasting blood vessels. *Nature*, 2004, **428**: 138–139.

[29] Han, C.I., G.R. Campbell, and J.H. Campbell, Circulating bone marrow cells can contribute to neointimal formation. *J. Vasc. Res.*, 2001, **38**: 113–119.

[30] Simper, D. et al., Smooth muscle progenitor cells in human blood. *Circulation*, 2002, **106**: 1199–1204.

[31] Sata, M. et al., Hematopoietic stem cells differentiate into vascular cells that participate in the pathogenesis of atherosclerosis. *Nat. Med.*, 2002, **8**: 403–409.

[32] Shimizu, K. et al., Host bone-marrow cells are a source of donor intimal smooth-muscle-like cells in murine aortic transplant arteriopathy. *Nat. Med.*, 2001, **7**: 738–741.

[33] Majka, S.M. et al., Distinct progenitor populations in skeletal muscle are bone marrow derived and exhibit different cell fates during vascular regeneration. *J. Clin. Invest.*, 2003, **111**: 71–79.

[34] Hirschi, K.K., S.A. Rohovsky, and P.A. D'Amore, PDGF, TGF-beta, and heterotypic cell–cell interactions mediate endothelial cell-induced recruitment of 10T1/2 cells and their differentiation to a smooth muscle fate. *J. Cell Biol.*, 1998, **141**: 805–814.

[35] Hirschi, K.K. et al., Endothelial cells modulate the proliferation of mural cell precursors via platelet-derived growth factor-BB and heterotypic cell contact. *Circ. Res.*, 1999, **84**: 298–305.

[36] Darland, D.C. and P.A. D'Amore, TGF beta is required for the formation of capillary-like structures in three-dimensional cocultures of 10T1/2 and endothelial cells. *Angiogenesis*, 2001, **4**: 11–20.

[37] Corti, S. et al., Chemotactic factors enhance myogenic cell migration across an endothelial monolayer. *Exp. Cell Res.*, 2001, **268**: 36–44.

[38] Adam, R.M. et al., Signaling through PI3K/Akt mediates stretch and PDGF-BB-dependent DNA synthesis in bladder smooth muscle cells. *J. Urol.*, 2003, **169**: 2388–2393.

[39] Iivanainen, E. et al., Angiopoietin-regulated recruitment of vascular smooth muscle cells by endothelial-derived heparin binding EGF-like growth factor. *FASEB J.*, 2003, **17**: 1609–1621.

[40] Stegemann, J.P. and R.M. Nerem, Phenotype modulation in vascular tissue engineering using biochemical and mechanical stimulation. *Ann. Biomed. Eng.*, 2003, **31**: 391–402.

[41] Hellstrom, M. et al., Role of PDGF-B and PDGFR-beta in recruitment of vascular smooth muscle cells and pericytes during embryonic blood vessel formation in the mouse. *Development*, 1999, **126**: 3047–3055.

[42] Stringa, E. et al., Collagen degradation and platelet-derived growth factor stimulate the migration of vascular smooth muscle cells. *J. Cell Sci.*, 2000, **113**(Pt 11): 2055–2064.

[43] Wrenn, R.W. et al., Transforming growth factor-beta: signal transduction via protein kinase C in cultured embryonic vascular smooth muscle cells. *In Vitro Cell Dev. Biol.*, 1993, **29A**: 73–78.

[44] Desmouliere, A. et al., Transforming growth factor-beta 1 induces alpha-smooth muscle actin expression in granulation tissue myofibroblasts and in quiescent and growing cultured fibroblasts. *J. Cell Biol.*, 1993, **122**: 103–111.

[45] Vaughan, M.B., E.W. Howard, and J.J. Tomasek, Transforming growth factor-beta1 promotes the morphological and functional differentiation of the myofibroblast. *Exp. Cell Res.*, 2000, **257**: 180–189.

[46] Mann, B.K., R.H. Schmedlen, and J.L. West, Tethered-TGF-beta increases extracellular matrix production of vascular smooth muscle cells. *Biomaterials*, 2001, **22**: 439–444.

[47] Hirschi, K.K. et al., Transforming growth factor-beta induction of smooth muscle cell phenotpye requires transcriptional and post-transcriptional control of serum response factor. *J. Biol. Chem.*, 2002, **277**: 6287–6295.

[48] Cutroneo, K.R., Gene therapy for tissue regeneration. *J. Cell Biochem.*, 2003, **88**: 418–425.

[49] Lobov, I.B., P.C. Brooks, and R.A. Lang, Angiopoietin-2 displays VEGF-dependent modulation of capillary structure and endothelial cell survival *in vivo*. *Proc. Natl Acad. Sci. USA*, 2002, **99**: 11205–11210.

[50] Du, L. et al., Signaling molecules in nonfamilial pulmonary hypertension. *N. Engl. J. Med.*, 2003, **348**: 500–509.

[51] Nishishita, T. and P.C. Lin, Angiopoietin 1, PDGF-B, and TGF-beta gene regulation in endothelial cell and smooth muscle cell interaction. *J. Cell Biochem.*, 2004, **91**: 584–593.

[52] Holland, T.A. et al., Transforming growth factor-beta1 release from oligo(poly(ethylene glycol) fumarate) hydrogels in conditions that model the cartilage wound healing environment. *J. Control. Release*, 2004, **94**: 101–114.

[53] Kim, B.S. et al., Cyclic mechanical strain regulates the development of engineered smooth muscle tissue. *Nat. Biotechnol.*, 1999, **17**: 979–983.

[54] Owens, G.K., Role of mechanical strain in regulation of differentiation of vascular smooth muscle cells. *Circ. Res.*, 1996, **79**: 1054–1055.

[55] Seliktar, D. et al., Dynamic mechanical conditioning of collagen-gel blood vessel constructs induces remodeling *in vitro*. *Ann. Biomed. Eng.*, 2000, **28**: 351–362.

[56] Kim, B.S. and D.J. Mooney, Scaffolds for engineering smooth muscle under cyclic mechanical strain conditions. *J. Biomech. Eng.*, 2000, **122**: 210–215.

[57] Niklason, L.E. et al., Functional arteries grown *in vitro*. *Science*, 1999, **284**: 489–493.

[58] Alberts, B., *Molecular Biology of the Cell*. 4th ed. New York: Garland Science, 2002.

[59] Davis, M.J. et al., Integrins and mechanotransduction of the vascular myogenic response. *Am. J. Physiol. Heart Circ. Physiol.*, 2001, **280**: H1427–H1433.

[60] Assoian, R.K. and M.A. Schwartz, Coordinate signaling by integrins and receptor tyrosine kinases in the regulation of G1 phase cell-cycle progression. *Curr. Opin. Genet. Dev.*, 2001, **11**: 48–53.

[61] Giancotti, F.G. and E. Ruoslahti, Integrin signaling. *Science*, 1999, **285**: 1028–1032.

[62] Schwartz, M.A. and M.H. Ginsberg, Networks and crosstalk: integrin signalling spreads. *Nat. Cell Biol.*, 2002, **4**: E65–E68.

[63] Howe, A.K., A.E. Aplin, and R.L. Juliano, Anchorage-dependent ERK signaling–mechanisms and consequences. *Curr. Opin. Genet. Dev.*, 2002, **12**: 30–35.

[64] Cunningham, J.J., J.J. Linderman, and D.J. Mooney, Externally applied cyclic strain regulates localization of focal contact components in cultured smooth muscle cells. *Ann. Biomed. Eng.*, 2002, **30**: 927–935.

[65] L'Heureux, N. et al., A completely biological tissue-engineered human blood vessel. *FASEB J.*, 1998, **12**: 47–56.

[66] Shin, H. et al., *In vivo* bone and soft tissue response to injectable, biodegradable oligo(poly(ethylene glycol) fumarate) hydrogels. *Biomaterials*, 2003, **24**: 3201–3211.

[67] Cao, Y. et al., Comparative study of the use of poly(glycolic acid), calcium alginate and pluronics in the engineering of autologous porcine cartilage. *J. Biomater. Sci. Polym. Ed.*, 1998, **9**: 475–487.

[68] Weinberg, C.B. and E. Bell, A blood vessel model constructed from collagen and cultured vascular cells. *Science*, 1986, **231**: 397–400.

[69] L'Heureux, N. et al., *In vitro* construction of a human blood vessel from cultured vascular cells: a morphologic study. *J. Vasc. Surg.*, 1993, **17**: 499–509.

[70] Thomson, R.C. et al., Fabrication of biodegradable polymer scaffolds to engineer trabecular bone. *J. Biomater. Sci. Polym. Ed.*, 1995, **7**: 23–38.

[71] Wong, W. and D. Mooney, Synthesis and properties of biodegradable polymers used in synthetic matrices for tissue engineering, In Atala, M.D., Ed., *Synthetic Biodegradable Polymer Scaffolds*, Birkhäuser: Boston. pp. 51–84, 1997.

[72] Mikos, A.G. et al., Laminated three-dimensional biodegradable foams for use in tissue engineering. *Biomaterials*, 1993, **14**: 323–330.

[73] Kim, B.S. and D.J. Mooney, Engineering smooth muscle tissue with a predefined structure. *J. Biomed. Mater. Res.*, 1998, **41**: 322–332.

[74] Peter, S.J. et al., Polymer concepts in tissue engineering. *J. Biomed. Mater. Res.*, 1998, **43**: 422–427.

[75] Niklason, L.E. and R.S. Langer, Advances in tissue engineering of blood vessels and other tissues. *Transpl. Immunol.*, 1997, **5**: 303–306.

[76] Mooney, D.J. et al., Stabilized polyglycolic acid fibre-based tubes for tissue engineering. *Biomaterials*, 1996, **17**: 115–124.

[77] Nikolovski, J. and D.J. Mooney, Smooth muscle cell adhesion to tissue engineering scaffolds. *Biomaterials*, 2000, **21**: 2025–2032.

[78] Mann, B.K. et al., Modification of surfaces with cell adhesion peptides alters extracellular matrix deposition. *Biomaterials*, 1999, **20**: 2281–2286.

[79] Gao, J., L. Niklason, and R. Langer, Surface hydrolysis of poly(glycolic acid) meshes increases the seeding density of vascular smooth muscle cells. *J. Biomed. Mater. Res.*, 1998, **42**: 417–424.

[80] Mooney, D.J. et al., Novel approach to fabricate porous sponges of poly(D,L-lactic-co-glycolic acid) without the use of organic solvents. *Biomaterials*, 1996, **17**: 1417–1422.

[81] Richardson, T.P. et al., Polymeric system for dual growth factor delivery. *Nat. Biotechnol.*, 2001, **19**: 1029–1034.

[82] Gombotz, W.R. et al., Protein adsorption to poly(ethylene oxide) surfaces. *J. Biomed. Mater. Res.*, 1991, **25**: 1547–1562.

[83] Mann, B.K. and J.L. West, Cell adhesion peptides alter smooth muscle cell adhesion, proliferation, migration, and matrix protein synthesis on modified surfaces and in polymer scaffolds. *J. Biomed. Mater. Res.*, 2002, **60**: 86–93.

[84] Tulis, D.A. et al., YC-1-mediated vascular protection through inhibition of smooth muscle cell proliferation and platelet function. *Biochem. Biophys. Res. Commun.*, 2002, **291**: 1014–1021.

[85] Gobin, A.S. and J.L. West, Cell migration through defined, synthetic ECM analogs. *FASEB J.*, 2002, **16**: 751–753.

[86] Drury, J.L. and D.J. Mooney, Hydrogels for tissue engineering: scaffold design variables and applications. *Biomaterials*, 2003, **24**: 4337–4351.

[87] Guan, J. et al., Synthesis, characterization, and cytocompatibility of elastomeric, biodegradable poly(ester-urethane)ureas based on poly(caprolactone) and putrescine. *J. Biomed. Mater. Res.*, 2002, **61**: 493–503.

[88] Lee, S.H. et al., Elastic biodegradable poly(glycolide-co-caprolactone) scaffold for tissue engineering. *J. Biomed. Mater. Res.*, 2003, **66A**: 29–37.

[89] Fromstein, J.D. and K.A. Woodhouse, Elastomeric biodegradable polyurethane blends for soft tissue applications. *J. Biomater. Sci. Polym. Ed.*, 2002, **13**: 391–406.

[90] Wang, Y. et al., A tough biodegradable elastomer. *Nat. Biotechnol.*, 2002, **20**: 602–6.

[91] Nakanishi, Y. et al., Tissue-engineered urinary bladder wall using PLGA mesh-collagen hybrid scaffolds: a comparison study of collagen sponge and gel as a scaffold. *J. Pediatr. Surg.*, 2003, **38**: 1781–1784.

[92] Pariente, J.L., B.S. Kim, and A. Atala, *In vitro* biocompatibility evaluation of naturally derived and synthetic biomaterials using normal human bladder smooth muscle cells. *J. Urol.*, 2002, **167**: 1867–1871.

[93] Cavallaro, J.F., P.D. Kemp, and K.H. Kraus, Collagen fabrics as biomaterials. *Biotechnol. Bioeng.*, 1994, **43**: 781–791.

[94] Li, S.T., Biologic biomaterials: tissue-derived biomaterials (collagen). In Brozino, J.D., Ed. *The Biomedical Engineering Handbook*. Boca Raton, FL: CRC Press, pp. 627–647, 1995.

[95] Kropp, B.P. et al., Regenerative urinary bladder augmentation using small intestinal submucosa: urodynamic and histopathologic assessment in long-term canine bladder augmentations. *J. Urol.*, 1996, **155**: 2098–2104.

[96] Badylak, S.F. et al., Comparison of the resistance to infection of intestinal submucosa arterial autografts versus polytetrafluoroethylene arterial prostheses in a dog model. *J. Vasc. Surg.*, 1994, **19**: 465–472.

[97] Zhang, Y. et al., Coculture of bladder urothelial and smooth muscle cells on small intestinal submucosa: potential applications for tissue engineering technology. *J. Urol.*, 2000, **164**: 928–934; discussion 934–935.

[98] Voytik-Harbin, S.L. et al., Identification of extractable growth factors from small intestinal submucosa. *J. Cell Biochem.*, 1997, **67**: 478–491.

[99] Chen, M.K. and S.F. Badylak, Small bowel tissue engineering using small intestinal submucosa as a scaffold. *J. Surg. Res.*, 2001, **99**: 352–358.

[100] Wang, Z.Q., Y. Watanabe, and A. Toki, Experimental assessment of small intestinal submucosa as a small bowel graft in a rat model. *J. Pediatr. Surg.*, 2003, **38**: 1596–1601.

[101] Yoo, J.J. et al., Bladder augmentation using allogenic bladder submucosa seeded with cells. *Urology*, 1998, **51**: 221–225.

[102] Huynh, T. et al., Remodeling of an acellular collagen graft into a physiologically responsive neovessel. *Nat. Biotechnol.*, 1999, **17**: 1083–6.

[103] Badylak, S.F. et al., Small intestinal submucosa as a large diameter vascular graft in the dog. *J. Surg. Res.*, 1989, **47**: 74–80.

[104] Falke, G., J. Caffaratti, and A. Atala, Tissue engineering of the bladder. *World J. Urol.*, 2000, **18**: 36–43.

[105] Tu, J.V. et al., Use of cardiac procedures and outcomes in elderly patients with myocardial infarction in the United States and Canada. *N. Engl. J. Med.*, 1997, **336**: 1500–1505.

[106] Seliktar, D., R.M. Nerem, and Z.S. Galis, Mechanical strain-stimulated remodeling of tissue-engineered blood vessel constructs. *Tissue Eng.*, 2003, **9**: 657–666.

[107] Reusch, P. et al., Mechanical strain increases smooth muscle and decreases nonmuscle myosin expression in rat vascular smooth muscle cells. *Circ. Res.*, 1996, **79**: 1046–1053.

[108] Probst, M. et al., Reproduction of functional smooth muscle tissue and partial bladder replacement. *Br. J. Urol.*, 1997, **79**: 505–515.

[109] Choi, R.S. and J.P. Vacanti, Preliminary studies of tissue-engineered intestine using isolated epithelial organoid units on tubular synthetic biodegradable scaffolds. *Transplant Proc.*, 1997, **29**: 848–851.

[110] Hori, Y. et al., Experimental study on tissue engineering of the small intestine by mesenchymal stem cell seeding. *J. Surg. Res.*, 2002, **102**: 156–160.

[111] Master, V.A. et al., Urothlelium facilitates the recruitment and trans-differentiation of fibroblasts into smooth muscle in acellular matrix. *J. Urol.*, 2003, **170**: 1628–1632.

[112] Vacanti, J.P., Tissue and organ engineering: can we build intestine and vital organs? *J. Gastrointest. Surg.*, 2003, **7**: 831–835.

25

Esophagus: A Tissue Engineering Challenge

B.D. Ratner
B.L. Beckstead
University of Washington

K.S. Chian
A.C. Ritchie
Nanyang Technological University

25.1 Medical Need/Clinical Problem

The esophagus, a muscular/mucosal tube connecting the mouth and pharynx to the stomach, is critical for life (and good quality of life) (Figure 25.1). This seemingly simple organ is surgically challenging to repair or replace. There are a number of conditions where surgical repair of the esophagus is indicated. These include accident and trauma, congenital defects such as esophageal atresia (incomplete formation of the esophagus) and tracheoesophageal fistulas and cancer. In 2003, roughly 14,000 people in the United States were diagnosed with esophageal cancer. The prevalence of esophageal cancer in the general population can be 10 to 100 times higher in Iran, China, Singapore, India, and South Africa. Worldwide, cancer of the esophagus is the seventh leading cause of cancer death.

Surgical removal of a section of the esophagus and reconnection with the stomach, the most common strategy for more advanced cancers, leads to complication rates as high as 40%. Strictures, dilation, leakage, and infection are often observed. Attempts to use synthetics such as polyethylene, polypropylene, teflon, or elastomers have also met with very limited success, problems being stenosis, leakage, infection,

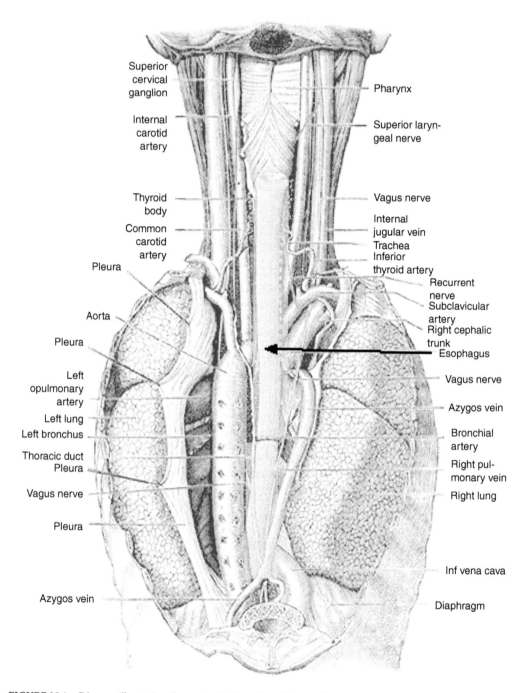

FIGURE 25.1   Diagram illustrating the anatomical location of the esophagus. (From *Gray's Anatomy*.)

scarring, ulceration, and migration [Leininger et al., 1970; Watanabe and Mark, 1971; Sato et al., 1997; Ure et al., 1998; Fuchs et al., 2001]. A living, nonimmunologic esophageal replacement could make a significant contribution to medical practice and patient treatment.

Tissue engineering, using healthy cells supplied by the patient, offers the possibility of a normal esophageal reconstruction after surgery, trauma, or for congenital repair. In recent years, the tissue engineering approach has been investigated as an alternative treatment of esophageal diseases [Natsume et al., 1993; Miki et al., 1999; Badylak et al., 2000; Yamamoto et al., 2000; Kajitani et al., 2001].

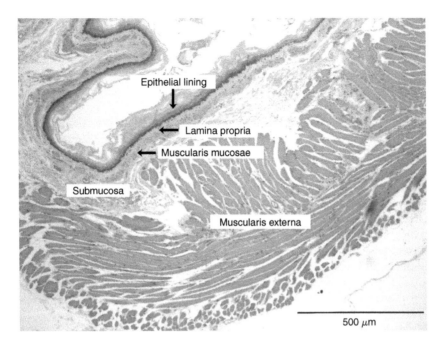

FIGURE 25.2 A cross-sectional histological section of the esophagus.

25.2 Anatomy and Physiology of the Esophagus

The esophagus is a thick-walled muscular tube extending from the pharynx to the stomach. It descends through the posterior mediasternum, passes through the diaphragm at the esophageal hiatus, and joins the stomach at the T9 level [Kumar, 1993]. Its length in adults is typically between 9 and 10 in. (225 to 250 mm). It consists of three main layers, the mucosa, submucosa, and muscularis. The esophagus has no serosa [Gray, 1995; Goyal and Sivarao, 1999; Ergun and Kahrilas, 1997; Wood, 1994].

Innervation: Sensory and motor function is supplied by branches of cranial nerves V, VII, IX, X, XI, and XII [Wood, 1994; Ergun and Kahrilas, 1997; Goyal and Sivarao, 1999]. Neural networks lie between the longitudinal and circular muscle layers (Auerbach's or myenteric plexus) and the circular muscle and submucosa (Meissner's or submucosal plexus) [Wood, 1994; Ergun and Kahrilas, 1997]. Swallowing and esophageal peristalsis come under the control of both the somatic and enteric nervous systems, with signal conduction primarily through the vagus nerve.

Vasculature: Blood supply to the esophagus is via shared vasculature. Along its length, branches of larger vessels including the thyroid artery, esophageal aortic arteries, and left gastric and splenic arteries supply blood to the arterial network around the esophageal lumen [Gershon et al., 1994; Ergun and Kahrilas, 1997]. Capillaries within the tissues drain into deep intrinsic and adventitial veins. Blood is transported back to the heart via the extrinsic serosal and periesophageal veins, which drain into the left gastric veins and azygos vein [Ergun and Kahrilas, 1997].

Structure: A cross-sectional diagram of a rat esophagus is shown in Figure 25.2. The layer closest to the lumen is typically composed of stratified squamous epithelium [Burkitt et al., 1993]. Acid reflux may cause Barrett's esophagus, where the stratified squamous epithelium is transformed into columnar epithelium [Goyal and Sivarao, 1999]. Lying beneath this layer is a thin lamina propria and a thin layer of smooth muscle, the muscularis mucosa [Burkitt et al., 1993].

The submucosa is a layer of highly vascularized and relatively loose connective tissue, which allows distension. Within the submucosa are small mucous glands for lubrication [Burkitt et al., 1993].

The muscularis layer is normally classified as two sublayers according to the orientation of the muscle cells. Closest to the submucosa is the circular muscle layer, where the myocytes are aligned tangentially to the esophageal lumen. The next layer is the longitudinal muscle, where the myocytes are aligned parallel to the esophageal axis [Burkitt et al., 1993; Ergun and Kahrilas, 1997].

The muscle type varies along the length of the esophagus. In the cervical esophagus, the muscularis layer is made up, almost exclusively, of skeletal muscle, and in the distal third (closest to the stomach), the muscularis layer consists of smooth muscle. The middle third is composed of a mixture of skeletal and smooth muscle.

Contraction of the skeletal muscle in the cervical esophagus may be initiated voluntarily or reflexively. The initial stage of swallowing occurs as the tongue pushes a bolus of masticated food into the oropharynx. This initiates the involuntary pharyngeal stage of swallowing, as the bolus of food stimulates receptors in the oropharynx, which then sends impulses to the deglutition center of the medulla oblongata and lower pons [Ergun and Kahrilas, 1997].

The bolus of masticated food is then pushed through the esophagus by peristaltic contraction of the muscularis layer. Peristaltic contraction may be initiated by either extrinsic or intrinsic neural pathways [Wingate, 1993]. Longitudinal muscles ahead of the bolus contract to widen the esophagus while the circular muscles behind the bolus of food contract to push it toward the stomach. Food normally passes through the esophagus within 10 sec [Ergun and Kahrilas, 1997]. The junction between the esophagus and the stomach is not a true sphincter and in certain conditions, matter may pass from the stomach into the esophagus [Kumar, 1993].

25.3   Criteria for a Tissue-Engineered Esophagus

Based upon the above anatomical description, and from an engineering and biology standpoint, what are the criteria that must be met before a practical and functional surgical alternative through tissue engineering is in place? The following demands challenge us in the development of a tissue-engineered living prosthesis:

Radially elastic/longitudinally rigid: The normal biomechanical behavior of the esophagus must be duplicated in a surgical replacement. First, this mechanically strong and compliant tube must expand radially to permit ingestion of a bolus of food or liquid, yet it should exhibit relatively little elasticity longitudinally. The organization of collagen and elastin within the structure, along with its convoluted luminal cross section can account for these mechanical properties.

Muscular: The esophagus is a highly muscular organ. Smooth muscle cells and skeletal muscle cells are present with proportions varying along the length of the tube. Muscle laminae in the walls of the esophagus are organized orthogonally to one another. The act of swallowing produces a peristaltic pulse along the esophagus to drive contents to the stomach. This peristaltic process must be replicated in a tissue-engineered construct.

Mucosal lining: The epithelial cell lining of the lumen of the esophagus generates a mucosal exterior layer that lubricates and also protects the air–tissue interface. This lining enables the esophagus able to deal with the range of "chemicals" and insults it must tolerate including hot foods, strong alcoholic beverages, abrasive foods, and acidic foods.

Innervation: A nerve network in the esophagus coordinates the peristaltic action.

Angiogenesis: The esophageal wall is a relatively thick tissue (> 1 mm) and requires its own blood vessel network to sustain the core muscle tissue and remove cell wastes.

Cell sources: Since we do not know how to deal with the immunological issues associated with allogeneic cell sources, a practical surgical prosthesis will probably be comprised of autologous cells. Cell harvesting from a biopsy should not be a problem. However, sterility issues, cell separation, expansion, and seeding on a scaffold are all challenges to address.

Scaffolds: A scaffold will be used to give anatomical shape and biological signals to the growing cells forming the new tissue. The criteria for scaffolds are many. Of course, it should be nontoxic (nominally "biocompatible"). The esophageal scaffold should have a shape and form similar to the organ to be replaced (including the puckered or invaginated form of the lumen). It should separate epithelial and muscular lamina, but allow biological communication between them. It should give cells the proper signals for attachment, growth, and orientation. This scaffold should be elastomeric as opposed to stiff. It should allow or even encourage angiogenesis and integration into the anatomical site. Finally, it should biodegrade without a trace after the living tissue has gained sufficient strength to support itself and function anatomically.

Bioreactor issues: The seeded cells will be cultured *in vitro* until the evolving tissue is adequate for transplantation into the patient. A bioreactor must sustain the cells with oxygen and nutrients, remove wastes, provide an appropriate mechanical environment to condition the cells, provide a sterile environment, and also create the air–tissue interface necessary for proper development of the epithelial cell–mucosal layer.

Surgical issues: Finally, the growing tissue engineered construct must be taken from the bioreactor and implanted. What is the optimal period of *in vitro* development before *in vivo* implantation? How should it be sutured? How should it be implanted to optimize angiogenesis? What patient management issues are needed pre and post surgery?

The following sections illustrate research efforts underway at the Nanyang Technological University and the University of Washington to take an engineering systems approach to the esophageal replacement problem and address the demanding criteria for a tissue-engineered esophagus.

25.4 Scaffold Possibilities

25.4.1 Background

Currently two main approaches to tissue/organ regeneration are *in vivo* and *in vitro* tissue engineering. *In vivo* tissue engineering uses noncell-seeded biomaterials, which include decellularized tissues such as the acellular small intestinal submucosa, amniotic membranes, and pig-heart valves [Badylak et al., 1995; Khan et al., 2001]. In contrast, *in vitro* approaches involve the manipulation of cells on biomaterial scaffolds *in vitro* prior to implantation. Despite these obvious differences, both approaches involve the use of biomaterial scaffold and rely on the body's ability to regenerate.

The scaffold plays a crucial role in tissue engineering of the esophagus. The growth of the anchorage-dependant esophageal cells requires a suitable scaffold for attachment in order to proliferate and function. These scaffolds are three-dimensional biodegradable structures that provide spatial cellular signaling environment necessary for the regenerative processes. These signaling processes are responsible for triggering the expression or repression of genes that regulate cell division, production of extracellular matrix (ECM), differentiation, proliferation, migration, and even apoptosis [Peters and Mooney, 1997; Bottaro and Heidaran, 2001]. In essence, a scaffold is a temporary biodegradable structure containing the appropriate cells that, through various biological remodeling processes, will eventually form vital tissues/organs.

A suitable tissue engineered scaffold for esophageal replacement must closely mimic the host tissue it replaces with respect to mechanical, surface, structural, and biological properties. Some of these considerations necessary for esophageal tissue engineering include (i) materials selection, (ii) scaffold design, and (iii) choice of fabrication techniques.

25.4.2 Materials Selection

The esophagus is a highly elastic and muscular organ, and one of the main considerations is to identify materials that are mechanically compatible. This criterion limits the choice to only polymeric biomaterials. These polymers must be biocompatible, biodegradable, mechanically compliant, and have suitable surface chemistry. In addition, the polymers must be amenable to fabrication and sterilization techniques without altering their biocompatibility and properties. Some of these biodegradable polymers used in tissue engineering applications have been comprehensively reviewed elsewhere [Pachence and Kohn, 2000; Langer and Tirrell, 2004].

The three groups of polymeric biomaterials commonly used in tissue engineering applications include (1) naturally derived polymers, that is, alginates, chitosan, hyaluronic acid; (2) biologically derived materials such as the decellularized tissues, that is, collagens, small intestinal submucosa, urinary bladder matrix, and amniotic membranes; and (3) synthetic polymers, that is, poly(lactic acid), poly(glycolic acid), and poly(lactic-co-glycolic acid), poly(hydroxybutyrate-valerate). Some of the recent developments in biomaterials for tissue engineering applications include self-assembly nanofibers [Huang et al., 2000; Hartgerink et al., 2002], and elastic protein-based polymer systems [Urry et al., 1991; McMillan and Conticello, 2000].

Many materials have been evaluated for esophagus repair and reconstruction. These include collagens [Natsume et al., 1993], poly(glycolic acid) [Shinhar et al., 1998, Miki et al., 1999], urinary bladder matrix [Badylak, et al. 2000]; elastin biomaterials obtained from porcine aorta [Kajitani et al., 2001], and Alloderm®[Isch et al., 2001]. All the above materials showed promise, especially the collagen and acellular matrices, but the problem of stenosis remained. The acellular matrices appear to show better cell–matrix interactions than synthetic ones. This may be due to the fact that these acellular matrices, being the ECM materials, contain a complex mixture of structural and functional proteins, glycoproteins, and proteoglycans arranged in a unique, tissue-specific three-dimensional ultrastructure [Badylak, 2002]. However, cell adhesion to degradable synthetic polymers can be improved by modifying the surfaces with RGD peptide for cell surface adhesion receptors [Glass et al., 1994; Cook et al., 1997; Schmedlen et al., 2002].

Our research group is currently evaluating the interactions between the esophageal epithelial and smooth muscle cells on various materials including chitosan, various blends of biodegradable polymers with chitosan, collagens; and decellularized porcine matrices such as urinary bladder matrix, small intestinal submucosa, and esophagus.

25.4.3 Scaffold Design

With the exception of the acellular matrices, all other scaffolds using synthetic polymers or pure collagen must be fabricated. As such, important structural features of the scaffold design must be considered. The ideal scaffold should direct the biological process of tissue formation and regeneration. One of the principal objectives in tissue engineering is to mimic the ECM in terms of their surface chemistry, mechanical properties and structure. In addition to the choice of biomaterials, to provide suitable surfaces for cell attachment and recognition, the physical structure of the scaffold plays an equally important role. The effects of pore size, morphology, microgeometry, and scaffold thickness are known to influence cellular adhesion, tissue organization, angiogenesis, and matrix deposition [Wake et al., 1994; Brauker et al., 1995; Zeltinger et al., 2001; Ward et al., 2002; Rosengren and Bjursten, 2003].

Pore size and total porosity, for example, are also known to influence fibrovascular tissue invasion and extent of fibrosis [Mikos et al., 1993]. In the case of the esophagus, fibrosis reaction must be minimized in order to maintain its mechanical performance. Conceptually, the scaffold for esophageal tissue should have a range of pore sizes. On the outer surface of the scaffold, the pore size should be large (ranging from 50 to 200 μm) to facilitate cell seeding, and transport of nutrients and waste. There should also be smaller pores (ranging from 35 to 70 μm) necessary to promote angiogenesis [Marshall et al., 2004]. The luminal surface, in order to mimic the basal membrane in the esophagus, should be dense (in the range of several microns in size). This barrier layer is to facilitate diffusion of signaling molecules and nutrients but prevents cell migration across the surface. An example of such a scaffold structure with varying pore sizes is shown in Figure 25.3a,b [Chian, 2003].

Another important aspect of the esophageal scaffold is the need for pores with specific orientation. The muscularis mucosa of the esophagus consists of a single layer of longitudinally oriented smooth muscle fibers, whereas the muscularis externa has an inner circular and outer longitudinal muscle layers. It is therefore advantageous to have channels in the scaffold that can provide directional guidance for these muscular tissues. Examples of such scaffolds with porous channels are shown in Figures 25.4a,b [Chian, 2003]. Our research effort is currently underway to study if these channels in the scaffold are effective in guiding these smooth muscle cells in culture.

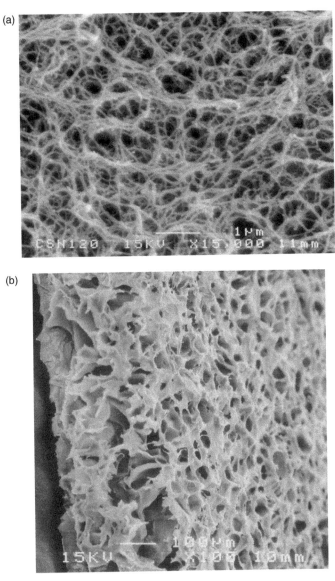

FIGURE 25.3 Chitosan scaffold showing a graded pore structure that may be suitable for replicating features found in the natural esophagus (a) Scaffold showing a dense basement membrane-like surface. (b) Scaffold showing highly interconnecting porous structure.

However, it must be noted that the features designed into the scaffold are only important in the initial stages of cell attachment and proliferation. As these scaffolds are biodegradable, the porous features and mechanical strengths are only transient. In an ideal situation, it is hoped that as the implanted cells interact suitably with the scaffold material, ECM produced by the cells will be laid down to replace the scaffold materials as they degrade. Therefore it is very important to select a suitable biomaterial as scaffold material that has a degradation timescale similar to the tissue forming process, which ranges from seconds to weeks.

25.5 Fabrication Processes

The methods of producing porous scaffolds for tissue engineering are well reviewed elsewhere [Thomson et al., 2000; Atala and Lanza, 2002]. Many of these processes used in scaffold fabrication are adapted from

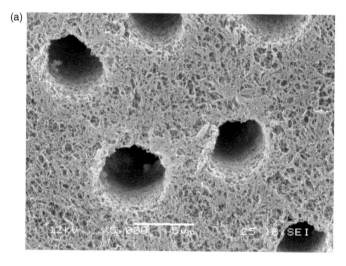

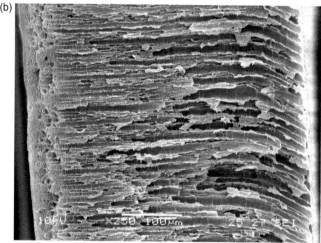

FIGURE 25.4 Examples of chitosan scaffolds with channel structures that may be useful for aligning muscle cells. (a) Porous surface of scaffold. (b) Cross section of a chitosan scaffold.

textile [Summanasinghe and King, 2003] and membrane technologies. Some of the common scaffold fabrication process that have been widely evaluated and reviewed [Sachlos and Czernuszka, 2003] include the following: Fiber bonding [Mooney et al., 1996], solvent casting and particulate leaching [Mikos et al., 1994; Wake et al., 1996], membrane lamination [Mikos et al., 1993], melt molding [Thomson et al., 1995], extrusion [Widmer et al., 1998], solid free-form methods [Giordano et al., 1996; Park et al., 1998], gas forming [Mooney et al., 1996], freeze drying [Whang et al., 1995], and phase inversion [Lo et al., 1995].

In this chapter, we will highlight some of the newer methods that are currently being explored for fabricating tissue engineering of tubular scaffolds. These include (i) electrostatic spinning, (ii) cryogenic molding, and (iii) rapid freeze prototyping.

25.5.1 Electrostatic Spinning

Electrostatic spinning is a well-established method for producing porous materials [Formhals, 1934; Amato, 1972; Bornat, 1982]. More recently, this technique has been adapted for producing biodegradable scaffolds from a range of polymers and collagens [Stitzel et al. 2000; Bowland et al., 2001; Matthews et al., 2002; Li et al., 2002; Wnek et al., 2003; Chu et al., 2004]. In an electrostatic spinning process, a high-voltage

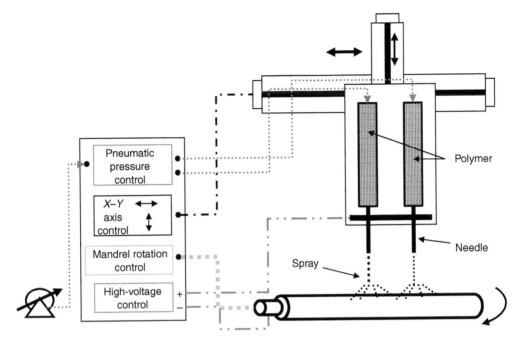

FIGURE 25.5 An electrostatic spinning system (a) apparatus for electrostatic spinning. (b) Porous nanofibers of PLA.

field is created between the polymer solution/melt and a collector. The polymer solution/melt is usually contained in a syringe and the needle is connected to an electrode. The oppositely charged electrode is connected to a collector, which can be either a stationary plate or a rotating mandrel. Typically, a high-voltage source of up to 30 kV is required for this process. Figure 25.5a shows a schematic diagram of a typical electrostatic spinning system. In order to form the electrostatic spray, the electric field between the end of the needle and the collector must increase until the mutual charge repulsion overcomes the surface tension of the polymer solution [Doshi and Reneker, 1995]. Increasing the electric field results in a charged stream of polymer fluid ejecting from the tip of the Taylor cone [Yarin et al., 2001]. The ejecting solution undergoes a whipping process [Shin et al., 2001], wherein the solvent evaporates leaving a charged polymer fiber randomly laid onto the grounded collector. The electrostatic spinning system offers many advantages over conventional methods of scaffold manufacture, and these include (i) the ability to produce varying fiber size, from nanometer to micron size, (ii) fabricating composite scaffolds, (iii) good porosity control, and (iv) the process is amenable to a wide range of synthetic and biological polymers. Figure 25.5b shows this use of this technique to produce nanofibers of polylactic acid with porous surfaces [Leong et al., 2004]. We are exploring further the potential of this method to fabricate scaffolds with various biodegradable polymers.

25.5.2 Cryogenic Molding

Another method for forming the esophageal scaffold that we are currently evaluating is the cryogenic molding process. In this process, the polymer solution is injected into a metal mould and allowed to freeze completely. The mould is then opened and the frozen polymer removed for freeze-drying or coagulated immediately. We have used this method successfully to produce a tubular scaffold made from chitosan solution. Figure 25.6 shows esophageal scaffolds that were made using this cryogenic molding process. This process offers the advantages of (i) reproducible scaffolds, (ii) it is amenable to a wide range of polymers, (iii) low cost, and (iv) it provides good porosity control, comparable to other phase separation methods commonly used in forming tissue engineering scaffolds.

FIGURE 25.6   A cryogenically molded tubular chitosan scaffold.

25.5.3 Rapid Freeze Prototyping

The various forms of rapid prototyping techniques have been successfully used in producing three-dimensional scaffolds for hard tissue implants [Giordano et al., 1996; Levy et al., 1997; Matsuda and Mizutani, 2002]. Attempts to produce scaffolds using the rapid prototyping technique for soft tissue engineering applications from agar hydrogels [Landers et al., 2002], fibrin hydrogels [Landers et al., 2002], chitosan, or chitosan-hydroxyapatite [Ang et al., 2002] have also been reported.

A new method was recently developed by our research group for producing scaffolds for soft tissue engineering application that is suitable for a wide range of polymers and biological materials. The method is adapted from the rapid freeze prototyping process that uses water to build ice prototypes. This process is capable of and has been successfully used to generate three-dimensional ice objects by depositing and rapidly freezing water layer by layer [Zhang et al., 2001; Chao et al., 2002]. However, in our adapted system, we used a robotic dispensing system to dispense chitosan solution onto a cold stage where it is allowed to freeze. The layers are built by repeatedly dispensing chitosan solution onto the previously frozen structure. When the required frozen structure is formed, it can be either freeze-dried or coagulated in alkaline solution to form the porous scaffold. Figure 25.7 shows samples of chitosan scaffolds fabricated using the adapted rapid freeze prototyping process.

The challenges in scaffold technology are many. The combination of selecting or developing a suitable material and utilizing a suitable fabrication method is often difficult. As cells have specific interactions with a substrate, a synthetic scaffold may eventually need to be a structure made from different materials, and with different pore size and surface chemistry. As we learn more about cell–material interactions, the closer we get to understanding, and enhancing our ability to mimic, the complex scaffold structure that nature can provide so readily. More research needs to be done in understanding the biological processes involved in tissue regeneration, and to develop novel and ingenious methods for fabricating scaffolds that replicate nature's ECM structures.

25.6   Cell Possibilities

25.6.1   Epithelial Characteristics

The epithelial lining of the esophagus is composed of stratified, squamous epithelial cells. In the human esophagus, these cells are nonkeratinizing, whereas in the rat they form a stratum corneum (see Figure 25.8) [Leeson and Leeson, 1981]. This epithelium is organized into distinct cellular layers, distinguished by appearance and protein expression. As the epithelial cells advance from the basal layer to the

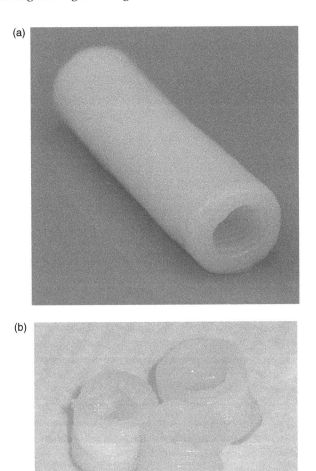

FIGURE 25.7 Samples of chitosan scaffolds fabricated using the rapid freeze prototyping process.

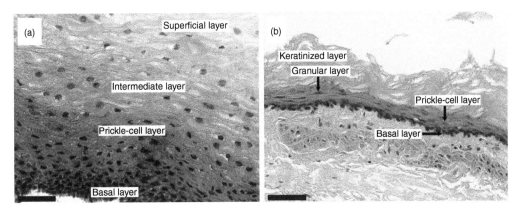

FIGURE 25.8 Esophageal epithelial morphology. (a) Nonkeratinizing human esophageal epithelium. (b) Keratinizing rat esophageal epithelium (H&E, bar = 50 μm).

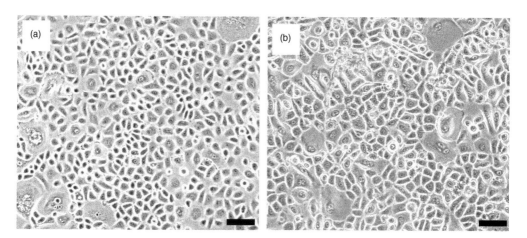

FIGURE 25.9 Esophageal epithelial cells in serum-free culture. (a) Human esophageal epithelial cells. (b) Rat esophageal epithelial cells (bar = 100 μm).

lumen, their appearance becomes flatter and more elongated. In nonkeratinizing epithelium, the various layers consist of the basal layer, prickle-cell layer, intermediate layer, and superficial layer (see Figure 25.8a) [Squier et al., 1976]. A keratinized esophagus shows a slightly different morphology and the epithelial layers are termed basal, prickle-cell, granular, and keratinized (see Figure 25.8b) [Squier et al., 1976]. As the cells progress toward the lumen, they become more differentiated and lose proliferative potential. In this manner, there is constant turnover and shedding of the epithelial cells. The turnover time for the epithelial lining is relatively short: 21 days in humans [Squier and Kremer, 2001] and 7 days in mice [Eastwood, 1977].

Epithelial stem cells reside in the basal layer of the esophagus and serve to replenish the lining [Seery and Watt, 2000]. Upon division of the stem cell, the daughter cells can either remain in their primitive state or enter the differentiation pathway [Seery and Watt, 2000]. This entrance into differentiation results in the development of a so-called transit-amplifying cell that is capable of further division, but quickly loses proliferative potential and becomes terminally differentiated [Seery and Watt, 2000]. The option between differentiation and proliferation has shown to be influenced by various factors, including calcium, phorbol esters, retinoic acid, vitamin A, air–liquid interface, and vitamin D3 [Fuchs and Green, 1981; Watt, 1984; Asselineau et al., 1985, 1989; Dotto, 1999]. The state of differentiation is spatially defined within the esophagus: cells residing in the basal layer are in a more primitive state, whereas differentiation progresses as the cell moves toward the lumen [Squier and Kremer, 2001]. Cytokeratin (intermediate cytoskeletal filaments) expression also varies spatially and provides a means to monitor differentiation of the epithelial cells. For example, within the human esophageal epithelial lining, cytokeratin 14 is expressed exclusively in the basal cells, whereas cytokeratin 13 is expressed in suprabasal cells [Takahasi et al., 1995]. Cytokeratin expression is closely tied to function and abnormal cytokeratin expression in other tissues has been associated with carcinomas and genetic diseases, such as epidermolysis bullosa [Takahasi et al. 1995; Fuchs, 1996].

The *in vitro* culture of esophageal epithelial cells has been well established [Compton et al., 1998; Oda et al., 1998; Okumura et al., 2003]. The majority of isolations involve an enzymatic digestion to allow for easy separation of the epithelium and underlying lamina propria. The epithelium is then trypsin-treated to obtain a single cell suspension. These cells can either be grown using a 3T3 feeder layer [Rheinwald and Green, 1975; Compton et al., 1998] or in a serum-free keratinocyte growth medium [Oda et al., 1998; Miki et al., 1999; Okumura et al., 2003], and propagated in culture for multiple passages (see Figure 25.9).

25.6.2 Epithelial Cell Source for Tissue Engineering

Attempts at engineering a replacement esophagus have exclusively employed the use of esophageal epithelial cells, rather than epithelial cells isolated from other tissues. Most likely, this stems from ease of isolation and physiological relevance. The most extensive research has been done by Kitajima and associates [Sato et al., 1993, 1994, 1997; Miki et al., 1999]. Originally, their work focused on seeding human esophageal epithelial cells on a collagen gel or poly(glycolic) acid (PGA) mesh embedded within a collagen gel. Using these scaffolds, they were able to obtain two to five layers of epithelial cells prior to implantation in the latissimus dorsi of a rat or mouse [Sato et al., 1993, 1994]. *In vivo*, the number of cell layers did increase and after 2 weeks, the epithelium was comparable to normal esophagus [Sato et al., 1993, 1994]. More recent studies have investigated the effect on epithelial cells of fibroblasts embedded in the collagen gels [Miki et al., 1999]. These fibroblasts serve a similar function as a 3T3 feeder layer used in monolayer culture. They reported a positive correlation between the number of dermal fibroblasts embedded in the collagen and the number of epithelial cell layers. At 8×10^5 fibroblasts/ml, the authors were able to obtain greater than 18 epithelial cell layers after 21 days of *in vitro* culture [Miki et al., 1999]. For *in vivo* studies, they seeded human esophageal epithelial cells with human esophageal fibroblasts in a collagen gel embedded within a PGA matrix. After implantation in the latissimus dorsi of a rat for 2 weeks, they observed approximately 20 layers of epithelial cells, with morphology similar to native esophagus [Miki et al., 1999]. However, no immunocytochemical staining was done to evaluate differentiation or cytokeratin expression of the epithelial lining and this construct has yet to be implanted in either a partial or a full circumferential esophageal defect.

The Vacanti research group has also reported creation of a tissue-engineered esophagus [Grikscheit et al., 2003]. For their epithelial cell source, they created "organoid units" (OUs), which are mesenchymal cells surrounded by epithelium. To create the OUs, a rat esophagus was harvested and digested using a dispase/collagenase type I solution. The suspension was washed, resuspended in culture media, and seeded immediately onto a PGA mesh for implantation. For *in vivo* implantation, the rat's omentum was wrapped around the construct before securing it in the peritoneum. After 4 weeks, the engineered construct was removed from the peritoneum and either used for histological analysis, implantation as an esophageal patch, or as an interposition graft. The epithelial lining of these grafts was similar to native rat esophagus with a stratified squamous keratinizing epithelium. The authors reported that the esophageal patch showed good integration and no stenosis, unlike the interposition graft, which showed stenosis at the upper anastomosis and dilation at the lower anastomosis [Grikscheit et al., 2003].

Our research group has also employed esophageal epithelial cells for use in esophageal tissue engineering. Both human and rat epithelial cells have been successfully isolated and cultured using previous published techniques [Oda et al., 1998]. We are currently investigating epithelial interactions with various synthetic and natural matrices.

25.6.3 Muscle Component of the Esophagus

In addition to the epithelial cells, another critical cellular component of a tissue-engineered esophagus is the muscle layers. The esophagus has two muscle layers: the muscularis mucosa and the muscularis externa [Leeson and Leeson, 1981]. The muscularis mucosa lies between the lamina propria and the submucosa. Its main function appears to be a support for the luminal lining during contraction of the muscularis externa [Squier and Kremer, 2001]. The muscularis externa is the outermost layer of the esophagus and serves to push the food down the esophagus through peristalsis [Leeson and Leeson, 1981]. In humans, the composition of the muscularis externa has traditionally been reported as striated muscle in the upper third of the esophagus, smooth muscle in the lower third, and mixed in the middle third [Leeson and Leeson, 1981]. However, this reported composition has been modified by other investigators. Meyer and Castell report that the inner circular layer of the muscularis externa is approximately 4% striated muscle, 35% mixed, and 62% smooth muscle [Meyer and Castell, 1983]. The outer longitudinal layer of the muscularis externa, follows similar trends with approximately 6% striated, 41% mixed, and 54% smooth muscle [Meyer and Castell, 1983]. In the rat esophagus, the compositions of the muscularis mucosa and externa are more defined. The muscularis mucosa is strictly smooth muscle, whereas the muscularis externa is striated muscle [Linnes, 2004].

25.6.4 Engineering the Muscularis Mucosa and Externa

The muscle layer of the esophagus has surprisingly received little attention in attempts to engineer a replacement esophagus. As stated above, Miki et al. [1999] attempted to improve epithelial differentiation through a coculture with fibroblasts. For *in vivo* studies, human esophageal fibroblasts were isolated from subepithelial tissues by mincing and enzymatic digestion [Miki et al., 1999]. Besides staining with an antihuman fibroblast antibody after implant harvest, no characterization of the cells was reported. It is unknown whether these cells can serve as an appropriate source for the muscularis mucosa or if they will populate the lamina propria or submucosa. For *in vivo* studies, the engineered tubes were implanted in the muscle flaps of the latissimus dorsi, which could theoretically serve to generate a muscularis externa [Miki et al., 1999]. However, no in-depth histological analysis of the muscularis mucosa or muscularis externa of these constructs has been reported.

Within the esophageal organoid units created by the Vacanti group are mesenchymal cells, which could serve as a smooth muscle cell source [Grikscheit et al., 2003]. As stated above, these organoid units were seeded onto a PGA mesh and wrapped with the omentum before suturing into the peritoneum. After 4 weeks, the construct was harvested and analyzed for alpha-smooth muscle actin expression [Grikscheit et al., 2003]. The staining did appear in the expected site of the muscularis mucosa, but was sparse and discontinuous. No mention of a muscularis externa was made. Thus, it is unclear whether the OU approach can lead to a morphological and functional equivalent to either muscle layer.

Clearly, engineering of the esophageal muscle layers is an area that requires increased attention. Since the muscle layer's composition varies from species to species, animal models will need to anticipate this variation. Studies investigating potential cell sources, matrix production, mechanical strength, and cellular alignment are needed to thoughtfully engineer the esophagus.

25.6.5 Esophageal Regeneration

As an alternative to the *in vitro* construction of a tissue-engineered esophagus, much research has been aimed at developing methods to stimulate *in vivo* regeneration. One widely employed approach has been to use a collagen-coated silicone tube to promote regeneration [Ike et al., 1989; Natsume et al., 1990, 1993; Takimoto et al., 1993, 1998; Yamamoto et al., 1999a, b, 2000; Hori et al., 2003]. In this approach, a silicone stent is coated with collagen types I and III, and the collagen is freeze-dried and lightly cross-linked around the stent. Both layers serve distinct purposes in the regeneration. The collagen provides a matrix for cell infiltration and tissue regeneration; the silicone stent provides mechanical integrity and protection from displacement, leakage, and infection [Natsume et al., 1990]. At the desired time after implantation, the silicone stent is dislodged endoscopically, leaving behind the neo-esophageal tissue. In one study, a 5-cm long prosthesis was implanted in the cervical esophagus of a canine animal model [Takimoto et al., 1998]. If the silicone stent was removed at 2 or 3 weeks, the neo-esophagus constricted rendering the dogs unable to swallow. However, if the stent was removed at 4 weeks, the regenerated tissue showed remarkable similarity to native esophagus. The regenerated tissue showed a stratified epithelium (8 to 10 layers) and both inner circular and outer longitudinal muscle layers [Takimoto et al., 1998]. The esophagus remained patent even after 12 months. Variations on this stent design have included preseeding with oral mucosal cells [Natsume et al., 1990], omental-pedicle wrapping [Yamamoto et al., 2000], and delivery of basic fibroblast growth factor from the collagen [Hori et al., 2003].

Extracellular matrix materials have also been studied as scaffolds for esophageal regeneration. Acellular human skin (AlloDerm®, LifeCell, Branchburg, NJ) was used as an esophageal patch by Isch and associates [Isch et al., 2001]. In this study, a section of canine esophagus was removed (2 × 1 cm) and replaced with AlloDerm®. While epithelial coverage was incomplete at 1 month, by 2 months, coverage was complete and vascularization was evident. However, elastin staining indicated that the AlloDerm® was still present in the wound site at 3 months. Also, there was no indication of smooth muscle cell repopulation [Isch et al., 2001].

Small intestinal submucosa (SIS) has also found use in esophageal repair. Badylak et al. reported using SIS for both partial (3 × 5 cm) and complete (5-cm length) circumferential defect repair in canines [Badylak et al., 2000]. By 35 days, epithelialization was complete and the repair site showed indications of striated muscle infiltration from adjacent esophagus. While remnants of the SIS could be seen at 35 days, by 50 days, the SIS appears to have been completely degraded. Stenosis was seen in the complete circumferential defect repair site, with an approximate 50% decrease in circumference. The authors hypothesize that this narrowing is caused by the lack of intraluminal pressure [Badylak et al., 2000].

Finally, Kajitani et al. [2001] employed acellular porcine aorta as an esophageal patch. A small half-circle defect (2-cm diameter) was made in the distal esophagus of a pig and the acellular aorta was used to repair the site. Complete epithelial coverage and evidence of muscle regeneration was seen in the defect site by 7 weeks. However, residual elastin fibers indicated incomplete degradation of the acellular patch [Kajitani et al., 2001].

Thus, the initial results for esophageal regeneration are promising. To repair circumferential defects, the collagen–silicone stent has shown the ability to form a neo-esophagus with morphological similarities to native esophagus [Takimoto et al., 1998]. For patch defects, use of ECM materials has resulted in esophageal regeneration, but these materials have yet to be successful as full circumferential replacements [Badylak, et al., 2000; Isch et al., 2001; Kajitani et al., 2001]. In addition, further studies must be performed to evaluate the effect of the residual ECM material that remains in the wound site after regeneration appears to be complete.

25.7 Bioreactors for Esophageal Tissue Engineering

25.7.1 Mechanical Conditioning of Smooth Muscle Tissue Constructs

The outcome of culture of cells on a scaffold has shown to be influenced by the underlying substrate in addition to the biochemical and biomechanical environment [Kanda et al., 1992; Kim and Mooney, 2000]. The effect of mechanical stimulation on vascular smooth muscle cells has been much studied as part of the ongoing research focus to develop bioartificial vascular prostheses [Nerem and Seliktar, 2001]. There is less data available in the literature on mechanical stimulation of visceral smooth muscle but the general principles and effects of mechanical stimulation are similar for the two myocyte types [Karim et al., 1992; Gooch and Tennant, 1997]. Mechanical stimulation of smooth muscle cells has been shown to have the following effects:

Cells align perpendicular to the stress direction in one-dimensional stress and exhibit morphological changes [Kanda et al., 1992; Gooch and Tennant, 1997]. Cyclic stretching significantly increases elastin production and promotes expression of contractile phenotype. Cells subjected to cyclic strain display prominent bundles of myofilaments while control cells (no strain) exhibited no such bundles, but conversely displayed significant amounts of rough endoplasmic reticulum (synthetic phenotype) [Kim and Mooney, 2000]. Mechanical stress causes elevated protein synthesis and gene expression in cells [Nerem and Seliktar, 2001], and has a favorable effect on cell proliferation [Stegemann and Nerem, 2001] and DNA synthesis [Gooch and Tennant, 1997; Karim et al., 1992].

In developing the bioartificial esophagus, the relationship between magnitude and frequency of stimulation and the optimal function of the smooth muscle cells has been examined *in vitro* using the apparatus shown schematically in Figure 25.10a. This apparatus is based on one-dimensional mechanical stimulators reported in the literature [Kanda et al., 1992; Kim and Mooney, 2000]. Apparatus for the conditioning of tissue constructs consists of a reservoir of culture medium, and a means for applying the mechanical stimulus, as shown schematically in Figure 25.10a,b. Simulation may be by means of a slider-crank mechanism [Kim and Mooney, 2000], lead screw [Kanda et al., 1992], or linear motor, as in the bioreactor in Figure 25.10b. One-dimensional stimulation as shown in Figure 25.10a,b has shown to influence the alignment of smooth muscle cells, so that they align perpendicular to the direction of principal strain [Kanda et al., 1992]. If this effect is not desired, a bioreactor as shown in Figure 25.11a may be used to subject the construct to a two-dimensional stress state [Gooch and Tennant, 1997]. The pressure difference

(a)

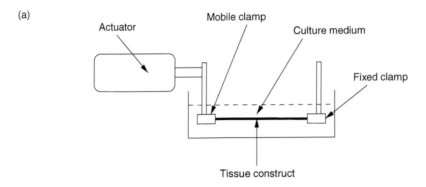

(b)

FIGURE 25.10   (a) Schematic of apparatus for tissue engineering. (b) Bioreactor developed at Nanyang Technological University for one-dimensional stimulation of smooth muscle and endothelial cells.

between the two chambers causes the scaffold to stretch, resulting in the bi-axial stress state as shown in Figure 25.11b. The culture chamber is maintained at atmospheric pressure (P1), while a pressure controller varies the pressure in the lower chamber [Gooch and Tennant, 1997]. Bioreactors based on this principle are available commercially, an example being the Flexcell Stage Flexer (Flexcell International Corp., Hillsborough, NC, USA).

Thicker tissue constructs have been fabricated from sheets incorporating a single layer of cells by rolling a matured, conditioned flat construct into a tubular structure [Nerem and Seliktar, 2001]. Alternatively, the tubular construct may also be mechanically conditioned, as shown in Figure 25.12. The construct is mounted between two tubular clamps and pressurized intermittently to provide radial and circumferential stress. This approach finds particular application in the tissue engineering of blood vessels although it also has application in the development of the bioartificial esophagus [Niklason et al., 1999; Nerem and Seliktar, 2001]. Mechanical conditioning has been found to significantly increase the burst strength of tissue-engineered blood vessels and may be expected to have a similar effect on the strength of an esophageal construct.

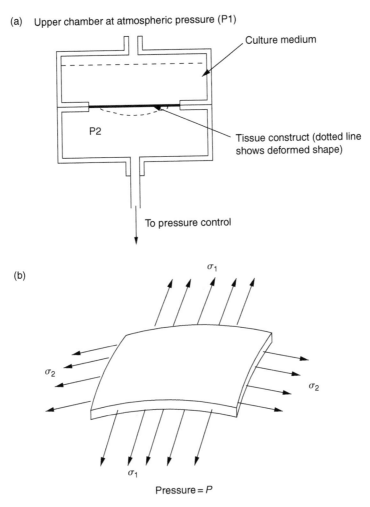

FIGURE 25.11 (a) Bioreactor for two-dimensional stress stimulation of cells. (b) Two-dimensional stress state in tissue construct.

As the esophagus is a thick-walled structure, necrosis of cells due to insufficient nutrition becomes a major issue. Myocytes are highly metabolically active and do not tolerate hypoxia well [Carrier et al., 1999; Radisic et al., 2003], and insufficient nutrition may result in necrosis of cells, particularly toward the center of the construct [Bethiaume and Yarmush, 2000]. A novel approach to this problem has been proposed [Kofidis et al., 2003] in which a section of native artery is used to provide a central vessel in a thick construct. The authors report viable cells in vascular constructs up to 8 mm in thickness, with a central vessel of mean diameter 2 mm.

The esophagus presents novel bioreactor challenges, and while much of the science generated in the development of bioartificial vascular constructs will influence application in other tissue types, a bioreactor must be developed to provide an analog of developmental conditions in the esophagus. A baby is able to swallow at birth: the neonatal esophagus is fully developed. We must therefore provide a means to replicate neonatal conditions *in vitro*. The final phase of development will be to build the bioreactor shown in Figure 25.13. It incorporates a tubular cell/scaffold construct with culture medium supplied to the outer jacket. The central canal of the esophagus can be drained from the bottom to provide the air/endothelium interface found in the esophagus. The central canal also incorporates a stimulator, consisting of a series of expandable chambers, which may be inflated in sequence by compressed air to simulate the passage of a bolus of masticated food through the esophagus.

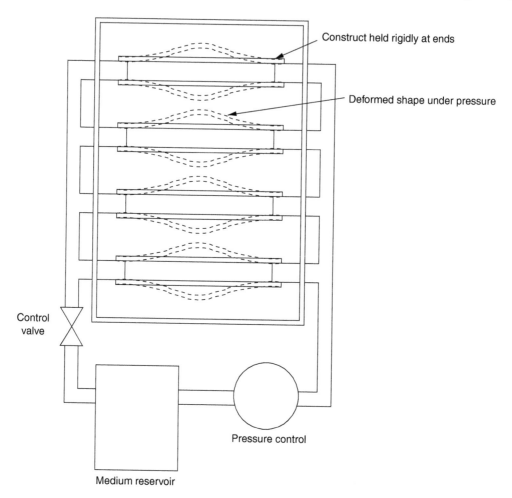

FIGURE 25.12 Bioreactor for mechanical stimulation of tubular constructs. Stress is induced by pressure difference between the inside of the construct and surroundings.

25.8 Conclusions and Prognostications

The potential to have significant impact on medicine and to make a positive contribution to the quality of life for esophagus surgery patients exists in the tissue engineering of esophagus. When this technology/biology triumph is realized, tissue engineering of other epithelial tissues such as stomach and colon will quickly follow.

The innovations that will make esophageal tissue engineering a routine procedure will come from interdisciplinary team efforts and an engineering systems approach. There are still many issues with regard to science and technology that must be resolved to develop the system for esophageal replacement. It is more than cell growth. It is more than surgery. It is more than chemistry. No one discipline can marshal all of the needed skills to make this happen. The engineered systems concept charts a path from science discoveries to products and generates a roadmap with needed team players, economic issues, milestones, and alternate strategies.

There are still significant technical challenges, challenges without clear solutions at this time. How we will go from an *in vitro* seeded cell construct to a vascularized, integrated tissue before hypoxic cell death occurs is not clear. How we will ultimately use allogeneic cells allowing an "off-the-shelf" surgical replacement challenges our understanding of cell-induced immune response. Matching (or exceeding) the mechanical properties of the natural tissue remains challenging. Many other sizable challenges remain.

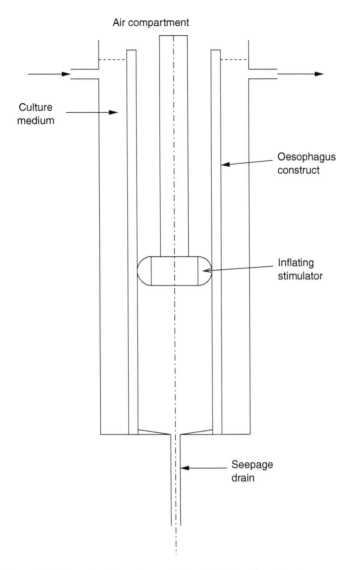

FIGURE 25.13 Schematic of bioreactor for environmental conditioning of constructs.

Acknowledgments

The authors acknowledge the generous support of the Singapore Agency for Science, Technology and Research (A*Star) to the Singapore-University of Washington Alliance (SUWA), and NSF support to the University of Washington Engineered Biomaterials (UWEB) program. SUWA is a close collaboration between UWEB and the Nanyang Technological University Biomedical Engineering Research Centre (BMERC).

References

Amato R. 1972. Textile machine. US Patent no. 3665695.
Ang T., Sultana F., Hutmacher D., Wong Y., Fuh J., Mo X., Loh H., Burdet E., and Teoh S. 2002. Fabrication of 3-d chitosan-hydroxyapatite scaffolds using a robotic dispensing system. *Mater. Sci. Eng.*, C20: 35.
Asselineau D., Bernard B., Bailly C., and Darmon M. 1989. Retinoic acid improves epidermal morphogenesis. *Dev. Biol.*, 133: 322–335.

Asselineau D., Bernhard B., Bailly C., and Darmon M. 1985. Epidermal morphogenesis and induction of the 67 kd keratin polypeptide by culture of human keratinocytes at the liquid–air interface. *Exp. Cell Res.*, 159: 536.

Atala A. and Lanza R. 2002. *Methods of Tissue Engineering.*, Academic Press, USA.

Badylak S. 2002. The extracellular matrix as a scaffold for tissue reconstruction. *Cell Dev. Biol.*, 13: 377.

Badylak S., Meurling S., Chen M., Spievack A., and Simmons-Byrd A. 2000. Resorbable bioscaffold for esophageal repair in a dog model. *J. Pediatr. Surg.*, 35: 1097.

Badylak S., Tullius R., Kokini K., Shelbourne K., Klootwyck T., Voytik S., Kraine M., and Simmons C. 1995. The use of xenographic small intestinal submucosa as a biomaterial for achilles tendon repair in a dog model. *J. Biomed. Mater. Res.*, 29: 977–985.

Bethiaume F. and Yarmush M. 2000. Tissue engineering. In: *The Biomedical Engineering Handbook*, Bronzino J., Ed., 2nd ed. CRC Press, Boca Raton, FL.

Bornat A. 1982. Electrostatic spinning of tubular products. US Patent no. 4323525.

Bottaro D. and Heidaran M. 2001. Engineered extracellular matrices: a biological solution for tissue repair, regeneration, and replacement. *e-Biomed*, 2: 9–12.

Bowland E., Wnek G., Simpson D., Pawlowski K., and Bowlin G. 2001. Tailoring tissue engineering scaffolds by employing electrostatic processing techniques: A study of poly(glycolic acid). *J. Macromol. Sci.*, 31: 1231–1243.

Brauker J., Carr-Brendel V., Martinson L., Crudele J., Johnston W., and Johnson R. 1995. Neovascularisation of synthetic membranes directed by membrane microarchitecture. *J. Biomed. Mater. Res.*, 29: 1517.

Burkitt H., Young B., and Heath J. 1993. *Wheater's Functional Histology.* 3rd ed. Longman, London.

Carrier R., Papadaki M., Rupnick M., Schoen F., Bursac N., Langer R., Freed L., and Vunjak-Novakovic G. 1999. Cardiac tissue engineering: cell seeding, cultivation parameters, and tissue construct characterization. *Biotechnol. Bioeng.*, 64: 580.

Chao F., Yan Y., and Zhang R. 2002. Comparison and analysis of continuously jetting and discretely jetting method in rapid ice prototype forming. *Mater. Des.*, 23: 77.

Chian K.S. 2003. Unpublished data. Nanyang Technological University.

Chu B., Hsiao B., Fang D., and Braithwaite C. 2004. Biodegradable and/or bioabsorbable fibrous articles and methods for using the article for medical applications. US Patent no. 6685956.

Compton C., Warland G., Nakagawa H., Opitz O., and Rustgi A. 1998. Cellular characterization and successful transfection of serially subcultured normal human esophageal keratinocytes. *J. Cell Physiol.*, 177: 274.

Cook A., Hrkach J., Gao N., Johnson I., Pajvani U., Cannizzaro S., and Langer R. 1997. Characterisation and development of rgd-peptide modified poly(lactic acid-co-lysine) as an interactive, resorbable biomaterial. *J. Biomed. Mater. Res.*, 35: 513.

Doshi J. and Reneker D. 1995. Electrospinning process and applications of electrospun fibers. *J. Electrostat.*, 35: 151.

Dotto G. 1999. Signal transduction pathways controlling the switch between keratinocyte growth and differentiation. *Crit. Rev. Oral Biol. Med.*, 10: 442.

Eastwood G. 1977. Gastrointestinal epithelial renewal. *Gastroenterology*, 72: 962.

Ergun G, and Kahrilas P. 1997. Esophageal anatomy and physiology. In: *Gastroenterology and Hepatology*, Orlando R, Ed., 3rd ed. Churchill Livingstone, Philadelphia.

Formhals A. 1934. Process and apparatus for preparing artificial threads. US Patent no. 1975504.

Fuchs E. 1996. The cytoskeleton and disease: genetic disorders of intermediate filaments. *Annu. Rev. Genet.*, 30: 197.

Fuchs E. and Green H. 1981. Regulation of terminal differentiation of cultured human keratinocytes by vitamin A. *Cell*, 25: 617.

Fuchs J., Nasseri B., and Vacanti J. 2001. Tissue engineering: a 21st century solution to surgical reconstruction. *Ann. Thorac. Surg.*, 72: 577.

Gershon M., Kirchgessner A., and Wade P. 1994. Functional anatomy of enteric nervous system. In: *Physiology of the Gastrointestinal Tract*, Johnson L., Ed., 3rd ed. Raven Press, New York.

Giordano R., Wu B., Borland S., Cima L., Sachs E., and Cima M. 1996. Mechanical properties of dense polylactic acid structures fabricated by three dimensional printing. *J. Biomat. Sci. Polym. Ed.*, 8: 63.

Glass J., Blevitt J., Dickerson K., Pierschbacher M., and Craig W. 1994. Cell attachment and motility on materials modified by surface-active rgd-containing peptides. *Ann. NY Acad. Sci.*, 745: 177.

Gooch K. and Tennant C. 1997. *Mechanical Forces: Their Effects on Cells and Tissues.* Spinger-Verlag and Landes Bioscience, Berlin.

Goyal R. and Sivarao D. 1999. Functional anatomy and physiology of swallowing and esophageal motility. In: *The Esophagus*, Castell D. and Richter J. Eds., 3rd ed. Lippincott Williams and Wilkins, Philadelphia.

Gray H. 1995. *Gray's Anatomy: The Anatomical Basis of Medicine and Surgery.* Churchill Livingstone, New York.

Grikscheit T., Ochoa E., Srinivasan A., Gaissert H., and Vacanti J. 2003. Tissue-engineered esophagus: experimental substitution by onlay patch or interposition. *J. Thorac. Cardiovasc. Surg.*, 126: 537.

Hartgerink J., Beniash E., and Stupp S. 2002. Peptide-amphiphile nanofibers: A versatile scaffold for the preparation of self-assembling materials. *Proc. Natl Acad. Sci. USA*, 99: 5133.

Hori Y., Nakamura T., Kimura D., Kaino K., Kurokawa Y., Satomi S., and Shimizu Y. 2003. Effect of basic fibroblast growth factor on vascularization in esophagus tissue engineering. *Int. J. Artif. Organs*, 26: 241.

Huang L., McMillan R., Apkarian R., Pourdeyhimi B., Conticello V., and Chaikof E. 2000. Generation of synthetic elastin-mimetic small diameter fibers and fiber networks. *Macromolecules*, 33: 2989.

Ike O., Shimizu Y., Okada T., Natsume T., Watanabe S., Ikada Y., and Hitomi S. 1989. Experimental studies on an artificial esophagus for the purpose of neoesophageal epithelization using a collagen-coated silicone tube. *ASAIO Trans.*, 35: 226.

Isch J., Engum S., Ruble C., Davis M., and Grosfeld J. 2001. Patch esophagoplasty using alloderm as a tissue scaffold. *J. Pediatr. Surg.*, 36: 266.

Kajitani M., Wadia Y., Hinds M., Teach J., Swartz K., and Gregory K. 2001. Successful repair of esophageal injury using an elastin based biomaterial patch. *J. Am. Soc. Artific. Internal Organs*, 47: 342.

Kanda K., Matsuda T., and Oka T. 1992. Two dimensional orientational response of smooth muscle cells to cyclic stretching. *J. Am. Soc. Artific. Internal Organs*, 38: M382.

Karim O., Pienta K., Seki N., and Mostwin J. 1992. Stretch-mediated visceral smooth muscle growth *in vitro. Am. J. Physiol.*, 262: R895–R992.

Khan S., Trento A., DeRoberts M., Kass R., Sandhu M., Czer L., Blanche C., Raissi S., Fontana G., Cheng W., Chaux A., and Matloff J. 2001. Twenty-year comparison of tissue and mechanical valve replacement. *J. Thorac. Cardiovasc. Surg.*, 122: 257–269.

Kim B., and Mooney D. 2000. Scaffolds for engineering smooth muscle under cyclic mechanical strain conditions. *Transac. ASME J. Biomech. Eng.*, 122: 210.

Kofidis T., Lenz A., Boublik J., Akhyari P., Wachsmann B., Stahl K., Haverich A., and Leyh R. 2003. Bioartificial grafts for transmural myocardial restoration: a new cardiovascular tissue culture concept. *Eur. J. Cardio-Thorac. Surg.*, 24: 906.

Kumar D. 1993. Morphology of the gastrointestinal motor system - gross morphology. In: *Gastrointestinal Motility*, Kumar D. and Wingate D., Eds., 2nd ed. Churchill Livingstone, Edinburgh.

Landers R., Pfister A., Hubner U., John H., Schmelzeisen R., and Mulhaupt R. 2002. Fabrication of soft tissue engineering scaffolds by means of rapid prototyping techniques. *J. Mater. Sci.*, 37: 3107.

Langer R. and Tirrell D. 2004. Designing materials for biology and medicine. *Nature*, 428: 487.

Leeson T. and Leeson C. 1981. *Histology.* 4th ed. W.B. Sauders, Philadelphia.

Leininger B., Peacock H., and Neville W. 1970. Esophageal mucosal regeneration following experimental prosthetic replacement of the esophagus. *Surgery*, 67: 468.

Leong M.F., Chian K.S., and Ritchie A.C. 2004. Unpublished data. Nanyang Technological University.

Levy R., Chu T., Halloran J., Feinberg S., and Hollister S. 1997. CT generated porous hydroxyapatite orbital floor prosthesis on a prototype bioimplant. *Am. J. Neuroradiol.*, 18: 1522.

Li W., Laurencin C., Caterson E., Tuan R., and Ko F. 2002. Electrospun nanofibrous structure: a novel scaffold for tissue engineering. *J. Biomed. Mater. Res.*, 60: 613.

Linnes M. 2004. Unpublished data. University of Washington.

Lo H., Ponticiello M., and Leong K. 1995. Fabrication of controlled release biodegradable foams by phase separation. *Tissue Eng.*, 1: 15.

Marshall A., Barker T., Sage E., Hauch K., and Ratner B. 2004. Pore size controls angiogenesis in subcutaneously implanted porous matrices. *Proceedings of the 7th World Biomaterials Congress in Sydney*, Australia.

Matsuda T., and Mizutani M. 2002. Liquid acrylate-endcapped biodegradable poly(e-caprolactone-co-trimethylene carbonate). ii. computer aided stereolithographic microarchitectural surface photoconstructs. *J. Biomed. Mater. Res.*, 62: 395.

Matthews J., Wnek G., Thomson D., and Bowlin G. 2002. Electrospinning of collagen nanofibers. *Biomolecules*, 3: 232.

McMillan R. and Conticello V. 2000. Synthesis and characterization of elastin-mimetic protein gels derived from a well-defined polypeptide precursor. *Macromolecules*, 33: 4809.

Meyer G. and Castell D. 1983. Anatomy and physiology of the esophageal body. In: *Esophageal Function in Health and Disease*, Castell D. and Johnson L., Eds., Elsevier Biomedical, New York.

Miki H., Ando N., Ozawa S., Sato M., Hayashi K., and Kitajima M. 1999. An artificial esophagus constructed of cultured human esophageal epithelial cells, fibroblasts, polyglycolic acid mesh and collagen. *J. Am. Soc. Artific. Internal Organs*, 45: 502.

Mikos A., Bao Y., Cima L., Ingber D., Vacanti J., and Langer R. 1993. Preparation of poly(glycolic acid) bonded fiber structures for cell attachment and transplantation. *J. Biomed. Mater. Res.*, 27: 183.

Mikos A., Lyman M., Freed L., and Langer R. 1994. Wetting of poly(l-lactic acid) and poly(dl-lactic-co-glycolic) foams for tissue culture. *Biomaterials*, 15: 55.

Mooney D., Baldwin D., Suh N., Vacanti J., and Langer R. 1996. Novel approach to fabricate porous sponges of poly(d,l-lactic-co-glycolic) without the use of organic solvents. *Biomaterials*, 17: 1417.

Natsume T., Ike O., Okada T., Shimizu Y., Ikada Y., and Tamura K. 1990. Experimental studies of a hybrid artificial esophagus combined with autologous mucosal cells. *ASAIO Trans.*, 36: M435.

Natsume T., Ike O., Okada T., Takimoto N., Shimizu Y., and Ikada Y. 1993. Porous collagen sponge for esophageal replacement. *J. Biomed. Mater. Res.*, 27: 867.

Nerem R. and Seliktar D. 2001. Vascular tissue engineering. *Annu. Rev. Biomed. Eng.*, 3: 225.

Niklason L., Gao J., Abbott W., Hirschi K., Houser S., Marini R., and Langer R. 1999. Functional arteries grown *in vitro*. *Science*, 284: 489.

Oda D., Savard C., Eng L., Sekijima J., Haigh G., and Lee S. 1998. Reconstituted human oral and esophageal mucosa in culture. *In vitro Cell Dev. Biol. Anim.*, 34: 46.

Okumura T., Shimada Y., Imamura M., and Yasumoto S. 2003. Neurotrophin receptor p75(ntr) characterizes human esophageal keratinocyte stem cells *in vitro*. *Oncogene*, 22: 4017.

Pachence J. and Kohn J. 2000. Biodegradable polymers. In: *Principles of Tissue Engineering*, Lanza R., Langer R., and Vacanti J., Eds., 2nd ed. Academic Press, New York.

Park A., Wu B., and Griffith L. 1998. Integration of surface modification and 3-d fabrication techniques to prepare patterned poly(l-lactide) substrates allowing regionally selective cell adhesion. *J. Biomater. Sci., Polym. Ed.*, 9: 89.

Peters M. and Mooney D. 1997. Synthetic extracellular matrices for cell transplantation. In: *Porous Materials for Tissue Engineering.*, Liu D. and Dixit V., Eds., volume 250 of *Materials Science Forum*. Trans Tech Publications, Enfield, pp. 43–52.

Radisic M., Euloth M., Yang L., Langer R., Freed L., and Vunjak-Novakovic G. 2003. High density seeding of myocyte cells for cardiac tissue engineering. *Biotechnol. Bioeng.*, 82: 403.

Rheinwald J. and Green H. 1975. Serial cultivation of strains of human epidermal keratinocytes: the formation of keratinizing colonies from single cells. *Cell*, 6: 331.

Rosengren A. and Bjursten L. 2003. Pore size in implanted polypropylene filters is critical for tissue organization. *J. Biomed. Mater. Res.*, 67A: 918.

Sachlos E. and Czernuszka J. 2003. Making tissue engineering scaffolds work. review on the application of solid freeform fabrication technology to the production of tissue engineering scaffolds. *Europ. Cells Mater.*, 5: 29.

Sato M., Ando N., Ozawa S., Miki H., Hayashi K., and Kitajima M. 1997. Artificial esophagus. In: *Porous Materials for Tissue Engineering.*, Liu D. and Dixit V., Eds., volume 250 of *Materials Science Forum*. Trans Tech Publications, Enfield.

Sato M., Ando N., Ozawa S., Miki H., and Kitajima M. 1994. An artificial esophagus consisting of cultured human esophageal epithelial cells, polyglycolic acid mesh, and collagen. *J. Am. Soc. Artific. Internal Organs*, 40: M389.

Sato M., Ando N., Ozawa S., Nagashima A., and Kitajima M. 1993. A hybrid artificial esophagus using cultured human esophageal epithelial cells. *J. Am. Soc. Artific. Internal Organs*, 39: M554.

Schmedlen R., Masters K., and West J. 2002. Photocrosslinkable polyvinyl alcohol hydrogels that can be modified with cell adhesion peptides for use in tissue engineering. *Biomaterials*, 23: 4325.

Seery J. and Watt F. 2000. Asymmetric stem-cell divisions define the architecture of human oesophageal epithelium. *Curr. Biol.*, 10: 1447.

Shin M., Hohman M., and Brenner M. 2001. Experimental characterization of electrospinning: the electrically forced jet and instabilities. *Polymer*, 42.

Shinhar D., Finaly R., Niska A., and Mares A. 1998. The use of collagen-coated vicryl mesh for reconstruction of the canine cervical esophagus. *Pediatr. Surg. Int.*, 13: 84.

Squier C., Johnson N., and Hopps R. 1976. *Human Oral Mucosa: Development, Structure, and Function.* Blackwell Scientific Publications, Oxford.

Squier C. and Kremer M. 2001. Biology of oral mucosa and esophagus. *J. Natl Cancer Inst. Monogr.*, 29: 7.

Stegemann J. and Nerem R. 2001. Effect of mechanical stimulation on smooth muscle cell proliferation and phenotype. *ASME Bioeng. Conf.*, BED 50: 609.

Stitzel J., Bowlin G., Mansfield K., Wnek G., and Simpson D. 2000. Electrospraying and electrospinning of polymers for biomedical applications: poly(lactic-co-glycolic acid) and poly(ethylene-co-vinyl acetate. In: *Proceedings of the 32nd Annual SAMPE Meeting.*

Summanasinghe R. and King M. 2003. New trends in biotextile — the challenges of tissue engineering. *J. Textile and Apparel, Technol. and Manag.*, 3: 1.

Takahashi H., Shikata N., Senzaki H., Shintaku M., and Tsubura A. 1995. Immunohistochemical staining patterns of keratins in normal oesophageal epithelium and carcinoma of the oesophagus. *Histopathology*, 26: 45.

Takimoto Y., Nakamura T., Yamamoto Y., Kiyotani T., Teramachi M., and Shimizu Y. 1998. The experimental replacement of a cervical esophageal segment with an artificial prosthesis with the use of collagen matrix and a silicone stent. *J. Thorac. Cardiovasc. Surg.*, 116: 98.

Takimoto Y., Okumura N., Nakamura T., Natsume T., and Shimizu Y. 1993. Long-term follow-up of the experimental replacement of the esophagus with a collagen–silicone composite tube. *J. Am. Soc. Artific. Internal Organs*, 39: M736.

Thomson R., Shung A., Yaszemski M., and Mikos A. 2000. Polymer scaffold processing. In: *Principles of Tissue Engineering*, Lanza R., Langer R., and Vacanti J., Eds., 2nd ed. Academic Press, New York.

Thomson R., Yaszemski M., Powers J., and Mikos A. 1995. Fabrication of biodegradable polymer scaffolds to engineer trabecular bone. *J. Biomater. Sci. Polymer Ed.*, 7: 23.

Ure B., Slany E., Eypasch E., Weiler K., Troidl H., and Holschneider A. 1998. Quality of life more than 20 years after repair of esophageal atresia. *J. Paediatr. Surg.*, 33: 511.

Urry D., Parker T., Reid M., and Gowda D. 1991. Biocompatibility of the bioelastic material poly(gvgvp) and its γ-irradiation crosslinked matrix. *J. Bioact. Compat. Polym.*, 3: 263.

Wake M., Gupta P., and Mikos A. 1996. Fabrication of pliable biodegradable polymer foams to engineer soft tissues. *Cell Transplant.*, 5: 465.

Wake M., Patrick C., and Mikos A. 1994. Pore morphology effects on the fribovascular tissue growth in porous polymer substrates. *Cell Transplant.*, 3: 339.

Ward W., Slobodzian E., Tiekotter K., and Wood M. 2002. The effect of microgeometry, implant thickness and polyurethane chemistry on the foreign body response to subcutaneous implants. *Biomaterials,* 23: 4185.

Watanabe K. and Mark J. 1971. Segmental replacement of the thoracic esophagus with silastic prosthesis. *Am. J. Surg.*, 121: 238.

Watt F. 1984. Selective migration of terminally differentiating cells from the basal layer of cultured human epidermis. *J. Cell Biol.*, 98: 16.

Whang K., Thomas H., and Healy K. 1995. A novel method to fabricate biodegradable scaffolds. *Polymer,* 36: 837.

Widmer M., Gupta P., Lu L., Meszlenyi R., Evans G., Brandt K., Savel T., Gurlek A., Patrick C., and Mikos A. 1998. Manufacture of porous biodegradable polymer conduits by an extrusion process for guided tissue regeneration. *Biomaterials*, 19: 1945.

Wingate D. 1993. Regulation of gastrointestinal motility — intrinsic and extrinsic neural control. In: *Gastrointestinal Motility*, Kumar D. and Wingate D., Eds., 2nd ed. Churchill Livingstone, Edinburgh.

Wnek G., Carr M., Simpson D., and Bowlin G. 2003. Electrospinning of nanofibrous fibrinogen structures. *Nano Lett.*, 3: 213.

Wood J. 1994. Physiology of enteric nervous system. In: *Physiology of the Gastrointestinal Tract*, Johnson L., Ed., 3rd ed. Raven Press, New York.

Yamamoto Y., Nakamura T., Shimizu Y., Matsumoto K., Takimoto Y., Kiyotani T., Sekine T., Ueda H., Liu Y., and Tamura N. 1999a. Intrathoracic esophageal replacement in the dog with the use of an artificial esophagus composed of a collagen sponge with a double-layered silicone tube. *J. Thorac. Cardiovasc. Surg.*, 118: 276.

Yamamoto Y., Nakamura T., Shimizu Y., Takimoto Y., Matsumoto K., Kiyotani T., Yu L., Ueda H., Sekine T., and Tamura N. 1999b. Experimental replacement of the thoracic esophagus with a bioabsorbable collagen sponge scaffold supported by a silicone stent in dogs. *J. Am. Soc. Artific. Internal Organs*, 45: 311.

Yamamoto Y., Nakamura T., Shimizu Y., Matsumoto K., Takimoto Y., Liu Y., Ueda H., Sekine T., and Tamura N. 2000. Intrathoracic esophageal replacement with a collagen sponge-silicone double layer tube: evaluation of omental-pedicle wrapping and prolonged placement of an inner stent. *J. Am. Soc. Artific. Internal Organs*, 46: 734.

Yarin A., Koombhongse S., and Reneker D. 2001. Taylor cone and jetting from liquid droplets in electrospinning of nanofibers. *J. Appl. Phys.*, 90: 4836–4846.

Zeltinger J., Sherwood J., Graham D., Meuller R., and Griffith L. 2001. Effect of pore size and void fraction on cellular adhesion, proliferation, and matrix deposition. *Tissue Eng.*, 7: 557.

Zhang W., Leu M., Ji Z., and Yan Y. 2001. Method and apparatus for rapid freeze prototyping. US Patent no. 6253116.

26

Tissue Engineered Vascular Grafts

Rachael H. Schmedlen
Wafa M. Elbjeirami
Andrea S. Gobin
Jennifer L. West
Rice University

26.1 Introduction

Cardiovascular disease remains the leading cause of death in the United States, claiming more lives each year than the next five leading causes of death combined (American Heart Association National Center for Health Statistics). Coronary heart disease caused more than 1 in every 5 American deaths in 2000 and required approximately 500,000 coronary artery bypass graft surgeries (CABGs) that year. Bypass grafting is also used in the treatment of aneurysmal disease or trauma. At present, surgeons use autologous tissue and synthetic biomaterials as vascular grafts. Transplantation of autologous tissue has the best outcome in small diameter applications such as CABG because synthetic grafts lack long-term patency for small diameter applications (<6 mm). However, autologous tissue is limited in supply. Recent advances in tissue engineering provide hope that new blood vessel substitutes may one day be fabricated for small diameter applications, such as CABG, where treatment options are often severely limited.

26.1.1 Current Vascular Grafts

Currently, occluded vessels with diameters <6 mm are bypassed with autologous native blood vessels such as the saphenous vein as a treatment. Autologous blood vessels were first employed in the beginning of the 20th century when Goyanes reported the use of a graft to replace an excised segment of artery by bridging the defect with the patient's own popliteal vein [Goyanes, 1906]. Since then, venous grafts have retained the best long-term patency rate, although limitations such as deterioration when exposed to increased flow and pressure still remain [Crawford et al., 1981; Szilagyi et al., 1973]. More recently, the internal mammary artery from the chest has shown to be a superior conduit to the saphenous vein with better long-term patency and has gained increasing use [Kock et al., 1996]. These native vessels generally display 50 to 70% graft patency over 10 years (VA study). Moreover, the superior patency of native vessels is attributed to their compliant natural tissue characteristics and to their antithrombogenic luminal endothelial lining [Richardson et al., 1980]. Unfortunately, ~60% of patients who require vascular bypass surgery do not have a suitable vessel for grafting [Moneta and Porter, 1995], either due to earlier procedures or advanced peripheral vascular disease. As a result, alternative conduits constructed with synthetic materials have been investigated.

The use of synthetic vascular prostheses for replacement of natural blood vessels has been explored since the 1950s [Vorhees et al., 1952]. The most widely employed materials include Dacron™ (polyethylene terepthalate; PET), Gore-Tex™ (expanded polytetrafluoroethylene; ePTFE), and compliant polyurethanes. A strong, yet flexible polymer, PET may be fabricated in woven, velour, or knitted fiber configurations [Coury et al., 1996]. ePTFE possesses a smooth surface and exhibits good durability and biocompatibility. Polyurethane has been increasingly studied because it has been developed to more closely match the compliance of native vessels than Gore-Tex™ or Dacron, which may reduce some clinical complications [Giudiceandrea et al., 1998]. These materials are readily available, relatively inexpensive, and have seen success clinically in applications with vessel diameters greater than 6 mm. However, synthetic grafts with a small diameter (<6 mm, e.g., coronary artery bypass and femoral-crural bypass) exhibit unsatisfactory long-term patency due to graft thrombogencity and neointima formation.

Platelet adhesion and aggregation on the graft surface often results in reocclusion of the vessel. This is caused by the adsorption of proteins that mediate platelet adhesion and coagulation processes on the relatively hydrophobic polymeric materials. The need for a nonthrombogenic surface has lead to the investigation of endothelial cell (EC) seeding on the lumen of the graft. Although EC seeding improves synthetic graft patency, EC retention under physiological shear conditions remains problematic [Thompson et al., 1994; Deutsch et al., 1999]. Furthermore, differences in elasticity between a synthetic material and the adjacent tissue create discrepancies in strain at the anastomoses, referred to as compliance mismatch, contributing to intimal hyperplasia [Geary et al., 1993; Greisler et al., 1993]. Moreover, synthetic grafts have been shown to carry an increased risk of infection [Clowes, 1993], further limiting their clinical efficacy.

26.1.2 Tissue Engineering

Most tissue engineering strategies attempt to create small caliber vascular grafts by closely mimicking the structure, function, and physiologic environment of native vessels. Normal arteries possess three distinct tissue layers (Figure 26.1): the intima, media, and adventitia. The intima consists of an EC

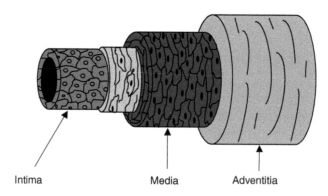

Intima   Media   Adventitia

FIGURE 26.1 The arterial wall is composed of three distinct layers: the intima, media, and adventitia. The intima, composed of endothelial cells, provides a nonthrombogenic surface. In the medial layer, smooth muscle cells and elastin fibers align circumferentially and provide mechanical integrity and contractility. The outer adventitia is a supportive connective tissue.

monolayer, which prevents platelet aggregation and regulates vessel permeability, vascular smooth muscle cell behavior, and homeostasis. Within the medial layer, smooth muscle cells (SMCs) and elastin fibers are aligned circumferentially, contributing the majority of the vessel's mechanical strength [Wight, 1996]. Dilation and constriction of the vessel are due to medial SMCs' response to external stimuli. Finally, the adventitial layer contains fibroblasts, connective tissue, the microvascular supply, and a neural network that regulates the vasotone of the blood vessel. The recreation of some or all of the vessel layers and their properties may result in the development of a patent, functional vascular graft. In all likelihood, an intima and media will be required to achieve any degree of success.

The general concept for creating a tissue-engineered vascular graft (TEVG) usually involves the harvest of desired cells, cell expansion in culture, cell seeding onto a scaffold, construct culture in an environment that induces tissue formation, and implantation of the construct back into the patient. Many options exist at each step of this process and each must be carefully considered. First, cell source and culture conditions must be determined. Due to concerns regarding immunogenicity, autologous cells are likely to be required, but these may be differentiated SMCs and ECs or a variety of types of vascular progenitors. Expansion *in vitro* is usually necessary, but *in vitro* culture time must also be minimized to avoid cell dedifferentiation. Genetic modification of cells may also be a means to improve TEVG performance. In addition to the cellular component, some type of material, or scaffolding, is generally required to provide mechanical support and integrity. Many varieties of scaffolds are used in tissue engineering, primarily composed of extracellular matrix (ECM) proteins and synthetic polymeric materials. Once cells are seeded onto scaffolds, another *in vitro* culture period is usually needed to allow for new tissue formation and development of appropriate mechanical and functional characteristics. During the culture period, the construct should receive appropriate chemical and mechanical signals for cells to synthesize proteins, remodel the tissue, and organize their environment such that the construct will develop into a functional graft with mechanical properties similar to native vessels. Therefore, a tissue-engineered construct may require several weeks of preparation before it can be implanted into the patient.

While efforts to create TEVGs remain in an early developmental stage, several problems potentially exist. Graft patency is still threatened by thrombosis, in all likelihood due to issues with retention of ECs after implantation or with alterations in EC function after culture *in vitro*. Also, the possibility of burst failure after implantation in the physiological flow environment raises concern since the consequences would be catastrophic. Mechanical properties of TEVGs are generally observed to be lower than those of native arteries, and thus a number of approaches have begun to address these issues. All of the strategies discussed above — cell source, genetic modification, scaffold materials, and culture conditions — will likely impact the fabrication of an optimal, clinically useful TEVG.

26.2 Cell Sources for Vascular Tissue Engineering

The development of a functional TEVG is likely to require the construction of an intima and media composed of ECs and SMCs. Limitations imposed by immunogenicity will probably require the use of autologous cells, so the majority of studies to date have utilized differentiated SMCs and ECs isolated from harvested blood vessels. Issues with donor site morbidity and the performance of these cells types in the engineered tissues have led to the consideration of alternative cell sources. Recent advances in stem cell biology offer hope for suitable progenitors that can be effectively differentiated into ECs and SMCs for use in vascular tissue engineering.

Circulating endothelial progenitor cells [Wijelath et al., 2004; Griese et al., 2003; He et al., 2003; Shirota et al., 2003] and smooth muscle progenitor cells [Simper et al., 2002] can be isolated from blood, offering a potential source of autologous cells for tissue engineering if they can be appropriately expanded in culture. Blood-derived endothelial progenitors have already been utilized as linings on synthetic vascular grafts in several studies [He et al., 2003; Shirota et al., 2003]. Grafts lined with these endothelial progenitors have been implanted in a canine carotid model, and after more than 30 days, 11 out of 12 grafts remained patent, and cells lining the surface appeared to be ECs [He et al., 2003]. Umbilical cord blood [Murga et al., 2004] and bone marrow [Hamilton et al., 2004] may be additional sources of autologous vascular progenitor cells.

26.3   Genetic Modification of Vascular Cells

The leading cause of vascular graft failure has been attributed to thrombosis. Genetic engineering of vascular cells *ex vivo* may provide an effective strategy to improve graft properties for tissue engineering applications. Investigators have reported transduction of vascular cells such as ECs [Wilson et al., 1989; Dunn et al., 1996], fibroblasts [Scharfmann et al., 1991], and SMCs [Lynch et al., 1992]. The seeding of small-diameter vascular graft constructs with either ECs or SMCs genetically engineered to secrete anti-thrombotic factors exists as a potential method of improving graft patency rates. As an example, baboon ECs have been genetically modified with a retroviral vector encoding antithrombotic factors, namely tissue plasminogen activator (tPA) and glycosylphosphatidylinositol-anchored urokinase-plasminogen activator (uPA) [Dichek et al., 1996]. The modified cells were seeded on the luminal surface of collagen-coated Dacron surfaces and introduced in arteriovenous shunts in baboons. A significant reduction in platelet and fibrin accumulation was observed in grafts containing modified ECs expressing either tPA or uPA. In another approach, the use of a platelet aggregation inhibitor, nitric oxide (NO), has shown to be promising. Using an *ex vivo* approach, bovine SMCs liposomally transfected with nitric oxide synthase III (NOS III) and GTP cyclohydrolase, which produces a cofactor essential for NOS activity, were grown as monolayers on plastic slides or biomaterials of interest and then placed in a parallel-plate flow chamber. Whole blood was introduced into the flow chamber to assess platelet adherence to the cell monolayers. The number of platelets that adhered to the NOS-transduced SMCs was significantly lower than those bound to mock-transduced SMCs and similar to the numbers adhered to cultured ECs. Moreover, the NOS-expressing SMCs exhibited decreased proliferative activity, which may reduce the incidence of intimal hyperplasia [Scott-Burden et al., 1996]. These studies demonstrated that an *ex vivo* genetic engineering approach may be useful for altering thrombogenicity of TEVG surfaces.

An alternative gene therapy approach to reducing the thrombogenicity of TEVG surfaces would utilize transfected VEGF. VEGF has shown to be mitogenic for ECs *in vitro* and stimulates angiogenesis *in vivo* [reviewed in Ahrendt et al., 1998]. More importantly, the mitogenic response associated with VEGF is restricted to ECs, which allows VEGF to be administered without concerns of SMC intimal hyperplasia. VEGF production in the TEVG may encourage proliferation of ECs seeded on the lumenal surface as well as stimulate endogenous host EC migration from the anastomoses. Transfecting SMCs with VEGF may allow for localized and prolonged treatment. Additionally, VEGF-producing SMCs have been found to promote EC proliferation and migration using *in vitro* models [Elbjeirami et al., 2004].

Improving the mechanical properties of TEVGs has also been a goal of numerous research efforts, and genetic modification of cells used to seed the TEVG may prove beneficial. The mechanical properties of a TEVG will be related in large part to the ECM that forms, both its composition and structure. ECM crosslinking can result from the enzymatic activity of lysyl oxidase (LO) or tissue transglutaminase [Aeschlimann et al., 1991] and may be a means to improve the mechanical properties of the TEVG. LO, a copper-dependent amine oxidase, forms lysine-derived crosslinks in connective tissue, particularly in collagen and elastin [Rucker et al., 1998]. Desmosine is produced in LO-mediated crosslinking of elastin and is commonly used as a biochemical marker of ECM crosslinking [Venturi et al., 1996]. The LO-catalyzed crosslinks are present in various connective tissues within the body including bone, cartilage, skin, and lung and are believed to be a major source of mechanical strength in tissues. Additionally, the LO-mediated enzymatic reaction renders crosslinked fibers less susceptible to proteolytic degradation [Vater et al., 1979]. A gene therapy strategy has demonstrated the enhancement of mechanical properties of tissue-engineered collagen constructs using vascular SMCs transfected with LO [Elbjeirami et al., 2003]. The elastic modulus and ultimate tensile strength of collagen gels seeded with LO-transfected SMCs nearly doubled as compared to gels seeded with mock-transfected SMCs. These enhanced mechanical properties resulted from increased ECM crosslinking rather than increased amounts of ECM, changes in the ECM composition, or increased cellularity. LO-mediated crosslinking, tissue transglutaminase, or glycation, may potentially be used in combination with mechanical conditioning, or biochemical factors, such as TGF-β, that increase the synthesis of ECM proteins to achieve synergistic effects. This strategy may ultimately enhance mechanical characteristics of TEVGs and minimize *in vitro* culture times prior to implantation.

Another effort in the use of genetic modification of vascular cells to improve TEVG performance has focused on regulation of SMC phenotype [Stegemann et al., 2004]. SMCs that were stably transfected with the gene for cyclic guanosine monophosphate-dependent protein kinase (PKG) were seeded into type I collagen constructs. PKG is an important regulator of SMC phenotype, and PKG-transfected cells showed substantially increased expression of smooth muscle α-actin, indicating a reduction in dedifferentiation during culture. Thus, TEVGs formed with such genetically modified cells may function more similarly to native tissues.

26.4   Scaffolds for Vascular Tissue Engineering

26.4.1   Natural Scaffold Materials

26.4.1.1   Collagen

One of the first TEVGs developed was based on a natural collagen type I scaffold supported by knitted Dacron [Weinberg and Bell, 1986]. This study provided early evidence of the feasibility of tissue engineering a blood vessel substitute by creating layers similar to those of a blood vessel using ECM components and vascular cells. Cultured bovine ECs, SMCs, and fibroblasts embedded within denatured bovine collagen were used to construct a multilayered vascular graft. Endothelium formation was confirmed by production of biosynthetic markers such as prostacylin and von Willebrand's factor. The burst strength of the engineered tissue was proportional to the collagen content and cell density. However, the resultant engineered graft did not achieve sufficient mechanical strength to withstand physiological conditions unless a supporting Dacron mesh sleeve was added. In fact, maximum burst strengths seen in this model ($\approx$325 mmHg) were significantly lower than that of native coronary artery ($\approx$ 5000 mmHg) or saphenous vein ($\approx$2000 mmHg).

Glycation, the nonenzymatic crosslinking of ECM proteins by reducing sugars, has been proposed as a strategy to enhance the stiffness and strength of collagen gel constructs seeded with SMCs [Girton et al., 1999, 2000]. These constructs were cultured for 10 weeks under high concentrations of glucose or ribose and then assessed for mechanical properties. The circumferential tensile stiffness of constructs incubated in medium containing 30 mM ribose showed a 16-fold increase over normally cultured constructs while tensile strength increased by 4-fold. Although the compliance of the construct compared favorably, the tensile strength and burst pressure still fell significantly below that of arteries.

Expanding upon work with collagen scaffolds, others have further explored the potential utility of collagen scaffolds by evaluating decellularized collagen matrices from porcine tissue [Hiles et al., 1993; Huynh et al., 1999]. Small intestine submucosa (SIS) has been isolated and chemically treated to remove the cellular component, leaving an intact matrix containing mainly collagen. The mechanical properties and remodeling ability of SIS have been previously investigated for numerous tissue-engineering applications [Sacks et al., 1999; Gloeckner et al., 2000]. For vascular grafts, SIS possesses acceptable mechanical properties and exhibits better compliance than currently employed vein grafts [Roeder et al., 1999]. In one effort, Huynh et al. prepared SIS tubes and minimally chemically crosslinked the collagen layers to provide mechanical strength while maintaining biocompatibility. The inner surface of the grafts were additionally treated with a heparin complex to inhibit thrombosis and then implanted into rabbits. For time periods up to 13 weeks, grafts remained patent, and SMCs and ECs were found to have migrated into the graft from the surrounding tissue.

26.4.1.2 Decellularized Vascular Matrix

Xenogenic acellular matrix conduits have also been investigated for the development of TEVG. In this approach, vessels are treated with reagents such as trypsin and EDTA [Bader et al., 2000] or sodium dodecyl sulfate [Schaner et al., 2004] to remove the cells, leaving an intact matrix that can be used as a scaffold. Decellularized porcine arteries have been seeded with human ECs and SMCs isolated from the saphenous vein [Teebken et al., 2000]. After exposure to pulsatile flow, the cell-seeded porcine matrix developed an endothelial monolayer. This study demonstrated the feasibility of generating vascular grafts *in vitro* from acellular, animal-derived, vascular matrices and human cells. Attempts to improve biocompatibility of xenograft matrices have included proteolysis of bovine or porcine carotid arteries [Moazami et al., 1998; Teebken et al., 2000]. This decellularized form provided an ECM with properly arranged collagen and elastin fibers that had minimal immune or inflammatory response [Teebken et al., 2000]. However, fibrosis eventually rendered these grafts unsuccessful.

26.4.1.3 Cell Sheets

L'Heureux and his collegues have applied a unique approach in the fabrication of a TEVG using biological materials, essentially using cells to generate the scaffold for the tissue-engineering construct [L'Heureux et al., 1998]. Umbilical vein SMCs and human skin fibroblasts were cultured for 5 weeks to form sheets containing superconfluent cells and cell-synthesized ECM. The cell sheets were sequentially rolled around a mandrel to form first the medial layer, then the adventitial layer. For 8 weeks, the construct was cultured in a bioreactor designed to provide perfusion of the culture medium and mechanical support. Following culture, the mandrel was removed, and the lumen was seeded with ECs, thus forming the three layers representative of a native artery. The burst strength of these grafts well exceeded that of native veins. Based on histological analysis, it appeared that SMCs were aligned circumferentially and produced a significant amount of new ECM. When implanted *in vivo*, these vascular substitutes had a 50% patency rate at one week after implantation. The construct consists entirely of human cells and human ECM proteins while exhibiting impressive mechanical strength, which makes this strategy attractive. However, much of the constructs' strength was attributed to the adventitial layer instead of the medial layer, as is normally observed in blood vessels. Some thrombus formation was also observed at the graft site following implantation. Furthermore, a 12-week time span was required to prepare a construct.

26.4.2 Synthetic Polymer Scaffolds

While the natural scaffold materials discussed above have achieved some success, concerns with disease transmission, difficulties with material processing, and often poor mechanical properties have led some groups to concentrate on the development of synthetic biomaterials as scaffolds. Bioabsorbable synthetic polymers may be designed to provide a transitional environment by providing a supporting structure to developing tissue. The degradation rate of these constructs may often be tailored to match the rate at which new tissue is formed, so that space is created for cell growth and matrix deposition. These materials serve as guides for tissue regeneration in three dimensions and offer the possibility to control structural variables and scaffold properties, such as the molecular structure, molecular weight, degradation properties, porosity, and mechanical properties [Ma et al., 1995].

26.4.2.1 Polyester Scaffolds

Polyglycolic acid (PGA), a biodegradable polyester, has demonstrated relatively good biocompatibility and has been extensively studied for numerous tissue-engineering applications [Freed et al., 1993; Kim et al., 1998; Mooney et al., 1996]. These well-characterized materials have been the first to illustrate the feasibility of polymeric scaffolds in vascular tissue engineering [Niklason et al., 1999]. TEVGs were fabricated using PGA mesh scaffolds, seeded with bovine aortic SMCs, then cultured in pulsatile flow bioreactors. After 8 weeks, these cultured vessels exhibited increased concentrations of collagen and enhanced mechanical properties. Burst pressure of these grafts compared well to that of typical vein grafts (human saphenous vein $= 1680 \pm 307$ mmHg vs. PGA graft $= 2150 \pm 700$ mmHg). When these grafts were seeded with autologous ECs on their luminal surface, continuously perfused for 3 days, and then implanted into pigs, they remained patent for up to 4 weeks.

26.4.2.2 Hydrogel Scaffolds

Derivatives of polyethylene glycol (PEG) currently are being studied as hydrogel scaffolds for blood vessel substitutes. These materials are hydrophilic, biocompatible, and intrinsically resistant to protein adsorption and cell adhesion [Merrill et al., 1983; Gombotz et al., 1991]. Thus, PEG essentially provides a "blank slate," devoid of biological interactions, upon which the desired biofunctionality can be built. Aqueous solutions of acrylated PEG can be rapidly photopolymerized in direct contact with cells and tissues [Sawhney et al., 1994; Hill-West et al., 1994]. Furthermore, PEG-based materials can be rendered bioactive by inclusion of proteolytically degradable peptides into the polymer backbone [West et al., 1999] and by grafting adhesion peptides [Hern et al., 1998] or growth factors [Mann et al., 2001a] into the hydrogel network during the photopolymerization process. Recently, PEG hydrogels that largely mimic the properties of collagen have been developed. PEG hydrogels grafted with a synthetic adhesive peptide RGDS and the collagenase-sensitive peptide sequence GGGLGPAGGK permitted cell migration [Gobin and West 2002; Mann et al., 2001b]. Following 7 days of incubation, approximately 70% as many cells migrated through collagenase-sensitive hydrogels as through collagen gels, with no statistically significant difference between the two groups. The elastin-derived peptide VAPG has shown to be specific for SMC adhesion, and PEG hydrogels modified with this adhesive peptide rather than RGDS supported adhesion and growth of vascular SMCs but not fibroblasts or platelets [Gobin et al., 2003]. Moreover, bioactive molecules like TGF-β may be covalently incorporated into scaffolds to induce protein synthesis by vascular SMCs. TGF-β has been reported to stimulate expression of several matrix components, including elastin, collagen, fibronectin, and proteoglycans [Amento et al., 1991; Lawrence et al., 1994; Tajima, 1996]. TGF-β covalently immobilized to PEG-based hydrogels significantly increased collagen production of vascular SMCs seeded within these scaffold materials [Mann et al., 2001]. Mechanical testing of these engineered tissues also determined that the elastic modulus was higher in TGF-β-tethered PEG scaffolds than PEG scaffolds without TGF-β, indicating that material properties for TEVGs may be improved using this technology. A cell-seeded graft formed from this biomimetic hydrogel scaffold is shown in Figure 26.2. These types of bioactive materials may allow one to capture the advantages of a natural scaffold, such as specific cell–material interactions and proteolytic remodeling in response to tissue formation, while also having the benefits of a synthetic material, namely the ease of processing and the ability to manipulate mechanical properties.

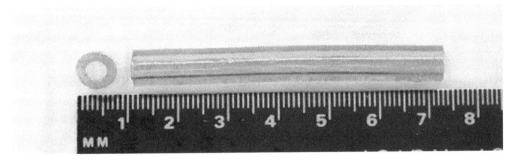

FIGURE 26.2 A PEG-based scaffold seeded with smooth muscle cells and endothelial cells ready for insertion into a bioreactor for *in vitro* culture of a TEVG (left, standing upright; right, laying on side). The cell-seeded scaffold is formed via photopolymerization, so the dimensions are easily tailored for a given application and cells are homogeneously seeded throughout the material. The scaffold is designed to degrade in response to cellular proteolytic activity during tissue formation.

26.5   Bioreactors for Mechanical Conditioning

In vivo, the pulsatile nature of blood flow imposes radial pressure upon the vessel wall, which subjects SMCs within the medial layer to cyclic strain. Thus, a great deal of research has examined SMC behavior in response to cyclic stretch and found such stimuli important in the fabrication of vascular tissue, particularly with respect to ECM synthesis and tissue organization. For example, SMCs seeded on purified elastin membranes and exposed to 2 days of cyclic stretching (10% beyond the resting length) have been shown to incorporate hydroxyproline into protein three to five times more rapidly than stationary controls, indicating increased collagen synthesis in response to strain [Leung et al., 1976]. Cyclic strain also increased the synthesis of collagen types I and III and chondroitin-6-sulfate without stimulating DNA synthesis. Another study also detected enhanced matrix production in collagen constructs seeded with rat aortic SMCs and subjected to 7% cyclic strain [Kim et al., 1999]. Over 20 weeks of culture under cyclic strain, SMCs upregulated expression of elastin and collagen type I. Elastin content from these SMCs increased 49% over unstretched controls. Furthermore, organization of the tissue was observed, as evidenced by perpendicular alignment of SMCs to the direction of the applied strain.

Similar results have been obtained in three-dimensional constructs. Tubular collagen constructs seeded with SMCs were cultured over thin-walled silicone sleeves and subsequently exposed to regulated intraluminal pressures to stretch the vessel in a repeatable fashion for up to 8 days. The 10% cyclic distension in diameter caused the scaffolds to contract, SMCs and bundles of collagen fibers to align circumferentially around the vessel, and improvement of the scaffold's mechanical properties [Seliktar et al., 2000]. Moreover, this model system was employed to investigate the remodeling capacity of these constructs via the activity of matrix metalloproteinases (MMPs) known to cleave solubilized type I collagen fragments [Seliktar et al., 2001]. Constructs mechanically conditioned for 4 days contained five times higher amounts of MMP-2 compared to static controls and increased MMP-2 activity. The increases in MMP-2 levels correlated favorably with improvements in mechanical strength and material modulus as a result of cyclic strain. When a nonspecific inhibitor of MMP-2 was added to the culture media, MMP-2 levels decreased and mechanical properties were reduced, negating the benefits of mechanical conditioning. These studies indicate that strain-mediated remodeling of collagen scaffolds is essential for improved construct of mechanical properties.

Because of the profound effects of cyclic strain on SMC orientation, ECM production, and tissue organization, preculture of vascular graft constructs in a pulsatile flow bioreactor system may help recreate the natural structure of native vessels and allow one to better achieve the mechanical properties required of the construct. A schematic of a typical pulsatile flow bioreactor system is shown in Figure 26.3. The mechanical stimuli from pulsatile flow could generate the cyclic strain necessary to alter ECM production, thereby creating a histologically organized, functional construct with satisfactory mechanical characteristics for implantation. To develop a blood vessel substitute, Niklason et al. [1999], cultured PGA constructs in a pulsatile blow bioreactor generating 165 beats per minute (bpm) and 5% radial strain. The pulse frequency of this system was chosen to mimic a fetal heart rate, believed to possibly provide optimal conditions for new tissue formation. However, most mechanical conditioning investigations mentioned above conducted strain studies at 60 bpm, more representative of an adult heart rate, with promising outcomes. Therefore, the optimal bioreactor culture conditions for the development of a TEVG remain to be elucidated. Nevertheless, such a system shows promise for the production of a blood vessel substitute with the necessary mechanical and biochemical components.

26.6   Conclusions

In the past couple of decades, a great deal of progress on TEVGs has been made. Still, many challenges remain and are currently being addressed, particularly with regard to the prevention of thrombosis and the improvement of graft mechanical properties. In order to develop a patent TEVG that grossly resembles native tissue, required culture times in most studies exceed 8 weeks. Even with further advances in

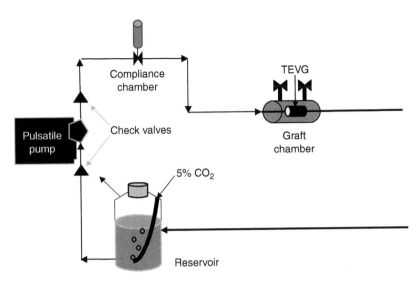

FIGURE 26.3 Diagram of a typical pulsatile flow bioreactor for culture of TEVGs.

the field, TEVGs will likely not be used in emergency situations because of the time necessary to allow for cell expansion, ECM production and organization, and attainment of desired mechanical strength. Furthermore, TEVGs will probably require the use of autologous tissue to prevent an immunogenic response, unless advances in immune acceptance render allogenic and xenogenic tissue use feasible. TEVGs have not yet been subjected to clinical trials, which will determine the efficacy of such grafts in the long term. Finally, off the shelf availability and cost will become the biggest hurdles in the development of a feasible TEVG product.

Although many obstacles still exist in the effort to develop a small-diameter TEVG, the potential benefits of such an achievement are exciting. In the near future, a nonthrombogenic TEVG with sufficient mechanical strength may be developed for clinical trials. Such a graft will have the minimum characteristics of biological tissue necessary to remain patent over a time period comparable to current vein graft therapies. As science and technology advance, TEVGs may evolve into complex blood vessel substitutes. TEVGs may become living grafts, capable of growing, remodeling, and responding to mechanical and biochemical stimuli in the surrounding environment. These blood vessel substitutes will closely resemble native vessels in almost every way, including structure, composition, mechanical properties, and function. They will possess vasoactive properties, able to dilate and constrict in response to stimuli. Close mimicry of native blood vessels may ultimately aid in the engineering of other tissues dependent upon vasculature to sustain function. With further understanding of the factors involved in cardiovascular development and function combined with the foundation of knowledge already in place, the development of TEVGs should one day lead to improved quality of life for those with vascular diseases and other life threatening conditions.

References

Aeschlimann, D. and Paulsson, M. Cross-linking of laminin-nidogen complexes by tissue transglutaminase. A novel mechanism for basement membrane stabilization. *J. Biol. Chem.* 1991; **266**: 15308–15317.

Ahrendt, G., Chickering, D.E., and Ranieri, J.P. Angiogenic growth factors: a review for tissue engineering. *Tissue Eng.* 1998; **4**: 117–130.

Amento, E.P., Ehsani, N., Palmer, H., and Libby, P. Cytokines and growth factors positively and negatively regulate interstitial collagen gene expression in human vascular smooth muscle cells. *Arterioscler. Thromb.* 1991; **11**: 1223–1230.

American Heart Association. Heart Disease and Stroke Statistics — 2003 Update. http://www.americanheart.org/presenter.jhtml?identifier=4439.

Anderson, J.M., Cima, L.G., Eskin, S.G., Graham, L.M., Greisler, H., Hubbell, J., Levy, R.J., Naughton, G., Northup, S.J., Ratner, B.D., Scott-Burden, T., Termin, P., and Didisheim, P. Tissue engineering in cardiovascular disease: a report. *J. Biomed. Mater. Res.* 1995; **29**:1473–1475.

Bader, A., Steinhoff, G., Strobl, K., Schilling, T., Brandes, G., Mertsching, H., Tsikas, D., Froelich, J., and Haverich, A. Engineering of human vascular aortic tissue based on a xenogenic starter matrix. *Transplantation* 2000; **70**: 7–14.

Clowes, A.W. Intimal hyperplasia and graft failure. *Cardiovasc. Pathol.* 1993; **2**: 179S–186S.

Coury, A.J., Levy, R.J., McMillin, C.R., Pathak, Y., Ratner, B.D., Schoen, F.J., Williams, D.F., and Williams, R. L. Degradation of materials in the biological environment. In *Biomaterials Science.* Ratner, B.D., Hoffman, A.S., Schoen, F.J., Lemons, J.E., Eds. Academic Press: San Diego, 1996.

Crawford, E.S., Bomberger, R.A., Glaeser, D.H., Saleh, S.A., and Russell, W.L. Aortoiliac occlusive disease: factors influencing survival and function following reconstructive operation over a twenty-five year period. *Surgery* 1981; **90**: 1055–1067.

Deutsch, M., Meinhart, J., Fischlein, T., Preiss, P., and Zilla, P. Clinical autologous *in vitro* endothelialization of infrainguinal ePTFE grafts in 100 patients: a 9-year experience. *Surgery* 1999; **126**: 847–855.

Dichek, D.A., Anderson, J., Kelly, A.B., Hanson, S.R., and Harker, L.A. Enhanced *in vivo* antithrombotic effects of endothelial cells expressing recombinant plasminogen activators transduced with retroviral vectors. *Circulation* 1996; **93**: 301–309.

Dunn, P.F., Newman, K.D., Jones, M., Yamada, I., Shayani, V., Virmani, R., and Dichek, D.A. Seeding of vascular grafts with genetically modified endothelial cells. *Circulation* 1996; **93**: 1439–1446.

Elbjeirami, W.M., Yonter, E.O., Starcher, B.C., and West, J.L. Enhancing mechanical properties of tissue engineered constructs via lysyl oxidase crosslinking activity. *J. Biomed. Mater. Res.* 2003; **66A**: 513–521.

Freed, L.E., Marquis, J.C., Nohia, A., Emmanual, J., Mikos, A.G., and Langer, R. Neocartilage formation *in vitro* and *in vivo* using cells cultured on synthetic biodegradable polymers. *J. Biomed. Mater. Res.* 1993; **27**: 11–23.

Geary, R.L, Kohler, T.R., Vergel, S., Kirkman, T.R., and Clowes, A.W. Time course of flow-induced smooth muscle cell proliferation and intimal thickening in endothelialized baboon vascular grafts. *Circulation Res.* 1993; **74**: 14–23.

Girton, T.S., Oegema, T.R., and Tranquillo, R.T. Exploiting glycation to stiffen and strengthen tissue equivalents for tissue engineering. *J. Biomed. Mater. Res.* 1999; **46**: 87–92.

Girton, T.S., Oegema, T.R., Grassel, E.D., Isenberg, B.C., and Tranquillo, R.T. Mechanisms of stiffening and strengthening in media-equivalents fabricated using glycation. *J. Biomech. Eng.* 2000; **122**: 216–223.

Giudiceandrea, A., Seifalian, A.M., Krijgsman, B., and Hamilton, G. Effect of prolonged pulsatile shear stress *in vitro* on endothelial cell seeded PTFE and compliant polyurethane vascular grafts. *Eur. J. Vasc. Endovasc. Surg.* 1998; **15**: 147–154.

Gloeckner, D.C., Sacks, M.S., Billiar, K.L., and Bachrach, N. Mechanical evaluation and design of a multilayered collagenous repair biomaterial. *J. Biomed. Mater. Res.* 2000; **52**: 365–373.

Gobin, A.S. and West, J.L. Cell migration through defined, synthetic extracellular matrix analogues. *FASEB J.* 2002.

Gobin, A.S. and West, J.L. Val-ala-pro-gly, an elastin-derived non-integrin ligand: smooth muscle cell adhesion and specificity, *J. Biomed. Mater. Res.* 2003; **67A**: 255–259.

Gombotz, W.R., Guanghui, W., Horbett, T.A., and Hoffman, A.S. Protein adsorption to poly(ethylene oxide) surfaces. *J. Biomed. Mater. Res.* 1991; **25**: 1547–1562.

Goyanes, J. Nuevos trabajos de cirugia vascular, substitucion plastica de les arterias por las venas or arterioplasia venosa, applicada, como nuevo metodo, al tratamiento de la aneurismas. *Siglo Med.* 1906; **53**: 546–549.

Griese, D.P., Ehsan, A., Melo, L.G., Kong, D, Zhang, L, Mann, M.J., Pratt, R.E., Mulligan, R.C., and Dzau, V.J. Isolation and transplantation of autologous circulating endothelial cell into denuded vessels and prosthetic grafts: implications for cell-based vascular therapy. *Circulation* 2003; **108**: 2710–2715.

Hamilton, D.W., Maul, T.M., and Vorp, D.A. Characterization of the response of bone marrow-derived progenitor cells to cyclic strain: implications for vascular tissue engineering applications. *Tissue Eng.* 2004; **10**: 361–369.

He, H., Shirota, T., and Matsuda, T. Canine endothelial progenitor cell-lined hybrid vascular graft with nonthrombogenic potential. *J. Thorac. Cardiovasc. Surg.* 2003; **126**: 455–464.

Hern, D.L. and Hubbell, J.A. Incorporation of adhesion peptides into nonadhesive hydrogels useful for tissue resurfacing. *J. Biomed. Mater. Res.* 1998; **39**: 266–276.

Hiles, M.C., Badylak, S.F., Geddes, L.A., Kokini, K., and Morff, R.J. Porosity of porcine small-intestinal submucosa for use as a vascular graft. *J. Biomed. Mater. Res.* 1993; **27**: 139–144.

Hill-West, J.L., Chowdhury, S.M., Sawhney, A.S., Pathak, C.P., Dunn, R.C., and Hubbell, J.A. Prevention of postoperative adhesions in the rat by *in situ* photopolymerization of bioresorbable hydrogel barriers. *Obstet. Gynecol.* 1994; **83**: 59–64.

Huynh, T., Abraham, G., Murray, J., Brockbank, K., Hagen, P.O., and Sullivan, S. Remodeling of an acellular collagen graft into a physiologically responsive neovessel. *Nat. Biotechnol.* 1999; **17**: 1083–1086.

Kim, B.S. and Mooney, D.J. Engineering smooth muscle tissue with a predefined structure. *J. Biomed. Mater. Res.* 1998; **41**: 322–332.

Kim, B.S., Nikolovski, J., Bonadio, J., and Mooney, D.J. Cyclic mechanical strain regulates the development of engineered smooth muscle tissue. *Nat. Biotech.* 1999; **17**: 979–983.

Kock, G., Gutschi, S., Pascher, O., Fruhwirth, J., and Hauser, H. Zur problematik des femoropoplitealen Gefassersatzes: vene ePTFE oder ovines Kollagen? *Zentralbl Chir* 1996; **121**: 761–767.

L'Heureux, N., Paquet, S., Labbe, R., Germain, L., and Auger, F.A. A completely biological tissue-engineered human blood vessel. *FASEB J.* 1998; **12**: 47–56.

Lawrence, R., Hartmann, D.J., and Sonenshein, G.E. Transforming growth factor β1 stimulates type V collagen expression in bovine vascular smooth muscle cells. *J. Biol. Chem.* 1994; **269**: 9603–9609.

Leung, D.Y.M., Glagov, S., and Mathews, M.B. Cyclic stretching stimulates synthesis of matrix components by arterial smooth muscle cells *in vitro*. *Science* 1976; **191**: 475–477.

Lynch, C.M., Clowes, M.M., Osborne, W.R.A., Clowes, A.W., and Miller, A.D. Long-term expression of human adenosine deaminase in vascular smooth muscle cells of rats: a model for gene therapy. *Proc. Natl Acad. Sci. USA* 1992; **89**: 1138–1142.

Ma, P.X. and Langer, R. Degradation, structure and properties of fibrous nonwoven poly(glycolic acid) scaffolds for tissue engineering. *Mat. Res. Soc. Symp. Proc.* 1995; **394**: 99–104.

Mann, B.K., Schmedlen, R.H., and West, J.L. Tethered-TGF-β increases extracellular matrix production of vascular smooth muscle cells. *Biomaterials* 2001; **22**: 439–444.

Mann, B.K., Gobin, A.S., Tsai, A.T., Schmedlen, R.H., and West, J.L. Smooth muscle cell growth in photopolymerized hydrogels with cell adhesive and proteolytically degradable domains: synthetic ECM analogs for tissue engineering. *Biomaterials* 2001; **22**: 3045–3051.

Merrill, E.A. and Salzman, E.W. Polyethylene oxide as a biomaterial. *ASAIO J.* 1983; **6**: 60–64.

Moazami, N., Argenziano, M., Williams, M., Cabreriza, S.B., Oz, M.C., and Nowygrod, R. Photo-oxidized bovine arterial grafts: short-term results. *ASAIO J.* 1998; **44**: 89–93.

Moneta, G.L. and Porter, J.M. Arterial substitutes in peripheral vascular surgery: a review. *J. Long-Term Effects Med. Implants* 1995; **5**: 47–67.

Mooney, D.J., Mazzoni, C.L., Breuer, C., McNamara, K., Hern, D., Vacanti, J.P., and Langer, R. Stabilized polyglycolic acid fibre-based tubes for tissue engineering. *Biomaterials* 1996; **17**: 115–124.

Murga, M., Yao, L., and Tosato, G. Derivation of endothelial cells from CD34-umbilical cord blood. *Stem Cells* 2004; **22**: 385–395.

Niklason, L.E., Gao, J., Abbott, W.M., Hirschi, K.K., Houser, S., Marini, R., and Langer, R. Functional arteries grown in *vitro*. *Science* 1999; **284**: 489–493.

Richardson, J.V., Wright, C.B., and Hiratzka, L.F. The role of endothelium in the patency of small venous substitutes. *J. Surg. Res.* 1980; **28**: 556–562.

Roeder, R., Wolfe, J., Lianakis, N., Hinson, T., Geddes, L.A., and Obermiller, J. Compliance, elasitic modulus, and burst pressure of small-intestine submucosa (SIS), small-diameter vascular grafts. *J. Biomed. Mater. Res.* 1999; **47**: 65–70.

Rucker, R.B., Kosonen, T., Clegg, M.S., Mitchell, A.E., Rucker, B.R., Uriu-Hare, J.Y., and Keen, C.L. Copper, lysyl oxidase, and extracellular matrix protein cross-linking. *Am. J. Clin. Nutri.* 1998; **67** Suppl: 996S–1002S.

Sacks, M.S. and Gloeckner, D.C. Quantification of the fiber architecture and biaxial mechanical behavior of porcine intestinal submucosa. *J. Biomed. Mater. Res.* 1999; **46**: 1–10.

Sawhney, A.S., Pathak, C.P., and Hubbell, J.A. Bioerodible hydrogels based on photopolymerized poly(ethylene glycol)-*co*-poly(α-hydroxy acid) diacrylate macromers. *Macromolecules* 1993; **26**: 581–587.

Schaner, P.J., Martin, N.D., Tulenko, T.N., Shapiro, I.M., Tarola, N.A., Leichter, R.F., Carabasi, R.A., and Dimuzio, P.J. Decellularized vein as a potential scaffold for vascular tissue engineering. *J. Vasc. Surg.* 2004; **40**: 146–153.

Scharfmann, R., Axelrod, J.H., and Verma, I.M. Long-term expression of retrovirus-mediated gene transfer in mouse fibroblast implants. *Proc. Natl Acad. Sci. USA* 1991; **88**: 4626–4630.

Scott-Burden, T., Tock, C.L., Schwarz, J.J., Casscells, S.W., and Engler, D.A. Genetically engineered smooth muscle cells as linings to improve the biocompatibility of cardiovascular prostheses. *Circulation* 1996; **94**[Suppl II]: II235–II238.

Seliktar, D., Black, R.A., Vito, R.P., and Nerem, R.M. Dynamic mechanical conditioning of collagen-gel blood vessel constructs induces remodeling *in vitro*. *Ann. Biomed. Eng.* 2000; **28**: 351–362.

Seliktar, D., Nerem, R.M., and Galis, Z.S. The role of matrix metalloproteinase-2 in the remodeling of cell-seeded vascular constructs subjected to cyclic strain. *Ann. Biomed. Eng.* 2001; **29**: 923–934.

Shirota, T., He, H., and Matsuda, T. Human endothelial progenitor cell-seeded hybrid graft: proliferative and antithrombogenic potentials *in vitro* and fabrication processing. *Tissue Eng.* 2003; **9**: 127–136.

Simper, D., Stalboerger, P.G., Panetta, C.J., Wang, S., and Caplice, N.M. Smooth muscle progenitor cells in human blood. *Circulation* 2002; **106**: 1199–1204.

Sinnaeve, P., Varenne, O., Collen, D., and Janssens, S. Gene therapy in the cardiovascular system: an update. *Cardiovasc. Res.* 1999; **44**: 498–506.

Stegemann, J.P., Dey, N.B., Lincoln, T.M., and Nerem, R.M. Genetic modification of smooth muscle cells to control phenotype and function in vascular tissue engineering. *Tissue Eng.* **10**: 189–199.

Szilagyi, D.E., Elliott, J.P., and Hageman, J.H. Biological fate of autologous implants as arterial substitutes: clinical angiographic and histopathological observations in femoropopliteal opertations for atherosclerosis. *Ann. Surg.* 1973; **178**: 232–246.

Tajima, S. Modulation of elastin expression and cell proliferation in vascular smooth muscle cells *in vitro*. *Keio J. Med.* 1996; **45**: 58–62.

Teebken, O.E., Bader, A., Steinhoff, G., and Haverich, A. Tissue engineering of vascular grafts: human cell seeding of decellularized porcine matrix. *Eur. J. Vasc. Endovasc. Surg.* 2000; **19**: 381–386.

Thompson, M.M., Budd, J.S., Eady, S.L., James, R.F.L., and Bell, P.R.F. Effect of pulsatile shear stress on endothelial attachment to native vascular surfaces. *Br. J. Surg.* 1994; **81**: 1121–1127.

VA Coronary Artery Bypass Surgery Cooperative Study Group. Eighteen-year follow-up in the veterans affairs cooperative study of coronary artery bypass surgery for stable angina. *Circulation* 1992; **86**: 121–130.

Vater, C.A., Harris, E.D., and Siegel, R.C. Native cross-links in collagen fibrils induce resistance to human synovial collagenase. *Biochem. J.* 1979; **181**: 639–645.

Venturi, M., Bonavina, L., Annoni, F., Colombo, L., Butera, C., Peracchia, A., and Mussini, E. Biochemical assay of collagen and elastin in the normal and varicose vein wall. *J. Surg. Res.* 1996; **60**: 245–248.

Vorhees, A.B., Jaretski, A., and Blakeore, A.H. Use of tubes constructed from vinyon "n" cloth in bridging arterial deficits. *Ann. Surg.* 1952; **135**: 332–338.

Weinberg, C.B. and Bell, E. A blood vessel model constructed from collagen and cultured vascular cells. *Science* 1986; **231**: 397–400.

West, J.L. and Hubbell, J.A. Polymeric biomaterials with degradation sites for proteases involved in cell migration. *Macromolecules* 1999; **32**: 241–244.

Wight, T.N. Arterial wall. In: *Extracellular Matrix*, Vol. 1. Howard Academic Publishers: Netherlands, 1996, pp. 175–202.

Wijelath, E.S., Rahman, S., Murray, J., Patel, Y., Savidge, G., and Sobel, M. Fibronectin promotes VEGF-induced CD34 cell differentiation into endothelial cells. *J. Vasc. Surg.* 2004; **39**: 655–660.

Wilson, J.M., Birinyi, L.K., Salomon, R.N., Libby, P., Callow, A.D., and Mulligan, R.C. Implantation of vascular grafts lined with genetically modified endothelial cells. *Science* 1989; **244**: 1344–1346.

27

Cardiac Tissue Engineering: Matching Native Architecture and Function to Develop Safe and Efficient Therapy

Nenad Bursac
Duke University

27.1 Introduction

After an acute myocardial infarction, lost cardiomyocytes are replaced by a noncontractile fibrous tissue. Although it is suggested that heart has a small regenerative potential via cell proliferation [1], or stem cell recruitment [2], the rate of renewal is insufficient to compensate for myocyte loss. As a result, altered workload of a surviving myocardium may ultimately lead to deterioration in contractile function and congestive heart failure (CHF). Besides traditional pharmacological therapies (diuretics, β-blockers, angiotensine, and aldosterone inhibitors) [3] or heart transplant [4], investigators are evaluating innovative approaches for treatment of CHF including mechanical assist devices [5], dynamic cardiomyoplasty [6], transmyocardial laser revascularization [7], and artificial heart [8]. Nevertheless, in end stage disease, heart transplant remains the only option with good long-term results [4]. However, inadequate availability of donor organs (~10% of current needs [9]) requires new strategies for treatment of increasing number of heart failure patients.

One promising approach is augmentation of the number of functional myocytes in the diseased heart using methodologies for cardiomyocyte cell cycle activation [10], adult stem cell mobilization [11], or cellular transplantation [12]. Cellular transplantation at the site of injury in the heart can be accomplished either by injecting isolated cells ("cellular cardiomyoplasty") [13], or by implanting a cardiac tissue patch engineered *in vitro* ("tissue cardiomyoplasty") [14].

27.1.1 Cellular Cardiomyoplasty

More than 10 years ago, pioneering studies in the laboratory of Lauren Field have shown feasibility of cell transplantation in the heart [15,16]. Since then, different investigators have used cardiac [15] or skeletal myoblast cell lines [16], fetal [17,18], neonatal [19], and adult [20] cardiac myocytes, autologous [21,22] and syngeneic [23] skeletal myoblasts, smooth muscle cells [24], endothelial cells [25], native [26] or genetically altered fibroblasts [27], embryonic [28], bone marrow [29], mesenchymal [30], or heart derived [31] stem cells, as potential donor cells. As a result, treated hearts have shown improvement in diastolic function, almost independent of transplanted cell type [26]. The improvement in the systolic function (generation of active force) on the other hand required use of cells with the myogenic (contractile) potential [26,32]. The possible therapeutic benefit of cellular cardiomyoplasty stems from structural remodeling of scar region [33], enhancement of myocardial revascularization [12], and direct structural and functional integration of donor cells with the host myocardium [18]. Presently, clinical studies in the United States, Europe, and Asia are under way to investigate feasibility and safety in using autologous bone marrow derived stem cells and skeletal myoblasts in treatment of postinfarction left ventricular disfunction [13,34,35]. Initial results are promising, but reveal risk for ventricular arrhythmias [34], limiting in some studies the pool of patients to only those that already have internal defibrillators. Since no systematic studies have been performed to assess the electrical performance of the heart postcardiomyoplasty, and little data exists on electrical interaction between donor and host cells *in vivo* or *in vitro* [36], the causes of the postoperative electrical instability are not known. Plausible explanations include inflammatory response and subsequent fibrosis at implantation site [35], possible electrical coupling in conjunction with different electrophysiology between implanted cells and cardiomyocytes [18,34], and possible stimulation of sympathetic nerve sprouting and overexpression of neurotransmitters after the cell transplantation [37].

27.1.2 Tissue Cardiomyoplasty

Some of the major hurdles in restoring heart function by cellular cardiomyoplasty include limited survival of injected cells in the region of scar tissue, and no architectural repair of the infarcted area. An alternative approach is tissue cardiomyoplasty, which involves *in vitro* cultivation of compact three-dimensional (3D) cardiac tissue patch, and subsequent implantation over or instead of the infracted scar tissue. Although surgically more challenging compared to cellular cardiomyoplasty, this methodology has a potential to improve efficiency and localization of tissue repair in larger size cardiac injury such as infarction of major coronary vessels, or congenital heart defects [38]. Ideally, based on the location, shape, and size of injury and architecture of surrounding tissue (assessed by ultrasound, MRI, or other noninvasive technique [39]), functional cardiac patch with needed geometry and 3D structure is engineered *in vitro* starting from selected cell type, natural or synthetic scaffold, and appropriate culturing vessel (bioreactor). Inside the bioreactor, cells attach to biocompatible (and possibly degradable) scaffold, interconnect, and assemble in three dimensions to reconstitute an *in vivo*-like cardiac tissue equivalent (construct). The combination of biochemical and physical stimuli during culture is designed to best mimic physiological state of tissue, and to support cell differentiation or transdifferentiation and desired 3D tissue architecture. At a proper time point, tissue construct is removed from the bioreactor and surgically implanted in the site of injury in order to restore or improve electrical and mechanical function of diseased heart.

In reality, however, the successful reconstitution of cardiac-like tissue patch *in vitro* starting from single cells is an extremely challenging problem due to limited proliferation potential and high metabolic demand of cardiac cells, as well as complex anisotropic architecture and electro-mechanical function of native cardiac tissue. While recent reviews on cardiac tissue engineering [14,40] have focused on scaffold biomaterials, bioreactors, and cultivation conditions, this chapter will provide emphasis on the electrophysiological considerations and role of tissue architecture in the development of functional cardiac patch. It is this author's view, that these factors will play an important role in the design of efficient and safe therapies, despite the fact that they are frequently neglected in current *in vitro* and *in vivo* studies.

27.2   Cardiac Architecture and Function

The crucial architectural and functional feature of cardiac muscle tissue (Figure 27.1) is anisotropy, that is, anatomically and biophysically, properties of cardiac muscle vary in different directions [41]. Microscopic structural anisotropy in cardiac tissue results from the spatial alignment of elongated cardiac myocytes (Figure 27.1a), and the preferential location of intercellular junctions (e.g., fascia adherence, gap junctions, desmosomes) in end-to-end vs. side-to-side cell connections [42]. Macroscopic anisotropy is a result of the presence of aligned cardiac muscle fibers and sheets that transmurally rotate inside the heart wall (180° rotation from endo- to epicardium) [43]. The unique anisotropic architecture of cardiac tissue enables an orderly sequence of electrical and mechanical activity, and efficient pumping of blood from the heart. Beside architecture, electrical membrane properties of cardiac myocytes also vary substantially depending on the location in the heart with distinct differences between atria and ventricles, endo- and epicardial regions, left and right heart, base and apex, etc. [44–46]. Moreover, the heart contains a large variety of nonmyocytes (e.g., fibroblasts, endothelial cells, smooth muscle cells, neural cells, leukocytes) with specific roles in the cardiac function that are still not fully elucidated.

Structural anisotropy and intercellular continuity of the excitable cardiac substrate have profound effect on electrical and mechanical functioning of the heart. For example, anatomical anisotropy in heart tissue causes a larger intracellular resistance per unit length in the transverse (across fiber) than longitudinal (along fiber) direction, resulting in smaller velocity but larger maximum slope of action potential upstroke and safer electrical propagation in the transverse direction [47,48]. As a consequence, electrical stimulation of a small region in the heart tissue results in development of elliptical rather than circular propagating wavefront [49] (Figure 27.1b). Directly related to anisotropy is evidence that cardiac impulse conduction at the microscopic level is discontinuous at sites of gap junctions, and even stochastic due to small local variations in ion channel function, gap junction distribution, and adjoining tissue architecture [50,51]. In contrast to these small physiological variations, larger variations of electrical properties (e.g., action potential duration, intercellular coupling, electrical load) at the cellular and tissue level may result in increased susceptibility to propagation slowing and conduction block [46,51]. Slow propagation velocity and unidirectional block are some of the main prerequisites for the initiation of reentrant cardiac arrhythmias [51,52].

The degree of anatomical and functional anisotropy depends on location in the heart and age of the individual, with main determinants being cell size and geometry, type, amount, and distribution of cell junctions in membrane, and macroscopic tissue architecture [53–55]. Electrical anisotropy changes in certain cardiac pathologies such as ischemia, infarction, and heart failure [56,57]. This change is a consequence of the altered gap junction distribution and expression, as well as formation of longitudinal collagenous septa between the cardiac fibers which result in discontinuous transverse propagation ("nonuniform anisotropy") and increased susceptibility to reentrant arrhythmias [58]. In particular, in canine hearts, "border zone" between infracted and normal tissue exhibits disarray of cardiac cells and gross change in anisotropy, resulting in conduction slowing, block, and reentrant "figure of eight" circuits

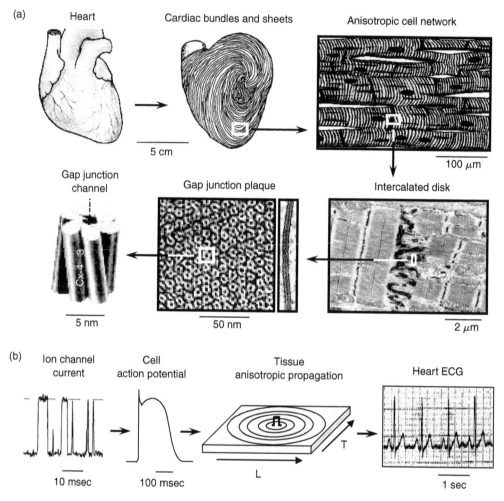

FIGURE 27.1 Levels of anatomical and electrophysiological organization in cardiac muscle. (a) Intercalated disk is specialized end-to-end connection between cardiac cells. Gap junction plaque is shown in cross-section and en face. Cx-43 is gap junction protein connexin-43. Note that structural complexity in heart spans many orders of magnitude from nanometer-size scale in single channels to centimeter-size scale in the heart. (b) Time constants of electrophysiological function in cardiac muscle range from nanoseconds for a single channel gating to seconds for heart beats. L and T denote longitudinal and transverse direction, respectively. Pulse sign denotes site of stimulus.

in the heart [59]. In addition, presence of noncontractile scar in heart milieu can cause locally increased stress gradients, which through mechano-electric feedback may yield in stretch-induced arrhythmias [60].

27.3   Current State of Cardiac Tissue Engineering

Over the last several years different strategies have been developed to design engineered cardiac tissues that could be used for pharmacological, genetic, and functional studies *in vitro* and possible implantation *in vivo*, as outlined in Table 27.1. These studies have shown that structure and function of cardiac tissue constructs depend on the animal species used for the cell dissociation [61–63], composition of seeded cells [14,64], initial cell seeding density [62,63,65,66], scaffold characteristics [40,67–70], composition of culture medium [69], type of bioreactor [63,65,69], and applied physical forces [61,71]. Most of these results are based on the evaluation of general histology, and assessment of cellular properties including cell

TABLE 27.1 Cardiac Tissue Engineering *In Vitro* and *In Vivo*

		In vitro		
Cell source	Scaffold	Bioreactor	Assessment	References
Neonatal rat ventricle	Microcarrier beads	HARV	Immunohistology, ultrastructure, pharmacology	[77,78]
Embryonic chick ventricle	Planar collagen gel with supplements	Static petri dish attachment to velcro	Immunohistology, pharmacology, ultrastructure, gene manipulation, mechanical contractile force	[79]
Neonatal rat ventricle	planar collagen gel with supplements	Static petri dish attachment to velcro	Immunohistology, pharmacology, ultrastructure, contractile force	[62]
Neonatal rat ventricle	Collagen gel ring with supplements	Static petri dish and cyclic stretch	Immunohistology, pharmacology, ultrastructure, contractile force	[72]
Rat smooth muscle, skin fibroblasts, fetal ventricle, human atria and ventricle	Rectangular gelatin mesh	Static petri dish	Histology, cell proliferation	[86]
Young human ventricle	Rectangular gelatin mesh	Cyclic stretch in dish	Histology, proliferation, mechanical	[71]
Neonatal rat ventricle	Rectangular collagen scaffold ("tissue fleece")	Static petri dish	RT-PCR, pharmacology, ultrastructure, mechanical	[73,87]
Neonatal rat ventricle	Fibrin glue and thick collagen gel around aorta	Unperfused and perfused through aorta	FDG-PET, Immunohistology	[88]
Neonatal rat ventricle	Cross-linked collagen mesh	HARV	Immunohistology, ultrastructure	[89]
Fetal and neonatal rat ventricle	Electrospun tubular collagen scaffold	HARV	Immunohistology, ultrastructure mechanical stress–strain curves	[74,90]
Neonatal rat ventricle	No scaffold	Static petri dish	Immunohistology, ultrastructure, electrical connectivity, mechanical, subcutaneous implantation	[75,92]
Embryonic chick and neonatal rat ventricle	Fibrous PGA disk	Static petri dish, spinner flask, HARV	Viability, metabolic activity, immunohistology, ultrastructure	[63]
Neonatal rat ventricle	Fibrous PGA disk	Spinner flask	Viability, metabolic activity, immunohistology, ultrastructure, tissue scale electrophysiology	[64]
Neonatal rat ventricle	Fibrous PGA disk	Perfusion cartridge	Viability, metabolic activity, immunohistology, ultrastructure	[94,95]
Neonatal rat ventricle C2C12 myoblasts	collagen sponge disk with matrigel	Orbitally mixed dish, Perfusion cartridge	Viability, metabolic activity, immunohistology, pharmacology, excitation threshold, capture rates	[65,96]
Neonatal rat ventricle	Surface hydrolyzed, laminin-coated PGA disk	Spinner flask, 3D gyrator, HARV	Immunohistology, immunoblotting, ultrastructure, viability, metabolic activity, tissue electrophysiology	[69]

Continued

TABLE 27.1   Continued

		In vitro		
Cell source	Scaffold	Bioreactor	Assessment	References
Neonatal rat ventricle	Surface hydrolyzed, laminin-coated PGA disk	HARV	Immunohistology, immunoblotting, pharmacology, cell electrophysiology	[76]
Neonatal rat ventricle	Fibronectin coated PLGA disk	HARV	Histology, ultrastructure, optical mapping of action potentials	[121,122]
Fetal rat ventricle	Alginate disk	Static petri dish	Viability, metabolic activity, histology	[66]

		In vivo implantation in heart		
Cell source	Scaffold	Site of implantation	Postoperative assessment	References
Fetal rat ventricle	Gelatin mesh cube	Infarct in cryoinjured left rat ventricle	Histology, ultrastructure, ventricular pressure in Langendorf preparation	[97]
Rat aortic smooth muscle	Gelatin mesh, PTFE patch, PGA, and PCLA sponge	Defect in right outflow ventricular tract in rat	Immunostaining, cell proliferation, morphometry	[67,98]
Rat aortic smooth muscle	PCLA sponge	Postinfarct aneurysm in rat left ventricular	Immunohistology, echocardiography, ventricular pressure	[99]
Fetal rat ventricle	Alginate disk	Rat coronary occlusion site	Immunohistology, echocardiography	[100]
Neonatal rat ventricle	Collagen gel ring with supplements	Perimeter of healthy rat ventricle	Immunohistology, ultrastructure, pharmacology, echocardiography	[85]

HARV — high-aspect-ratio-vessel, HFDG-PET — Fluor-Deoxy-Glucose-Positron-Emission-Tomography, PLGA — poly(lactic-co-glycolic) acid, PTFE — polytetrafluoroethylene, PCLA — ε-caprolactone-co-L-lactide reinforced with knitted poly-L-lactide fabric.

number, viability, metabolic activity, expression of cardiac-specific proteins, and ultrastructural features. Few groups also focused on measurements of contractile force at tissue scale [62,72–75], while only one group has studied in detail microscopic and macroscopic electrical properties of tissue constructs [64,69,76]. The following paragraphs will give an overview of existing *in vitro* and *in vivo* efforts in the emerging field of cardiac tissue engineering.

27.3.1   Cardiogenesis *In Vitro*

Akins et al. [77,78] have shown that neonatal rat ventricular myocytes can form multilayered interconnected structures when cultivated on fibronectin-coated polystyrene beads or collagen fibers inside high-aspect-ratio-vessel (HARV) bioreactors. After 6 days in culture, cardiac cells formed small, several layers thick clusters in the regions between the beads, exhibited presence of sarcomeres and gap junctions, and rhythmically contracted at rates that were slower in the presence of propranolol. The nonmyocytes were distributed throughout the tissue clusters, with most of the endothelial cells lining on the interface between the cluster and culture medium.

Group of Eschenhagen has done some of the most comprehensive work in the field, using mixtures of embryonic chick [79] or neonatal rat cardiac cells [62,72] and gels made of collagen type I supplemented with matrigel, chick embryo extract, and horse serum. Their initial work was based on Vandenburgh's approach for engineering of skeletal muscle [80], where cell–gel mixture was cast in the thin planar geometry between two parallel Velcro-coated glass tubes. Firm attachment to Velcro-imposed static stress on free edges of the gel resulting in thin biconcave tissue construct ($8 \times 15 \times 0.18$ mm^3) with loose, aligned cardiac cell network formed mostly along the construct edges [79]. The alignment and density of this network was improved by use of the chronic cyclic stretch during cultivation [61]. In their current approach [72], cardiac constructs termed engineered heart tissues (EHTs) are made by embedding neonatal rat ventricular cells in circularly molded collagen gels, which are subsequently cultivated in static conditions for 7 days and subjected to chronic cyclic stretch (10%, 2 Hz) for additional 7 days. Resulting submillimeter thick rings of tissue contain aligned cardiomyocytes organized in loose but uniform tissue-like network with frequently forming 20 to 50 μm thick cardiac fibers [72]. Myocytes in this network spontaneously contract at steady rates of ~2 Hz, and exhibit differentiated cardiac-specific ultrastructure including parallel sarcomeres, T-tubules, SR vesicles, formed dyads, and basement membrane [72]. The initial seeding of unpurified cell mixture (no differential preplating) result in the presence of microphages and abundant fibroblasts, scattered throughout the EHT, as well as endothelial and smooth muscle cells, packed more densely in the outer compared to inner region. When electrically and pharmacologically stimulated, EHTs exhibit cardiac-specific mechanical properties including Frank–Starling behavior, a positive inotropic response to extracellular calcium and isoprenaline, and negative inotropic effect to carbachol. Although recorded twitch amplitudes of 1 to 2 mN/mm^2 are an order of magnitude lower than those found in native cardiac tissues [81], the twitch to resting tension ratio is larger than 1, similar to native muscle. The use of rat cells, horse serum, chick embryo extract, matrigel, and unpurified cell seeding mixture are all found to increase the maximum developed force and mechanical integrity of EHTs, while increase in collagen content seems to decrease twitch tension [14,70]. Up to now, EHTs have been used for studying the effect of genetic and pharmacological manipulations on cardiac contractile function [62,82–84], and were also implanted *in vivo* (see work by Zimmermann et al. [85]).

Group of Li [86] seeded biodegradable gelatin meshes with different cell types including stomach smooth muscle cells, skin fibroblasts and fetal ventricular myocytes from rat, and adult atrial and ventricular myocytes from humans. Rat cells and human atrial, but not ventricular, cells proliferated over 3 to 4 weeks in culture. All cells migrated in a 300 to 500 μm thick outside layer of gelatin scaffold, which slowly degraded with the highest degradation rate found in the presence of fibroblasts. In separate *in vitro* study [71], the same group showed that 2 weeks of cyclic mechanical stretch improved cell proliferation, distribution, and mechanical strength of tissue constructs made using gelatin scaffolds and heart cells isolated from children who underwent repair of Tetralogy of Fallot.

Kofidis et al. [73,87], used $20 \times 15 \times 2$ mm^3 commercially available collagen-based scaffolds ("tissue fleece") that were inoculated with neonatal rat cardiac cells and cultured in petri dishes. The randomly distributed cells formed sparse synchronously contractile networks, and exhibited cardiac specific mechanical responses to stretch, extracellular calcium, and epinephrine. In an attempt to increase the thickness of the engineered cardiac tissue, the same group recently embedded a rat aorta in the 8.5 mm thick mixture of collagen gel and cardiac cells, and used pulsatile flow through the aorta for 2 weeks as a vehicle for nutrition and oxygen delivery [88]. The aorta remained patent throughout the culture and viability was increased compared to unperfused controls.

van Luyn et al. [89] have also used neonatal rat cells and commercially available cross-linked collagen I bovine matrices, and cultured them in HARV bioreactors for up to 3 weeks. Spatially scattered cells exhibited immature sarcomeres, gap junctions, and stained for troponin-T.

In recent studies, Evans et al. [90] and Yost et al. [74] cultured embryonic and neonatal rat cardiac cells on fibronectin coated aligned tubular scaffolds (15 mm long, 4 mm inner, 5 mm outer diameter) made from extruded collagen I fibers [91]. After 3 to 6 weeks in HARV bioreactors, cardiac cells aligned, contracted spontaneously, formed few interconnected cell layers (with total thickness of ~20 μm) on the inside and outside lumen of the tube, and exhibited registered sarcomeres and randomly distributed gap junctions. Tubular collagen scaffolds exhibited viscoelastic properties qualitatively resembling those of native papillary muscle [74] only when seeded with cardiac cells, as inferred from the shape of the stress–strain hysteresis loops.

Very elegant studies by Shimizu et al. [75,92,93] have demonstrated that cardiac cells can form 3D multilayer tissue-like structures without the use of any type of scaffold. Isotropic monolayers of purified cardiac cells were cultured to confluence on the surfaces made of temperature responsive polymer poly(N-isopropylacrylamide). This polymer is slightly hydrophobic and cell adhesive at 37°C and becomes hydrophilic and cell repellent when cooled below 32°C. After 4 days, up to four cardiac sheets were detached (together with secreted extracellular matrix) from polymer surface by cooling to 20°C and overlaid using pipette or polyethylene mesh. Overlaid sheets exhibited uniform gap junction distribution, connected electrically, and formed compact multilayered spontaneously contractile cardiac constructs with area of 1 cm^2 and thickness of up to 50 μm. After subcutaneous implantation, cardiac constructs survived up to 12 weeks, spontaneously contracted, appeared vascularized and at 3 weeks exhibited twitch tension of 1.2 mN [75].

Group of Freed and Vunjak-Novakovic has utilized various approaches to engineering of cardiac tissue based on the use of biodegradable polymer scaffolds and different tissue culture bioreactors. Initial studies of Bursac et al. [64], and Carrier et al. [63] have demonstrated that cardiac cells formed tissue-like constructs when seeded on 5 mm diameter × 2 mm thick fibrous poly(glycolic) acid (PGA) disks inside spinner flask bioreactors. Cells in the outer 50 to 70 μm thick region were randomly oriented and connected in the relatively dense multilayer network. These cells expressed cardiac specific proteins (α-sarcomeric actin and troponin-T), end ultrastructural features characteristic of cardiac myocytes (parallel sarcomeres, all types of specialized junctions, dense mitochondria, glycogen granules). The cells in the interior of these tissue constructs were sparsely distributed and often necrotic. Spontaneous macroscopic contractions were observed at days 2 to 4 of culture and generally ceased thereafter, with occasional activity at rates of less then 1 Hz on culture day 7. Action potential propagation and electrical excitability were studied using linear array of eight metal microelectrodes with 500 μm spatial resolution (Figure 27.2a). Constructs were electrically excitable and exhibited isotropic, macroscopically continuous electrical propagation with velocities as high as 60% of those found in native ventricles [64]. Use of purified cell mixture (after differential preplating) for seeding resulted in superior electrophysiological properties including higher velocity of propagation, increased maximum rates of tissue capture and lower excitation threshold compared with use of unpurified (no preplating) cell mixture. In addition, use of neonatal vs. chick cardiac cells, dynamic seeding and cultivation in bioreactors vs. static petri dishes, and increase in the number of seeded cells up to 8 × 10^6 cells per scaffold have all improved cell packing density, metabolic activity, and electrophysiological properties of constructs [63,64]. In further studies, Carrier et al. [94] looked in the use of perfusion through tissue construct as means to improve the cellularity and tissue architecture, and studied effect of oxygen deprivation on engineered cardiac muscle [95]. In the most recent studies from the same group, Radisic et al. [65,96] used cell–matrigel mixture to densely inoculate cardiac cells inside collagen sponge scaffolds, and employed similar perfusion bioreactor for construct cultivation. The viability, metabolic activity, and cellular density through ~1 mm thick region were higher than in constructs cultured in orbital shakers and those from studies by Carrier et al. In a different study, Papadaki et al. [69] have shown significant improvements in structure and function of cardiac constructs when PGA scaffolds were hydrophilized and coated with laminin, percentage of serum in culture medium reduced after 2 days of cultivation, constructs seeded with concentrated cell suspension in rotating gyrators, and cultivation performed in HARV bioreactors. Compared to previous studies, tissue constructs exhibited better cell viability yielding thicker (120 to 160 μm) cardiac-like outer region, and higher cellular expression of differentiation marker proteins including creatine kinase-MM (involved in metabolism), sarcomeric myosin heavy chain (involved in contractile function), and gap junction protein Connexin-43. Tissue scale electrical properties approached those found in native muscle with conduction velocity at basic stimulation rate of 1 Hz and maximum capture rate reaching 90 and 70% of those found in donor neonatal ventricles, respectively (Figure 27.2a). Macroscopic electrical propagation was effectively isotropic due to random cell orientation and uniform gap junction distribution. Further electrophysiological and pharmacological studies at microscopic scale by Bursac et al. [76] demonstrated that action potentials in cardiac cells from 7-day constructs were comparable to those in 2-day old donor ventricles with respect to depolarization upstroke, amplitude, and resting potential. The major difference was prolonged action potential duration

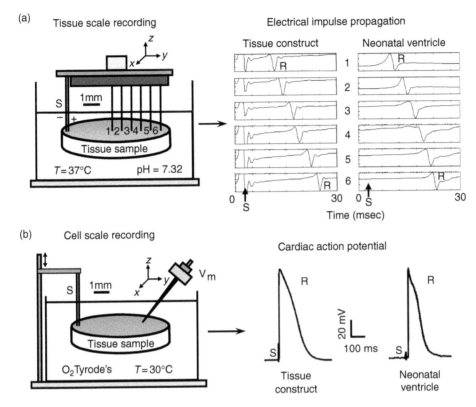

FIGURE 27.2 Tissue and cell scale electrophysiological recordings in 7-day old cardiac tissue constructs and 2-day old neonatal ventricles. (a) Custom-built linear array of two stimulating and eight recording microelectrodes (only six are shown) was used for assessment of macroscopic impulse propagation. S and R in two tracings denote stimulus artifact and responses, respectively. Note relatively smooth biphasic shapes of recorded extracellular waveforms in constructs and ventricles. The response amplitude in constructs is an order of magnitude lower than in ventricles due to smaller number of cardiac cell layers present. Propagation velocities in constructs and ventricles are comparable (i.e., times for propagation from electrode 1 to 6 are similar). (b) Glass capillary microelectrodes are used for action potential recordings from single cells in the tissue sample. Note fast upstroke and similar resting potential, but longer action potential in constructs than in ventricles. S and R denote stimulus artifact and response, respectively.

(Figure 27.2b) and absence of early repolarization notch due to downregulation of transient outward potassium current. In addition, cardiac cells cultured in 3D tissue constructs maintained more differentiated phenotype (higher expression of marker proteins) and more *in-vivo* like action potential features compared to those cultured in 2D monolayers under similar cultivation conditions.

27.3.2 *In Vivo* Implantation for Cardiac Repair

By April 2004, *in vivo* implantation of engineered cardiac tissue in infracted heart was attempted by only three groups.

Li et al. [97] cultured fetal rat myocytes on biodegradable gelatin meshes ($15 \times 15 \times 5$ mm^3) for 7 days and implanted cardiac constructs over the scar area in cryoinjured syngeneic rat hearts. Cells populated sparse interstices of gelatin meshes (see *in vitro* work by Li's group) and continued to proliferate and spontaneously contract *in vitro* for at least 26 days. Epicardially implanted grafts survived for 5 weeks, exhibited increased cellularity, slight degradation, and moderate degree of vascularization. Left ventricular developed pressure, showed no improvement over the control animals. In other studies [67,98], the same group evaluated use of various scaffold materials seeded with aortic smooth muscle cells (used to presumably increase elasticity of the patch) for repair of defect in the right ventricular outflow tract in syngeneic rats. Eight weeks postimplantation constructs made of ε-caprolactone-co-L-lactide reinforced with knitted poly-L-lactide fabric (PCLA) outperformed those made of gelatin, PGA, and polytetrafluoroethylene (PTFE) with respect to cellularity, elastin content, and preserved thickness. In the next study [99], grafts made of PCLA and smooth muscle cells were used to repair left ventricular aneurysm in the rat hearts after transmural infraction. Cell-seeded grafts reduced abnormal chamber distensibility and improved ventricular function compared with implanted cell-free grafts, as assessed by echocardiography and constant pressure measurements in Langendorff preparation.

Leor et al. [100] cultured fetal rat myocytes on porous biodegradable alginate disks (6 mm diameter, 1 mm thick) inside 96-well plates for 4 days, and implanted cardiac tissue constructs over infracted region in rat hearts 7 days after permanent occlusion of left main coronary artery. Nine weeks postimplantation, cardiac constructs survived, while alginate scaffold substantially degraded. Cardiac grafts were neovascularized, contained infiltrated macrophages and lymphocytes due to use of allogenic cells and no immunosuppression, and exhibited small number of sparsely distributed cardiac cells presumably due to low initial seeding density. Echocardiography revealed attenuated left ventricular dilatation and maintained contractile function, although it was not clear if implanted cell-free scaffolds would have produced similar results. Further *in vitro* study from the same group [66] focused on methods to increase cell density in alginate scaffolds by applying moderate centrifugal forces during seeding.

Zimmermann et al. [85] implanted 12-day old ring-shaped EHTs (see *in vitro* work from Eschehagen's group) around the circumference of healthy syngeneic rat hearts. Two weeks after implantation, EHTs were vascularized, innervated, expressed differentiated cardiac phenotype, and did not alter left ventricular function compared to preoperative state, as assessed by echocardiography. Spontaneous contractions were preserved *in vivo*, but no intercellular coupling of EHTs and host tissue could be demonstrated. Despite the syngeneic approach, EHTs were completely degraded in the absence of immunosuppression, presumably due to presence of allogenic components in reconstitution mixture (e.g., matrigel, horse serum, chick embryo extract).

In all of the described *in vivo* attempts no electrophysiological studies of engineered patch were done pre- or postimplantation.

27.4   Design Considerations

Ultimate success of cardiac tissue repair with an implanted cardiac patch depends on thorough understanding of the key parameters of tissue design *in vitro*, and careful definition of the desired tissue engineering outcomes.

27.4.1   Cell Source and Immunology

One of the crucial aspects for successful engineering of cardiac tissue is a choice of implanted cells. Experiences from cellular cardiomyoplasty show that for the improvement of heart systolic function implanted cells need to be (or be capable of becoming) contractile [26]. Although fetal and neonatal cardiac cells are clearly shown to functionally incorporate in the myocardium [17–19], they are not cells of choice due to limited proliferation potential, immunological, and ethical issues. For this reason, their use will probably stay limited to *in vitro* model systems and proof-of-concept *in vivo* studies. Possible "ideal" cell source may be human embryonic or adult stem cells. Although human embryonic stem cells are shown to differentiate into cardiac myocytes [101,102], their immunogenic and tumorogenic nature, and low efficiency and specificity of differentiation ($<1\%$ of cells differentiate into mixture of atrial, ventricular, and nodal cells), as well as ethical issues, may finally preclude their clinical use. Some hope lies in the nuclear transfer technology ("therapeutic cloning") [103], and genetic knock-out of major histocompatibility complexes [104], which may offer strategies for preventing immune rejection. Autologous adult stem cells from skeletal muscle, peripheral blood, or bone marrow appear as better choice for cell transplantation than embryonic stem cells. For example, autologous skeletal myoblasts are easy to proliferate *in vitro* and implant *in vivo* and cause no immune response, which currently makes them one of the cell types used in clinical trials [33]. Unfortunately, they do not express gap junction proteins and are still not shown to functionally couple with host cardiac tissue when implanted [105,106], although this may be resolved with stable transgene expression of connexin molecules [107]. Mesenchymal or hematopoietic stem cells derived from bone marrow may represent an ideal cell source [29,108]. However, the efficiency of their transdifferentiation into cardiac myocytes remains controversial in light of recent findings that question techniques used to quantify the number of transdifferentated cells in heart [109,110]. Still their beneficial effect may come from induced neovascularization in implantation sites. Current research efforts are focused on increase in percentage of embryonic or adult stem cells that commit to cardiac phenotype by use of different growth factors or media compositions, introduction of early cardiac genes in undifferentiated stem cells *in vitro*, or coculture of stem cells with differentiated cardiac myocytes. Another promising alternative is *in vitro* genetic reprogramming of differentiated somatic cells (for review see Reference 111).

27.4.2 Cellular Composition

The heart is composed of different types of cells. Engineering of functional tissue patch requires selection of the appropriate composition of seeded cell mixture. For example, if an implant were to be localized in the ventricle, cardiac myocytes that comprise a main fraction of seeded cells should be of ventricular origin or with ventricular characteristics. The percentage of nonmyocytes in the seeding mixture is a parameter that can be varied. Eschenhagen's group found that it was necessary to use higher percentage of nonmyocytes (no preplating after cell isolation) to improve mechanical integrity and twitch tension of EHTs [14]. In contrast, scaffold-free constructs by Shimizu et al. [75] developed similar twitch tensions despite the fact that they were made of purified cardiac cell mixture. This suggests that the percentage of "needed" nonmyocytes may actually depend on the presence and type of scaffold. It is possible that in EHTs, higher number of fibroblasts contributed initial contraction of collagen gel [112], which in turn increased the proximity and intercellular connectivity between myocytes, yielding increased contractile force. In addition, Bursac et al. [64] have shown that use of unpurified cardiac cell mixture decreased propagation velocity and compromised electrical properties in cardiac constructs. Therefore, for a given type of scaffold, the percentage of seeded nonmyocytes should be selected to optimize for both mechanical and electrical function of cardiac constructs.

27.4.3 Tissue Architecture

The anisotropic architecture and dense cell packing of native cardiac tissue impose important design rules in engineering of functional cardiac patch. For instance, velocity of electrical propagation and mechanical stiffness in healthy adult human ventricles are on average 2 to 3 times larger in longitudinal (fiber) than in transverse (cross-fiber) direction [113,114]. This can vary widely with location in the heart, age of individual, and heart disease. Therefore, only a cardiac patch with dense 3D network of elongated and aligned cardiac myocytes that mimic architecture of native tissue can generate desirable spatio-temporal distribution of electrical and mechanical activity, and result in efficient and safe therapy. Aligned growth of cardiac cells can be induced by static or dynamic stretch [61,115,116], presence of free tissue boundaries [117], or by cell guidance with oriented surface topography [49,118,119] and anisotropic distribution of chemical cues for cell attachment [49,120]. Eschenhagen's group applied cyclic stretch and ring geometry creating a sparse network of oriented cardiac cells in the form of a thin (submillimeter diameter) cardiac cable. Repair of larger injured area in the heart will, however, require engineering of an anisotropic slab of 3D cardiac-like tissue with controllable shape, size, and geometry.

Bursac et al. [49] have recently shown that anisotropic monolayers of cardiac cells (that mimic longitudinal sections of native cardiac tissue) can be designed with highly controllable architecture using surface microabrasion or micropatterning of extracellular matrix proteins (e.g., fibronectin) (Figure 27.3b). These

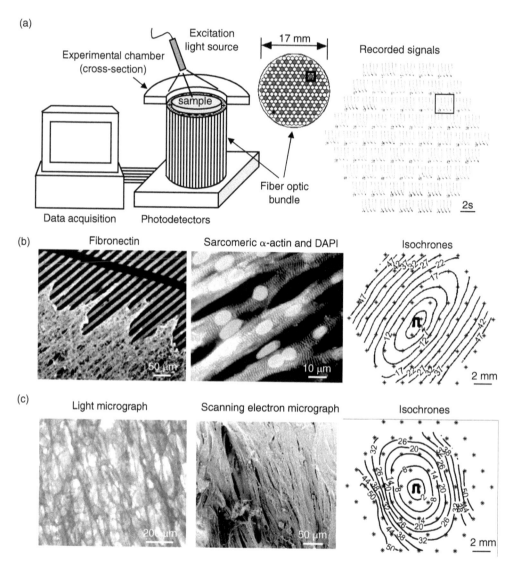

FIGURE 27.3 Architecture and impulse propagation in 2D and 3D anisotropic cultures of neonatal rat ventricular cells. (a) Optical mapping setup. Tissue samples stained with voltage sensitive dye were positioned over 61 hexagonally arranged fibers and transilluminated with excitation light. Two seconds of optically recorded action potentials from a cardiac cell monolayer are shown on the right. Gray circles denote active fibers. Square denotes a recording site in the bundle and corresponding voltage trace (for details see Reference 49). (b) Anisotropic monolayer of cardiac cells. Cells were cultured on micropatterned lines of fibronectin. On the left, culture is deliberately scratched to expose patterned lines and cell-deposited fibronectin. Aligned cells exhibit prominent sarcomeres and elongated nuclei (middle). Elliptical isochrones demonstrate anisotropic propagation (right). * denotes recording sites. Pulse symbol denotes site of stimulus. The degree of anisotropy can be systematically varied by controlling the amount of intercellular clefts and cell co-alignment [49]. (c) Anisotropic tissue construct. Oriented fibrous architecture in PLGA scaffolds (left) was accomplished by leaching sucrose from a polymer-coated template made of aligned sucrose fibers [121]. Cardiac cells were aligned in numerous regions along the direction of PLGA fibers (middle). Macroscopic propagation was anisotropic (albeit moderately), as assessed by optical mapping of transmembrane voltage (right) [122].

methodologies enabled systematic control over the degree of anisotropy (longitudinal-to-transverse velocity anisotropy ratios from 1 to 5.6), fiber direction, and amount of longitudinal intercellular clefts in the 2D cardiac monolayers. Optically recorded action potentials with voltage sensitive dye RH237 (at 61 sites over 2 cm^2 area) were used to map propagation of electrical activity and degree of functional anisotropy (Figure 27.3a). It would be ideal if techniques for 3D cardiac tissue culture could be developed with a similar level of architectural control as those used for 2D cell culture. In most recent studies Bursac et al. [121,122] extruded and baked sucrose to form 3D aligned fibrous templates in an attempt to induce anisotropic architecture in poly(lactic-co-glycolic) acid (PLGA) scaffold disks (Figure 27.3c). After 2 weeks of culture in HARV bioreactors, 12 mm diameter, 0.7 mm thick cardiac constructs exhibited regions with aligned cells, and moderate degrees of functional anisotropy (i.e., longitudinal-to-transverse velocity ratio of up to 2), as assessed by optical mapping of electrical propagation (Figure 27.3c). Using similar techniques to align elastomeric polymers [123] instead of PLGA in conjunction with chronic mechanical stimulation may yield 3D cardiac patches with physiological degrees of structural and functional anisotropy. An alternative approach is the use of electrospun scaffolds [74,91,124] providing that obtained degrees of porosity can support deeper cell penetration during seeding than achievable by current methodologies.

27.4.4 Tissue Thickness

Another important parameter in the design of a functional cardiac patch is thickness of the tissue construct. Compared to other cells in the body, cardiac myocytes require high oxygen and nutrient supply due to continuous contractile activity. High metabolic demand in myocardium is sustained by dense vascularization with average arterial intercapillary distances that range from 15 to 50 μm depending on the size of the heart and basal heart rate [125,126]. Absence of vasculature in tissue constructs is a limiting factor for engineering thicker ($>$200 to 300 μm) cardiac patches with physiological cell packing densities. Nonetheless, high cell density is necessary for establishment of proper intercellular communication, which in turn enables efficient generation of mechanical force and fast, electrically safe impulse propagation. Mixing and perfusion of culture medium, and/or cyclic mechanical stimulation are some of the methods that can alleviate diffusional limits of oxygen and nutrient supply, but only to a certain extent. Nevertheless, thin but dense engineered cardiac tissue is still a good *in vitro* approximation of a viable portion of explanted ventricular slice after several hours of superfusion [127], and thus could be used for different tissue-scale functional studies *in vitro*. These studies would be more versatile and technically easier than studies in monolayer cultures (e.g., they would enable variety of force measurements, and yield substantially higher signal-to-noise ratio during extracellular and optical recordings of electrical propagation). Moreover, implantation of even 10 to 20 well-coupled anisotropic cardiomyocyte layers over the infracted area may still have a significant therapeutic value regarding the facts that after epicardial infarction only several disarrayed cell layers may survive above the scar, and that increased thickness of survived layers is directly correlated with decreased incidence of arrhythmias [128]. This is one of the reasons why approach by Shimizu et al. [75] with scaffold-free cardiac multilayers deserves close attention. On the other hand, engineering of thicker cardiac slices will depend upon the development of externally perfusable, patent microcapillary-like networks inside the tissue constructs. Recent work by Vacanti's group [129] represents an interesting approach to this problem.

27.4.5 Electrical Function and Safety

One of the most important criteria for successful design of cardiac patch is the issue of electrical and mechanical safety. It is important to understand that haphazardly adding donor cells or transplanting a poorly designed tissue patch into an already compromised heart may only increase the likelihood for aneurisms, tissue rupture, or arrhythmias. In general, the injection of donor cells or implantation of a tissue patch introduces structural and functional heterogeneity in the cardiac milieu that depends on (1) electromechanical characteristics of donor cells, (2) the density, coupling, and geometrical arrangement of donor cells within the implant, (3) the degree of interaction between the donor and host cells, and

(4) architectural differences and relative position between implant and host tissue. Understanding how normal functioning of the cardiac cell network changes with the presence of different types of cellular, structural, and functional heterogeneities is a necessary step in the design of safer and more efficient cell and tissue transplantation therapies.

For example, it is already known from computational and experimental studies that contact between two cells with different resting potentials (e.g., injured and healthy cell [130]) or action potential durations (e.g., purkinje and ventricular cell [131], epicardial and M cell [132]), may trigger afterdepolarizations or create conduction block depending on the degree of their electrical coupling. It is also known that partial decoupling between the cells may actually increase the safety of propagation [133,134] as is the case in transverse vs. longitudinal propagation in healthy cardiac muscle [47]. On the other hand, excessive gap junction decoupling [134], presence of fibrotic regions [135], geometrical expansions [136], and repetitive tissue branching [137] (e.g., sparse cardiac network in ischemic or infarcted area [138]) can yield extreme slowing of impulse propagation and occurence of conduction block and introduce the susceptibility of cardiac tissue to formation of micro- or macroreentrant circuits. In the ideal world, the safest and most efficient therapy would be to engineer a cardiac patch that exactly mimics the geometry, architecture, and function of the "missing region" in the heart, and to perfectly couple it to the host tissue, as "a piece of the jigsaw puzzle." In the "semi-real" world, even if one is to produce a patch with perfect anisotropic architecture, its optimal orientation relative to host tissue when implanted may change depending on the electrical differences between the donor and host cells, or capability of patch to couple with the host tissue. In the real world, things are very complex due to interplay of many factors. For example, the use of skeletal myoblasts or stem cell-derived cardiomyocytes may demand different patch design and different implantation strategy as determined by the functional differences between these cell types (e.g., shorter vs. longer action potential, different resting potentials, none vs. significant expression of gap junctions, etc.). The simulation may be further complicated by the presence of a fibrous capsule at the implantation site.

Due to the complexity of the problem, carefully designed studies on possible implantation scenarios will be crucial as a prescreening tool before the actual implantation. The author's view is that there are at least two simplified settings that could be used to systematically and reproducibly study the factors affecting the likelihood of arrhythmia arising from cell and tissue cardiomyoplasty. One is the use of micropatterned cocultures [139,140] of host cardiomyocytes and different types of donor cells or mixtures of donor cells, with the possibility for well-controlled studies based on (1) simplified cellular composition and tissue architecture compared to the 3D heart, (2) precise control of the cell microenvironment, cellular geometry and distribution, and geometry of cellular interactions between donor, host, and donor and host cells, and (3) the possibility to optically assess and exactly correlate electrical/mechanical activity and underlying tissue architecture at microscopic and macroscopic spatial scales. Optical mapping of transmembrane potentials and intracellular calcium [141] are well-suited not only for these studies, but also for electrophysiological evaluation of engineered cardiac patch pre and postimplantation *ex vivo*. The other setting is the use of computer models that incorporate cell-specific membrane properties, cell geometry, distribution of intercellular connections, and discrete tissue microarchitecture. With increase in computing efficiency, these detailed models could be used as counterparts to different *in vitro* or *in vivo* implantation scenarios to help interpretation and design of experiments, and eventually yield safe and efficient therapies.

27.4.6 Spontaneous Activity

Another electrophysiological parameter for consideration is a somewhat misinterpreted presence of spontaneous contractile activity in cardiac cell cultures. It is important to understand that although spontaneous contractions in cardiac cultures represent convenient way to visually identify cardiac cells, they are by no means a physiologically normal state for adult or neonatal ventricular muscle (which is most often the source tissue for cell dissociation). While early embryonic ventricular cells still spontaneously depolarize and contract [142], their resting potential hyperpolarizes with maturation such that neonatal ventricular cells already exhibit steady resting potentials (less than -70 mV), fast action potential upstroke (more than 100 V/sec), and no spontaneous activity [76,142,143]. Rather, spontaneous activity of cardiac cells observed *in vitro* is an artifact of the cell culture possibly caused by (1) membrane depolarization due to cell injury during or after dissociation, (2) dedifferentiation of cultured cells to more embryonic state due to presence of high serum and inadequate media formulation, (3) large presence of nodal or other pacemaking cells due to unselective cell dissociation, and/or (4) decreased intercellular coupling compared to native tissue due to 2D or sparse 3D arrangement of ventricular cells. The first three causes can be alleviated with time in culture, proper cultivation conditions, and careful cell dissociation. However, the sparse cardiac network, if present, will always result in less electrotonic load "seen" by a cell and facilitate propagation from single or small group of pacemaking cells into the quiescent tissue, favoring spontaneous activity [144,145]. Our observations that spontaneous contractions in cardiac constructs can be more readily observed when cells are less coupled (e.g., at lower seeding density, in the beginning of cultivation, or on the perturbing polymer fibers at the edges of the tissue construct) are in agreement with this reasoning. After 5 to 7 days in culture, tissue constructs in our studies usually become quiescent, presumably due to increased packing density and connectivity of cardiac cells. The low percentage of nonmyocytes in tissue constructs as assessed by immunostaining, and increase in conduction velocity between culture days 4 and 7, exclude fibroblast overgrowth as a possible cause of cardiac cells quiescence.

Although reported spontaneous beating rates in cardiac constructs are usually lower than regular heart rate, presence of paracrine chronotropic factors (e.g., epinephrine, adrenaline) after implantation in the heart may result in faster activity and possible arrhythmogenic hazard. For electrical safety reasons, a nonstimulated ventricular cardiac patch should be electrically quiescent, but readily excitable and fast conducting, similar to native ventricular muscle. Moreover, a quiescent avascular patch may have lower metabolic demand and increased chance for survival after implantation compared to a spontaneously contractile patch. However, persistent contractions and presence of mechanical load are essential for maintenance of ventricular mass *in vivo* [146], and for cell hypertrophy [147], spreading to confluence [148], and establishment of cell contacts *in vitro* [149,150]. (Our experience is that if cardiac cell culture is noncontracting for more than 3 days, cells start to atrophy and loose cell contacts.) Therefore, spontaneously quiescent engineered cardiac constructs still need to be maintained mechanically active in culture either by use of electrical, mechanical, or chemical (e.g., small concentrations of epinephrine) stimulation [150], or by coculture with spatially distinct population of pacemaking cells, which can be easily dissected before the implantation in the ventricle.

In any case, thorough evaluation of electrical properties and susceptibility to conduction block and arrhythmic behavior should be routinely done in cultured cardiac constructs and used as one of the tissue design criteria. From our experience, it is important to assess the properties of engineered tissue in challenging regimes such as fast or premature stimulation, similar to standard clinical pacing protocols for testing the susceptibility to arrhythmias [151]. For example, a tissue construct can support macroscopically continuous and relatively fast electrical propagation at low pacing rates, but yield conduction blocks and high incidence of reentrant arrhythmias [122] when paced rapidly (Figure 27.4) due to large spatial dispersion of refractoriness, or abundance of microscopic anatomic heterogeneities.

27.5 Future Work

The crucial technical aspect of this and all other cell-based therapies will be a choice of appropriate cell source, which is to be primarily determined by developmental biologists, geneticists, and immunologists. Engineering of a functional cardiac tissue patch using embryonic or adult stem cells and subsequent implantation studies are still to be done. Simultaneous efforts on design of appropriate scaffolds and control over the cellular connectivity and tissue architecture in three dimensions will be essential to the development of functional cardiac patch for use in laboratory and clinics. Systematic *in vitro* and *in vivo* studies on the role of heterogeneities on structure and function of cardiac tissue will help in designing more efficient and safer therapies. Possible engineering of the patent capillary- and microcapillary-like networks inside the tissue constructs may enable culture of thick tissue slices, and facilitate immediate perfusion

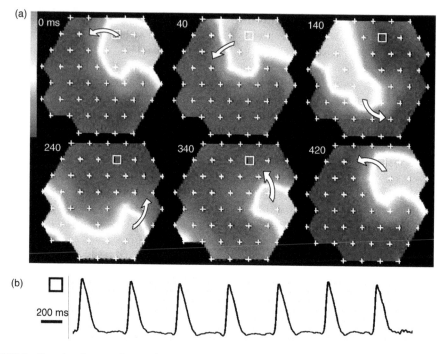

FIGURE 27.4 Functional reentry in a cardiac tissue construct induced by rapid point pacing, and assessed by optical mapping of transmembrane potential. (a) Single rotation of a counterclockwise reentrant wave shown through a series of voltage snapshots in time. Frames progress left to right, top to bottom. Time in milliseconds is marked at the top left corner of each frame. Arrows denote the direction of wave front. + denotes 2-mm spaced recording sites. Gray scale bar next to the first frame corresponds to a normalized transmembrane voltage with bottom and top denoting rest and peak of an action potential, respectively. (b) Recording of transmembrane voltage from the site marked by a white square in the frames of panel A. Reentrant activity appears stationary and periodic.

during implantation by attachment to one of the host arteries. The success will only be accomplished by a profound understanding of a number of complex topics, and achieved through a joint effort among basic scientists, engineers, and clinicians.

27.6 Conclusions

Cardiac tissue engineering is a new and exciting field with many obstacles to be surmounted before start of clinical trials. With technical aspects resolved, the main advantage of implanted cardiac tissue patch over the injected cells will be a structural and functional repair of large tissue defects. Given that implantation of cardiac patch will repair a centimeter-size tissue region, inference about the quality of tissue construct sprior to implantation should be based not only upon different assessments at cellular or subcellular level, but also upon detailed evaluation of both electrical and mechanical function at centimeter-size scale (i.e., measurements of impulse propagation and contractile force in constructs relative to those in native cardiac muscle). Similarly, only systematic micro- and macroscopic assessments of heart structure and electromechanical function postimplantation will provide necessary selectivity towards the design of safe and efficient clinical therapies.

Acknowledgment

The author would like to thank P. Bursac and K. Cahill for reviewing the manuscript. This manuscript was submitted for press on May 1, 2004. Due to rapid developments in the field, some of the most contemporary work will not be reviewed, and certain statements may be outdated at the time of press. Author apologizes in advance if this is the case.

References

[1] Anversa P. and Kajstura J. Ventricular myocytes are not terminally differentiated in the adult mammalian heart. *Circ. Res.* 1998; 83: 1–14.

[2] Laflamme M.A., Myerson D., Saffitz J.E., and Murry C.E. Evidence for cardiomyocyte repopulation by extracardiac progenitors in transplanted human hearts. *Circ. Res.* 2002; 90: 634–640.

[3] Guyatt G.H. and Devereaux P.J. A review of heart failure treatment. *Mt. Sinai. J. Med.* 2004;71: 47–54.

[4] Miniati D.N. and Robbinson R.C. Heart transplantation: a thirty year perspective. *Annu. Rev. Med.* 2002; 53: 189–205.

[5] Boehmer J.P. Device therapy for heart failure. *Am. J. Cardiol.* 2003; 91: 53D–59D.

[6] Moreira L.F. and Stolf N.A. Dynamic cardiomyoplasty as a therapeutic alternative: current status. *Heart Fail. Rev.* 2001; 6: 201–212.

[7] Szatkowski A., Ndubuka-Irobunda C., Oesterle S.N., and Burkhoff D. Transmyocardial laser revascularization: a review of basic and clinical aspects. *Am. J. Cardiovasc. Drugs* 2002; 2: 255–266.

[8] Sorelle R. Cardiovascular news: totally contained AbioCor artificial heart implanted July 3, 2001. *Circulation* 2001; 104: E9005–E9006.

[9] Association A.H. *2000 Heart and Stroke Statistical Update.* Dallas, TX: American Heart Association, 1999.

[10] Pasumarthi K.B.S. and Field L.J. Cardiomyocyte cell cycle regulation. *Circ. Res.* 2002; 90: 1044–1054.

[11] Orlic D., Kajstura J., Chimenti S., Limana F., Jakoniuk I., Quaini F., Nadal-Ginard B., Bodine D.M., Leri A., and Anversa P. Mobilized bone marrow cells repair the infarcted heart, improving function and survival. *Proc. Natl Acad. Sci. USA* 2001; 98: 10344–10349.

[12] Dowell J.D., Rubart M., Pasumarthi K.B., Soonpaa M.H., and Field L.J. Myocyte and myogenic stem cell transplantation in the heart. *Cardiovasc. Res.* 2003; 58: 336–350.

[13] Chachques J.C., Acar C., Herreros J., Trainini J.C., Prosper F., D'Attellis N., Fabiani J.N., and Carpentier A.F. Cellular cardiomyoplasty: clinical application. *Ann. Thorac. Surg.* 2004; 77: 1121–1130.

[14] Zimmermann W.H. and Eschenhagen T. Cardiac tissue engineering for replacement therapy. *Heart Fail. Rev.* 2003; 8: 259–269.

[15] Koh G., Soonpaa M., Klug M., and Field L. Long-term survival of AT-1 cardyomyocyte grafts in syngeneic myocardium. *Am. J. Physiol.* 1993; 264: H1727–H1733.

[16] Koh G., Klug M., Soonpaa M., and Field L. Differentiation and long term survival of C2C12 myoblasts graft in heart. *J. Clin. Invest.* 1993; 92: 1548–1554.

[17] Soonpa M.H., Koh G.Y., Klug M.G., and Field L.J. Formation of nascent intercalated disks between grafted fetal cardiomyocytes and host myocardium. *Science* 1994; 264: 98–101.

[18] Rubart M., Pasumarthi K.B., Nakajima H., Soonpaa M.H., Nakajima H.O., and Field L.J. Physiological coupling of donor and host cardiomyocytes after cellular transplantation. *Circ. Res.* 2003; 92: 1217–1224.

[19] Reinecke H., Zhang M., Bartosek T., and Murry C.E. Survival, integration, and differentiation of cardiomyocyte grafts: a study in normal and injured rat hearts. *Circulation* 1999; 100: 193–202.

[20] Sakai T., Li R.K., Weisel R.D., Mickle D.A., Kim E.J., Tomita S., Jia Z.Q., and Yau T.M. Autologous heart cell transplantation improves cardiac function after myocardial injury. *Ann. Thorac. Surg.* 1999; 68: 2074–2080; discussion 2080–2081.

[21] Taylor D., Atkins B., Hungspreugs P., Jones T., Reedy M., Hutcheson K., Glower D., and Kraus W. Regenerating functional myocardium: improved performance after skeletal myoblast transplantation. *Nat. Med.* 1998; 4: 929–933.

[22] Kessler P.D. and Byrne B.J. Myoblast cell grafting into heart muscle: cellular biology and potential applications. *Annu. Rev. Physiol.* 1999; 61: 219–242.

[23] Murry C., Wiseman R., Schwartz S., and Hauschka S. Skeletal myoblast transplantation for repair of myocardial necrosis. *J. Clin. Invest.* 1996; 98: 2512–2523.

[24] Li G., Ouyang Q., Petrov V.V., and Swinney H.L. Transition from simple rotating chemical spirals to meandering and traveling spirals. *Phys. Rev. Lett.* 1996; 77: 2105–2108.

[25] Kim E.J., Li R.K., Weisel R.D., Mickle D.A., Jia Z.Q., Tomita S., Sakai T., and Yau T.M. Angiogenesis by endothelial cell transplantation. *J. Thorac. Cardiovasc. Surg.* 2001; 122: 963–971.

[26] Hutcheson K.A., Atkins B.Z., Hueman M.T., Hopkins M.B., Glower D.D., and Taylor D.A. Comparison of benefits on myocardial performance of cellular cardiomyoplasty with skeletal myoblasts and fibroblasts. *Cell Transplant.* 2000; 9: 359–368.

[27] Etzion S., Barbash I.M., Feinberg M.S., Zarin P., Miller L., Guetta E., Holbova R., Kloner R.A., Kedes L.H., and Leor J. Cellular cardiomyoplasty of cardiac fibroblasts by adenoviral delivery of MyoD *ex vivo*: an unlimited source of cells for myocardial repair. *Circulation* 2002; 106: I125–I130.

[28] Klug M.G., Soonpaa M.H., Koh G.Y., and Field L.J. Genetically selected cardiomyocytes from differentiating embronic stem cells form stable intracardiac grafts. *J. Clin. Invest.* 1996; 98: 216–224.

[29] Orlic D., Kajstura J., Chimenti S., Jakoniuk I., Anderson S.M., Li B., Pickel J., McKay R., Nadal-Ginard B., Bodine D.M., Leri A., and Anversa P. Bone marrow cells regenerate infarcted myocardium. *Nature* 2001; 410: 701–705.

[30] Vulliet P.R., Greeley M., Halloran S.M., MacDonald K.A., and Kittleson M.D. Intra-coronary arterial injection of mesenchymal stromal cells and microinfarction in dogs. *Lancet* 2004; 363: 783–784.

[31] Beltrami A.P., Barlucchi L., Torella D., Baker M., Limana F., Chimenti S., Kasahara H., Rota M., Musso E., Urbanek K., Leri A., Kajstura J., Nadal-Ginard B., and Anversa P. Adult cardiac stem cells are multipotent and support myocardial regeneration. *Cell* 2003; 114: 763–776.

[32] Sakai T., Li R.K., Weisel R.D., Mickle D.A., Jia Z.Q., Tomita S., Kim E.J., and Yau T.M. Fetal cell transplantation: a comparison of three cell types. *J. Thorac. Cardiovasc. Surg.* 1999; 118: 715–724.

[33] Menasche P. Myoblast-based cell transplantation. *Heart Fail. Rev.* 2003; 8: 221–227.

[34] Menasche P., Hagege A.A., Vilquin J.T., Desnos M., Abergel E., Pouzet B., Bel A., Sarateanu S., Scorsin M., Schwartz K., Bruneval P., Benbunan M., Marolleau J.P., and Duboc D. Autologous skeletal myoblast transplantation for severe postinfarction left ventricular dysfunction. *J. Am. Coll. Cardiol.* 2003; 41: 1078–1083.

[35] Menasche P. Cellular transplantation: hurdles remaining before widespread clinical use. *Curr. Opin. Cardiol.* 2004; 19: 154–161.

[36] Reinecke H., MacDonald G.H., Hauschka S.D., and Murry C.E. Electromechanical coupling between skeletal and cardiac muscle. Implications for infarct repair. *J. Cell Biol.* 2000; 149: 731–740.

[37] Pak H.N., Qayyum M., Kim D.T., Hamabe A., Miyauchi Y., Lill M.C., Frantzen M., Takizawa K., Chen L.S., Fishbein M.C., Sharifi B.G., Chen P.S., and Makkar R. Mesenchymal stem cell injection induces cardiac nerve sprouting and increased tenascin expression in a Swine model of myocardial infarction. *J. Cardiovasc. Electrophysiol.* 2003; 14: 841–848.

[38] Hoffman J.I., Kaplan S., and Liberthson R.R. Prevalence of congenital heart disease. *Am. Heart J.* 2004; 147: 425–439.

[39] O'Dell W.G. and McCulloch A.D. Imaging three-dimensional cardiac function. *Annu. Rev. Biomed. Eng.* 2000; 2: 431–456.

[40] Shachar M. and Cohen S. Cardiac tissue engineering, *ex-vivo*: design principles in biomaterials and bioreactors. *Heart Fail. Rev.* 2003; 8: 271–276.

[41] Spach M. Anisotropy of cardiac tissue: a major determinant of conduction. *J. Cardiovasc. Electrophysiol.* 1999; 10: 887–890.

[42] Saffitz J., Kanter H., Green K., Tolley T., and Beyer E. Tissue-specific determinants of anisotropic conduction velocity in canine atrial and ventricular myocardium. *Circ. Res.* 1994; 74: 1065–1070.

[43] Arts T., Costa K.D., Covell J.W., and McCulloch A.D. Relating myocardial laminar architecture to shear strain and muscle fiber orientation. *Am. J. Physiol. Heart Circ. Physiol.* 2001; 280: H2222–H2229.

[44] Antzelevitch C., Yan G.-X., Shimizu W., and Burashnikov A. Electrical heterogeneity, the ECG, and cardiac arrhythmias. In Zipes D.P. and Jalife J., Eds. *Cardiac Electrophysiology — From Cell to Bedside.* 3rd ed. Philadelphia, PA: W. B. Saunders Co., 2000, pp. 222–238.

[45] de Bakker J.M. and OptHof T. Is the apico-basal gradient larger than the transmural gradient? *J. Cardiovasc. Pharmacol.* 2002; 39: 328–331.

[46] Wolk R., Cobbe S.M., Hicks M.N., and Kane K.A. Functional, structural, and dynamic basis of electrical heterogeneity in healthy and diseased cardiac muscle: implications for arrhythmogenesis and anti-arrhythmic drug therapy. *Pharmacol. Ther.* 1999; 84: 207–231.

[47] Spach M.S., Dolber P.C., Heidlage J.F., Kootsey J.M., and Johnson E.A. Propagating depolarization in anisotropic human and canine cardiac muscle: apparent directional differences in membrane capacitance. A simplified model for selective directional effects of modifying the sodium conductance on Vmax, tau foot, and the propagation safety factor. *Circ. Res.* 1987; 60: 206–219.

[48] Roberge F.A. and Leon L.J. Propagation of activation in cardiac muscle. *J. Electrocardiol.* 1992; 25 Suppl: 69–79.

[49] Bursac N., Parker K.K., Iravanian S., and Tung L. Cardiomyocyte cultures with controlled macroscopic anisotropy. A model for functional electrophysiological studies of cardiac muscle. *Circ. Res.* 2002; 91: e45–e54.

[50] Spach M.S. and Heidlage J.F. The stochastic nature of cardiac propagation at a microscopic level. Electrical description of myocardial architecture and its application to conduction. *Circ. Res.* 1995; 76: 366–380.

[51] Kleber A.G. and Rudy Y. Basic mechanisms of cardiac impulse propagation and associated arrhythmias. *Physiol. Rev.* 2004; 84: 431–488.

[52] Antzelevitch C. Basic mechanisms of reentrant arrhythmias. *Curr. Opin. Cardiol.* 2001; 16: 1–7.

[53] Spach M.S. and Dolber P.C. Relating extracellular potentials and their derivatives to anisotropic propagation at a microscopic level in human cardiac muscle. Evidence for electrical uncoupling of side-to-side fiber connections with increasing age. *Circ. Res.* 1986; 58: 356–371.

[54] Koura T., Hara M., Takeuchi S., Ota K., Okada Y., Miyoshi S., Watanabe A., Shiraiwa K., Mitamura H., Kodama I., and Ogawa S. Anisotropic conduction properties in canine atria analyzed by high-resolution optical mapping: preferential direction of conduction block changes from longitudinal to transverse with increasing age. *Circulation* 2002; 105: 2092–2098.

[55] McCulloch A., Bassingthwaighte J., Hunter P., and Noble D. Computational biology of the heart: from structure to function. *Prog. Biophys. Mol. Biol.* 1998; 69: 153–155.

[56] de Bakker J.M., van Capelle F.J., Janse M.J., Tasseron S., Vermeulen J.T., de Jonge N., and Lahpor J.R. Fractionated electrograms in dilated cardiomyopathy: origin and relation to abnormal conduction. *J. Am. Coll. Cardiol.* 1996; 27: 1071–1078.

[57] De Bakker J.M.T and Janse M.J. Pathophysiological correlates of ventricular tachycardia in hearts with a healed infarct. In Zipes D.P. and Jalife J., Eds. *Cardiac Electrophysiology — From Cell to Bedside.* 3rd ed. Philadelphia, PA: W.B. Saunders Co., 2000, pp. 415–422.

[58] Spach M.S. and Josephson M.E. Initiating reentry: the role of nonuniform anisotropy in small circuits. *J. Cardiovasc. Electrophysiol.* 1994; 5: 182–209.

[59] Wit A.L., Dillon S.M., and Coromilas J. Anisotropic reentry as a cause of ventricular tachyarrhythmias in myocardial infarction. In Zipes D.P. and Jalife J., Eds. *Cardiac Electrophysiology — From Cell to Bedside.* Philadelphia, PA: W.B. Saunders Co., 1995, pp. 511–526.

[60] Ravens U. Mechano-electric feedback and arrhythmias. *Prog. Biophys. Mol. Biol.* 2003; 82: 255–266.

[61] Fink C., Ergun S., Kralisch D., Remmers U., Weil J., and Eschenhagen T. Chronic stretch of engineered heart tissue induces hypertrophy and functional improvement. *FASEB J.* 2000; 14: 669–679.

[62] Zimmermann W., Fink C., Kralisch D., Remmers U., Weil J., and Eschenhagen T. Three-dimensional engineered heart tissue from neonatal rat cardiac cells. *Biotechnol. Bioeng.* 2000; 68: 106–114.

[63] Carrier R., Papadaki M., Rupnick M., Schoen F., Bursac N., Langer R., Freed L., and Vunjak-Novakovic G. Development of a model system for cardiac tissue engineering: investigation of key parameters. *Biotechnol. Bioeng.* 1999; 64: 580–589.

[64] Bursac N., Papadaki M., Cohen R., Schoen F., Eisenberg S., Carrier R., Vunjak-Novakovic G., and Freed L. Cardiac muscle tissue engineering: towards an *in vitro* model for electrophysiological studies. *Am. J. Physiol. (Heart Circ. Physiol. 46)* 1999; 277: H433–H444.

[65] Radisic M., Euloth M., Yang L., Langer R., Freed L.E., and Vunjak-Novakovic G. High-density seeding of myocyte cells for cardiac tissue engineering. *Biotechnol. Bioeng.* 2003; 82: 403–414.

[66] Dar A., Shachar M., Leor J., and Cohen S. Optimization of cardiac cell seeding and distribution in 3D porous alginate scaffolds. *Biotechnol. Bioeng.* 2002; 80: 305–312.

[67] Ozawa T., Mickle D.A., Weisel R.D., Koyama N., Ozawa S., and Li R.K. Optimal biomaterial for creation of autologous cardiac grafts. *Circulation* 2002; 106: I176–I182.

[68] Pego A.P., Poot A.A., Grijpma D.W., and Feijen J. Biodegradable elastomeric scaffolds for soft tissue engineering. *J. Control. Release* 2003; 87: 69–79.

[69] Papadaki M., Bursac N., Langer R., Merok J., Vunjak-Novakovic G., and Freed L. Tissue engineering of functional cardiac muscle: molecular, structural and electrophysiological studies. *Am. J. Physiol. (Heart Circ. Physiol.)* 2001; 280: H168–H178.

[70] Zimmermann W.H., Melnychenko I., and Eschenhagen T. Engineered heart tissue for regeneration of diseased hearts. *Biomaterials* 2004; 25: 1639–1647.

[71] Akhyari P., Fedak P.W., Weisel R.D., Lee T.Y., Verma S., Mickle D.A., and Li R.K. Mechanical stretch regimen enhances the formation of bioengineered autologous cardiac muscle grafts. *Circulation* 2002; 106: I137–I142.

[72] Zimmermann W., Schneiderbanger K., Schubert P., Didie M., Munzel F., Heubach J., Kostin S., Nehuber W., and Eschenhagen T. Tissue engineering of a differentiated cardiac muscle construct. *Circ. Res.* 2002; 90: 223–230.

[73] Kofidis T., Akhyari P., Wachsmann B., Mueller-Stahl K., Boublik J., Ruhparwar A., Mertsching H., Balsam L., Robbins R., and Haverich A. Clinically established hemostatic scaffold (tissue fleece) as biomatrix in tissue- and organ-engineering research. *Tissue Eng.* 2003; 9: 517–523.

[74] Yost M.J., Baicu C.F., Stonerock C.E., Goodwin R.L., Price R.L., Davis J.M., Evans H., Watson P.D., Gore C.M., Sweet J., Creech L., Zile M.R., and Terracio L. A novel tubular scaffold for cardiovascular tissue engineering. *Tissue Eng.* 2004; 10: 273–284.

[75] Shimizu T., Yamato M., Isoi Y., Akutsu T., Setomaru T., Abe K., Kikuchi A., Umezu M., and Okano T. Fabrication of pulsatile cardiac tissue grafts using a novel 3-dimensional cell sheet manipulation technique and temperature-responsive cell culture surfaces. *Circ. Res.* 2002; 90: e40.

[76] Bursac N., Papadaki M., White J.A., Eisenberg S.R., Vunjak-Novakovic G., and Freed L.E. Cultivation in rotating bioreactors promotes maintenance of cardiac myocyte electrophysiology and molecular properties. *Tissue Eng.* 2003; 9: 1243–1253.

[77] Akins R.E., Schroedl N.A., Gonda S.R., and Hartzell C.R. Neonatal rat heart cells cultured in simulated microgravity. *In Vitro Cell Dev. Biol.* 1997; 337–343.

[78] Akins R.E., Boyce R.A., Madonna M.L., Schroedl N.A., Gonda S.R., McLaughlin T.A., and Hartzell C.R. Cardiac organogenesis in vitro: reestablishment of three-dimensional tissue architecture by dissociated neonatal rat ventricular cells. *Tissue Eng.* 1999; 5: 103–118.

[79] Eschenhagen T., Fink C., Remmers U., Scholz H., Wattchow J., Weil J., Zimmerman W., Dohmen H., Schafer H., Bishopric N., Wakatsuki T., and Elson E. Three-dimensional reconstitution of embryonic cardimyocytes in a collagen matrix: a new heart muscle model system. *FASEB J.* 1997; 11: 683–694.

[80] Vandenburgh H., Del Tatto M, Shansky J., Lemaire J., Chang A., Payumo F., Lee P., Goodyear A., and Raven L. Tissue-engineered skeletal muscle organoids for reversible gene therapy. *Hum. Gene Ther.* 1996; 7: 2195–2200.

[81] Holubarsch C., Ruf T., Goldstein D.J., Ashton R.C., Nickl W., Pieske B., Pioch K., Ludemann J., Wiesner S., Hasenfuss G., Posival H., Just H., and Burkhoff D. Existence of the Frank–Starling mechanism in the failing human heart. Investigations on the organ, tissue, and sarcomere levels. *Circulation* 1996; 94: 683–689.

[82] Rau T., Nose M., Remmers U., Weil J., Weissmuller A., Davia K., Harding S., Peppel K., Koch W.J., and Eschenhagen T. Overexpression of wild-type Galpha(i)-2 suppresses beta-adrenergic signaling in cardiac myocytes. *FASEB J.* 2003; 17: 523–525.

[83] El-Armouche A., Rau T., Zolk O., Ditz D., Pamminger T., Zimmermann W.H., Jackel E., Harding S.E., Boknik P., Neumann J., and Eschenhagen T. Evidence for protein phosphatase inhibitor-1 playing an amplifier role in beta-adrenergic signaling in cardiac myocytes. *FASEB J.* 2003; 17: 437–439.

[84] Zolk O., Munzel F., and Eschenhagen T. Effects of chronic endothelin-1 stimulation on cardiac myocyte contractile function. *Am. J. Physiol. Heart Circ. Physiol.* 2004; 286: H1248–H1257.

[85] Zimmermann W.H., Didie M., Wasmeier G.H., Nixdorff U., Hess A., Melnychenko I., Boy O., Neuhuber W.L., Weyand M., and Eschenhagen T. Cardiac grafting of engineered heart tissue in syngenic rats. *Circulation* 2002; 106: I151–I157.

[86] Li R.K., Yau T.M., Weisel R.D., Mickle D.A., Sakai T., Choi A., and Jia Z.Q. Construction of a bioengineered cardiac graft. *J. Thorac. Cardiovasc. Surg.* 2000; 119: 368–375.

[87] Kofidis T., Akhyari P., Wachsmann B., Boublik J., Mueller-Stahl K., Leyh R., Fischer S., and Haverich A. A novel bioartificial myocardial tissue and its prospective use in cardiac surgery. *Eur. J. Cardiothorac. Surg.* 2002; 22: 238–243.

[88] Kofidis T., Lenz A., Boublik J., Akhyari P., Wachsmann B., Mueller-Stahl K., Hofmann M., and Haverich A. Pulsatile perfusion and cardiomyocyte viability in a solid three-dimensional matrix. *Biomaterials* 2003; 24: 5009–5014.

[89] van Luyn M.J., Tio R.A., Gallego y van Seijen X.J., Plantinga J.A., de Leij L.F., DeJongste M.J., and van Wachem P.B. Cardiac tissue engineering: characteristics of in unison contracting two- and three-dimensional neonatal rat ventricle cell (co)-cultures. *Biomaterials* 2002; 23: 4793–4801.

[90] Evans H.J., Sweet J.K., Price R.L., Yost M., and Goodwin R.L. Novel 3D culture system for study of cardiac myocyte development. *Am J. Physiol. Heart Circ. Physiol.* 2003; 285: H570–H578.

[91] Matthews J.A., Wnek G.E., Simpson D.G., and Bowlin G.L. Electrospinning of collagen nanofibers. *Biomacromolecules* 2002; 3: 232–238.

[92] Shimizu T., Yamato M., Akihiko K., and Okano T. Two-dimensional manipulation of cardiac myocyte sheets utilizing temperature-responsive culture dishes augments pulsatile amplitude. *Tissue Eng.* 2001; 7: 141–151.

[93] Shimizu T., Yamato M., Akutsu T., Shibata T., Isoi Y., Kikuchi A., Umezu M., and Okano T. Electrically communicating three-dimensional cardiac tissue mimic fabricated by layered cultured cardiomyocyte sheets. *J. Biomed. Mater. Res.* 2002; 60: 110–117.

[94] Carrier R.L., Rupnick M., Langer R., Schoen F.J., Freed L.E., and Vunjak-Novakovic G. Perfusion improves tissue architecture of engineered cardiac muscle. *Tissue Eng.* 2002; 8: 175–188.

[95] Carrier R.L., Rupnick M., Langer R., Schoen F.J., Freed L.E., and Vunjak-Novakovic G. Effects of oxygen on engineered cardiac muscle. *Biotechnol. Bioeng.* 2002; 78: 617–625.

[96] Radisic M., Yang L., Boublik J., Cohen R.J., Langer R., Freed L.E., and Vunjak-Novakovic G. Medium perfusion enables engineering of compact and contractile cardiac tissue. *Am. J. Physiol. Heart Circ. Physiol.* 2004; 286: H507–H516.

[97] Li R.K., Jia Z.Q., Weisel R.D., Mickle D.A., Choi A., and Yau T.M. Survival and function of bioengineered cardiac grafts. *Circulation* 1999; 100: II63–II69.

[98] Ozawa T., Mickle D.A., Weisel R.D., Koyama N., Wong H., Ozawa S., and Li R.K. Histologic changes of nonbiodegradable and biodegradable biomaterials used to repair right ventricular heart defects in rats. *J. Thorac. Cardiovasc. Surg.* 2002; 124: 1157–1164.

[99] Matsubayashi K., Fedak P.W., Mickle D.A., Weisel R.D., Ozawa T., and Li R.K. Improved left ventricular aneurysm repair with bioengineered vascular smooth muscle grafts. *Circulation* 2003; 108 Suppl 1: II219–II225.

[100] Leor J., Aboulafia-Etzion S., Dar A., Shapiro L., Barbash I., Battler A., Granot Y., and Cohen S. Bioengineered cardiac grafts: a new approach to repair the infarcted myocardium? *Circulation* 2000; 102: III56–III61.

[101] Kehat I., Kenyagin-Karsenti D., Snir M., Segev H., Amit M., Gepstein A., Livne E., Binah O., Itskovitz-Eldor J., and Gepstein L. Human embryonic stem cells can differentiate into myocytes with structural and functional properties of cardiomyocytes. *J. Clin. Invest.* 2001; 108: 407–414.

[102] Xu C., Police S., Rao N., and Carpenter M.K. Characterization and enrichment of cardiomyocytes derived from human embryonic stem cells. *Circ. Res.* 2002; 91: 501–508.

[103] Lanza R., Moore M.A., Wakayama T., Perry A.C., Shieh J.H., Hendrikx J., Leri A., Chimenti S., Monsen A., Nurzynska D., West M.D., Kajstura J., and Anversa P. Regeneration of the infarcted heart with stem cells derived by nuclear transplantation. *Circ. Res.* 2004; 94: 820–827.

[104] Grusby M.J., Auchincloss H., Jr., Lee R., Johnson R.S., Spencer J.P., Zijlstra M., Jaenisch R., Papaioannou V.E., and Glimcher L.H. Mice lacking major histocompatibility complex class I and class II molecules. *Proc. Natl Acad. Sci. USA* 1993; 90: 3913–3917.

[105] Leobon B., Garcin I., Menasche P., Vilquin J.T., Audinat E., and Charpak S. Myoblasts transplanted into rat infarcted myocardium are functionally isolated from their host. *Proc. Natl Acad. Sci. USA* 2003; 100: 7808–7811.

[106] Reinecke H., Poppa V., and Murry C.E. Skeletal muscle stem cells do not transdifferentiate into cardiomyocytes after cardiac grafting. *J. Mol. Cell Cardiol.* 2002; 34: 241–249.

[107] Suzuki K., Brand N.J., Allen S., Khan M.A., Farrell A.O., Murtuza B., Oakley R.E., and Yacoub M.H. Overexpression of connexin 43 in skeletal myoblasts: relevance to cell transplantation to the heart. *J. Thorac. Cardiovasc. Surg.* 2001; 122: 759–766.

[108] Thompson R.B., Emani S.M., Davis B.H., van den Bos E.J., Morimoto Y., Craig D., Glower D., and Taylor D.A. Comparison of intracardiac cell transplantation: autologous skeletal myoblasts versus bone marrow cells. *Circulation* 2003; 108 Suppl 1: II264–II271.

[109] Murry C.E., Soonpaa M.H., Reinecke H., Nakajima H., Nakajima H.O., Rubart M., Pasumarthi K.B., Virag J.I., Bartelmez S.H., Poppa V., Bradford G., Dowell J.D., Williams D.A., and Field L.J. Haematopoietic stem cells do not transdifferentiate into cardiac myocytes in myocardial infarcts. *Nature* 2004; 428: 664–668.

[110] Balsam L.B., Wagers A.J., Christensen J.L., Kofidis T., Weissman I.L., and Robbins R.C. Haematopoietic stem cells adopt mature haematopoietic fates in ischaemic myocardium. *Nature* 2004; 428: 668–673.

[111] Leor J., Battler A., Kloner R.A., and Etzion S. Reprogramming cells for transplantation. *Heart Fail. Rev.* 2003; 8: 285–292.

[112] Lijnen P., Petrov V. and Fagard R. *In vitro* assay of collagen gel contraction by cardiac fibroblasts in serum-free conditions. *Meth. Find. Exp. Clin. Pharmacol.* 2001; 23: 377–382.

[113] Taggart P., Sutton P.M., Opthof T., Coronel R., Trimlett R., Pugsley W., and Kallis P. Inhomogeneous transmural conduction during early ischaemia in patients with coronary artery disease. *J. Mol. Cell Cardiol.* 2000; 32: 621–630.

[114] Hunter P.J., McCulloch A.D., and ter Keurs H.E. Modelling the mechanical properties of cardiac muscle. *Prog. Biophys. Mol. Biol.* 1998; 69: 289–331.

[115] Terracio L., Miller B., and Borg T.K. Effects of cyclic mechanical stimulation of the cellular components of the heart: *in vitro*. *In Vitro Cell Dev. Biol.* 1988; 24: 53–58.

[116] Simpson D., Majeski M., Borg T., and Terracio L. Regulation of cardiac myocyte protein turnover and myofibrillar structure *in vitro* by specific directions of stretch. *Circ. Res.* 1999; 85: e59–e69.

[117] Costa K.D., Lee E.J., and Holmes J.W. Creating alignment and anisotropy in engineered heart tissue: role of boundary conditions in a model three-dimensional culture system. *Tissue Eng.* 2003; 9: 567–577.

[118] Simpson D., Terracio L., Terracio M., Price R., Turner D., and Borg T. Modulation of cardiac myocyte phenotype *in vitro* by the composition and orientation of the extracellular matrix. *J. Cell Physiol.* 1994; 161: 89–105.

[119] Bien H., Yin L., and Entcheva E. Cardiac cell networks on elastic microgrooved scaffolds. *IEEE Eng. Med. Biol. Mag.* 2003; 22: 108–112.

[120] McDevitt T., Angello J., Whitney M., Reinecke H., Hauschka S., Murry C., and Stayton P. *In vitro* generation of differentiated cardiac myofibers on micropatterned laminin surfaces. *J. Biomed. Mater. Res.* 2002; 60: 472–479.

[121] Bursac N., Loo Y., Irby M.E., Leong K., and Tung L. Polymer scaffolds for anisotropic growth of engineered cardiac tissue. In Vossoughi J., Ed. *Southern Biomedical Engineering Conference.* Washington, DC: 2002, pp. 141–142.

[122] Bursac N., Loo Y., Irby M.E., Leong K., and Tung L. Electrophysiological studies in anisotropic 3D cardiac cultures. In Pace, Ed. *NASPE.* Washington, DC: Futura, 2003, p. 1045.

[123] Pego A.P., Siebum B., Van Luyn M.J., Gallego y Van Seijen X.J., Poot A.A., Grijpma D.W., and Feijen J. Preparation of degradable porous structures based on 1,3-trimethylene carbonate and D,L-lactide (co)polymers for heart tissue engineering. *Tissue Eng.* 2003; 9: 981–994.

[124] Li W.J., Laurencin C.T., Caterson E.J., Tuan R.S., and Ko F.K. Electrospun nanofibrous structure: a novel scaffold for tissue engineering. *J. Biomed. Mater. Res.* 2002; 60: 613–621.

[125] Stoker M.E., Gerdes A.M., and May J.F. Regional differences in capillary density and myocyte size in the normal human heart. *Anat. Rec.* 1982; 202: 187–191.

[126] Xie Z., Gao M., Batra S., and Koyama T. The capillarity of left ventricular tissue of rats subjected to coronary artery occlusion. *Cardiovasc. Res.* 1997; 33: 671–676.

[127] Spach M., Dolber P., and Heidlage J. Properties of discontinuous anisotropic propagation at a microscopic level. *Ann. N Y Acad. Sci.* 1990; 591: 62–74.

[128] Peters N.S., Coromilas J., Severs N.J., and Wit A.L. Disturbed connexin43 gap junction distribution correlates with the location of reentrant circuits in the epicardial border zone of healing canine infarcts that cause ventricular tachycardia. *Circulation* 1997; 95: 988–996.

[129] Kaihara S., Borenstein J., Koka R., Lalan S., Ochoa E.R., Ravens M., Pien H., Cunningham B., and Vacanti J.P. Silicon micromachining to tissue engineer branched vascular channels for liver fabrication. *Tissue Eng.* 2000; 6: 105–117.

[130] Arutunyan A., Swift L.M., and Sarvazyan N. Initiation and propagation of ectopic waves: insights from an in vitro model of ischemia-reperfusion injury. *Am. J. Physiol. Heart Circ. Physiol.* 2002; 283: H741–H749.

[131] Huelsing D.J., Spitzer K.W., Cordeiro J.M., and Pollard A.E. Conduction between isolated rabbit Purkinje and ventricular myocytes coupled by a variable resistance. *Am. J. Physiol.* 1998; 274: H1163–H1173.

[132] Antzelevitch C., Shimizu W., Yan G.X., Sicouri S., Weissenburger J., Nesterenko V.V., Burashnikov A., Di Diego J., Saffitz J., and Thomas G.P. The M cell: its contribution to the ECG and to normal and abnormal electrical function of the heart. *J. Cardiovasc. Electrophysiol.* 1999; 10: 1124–1152.

[133] Rohr S., Kucera J., Fast V., and Kleber A. Paradoxical improvement of impulse conduction in cardiac tissue by partial cellular uncoupling. *Science* 1997; 275: 841–844.

[134] Shaw R. and Rudy Y. Ionic mechanisms of propagation in cardiac tissue. roles of the sodium and L-type calcium currents during reduced excitability and decreased gap junctional coupling. *Circ. Res.* 1997; 81: 727–741.

[135] Gaudesius G., Miragoli M., Thomas S.P., and Rohr S. Coupling of cardiac electrical activity over extended distances by fibroblasts of cardiac origin. *Circ. Res.* 2003; 93: 421–428.

[136] Rohr S. and Salzberg B. Characterization of impulse propagation at the microscopic level across geometrically defined expansions of excitable tissue: multiple site optical recording of transmembrane voltage (MSORTV) in patterned growth heart cell cultures. *J. Gen. Physiol.* 1994; 104: 287–309.

[137] Kucera J., Kleber A., and Rohr S. Slow conduction in cardiac tissue, II. Effects of branching tissue geometry. *Circ. Res.* 1998; 83: 795–805.

[138] de Bakker J.M., van Capelle F.J., Janse M.J., Tasseron S., Vermeulen J.T., de Jonge N., and Lahpor J.R. Slow conduction in the infarcted human heart. 'Zigzag' course of activation. *Circulation* 1993; 88: 915–926.

[139] Folch A. and Toner M. Microengineering of cellular interactions. *Annu. Rev. Biomed. Eng.* 2000; 2: 227–256.

[140] Khademhosseini A., Suh K.Y., Yang J.M., Eng G., Yeh J., Levenberg S., and Langer R. Layer-by-layer deposition of hyaluronic acid and poly-L-lysine for patterned cell co-cultures. *Biomaterials* 2004; 25: 3583–3592.

[141] Fast V.G. and Ideker R.E. Simultaneous optical mapping of transmembrane potential and intracellular calcium in myocyte cultures. *J. Cardiovasc. Electrophysiol.* 2000; 11: 547–556.

[142] Sperelakis N. and Haddad G. Developmental changes in membrane electrical properties of the heart. In Sperelakis N., Ed. *Physiology and the Pathophysiology of the Heart.* 3rd ed. Norwell, MA: Kluwer Academic Publishers, 1995, pp. 669–700.

[143] Athias P., Frelin C., Groz B., Dumas J., Klepping J., and Padieu P. Myocardial electophysiology: intracellular studies on heart cell cultures from newborn rats. *Path. Biol.* 1979; 27: 13–19.

[144] Joyner R. Interactions between spontaneously pacing and quiescent but excitable heart cells. *Can. J. Cardiol.* 1997; 13: 1085–1092.

[145] Wagner M., Golod D., Wilders R., Verheijck E., Joyner R., Kumar R., Jonsma H., Van Ginneken A., and Goolsby W. Modulation of propagation from an ectopic focus by electrical load and by extracellular potassium. *Am. J. Physiol. (Heart Circ. Physiol.)* 1997; 272: H1759–H1769.

[146] Antonutto G. and di Prampero P.E. Cardiovascular deconditioning in microgravity: some possible countermeasures. *Eur. J. Appl. Physiol.* 2003; 90: 283–291.

[147] Decker M.L., Simpson D.G., Behnke M., Cook M.G., and Decker R.S. Morphological analysis of contracting and quiescent adult rabbit cardiac myocytes in long-term culture. *Anat. Rec.* 1990; 227: 285–299.

[148] Clark W.A., Decker M.L., Behnke-Barclay M., Janes D.M., and Decker R.S. Cell contact as an independent factor modulating cardiac myocyte hypertrophy and survival in long-term primary culture. *J. Mol. Cell. Cardiol.* 1998; 30: 139–155.

[149] Simpson D.G., Decker M.L., Clark W.A., and Decker R.S. Contractile activity and cell–cell contact regulate myofibrillar organization in cultured cardiac myocytes. *J. Cell Biol.* 1993; 123: 323–336.

[150] Clark W.A., Rudnick S.J., LaPres J.J., Andersen L.C., and LaPointe M.C. Regulation of hypertrophy and atrophy in cultured adult heart cells. *Circ. Res.* 1993; 73: 1163–1176.

[151] Ferrick K.J., Maher M., Roth J.A., Kim S.G., and Fisher J.D. Reproducibility of electrophysiological testing during antiarrhythmic therapy for ventricular arrhythmias unrelated to coronary artery disease. *Pacing Clin. Electrophysiol.* 1995; 18: 1395–1400.

28

Tissue Engineering of Heart Valves

K. Jane Grande-Allen
Rice University

Heart valves are essential to the normal function of the heart and cardiovascular/cardiopulmonary systems. When functioning properly, the heart valves allow unrestricted, unidirectional blood flow through the heart for subsequent distribution throughout the body. Consequently, valve disease or dysfunction can result in significant harm, as the reduction in the forward flow of blood limits the oxygenation of the tissues and can induce cardiac, cardiovascular, or cardiopulmonary compensation. Valve disease is prevalent in our society, with valve replacement or repair in approximately 90,000 people in the United States in 2001 [1] (275,000 worldwide [2]). Moreover, valve disease can be either congenital or acquired. For example, approximately 9 to 14 of every 10,000 children born are affected with the **Tetralogy of Fallot** [3,4], a congenital heart disorder characterized by a narrowing of the pulmonary valve among other anomalies. Acquired valve disease can affect people of all ages and may be due to an infectious agent (rheumatic heart disease, endocarditis), systemic diseases (lupus, carcinoid syndrome), other cardiac disease, trauma, pharmacologic agents, aging-related changes, or many other causes, some of which remain unknown [5].

The majority of current treatments for heart valve disease involve elective surgical replacement of the valve with a mechanical, bioprosthetic, or cryopreserved allograft (homograft) valve. The allograft is the treatment of choice for children, because bioprosthetic valves will calcify rapidly in children and mechanical valves cannot grow with the child [6]. Aortic and pulmonary allografts have also been used very successfully in adults, with the pulmonary conduit having a 90% freedom from replacement at 20 years [7], but the vascular remnant of these allografts eventually calcifies. Unfortunately, allografts, much like other donated organs, are in scarce supply. Moreover, allografts needs to be matched to the recipient tissue type to prevent immunological rejection [7], which narrows the diminishing pool of donated organs even further. Alternative options such as mechanical or bioprosthetic heart valves can be used in many situations, and are widely available, but have their own limitations [5]. Mechanical heart valves require anticoagulation therapy, which some patients cannot tolerate. Bioprosthetic valves do not require any anticoagulation, but do not contain any living tissues, and they undergo stiffening, calcification, and structural deterioration *in vivo* as a result of their glutaraldehyde fixation during manufacturing [8]. Bioprosthetic valves demonstrate a freedom from structural deterioration of 49% at 10 years and only 32% at 15 years, [9] and eventually require another surgical replacement. Overall, there is great need for a living, unfixed tissue-engineered heart valve (TEHV) or valved conduit in adults who require valve replacements. A TEHV with the potential for growth would also provide pediatric patients with a superior alternative for the treatment of valve defects.

28.1 The Native Heart Valve as a Design Goal

28.1.1 Anatomy and Terminology

The aortic valve is one of four valves in the heart, but it is replaced most frequently, and therefore will be discussed in greater detail. The aortic valve consists of three pieces of connective tissue (the right, left, and noncoronary leaflets) that are attached to the aorta at one edge, and are free to move at the other edge. These free edges meet centrally to close the valve and keep the blood from reentering the left ventricle. The valve is located in the bulbous base of the aorta, which is known as the aortic root (an anatomic recreation for a TEHV is shown in Figure 28.1). There are several distinct anatomic regions of the valve leaflet itself (Figure 28.2). The leaflet attachment edge inserts into the aortic root wall at the crown-shaped annulus [10]. The common region where two leaflets insert into the root wall is their commissure. The leaflet belly, or body, is the main portion of the leaflet (0.4 mm thick in humans [11]), and bears the majority of the pressure load when the valve is closed [12]. The **coaptation** area is the 0.5 to 0.6 mm thick [11] region of the leaflet that is in contact with the two other leaflets when the valve is closed. The free margin is the unattached edge of the leaflet, and suspends the leaflet between the tops of the commissures, much like cables of a suspension bridge [10,13]. Finally, the central portion of the edge of the valve leaflet is the Nodule of Arantius, a thickened area (0.95 to 1.2 mm [11]) that helps maintain valve closure [14].

The pulmonary valve (the other "**semilunar**" valve) is located in the pulmonary root between the right ventricle and pulmonary artery, and is thinner and more delicate than but otherwise almost identical to the aortic valve. The semilunar valves are quite different structurally from the "**atrioventricular**" valves. The mitral valve consists of two differently shaped leaflets attached at their outer border to the junction between the left atrium and left ventricle. The free edges and ventricular surfaces of the leaflets are connected to the papillary muscles of the left ventricle by numerous chordae tendineae. Likewise, the tricuspid valve is located between the right atrium and right ventricle. The tricuspid valve also contains chordae, but has three differently shaped leaflets as opposed to two. The tricuspid leaflets and chordae are thinner, shorter, and more delicate than those in the mitral valve.

28.1.2 Microstructure and Material Behavior

The semilunar valve leaflets consist of three histologically defined layers: the ventricularis forms the lower surface, the fibrosa forms the upper surface, and the spongiosa layer lies in between [15] (Figure 28.3 and Figure 28.4). The ventricularis contains a meshwork of elastic fibers along with loosely scattered collagen fibers [15]. The predominant elastic makeup allows this layer to expand in response to tension in the closed state of the valve, and then retract when the valve opens in response to ventricular ejection [15]. The fibrosa contains collagen fibers (predominantly type I), which are aligned largely circumferentially, although radially aligned fibers are found near the root-valve annulus [15]. The collagen fiber bundles in the fibrosa serve as the main source of strength for the diastolic pressure. The ridged appearance of the fibrosa is attributed to a corrugation of that tissue layer, in addition to the collagen bundles [16]. The spongiosa is a gelatinous layer containing loose connective tissue that is rich in proteoglycans

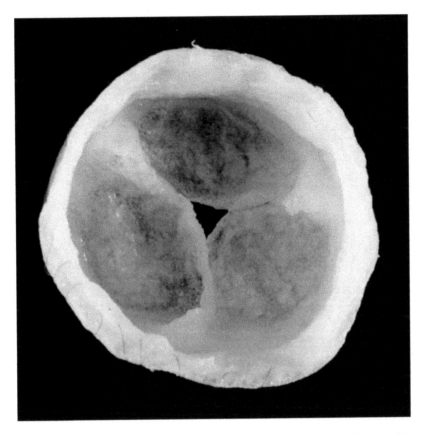

FIGURE 28.1 PGA/P4HB scaffold after cell seeding and 2 weeks of bioreactor conditioning. (Reprinted from Hoerstrup, S.P., Sodian, R., Daebritz, S., Wang, J., Bacha, E.A., Martin, D.P., Moran, A.M., Guleserian, K.J., Sperling, J. S., Kaushal, S., Vacanti, J.P., Schoen, F.J., Mayer, J.E., *Circulation*, 102, III-46, Copyright 2000, with permission from Lippincott Williams & Wilkins.)

(PGs), and serves as a mechanism for compressive resistance [17] and shear between the fibrosa and ventricularis [16].

The mitral and tricuspid valves have a similar laminated structure, except the respective outer layers are "upside down" from the arrangement shown in Figure 28.3. In these valves, the thick, heavily collagenous layer is located on the ventricular side, whereas the thin, predominantly elastic layer is found on the atrial side; these layers are also separated by a spongiosa. These similarities, which may be beneficial in future tissue engineering of atrioventricular valves, end with the chordae tendineae, which are not found in semilunar valves. The chordae are strong, thin, cable-like structures that contain a core of highly aligned collagen inside a thin outer sheath of elastic fibers and endothelial cells.

The interaction between the extracellular matrix (ECM) constituents within the valve microstructure allows distensibility, strength, elastic recovery, viscoelasticity, and an even distribution of deformation over a wide range of loading [18]. Like many other biological soft tissues, the stress–strain and load elongation curves of heart valve tissues are characterized by a low pretransition elastic modulus at initial strain (due to elastic fibers), followed by a transition zone to a higher posttransition elastic modulus at higher strains (due to the uncrimped collagen) [18]. The unique collagen and elastic fiber arrangements in the different layers, however, bestow the leaflet with anisotropic behavior (Figure 28.5). The greater circumferential stiffness (due to collagen) contributes to normal aortic valve function by restricting downward leaflet motion, while the lower radial stiffness permits the inward motion toward leaflet coaptation. This properly closing aortic valve will allow blood flow from the left ventricle into the ascending aorta, and prevent reverse flow. During this functional cycle, the leaflets interact with complex patterns of blood flow [5], and are

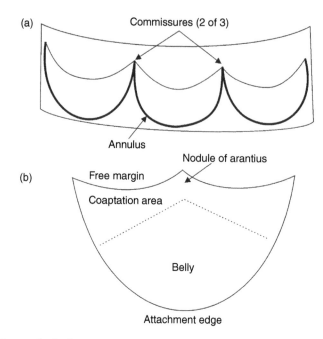

FIGURE 28.2   Semilunar valve leaflet anatomy. (a) Illustration of aortic valve leaflets within an opened aortic root. (b) Single valve leaflet.

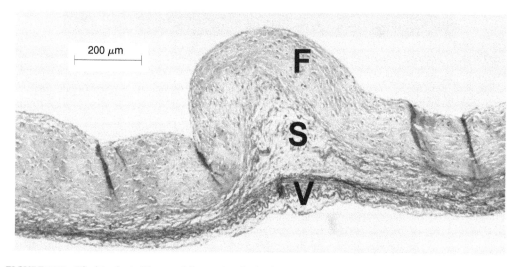

FIGURE 28.3   The histological layers of the aortic valve leaflet. Movat's Pentachrome stain. F, fibrosa; S, spongiosa; V, ventricularis.

subjected to transvalvular pressures as high as 120 mmHg [19] and shear stresses as high as 7.9 Pa [20]. These magnitudes of load are lower in the pulmonary circulation, where the transvalvular pressure across the pulmonary valve is only 25 mmHg [19].

28.1.3   Valvular Cells

Heart valves contain both endothelial and interstitial cells [21–23]. The valvular endothelial cells (VECs) populate the outer surfaces of the valves, whereas "interstitial cells" are all the cells that populate the inside

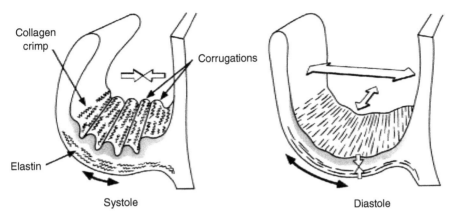

FIGURE 28.4 Illustration of the histological layers of the valve during systole and diastole. (Reprinted from Schoen, F. J., *J. Heart Valve Dis.*, 6, 2, Copyright 1997, with permission from ICR Publishers.)

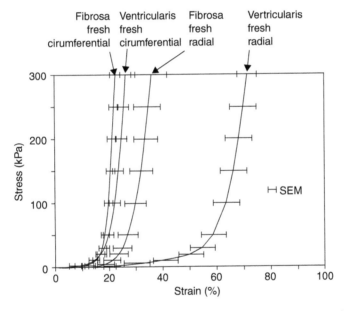

FIGURE 28.5 Radial and circumferential stress–strain curves of the ventricularis and fibrosa of aortic valve leaflets. (Reprinted from Vesely, I. and Noseworthy, R., Micromechanics of the fibrosa and the ventricularis in aortic valve leaflets, *J. Biomech.*, 25, 107, Copyright 1992, with permission from Elsevier.)

of the leaflets. Although it is presumed that the VECs provide the tissue with a nonthrombogenic surface, their functions otherwise are only beginning to be explored [23,24]. The **valvular interstitial cells** (VICs), which are only slightly better understood, are responsible for the synthesis of extracellular matrix components, including collagen, elastin, proteoglycans, and hyaluronan (for reviews see References 25 and 26). A key characteristic of VICs is that this group of cells exhibits a mixed phenotype of both fibroblastic and smooth muscle cell characteristics and are yet uniquely different from both these cell types [21,22,27,28]. It remains unclear whether this dual phenotype is caused by a single population of cells that express both features simultaneously [22,27], a single population of cells that can switch between these two phenotypes [21], or a population of several types of cells [29,30]. The different phenotypes of VICs are typically distinguished by their morphological appearance and immunohistochemical staining [21,27,28]. The fibroblastic phenotype is marked by elongated cells that contain numerous organelles for matrix synthesis, and stain for prolyl-4-hydroxlyase. The smooth muscle cell phenotype is denoted by cobblestone cells that stain for smooth muscle α-actin and stress fibers. VICs display this dual phenotype consistently throughout passaging [22,27]. Although in native valves these different cell phenotypes are slightly segregated [29,30], cells harvested from different regions of the valve had a consistent dual phenotypic appearance and growth characteristics [22]. It has also been difficult to separate these phenotypes in culture [28].

28.2   Approaches to the Tissue Engineered Heart Valve

The basic approach to constructing a TEHV, as with many other engineered tissues, is first to seed an appropriate cell type onto or within a suitable scaffold, and then to have a period of incubation during which the cells remodel or otherwise become integrated with the scaffold, and form a neotissue. This definition is intentionally ambiguous because a wide range of cells, scaffolds, and incubation environments have been used. All of the proposed TEHVs, however, have been designed to have as many of the ideal features of a heart valve substitute as possible [31,32] (i) maintain normal structural and biological function over the patient's lifetime, (ii) not elicit any inflammatory, foreign body, or immunologic responses, (iii) have antithrombotic surfaces and the potential for growth and self-repair, (iv) manufactured for each individual, (v) easy to implant with little technical variability, and (vi) available in an unlimited supply [31,32]. The majority of research of TEHVs has focused upon the structural, antithrombotic, immunologic, and availability aspects of this lofty goal. Although the studies described here do not represent an exhaustive discussion of TEHVs, several reviews provide more thorough detail [2,26,33].

28.2.1   Biodegradable Polymeric Scaffolds

Almost half of the proposed designs for TEHVs involve seeding cells on or within a polymeric biodegradable scaffold. The purpose in using a biodegradable scaffold is to anchor the seeded cells within an environment that is originally strong enough to withstand the *in vivo* mechanical forces, yet will subsequently degrade slowly, thereby transferring the function of load-bearing to the nascent ECM produced by the cells. Ideally, scaffold degradation rate and cellular synthesis rate should be balanced so that the scaffold has been completely degraded when the seeded cells have generated an amount of ECM comparable to native heart valves. The scaffold should also have initial material properties that are comparable to native valves.

The first such scaffold used to generate a TEHV was a woven mesh of 90% poly(glycolic acid) (PGA)/10% poly(lactic acid) (PLA) sandwiched between nonwoven PGA mesh, which was seeded with ovine vascular myofibroblasts and endothelial cells and used to replace the right leaflet of the pulmonary valve in a lamb model [34]. Although the resulting pulmonary valve was functional in the short term (3 weeks), this scaffold was found to be too stiff and thick for long-term use [35,36]. Conversely, seeding cells in PGA mesh alone produced a neotissue that was too delicate to handle [37], although the high porosity of this scaffold (95%) encouraged high seeding efficiencies and subsequent ECM production [35].

To avoid the mechanical limitations of the previous scaffolds, Sodian et al. [38] developed a new TEHV scaffold using poly(hydroxyalkanoate) (PHA), a thermoplastic, easily moldable polymer. The polymer was cast into a valved-conduit-shaped mold, made porous through a salt leaching process, and seeded with **autologous** ovine vascular myofibroblasts and endothelial cells. Although this valved conduit functioned normally in a sheep model for up to 17 weeks, the PHA scaffold material did not degrade fully by that time, and the developing neotissue did not contain any histologically detectable elastin or have an endothelial cell coating [38]. Moreover, PHA has a high echocardiographic density, which prevented the TEHV performance from being evaluated by Doppler echocardiography.

Because the PHA did not degrade rapidly enough, valved conduits were next assembled from nonwoven PGA mesh coated with a thin layer of poly-4-hydroxybutyrate (P4HB), a biodegradable thermoplastic moldable polymer that provided the nonwoven PGA with additional strength [39]. After seeding with

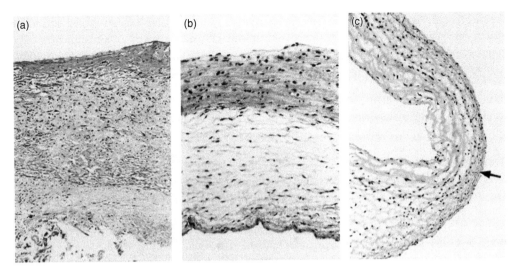

FIGURE 28.6 After 6 weeks *in vivo* (a), the TEHV from Figure 28.5 demonstrates preliminary organization (50×). After 16 weeks (b) and 20 weeks (c), the TEHV leaflet demonstrates organized, dense collagen on the outflow surface, elastic fibers on the inflow surface (arrow), and spongy organization within (both 100×). (Reprinted from Hoerstrup, S. P., Sodian, R., Daebritz, S., Wang, J., Bacha, E.A., Martin, D.P., Moran, A.M., Guleserian, K.J., Sperling, J. S., Kaushal, S., Vacanti, J.P., Schoen, F.J., Mayer, J.E., *Circulation*, 102, III-48, Copyright 2000, with permission from Lippincott Williams & Wilkins.)

autologous ovine vascular cells and 2 weeks of dynamic conditioning in a bioreactor (Figure 28.1), these TEHVs were implanted in the pulmonary position of sheep for 20 weeks. These TEHVs functioned well, with only mild to moderate pulmonary regurgitation at 16 and 20 weeks. Upon explant, the TEHV leaflets were found to have the normal three-layered structure of native heart valves (Figure 28.6), although their biochemically measured concentrations of collagen, elastin, and GAGs, as well as their ultimate tensile strength, were significantly higher than normal native pulmonary leaflets. Overall, the PGA/P4HB has been considered a very successful scaffold for TEHV development and is still being investigated [40,41]. Other biodegradable scaffolds that have been explored include biodegradable polyurethane, which was found to have not degraded completely at 6 weeks [42]. Finally, new classes of biodegradable scaffolds are being designed, such as a combination of poly(vinyl alcohol) (PVA) with brush groups of modified PLA [43]. This novel scaffold should combine the advantages of PVA, a high water content hydrogel with high elasticity and the ability to incorporate biologically active molecules on its hydroxyl groups, with the features of PLA, which is biodegradable, can be crosslinked, and is hydrophobically attractive to cells.

28.2.2 Decellularized Leaflet Scaffolds

Although polymeric biodegradable scaffolds have a long history in TEHV designs, another early approach that is still in active development today is the use of decellularized semilunar valve leaflets as a scaffold, with the rationale that they would provide the requisite strength [44], and already contain ECM in the correct microstructural arrangement [45]. Unlike polymeric designs, there is no need to fabricate or mold these scaffolds. Moreover, removing the cells would presumably eliminate the most antigenic elements, thereby avoiding any immunological response in the recipient. This reasoning has enabled the development of decellularized scaffolds from not only human heart valves (from donated allograft organs [45–47] or cadavers [48]), but also porcine heart valves. The predominant matrix element in these scaffolds, collagen, is highly conserved between species and is thus considered minimally antigenic [49,50], although there are concerns about the potential for transmission of **xenogenic** diseases [51].

A major area of research in the development of this scaffold is determining the best method to remove the cells from the original valve leaflet. Several different methods, involving ionic and nonionic detergents, solutions that are hypotonic or hypertonic, and enzyme treatments, have been attempted, abandoned, debated, and revisited, with only few direct comparisons [52]. In early studies, a treatment involving hypotonic saline to lyse the cells followed by two washes with the nonionic detergent Triton X-100 and one enzymatic soak in DNAse and RNAse was effective in the removal of all cells and cell debris [44,53]. This method preserved the majority of the thermal, physical, and material properties with the exception of a slight swelling and a slight increase in the stress relaxation of the tissue. This successful treatment was in contrast to their previous experimentation with the ionic detergent sodium dodecyl sulfate (SDS), in which the leaflet matrix swelled up to three times and had significant thermal denaturation [53]. On a single wash basis, however, SDS appears to remove more cells than Triton X-100 [54]. A combination of 0.5% trypsin and 0.2% EDTA was also successful in removing cells from human and porcine valves [48,55], as was a solution of 1% deoxycholic acid [56]. A solution of 0.1% N-cetylpyridinium chloride was shown to remove cells effectively and to preserve the tissue's microstructure and mechanics but this treatment induced calcification when the decellularized leaflet was tested in a rat subcutaneous dermal model [57]. Booth et al. [52] found that solutions of 0.03 to 0.1% SDS and 0.5% Na deoxycholate in hypotonic solutions worked best (with SDS causing a slight increase in tissue extensibility [58]), which they attribute to better protease inhibition than in the previous studies that implicated SDS in fiber damage. Although a few studies reported using protease inhibitors [48,53,55], such as phenylmethyl sulfonyl fluoride (PMSF) and EDTA, to block the endogenous lysosomal proteases released during cell lysis and to prevent degradation of the matrix scaffold, certain protease inhibitors (including PMSF) are short-lived in aqueous solutions and a more stable compound such as aprotinin may be preferable [52]. Many studies did not report any use of protease inhibitors, which could result in partial degradation of the collagen and elastic matrix components of the scaffold. The partial degradation of elastin in these scaffolds is considered particularly risky given that decellularized aortic wall, found to contain an abundance of partially degraded elastin, was prone to calcification in a rat subcutaneous dermal model [59].

Once prepared, the decellularized leaflet scaffold is almost always reseeded with autologous cells derived from the same host animal to be used in the TEHV study. Several of these scaffolds have been reseeded with endothelial cells only [56,60]. The main intent of this seeding is to form an antithrombotic coating around the bare collagen [60], and is an especially important consideration in planning for human TEHV use, because humans may have a more difficult time endothelizing structures than do the sheep models used in most of these studies [61]. Many other decellularized scaffolds, however, have also received a preliminary reseeding with vascular myofibroblasts in attempts to accelerate the eventual remodeling of the matrix [51,55,62–64]. Despite the preliminary seeding of the decellularized scaffolds, dispersing the cells within the existing matrix has proven difficult, with some tendency for the cells to remain on the surface or to merely line the largest pores [65]. In addition, Steinhoff et al. [62] found that the seeded cells tended to make new matrix on top of as opposed to within the existing scaffold matrix.

The Synergraft™ valve (Cryolife, Inc., Kennesaw, Georgia) consists of a decellularized porcine pulmonary root or composite aortic root (constructed from three noncoronary root-valve segments); decellularized human allografts are also available [46,66]. In contrast to the other approaches, the Synergraft valved conduits were not reseeded with any cells before implantation in sheep models, but became entirely repopulated with host cells and were completely functional for one year [45]. Although these scaffolds were developed with many techniques similar to those used for other TEHVs, they are not universally considered tissue-engineered structures because they are not reseeded with cells before implantation.

28.2.3 Cell Seeding

The methods used to seed the cells on and within the polymeric biodegradable scaffolds and the decellularized valve leaflets tend to be very straightforward: a concentrated solution containing the cells is dripped onto the scaffold surface and the cells disperse by gravity [67,68]. This seeding dispersion was encouraged by gentle agitation in some studies [30], but there was no report of any improved seeding efficiency due to this method. Many TEHVs have been developed by seeding first with myofibroblasts (several million cells), incubating 10 to 14 days, and then seeding with endothelial cells [34,37,38,62,69–72]. In preparation for this staged seeding process, a mixed population of vascular cells can separated into endothelial and nonendothelial cells using fluorescence-activated cell sorting (FACS)-based binding to acetylated low-density lipoprotein (positive binding in endothelial cells [73]). Staged seeding is not necessarily required [74]; mixed vascular cells seeded onto PGA/PLA–PGA sandwich scaffolds tended to segregate during incubation and *in vivo* implantation, forming neotissue with endothelial cells (staining for Factor XIII) on the outside and myofibroblasts (negative for Factor XIII) within. Other factors shown to improve seeding efficiency include 24 or 36 h seeding intervals (as opposed to 2 or 12 h intervals [68]), using polymeric scaffolds with high porosity such in PGA [67,68], mixing soluble collagen into the cell seeding solution [70], and coating the scaffold with Matrigel before seeding [75].

28.2.4 Natural Materials

Although the majority of TEHVs to date have been constructed from either biodegradable polymeric or decellularized leaflet scaffolds, there are a number of alternative scaffolds that have been developed using natural polymers such as collagen. The rationale behind using these natural materials is that the synthetic biodegradable polymers may have cytotoxic degradation products such as lactic acid, which can lower the pH of the culture medium [76]. A very early approach, reported by Carpentier et al. [77], was to inject solubilized collagen into a leaflet shaped mold. The resulting valve was implanted in sheep in the tricuspid or mitral position and functioned for up to 10 months without incompetence. Upon explant, fibroblasts were found within the collagen leaflets, but the leaflets were determined to have thickened slightly *in vivo*. More recently, valvular interstitial cells seeded within collagen sponge scaffolds were found to demonstrate phenotypic characteristics very similar to cells within intact valve leaflets [78]. Rothenburger et al. [71,75] grew human and porcine vascular and valvular cells within a freeze-dried porous type I collagen matrix *in vitro* and found that the cells produced the large hydrating proteoglycan (PG) versican, the small collagen-binding PG decorin, fibronectin, thrombospondin, and a medium-sized heparan sulfate PG (possibly perlecan or syndecan). The production of these PGs and glycoproteins indicate that the cells were interacting with and organizing the nascent ECM. Another natural material, fibrin, is being explored as a natural scaffold because fibrin gels can be made autologously from a patient's own blood and thereby prevent an immunologic reaction [67,68,76]. Fibrin gel components can also be coupled with exogenous biologically functional groups, such as growth factors, for improved neotissue formation. In addition, the use of injection molding to cast a cell-seeded fibrin gel ensures that the cells will be evenly distributed throughout the TEHV. This initial even distribution of cells is an advantage over the seeding of fibrous or sponge structures, where seeding dispersion is due to gravity and barely 50% of cells attach [67]. The disadvantage with fibrin gels is their initial weakness, which needs to be improved before *in vivo* studies can be performed. Yet another natural polymer that has been proposed is chitosan, a polysaccharide derivative that has been used for other tissue engineering applications [23]. Although three-dimensional chitosan scaffolds have yet to be tested in a TEHV, monolayer cultures of VECs adhered better to chitosan surfaces — and chitosan/collagen IV combinations in particular — than to PHA surfaces.

28.2.5 Building Block Approach

Almost without exception, the approach to TEHV development has been to use scaffolds that are destined for degradation or remodeling by the cells that will populate these constructs. The exception to this paradigm is the approach by Vesely et al. [79], in which the different ECM structural components are derived and then assembled into an approximation of the native aortic valve microstructure *in vitro*. The collagen bundles that provide the leaflet strength are replicated by neonatal rat aortic smooth muscle cells (NRASMCs) seeded within a type I collagen gel and anchored to promote uniaxial or branched contraction. The lubricating glycosaminoglycan component will be represented by cross-linked hyaluronan (also seeded with NRASMCs), and the network and sheets of elastic fibers are synthesized by the NRASMCs atop the crosslinked hyaluronan and around the collagen bundles. The final valve structure will be assembled by stacking alternating layers of hyaluronan/elastic sheets with layers of collagen bundles/elastic sheaths that are oriented as they would be in the native aortic valve (predominantly circumferential). This novel approach is still in the *in vitro* developmental stages, but it is envisaged that the multiple layers would compact, and the embedded cells would synthesize additional ECM, during a period of *in vitro* dynamic conditioning.

28.2.6 Cell Origin

It is generally agreed that the source of cells for a TEHV should be autologously derived from the intended valve recipient. This is particularly feasible given that most valve surgeries are elective, with criteria for the decisions of surgical timing being continuously reevaluated [5,80]. Autologous cells may perform best in pediatric cases, given that adult cells may have diminished proliferative ability (cell sources are discussed extensively in Reference 2). The anatomic source of these autologous cells, however, has not been firmly established. Shinoka et al. [74] determined that arterial myofibroblasts were superior to dermal fibroblasts in generating a strong, collagen-rich, and organized neotissue; Schnell et al. [42] found that venous cells were better yet. Given that saphenous veins are more easily harvested than arteries, the use of venous cells was therefore considered a promising alternative to arterial cells for seeding a TEHV scaffold. Although many studies continue to explore the use of cells derived from vascular sources, there were two recent reports that used either the stromal cells [40] or mesenchymal stem cells [81] from bone marrow. In the first case, human bone marrow stromal cells (obtained from the sternum) were seeded on a PGA/P4HB valved conduit, then conditioned in a bioreactor. After 14 days, the cells showed phenotypic characteristics of myofibroblasts (smooth muscle α-actin and vimentin), but did not show markers for muscle, osteogenic, or endothelial differentiation. In the second case, ovine bone marrow mesenchymal stem cells were seeded on the same type of valved conduit, then conditioned in bioreactor. Although the differentiation characteristics such as cell phenotype were not reported, the cells were able to form a functional neotissue *in vitro*.

Very few research groups have actually attempted to use VECs and VICs in the design of TEHVs [30,75,78]. It appears to have been assumed, if not often explicitly postulated, that whatever cell type is seeded within the TEHV will soon differentiate and begin to express the phenotypic characteristics of valvular cells. With rare exception [82], this assumption has not been tested in any depth, even though recent gene expression work has shown that VECs have many transcriptional differences from vascular endothelial cells [24]. VECs were also four times more proliferative in culture than were vascular endothelial cells [24], which may explain why vascular endothelial cells have frequently failed to form a continuous endothelial coating of the TEHV scaffolds *in vitro* [71] or *in vivo* [38,44,63]. VICs seeded in decellularized porcine leaflets differentiated and segregated into regionally specific myofibroblasts, fibroblasts, endothelial cells, and smooth muscle cells after 15 days *in vitro* [30]. After seeding in porous collagen sponge-like structures, VICs demonstrated phenotypic similarity to cells in native valve leaflets [78] and produced many of the PGs normally found in valves [75] A potential source for autologous valvular cells was proposed by Maish et al. [83], who found that valvular cells could be harvested from an ovine tricuspid valve without compromising the function of the tricuspid valve in the donor animal. This option may be important if the restoration of the VIC and VEC phenotype proves necessary and cannot be accomplished using an alternative cell source.

28.2.7 Bioreactors for Conditioning and Proof of Concept

Bioreactors have been widely used to provide the developing TEHV with strength-building mechanical stimulation prior to implantation within the animal model. Compared with static controls, bioreactor-conditioned TEHVs have demonstrated greater production of collagen, greater cell densities, improved mechanical strength, and more compact matrix organization [39,40,55]. It has also been proposed that bioreactor conditioning maintains the synthesis of ECM and DNA at levels normally found in the valve leaflets [84]. Bioreactors can also accelerate the turnover of the biodegradable matrix. Incubation in

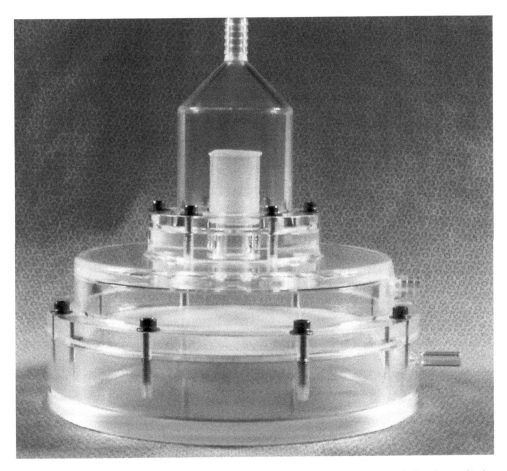

FIGURE 28.7 This bioreactor design creates pulsatile flow by cyclically pumping air beneath the silicone diaphragm separating the medium and air chambers. This action draws medium from a reservoir and propels it through the TEHV fixed to the silicone tubing above. (Reprinted from Hoerstrup, S.P., Sodian R., Sperling, J.S., Vacanti, J.P., Mayer J. E., *Tissue Eng.*, 6, 1, Copyright 2000, with permission from Mary Ann Liebert, Inc.)

a dynamic flexure bioreactor also augmented the bulk degradation of PGA/P4HB and PGA/PLA/P4HB scaffolds, resulting in significantly less flexural stiffness at 3 weeks compared to unflexed controls [41]. The types of bioreactors developed for TEHV studies can be generally classified into one of two categories. The first category, which has a continuously circulating supply of tissue culture medium, is predominantly used for the conditioning of intact TEHV and valved conduits. These bioreactors generally consist of a medium reservoir, a chamber to hold the TEHV, a pressure source, ports for gas exchange, optional pressure and flow transducers, and often have components to mimic vascular compliance and resistance [85,86] (Figure 28.7 and Figure 28.8). A similar setup was used in a study of the synthesis of collagen, GAGs, and DNA by valvular cells [84]. Because these bioreactor systems are capable of producing pulsatile flows ranging from 50 to 2000 ml/min [85] or even 20 l/min [84], and transvalvular pressures of 10 to 240 mmHg, they can easily be tailored to the conditioning or testing of aortic or pulmonary TEHVs. Feedback-driven tensioning systems have also been incorporated into these bioreactors [87]. The second category of bioreactor tends to be smaller and scalable, and is used more for proof of concept studies that examine how mechanical stimulation can improve the TEHV characteristics. An example of this was developed by Engelmayr et al. [41], in which scaffolds identical to those used in certain TEHVs were subjected to flexural conditioning, but were not bathed in circulating medium.

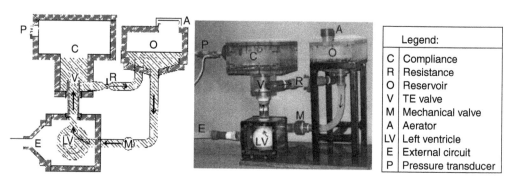

FIGURE 28.8 A bioreactor flow loop that includes components for variable compliance, resistance, and pressure measurements. (Reprinted from Dumont, K., Yperman, J., Verbeken, E., Segers, P., Mauris, B., Vandenberghe, S., Flameng, W., Verdonck, P.R., *Artif. Organs*, 26, 711, Copyright 2002, with permission from Blackwell Publishing.)

28.2.8 *In Vivo* Testing in Animal Models

Although many potential TEHV designs have been successfully compared using *in vitro* and bioreactor analyses, evaluation of the inflammatory, immunologic, and calcific responses requires an *in vivo* model. Canines have been used at least twice [44,88], but the majority of TEHVs have been tested *in vivo* in the widely accepted ovine models (either lamb or mature sheep) [33]. TEHVs have been implanted into the atrioventricular position in the past [77], and a **heterotopic** implantation in the abdominal aorta was recently reported [88], but most are tested in the pulmonary position, either as an **orthotopic** replacement of the pulmonary valve [38,39,45,51,56,72] or heterotopically interposed between the main and left pulmonary artery [44]. Single [34,69] and double [36] leaflet replacements within the pulmonary valve have been attempted with more success in the single leaflet replacement. The double leaflet replacement involved two identical PGA/PLA–PGA sandwich scaffolds: one seeded TEHV leaflet and one unseeded control. Unfortunately, this otherwise ingenious approach did not work because the stiff, thick scaffolds were unable to coapt normally and cause pulmonary insufficiency [36]. More recent TEHV designs are often tested using a valved conduit, containing all three leaflets, in the orthotopic pulmonary position.

These *in vivo* studies have often demonstrated an early but very mild inflammatory response to the TEHV [47], although this response was stronger when **allogeneic** as opposed to autologous cells were seeded [34]. A late inflammatory response (9 to 12 weeks) was reported with a reseeded decellularized porcine TEHV [63]. The remnant muscular shelf of decellularized leaflets has also been shown to invoke a mild inflammatory and calcific response [62]. Otherwise, reports of calcification of the scaffolds have been with mixed; there was heavy calcification of reseeded decellularized scaffolds after 12 and 24 weeks *in vivo* [64], while a Synergraft decellularized porcine scaffold (not reseeded) showed no calcification after 21 weeks [66].

28.2.9 Clinical Experience

The majority of TEHVs have only been tested *in vitro* or in animal models; no TEHV has been approved for clinical use in the United States. The porcine and human Synergrafts, which are technically not TEHVs since they are not reseeded with cells, have been approved in the United States [46]. As of 2003, these decellularized human valved conduits have been implanted in more than 150 patients in the United States and more than 1000 patients worldwide, approximately 84% of which have been in the pulmonary position (16% aortic). The distribution of patient age in the United States has been approximately 1/3 children and 2/3 adults, with a mean follow-up time of 231 days (range 0 to 788 days). In the two cases where these pulmonary allograft conduits were explanted (due to external fibrosis at 14 months and unrelated infection at 3 months), the leaflets and conduit had been repopulated with fibroblasts and myofibroblasts and demonstrated an endothelial lining and normal leaflet architecture. Function of the pulmonary Synergraft allografts have been comparable to cryopreserved allografts [46,61], whereas the aortic Synergraft allografts have evoked humoral antibodies in 25% of patients, but have shown normal aortic valve function with only trivial regurgitation [46]. The porcine pulmonary Synergrafts have been implanted in 20 patients [46], with 3 deaths and conduit pseudoaneurysm formation in 3 patients, but 14 other patients have had satisfactory valve performance for a maximum of one year. Unfortunately, the porcine Synergraft has not been as successful in the pediatric population, with a massive inflammatory response leading to leaflet degeneration and rupture in at least seven children in Austria and Norway [61]. One Synergraft that was examined after being implanted for one year showed no host recellularization of the implant at all and the formation of a fibrous capsule around the porcine tissue, which was very different from the results in a sheep model [45], in which the entire leaflet was repopulated with cells after one year.

In Germany, decellularized human cryopreserved pulmonary allografts that were reseeded with autologous venous endothelial cells were implanted in at least 6 **Ross procedure** patients with a mean age of 44 years [60]. All patients demonstrated an improvement of **ejection fraction** at 3 months. One patient was reported to be excellent condition after 12 months, and had not demonstrated any fever during this time. This absence of fever was interpreted as an absence of any antigenic inflammatory or immunologic response.

28.3 Conclusions and Future Challenges

There are many challenges ahead for the future of TEHVs. The first challenge appears to be the recapitulation of the many aspects of valvular biology in a living TEHV, particularly in designs involving nonvalvular cells. Quite frankly, it remains to be determined if such recapitulation would even be necessary for TEHV function, even though that finding could help direct future TEHV research. Moreover, the nascent field of valvular biology lags far behind vascular biology. Regardless, future TEHV research may wish to build upon the many studies over the last several years that have examined the contractile, synthetic, signaling, interactive, and diffusion characteristics of heart valves at the tissue, cellular, and molecular level. Normal aortic valve leaflets and VICs, for example, will contract slowly in response to stimulation with vasoactive agents such as serotonin, epinephrine, and endothelin-1 [28,89], potentially to maintain a baseline resting tension in the normal aortic valve [89]. More recently, the proliferative, migratory, synthetic, signaling, and phenotypic transdifferentiation responses of VICs to members of the renin–angiotensin system (including TGF-β) and vasoactive agents have been assessed [90–92]. Furthermore, the receptors and/or mRNA for serotonin, angiotensin II, bradykinin, and angiotensin-converting enzyme have been identified in VICs [92,93]. In fact, many of the transcriptional and proliferative characteristics of valvular cells have been shown to be different from those expressed by vascular cells [24,93]. Cell–matrix interactions have been demonstrated at the molecular level by the characterization of several integrins in bovine and baboon VICs [94]. Finally, oxygen diffusion within young porcine aortic valves was recently measured [95] and it was observed that blood vessels were present in most regions of the valve thicker than 0.5 mm. This finding highlights the differences between pediatric heart valves, which are vascularized [96,97] and even innervated [98], and adult heart valves, which have far less nerves and vessels, but contain viable, synthetic cells nonetheless. Exactly how well oxygen and nutrients diffuse to these adult VICs, and likewise to the cells within a TEHV, is a fertile topic for further research.

Another challenge for the future is the development of TEHV replacements for the aortic, atrioventricular, and venous valves. The highest clinical need is for the aortic position. To meet the higher pressure and flow requirements of the systemic circulation, the current pulmonary TEHV designs will need to be adapted to produce a stronger neotissue. In a recent proof of concept study that explored this need, a PGA/P4HB scaffold seeded with human venous myofibroblasts became more than three times stiffer when the maximum cyclic strain was raised from 7 to 10% during bioreactor conditioning [99]. Given that the atrioventricular valves are more commonly repaired than replaced [5], recent TEHV work in that regard has focused on replacement components, as opposed to the entire valve. The chordae, in particular, frequently rupture in prolapsed mitral valves [80]; torn chordae are currently replaced with Gore-Tex™ sutures. Because these sutures are several times stiffer than normal chordae [100], Vesely et al. [79] have been developing tissue-engineered chordae (large collagen fiber bundles) by seeding NRASMCs in a type 1 collagen gel and allowing the cell-seeded gel to contract uniaxially. Although these tissue-engineered chordae have not yet been tested *in vivo*, and are an order of magnitude less stiff than normal native human chordae, they have approximately normal extensibility [79]. As this avenue of TEHV research is very young, these preliminary results are likely to improve substantially in the years to come. Tissue engineered venous valves have been developed in a sheep model using allogeneic decellularized jugular veins seeded with autologous venous myofibroblasts and endothelial cells; these valves were still patent at 12 weeks [101].

The final challenge, particularly given the clinical reports of failure of the Synergraft valves in pediatric patients [61], are the ethical considerations involved as these research findings are eventually translated to clinical practice [102]. Although there are no current governmental regulatory divisions that are focusing exclusively on engineered tissues, U.S. and European regulatory agencies are planning centers that will meet this need. Beyond that goal lies the problem of developing standards to which engineered tissues such as the TEHV would be held. Furthermore, we are reminded by Sutherland and Mayer in their excellent review of this topic, "compliance with standards alone carries no guarantee that a given device will be free from risk" [102]. Applying a risk management approach, with identification and severity grading of potential hazards, is an appropriate first step along this path. Hazards particularly associated with the current characteristics of the TEHV described in this chapter could be scaffold-related, such as adverse reactions to the degradation products of the biodegradable scaffolds, or the potential for transmission of porcine endogenous retrovirus from the decellularized porcine valve scaffold to the TEHV recipient [51,102]. Other hazards could be related to the cells involved, particularly if they are not autologous, or to the performance of the TEHV *in vivo*.

The field of TEHV research is growing exponentially and has demonstrated many recent advances, particularly in the development and application of bioreactor technologies. New research in valvular biology and targeted scaffold development, together with frequent reflection and review [2,26,33], are certain to drive this field ahead. Although there are many hurdles to overcome before the many promises of TEHVs become fully realized, the potential to create a living, healing, growing valve continues to inspire significant effort in this exciting field.

Defining Terms

Allogeneic: Taken from a different individual of the same species
Atrioventricular: Referring to the position between the atrium and the ventricle. The mitral and tricuspid valves are atrioventricular valves.
Autologous: Taken or derived from the same individual.
Coaptation: Referring to contact between different heart valve leaflets, or regions of the leaflets where this contact occurs when the valve is closed.
Ejection fraction: The percentage of blood pumped out of an individual's left ventricle.
Heterotopic: Referring to a region of the body where a structure is not normally located.
Incompetence and insufficiency: Inability of the heart valve to close property.
Orthotopic: Referring to a region of the body where a structure is normallt located.
Ross Procedure: Surgical replacement of a patient's deficient aortic valve with a pulmonary autograft, followed by implantation of a cryopreserved pulmonary allograft in the pulmonary position.
Semilunar: Referring to the half-moon shape of the leaflets of the pulmonary and aortic valves.
Tetralogy of Fallot: A congenital heart defect with four predominant symptoms (ventricular septal defect, pulmonary valve stenosis, right ventricular hypertrophy, and malposition of the aorta over both ventricles.

Valvular interstitial cells: An all-encompassing term for the nonendothelial cells that are contained within the heart valve leaflets. This definition includes cells of variable phenotype.

Xenogenic: Taken from a different species.

References

[1] American Heart Association, *Heart Disease and Stroke Statistics — 2004 Update*, American Heart Association, Dallas, 2004.

[2] Rabkin, E. and Schoen, F.J., Cardiovascular tissue engineering, *Cardiovasc. Pathol.* 11, 305, 2002.

[3] American Heart Association, *Youth and Cardiovascular Diseases — Statistical Fact Sheet*, American Heart Association, Dallas, 2004.

[4] Topol, E.J., *Cleveland Clinic Heart Book*, Hyperion, New York, 2000.

[5] Otto, C.M., *Valvuar Heart Disease*, 2nd ed., Saunders, Philadephia, 2004.

[6] Mayer, J.E. Jr., In search of the ideal valve replacement device, *J. Thorac. Cardiovasc. Surg.* 122, 8, 2001.

[7] Ross, D.N., Options for right ventricular outflow tract reconstruction, *J. Cardiovasc. Surg.* 13, 186, 1998.

[8] Vesely, I., Barber, J.E., and Ratliff, N.B., Tissue damage and calcification may be independent mechanisms of bioprosthetic heart valve failure, *J. Heart Valve Dis.* 10, 471, 2001.

[9] Fann, J.I. et al., Twenty-year clinical experience with porcine bioprostheses, *Ann. Thorac. Surg.* 62, 1301, 1996.

[10] Mercer, L., Benedicty, M., and Bahnson, H., The geometry and construction of the aortic leaflet, *J. Thorac. Cardiovasc. Surg.* 65, 511, 1973.

[11] Sahasakul, Y. et al., Age-related changes in aortic and mitral valve thickness: implications for two-dimensional echocardiography based on an autopsy study of 200 normal human hearts, *Am. J. Cardiol.* 62, 424, 1988.

[12] Christie, G.W. and Stephenson, R.A., Modelling the mechanical role of the fibrosa and the ventricularis in the porcine bioprosthesis, in *Surgery for Heart Valve Disease*, Bodnar, E., Ed., Heart Valve Diseases, London, 1989, p. 815.

[13] Sutton, J.P. 3rd, Ho, S.Y., and Anderson, R.H., The forgotten interleaflet triangles: a review of the surgical anatomy of the aortic valve, *Ann. Thorac. Surg.* 59, 419, 1995.

[14] Clark, R.E. and Finke, E.H., Scanning and light microscopy of human aortic leaflets in stressed and relaxed states, *J. Thorac. Cardiovasc. Surg.* 67, 792, 1974.

[15] Sauren, A.A. et al., Aortic valve histology and its relation with mechanics — preliminary report, *J. Biomech.* 13, 97, 1980.

[16] Vesely, I. and Noseworthy, R., Micromechanics of the fibrosa and the ventricularis in aortic valve leaflets, *J. Biomech.* 25, 101, 1992.

[17] Grande-Allen, K.J. et al., Glycosaminoglycans and proteoglycans in normal mitral valve leaflets and chordae: association with regions of tensile and compressive loading, *Glycobiology* 14, 621, 2004.

[18] Sauren, A.A. et al., The mechanical properties of porcine aortic valve tissues, *J. Biomech.* 16, 327, 1983.

[19] Patton, H.D. et al., *Textbook of Physiology. Circulation, Respiration, Body Fluids, Metabolism, and Endocrinology*, W.B. Saunders, Philadelphia, 1989.

[20] Weston, M.W., LaBorde, D.V., and Yoganathan, A.P., Estimation of the shear stress on the surface of an aortic valve leaflet, *Ann. Biomed. Eng.* 27, 572, 1999.

[21] Taylor, P., Allen, S., and Yacoub, M., Phenotypic and functional characterization of interstitial cells from human heart valves, pericardium and skin, *J. Heart Valve Dis.* 9, 150, 2000.

[22] Johnson, C.M., Hanson, M.N., and Helgeson, S.C., Porcine cardiac valvular subendothelial cells in culture: cell isolation and growth characteristics, *J. Mol. Cell. Cardiol.* 19, 1185, 1987.

[23] Cuy, J.L. et al., Adhesive protein interactions with chitosan: consequences for valve endothelial cell growth on tissue-engineering materials, *J. Biomed. Mater. Res.* 67A, 538, 2003.

[24] Farivar, R.S. et al., Transcriptional profiling and growth kinetics of endothelium reveals differences between cells derived from porcine aorta versus aortic valve, *Eur. J. Cardiothorac. Surg.* 24, 527, 2003.

[25] Taylor, P.M. et al., The cardiac valve interstitial cell, *Int. J. Biochem. Cell. Biol.* 35, 113, 2003.

[26] Flanagan, T.C. and Pandit, A., Living artificial heart valve alternatives: a review, *Eur. Cell. Mater.* 6, 28, 2003.

[27] Lester, W.M. and Gotlieb, A.I., *In vitro* repair of the wounded porcine mitral valve, *Circ. Res.* 62, 833, 1988.

[28] Messier, R.H. et al., Dual structural and functional phenotypes of the porcine aortic valve interstitial population: characteristics of the leaflet myofibroblast, *J. Surg. Res.* 57, 1, 1994.

[29] Della Rocca, F. et al., Cell comparison of the human pulmonary valve: a comparison study with the aortic valve — the VESALIO project, *Ann. Thorac. Surg.* 70, 1594, 2000.

[30] Bertipaglia, B. et al., Cell characterization of porcine aortic valve and decellularized leaflets repopulated with aortic valve interstitial cells: the VESALIO project (Vitalitate Exornatum Succedaneum Aorticum Labore Ingenioso Obtenibitur), *Ann. Thorac. Surg.* 75, 1274, 2003.

[31] Cohn, L.H. and Lipson, W., Selection and complications of cardiac valvular prostheses, in *Glenn's Thoracic and Cardiovascular Surgery*, Baue, A.E. et al., Eds., Appleton & Lange, Stamford, CT, 1996, p. 2043.

[32] Harken, D.E. and Curtis, L.E., Heart surgery — legend and a long look, *Am. J. Cardiol.* 19, 393, 1967.

[33] Stock, U.A. et al., Tissue engineering of heart valves — current aspects, *Thorac. Cardiovasc. Surg.* 50, 184, 2002.

[34] Shinoka, T. et al., Tissue engineering heart valves: valve leaflet replacement study in a lamb model, *Ann. Thorac. Surg.* 60, S513, 1995.

[35] Sodian, R. et al., Evaluation of biodegradable, three-dimensional matrices for tissue engineering of heart valves, *ASAIO J.* 46, 107, 2000.

[36] Kim, W.G. et al., Tissue-engineered heart valve leaflets: an animal study, *Int. J. Artif. Organs* 24, 642, 2001.

[37] Zund, G. et al., The *in vitro* construction of a tissue engineered bioprosthetic heart valve, *Eur. J. Cardiothorac. Surg.* 11, 493, 1997.

[38] Sodian, R. et al., Early *in vivo* experience with tissue-engineered trileaflet heart valves, *Circulation* 102, III22, 2000.

[39] Hoerstrup, S.P. et al., Functional living trileaflet heart valves grown *in vitro*, *Circulation* 102, III44, 2000.

[40] Hoerstrup, S.P. et al., Tissue engineering of functional trileaflet heart valves from human marrow stromal cells, *Circulation* 106, I143, 2002.

[41] Engelmayr, G.C. Jr. et al., A novel bioreactor for the dynamic flexural stimulation of tissue engineered heart valve biomaterials, *Biomaterials* 24, 2523, 2003.

[42] Schnell, A.M. et al., Optimal cell source for cardiovascular tissue engineering: venous vs. aortic human myofibroblasts, *Thorac. Cardiovasc. Surg.* 49, 221, 2001.

[43] Nuttelman, C.R., Henry, S.M., and Anseth, K.S., Synthesis and characterization of photocross-linkable, degradable poly(vinyl alcohol)-based tissue engineering scaffolds, *Biomaterials* 23, 3617, 2002.

[44] Wilson, G.J. et al., Acellular matrix: a biomaterials approach for coronary artery bypass and heart valve replacement, *Ann. Thorac. Surg.* 60, S353, 1995.

[45] Goldstein, S. et al., Transpecies heart valve transplant: advanced studies of a bioengineered xeno-autograft, *Ann. Thorac. Surg.* 70, 1962, 2000.

[46] Elkins, R.C., Is tissue-engineered heart valve replacement clinically applicable? *Curr. Cardiol. Rep.* 5, 125, 2003.

[47] Elkins, R.C. et al., Decellularized human valve allografts, *Ann. Thorac. Surg.* 71, S428, 2001.

[48] Cebotari, S. et al., Construction of autologous human heart valves based on an acellular allograft matrix, *Circulation* 106, I63, 2002.

[49] Yannas, I.V., Natural materials, in *Biomaterials Science. An Introduction to Materials in Medicine*, Ratner, B.D. et al., Eds., Academic Press, San Diego, 1996, p. 84.

[50] Miller, E.J., Chemistry of the collagens and their distribution, in *Extracellular Matrix Biochemistry*, Piez, K.A. and Reddi, A.H., Eds., Elsevier, New York, 1984, p. 41.

[51] Leyh, R.G. et al., Acellularized porcine heart valve scaffolds for heart valve tissue engineering and the risk of cross-species transmission of porcine endogenous retrovirus, *J. Thorac. Cardiovasc. Surg.* 126, 1000, 2003.

[52] Booth, C. et al., Tissue engineering of cardiac valve prostheses I: development and histological characterization of an acellular porcine scaffold, *J. Heart Valve Dis.* 11, 457, 2002.

[53] Courtman, D.W. et al., Development of a pericardial acellular matrix biomaterial: biochemical and mechanical effects of cell extraction, *J. Biomed. Mater. Res.* 28, 655, 1994.

[54] Kim, W.G., Park, J.K., and Lee, W.Y., Tissue-engineered heart valve leaflets: an effective method of obtaining acellularized valve xenografts, *Int. J. Artif. Organs* 25, 791, 2002.

[55] Schenke-Layland, K. et al., Complete dynamic repopulation of decellularized heart valves by application of defined physical signals — an *in vitro* study, *Cardiovasc. Res.* 60, 497, 2003.

[56] Dohmen, P.M. et al., Ross operation with a tissue-engineered heart valve, *Ann. Thorac. Surg.* 74, 1438, 2002.

[57] Spina, M. et al., Isolation of intact aortic valve scaffolds for heart-valve bioprostheses: extracellular matrix structure, prevention from calcification, and cell repopulation features, *J. Biomed. Mater. Res.* 67A, 1338, 2003.

[58] Korossis, S.A., Fisher, J., and Ingham, E., Cardiac valve replacement: a bioengineering approach, *Biomed. Mater. Eng.* 10, 83, 2000.

[59] Bailey, M.T. et al., Role of elastin in pathologic calcification of xenograft heart valves, *J. Biomed. Mater. Res.* 66A, 93, 2003.

[60] Dohmen, P.M. et al., Tissue engineering of an auto-xenograft pulmonary heart valve, *Asian Cardiovasc. Thorac. Ann.* 10, 25, 2002.

[61] Simon, P. et al., Early failure of the tissue engineered porcine heart valve SYNERGRAFT in pediatric patients, *Eur. J. Cardiothorac. Surg.* 23, 1002, 2003.

[62] Steinhoff, G. et al., Tissue engineering of pulmonary heart valves on allogenic acellular matrix conduits: *in vivo* restoration of valve tissue, *Circulation* 102, III50, 2000.

[63] Wilhelmi, M.H. et al., Role of inflammation and ischemia after implantation of xenogeneic pulmonary valve conduits: histological evaluation after 6 to 12 months in sheep, *Int. J. Artif. Organs* 26, 411, 2003.

[64] Leyh, R.G. et al., *In vivo* repopulation of xenogeneic and allogeneic acellular valve matrix conduits in the pulmonary circulation, *Ann. Thorac. Surg.* 75, 1457, 2003.

[65] Curtil, A., Pegg, D.E., and Wilson, A., Repopulation of freeze-dried porcine valves with human fibroblasts and endothelial cells, *J. Heart Valve Dis.* 6, 296, 1997.

[66] O'Brien, M.F. et al., The SynerGraft valve: a new acellular (nonglutaraldehyde-fixed) tissue heart valve for autologous recellularization first experimental studies before clinical implantation, *Semin. Thorac. Cardiovasc. Surg.* 11, 194, 1999.

[67] Ye, Q. et al., Scaffold precoating with human autologous extracellular matrix for improved cell attachment in cardiovascular tissue engineering, *ASAIO J.* 46, 730, 2000.

[68] Zund, G. et al., Tissue engineering in cardiovascular surgery: MTT, a rapid and reliable quantitative method to assess the optimal human cell seeding on polymeric meshes, *Eur. J. Cardiothorac. Surg.* 15, 519, 1999.

[69] Shinoka, T. et al., Tissue-engineered heart valves. Autologous valve leaflet replacement study in a lamb model, *Circulation* 94, II164, 1996.

[70] Kim, W.G. et al., Tissue-engineered heart valve leaflets: an effective method for seeding autologous cells on scaffolds, *Int. J. Artif. Organs* 23, 624, 2000.

[71] Rothenburger, M. et al., *In vitro* modelling of tissue using isolated vascular cells on a synthetic collagen matrix as a substitute for heart valves, *Thorac. Cardiovasc. Surg.* 49, 204, 2001.

[72] Stock, U.A. et al., Tissue-engineered valved conduits in the pulmonary circulation, *J. Thorac. Cardiovasc. Surg.* 119, 732, 2000.

[73] Hoerstrup, S.P. et al., Fluorescence activated cell sorting: a reliable method in tissue engineering of a bioprosthetic heart valve, *Ann. Thorac. Surg.* 66, 1653, 1998.

[74] Shinoka, T. et al., Tissue-engineered heart valve leaflets: does cell origin affect outcome? *Circulation* 96, II, 1997.

[75] Rothenburger, M. et al., Tissue engineering of heart valves: formation of a three-dimensional tissue using porcine heart valve cells, *ASAIO J.* 48, 586, 2002.

[76] Jockenhoevel, S. et al., Fibrin gel — advantages of a new scaffold in cardiovascular tissue engineering, *Eur. J. Cardiothorac. Surg.* 19, 424, 2001.

[77] Carpentier, A. et al., Collagen-derived heart valves: concept and experimental results, *J. Thorac. Cardiovasc. Surg.* 62, 707, 1971.

[78] Taylor, P.M. et al., Human cardiac valve interstitial cells in collagen sponge: a biological three-dimensional matrix for tissue engineering, *J. Heart Valve Dis.* 11, 298, 2002.

[79] Shi, Y., Ramamurthi, A., and Vesely, I., Towards tissue engineering of a composite aortic valve, *Biomed. Sci. Instrum.* 38, 35, 2002.

[80] Cosgrove, D.M. and Stewart, W.J., Mitral valvuloplasty, *Curr. Probl. Cardiol.* 14, 359, 1989.

[81] Perry, T.E. et al., Thoracic Surgery Directors Association Award. Bone marrow as a cell source for tissue engineering heart valves, *Ann. Thorac. Surg.* 75, 761, 2003.

[82] Narine, K. et al., Transforming growth factor-beta-induced transition of fibroblasts: a model for myofibroblast procurement in tissue valve engineering, *J. Heart Valve Dis.* 13, 281, 2004.

[83] Maish, M.S. et al., Tricuspid valve biopsy: a potential source of cardiac myofibroblast cells for tissue-engineered cardiac valves, *J. Heart Valve Dis.* 12, 264, 2003.

[84] Weston, M.W. and Yoganathan, A.P., Biosynthetic activity in heart valve leaflets in response to *in vitro* flow environments, *Ann. Biomed. Eng.* 29, 752, 2001.

[85] Hoerstrup, S.P. et al., New pulsatile bioreactor for *in vitro* formation of tissue engineered heart valves, *Tissue Eng.* 6, 75, 2000.

[86] Dumont, K. et al., Design of a new pulsatile bioreactor for tissue engineered aortic heart valve formation, *Artif. Organs* 26, 710, 2002.

[87] Sarraf, C.E. et al., Heart valve and arterial tissue engineering, *Cell. Prolif.* 36, 241, 2003.

[88] Zhao, D.E. et al., Tissue-engineered heart valve on acellular aortic valve scaffold: *in-vivo* study, *Asian Cardiovasc. Thorac. Ann.* 11, 153, 2003.

[89] Chester, A.H., Misfeld, M., and Yacoub, M.H., Receptor-mediated contraction of aortic valve leaflets, *J. Heart Valve Dis.* 9, 250, 2000.

[90] Hafizi, S. et al., Mitogenic and secretory responses of human valve interstitial cells to vasoactive agents, *J. Heart Valve Dis.* 9, 454, 2000.

[91] Jian, B. et al., Serotonin mechanisms in heart valve disease I: serotonin-induced up-regulation of transforming growth factor-beta1 via G-protein signal transduction in aortic valve interstitial cells, *Am. J. Pathol.* 161, 2111, 2002.

[92] Katwa, L.C. et al., Angiotensin converting enzyme and kininase-II-like activities in cultured valvular interstitial cells of the rat heart, *Cardiovasc. Res.* 29, 57, 1995.

[93] Roy, A., Brand, N.J., and Yacoub, M.H., Expression of 5-hydroxytryptamine receptor subtype messenger RNA in interstitial cells from human heart valves, *J. Heart Valve Dis.* 9, 256, 2000.

[94] Wiester, L.M. and Giachelli, C.M., Expression and function of the integrin alpha9beta1 in bovine aortic valve interstitial cells, *J. Heart Valve Dis.* 12, 605, 2003.

[95] Weind, K.L. et al., Oxygen diffusion and consumption of aortic valve cusps, *Am. J. Physiol. Heart Circ. Physiol.* 281, H2604, 2001.

[96] Weind, K.L., Ellis, C.G., and Boughner, D.R., Aortic valve cusp vessel density: relationship with tissue thickness, *J. Thorac. Cardiovasc. Surg.* 123, 333, 2002.

[97] Duran, C.M. and Gunning, A.J., The vascularization of the heart valves: a comparative study, *Cardiovasc. Res.* 2, 290, 1968.

[98] Marron, K. et al., Innervation of human atrioventricular and arterial valves, *Circulation* 94, 368, 1996.

[99] Mol, A. et al., The relevance of large strains in functional tissue engineering of heart valves, *Thorac. Cardiovasc. Surg.* 51, 78, 2003.

[100] Cochran, R.P. and Kunzelman, K.S., Comparison of viscoelastic properties of suture versus porcine mitral valve chordae tendineae, *J. Cardiovasc. Surg.* 6, 508, 1991.

[101] Teebken, O.E. et al., Tissue-engineered bioprosthetic venous valve: a long-term study in sheep, *Eur. J. Vasc. Endovasc. Surg.* 25, 305, 2003.

[102] Sutherland, F.W. and Mayer, J.E. Jr., Ethical and regulatory issues concerning engineered tissues for congenital heart repair, *Semin. Thorac. Cardiovasc. Surg. Pediatr. Card. Surg. Annu.* 6, 152, 2003.

29

Tissue Engineering, Stem Cells and Cloning for the Regeneration of Urologic Organs

J. Daniell Rackley
Anthony Atala
Wake Forest University
Baptist Medical Center

The genitourinary system is exposed to a variety of possible injuries from the time the fetus develops. Aside from congenital abnormalities, individuals may also suffer from other disorders such as cancer, trauma, infection, inflammation, iatrogenic injuries, or other conditions that may lead to genitourinary organ damage or loss, requiring eventual reconstruction. The type of tissue chosen for replacement depends on which organ requires reconstruction. Bladder and ureteral reconstruction may be performed with gastrointestinal tissues. Urethral reconstruction is performed with skin, mucosal grafts from the bladder, rectum, or oral cavity. Vaginas can be reconstructed with skin, small bowel, sigmoid colon, and rectum. However, a shortage of donor tissue may limit these types of reconstructions and there is a degree of morbidity associated with the harvest procedure. In addition, these approaches rarely replace the entire function of the original organ. The tissues used for reconstruction may lead to complications due to their inherently different functional parameters. In most cases, the replacement of lost or deficient tissues with functionally equivalent tissues would improve the outcome for these patients. This goal may be attainable with the use of tissue engineering techniques.

29.1 Cell Growth

One of the initial limitations of applying cell-based tissue engineering techniques to urologic organs had been the previously encountered inherent difficulty of growing genitourinary associated cells in large quantities. In the past, it was believed that urothelial cells had a natural senescence, which was hard to overcome. Normal urothelial cells could be grown in the laboratory setting, but with limited expansion. Several protocols were developed over the last two decades which improved urothelial growth and expansion [1–4]. A system of urothelial cell harvest was developed, which does not use any enzymes or serum and has a large expansion potential. Using these methods of cell culture, it is possible to expand a urothelial strain from a single specimen, which initially covers a surface area of 1 cm^2 to one covering a surface area of 4202 m^2 (the equivalent area of one football field) within 8 weeks [1]. These studies indicated that it should be possible to collect autologous urothelial cells from human patients, expand them in culture, and return them to the human donor in sufficient quantities for reconstructive purposes. Bladder, ureter, and renal pelvis cells can be equally harvested, cultured, and expanded in a similar fashion. Normal human bladder epithelial and muscle cells can be efficiently harvested from surgical material, extensively expanded in culture, and their differentiation characteristics, growth requirements, and other biological properties studied [1,3–13].

29.2 Biomaterials for Genitourinary Tissue Engineering

Biomaterials provide a cell–adhesion substrate and can be used to achieve cell delivery with high loading and efficiency to specific sites in the body. The configuration of the biomaterials can guide the structure of an engineered tissue. The biomaterials provide mechanical support against *in vivo* forces, thus maintaining a predefined structure during the process of tissue development. The biomaterials can be loaded with bioactive signals, such as cell–adhesion peptides and growth factors, which can regulate cellular function. The design and selection of the biomaterial is critical in the development of engineered genitourinary tissues. The biomaterial must be capable of controlling the structure and function of the engineered tissue in a predesigned manner by interacting with transplanted cells or the host cells. Generally, the ideal biomaterial should be biocompatible, promote cellular interaction and tissue development, and possess proper mechanical and physical properties.

The selected biomaterial should be biodegradable and bioresorbable to support the reconstruction of a completely normal tissue without inflammation. The degradation products should not provoke inflammation or toxicity and must be removed from the body via metabolic pathways. The degradation rate and the concentration of degradation products in the tissues surrounding the implant must be at a tolerable level [14]. The mechanical support of the biomaterials should be maintained until the engineered tissue has sufficient mechanical integrity to support itself [15].

This can be potentially achieved by an appropriate choice of mechanical and degradative properties of the biomaterials [16].

29.2.1 Types of Biomaterials

Generally, three classes of biomaterials have been utilized for engineering genitourinary tissues: naturally derived materials (e.g., collagen and alginate), acellular tissue matrices (e.g., bladder submucosa and small intestinal submucosa), and synthetic polymers (e.g., polyglycolic acid (PGA), polylactic acid (PLA), and poly(lactic-co-glycolic acid) (PLGA). These classes of biomaterials have been tested in respect to their biocompatibility with primary human urothelial and bladder muscle cells [17,18]. Naturally derived materials and acellular tissue matrices have the potential advantage of biological recognition. Synthetic polymers can be produced reproducibly on a large scale with controlled properties of their strength, degradation rate, and microstructure.

Collagen is the most abundant and ubiquitous structural protein in the body, and may be readily purified from both animal and human tissues with an enzyme treatment and salt/acid extraction [19]. Collagen implants degrade through a sequential attack by lysosomal enzymes. The *in vivo* resorption rate can be regulated by controlling the density of the implant and the extent of intermolecular crosslinking. The lower the density, the greater the interstitial space and generally the larger the pores for cell infiltration, leading to a higher rate of implant degradation. Collagen contains cell-adhesion domain sequences (e.g., RGD), which exhibit specific cellular interactions. This may assist in retaining the phenotype and activity of many types of cells, including fibroblasts [20] and chondrocytes [21].

Alginate, a polysaccharide isolated from sea weed, has been used as an injectable cell delivery vehicle [22] and a cell immobilization matrix [23] owing to its gentle gelling properties in the presence of divalent ions such as calcium. Alginate is relatively biocompatible and approved by the FDA for human use as wound dressing material. Alginate is a family of copolymers of D-mannuronate and L-guluronate. The physical and mechanical properties of alginate gel are strongly correlated with the proportion and length of polyguluronate block in the alginate chains [22].

Acellular tissue matrices are collagen-rich matrices prepared by removing cellular components from tissues. The matrices are often prepared by mechanical and chemical manipulation of a segment of tissue [24–27]. The matrices slowly degrade upon implantation, and are replaced and remodeled by ECM proteins synthesized and secreted by transplanted or ingrowing cells.

Polyesters of naturally occurring α-hydroxy acids, including PGA, PLA, and PLGA, are widely used in tissue engineering. These polymers have gained FDA approval for human use in a variety of applications, including sutures [28]. The ester bonds in these polymers are hydrolytically labile, and these polymers degrade by nonenzymatic hydrolysis. The degradation products of PGA, PLA, and PLGA are nontoxic, natural metabolites and are eventually eliminated from the body in the form of carbon dioxide and water [28]. The degradation rate of these polymers can be tailored from several weeks to several years by altering crystallinity, initial molecular weight, and the copolymer ratio of lactic to glycolic acid. Since these polymers are thermoplastics, they can be easily formed into a three-dimensional scaffold with a desired microstructure, gross shape and dimension by various techniques, including molding, extrusion [29], solvent casting [30], phase separation techniques, and gas foaming techniques [31]. Many applications in genitourinary tissue engineering often require a scaffold with high porosity and ratio of surface area to volume. Other biodegradable synthetic polymers, including poly(anhydrides) and poly(ortho-esters), can also be used to fabricate scaffolds for genitourinary tissue engineering with controlled properties [32].

29.3 Tissue Engineering of Urologic Structures

29.3.1 Urethra

Various biomaterials without cells have been used experimentally (in animal models) for the regeneration of urethral tissue, including PGA, and acellular collagen based matrices from small intestine, bladder, and skin [27,33–37]. Some of these biomaterials, like acellular collagen matrices derived from bladder submucosa, have also been seeded with autologous cells for urethral reconstruction. Our laboratory has been able to replace tubularized urethral segments with cell-seeded collagen matrices.

A cellular collagen matrices derived from bladder submucosa by our laboratory have been used experimentally and clinically. In animal studies, segments of the urethra were resected and replaced with acellular matrix grafts in an onlay fashion. Histological examination showed complete epithelialization and progressive vessel and muscle infiltration. The animals were able to void through the neourethras [27]. These results were confirmed clinically in a series of patients with hypospadias and urethral stricture disease [38,39]. Cadaveric bladders were microdissected and the submucosal layers were isolated. The submucosa was washed and decellularized. The matrix was used for urethral repair in patients with stricture disease

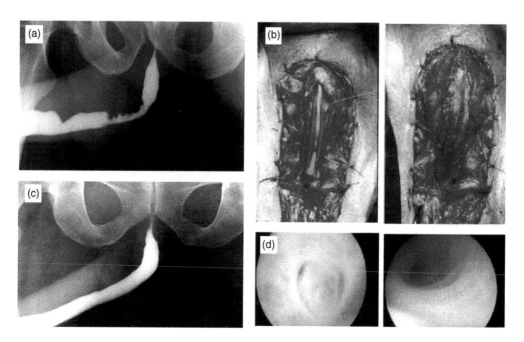

FIGURE 29.1 Representative case of a patient with a bulbar stricture repaired with a collagen matrix. (a) Preoperative urethrogram. (b) Urethral repair. Structured tissue is excised, preserving urethral plate on left side and matrix is anastomosed to urethral plate in an onlay fashion on right side. (c) Urethrogram 6 months after repair. (d) Cystoscopic view of urethra preoperatively on left side and 4 months after repair on right side.

($n = 33$; 28 adults, 5 children) and hypospadias ($n = 7$ children). The matrices were trimmed to size and the neourethras were created by anastomosing the matrix in an onlay fashion to the urethral plate. The size of the neourethras ranged from 2 to 16 cm. Voiding histories, physical examination, retrograde urethrography, uroflowmetry, and cystoscopies were performed serially, pre- and postoperatively, with up to a 7-year follow-up. After a 4- to 7-year follow-up, 34 of the 40 patients had a successful outcome. Six patients with a urethral stricture had a recurrence, and one patient with hypospadias developed a fistula. The mean maximum urine flow rate significantly increased postoperatively. Cystoscopic studies showed adequate caliber conduits. Histologic examination of the biopsies showed the typical urethral epithelium. The use of an off-the-shelf matrix appears to be beneficial for patients with abnormal urethral conditions, and obviates the need for obtaining autologous grafts, thus decreasing operative time and eliminating donor site morbidity (Figure 29.1).

Unfortunately, the above techniques are not applicable for tubularized urethral repairs. The collagen matrices are able to replace urethral segments when used in an onlay fashion. However, if a tubularized repair is needed, the collagen matrices need to be seeded with autologous cells [40,41]. Autologous bladder epithelial and smooth muscle cells from male rabbits were grown and seeded onto preconfigured tubular matrices. The entire anterior urethra was resected and urethroplasties were performed with tubularized collagen matrices seeded with cells in nine animals, and without cells in six animals. Serial urethrograms showed a wide urethral caliber without strictures in the animals implanted with the cell seeded matrices, and collapsed urethral segments with strictures within the unseeded scaffolds. Gross examination of the urethral implants seeded with cells showed normal appearing tissue without any evidence of fibrosis. Histologically, a transitional cell layer surrounded by muscle cell fiber bundles with increasing cellular organization over time were observed on the cell seeded constructs. The epithelial and muscle phenotypes were confirmed with pAE1/AE3 and smooth muscle specific alpha actin antibodies. A transitional cell layer with scant unorganized muscle fiber bundles and large areas of fibrosis were present at the anastomotic sites on the unseeded constructs. Therefore, tubularized collagen matrices seeded with autologous cells can be used successfully for total penile urethra replacement; whereas, tubularized collagen matrices without cells lead to poor tissue development and stricture formation. The cell seeded collagen matrices form new tissue which is histologically similar to native urethra. This technology may be applicable to patients requiring tubularized urethral repair.

29.3.2 Bladder

Currently, gastrointestinal segments are commonly used as tissues for bladder replacement or repair. However, gastrointestinal tissues are designed to absorb specific solutes, whereas bladder tissue is designed for the excretion of solutes. Due to the problems encountered with the use of gastrointestinal segments, numerous investigators have attempted alternative materials and tissues for bladder replacement or repair.

Over the last few decades, several bladder wall substitutes have been attempted with both synthetic and organic materials. The first application of a free tissue graft for bladder replacement was reported by Neuhoff in 1917, when fascia was used to augment bladders in dogs [42]. Since that first report, multiple other free graft materials have been used experimentally and clinically, including bladder allografts, SIS, pericardium, dura, and placenta [24,25,43–50]. In multiple studies using different materials as an acellular graft for cystoplasty, the urothelial layer was able to regenerate normally, but the muscle layer, although present, was not fully developed [24,48,51,52]. When using cell-free collagen matrices, scarring and graft contracture may occur over time [53–58]. Synthetic materials, which have been tried previously in experimental and clinical settings, include polyvinyl sponge, tetrafluoroethylene (Teflon), collagen matrices, vicryl matrices, and silicone [59–62]. Most of the above attempts have usually failed due to either mechanical, structural, functional, or biocompatibility problems. Usually, permanent synthetic materials used for bladder reconstruction succumb to mechanical failure and urinary stone formation and degradable materials lead to fibroblast deposition, scarring, graft contracture, and a reduced reservoir volume over time.

Engineering tissue using selective cell transplantation may provide a means to create functional new bladder segments [63]. The success of using cell transplantation strategies for bladder reconstruction depends on the ability to use donor tissue efficiently and to provide the right conditions for long-term survival, differentiation, and growth. Urothelial and muscle cells can be expanded *in vitro*, seeded onto the polymer scaffold, and allowed to attach and form sheets of cells. The cell-polymer scaffold can then be implanted *in vivo*. A series of *in vivo* urologic associated cell-polymer experiments were performed. Histologic analysis of human urothelial, bladder muscle, and composite urothelial and bladder muscle-polymer scaffolds, implanted in athymic mice and retrieved at different time points, indicated that viable cells were evident in all three experimental groups [64]. Implanted cells oriented themselves spatially along the polymer surfaces. The cell populations appeared to expand from one layer to several layers of thickness with progressive cell organization with extended implantation times. Cell-polymer composite implants of urothelial and muscle cells, retrieved at extended times (50 days), showed extensive formation of multilayered sheet-like structures and well-defined muscle layers. Polymers seeded with cells and manipulated into a tubular configuration showed layers of muscle cells lining the multilayered epithelial sheets. Cellular debris appeared reproducibly in the luminal spaces, suggesting that epithelial cells lining the lumina are sloughed into the luminal space. Cell polymers implanted with human bladder muscle cells alone showed almost complete replacement of the polymer with sheets of smooth muscle at 50 days. This experiment demonstrated, for the first time, that composite tissue engineered structures could be created *de novo*. Prior to this study, only single cell type tissue engineered structures had been created.

29.3.2.1 Formation of Bladder Tissue *Ex-Situ*

In order to determine the effects of implanting engineered tissues in continuity with the urinary tract, an animal model of bladder augmentation was utilized [24]. Partial cystectomies, which involved removing approximately 50% of the native bladders, were performed in 10 beagles. In five, the retrieved bladder tissue was microdissected and the mucosal and muscular layers separated. The bladder urothelial and muscle cells were cultured using the techniques described above. Both urothelial and smooth muscle cells were harvested and expanded separately. A collagen-based matrix, derived from allogeneic bladder submucosa, was used for cell delivery. This material was chosen for these experiments due to its native elasticity. Within 6 weeks, the expanded urothelial cells were collected as a pellet. The cells were seeded on the luminal surface of the allogeneic bladder submucosa and incubated in serum-free keratinocyte growth medium for 5 days. Muscle cells were seeded on the opposite side of the bladder submucosa and subsequently placed in DMEM supplemented with 10% fetal calf serum for an additional 5 days. The seeding density on the allogeneic bladder submucosa was approximately 1×10^7 cells/cm^2.

Preoperative fluoroscopic cystography and urodynamic studies were performed in all animals. Augmentation cystoplasty was performed with the matrix with cells in one group, and with the matrix without cells in the second group. The augmented bladders were covered with omentum in order to facilitate angiogenesis to the implant. Cystostomy catheters were used for urinary diversion for 10 to 14 days. Urodynamic studies and fluoroscopic cystography were performed at 1, 2, and 3 months postoperatively. Augmented bladders were retrieved 2 ($n = 6$) and 3 ($n = 4$) months after surgery and examined grossly, histologically, and immunocytochemically.

Bladders augmented with the matrix seeded with cells showed a 99% increase in capacity compared to bladders augmented with the cell-free matrix, which showed only a 30% increase in capacity. Functionally, all animals showed a normal bladder compliance as evidenced by urodynamic studies, however, the remaining native bladder tissue may have accounted for these results. Histologically, the retrieved engineered bladders contained a cellular organization consisting of a urothelial lined lumen surrounded by submucosal tissue and smooth muscle. However, the muscular layer was markedly more prominent in the cell reconstituted scaffold [24].

Most of the free grafts (without cells) utilized for bladder replacement in the past have been able to show adequate histology in terms of a well-developed urothelial layer, however they have been associated with an abnormal muscular layer that varies in terms of its full development [15,65]. It has been well established for decades that the bladder is able to regenerate generously over free grafts. Urothelium is associated with a high reparative capacity [66]. Bladder muscle tissue is less likely to regenerate in a normal fashion. Both urothelial and muscle ingrowth are believed to be initiated from the edges of the normal bladder toward the region of the free graft [67,68]. Usually, however, contracture or resorption of the graft has been evident. The inflammatory response toward the matrix may contribute to the resorption of the free graft.

It was hypothesized that building the three-dimensional structure constructs *in vitro*, prior to implantation, would facilitate the eventual terminal differentiation of the cells after implantation *in vivo*, and would minimize the inflammatory response towards the matrix, thus avoiding graft contracture and shrinkage. This study demonstrated that there was a major difference evident between matrices used with autologous cells (tissue engineered) and matrices used without cells [24]. Matrices implanted with cells for bladder augmentation retained most of their implanted diameter, as opposed to matrices implanted without cells for bladder augmentation, wherein graft contraction and shrinkage occurred. The histomorphology demonstrated a marked paucity of muscle cells and a more aggressive inflammatory reaction in the matrices implanted without cells. Of interest is that the urothelial cell layers appeared normal, even though their underlying matrix was significantly inflamed. It was further hypothesized, that having an adequate urothelial layer from the outset would limit the amount of urine contact with the matrix, and would therefore decrease the inflammatory response, and that the muscle cells were also necessary for bioengineering, being that native muscle cells are less likely to regenerate over the free grafts. Further studies confirmed this hypothesis [69]. Thus, it appears that the presence of both urothelial and muscle cells on the matrices used for bladder replacement appear to be important for successful tissue bioengineering.

29.3.2.2   Bladder Replacement Using Tissue Engineering

The results of initial studies showed that the creation of artificial bladders may be achieved *in vivo*, however, it could not be determined whether the functional parameters noted were due to the augmented segment or the intact native bladder tissue. In order to better address the functional parameters of tissue engineered bladders, an animal model was designed, which required a subtotal cystectomy with subsequent replacement by a tissue engineered organ [69].

  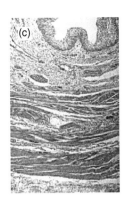

FIGURE 29.2 Hematoxylin and Eosin histological results 6 months after surgery (original magnification: ×250). (a) Normal canine bladder. (b) The bladder dome of the cell-free polymer reconstructed bladder consists of a thickened layer of collagen and fibrotic tissue. (c) The tissue engineered neo-organ shows a histo-morphologically normal appearance. A trilayered architecture consisting of urothelium, submucosa, and smooth muscle is evident.

A total of 14 beagle dogs underwent a trigone-sparing cystectomy. The animals were randomly assigned to one of three groups. Group A ($n = 2$) underwent closure of the trigone without a reconstructive procedure. Group B ($n = 6$) underwent reconstruction with a cell-free bladder shaped biodegradable polymer. Group C ($n = 6$) underwent reconstruction using a bladder shaped biodegradable polymer that delivered autologous urothelial cells and smooth muscle cells. The cell populations had been separately expanded from a previously harvested autologous bladder biopsy. Preoperative and postoperative urodynamic and radiographic studies were performed serially. Animals were sacrificed at 1, 2, 3, 4, 6, and 11 months postoperatively. Gross, histological, and immunocytochemical analyses were performed [69].

Cystectomy only controls and polymer only grafts maintained average capacities of 22 and 46% of preoperative values, respectively. An average bladder capacity of 95% of the original precystectomy volume was achieved in the tissue engineered bladder replacements. These findings were confirmed radiographically. The subtotal cystectomy reservoirs, which were not reconstructed and polymer only reconstructed bladders showed a marked decrease in bladder compliance (10 and 42%). The compliance of the tissue engineered bladders showed almost no difference from preoperative values that were measured when the native bladder was present (106%). Histologically, the polymer only bladders presented a pattern of normal urothelial cells with a thickened fibrotic submucosa and a thin layer of muscle fibers. The retrieved tissue engineered bladders showed a normal cellular organization, consisting of a trilayer of urothelium, submucosa, and muscle (Figure 29.2). Immunocytochemical analyses for desmin, alpha actin, cytokeratin 7, pancytokeratins AE1/AE3 and uroplakin III confirmed the muscle and urothelial phenotype. S-100 staining indicated the presence of neural structures. The results from this study showed that it is possible to tissue engineer bladders, which are anatomically and functionally normal [69]. Clinical trials for the application of this technology are currently being arranged.

29.3.3 Genital Tissues

Reconstructive surgery is required for a wide variety of pathologic penile conditions, such as penile carcinoma, trauma, severe erectile dysfunction, and congenital conditions like ambiguous genitalia, hypospadias, and epispadias. One of the major limitations of phallic reconstructive surgery is the availability of sufficient autologous tissue. Phallic reconstruction using autologous tissue, derived from the patient's own cells, may be preferable in selected cases.

29.3.4 Reconstruction of Corporal Tissues

One of the major components of the phallus is corporal smooth muscle. The creation of autologous functional and structural corporal tissue *de novo* would be beneficial.

Initial experiments were performed in order to determine the feasibility of creating corporal tissue *in vivo* using cultured human corporal smooth muscle cells seeded onto biodegradable polymers [70]. Primary normal human corpus cavernosal smooth muscle cells were isolated from normal young adult patients after informed consent during routine penile surgery. Muscle cells were maintained in culture, seeded onto biodegradable polymer scaffolds, and implanted subcutaneously in athymic mice. Implants were retrieved at 7, 14, and 24 days after surgery for analyses. Corporal smooth muscle tissue was identified grossly and histologically. Intact smooth muscle cell multilayers were observed growing along the surface of the polymers throughout all time points. Early vascular ingrowth at the periphery of the implants was evident by 7 days. By 24 days, there was evidence of polymer degradation. Smooth muscle phenotype was confirmed immunocytochemically and by Western blot analyses with antibodies to alpha smooth muscle actin.

In order to engineer functional corpus cavernosum, both smooth muscle and sinusoidal endothelial cells are essential. However, penile sinusoidal endothelial cells had not been extensively cultured in the past, and had not been fully characterized. A method of isolation and expansion of sinusoidal endothelial cells from corpora cavernosa was devised, and cell function and gene expression were characterized.

When grown on collagen, corporal cavernosal endothelial cells formed capillary structures, which created a complex three-dimensional capillary network. The possibility of developing human corporal tissue *in vivo* by combining smooth muscle and endothelial cells was investigated [71]. Primary normal human corpus cavernosal smooth muscle cells and ECV 304 human endothelial cells were seeded on biodegradable polymers and implanted in the subcutaneous space of athymic mice. At retrieval all polymer scaffolds seeded with cells had formed distinct tissue structures and maintained their preimplantation size. The control scaffolds without cells had decreased in size with increasing time. Histologically, all of the retrieved polymers seeded with corporal smooth muscle and endothelial cells showed the survival of the implanted cells. The presence of penetrating native vasculature was observed 5 days after implantation. The formation of multilayered strips of smooth muscle adjacent to endothelium was evident by 7 days after implantation. Increased smooth muscle organization and accumulation of endothelium lining the luminal structures were evident 14 days after implantation. A well-organized construct, consisting of muscle and endothelial cells, was noted at 28 and 42 days after implantation. A marked degradation of the polymer fibers was observed by 28 days. There was no evidence of tissue formation in the controls (polymers without cells). The results of these studies suggested that the creation of well-vascularized autologous corporal-like tissue, consisting of smooth muscle and endothelial cells, may be possible.

The aim of phallic reconstruction is to achieve structurally and functionally normal genitalia. It had been shown that human cavernosal smooth muscle and endothelial cells seeded on polymers would form tissue composed of corporal cells when implanted *in vivo*. However, corporal tissue structurally identical to the native corpus cavernosum was not achieved, due to the type of polymers used. Therefore, a naturally derived acellular corporal tissue matrix that possesses the same architecture as native corpora was developed. The feasibility of developing corporal tissue, consisting of human cavernosal smooth muscle and endothelial cells *in vivo*, using an acellular corporal tissue matrix as a cell delivery vehicle was explored [72]. Acellular collagen matrices were derived from processed donor rabbit corpora using cell lysis techniques. Human corpus cavernosal muscle and endothelial cells were derived from donor penile tissue, the cells were expanded *in vitro*, seeded on the acellular matrices, and implanted subcutaneously in athymic mice. Western blot analysis detected alpha actin, myosin and tropomyosin proteins from human corporal smooth muscle cells. Expression of muscarinic acetylcholine receptor (mAChR) subtype m4 mRNA was demonstrated by RT-PCR from corporal muscle cells 8 weeks prior to and after seeding. The implanted matrices showed neovascularity into the sinusoidal spaces by 1 week after implantation. Increasing organization of smooth muscle and endothelial cells lining the sinusoidal walls was observed at 2 weeks and continued with time. The matrices were covered with the appropriate cell architecture 4 weeks after implantation. The matrices showed a stable collagen concentration over 8 weeks, as determined by hydroxy-proline quantification. Immunocytochemical studies using alpha actin and Factor VIII antibodies confirmed the presence of corporal smooth muscle and endothelial cells, both *in vitro* and *in vivo*, at all time points. There was no evidence of cellular organization in the control matrices.

In another study, we attempted to replace entire crossectional segments of both corporal bodies of penis *in vivo* by interposing engineered tissue in rabbits and investigated their structural and functional integrity [73]. Autologous cavernosal smooth muscle and endothelial cells were harvested, expanded, and seeded on acellular collagen matrices. The entire cross section of the protruding rabbit phallus (~0.7 cm long; 1/3 of penile shaft) was excised, leaving the urethra intact. Matrices with and without cells were interposed into the excised corporal space. Additional rabbits, without surgical intervention, served as controls. The experimental corporal bodies demonstrated adequate structural and functional integrity by cavernosography and cavernosometry. Mating activity in the animals with the engineered corpora normalized by 3 months. The presence of sperm was confirmed during mating, and was present in all the rabbits with the engineered corpora. Grossly, the corporal implants with cells showed continuous integration of the graft into native tissue. Histologically, sinusoidal spaces and walls, lined with endothelium and smooth muscle, were observed in the engineered grafts. Grafts without cells contained fibrotic tissue and calcifications with sparse corporal elements. Each cell type was identified immunohistochemically and by Western blot analyses. These studies demonstrate that it is possible to engineer autologous functional penile tissue. Our laboratory is currently working on increasing the size of the engineered constructs.

29.4   Engineered Penile Prostheses

Although silicone is an accepted biomaterial for penile prostheses, biocompatibility is a concern [74,75]. The use of a natural prosthesis composed of autologous cells may be advantageous. A feasibility study for creating natural penile prostheses made of cartilage was performed initially [76].

Cartilages, harvested from the articular surface of calf shoulders, were isolated, grown, and expanded in culture. The cells were seeded onto preformed cylindrical polyglycolic acid polymer rods (1 cm in diameter and 3 cm in length). The cell-polymer scaffolds were implanted in the subcutaneous space of 20 athymic mice. Each animal had two implantation sites consisting of a polymer scaffold seeded with chondrocytes and a control (polymer alone). The rods were retrieved at 1, 2, 4, and 6 months post-implantation. Biomechanical properties, including compression, tension, and bending were measured on the retrieved structures. Histological analyses were performed to confirm the cellular composition. At retrieval, all of the polymer scaffolds seeded with cells formed milky-white rod-shaped solid cartilaginous structures, maintaining their preimplantation size and shape. The control scaffolds without cells failed to form cartilage. There was no evidence of erosion, inflammation, or infection in any of the implanted cartilage rods.

The compression, tension, and bending studies showed that the cartilage structures were readily elastic and could withstand high degrees of pressure. Biomechanical analyses showed that the engineered cartilage rods possessed the mechanical properties required to maintain penile rigidity. The compression studies showed that the cartilage rods were able to withstand high degrees of pressure. A ramp compression speed of 200 μm/sec, applied to each cartilage rod up to 2000 μm in distance, resulted in 3.8 kg of resistance. The tension relaxation studies demonstrated that the retrieved cartilage rods were able to withstand stress and were able to return to their initial state while maintaining their biomechanical properties. A ramp tension speed of 200 μm/sec applied to each cartilage rod created a tensile strength of 2.2 kg, which physically lengthened the rods an average of 0.48 cm. Relaxation of tension at the same speed resulted in retraction of the cartilage rods to their initial state. The bending studies performed at two different speeds showed that the engineered cartilage rods were durable, malleable, and were able to retain their mechanical properties. Cyclic compression, performed at rates of 500 and 20,000 μm/sec, demonstrated that the cartilage rods could withstand up to 3.5 kg of pressure at a predetermined distance of 5000 μm. The relaxation phase of the cyclic compression studies showed that the engineered rods were able to maintain their tensile strength. None of the rods were ruptured during the biomechanical stress relaxation studies.

Histological examination with hematoxylin and eosin showed the presence of mature and well-formed cartilage in all the chondrocyte-polymer implants. The polymer fibers were progressively replaced by cartilage with time progression. Undegraded polymer fibers were observed at 1 and 2 months after implantation. However, remnants of polymer scaffolds were not present in the cartilage rods at 6 months. Aldehyde fuschin-alcian blue and toluidine blue staining demonstrated the presence of highly sulfated mucopolysaccharides, which are differentiated products of chondrocytes. There was no evidence of cartilage formation in the controls.

In a subsequent study using an autologous system, the feasibility of applying the engineered cartilage rods *in situ* was investigated [77]. Autologous chondrocytes harvested from rabbit ear were grown and expanded in culture. The cells were seeded onto biodegradable poly-L-lactic acid coated polyglycolic acid polymer rods at a concentration of 50×10^6 chondrocytes/cm^3. Eighteen chondrocyte-polymer scaffolds were implanted into the corporal spaces of 10 rabbits. As controls, two corpora, one each in 2 rabbits, were not implanted. The animals were sacrificed at 1, 2, 3, and 6 months after implantation. Histological analyses were performed with hematoxylin and eosin, aldehyde fuschin-alcian blue, and toluidine blue staining. All animals tolerated the implants for the duration of the study without any complications. Gross examination at retrieval showed the presence of well-formed milky-white cartilage structures within the corpora at 1 month. All polymers were fully degraded by 2 months. There was no evidence of erosion or infection in any of the implant sites. Histological analyses with alcian blue and toluidine blue staining demonstrated the presence of mature and well-formed chondrocytes in the retrieved implants. Subsequent studies were performed assessing the functionality of the cartilage penile rods *in vivo* long term. To date, the animals have done well, and can copulate and impregnate their female partners without problems. Further functional studies need to be completed before applying this technology to the clinical setting.

29.5   Other Applications of Genitourinary Tissue Engineering

29.5.1   Injectable Therapies

Both urinary incontinence and vesicoureteral reflux are common conditions affecting the genitourinary system, wherein injectable bulking agents can be used for treatment. There are definite advantages in treating urinary incontinence and vesicoureteral reflux endoscopically. The method is simple and can be completed in less than 15 min, it has a low morbidity and it can be performed on an outpatient basis. The goal of several investigators has been to find alternate implant materials, which would be safe for human use [78].

The ideal substance for the endoscopic treatment of reflux and incontinence should be injectable, nonantigenic, nonmigratory, volume stable, and safe for human use. Toward this goal long-term studies were conducted to determine the effect of injectable chondrocytes *in vivo* [79]. It was initially determined that alginate, a liquid solution of gluronic and mannuronic acid, embedded with chondrocytes, could serve as a synthetic substrate for the injectable delivery and maintenance of cartilage architecture *in vivo*. Alginate undergoes hydrolytic biodegradation and its degradation time can be varied depending on the concentration of each of the polysaccharides. The use of autologous cartilage for the treatment of vesicoureteral reflux in humans would satisfy all the requirements of an ideal injectable substance. A biopsy of the ear could be easily and quickly performed, followed by chondrocyte processing and endoscopic injection of the autologous chondrocyte suspension for the treatment reflux.

Chondrocytes can be readily grown and expanded in culture. Neocartilage formation can be achieved *in vitro* and *in vivo* using chondrocytes cultured on synthetic biodegradable polymers [79]. In these experiments, the cartilage matrix replaced the alginate as the polysaccharide polymer underwent biodegradation. This system was adapted for the treatment of vesicoureteral reflux in a porcine model [80].

Six mini swine underwent bilateral creation of reflux. All six were found to have bilateral reflux without evidence of obstruction at 3 months following the procedure. Chondrocytes were harvested from the left auricular surface of each mini swine and expanded with a final concentration of $50-150 \times 10^6$ viable cells per animal. The animals underwent endoscopic repair of reflux with the injectable autologous chondrocyte solution on the right side only. Serial cystograms showed no evidence of reflux on the treated side and persistent reflux in the uncorrected control ureter in all animals. All animals had a successful cure of reflux in the repaired ureter without evidence of hydronephrosis on excretory urography. The harvested ears had evidence of cartilage regrowth within 1 month of chondrocyte retrieval.

At the time of sacrifice, gross examination of the bladder injection site showed a well-defined rubbery to hard cartilage structure in the subureteral region. Histologic examination of these specimens showed evidence of normal cartilage formation. The polymer gels were progressively replaced by cartilage with increasing time. Aldehyde fuschin-alcian blue staining suggested the presence of chondroitin sulfate. Microscopic analyses of the tissues surrounding the injection site showed no inflammation. Tissue sections from the bladder, ureters, lymph nodes, kidneys, lungs, liver, and spleen showed no evidence of chondrocyte or alginate migration, or granuloma formation. These studies showed that chondrocytes can be easily harvested and combined with alginate *in vitro*, the suspension can be easily injected cystoscopically and the elastic cartilage tissue formed is able to correct vesicoureteral reflux without any evidence of obstruction [80].

Two multicenter clinical trials were conducrted using the above engineered chondrocyte technology. Patients with vesicoureteral reflux were treated at 10 centers throughout the United States. The patients had a similar success rate as with other injectable substances in terms of cure (Figure 29.3). Chondrocyte formation was not noted in patients who had treatment failure. The patients who were cured would supposedly have a biocompatible region of engineered autologous tissue present, rather than a foreign material [81]. Patients with urinary incontinence were also treated endoscopically with injected chondrocytes at three different medical centers. Phase 1 trials showed an approximate success rate of 80% at both 3 and 12 months postoperatively [82].

The potential use of injectable, cultured myoblasts for the treatment of stress urinary incontinence has been investigated [83,84]. Primary myoblasts obtained from mouse skeletal muscle were transduced *in vitro* to carry the β-galactosidase reporter gene and were then incubated with fluorescent microspheres, which would serve as markers for the original cell population. Cells were then directly injected into the proximal urethra and lateral bladder walls of nude mice with a micro-syringe in an open surgical procedure. Tissue was harvested up to 35 days postinjection, analyzed histologically, and assayed for β-galactosidase expression. Myoblasts expressing B-galactosidase and containing fluorescent microspheres were found at each of the retrieved time points. In addition, regenerative myofibers expressing β-galactosidase were identified within the bladder wall. By 35 days postinjection, some of the injected cells expressed the contractile filament α-smooth muscle actin, suggesting the possibility of myoblastic differentiation into smooth muscle. The authors reported that a significant portion of the injected myoblast population persisted *in vivo*. Similar techniques of sphincteric derived muscle cells have been used for the treatment of urinary incontinence. Strasser from Innsbuck, Austria harvested muscle samples from pigs, dissociated the cells, and injected autologous pure clones of myoblasts into the urethral wall of pigs under sonographic visualization. Postoperatively maximal urethral closure pressures were increased markedly in most pigs and the zone of higher urethral closure pressure was lengthened compared to preoperative measurements [85]. The fact that myoblasts can be transfected, survive after injection, and begin the process of myogenic differentiation, further supports the feasibility of using cultured cells of muscular origin as an injectable bioimplant.

29.6   Stem Cells for Tissue Engineering

Most current strategies for engineering urologic tissues involve harvesting of autologous cells from the host diseased organ. However, in situations wherein extensive end stage organ failure is present, a tissue biopsy may not yield enough normal cells for expansion. Under these circumstances, the availability of pluripotent stem cells may be beneficial. Pluripotent embryonic stem cells are known to form teratomas *in vivo*, which are composed of a variety of differentiated cells. However, these cells are immunocompetent, and would require immunosuppression if used clinically.

The possibility of deriving stem cells from postnatal mesenchymal tissue from the same host, and inducing their differentiation *in vitro* and *in vivo*, was investigasted. Stem cells were isolated from human

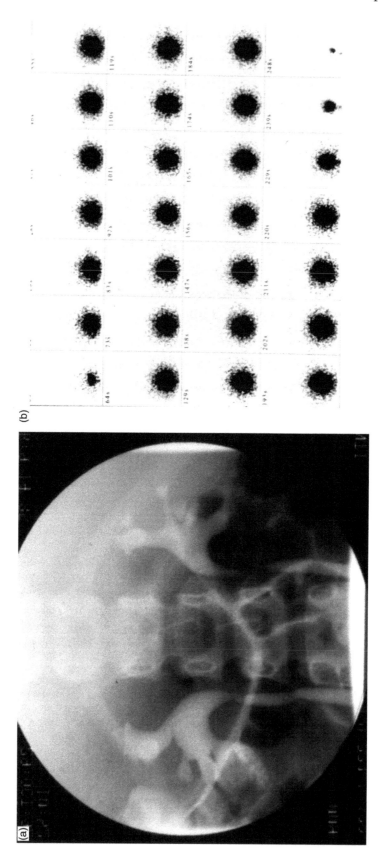

FIGURE 29.3 (a) Preoperative voiding cystourethrogram of a patient showing bilateral reflux; (b) Postoperative radionuclide cystogram of the same patient 6 months after the injection of autologous chondrocytes.

foreskin derived fibroblasts. Stem cell derived chondrocytes were obtained through a chondrogenic lineage process. The cells were grown, expanded, seeded onto biodegradable scaffolds, and implanted *in vivo*, where they formed mature cartilage structures. This was the first demonstration that stem cells can be derived from postnatal connective tissue and can be used for engineering tissues *in vivo ex situ* [86].

A second approach which has been pursued for stem lineage isolation involves the isolation of stem cells from individual organs. For example, daily female hormone supplementation is used widely, most commonly in postmenopausal women. A continuous and unlimited hormonal supply produced from ovarian granulosa cells would be an attractive alternative. The feasibility of isolating functional human ovarian granulosa stem cells, which, unlike primary cells, may have the ability to proliferate and function indefinitely, was investigated.

Granulosa stem cells were selectively isolated from postmenopausal human ovaries and their phenotype was confirmed with the stem cell marker antibodies, CD 34, CD 105, and CD 90. The granulosa stem cells in culture showed steady state progesterone (5 to 7 ng/ml) and estradiol (2500 to 3000 pg/ml) production either with or without hCG stimulation [87].

29.7 Therapeutic Cloning

Nuclear transplantation ("therapeutic cloning") could theoretically provide a limitless source of cells for regenerative therapy. According to data from the Centers for Disease Control, as many as 3000 Americans die every day from diseases that in the future may be treatable with embryonic stem (ES)-derived tissues [88]. In addition to generating functional replacement cells such as cardiomyocytes and neurons, there is also the possibility that these cells could be used to reconstitute more complex tissues and organs, including kidneys [89–91]. Somatic cell nuclear transfer (SCNT) has the potential to eliminate immune responses associated with the transplantation of these various tissues, and thus the requirement for immunosuppressive drugs or immunomodulatory protocols that carry the risk of a wide variety of serious and potentially life-threatening complications (Figure 29.4) [92].

Although the goal of "therapeutic" cloning is to generate replacement cells and tissues that are genetically identical with the donor, numerous studies have shown that animals produced by the SCNT technique

Therapeutic Cloning Strategies

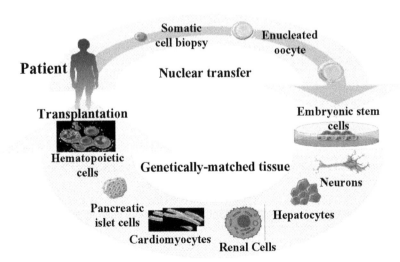

FIGURE 29.4 Therapeutic cloning strategy and its application to the engineering of tissues and organs.

inherit their mitochondria entirely or in part from the recipient oocyte and not the donor cell [93–95]. This raises the question of whether nonself mitochondrial proteins in cells could lead to immunogenicity after transplantation and defeat the main objective of the procedure.

We tested the histocompatibility of nuclear-transfer-generated cells and engineered tissues in a large animal model, the cow (*Bos taurus*). Cloned muscle cell implants were not rejected, and they remained viable after being transplanted back into the nuclear donor animal despite expressing a different mtDNA haplotype. We also showed that nuclear transplantation can be used to generate functional renal structures. Owing to its complex structure and function [95], the kidney is one of the most challenging organs to reconstruct in the body. Previous efforts at tissue engineering the kidney have been directed toward development of an extracorporeal renal support system comprising both biologic and synthetic components [97–99]. This approach was first described by Aebischer et al. [100,101] and is being focused toward the treatment of acute rather than chronic renal failure. Humes et al. [97] have shown that the combination of hemofiltration and a renal-assist device containing tubule cells can replace certain physiologic functions of the kidney when they are connected in an extravascular perfusion circuit in uremic dogs. Heat exchangers, flow and pressure monitors, and multiple pumps are required for optimal functioning of this device [102,103]. Although *ex vivo* organ substitution therapy would be life-sustaining, there would be obvious benefits for patients if such devices could be implanted long term without the need for an extracorporeal perfusion circuit or immunosuppressive drugs or immune modulatory protocols. While synthetic, selectively permeable barriers can be used *ex vivo* to separate transplanted cells from the immune system of the body, the implantation of such immunoisolation systems would pose significant difficulties in both the long and short term [104–107]. We demonstrated that it may be feasible to use therapeutic cloning to generate functional immune-compatible renal tissues [108].

Dermal fibroblasts were isolated from adult Holstein steers by ear notch. Bovine oocytes were obtained from abattoir-derived ovaries. The oocytes were mechanically enucleated at 18 to 22 h postmaturation, and complete enucleation of the metaphase plate was confirmed with *bis*Benzimide dye under fluorescence microscopy. A suspension of actively dividing cells was prepared immediately prior to nuclear transfer. Single donor cells were selected and transferred into the perivitelline space of the enucleated oocytes. Fusion of the cell–oocyte complexes was accomplished by applying a single pulse of 2.4 kV/cm for 15 μsec. Nuclear transfer embryos were activated with exposure to Ionomycin. The resulting blastocysts were nonsurgically transferred into progestrin-synchronized recipients. The cloned renal and muscle cells were isolated and expanded *in vitro* after 12 weeks. The expanded cloned renal cells were successfully seeded onto renal units, and implanted back into the nuclear donor organism without immune destruction. The cells organized into glomeruli- and tubule-like structures with the ability to excrete toxic metabolic waste products through a urine-like fluid.

29.7.1 Muscle

Tissue engineered constructs containing bovine muscle cells seeded onto PGA matrices were transplanted subcutaneously and retrieved 6 weeks after implantation. After retrieval of the first-set implants, a second set of constructs from the same donor were transplanted for a further 12 weeks. On a histological level, the cloned muscle tissue appeared intact, and showed a well-organized cellular orientation with spindle-shaped nuclei. Immunohistochemical analysis identified muscle fibers within the implanted constructs. In contrast to the cloned implants, the allogeneic, control cell implants failed to form muscle bundles, and showed an increased number of inflammatory cells, fibrosis, and necrotic debris consistent with acute rejection. Semiquantitative RT-PCR and Western blot analysis confirmed the expression of muscle specific mRNA and proteins in the retrieved tissues despite the presence of allogeneic mitochondria. In contrast, expression intensities were significantly lower or absent in constructs generated from genetically unrelated cattle.

Immunocytochemical analysis using CD4- and CD8-specific antibodies identified an approximately twofold increase in CD4+ and CD8+ T cells within the explanted first and second set control vs. cloned constructs. Importantly, first and second set cloned constructs exhibited comparable levels of CD4 and

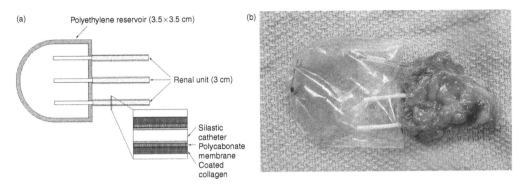

FIGURE 29.5   Tissue-engineered renal units. Illustration of renal unit (a) and unit seeded with cloned cells, showing the accumulation of urinelike fluid, retrieved 3 months after implantation (b).

CD8 expression, arguing against the presence of an enhanced second set reaction as would be expected if mtDNA-encoded minor antigen differences were present. Western blot analysis of the first-set explants indicated an approximately sixfold increase in expression intensity of CD4 in the control vs. cloned constructs at 6 weeks, confirming a primary immune response to the control grafts. There was also a significant increase in the mean expression intensities of CD8 in the control vs. cloned constructs at 6 weeks. Twelve weeks after second-set implantation, mean expression intensities of CD4 and CD8 continued to remain significantly elevated in the control vs. cloned constructs.

29.7.2 Kidney

The renal cells obtained through nuclear transfer demonstrated immunochemically the expression of renal specific proteins, including synaptopodin (produced by podocytes), aquaporin 1 (AQP1, produced by proximal tubules and the descending limb of the loop of Henle), aquaporin 2 (AQP2, produced by collecting ducts), Tamm–Horsfall protein (produced by the ascending limb of the loop of Henle), and factor VIII (produced by endothelial cells). Synaptopodin and AQP1 & 2 expressing cells exhibited circular and linear patterns in two-dimensional culture, respectively. After expansion, the renal cells were shown to produce both erythropoietin and 1,25-dihydroxyvitamin D_3, a key endocrinologic metabolite. The cloned cells produced erythropoietin and were responsive to hypoxic stimulation.

The cloned renal cells were seeded onto collagen-coated cylindrical polycarbonate membranes. Renal devices with collecting systems were constructed by connecting the ends of three membranes with catheters that terminated in a reservoir (Figure 29.5). Thirty-one units ($n = 19$ with cloned cells, $n = 6$ without cells, and $n = 6$ with cells from an allogeneic control fetus) were transplanted subcutaneously and retrieved 12 weeks after implantation back into the nuclear donor animal.

On gross examination, the explanted units appeared intact, and straw-yellow colored fluid could be observed in the reservoirs of the cloned group. There was a sixfold increase in volume in the experimental group vs. the control groups. Chemical analysis of the fluid suggested unidirectional secretion and concentration of urea nitrogen and creatinine.

Physiological function of the implanted units was further evidenced by analysis of the electrolyte levels in the collected fluid as well as specific gravity and glucose concentrations. The electrolyte levels detected in the fluid of the experimental group were significantly different from plasma or the controls. These findings indicate that the implanted renal cells possess filtration, reabsorption, and secretory functions. Urine specific gravity is an indicator of kidney function and reflects the action of the tubules and collecting ducts on the glomerular filtrate by furnishing an estimate of the number of particles dissolved in the urine. The urine-specific gravity of cattle is reported as approximately 1.025 (vs. 1.027 ± 0.001 for the fluid that was produced by the cloned renal units), and normally ranges from 1.020 to 1.040 (vs. approximately 1.010 in normal bovine serum) [16,17]. The normal range of urine pH for adult herbivores is alkaline, with values

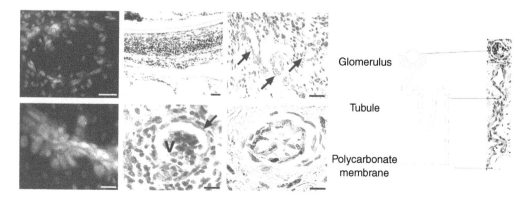

FIGURE 29.6   Characterization of renal explants. Cloned cells stained positively with synaptopodin antibody (a) and AQP1 antibody (b). The allogeneic controls displayed a foreign body reaction with necrosis (c). Cloned explant shows organized glomeruli (d) and tubule-like structures (e). H&E, reduced from 400×. Immunohistochemical analysis using factor VIII antibodies identifies vascular structure within d (f). Reduced from ×400. There was a clear unidirectional continuity between the mature glomeruli, their tubules, and the polycarbonate membrane (g).

ranging from 7.0 to 9.0 [17] (the pH of the fluid from the cloned renal units was 8.1±0.20). Glucose is reabsorbed in the proximal tubules, and is seldom present in the urine of cattle. Glucose was undetectable (<10 mg/dl) in the cloned renal fluid (vs. blood glucose concentrations of 76.6 ± 0.04 mg/dl).

The retrieved implants demonstrated extensive vascularization, and had self-assembled into glomeruli and tubule-like structures. The latter were lined with cuboid epithelial cells with large, spherical, and pale-stained nuclei, whereas the glomeruli structures exhibited a variety of cell types with abundant red blood cells. There was a clear continuity between the mature glomeruli, their tubules, and the polycarbonate membrane. The renal tissues were integrally connected in a unidirectional manner to the reservoirs, resulting in the excretion of dilute urine into the collecting systems.

Immunohistochemical analysis confirmed expression of renal specific proteins, including AQP1, AQP2, synaptopodin, and factor VIII. Antibodies for AQP1, AQP2, and synaptopodin identified tubular, collecting tubule, and glomerular segments within the constructs, respectively. In contrast, the allogeneic controls displayed a foreign body reaction with necrosis, consistent with the finding of acute rejection. RT-PCR analysis confirmed the transcription of AQP1, AQP2, synaptopodin, and Tamm–Horsfall genes exclusively in the cloned group. Cultured and cloned cells also expressed high protein levels of AQP1, AQP2, synaptopodin, and Tamm–Horsfall protein as determined by Western blot analysis. Expression intensity of CD4 and CD8, markers for inflammation and rejection, were also significantly higher in the control vs. cloned group (Figure 29.6).

29.7.2.1   Mitochondrial DNA analysis

Previous studies showed that bovine clones harbor the oocyte mtDNA [93–95,109]. Differences in mtDNA-encoded proteins expressed by clone cells could stimulate a T cell response specific for mtDNA-encoded minor histocompatibility antigens (miHA) [110] when clone cells are transplanted back to the original nuclear donor. The most straightforward approach to resolve the question of miHA involvement is the identification of potential antigens by nucleotide sequencing of the mtDNA genomes of the clone and fibroblast nuclear donor. The contiguous segments of mtDNA that encode 13 mitochondrial proteins and tRNAs were amplified by PCR from total cell DNA in five overlapping segments. These amplicons were directly sequenced on one strand with a panel of sequencing primers spaced at 500 bp intervals.

The resulting nucleotide sequences (13,210 bp) revealed nine nucleotide substitutions for the first donor : recipient combination (muscle constructs). One substitution was in the tRNA-Gly segment and five substitutions were synonymous. The sixth substitution, in the ND1 gene, was heteroplasmic in the nuclear donor where one of the two alternative nucleotides was shared with the clone. A Leu or Arg would be translated at this position in ND1. The eighth and ninth substitutions resulted in amino acid (AA) interchanges of Asn > Ser and Val > Ala in the ATPase6 and ND4L genes, respectively. For the second donor : recipient combination (renal constructs), we obtained 12,785 bp from both the clone and nuclear donor animal. The resulting sequences revealed six nucleotide substitutions. One substitution was in the tRNA-Arg segment and three substitutions were synonymous. The fifth and sixth substitutions resulted in AA interchanges of Ile>Thr and Thr>Ile in the ND2 and ND5 genes, respectively. The identification of two AA substitutions that distinguish the clone and the nuclear donor confirm that a maximum of only two miHA peptides could be defined by the second donor : recipient combination. Given the lack of knowledge concerning peptide binding motifs for bovine MHC class I molecules, there is no reliable method to predict the impact of these AA substitutions on the ability of mtDNA-encoded peptides to either bind to bovine class I molecules or activate CD8+CTLs.

Although the cloned renal cells derived their nuclear genome from the original fibroblast donor, their *mt*DNA was derived from the original recipient oocyte. A relatively limited number of *mt*DNA polymorphisms have been shown to define maternally transmitted miHA in mice [110]. This class of miHA has been shown to stimulate both skin allograft rejection *in vivo* and expansion of cytotoxic T lymphocytes (CTL) *in vitro* [110], and could constitute a barrier to successful clinical use of such cloned devices as hypothesized for chronic rejection of MHC-matched human renal transplants [111,112]. We chose to investigate a possible anti-miHA T cell response to the cloned renal devices through both delayed-type hypersensitivity (DTH) testing *in vivo* and Elispot analysis of IFNg-secreting T cells *in vitro*. An *in vivo* assay of anti-miHA immunity was chosen based on the ability skin allograft rejection to detect a wide range of miHA in mice with survival times exceeding 10 weeks [113] and the relative insensitivity of *in vitro* assays in detecting miHA incompatibility, highlighted by the requirement for *in vivo* priming to generate CTL [114]. We were unable to discern an immunological response directed against the cloned cells by DTH testing *in vivo*. Cloned and control allogeneic cells were intradermally injected back into the nuclear donor animal 80 days after the initial transplantation. A positive DTH response was observed after 48 h for the allogeneic control cells but not the cloned cells.

The results of DTH analysis were mirrored by Elispot-derived estimates of the frequencies of T cells that secreted IFN-gamma following *in vitro* stimulation. PBLs were harvested from the transplanted recipient 1 month after retrieval of the devices. These PBLs were stimulated in primary mixed lymphocyte cultures (MLCs) with allogeneic renal cells, cloned renal cells, and nuclear donor fibroblasts. Surviving T cells were restimulated in anti-IFN-gamma-coated wells with either nuclear donor fibroblasts (autologous control) or the respective stimulators used in the primary MLCs. Elispot analysis revealed a relatively strong T cell response to allogeneic renal stimulator cells relative to the responses to either cloned renal cells or nuclear donor fibroblasts. These results corroborate both the relative CD4 and CD8 expression in Western blots as well as the results of *in vivo* DTH testing to support the conclusion that there was no detectable rejection response that was specific for cloned renal cells following either primary or secondary challenge. Our results suggest that cloned cells and tissues can be grafted back into the nuclear donor organism without immune destruction. These were the first proof-of-principle studies to demonstrate that therapeutic cloning is feasible.

29.8   Conclusion

Tissue engineering efforts are currently being undertaken for every type of tissue and organ within the urinary system. Most of the effort expended to engineer genitourinary tissues has occurred within the last decade. Tissue engineering techniques require a cell culture, facility designed for human application. Personnel who have mastered the techniques of cell harvest, culture, and expansion as well as polymer design are essential for the successful application of this technology. Various engineered genitourinary tissues are at different stages of development, with some already being used clinically, a few in preclinical trials, and some in the discovery stage. Recent progress suggests that engineered urologic tissues may have an expanded clinical applicability in the future.

References

[1] Cilento, B.G., Freeman, M.R., Schneck, F.X., Retik, A.B., and Atala, A. Phenotypic and cytogenetic characterization of human bladder urothelia expanded *in vitro. J. Urol.* 1994; 152: 655.

[2] Scriven, S.D., Booth, C., Thomas, D.F., Trejdosiewicz, L.K., and Southgate, J. Reconstitution of human urothelium from monolayer cultures. *J. Urol.* 1997; 158: 1147–52.

[3] Liebert, M., Hubbel, A., Chung, M., Wedemeyer, G., Lomax, M.I., Hegeman, A., Yuan, T.Y., Brozovich, M., Wheelock, M.J., and Grossman, H.B. Expression of mal is associated with urothelial differentiation in vitro: identification by differential display reverse-transcriptase polymerase chain reaction. *Differentiation* 1997; 61: 177–85.

[4] Puthenveettil, J.A., Burger, M.S., and Reznikoff, C.A. Replicative senescence in human uroepithelial cells. *Adv. Exp. Med. Biol.* 1999; 462: 83–91.

[5] Fauza, D.O., Fishman, S., Mehegan, K., and Atala, A. Videofetoscopically assisted fetal tissue engineering: bladder augmentation. *J. Ped. Surg.* 1998; 33: 7–12.

[6] Liebert, M., Wedemeyer, G., Abruzzo, L.V., Kunkel, S.L., Hammerberg, C., Cooper, K.D., and Grossman, H.B. Stimulated urothelial cells produce cytokines and express an activated cell surface antigenic phenotype. *Semin. Urol.* 1991; 9: 124–30.

[7] Tobin, M.S., Freeman, M.R., and Atala, A. Maturational response of normal human urothelial cells in culture is dependent on extracellular matrix and serum additives. *Surgical Forum* 1994; 45: 786.

[8] Freeman, M.R., Yoo, J.J., Raab, G., Soker, S., Adam, R.M., Schneck, F.X., Renshaw, A.A., Klagsbrun, M., and Atala, A. Heparin-binding EGF-like growth factor is an autocrine factor for human urothelial cells and is synthesized by epithelial and smooth muscle cells in the human bladder. *J. Clin. Invest.* 1997; 99: 1028.

[9] Nguyen, H.T., Park, J.M., Peters, C.A., Adam, R.A., Orsola, A., Atala, A., and Freeman, M.R. Cell-specific activation of the HB-EGF and ErbB1 genes by stretch in primary human bladder cells. *In Vitro Cell Dev. Biol.* 1999; 35: 371–375.

[10] Harriss, D.R. Smooth muscle cell culture: a new approach to the study of human detrusor physiology and pathophysiology. *Br. J. Urol.* 1995; 75(Suppl 1): 18–26.

[11] Solomon, L.Z., Jennings, A.M., Sharpe, P., Cooper, A.J., and Malone, P.S. Effects of short-chain fatty acids on primary urothelial cells in culture: implications for intravesical use in enterocystoplasties. *J. Lab. Clin. Med.* 1998; 132: 279–83.

[12] Lobban, E.D., Smith, B.A., Hall, G.D., Harnden, P., Roberts, P., Selby, P.J., Trejdosiewicz, L.K., and Southgate, J. Uroplakin gene expression by normal and neoplastic human urothelium. *Am. J. Pathol.* 1998; 153: 1957–67.

[13] Rackley, R.R., Bandyopadhyay, S.K., Fazeli-Matin, S., Shin, M.S., and Appell, R. Immunoregulatory potential of urothelium: characterization of NF-kappaB signal transduction. *J. Urol.* 1999; 162: 1812–6.

[14] Bergsma, J.E., Rozema, F.R., Bos, R.R.M, van Rozendaal, A.W.M, de Jong, W.H., Teppema, J.S., and Joziasse, C.A.P. Biocompatibility and degradatin mechanism of predegraded and non-degraded poly(lactide) implants: an animal study. *Mater. Med.* 1995; 6: 715–24.

[15] Atala, A. Autologous cell transplantation for urologic reconstruction. *J. Urol.* 1998; 159: 2–3.

[16] Kim, B.S. and Mooney, D.J. Development of biocompatible synthetic extracellular matrices for tissue engineering. *Trend Biotechnol.* 1998; 16: 224–30.

[17] Pariente, J.L., Kim, B.S., and Atala, A. In vitro biocompatibility assessment of naturally-derived and sythetic biomaterials using normal human urothelial cells. *J. Biomed. Mater. Res.* 2001; 55: 33–39.

[18] Pariente, J.L., Kim, B.S., and Atala, A. *In vitro* biocompatibility evaluation of naturally derived and synthetic biomaterials using normal human bladder smooth muscle. *J. Urol.* 2002; 167: 1867–71.

[19] Li, S.T. Biologic biomaterials: tissue-derived biomaterials (collagen). In: Brozino, J.D., Ed. *The Biomedical Engineering Handbook.* Boca Ranton, FL: CRC Press, 1995, pp. 627–647.

[20] Silver, F.H. and Pins, G. Cell growth on collagen: a review of tissue engineering using scaffolds containing extracellular matrix. *J. Long-Term Effects Med. Implants* 1992; 2: 67–80.

[21] Sam, A.E. and Nixon, A.J. Chondrocyte-laden collagen scaffolds for resurfacing extensive articular cartilage defects. *Osteoarthritis Cartil.* 1995; 3: 47–59.

[22] Smidsrød, O. and Skjåk-Bræk, G. Alginate as an immobilization matrix for cells. *Trend Biotechnol.* 1990; 8: 71–78.

[23] Lim, F. and Sun, A.M. Microencapsulated islets as bioartificial endocrine pancreas. *Science* 1980; 210: 908–910.

[24] Yoo, J.J., Meng, J., Oberpennin, F., and Atala, A. Bladder augmentation using allogenic bladder submucosa seeded with cells. *J. Urol.* 1998; 51: 221.

[25] Piechota, H.J., Dahms, S.E., Nunes, L.S., Dahiya, R., Lue, T.F., and Tanagho, E.A. In vitro functional properties of the rat bladder regenerated by the bladder acellular matrix graft. *J. Urol.* 1998; 159: 1717–24.

[26] Dahms, S.E., Piechota, H.J., Dahiya, R., Lue, T.F., and Tanagho, E.A. Composition and biochemical properties of the bladder acellular matrix graft: comparative analysis in rat, pig and human. *Br. J. Urol.* 1998; 82: 411–19.

[27] Chen, F., Yoo, J.J., and Atala, A. Acellular collagen matrix as a possible "off the shelf" biomaterial for urethral repair. *Urology* 1999; 54: 407–10.

[28] Gilding, D.K. Biodegradable polymers. In: Williams, D.F., Ed. *Biocompatibility of Clinical Implant Materials*. Boca Raton, FL: CRC Press, 1981, pp. 209–32.

[29] Freed, L.E., Vunjak-Novakovic, G., Biron, R.J., Eagles, D.B., Lesnoy, D.C., Barlow, S.K., and Langer, R. Biodegradable polymer scaffolds for tissue engineering. *BioTechnology* 1994; 12: 689–93.

[30] Mikos, A.G., Thorsen, A.J., Czerwonka, L.A., Bao, Y., Langer, R., Winslow, D.N., and Vacanti, J.P. Preparation and characterization of poly(L-lactic acid) foams. *Polymer* 1994; 35: 1068–77.

[31] Harris, L.D., Kim, B.S., and Mooney, D.J. Open pore biodegradable matrices formed with gas foaming. *J. Biomed. Mater. Res.* 1998; 42: 396–402.

[32] Peppas, N.A. and Langer, R. New challenges in biomaterials. *Science* 1994; 263: 1715–20.

[33] Bazeed, M.A., Thüroff, J.W., Schmidt, R.A., and Tanagho, E.A. New treatment for urethral strictures. *Urology* 1983; 21: 53–57.

[34] Atala, A., Vacanti, J.P., Peters, C.A., Mandell, J., Retik, A.B., and Freeman, M.R. Formation of urothelial structures in vivo from dissociated cells attached to biodegradable polymer scaffolds *in vitro*. *J. Urol.* 1992; 148: 658.

[35] Olsen, L., Bowald, S., Busch, C. et al. Urethral reconstruction with a new synthetic absorbable device. *Scan. J. Urol. Nephrol.* 1992; 26: 323–6.

[36] Kropp, B.P., Ludlow, J.K., Spicer, D., Rippy, M.K., Badylak, S.F., Adams, M.C., Keating, M.A., Rink, R.C., Birhle, R., and Thor, K.B. Rabbit urethral regeneration using small intestinal submucosa onlay grafts. *Urology* 1998; 52: 138–42.

[37] Sievert, K.D., Bakircioglu, M.E., Nunes, L., Tu, R., Dahiya, R., and Tanagho, E.A. Homologous acellular matrix graft for urethral reconstruction in the rabbit: histological and functional evaluation. *J. Urol.* 2000; 163: 1958–65.

[38] Atala, A. Future perspectives in reconstructive surgery using tissue engineering. *Urol. Clin.* 1999; 26: 157–65.

[39] ElKassaby, A.W., Retik, A.B., Yoo, J.J., and Atala, A. Urethral stricture repair with an "off the shelf" collagen matrix. *J. Urol.* 2003; 169.

[40] DeFilippo, R.E., Yoo, J.J., Chen, F., and Atala, A. Urethral replacement using cell-seeded tubularized collagen matrices. *J. Urol.* 2002; 168: 1789–93.

[41] DeFilippo, R.E., Pohl, H.G., Yoo, J.J., and Atala, A. Total penile urethra replacement with autologous cell-seeded collagen matrices [abstract]. *J. Urol.* 2002; 167(Suppl): 152–3.

[42] Neuhof, H. Fascial transplantation into visceral defects: An experimental and clinical study. *Surg. Gynecol Obstet.* 1917; 25: 383.

[43] Tsuji, I., Ishida, H., and Fujieda, J. Experimental cystoplasty using preserved bladder graft. *J. Urol.* 1961; 85: 42.

[44] Kambic, H., Kay, R., Chen, J.F., Matsushita, M., Harasaki, H., and Zilber, S. Biodegradable pericardial implants for bladder augmentation: a 2.5-year study in dogs. *J. Urol.* 1992; 148: 53943.

[45] Kelami, A., Ludtke-Handjery, A., Korb, G., Roll, J., Schnell, J., and Danigel, K.H. Alloplastic replacement of the urinary bladder wall with lyophilized human dura. *Eur. Surg. Res.* 1970; 2: 195.

[46] Fishman, I.J., Flores, F.N., Scott, B., Spjut, H.J., and Morrow, B. Use of fresh placental membranes for bladder reconstruction. *J. Urol.* 1987; 138: 1291.

[47] Probst, M., Dahiya, R., Carrier, S., and Tanagho, E.A. Reproduction of functional smooth muscle tissue and partial bladder replacement. *Br. J. Urol.* 1997; 79: 505.

[48] Sutherland, R.S., Baskin, L.S., Hayward, S.W., and Cunha, G.R. Regeneration of bladder urothelium, smooth muscle, blood vessels, and nerves into an acellular tissue matrix. *J. Urol.* 1996; 156: 571–7.

[49] Kropp, B.P., Sawyer, B.D., Shannon, H.E., Rippy, M.K., Badylak, S.F., Adams, M.C., Keating, M.A., Rink, R.C., and Thor, K.B. Characterization of using small intestine submucosa regenerated canine detrusor: assessment of reinnervaation, in vitro compliance and contractility. *J. Urol.* 1996; 156: 599–607.

[50] Vaught, J.D., Kroop, B.P., Sawyer, B.D., Rippy, M.K., Badylak, S.F., Shannon, H.E., and Thor, K.B. Detrusor regeneration in the rat using porcine small intestine submucosal grafts: functional innervation and receptor expression. *J. Urol.* 1996; 155: 374–8.

[51] Probst, M., Dahiya, R., Carrier, S., and Tanagho, E.A. Reproduction of functional smooth muscle tissue and partial bladder replacement. *Br. J. Urol.* 1997; 79: 505–15.

[52] Kropp, B.P., Rippy, M.K., Badylak, S.F., Adams, M.C., Keating, M.A., Rink, R.C., and Thor, K.B. Small intestinal submucosa: urodynamic and histopahtologic evaluation in long term canine bladder augmentations. *J. Urol.* 1996; 155: 2098–2104.

[53] Lai, J.Y., Yoo, J.J., Wulf, T., and Atala, A. Bladder augmentation using small intestinal submucosa seeded with cells. [abstract]. *J. Urol.* 2002; 167(Suppl): 257.

[54] Brown, A.L., Farhat, W., Merguerian, P.A., Wilson, G.J., Khoury, A.E., and Woodhouse, K.A. 22 week assessment of bladder accellular matrix as a bladder augmentation material in a porcine model. *Biomaterials* 2002; 23: 2179–90.

[55] Reddy, P.P., Barrieras, D.J., Wilson, G., Bagli, D.J., McLorie, G.A., Khoury, A.E., and Merguerian, P.A. Regeneration of functional bladder substitutes using large segment acellular matrix allografts in a porcine model. *J. Urol.* 2000; 164: 936–41.

[56] Merguerian, P.A., Reddy, P.P., Barrieras, D.J., Wilson, G.J., Woodhouse, K., Bagli, D.J., McLorie, G.A., and Khoury, A.E. Acellular bladder matrix allografts in the regeneration of functional bladders: evaluation of large-segment (>24 cm) substitution in a porcine model. *BJU Int.* 2000; 85: 894–8.

[57] Portis, A.J., Elbahnasy, A.M., Shalhav, A.L., Brewer, A., Humphrey, P., McDougall, E.M., and Clayman, R.V. Laparascopic augmentation cystoplasty with different biodegradable grafts in an animal model. *J. Urol.* 2000; 164: 1405–11.

[58] Portis, A.J., Elbahnasy, A.M., Shalhav, A.L., Brewer, A.V., Olweny, E., Humphrey, P.A., McDougall, E.M., and Clayman, R.V. Laparoscopic midsagittal hemicystectomy and replacement of bladder wall with small intestinal submucosa and reimplantation of ureter into graft. *J. Endourol.* 2000; 14: 203–11.

[59] Gleeson, M.J. and Griffith, D.P. The use of aloplastic biomaterials in bladder substitution. *J. Urol.* 1992; 148: 1377.

[60] Bona, A.V. and De Gresti, A. Partial substitution of urinary bladder with Teflon prothesis. *Minerva Urol.* 1966; 18: 43.

[61] Monsour, M.J., Mohammed, R., Gorham, S.D., French, D.A., and Scott, R. An assessment of a collagen/vicryl composite memebrane to repair defects of the urinary bladder in rabbits. *Urology* 1987; 15: 235.

[62] Rohrmann, D., Albrecht, D., Hannappel, J., Gerlach, R., Schwarzkopp, G., and Lutzeyer, W. Alloplastic replacement of the urinary bladder. *J. Urol.* 1996; 156: 2094.

Atala, A. Tissue engineering in the genitourinary system. In: Atala, Moone, D. (Eds.) *Tissue Engineering* Boston: Birkhouser Press, 1997, p. 149.

[63] Atala, A., Freeman, M.R., Vacanti, J.P., Shepard, J., and Retik, A.B. Implantation *in vivo* and retrieval of artificial structures consisting of rabbit and human urothelium and human bladder muscle. *J. Urol.* 1993; 150: 608–12.

[64] Atala, A. Commentary on the replacement of urologic associated mucosa. *J. Urol.* 1995; 156: 338.

[65] De Boer, W.I., Schuller, A.G., Vermay, M., and van der Kwast, T.H. Expression of growth factors and receptors during specific phases in regenerating urothelium after acute injury *in vivo*. *Am. J. Pathol.* 1994; 145: 1199.

[66] Baker, R., Kelly, T., Tehan, T., Putman, C., and Beaugard, E. Subtotal cystectomy and total bladder regeneration in treatment of bladder cancer. *J. Am. Med. Assoc.* 1955; 168: 1178.

[67] Gorham, S.D., French, D.A., Shivas, A.A., and Scott, R. Some observations on the regeneration of smooth muscle in the repaired urinary bladder of the rabbit. *Eur. Urol.* 1989; 16: 440.

[68] Oberpenning, F.O., Meng, J., Yoo, J., and Atala, A. *De novo* reconstitution of a functional urinary bladder by tissue engineering. *Nat. Biotechnol.* 1999; 17: 2.

[69] Kershen, R.T., Yoo, J.J., Moreland, R.B., Krane, R.J., and Atala, A. Novel system for the formation of human corpus cavernosum smooth muscle tissue *in vivo*. *J. Urol.* 1998; 159(Suppl): 156.

[70] Park, H.J., Kershen, R., Yoo, J., and Atala, A. Reconstitution of human corporal smooth muscle and endothelial cells *in vivo*. *J. Urol.* 1999; 162: 1106–9.

[71] Falke, G., Yoo, J., Machado, M., Moreland, R., and Atala, A. Formation of corporal tissue *in vivo* using human cavernosal muscle and endothelium cells seeded on collagen matrices. *Tissue Eng.* 2003; 9: 871–9.

[72] Kwon, T.G., Yoo, J.J., and Atala, A. Autologous penile corpora cavernosa replacement using tissue engineering techniques. *J. Urol.* 2002; 168: 1754–58.

[73] Nukui, F., Okamoto, S., Nagata, M., Kurokawa, J., and Fukui, J. Complications and reimplantation of penile implants. *Int. J. Urol.* 1997; 4: 52.

[74] Thomalla, J.V., Thompson, S.T., Rowland, R.G., and Mulcahy, J.J. Infectious complications of penile prosthetic implants. *J. Urol.* 1987; 138: 65–7.

[75] Yoo, J.J., Lee, I., and Atala, A. Cartilage rods as a potential material for penile reconstruction. *J. Urol.* 1998; 160: 1164.

[76] Yoo, J.J., Park, H., Lee, I., and Atala, A. Autologous engineered cartilage rods for penile reconstruction. *J. Urol.* 1999; 162: 1119–21.

[77] Kershen, R.T. and Atala, A. Advances in injectable therapies for the treatment of incontinence and vesicoureteral reflux. *Urol. Clin.* 1999; 26: 81–94.

[78] Atala, A., Cima, L.G., Kim, W., Paige, K.T., Vacanti, J.P., Retik, A.B., and Vacanti, C.A. Injectable alginate seeded with chondrocytes as a potential treatment for vesicoureteral reflux. *J. Urol.* 1993; 150: 745–5.

[79] Atala, A., Kim, W., Paige, K.T., Vacanti, C.A., and Retik, A.B. Endoscopic treatment of vesicoureteral reflux with chondrocyte-alginate suspension. *J. Urol.* 1994; 152: 641.

[80] Diamond, D.A., and Caldamone, A.A. Endoscopic correction of vesicoureteral reflux in children using autologous chondrocytes: preliminary results. *J. Urol.* 1999; 162: 1185.

[81] Bent, A., Tutrone, R., McLennan, M., Lloyd, K., Kennelly, M., and Badlani, G. Treatment of intrinsic sphincter deficiency using autologous ear chondrocytes as a bulking agent. *Neurourol. Urodynam* 2001; 20: 157–65.

[82] Yokoyama, T., Chancellor, M.B., Watanabe, T., Ozawa, H., Yoshimura, N., de Groat, W.C., Qu, Z., and Huard, J. Primary myoblasts injection into the urethra and bladder as a potential treatment of stress urinary incontinence and impaired detrusor contractility ; long term survival without significant cytotoxicity. *J. Urol.* 1999; 161: 307.

[83] Chancellor, M.B., Yokoyama, T., Tirney, S., Mattes, C.E., Ozawa, H., Yoshimura, N., de Groat, W.C., and Huard, J. Preliminary results of myoblast injection into the urethra and bladder wall: a possible method for the treatment of stree urinary incontinence and impaired detrusor contractility. *Neurourol. Urodyn.* 2000; 19: 279–87.

[84] Strasser, H., Marksteiner, R., Eva, M., Stanislav, B., Guenther, K., Helga, F. et al. Transurethral Ultrasound Guided Injection of Clonally Cultured Autologous Myoblasts and Fibroblasts: Experimental Results. In: *Proceedings of the 2003 International Bladder Symposium*, 2003 March 6–9; Arlington, VA. Ridgefield, CT: National Bladder Foundation, 2003, p. 6.

[85] Bartsch, G., Yoo, J., Kim, B., and Atala, A. Stem cells in tissue engineering applications for incontinence. *J. Urol.* 2000; 1009S: 227.

[86] Raya-Rivera, A., Yoo, J., and Atala, A. Hormone producing granulosa stem cells for intersex disorders [abstract 29]. American Academy of Pediatrics Meeting, Urology section, Chicago, 2000.

[87] Lanza, R.P. et al. The ethical reasons for stem cell research. *Science* 2001; 293: 1299.

[88] Machluf, M. and Atala, A. Emerging concepts for tissue and organ transplantation. *Graft* 1998; 1: 31.

[89] Atala, A. and Lanza, R.P. *Methods of Tissue Engineering*. San Diego, CA: Academic Press, 2001.

[90] Atala, A. and Mooney, D. *Synthetic Biodegradable Polymer Scaffolds*. Boston, MA: Birkhaüser, 1997.

[91] Lanza, R.P., Cibelli, J.B., and West, M.D. Prospects for the use of nuclear transfer in human transplantation. *Nat. Biotechnol.* 1999; 17: 1171–4.

[92] Evans, M.J. et al. Mitochondrial, DNA genotypes in nuclear transfer-derived cloned sheep. *Nat. Genet.* 1999; 23: 90–3.

[93] Hiendleder, S., Schmutz, S.M., Erhardt, G., Green, R.D., and Plante, Y. Transmitochondrial differences and varying levels of heteroplasmy in nuclear transfer cloned cattle. *Mol. Reprod. Dev.* 1999; 54: 24–31.

[94] Steinborn, R., et al. Mitochondrial DNA heteroplasmy in cloned cattle produced by fetal and adult cell cloning. *Nat. Genet.* 2000; 25: 255–7.

[95] Amiel, G.E. and Atala, A. Current and future modalities for functional renal replacement. *Urol. Clin.* 1999; 26.

[96] Humes, H.D., Buffington, D.A., MacKay, S.M., Funke, A.J., and Weitzel, W.F. Replacement of renal function in uremic animals with a tissue-engineered kidney. *Nat. Biotechnol.* 1999; 17: 451455.

[97] Cieslinsk, D.A. and Humes, H.D. Tissue engineering of a bioartificial kidney. *Biotechnol. Bioeng.* 1994; 43: 781–91.

[98] MacKay, S.M., Kunke, A.J., Buffington, D.A., and Humes, H.D. Tissue engineering of a bioartificial renal tubule *ASAIO J.* 1998; 44: 179–83.

[99] Aebischer, P., Ip, T.K., Panol, G., and Galletti, P.M. The bioartificial kidney: progress towards an ultrafiltration device with renal epithelial cells processing. *Life Support Syst.* 1987; 5: 159–68.

[100] Ip, T., Aebischer, P., and Galletti, P.M. Cellular control of membrane permeability. Implications for a bioartificial renal tubule. *ASAIO Trans.* 1988; 34: 351–55.

[101] Humes, H.D. Renal replacement devices. In: Lanza, R.P., Langer, R., and Vacanti, J. Eds. *Principles of Tissue Engineering*, 2nd ed. San Diego: Academic Press, 2000, pp. 645–653.

[102] Amiel, A., Yoo, J., and Atala, A. Renal therapy using tissue engineered constructs and gene delivery. *J. Urol.* 2000; 18: 71–9.

[103] Lanz, R.P., Hayes, J.L., and Chick, W.L. Encapsulated cell technology. *Nat. Biotechol.* 1996; 14: 1107–11.

[104] Kuhtreiber, W.M., Lanza, R.P., and Chick, W.L. (Eds). *Cell Encapsulation Technology and Therapeutics*. Boston: Birkhauser, 1998.

[105] Lanza, R.P. and Chick, W.L. (Eds.) *Immunoisolation of Pancreatic Islets*. Austin, TX: R.G. Landes, 1994.

[106] Joki, T., Machluf, M., Atala, A., Zhu, J., Seyfried, N.T., Dunn, I.F., Abe, T., Carroll, R.S., and Black, P.M. Continuous release of endostatin from microencapsulated engineered cells for tumor therapy. *Nat. Biotechnol.* 2001; 19: 35–9.

[107] Lanza, R.P., Chung, H.Y., Yoo, J.J., Wettstein, P., Blackwell, C., Atala, A. et al. Generation of histocompatible tissues using nuclear transplantation. *Nat. Biotechnol.* 2002; 20: 689.

[108] Lanza, R.P. et al. Cloning of an endangered species (Bos gaurus) using interspecies nuclear transfer. *Cloning* 2000; 2: 79–90.

[109] Fischer Lindahl, K., Hermel, E., Loveland, B.E., and Wang, C.R. Maternally transmitted antigen of mice. *Ann. Rev. Immunol.* 1991; 9: 351–72.

[110] Hadley, G.A., Linders, B., and Mohanakumar, T. Immunogenicity of MHC class I alloantigens expressed on parenchymal cells in the human kidney. *Transplantation* 1992; 54: 537–42.

[111] Yard, B.A. et al. Analysis of T cell lines from rejecting renal allografts. *Kidney Int.* 1993; 43: S133–8.

[112] Bailey, D.W. Genetics of histocompatibility in mice. I. New loci and congenic lines. *Immunogenetics* 1975; 2: 249–56.

[113] Mohanakumar, T. The role of MHC and non-MHC antigens in allograft immunity. Austin, TX: R.G. Landes Company, 1994, pp. 1–115.

30

Hepatic Tissue Engineering for Adjunct and Temporary Liver Support

François Berthiaume
Arno W. Tilles
Mehmet Toner
Martin L. Yarmush
Harvard Medical School
Shriners Hospital for Children

Christina Chan
Harvard Medical School
Shriners Hospital for Children
Michigan State University

30.1 Introduction

Approximately 30,000 patients die each year from end-stage liver disease in the United States. About 80% of these patients have decompensated chronic liver disease, typically caused by alcoholism or chronic hepatitic C infection, and less commonly by a genetic — hepatocellular or anatomic — defect of liver function, or cancer. The other 20% die of acute liver failure (without preexisting chronic liver disease), which has various etiologies, including ischemia–reperfusion injury during liver surgery, acetaminophen poisoning, viral hepatitis, severe sepsis, idiosyncratic drug reactions, etc. Acute liver failure symptoms develop over a period of 6 weeks to 6 months and lead to death in over 80% of the cases, usually from cerebral edema, complications due to coagulopathy, and renal dysfunction. A more severe form of acute liver failure — fulminant hepatic failure — is characterized by a more rapid evolution (2 to 6 weeks).

Orthotopic liver transplantation (OLT) is the only clinically proven effective treatment for patients with end-stage liver disease. The majority of donor livers are obtained from brain-dead cadavers that still possess respiratory and circulatory functions at the time of organ retrieval. Expansion of the donor pool to include living donors, marginal and domino livers, as well as using split livers has been a major focus of transplant surgeons in the past few years. Although this may alleviate the donor organ shortage, there still remains a great potential benefit to developing alternatives that could be more cost-effective and less invasive, such as adjunct and temporary liver support. These approaches may find applications in the treatment of acute liver failure by allowing endogenous liver regeneration, as well as in chronic liver failure by ameliorating complications arising from the disease. Temporary liver support may also serve as a bridge to OLT by allowing more time to find a better match between donor and recipient or stabilize the patient prior to surgery.

30.2   Adjunct Internal Liver Support

A situation in which the native liver retains some functional capabilities is most amenable to adjunct liver support. The concept of adjunct liver support has been validated by the success of auxiliary partial liver transplantation [1,2]. However, primary nonfunction and vascular complications, for example, portal vein thrombosis, are more frequent with auxiliary partial liver transplantation than with whole liver transplantation. On the other hand, in certain situations, for example, fulminant hepatic failure (FHF), the native liver recovered and the patients could be safely removed from immunosuppressive drug therapy. Hepatocyte transplantation and hepatocyte-based implantable devices are an appealing alternative to auxiliary partial liver transplantation for several reasons (a) several patients could be treated with one single donor liver; (b) the implantation procedure could be performed using less invasive surgery; (c) isolated liver cells can be cryopreserved for long times; and (d) the liver cells could be genetically engineered *in vitro* to upregulate specific functions.

30.2.1   Hepatocyte Transplantation

Hepatocyte transplantation is the simplest form of adjunct internal liver support and has been investigated for over 25 years. In general, the efficiency of engraftment has been found to be quite low and a lag time, which may be as much as 48 h, is necessary before any clinical benefit occurs [3]. Thus, this approach offers an attractive prospect for correcting mostly nonemergency conditions such as inherited metabolic defects of the liver [4]. In early studies the choice of the transplantation site was dictated by accessibility and ease of procedure, as well as by spatial considerations: the pulmonary vascular bed, dorsal and inguinal fat pads, and peritoneal cavity. However, expression of liver-specific functions by transplanted hepatocytes could not be achieved in most of these ectopic sites. A microenvironment resembling that of liver, including a basement substrate to promote hepatocyte anchorage and a venous blood supply mimicking the mechanical and biochemical environment of the hepatic sinusoid is required [5]. The splenic pulp and the host liver itself are now the preferred sites for transplantation of hepatocytes [6]. When implanted into the spleen, hepatocytes may engraft locally or migrate into the liver. Some of the successes with hepatocyte transplantation in experimental animals, although often not very dramatic, have prompted clinical studies. The best results have been obtained in the treatment of specific metabolic disorders; however, except for one case with Crigler–Najjar syndrome type I, there was no detectable long-term function of transplanted human hepatocytes [7].

30.2.2   Implantable Devices

To improve the survival and function of implanted hepatocytes, the latter have been incorporated into biocompatible support materials, effectively constituting an implantable device. There are two major types of implantables devices (a) hepatocytes in open matrices that allow tissue — especially blood vessels — ingrowth from the host, thus leading to integration with the surrounding tissue, and (b) hepatocytes isolated from the surrounding environment in the host by a selective membrane barrier.

In early studies, isolated hepatocytes attached to collagen-coated dextran microcarriers were transplanted by intraperitoneal injection in two rat models of liver dysfunction (a) the Nagase analbuminemic rat, and (b) the Gunn rat, which has and inherited deficiency of bilirubin–uridine disphosphate glucuronosyltransferase activity causing a lack of conjugated bilirubin in the bile [8]. In both models, microcarriers promoted cell attachment, survival, and function of the transplanted hepatocytes.

Prevascularizing the cell polymer devices in combination with hepatotrophic stimulation have been used to encourage liver tissue regeneration around the implant [9]. Furthermore, materials that biodegrade at controlled rates *in vivo* (such as collagen and poly-lactic-glycolic co-polymers) can be used [10,11], and novel techniques, such as solid freeform fabrication can be used to reproducibly manufacture three-dimensional porous materials of well-defined pore size, distribution and interconnectivity [12,13]. Recently, novel biomaterials that are bioactive as well as resorbable have been developed [14]. For example, biomaterials are being designed to stimulate tissue repair through the release of factors that elicit specific cellular responses, such as cell proliferation, differentiation, and synthesis of extracellular matrix. Thus far, one of the more common approaches is to incorporate growth factors into tissue-engineering scaffolds [15,16]. There is also an interest in using "smart" materials consisting of stimuli-responsive polymers that change their properties in response to changes in the external environment [17].

30.2.3 Encapsulated Hepatocytes

A limitation of hepatocyte-based devices using open matrices is the need to use the host's own cells or at the very least an allogeneic cell source, both of which are very difficult to obtain, which seriously limits the usefulness of these devices. To circumvent this problem, there is great interest in using hepatocytes from xenogeneic sources. Since there is no immunosuppressive regimen that currently exists to prevent rejection of xenografts, hepatocytes have been encapsulated into small microspheres as well as into hollow fibers. In theory, encapsulating with a synthetic, permeable membrane provides a physical separation which protects the cells from the immune system of the host by excluding high molecular weight immunocompetent proteins (e.g., antibodies and complement) as well as leukocytes, while allowing free exchange of nutrients and oxygen. Nevertheless, if the microcapsule causes complement activation after implantation, the breakdown complement products could be small enough to enter the microcapsules and damage the transplanted cells. Initial applications of semipermeable microcapsules contained hemoglobin as blood substitute, enzymes to treat inborn errors of metabolism or absorbents to treat drug overdoses [18]. With advances in genetically engineered cells, microencapsulated cells have been used to remove ammonia in liver failure and amino acids such as phenylalanine in phenylketonuria [19]. Numerous studies have been performed with encapsulated hepatocytes without immunosuppressive drugs. Transplantation of microencapsulated xenogeneic hepatocytes into Gunn rats without immunosuppression reduced serum bilirubin levels for up to 9 weeks before returning to control rat levels, possibly due to the deterioration of the biomaterial [20]. The viability and function of encapsulated hepatocytes is highly dependent on the composition of the hollow fiber material [21]. Better results may be possible if angiogenesis near the capsule surface can be promoted [22], and the formation of a fibrotic layer around the capsule can be avoided [23].

30.3 Extracorporeal Temporary Liver Support

Extracorporeal temporary liver support systems are life-support systems that are analogous in concept to kidney dialysis machines, but specifically designed for liver failure patients. Since the liver has the ability to regenerate, temporary liver support may be sufficient to prevent patient death during the most severe phase of the illness, and allow regeneration of the host liver. The other main purpose of liver support systems is to provide a bridge to transplantation while awaiting a suitable donor. Table 30.1 provides a listing of the various techniques and systems currently being tested in a clinical setting.

30.3.1 Extracorporeal Whole Liver Perfusion

This technique was first used in humans in 1964 [24]. Xenogeneic (pig) livers were used for the first time in human studies in 1965 [25]. Although it fell out of favor due to the development of OLT, extracorporeal

TABLE 30.1 Clinical Trials for Temporary Extracorporeal Liver Support

Device[a]	Configuration	Cell mass and source	Perfusate and treatment protocol	Trial phase	Refs.
Whole liver perfusion					
Whole pig, baboon, or human liver			Whole blood, 5 h median perfusion time, most patients received 1 or 2 perfusions	I/II	[26,36]
Dialysis and filtration systems					
MARS (Teraklin AG, Rostock, Germany)	Albumin-loaded hemofilter, 60-kDa cut-off	None	Whole blood, 12–132 h	I/II	[123,124]
Liver Dialysis Unit (HemoCleanse Technologies, West Lafayette, IN)	Hemodiabsorption across 5-kDa cut-off cellulosic membranes	None	Whole blood, 6 h/day; up to 5 days	FDA approved	[38,39]
Prometheus (Fresenius Medical Care AG, Bad Homburg, Germany)	Hemofilter, 250-kD a cut-off, connected to two adsorber cartridges, in series with conventional dialyser	None	Whole blood, up to 12 h divided into 2 treatments over 2 days	I	[40,125]
Bioartificial livers					
HepatAssist (Circe Biomedical, Lexington, MA)	Hollow-fiber, polysulphone, 0.15–0.20 μm pore size	50 g cryopreserved primary porcine hepatocytes on microcarrier beads[b]	Plasma, 6 h/session; up to 14 sessions	II/III	[126–128]
BLSS (Excorp Medical, Oakdale, MN)	Hollow-fiber, cellulose acetate, 100-kDa cut-off	70–100 g primary porcine hepatocytes	Whole blood, 12 h/session; up to 2 sessions	I/II	[129,130]
MELS (Charité Virchow, Berlin, Germany)	Hollow-fiber (interwoven, multi-compartment)	250–500 g primary porcine or human hepatocytes	Plasma, continuous up to 3 days	I/II	[131]
ELAD (Vital Therapies, La Jolla, CA)	Hollow-fiber, cellulose acetate, 120-kDa cut-off	100 g human hepatoblastoma C3A cells per cartridge, up to 4 cartridges/device	Plasma, continuous up to 107 h	II (due to resume in 2004)	[132]
AMC-BAL (Hep-Art Medical Devices, B.V., Amsterdam, The Netherlands)	Spirally wound, nonwoven polyester matrix, no membrane	70–150 g[b] primary porcine hepatocytes	Plasma, up to 18 h/session; up to 2 sessions	I	[99,133]
Radial-flow bioreactor (Sant'Anna University Hospital, Italy)	Radial-flow bioreactor	230 g primary porcine hepatocytes	Plasma, 6–24 h treatments, mostly in one session	I/II	[134]
LiverX-2000 (Algenix, Inc., Minneapolis, MN)	Cells embedded in collagen matrix within hollow-fibers	40 g primary porcine hepatocytes per cartridge, 2 cartridges/device	Blood	I/II	[135][c]
Hybrid bioartificial liver (Hepatobiliary Institute of Nanjing University, China)	Polysulfone hollow-fiber cartridge with 100-kDa cut-off combined with adsorption column	100 g primary porcine hepatocytes[b]	Plasma, one 6 h treatment, except one patient with 2 × 6 h treatments	I	[136]

[a] ELAD, Extracorporeal liver assist device; BLSS, bioartificial liver support system; MELS, modular extracorporeal liver system; AMC–BAL, Amsterdam bioartificial liver system; MARS, molecular adsorbent recycling system; FDA, Food and Drug Administration.

[b] Based on 10^8 hepatocytes/g liver tissue.

[c] Clinical data not yet published, reference describes device design.

whole liver perfusion has experienced renewed interest in recent years. Pascher et al. [26] analyzed data from nearly 200 patients from studies conducted from 1964 to 2000, and overall the study concluded that extracorporeal whole liver perfusion was not superior to conventional intensive care treatment approaches.

Early on, a major challenge of this technique was the relatively poor and unstable function of the extracorporeal liver and hemodynamic instability of the patient, which have been improved by using dual vessel (hepatic artery and portal vein) perfusion [27]. Studies in the period ranging from 1990 to 2000 have shown that patients treated with an extracorporeal liver and that survived all eventually received OLT; thus, the extracorporeal liver was potentially effective as a bridge to transplantation, but not as a substitute.

The shortage of human donors provides a strong motivation for the use of xenogeneic livers. Transgenic pigs and immunoabsorption techniques have been used to reduce the effects of hyperacute rejection [28–30]. Given that immune function is severely depressed in acute liver failure patients, this may not be necessary for perfusions lasting 24 to 36 h [31–33]. Borie et al. [34] described an alternative approach to isolate the xenogeneic liver from the host whereby a pig liver is perfused with pig blood in a secondary circuit which is separated from the host's blood by a microporous membrane. Although it appears that immune incompatibilities could be addressed, some of the data suggest that baboon and human livers may be more effective than livers from other species, suggesting an important role for proper matching of the metabolic and physiological activities of the extracorporeal liver and host [35,36].

30.3.2 Dialysis and Filtration Systems

The first attempts at developing devices for temporary and adjunct liver support consisted of nonbiological devices incorporating hemodialysis, hemofiltration, and/or plasma exchange units aimed at removing toxins accumulating in the patient's blood. Charcoal perfusion, the most extensively characterized nonbiological method, showed benefits in various animal models, but no survival benefit was reported in the only one reported randomized clinical trial [37]. Recently, there has been renewed interest in further refining these approaches, with three different systems at various stages of clinical assessment.

The Liver Dialysis Unit (Hemocleanse Technologies, West Lafayette, IN), an approved device in the United States since 1996, is a modified dialysis machine wherein blood is dialyzed against a solution that is continuously recycled through a mixture of sorbents including activated charcoal and an ion-exchange resin. In several small randomized prospectively controlled trials carried out from 1992 to 1998, patients were treated 6 h/day for 1 to 5 days, and the results showed a better outcome with patients with acute on chronic liver failure, although there was no benefit for patients with FHF [38]. The lack of benefit in FHF patients was attributed to the inability to clear strongly protein- or lipid-bound toxins, including bilirubin, endotoxin, and inflammatory cytokines, that are too big to go across the 5 kDa molecular weight cut-off of the dialysis module. A modified version of the device includes a plasma filter module wherein plasma interacts directly with the sorbent particles, thus eliminating this barrier. In preliminary clinical studies, including the plasma filter module resulted in decreases in bilirubin, aromatic amino acids, ammonium, creatinine, and inflammatory mediators such as interleukin-1β [39], but not enough information was available to make conclusions on the overall clinical benefit. A "second-generation" system is currently being designed as kit to convert an existing kidney dialysis into a liver dialysis system, and being touted as a more cost-effective system than its predecessor.

The Prometheus System (Fresenius Medical Care AG, Bad Homburg, Germany) is conceptually very similar to the competing system described above, in that it also includes of two separate modules (a) a high-flux dialyzer that removes water soluble toxins, and (b) a plasma filter module. The latter consists of a large pore (250 kDa) hollow-fiber module that enables albumin along with hydrophobic toxins bound to it to enter a closed loop circuit that contains sorbent materials that strip off the toxins and free up the binding sites on albumin before it is returned to the blood stream. A clinical study in patients with acute-on-chronic liver failure with accompanying hepatorenal syndrome shows that treatment decreased circulating levels of many toxins, such as ammonia, bilirubin, bile acids, etc. although there was no improvement in the hepatic encephalopathy score [40]. More information on clinical efficacy awaits prospective controlled studies with longer treatment periods.

The Molecular Adsorbent Recirculating System, also known as "MARS" (Teraklin, Rostock, Germany) is a device wherein the patient's blood is dialyzed against an albumin solution, the latter of which is recycled continuously over stripping columns containing various sorbent materials, including activated charcoal [41]. The dialysis membrane has a pore size of 50 kDa, which in principle allows small water-soluble toxins (such as ammonia) to escape, and a hydrophobic coating which allows albumin-bound liposoluble substances in the blood (such as bilirubin and benzodiazepines) to cross the membrane and be picked up by the albumin in the dialysate. In this design, a single module can remove both water-soluble and lipid-soluble toxins. Furthermore, the patient's blood contacts the biocompatible membrane only, and never comes into direct contact with the sorbent materials. Over 3000 patients with various etiologies of liver dysfunction have been treated with this device, and generally show neurological improvement, hemodynamic stabilization, and better hepatic and kidney functions following treatment [42,43]. The small number of controlled trials available for acute liver failure also suggest increased survival in MARS-treated patients [44–47]. Larger multicenter trials in the United States and Europe are currently under way to confirm these very encouraging, yet preliminary, findings. Evidence shows that markers of oxidative stress and systemic inflammation are also reduced after MARS treatment [48].

A meta-analysis published early in 2003 suggests that, overall, artificial liver support systems containing no cells significantly reduced mortality in acute-on-chronic liver failure, although not acute liver failure [49]. Preliminary economic evaluations of such treatments have been performed. In one study with cirrhosis patients undergoing superimposed acute liver injury, cost savings due to reduced liver-disease-related complications more than offset the additional cost of MARS treatment relative to conventional therapy [50]. In another study with patients with acute-on-chronic liver failure due to alcoholic liver disease, cumulative costs per patient in the first year were much higher in the MARS-treated group, although the main explanation appears to be an increase in mean survival time of the patients [51].

30.3.3 Bioartificial Livers

Although such devices are in principle more complex than dialysis and filtration systems, they could provide biochemical and synthetic functions that are not available in the systems containing no cells [52]. The mechanisms of liver failure are not yet well understood and the most critical hepatic functions in patients undergoing liver failure not known; therefore, it is yet unclear whether dialysis and filtration systems, which are likely to be cheaper, will supplant hepatocyte- or cell-based bioartificial livers.

30.3.3.1 Long-Term Hepatocyte Culture Systems

The availability of stable long-term liver cell culture systems that express high levels of liver-specific functions is an essential step in the development of liver-assist devices using hepatocytes. Three types of long-term culture techniques for adult hepatocytes have been used for bioartificial liver development: (a) co-culture of hepatocytes with a "feeder" cell line, such as fibroblasts, (b) three-dimensional network of collagen or other matrix, and (c) hepatocyte aggregates or spheroids. Hepatoma cell lines, which do not require specific substrate configurations, have been used as well. Some of these techniques can be combined; for example, Takezawa et al. [53,54] used thermally responsive polymer substrates to develop multicellular spheroids of fibroblasts and hepatocytes.

30.3.3.1.1 Introducing Non-Parenchymal Cells

Approximately 20 years ago, it was discovered that hepatocytes could be cultured on "feeder or supportive cells" to maintain their viability and function [55]. More recent studies showed that nonhepatic cells, even from other species, could be used. In these culture systems, cell–cell interactions among hepatocytes and cells of another type (rat liver epithelial cells, liver sinusoidal endothelial cells, or mouse embryonic fibroblasts), or "heterotypic interactions," are critical for the expression of hepatocellular functions. The disadvantages of co-culture systems include the potential variability in the cell line used, and the additional work needed to propagate that cell line in addition to attending to the isolation of hepatocytes.

It may be desirable to optimize heterotypic cell–cell interactions in order to maximize the expression of liver-specific functions of the co-cultures. Keeping in mind that cells cultured on surfaces do not usually layer onto each other (except for malignant cancer cell lines), random seeding using a low ratio of parenchymal cells to feeder cells will achieve this goal, but at the expense of using a lot of the available surface for fibroblasts, which do not provide the desired metabolic activity. On the other hand, micropatterning techniques enable the optimization of the seeding pattern of both cell types so as to ensure that each hepatocyte is near a feeder cell while minimizing the number of feeder cells [56]. As a result, metabolic function per area of culture is increased and the ultimate size of a BAL with the required functional capacity is reduced. In prior studies using circular micropatterns, function per hepatocyte increased when the hepatocyte circle diameter decreased, and function per unit area of culture increased when the space occupied by fibroblasts in-between the hepatocyte islands decreased (for a constant cell number ratio of the two cell types).

Various methods for patterning the deposition of extracellular matrix or other cell attachment factors onto surfaces have been developed [57]. Photolithography involves spin-coating a surface (typically silicon or glass) with a ~1 μm thick layer of photo-resist material, exposing the coated material to ultraviolet light through a mask which contains the pattern of interest, and treating the surface with a developer solution which dissolves the exposed regions of photo-resist only. This process leaves photo-resist in previously unexposed areas of the substrate. The exposed areas of substrate can be chemically modified for attaching proteins, etc., or treated with hydrofluoric acid to etch the material. The etching time controls the depth of the channels created. The etched surfaces produced by photolithography can be used to micromold various shapes in a polymer called poly(dimethylsiloxane) (PDMS). The PDMS cast faithfully reproduces the shape of the silicon or glass mold to the micrometer scale, and can be used in various "soft lithography" techniques, including microstamping, microfluidic patterning, and stencil patterning. An infinite number of identical PDMS casts can be generated from a single master mold, which makes the technique very inexpensive. Soft lithography methods can be used on virtually any type of surface, including curved surfaces, owing to the flexibility of PDMS.

30.3.3.1.2 Hepatocyte Functional Heterogeneity
In the hepatic lobule, blood flows from the periportal outer region towards the central hepatic vein. Hepatocytes in the periportal, intermediate or centrilobular, and perivenous zones exhibit different morphological and functional characteristics. Spatial heterogeneity in the hepatic lobule is clearly important for some aspects of hepatic function (Figure 30.1a). For example, urea synthesis is a process with high capacity to metabolize ammonia but low affinity for the substrate. Ammonia removal by glutamine synthesis is a high affinity process which removes traces of ammonia which cannot be metabolized by the urea cycle [58]. Co-expression of both enzyme systems would not be productive because the higher affinity process (glutamine synthesis) would be saturated under most operating conditions, leading to a reduced efficiency in ammonia extraction. On the other hand, replicating the functional heterogeneity of hepatocytes in the lobule would likely enhance the performance at the tissue level.

Functional heterogeneity also has important implications in the metabolism of hepatotoxins such as acetaminophen. Acetaminophen is normally degraded by glucuronidation and sulfation reactions which are uniformly distributed along the acinus. After acetaminophen overdose, these processes are saturated and cytochrome P450 activities primarily located in the centrilobular region metabolize significant amounts to toxic metabolites causing oxidative stress and protein cross-linking. Although these metabolites can be detoxified by glutathione-dependent reactions, centrilobular hepatocytes do not have an efficient glutathione recycling system, and as a result are the main target of acetaminophen-induced hepatotoxicity. Repeat exposure to incremental doses of acetaminophen increases the tolerance to hepatic damage by partially shifting the expression of cytochrome P450 towards the periportal region [59], which has the most active glutathione recycling metabolism in the liver [60].

The maintenance of functional heterogeneity in the liver is dependent on several factors, including gradients of hormones, substrates, oxygen, and extracellular matrix composition, although the relative importance of each one of these factors is currently unknown. In one study where hepatocytes were

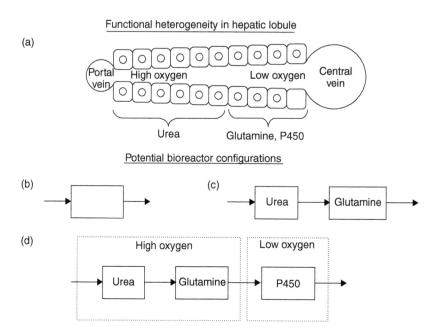

FIGURE 30.1 Potential bioreactor configurations in a bioartificial liver. (a) *In vivo* distribution of hepatocellular functions. (b) Single unit optimized to perform all functions. (c) Two subunits, the first one optimized for ammonia conversion to urea, a high capacity but low affinity process, and the second one optimized for ammonia conversion to glutamine, a high affinity process that scavenges ammonia not metabolized in the first subunit. (d) Three subunits, the first two designed to clear ammonia under high oxygen tension, and a third subunit operating at lower oxygen tension that is optimized for efficient P450 detoxification pathways.

chronically exposed to increasing oxygen tensions within the physiological range of about 5 mmHg (perivenous) to 85 mmHg (periportal), urea synthesis increased about 10 fold, while P450IA1 activity decreased slightly and albumin secretion was unchanged [61]. These data suggest that by creating environmental conditions which emulate certain parts of the liver sinusoid, it is possible to modulate hepatocyte metabolism in a way that is consistent with *in vivo* behavior. Spatial control of the layout of the cells in the device may be achieved using micropatterning and microfabrication techniques [62,63], or using separate bioreactor modules that are optimized to perform a subset of hepatocellular functions, as illustrated in Figure 30.1b–d.

It is also possible to profoundly affect the expression of liver-specific functions by hepatocytes by changing a number of environmental conditions in the bioreactor environment. For example, urea synthesis dramatically increases with increasing oxygen tension while cytochrome P450 decreases [64]. Amino acid supplementation to human plasma increases urea and albumin synthesis, as well as cytochrome P450 activities [65]. Co-culture with mouse 3T3 cells also increases albumin and urea secretion to levels which exceed *in vivo* rates severalfold [66,67]. While albumin and urea secretion decrease at higher fluid shear rates, the latter tend to increase cytochrome P450 detoxification rates, at least in the short term [68]. Sophisticated optimization techniques that can tackle the large number of adjustable environmental variables may be helpful for optimizing the bioreactor environment [69,70]. Since varying one specific environmental condition increases the expression of liver-specific functions many times, it is reasonable to assume that optimization of several such parameters simultaneously may yield an order of magnitude or more in improvement.

30.3.3.1.3 Pre-Conditioning Hepatocytes Prior to Plasma Exposure
Rat hepatocytes which are seeded and maintained in standard hepatocyte culture medium and then exposed to either rat or human plasma become severely fatty within 24 h with a concomitant reduction in

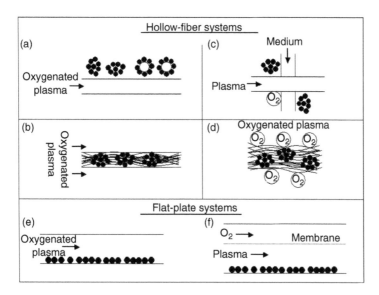

FIGURE 30.2   Most common bioreactor design for bioartificial livers. (a) Hepatocyte aggregates or seeded on microcarriers are placed on the outside of hollow fibers. Oxygenated plasma is flown through the hollow fibers. (b) Hepatocyte aggregates in a supporting matrix are inside hollow fibers and oxygenated plasma is flown outside the hollow fibers. (c) Similar to panel A, although separate hollow fibers are used to deliver hepatocyte culture medium and oxygen into the system. Circle with O_2 is a hollow fiber perpendicular to the plane of the paper. (d) Hepatocyte aggregates are in a supporting matrix next to hollow fibers that deliver oxygen. Oxygenated plasma is flown in the space outside of the hollow fibers and percolates through the matrix–hepatocyte network. (e) Hepatocytes are seeded as a monolayer on the bottom surface of a flat plate and placed within a parallel-plate flow chamber. Oxygenated plasma is flown directly above the cells. (f) System is similar to panel E, except that oxygen is delivered through a permeable membrane directly above the flow channel with the hepatocytes.

liver-specific functions. Thus, plasma appears to be a rather inhospitable environment to the hepatocytes, yet it is clear that hepatocytes must be made to tolerate it for the concept of bioartificial liver to become reality. Supplementation of human anticoagulated plasma with hormones and amino acids (to bring those metabolites to levels similar to that found in standard hepatocyte culture medium) eliminate intracellular lipid accumulation and restores albumin and urea synthesis as well as P450-dependent detoxification [71,72]. However, direct supplementation of plasma, especially with respect to the high levels of hormones used, would be very costly and pose a health risk to the patient.

In prior studies, we have shown that the culture conditions used prior to placing the hepatocytes in contact with human plasma as well as during plasma exposure, can dramatically affect hepatocellular metabolism. For example, hepatocytes cultured in standard hepatocyte culture medium containing supra-physiological levels of insulin become fatty once they are exposed to plasma, and this can be prevented by "preconditioning" the cells in a medium containing physiological levels of insulin [64]. Direct amino acid supplementation to the plasma also increased both urea and albumin secretion rates by the hepatocytes. Thus, a combination of preconditioning and plasma supplementation can be used to upregulate liver-specific functions of hepatocytes during plasma exposure.

30.3.3.2 Hepatocyte Bioreactor Designs

The most popular bioreactor designs are shown in Figure 30.2 and discussed in greater detail below. Most devices tested clinically consist of hollow fiber cartridges containing either porcine hepatocytes or human hepatoblastoma cells. In most cases, cells are loaded into the extraluminal compartment and patient plasma or blood is perfused through the fiber lumens [73–75]. Similar hollow fiber cartridges have also been used in animal studies with hepatocytes seeded inside the fibers and the plasma flowing over the outer surface of the fibers [76,77]. Because of the relatively large diameter of the fibers as well as transport limitations associated with the fiber wall, these systems are prone to substrate transport limitations [78].

30.3.3.2.1 Minimum Cell Mass and Functional Capacity

The cell mass required to support an animal model of hepatic failure has not been systematically determined. Prior studies have shown significant improvements in various parameters using as low as 2 to 3% of the normal liver mass of the animal [79,80]. Devices that have undergone clinical testing have used 6×10^9 to 1×10^{11} porcine hepatocytes [81,82] or 4×10^{10} C3A cells [83]. Recently, in an experimental pig model of hepatic failure, treatment with a bioartificial liver containing 6×10^8 pig hepatocytes (about 3 to 5% of the liver mass) significantly improved survival [84]. Recently, there have been efforts to improve cell viability in large-scale devices. Hepatocytes have been transfected with an antiapoptotic gene (nitric oxide synthase) or exposed to an antiapoptotic drug (ZVAD-fmk) to increase their resistance to what appears to be mainly hypoxic injury [85,86].

Clinical improvements have also been seen in hepatocyte transplantation studies using less than 10% of the host's liver mass. Intrasplenic transplantation of 2.5×10^7 allogeneic rat hepatocytes (about 5% of the rat liver mass) prolonged the survival, improved blood chemistry, and lowered blood TGF-β_1 (an inhibitor of hepatocyte growth) levels in anhepatic rats [87]. In another study using reversibly transformed human hepatocytes, 50×10^6 cells were injected intra-splenically into rats subjected to a 90% hepatectomy [88]. In a recent study on humans with acute liver failure, intrasplenic and intra-arterial injections of human hepatocytes (ranging from 10^9 to 4×10^{10} per patient, i.e., 1 to 10% of the total liver mass) transiently improved several blood chemistry parameters and brain function after a lag time of about 48 h, but did not improve survival [3]. The lag time before any benefit is observed may reflect the time required for the engraftment of the cells in the host liver. Better survival of the injected cells may be possible if the cells are seeded in prevascularized polymeric scaffolds [89]. The relatively low number of hepatocytes needed to effect a therapeutic benefit may be due, in part, to the fact that the exogenously supplied hepatocytes may aid the regeneration of the native liver [79].

Assuming that the minimum cell mass necessary to support a patient undergoing acute liver failure is about 5 to 10% of the total liver weight, this yields a bioartificial liver containing about 10^{10} cells. Designing this system with a priming volume not exceeding about 1 l is still a daunting challenge. Knowing which functions are most critical would help to rationally improve the efficacy of bioartificial liver systems and dramatically reduce the minimum therapeutic cell mass. For example, it is well known that hepatocytes exhibit a metabolic zonation along the acinus [61]. Periportal and centrilobular hepatocytes express high levels of urea cycle enzymes and low levels of glutamine synthetase while pericentral hepatocytes are the opposite [58]. Another example is the reduction in albumin synthesis during the acute phase response, a process which may help sustain the increased level of acute phase proteins [90].

30.3.3.2.2 Oxygen Transport Issues

In a normal liver, no hepatocyte is further than a few micrometers from circulating blood; thus, transport by diffusion only has to occur over very short distances. Although oxygen diffusivity is an order of magnitude greater that that of many other small metabolites (e.g., glucose and amino acids), it has a very low solubility in physiological fluids deprived of oxygen carriers. Thus, it is not possible to create large concentration gradients which would provide the driving force for rapid oxygen transport over long distances. This, in addition to the fact that hepatocytes have a relatively high oxygen uptake rate [91,92], makes oxygen transport the most constraining parameter in the design of BAL devices.

Oxygen transport and uptake of hepatocytes has been extensively studied in the sandwich culture configuration in order to obtain the essential oxygen uptake parameters needed in the design of bioreactor configurations [93]. The maximum oxygen uptake rate of cultured rat hepatocytes was measured to be about 13.5 pmol/sec/μg DNA, which is fairly stable after the first day in culture and for up to 2 weeks. Interestingly, the oxygen uptake was about twice in the first day after cell seeding, presumably because of the increased energy requirement for cell attachment and spreading. This may need to be taken into account when seeding hepatocytes into a BAL. Oxygen uptake was not sensitive to the oxygen tension in the vicinity of the hepatocytes up to a lower limit of about 0.5 mmHg, below which oxygen uptake decreased, suggesting that it becomes a limiting substrate for intercellular hepatocyte metabolism. Since oxygen is essential for hepatic ATP synthesis, a reasonable design criterion is that the oxygen tension should remain above ~0.5 mmHg.

Based on these parameters, it is possible to estimate oxygen concentration profiles in various bioreactor configurations based on a simple diffusion–reaction models assuming Michaelis–Menten kinetics. Generally, one can estimate that the maximum thickness of a static layer of aqueous medium on the surface of a confluent single hepatocyte layer is about 400 μm [94]. Calculations on oxygen transport through hepatocyte aggregates suggest that even a relatively low density of cells (10^7 cells/cm^3) cannot have a thickness exceeding about 300 to 500 μm. At cell densities of 10^8 cells/cm^3, which is similar to that found in normal liver, that thickness is only 100 to 200 μm.

30.3.3.2.3 Hollow-Fiber Systems
The hollow fiber system has been the most widely used type of bioreactor in BAL development [77,95]. The hollow fiber cartridge consists of a shell traversed by a large number of small diameter tubes. The cells may be placed within the fibers in the intracapillary space or on the shell side in the extracapillary space. The compartment which does not contain the cells is generally perfused with culture medium or the patient's plasma or blood. The fiber walls may provide the attaching surface for the cells and/or act as barrier against the immune system of the host. Microcarriers have also been used as a way to provide an attachment surface for anchorage-dependent cells introduced in the shell side of hollow fiber devices. There are many studies on how to determine optimal fiber dimensions, spacing, and reactor length based on oxygen transport considerations [78].

One difficulty with the hollow fiber configuration is that interfiber distances, and consequently transport properties within the shell space, are not well controlled. Thus, it may be advantageous to place cells in the lumen of small fibers because the diffusional distance between the shell (where the nutrient supply would be) and the cells is essentially equal to the fiber thickness. In one configuration, hepatocytes have been suspended in a collagen solution and injected into the lumen of fibers where the collagen is allowed to gel. Contraction of the collagen lattice by the cells even creates a void in the intraluminal space, which can theoretically be perfused with hormonal supplements, etc. to enhance the viability and function of the cells, while the patient's plasma flows on the shell side. Because of the relatively large diameter of the fibers used as well as transport limitations associated with the fiber wall, these systems have been prone to substrate transport limitations.

To improve oxygen delivery, novel designs using additional fibers which carry gaseous oxygen straight into the device have been used [96,97]. Using this approach, Gerlach et al. [82] were able to demonstrate that hepatocytes could express differentiated functions over several weeks. Using a device consisting of hepatocytes seeded onto a woven polyester substrate with integrated hollow fibers for oxygen supply, Flendrig et al. [96,98] showed that the survival time of pigs undergoing total hepatic ischemia was significantly increased over the control group; more recently, this device was successfully used to treat seven acute liver failure patients, of which six were bridged to a transplant and one spontaneously recovered [99].

30.3.3.2.4 Parallel Plate Systems
An alternative bioreactor configuration is based on a flat surface geometry [80,93,100,101] where it is easier to control the internal flow distribution and ensure that all cells are adequately perfused. Its main drawback is that it is difficult to build a system which contains a sufficient cell concentration (Figure 30.3). For example, a channel height of 1 mm would result in a 10 l reactor to support 20×10^9 hepatocytes cultured on an area of 10 m^2. For a liver failure patient who is probably hemodynamically unstable, it is generally accepted that the priming volume of the system should not exceed 1 l.

The volume of the device in the flat-plate geometry can be decreased by reducing the channel height (Figure 30.3). However, this forces the fluid to move through a smaller gap, which rapidly increases the drag force (shear stress) imparted by the flow on the cells. Recent data suggest that hepatocyte function decreases significantly at shear stresses >5 dyn/cm^2 [102]. To reduce the deleterious effects of high shear,

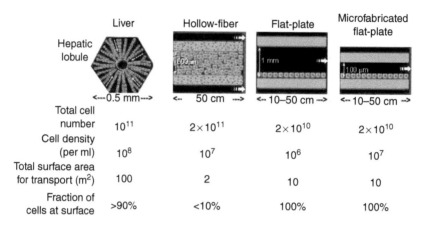

FIGURE 30.3 Comparison between popular hepatocyte bioreactor configurations.

it may be possible to use grooved surfaces where cells lodge and are less exposed to the shear stress, as previously done for blood cells [103,104]. Cells lodge inside the grooves where they are less exposed to the shear stress, which allows for faster flow without causing cell damage. The grooves may on the other hand significantly increase the fluid hold-up volume [105].

In an attempt to provide to cells adequate oxygenation and protection from shear in perfused bioreactors, gas-permeable membranes as well as membranes separating cells from plasma have been incorporated into the flat-plate geometry. Recently, a flat-plate microchannel bioreactor where cells directly contact the circulating medium was developed [93]. The channel is closed by a gas-permeable membrane on one surface, which decouples oxygen transport from the flow rate in the device. Comparing this with a similar flat-plate design where a nonpermeable glass surface is substituted to the membrane, internal membrane oxygenation removed the oxygen limitations that occur at low volumetric flow rates [105]. De Bartolo et al. [80] incorporated two membranes into the flat-plate geometry. The first membrane is gas-permeable and minimizes the oxygen transport limitations in the system. The second membrane separates cells from plasma and adds a significant barrier to the transport of protein-bound toxins that need to be processed by the cells [78]. Others have reported the design of a radial flow bioreactor with an internal membrane oxygenator for the culture of hematopoietic cells [106]. Based on a theoretical analysis, the proposed design would have removed oxygen transport limitations in the bioreactor, but no experimental data were shown.

30.3.3.3 Potential Sources of Cells for Bioartificial Livers

Although several technical difficulties remain to be addressed with respect to the design of implantable and extracorporeal liver-assist devices, clearly a major hurdle for both approaches is the procurement of a sufficient number of cells that are required to achieve a therapeutic effect. Human hepatocytes appear to be the "natural" choice for hepatocyte transplantation, internal and external liver assist devices, however, they are scarce due to a competing demand of OLT. Whether adult human hepatocytes can be induced to replicate *in vitro* and the daughter cells express high levels of liver-specific functions remains to be shown. Human hepatocyte cell lines have been developed via spontaneous transformation [107], as well as via retroviral transfection of the simian virus 40 large T antigen [108]. Recently, a novel technology which uses a reversible transformation strategy with the SV40 T antigen and Cre–Lox recombination was used to grow human hepatocytes *in vitro* [88]. These cells, when transplanted into the spleen of 90% hepatectomized rats, improved biochemical and clinical parameters. In bioartificial liver devices tested so far, the only human cells used have been the cancer-derived C3A line [74,83]. However, one study suggests that C3A cells have lower levels of P450IA1 activity, ammonia removal, and amino acid metabolism that adult porcine hepatocytes [109]. Furthermore, when using immortalized human cell lines, there are concerns with the possibility of transmission of tumorigenic products into the patient. Xenogeneic hepatocytes offer no risk of transmitting malignancies to the patient, but pose other problems, including the risk of hyperacute rejection [110], transmission of zoonoses [111], and potential mismatch between xenogeneic and human liver functions. The first two could be addressed by dedicated breeding programs of transgenic animals. On the other hand, little is known about the third factor.

Although no one has achieved the goal of generating a safe, fully functional yet clonal, immortalized, or genetically engineered human cell that can be substituted for primary hepatocytes, a new promising avenue is the discovery of liver stem cells. The existence of hepatic stem cells was hypothesized over 40 years ago [112], and recent data suggest that there are stem cells present within [113–115] as well as outside the liver [116], which can differentiate into fully mature hepatocytes. *In vitro* studies suggest the presence of a subpopulation of small hepatocytes in rat liver with a high proliferative potential [117]. Three independent studies in rats, mice, and humans have shown that a major extrahepatic source is stem cells of the bone marrow which may take part in normal tissue renewal as well as in liver regeneration after severe experimentally induced hepatic injury [116,118,119].

30.3.3.4 Techniques for Preservation of Hepatocytes and Liver Cells

The development of optimal preservation protocols for hepatocytes that enable the storage and ready availability of cells for BALs, has been the subject of several studies. Hepatocytes have been cryopreserved shortly after isolation as well as after culture for several days. Compared to isolated cells, cultured hepatocytes exhibit greater resistance to high concentrations of the cryoprotective agent dimethyl sulfoxide, as evidenced by preservation of cell viability, cytoskeleton, and function. Based on experimental and theoretical studies, cooling rates between 5 and 10°C/min caused no significant decrease in albumin secretion rate compared to control, unfrozen, cultures [120,121]. There have been attempts to store hepatocyte cultures in various solutions used for cold storage of whole donor livers. One study showed that cultured hepatocytes maintained at 4°C lose significant viability after a few hours of cold storage, but that addition of polyethylene glycol significantly extends functionality and survival [122]. Interestingly, the use of the University of Wisconsin (UW) solution, currently the most widely used solution for cold organ storage, has not performed better than leaving the cells in standard hepatocyte culture medium. It is conceivable that the UW solution mediates its effect by prolonging the survival of nonparenchymal cells. It is hoped that further improvements in preservation solutions will enable the storage of BAL systems, as well as lengthen the useful cold storage time of whole livers for transplantation.

30.4 Summary

The severe donor liver shortage, high cost, and complexity of orthotopic liver transplantation have prompted the search for alternative treatment strategies for end-stage liver disease that would require less donor material, be cheaper, and less invasive. Adjunct internal liver support, which may be provided via auxiliary partial liver transplantation or hepatocyte transplantation, is most suitable for cases where the native liver retains some functional capabilities, and may be a cure for patients who suffer from specific metabolic disorders. Acute liver failure patients will benefit most from extracorporeal temporary liver support, which can be used as a bridge to transplantation, or as a means to support the patient until its own liver regenerates. Currently, there are three approaches for extracorporeal temporary liver support: extracorporeal liver perfusion, dialysis and filtration systems containing no cells, and bioartificial livers. Dialysis and filtration systems, which do not contain any living cells, are ahead with respect to clinical testing and gaining regulatory approval. A concern with such systems is that their efficacy may be limited due to the lack of metabolic and protein synthetic activities which are normally present in the liver. Bioartificial livers containing liver cells would overcome this limitation, and have passed the "proof of principle" test in preclinical and clinical studies, although tangible clinical benefits have not yet been demonstrated. Important unresolved issues for bioartificial livers and extracorporeal liver perfusion are the identification of a reliable cell/tissue source and a better understanding of metabolic and immune incompatibilities arising from the use of allogeneic and xenogeneic liver cells. Ultimately, several temporary and adjunct treatment approaches may be available, and the best choice may depend on the etiology of liver failure in each individual patient.

Acknowledgments

This work was partially supported by the NIH grant R01 DK37743 and the Shriners Hospitals for Children.

References

[1] Boudjema, K. et al., Auxiliary liver transplantation and bioartificial bridging procedures in treatment of acute liver failure, *World J. Surg.*, 26, 264, 2002.

[2] Azoulay, D. et al., Auxiliary partial orthotopic versus standard orthotopic whole liver transplantation for acute liver failure: a reappraisal from a single center by a case-control study, *Ann. Surg.*, 234, 723, 2001.

[3] Bilir, B.M. et al., Hepatocyte transplantation in acute liver failure, *Liver Transplant.*, 6, 32, 2000.

[4] Fox, I.J. et al., Treatment of the Crigler-Najjar syndrome type I with hepatocyte transplantation, *N. Engl. J. Med.*, 338, 1422, 1998.

[5] Fox, I.J. and Roy-Chowdhury, J., Hepatocyte transplantation, *J. Hepatol.*, 40, 878, 2004.

[6] Gupta, S., Bhargava, K.K., and Novikoff, P.M., Mechanisms of cell engraftment during liver repopulation with hepatocyte transplantation, *Semin. Liver Dis.*, 19, 15, 1999.

[7] Fox, I.J. and Chowdhury, J.R., Hepatocyte transplantation, *Am. J. Transpl.*, 4, 7, 2004.

[8] Demetriou, A.A. et al., Survival, organization, and function of microcarrier-attached hepatocytes transplanted in rats, *Proc. Natl Acad. Sci. USA*, 83, 7475, 1986.

[9] Fontaine, M. et al., Human hepatocyte isolation and transplantation into an athymic rat, using prevascularized cell polymer constructs, *J. Pediatr. Surg.*, 30, 56, 1995.

[10] Hasirci, V. et al., Expression of liver-specific functions by rat hepatocytes seeded in treated poly(lactic-co-glycolic) acid biodegradable foams, *Tissue Eng.*, 7, 385, 2001.

[11] Ranucci, C.S. et al., Control of hepatocyte function on collagen foams: sizing matrix pores toward selective induction of 2-D and 3-D cellular morphogenesis, *Biomaterials*, 21, 783, 2000.

[12] Sachols, E. and Czernuszka, J.T., Making tissue engineering scaffolds work. Review on the application of solid freeform fabrication technology to the production of tissue engineering scaffolds, *Eur. Cells Mater.*, 5, 29, 2003.

[13] Leong, K.F., Cheah, C.M., and Chua, C.K., Solid freeform fabrication of three-dimensional scaffolds for engineering replacement tissues and organs, *Biomaterials*, 295, 2363, 2003.

[14] Hench, L.L. and Polak, J.M., Third-generation biomedical materials, *Science*, 295, 1014, 2002.

[15] Whitaker, M.J. et al., Growth factor release from tissue engineering scaffolds, *J. Phar. Pharmacol.*, 53, 1427, 2001.

[16] Sakai, Y. et al., *In vitro* organization of biohybrid rat liver tissue incorporating growth factor- and hormone-releasing biodegradable polymer microcapsules, *Cell Transplant.*, 10, 479, 2001.

[17] Jeong, B. and Gutowska, A., Lessons from nature: stimuli-responsive polymers and their biomedical applications, *Trends Biotechnol.*, 20, 305, 2002.

[18] Chang, T.M. and Prakash, S., Therapeutic uses of microencapsulated genetically engineered cells, *Mol. Med. Today*, 4, 221, 1998.

[19] Lanza, R.P., Hayes, J.L., and Chick, W.L., Encapsulated cell technology, *Nat. Biotechnol.*, 14, 1107, 1996.

[20] Gomez, N. et al., Evidence for survival and metabolic activity of encapsulated xenogeneic hepatocytes transplanted without immunosuppression in Gunn rats, *Transplantation*, 63, 1718, 1997.

[21] Yang, M.B., Vacanti, J.P., and Ingber, D.E., Hollow fibers for hepatocyte encapsulation and transplantation: studies of survival and function in rats, *Cell Transplant.*, 3, 373, 1994.

[22] Yoon, J.J. et al., Surface immobilization of galactose onto aliphatic biodegradable polymers for hepatocyte culture, *Biotechnol. Bioeng.*, 78, 1, 2002.

[23] Granicka, L.H. et al., Polypropylene hollow fiber for cells isolation: methods for evaluation of diffusive transport and quality of cells encapsulation, *Artif. Cells Blood Substit. Immobil. Biotechnol.*, 31, 249, 2003.

[24] Sen, P.K. et al., Use of isolated perfused cadaveric liver in the management of hepatic failure, *Surgery*, 59, 774, 1966.

[25] Eiseman, B.,Liem, D.S., and Raffucci, F., Heterologous liver perfusion in treatment of hepatic failure, *Ann. Surg.*, 162, 329, 1965.

[26] Pascher, A. et al., Extracorporeal liver perfusion as hepatic assist in acute liver failure: a review of world experience, *Xenotransplantation*, 9, 309, 2002.

[27] Mora, N. et al., Single vs. dual vessel porcine extracorporeal liver perfusion, *J. Surg. Res.*, 103, 228, 2002.

[28] Pascher, A. et al., Immunopathological observations after xenogeneic liver perfusions using donor pigs transgenic for human decay-accelerating factor, *Transplantation*, 64, 384, 1997.

[29] Pascher, A. et al., Application of immunoapheresis for delaying hyperacute rejection during isolated xenogeneic pig liver perfusion, *Transplantation*, 63, 867, 1997.

[30] Pascher, A. et al., Impact of immunoadsorption on complement activation, immunopathology, and hepatic perfusion during xenogeneic pig liver perfusion, *Transplantation*, 65, 737, 1998.

[31] Tector, A.J. et al., Mechanisms of resistance to injury in pig livers perfused with blood from patients in liver failure, *Transplant. Proc.*, 29, 966, 1997.

[32] Horslen, S.P. et al., Extracorporeal liver perfusion using human and pig livers for acute liver failure, *Transplantation*, 70, 1472, 2000.

[33] Collins, B.H. et al., Immunopathology of porcine livers perfused with blood of humans with fulminant hepatic failure, *Transplant. Proc.*, 27, 280, 1995.

[34] Borie, D.C. et al., Functional metabolic characteristics of intact pig livers during prolonged extracorporeal perfusion: potential for a unique biological liver-assist device, *Transplantation*, 72, 393, 2001.

[35] Foley, D.P. et al., Bile acids in xenogeneic *ex-vivo* liver perfusion: function of xenoperfused livers and compatibility with human bile salts and porcine livers, *Transplantation*, 69, 242, 2000.

[36] Pascher, A., Sauer, I.M., and Neuhaus, P., Analysis of allogeneic versus xenogeneic auxiliary organ perfusion in liver failure reveals superior efficacy of human livers, *Int. J. Artif. Organs*, 25, 1006, 2002.

[37] O'Grady, J.G. et al., Controlled trials of charcoal haemoperfusion and prognostic factors in fulminant hepatic failure, *Gastroenterology*, 94, 1186, 1998.

[38] Ash, S.R. et al., Push-pull sorbent-based pheresis and hemodiabsorption in the treatment of hepatic failure: preliminary results of a clinical trial with the BioLogic-DTPF System, *Ther. Apher.*, 4, 218, 2000.

[39] Ash, S.R. et al., Treatment of acetaminophen-induced hepatitis and fulminant hepatic failure with extracorporeal sorbent-based devices, *Adv. Ren. Replace Ther.*, 9, 42, 2002.

[40] Rifai, K. et al., Prometheus((R)) — a new extracorporeal system for the treatment of liver failure (small star, filled), *J. Hepatol.*, 39, 984, 2003.

[41] Stange, J. et al., Molecular adsorbent recycling system (MARS): clinical results of a new membrane-based blood purification system for bioartificial liver support, *Artif. Organs*, 23, 319, 1999.

[42] Butterworth, R.F., Role of circulating neurotoxins in the pathogenesis of hepatic encephalopathy: potential for improvement following their removal by liver assist devices, *Liver Int.*, 23, 5, 2003.

[43] Mitzner, S. et al., Improvement in central nervous system functions during treatment of liver failure with albumin dialysis MARS — a review of clinical, biochemical, and electrophysiological data, *Metab. Brain Dis.*, 17, 463, 2002.

[44] Heemann, U. et al., Albumin dialysis in cirrhosis with superimposed acute liver injury: a prospective, controlled study, *Hepatology*, 36, 949, 2002.

[45] Chen, S. et al., Molecular adsorbent recirculating system: clinical experience in patients with liver failure based on hepatitis B in China, *Liver*, 22, 48, 2002.

[46] Schmidt, L.E. et al., Systemic hemodynamic effects of treatment with the molecular adsorbents recirculating system in patients with hyperacute liver failure: a prospective controlled trial, *Liver Transpl.*, 9, 290, 2003.

[47] Mitzner, S.R. et al., Improvement of hepatorenal syndrome with extracorporeal albumin dialysis MARS: results of a prospective, randomized, controlled clinical trial, *Liver Transplant.*, 6, 277, 2000.

[48] Sen, S., Jalan, R., and Williams, R., Liver failure: basis of benefit of therapy with the molecular adsorbents recirculating system, *Int. J. Biochem. Cell Biol.*, 35, 1306, 2003.

[49] Kjaergard, L.L. et al., Artificial and bioartificial support systems for acute and acute-on-chronic liver failure: a systematic review, *JAMA*, 289, 217, 2003.

[50] Hassanein, T. et al., Albumin dialysis in cirrhosis with superimposed acute liver injury: possible impact of albumin dialysis on hospitalization costs, *Liver Int.*, 23, 61, 2003.

[51] Hessel, F.P. et al., Economic evaluation and 1-year survival analysis of MARS in patients with alcoholic liver disease, *Liver Int.*, 23, 66, 2003.

[52] Chamuleau, R.A., Artificial liver support in the third millennium, *Artif. Cells Blood Substit. Immobil. Biotechnol.*, 31, 117, 2003.

[53] Takezawa, T. et al., Morphological and immuno-cytochemical characterization of a hetero-spheroid composed of fibroblasts and hepatocytes, *J. Cell Sci.*, 101, 495, 1992.

[54] Takezawa, T. et al., Characterization of morphology and cellular metabolism during the spheroid formation by fibroblasts, *Exp. Cell Res.*, 208, 430, 1993.

[55] Guguen-Guillouzo, C. et al., Maintenance and reversibility of active albumin secretion by adult rat hepatocytes co-cultured with another liver epithelial cell type, *Exp. Cell Res.*, 143, 47, 1983.

[56] Bhatia, S.N. et al., Effect of cell–cell interactions in preservation of cellular phenotype: cocultivation of hepatocytes and nonparenchymal cells, *FASEB J.*, 13, 1883, 1999.

[57] Folch, A. and Toner, M., Microengineering of cellular interactions, *Annu. Rev. Biomed. Eng.*, 2, 227, 2000.

[58] Häussinger, D., Gerok, W., and Sies, H., Regulation of flux through glutaminase and glutamine synthetase in isolated perfused rat liver, *Biochim. Biophys. Acta*, 755, 272, 1983.

[59] Shayiq, R.M. et al., Repeat exposure to incremental doses of acetaminophen provides protection against acetaminophen-induced lethality in mice: an explanation for high acetaminophen dosage in humans without hepatic injury, *Hepatology*, 29, 451, 1999.

[60] Kera, Y., Penttila, K.E., and Lindros, K.O., Glutathione replenishment capacity is lower in isolated perivenous than in periportal hepatocytes, *Biochem. J.*, 254, 411, 1988.

[61] Bhatia, S.N. et al., Zonal liver cell heterogeneity: effects of oxygen on metabolic functions of hepatocytes, *Cell Eng.*, 1, 125, 1996.

[62] Bhatia, S.N. et al., Selective adhesion of hepatocytes on patterned surfaces, *Ann. NY Acad. Sci. USA*, 745, 187, 1994.

[63] Bhatia, S.N., Yarmush, M.L., and Toner, M., Controlling cell interactions by micropatterning in co-cultures: hepatocytes and 3T3 fibroblasts, *J. Biomed. Mater. Res.*, 34, 189, 1997.

[64] Chan, C. et al., Metabolic flux analysis of hepatocyte function in hormone- and amino acid-supplemented plasma, *Metab. Eng.*, 5, 1, 2003.

[65] Washizu, J. et al., Optimization of rat hepatocyte culture in citrated human plasma, *J. Surg. Res.*, 93, 237, 2000.

[66] Bhatia, S.N. et al., Probing heterotypic cell interactions: hepatocyte function in microfabricated co-cultures, *J. Biomater. Sci. Polym. Ed.*, 9, 1137, 1998.

[67] Bhatia, S.N. et al., Microfabrication of hepatocyte/fibroblast co-cultures: role of homotypic cell interactions, *Biotechnol. Prog.*, 14, 378, 1998.

[68] Roy, P. et al., Effect of flow on the detoxification function of rat hepatocytes in a bioartificial liver reactor, *Cell Transplant.*, 10, 609, 2001.

[69] Chan, C. et al., Application of multivariate analysis to optimize function of cultured hepatocytes, *Biotechnol. Prog.*, 19, 580, 2003.

[70] Chan, C. et al., Metabolic flux analysis of cultured hepatocytes exposed to plasma, *Biotechnol. Bioeng.*, 81, 33, 2003.

[71] Washizu, J. et al., Amino acid supplementation improves cell-specific functions of the rat hepatocytes exposed to human plasma, *Tissue Eng.*, 6, 497, 2000.

[72] Washizu, J. et al., Long-term maintenance of cytochrome P450 activities by rat hepatocyte/3T3 cell co-cultures in heparinized human plasma, *Tissue Eng.*, 7, 691, 2001.

[73] Watanabe, F.D. et al., Clinical experience with a bioartificial liver in the treatment of severe liver failure. A phase I clinical trial, *Ann. Surg.*, 225, 484, 1997.

[74] Kamohara, Y.,Rozga, J., and Demetriou, A.A., Artificial liver: review and Cedars–Sinai experience, *J. Hepatobil. Pancreat. Surg.*, 5, 273, 1998.

[75] Ellis, A.J. et al., Pilot-controlled trial of the extracorporeal liver assist device in acute liver failure, *Hepatology*, 24, 1446, 1996.

[76] Hu, W.-S. et al., Development of a bioartificial liver employing xenogeneic hepatocytes, *Cytotechnology*, 23, 29, 1997.

[77] Tzanakakis, E.S. et al., Extracorporeal tissue engineered liver-assist devices, *Annu. Rev. Biomed. Eng.*, 2, 607, 2000.

[78] Catapano, G., Mass transfer limitations to the performance of membrane bioartificial liver support devices, *Int. J. Artif. Organs*, 19, 18, 1996.

[79] Eguchi, S. et al., Loss and recovery of liver regeneration in rats with fulminant hepatic failure, *J. Surg. Res.*, 72, 112, 1997.

[80] De Bartolo, L. et al., A novel full-scale flat membrane bioreactor utilizing porcine hepatocytes: cell viability and tissue-specific functions, *Biotechnol. Prog.*, 16, 102, 2000.

[81] Rozga, J. et al., Development of a hybrid bioartificial liver, *Ann. Surg.*, 217, 502, 1993.

[82] Gerlach, J.C. et al., Bioreactor for a larger scale hepatocyte *in vitro* perfusion, *Transplantation*, 58, 984, 1994.

[83] Sussman, N.L. et al., The Hepatix extracorporeal liver assist device: initial clinical experience, *Artif. Organs*, 18, 390, 1994.

[84] Cuervas-Mons, V. et al., *In vivo* efficacy of a bioartificial liver in improving spontaneous recovery from fulminant hepatic failure: a controlled study in pigs, *Transplantation*, 69, 337, 2000.

[85] Tzeng, E. et al., Adenovirus-mediated inducible nitric oxide synthase gene transfer inhibits hepatocyte apoptosis, *Surgery*, 124, 278, 1998.

[86] Nyberg, S.L. et al., Cytoprotective influence of ZVAD-fmk and glycine on gel-entrapped rat hepatocytes in a bioartificial liver, *Surgery*, 127, 447, 2000.

[87] Arkadopoulos, N. et al., Intrasplenic transplantation of allogeneic hepatocytes prolongs survival in anhepatic rats, *Hepatology*, 28, 1365, 1998.

[88] Kobayashi, N. et al., Prevention of acute liver failure in rats with reversibly immortalized human hepatocytes, *Science*, 287, 1258, 2000.

[89] Kim, S.S. et al., Survival and function of hepatocytes on a novel three-dimensional synthetic polymer scaffold with an intrinsic network of channels, *Ann. Surg.*, 228, 8, 1998.

[90] Baumann, H. and Gauldie, J., The acute phase response, *Immunol. Today*, 15, 74, 1994.

[91] Rotem, A. et al., Oxygen uptake rates in cultured hepatocytes, *Biotechnol. Bioeng.*, 40, 1286, 1992.

[92] Rotem, A. et al., Oxygen is a factor determining *in vitro* tissue assembly: effects on attachment and spreading of hepatocytes, *Biotechnol. Bioeng.*, 43, 654, 1994.

[93] Tilles, A.W. et al., Critical issues in bioartificial liver development, *Technol. Health Care*, 10, 177, 2002.

[94] Yarmush, M.L. et al., Hepatic tissue engineering. Development of critical technologies, *Ann. N. Y. Acad. Sci. USA*, 665, 238, 1992.

[95] Allen, J.W., Hassanein, T., and Bhatia, S.N., Advances in bioartificial liver devices, *Hepatology*, 34, 447, 2001.

[96] Flendrig, L.M. et al., *In vitro* evaluation of a novel bioreactor based on an integral oxygenator and a spirally wound nonwoven polyester matrix for hepatocyte culture as small aggregates, *J. Hepatol.*, 26, 1379, 1997.

[97] Sauer, I.M. et al., Development of a hybrid liver support system, *Ann. NY Acad. Sci.*, 944, 308, 2001.

[98] Flendrig, L.M. et al., Significantly improved survival time in pigs with complete liver ischemia treated with a novel bioartificial liver, *Int. J. Artif. Organs*, 22, 701, 1999.

[99] van de Kerkhove, M.P. et al., Phase I clinical trial with the AMC-bioartificial liver, *Int. J. Artif. Organs*, 25, 950, 2002.

[100] Kan, P. et al., Effects of shear stress on metabolic function of the co-culture system of hepatocyte/nonparenchymal cells for a bioartificial liver, *ASAIO J.*, 44, M441, 1998.

[101] Taguchi, K. et al., Development of a bioartificial liver with sandwiched-cultured hepatocytes between two collagen gel layers, *Artif. Organs*, 20, 178, 1996.

[102] Tilles, A.W. et al., Effects of oxygenation and flow on the viability and function of rat hepatocytes cocultured in a microchannel flat-plate bioreactor, *Biotechnol. Bioeng.*, 73, 379, 2001.

[103] Sandstrom, C.E. et al., Development of novel perfusion chamber to retain nonadherent cells and its use for comparison of human "mobilized" peripheral blood mononuclear cell cultures with and without irradiated bone marrow stroma, *Biotechnol. Bioeng.*, 50, 493, 1996.

[104] Horner, M. et al., Transport in a grooved perfusion flat-bed bioreactor for cell therapy applications, *Biotechnol. Prog.*, 14, 689, 1998.

[105] Roy, P. et al., Analysis of oxygen transport to hepatocytes in a flat-plate microchannel bioreactor, *Ann. Biomed. Eng.*, 29, 947, 2001.

[106] Peng, C.A. and Palsson, B.O., Determination of specific oxygen uptake rates in human hematopoietic cultures and implications for bioreactor design, *Ann. Biomed. Eng.*, 24, 373, 1996.

[107] Roberts, E.A. et al., Characterization of human hepatocyte lines derived from normal liver tissue, *Hepatology*, 19, 1390, 1994.

[108] Kobayashi, N. et al., Transplantation of highly differentiated immortalized human hepatocytes to treat acute liver failure, *Transplantation*, 69, 202, 2000.

[109] Wang, L. et al., Comparison of porcine hepatocytes with human hepatoma (C3A) cells for use in a bioartificial liver support system, *Cell Transplant.*, 7, 459, 1998.

[110] Butler, D., Last chance to stop and think on risks of xenotransplants, *Nature*, 391, 320, 1998.

[111] Le Tissier, P. et al., Two sets of human-tropic pig retrovirus, *Nature*, 389, 681, 1997.

[112] Wilson, J.W. and Leduc, R.H., Role of cholangioles in restoration of the liver of the mouse after dietary injury, *J. Pathol. Bacteriol.*, 76, 441, 1958.

[113] Sigal, S.H. et al., The liver as a stem cell and lineage system, *Am. J. Physiol.*, 263, G139, 1992.

[114] Thorgeirsson, S.S., Hepatic stem cells in liver regeneration, *FASEB J.*, 10, 1249, 1996.

[115] Theise, N.D. et al., The canals of Hering and hepatic stem cells in humans, *Hepatology*, 30, 1425, 1999.

[116] Theise, N.D. et al., Liver from bone marrow in humans, *Hepatology*, 32, 11, 2000.

[117] Tateno, C. et al., Heterogeneity of growth potential of adult rat hepatocytes *in vitro*, *Hepatology*, 31, 65, 2000.

[118] Theise, N.D. et al., Derivation of hepatocytes from bone marrow cells in mice after radiation-induced myeloablation, *Hepatology*, 31, 235, 2000.

[119] Petersen, B.E. et al., Bone marrow as a potential source of hepatic oval cells, *Science*, 284, 1168, 1999.

[120] Borel-Rinkes, I.H.M. et al., Nucleation and growth of ice crystals insude cultured hepatocytes during freezing in the presence of dimethylsulfoxide, *Biophys. J.*, 65, 2524, 1993.

[121] Karlsson, J.O.M. et al., Long-term functional recovery of hepatocytes after cryopreservation in a three-dimensional culture configuration, *Cell Transplant.*, 1, 281, 1992.

[122] Stefanovich, P. et al., Effects of hypothermia on the function, membrane integrity, and cytoskeletal structure of hepatocytes, *Cryobiology*, 23, 389, 1995.

[123] Stange, J. et al., The molecular adsorbents recycling system as a liver support system based on albumin dialysis: a summary of preclinical investigations, prospective, randomized, controlled clinical trial, and clinical experience from 19 centers, *Artif. Organs*, 26, 103, 2002.

[124] Chang, M.-H. et al., Albumin dialysis MARS 2003, *Liver Int.*, 23, 3, 2003.

[125] Kramer, L. et al., Successful treatment of refractory cerebral oedema in ecstasy/cocaine-induced fulminant hepatic failure using a new high-efficacy liver detoxification device (FPSA-Prometheus), *Wien. Klin. Wochenschr.*, 115, 599, 2003.

[126] Samuel, D. et al., Neurological improvement during bioartificial liver sessions in patients with acute liver failure awaiting transplantation, *Transplantation*, 73, 257, 2002.

[127] Pazzi, P. et al., Serum bile acids in patients with liver failure supported with a bioartificial liver, *Aliment. Pharmacol. Ther.*, 16, 1547, 2002.

[128] Demetriou, A.A. et al., Prospective, randomized, multicenter, controlled trial of a bioartificial liver in treating acute liver failure, *Ann. Surg.*, 239, 660, 2004.

[129] Patzer II, J. et al., Bioartificial liver assist devices in support of patients with liver failure, *Hepatobiliary Pancreat. Dis. Int.*, 1, 18, 2002.

[130] Mazariegos, G.V. et al., Safety observations in phase I clinical evaluation of the Excorp Medical Bioartificial Liver Support System after the first four patients, *ASAIO J.*, 47, 471, 2001.

[131] Sauer, I.M. et al., Clinical extracorporeal hybrid liver support — phase I study with primary porcine liver cells, *Xenotransplantation*, 10, 460, 2003.

[132] Millis, J.M. et al., Initial experience with the modified extracorporeal liver-assist device for patients with fulminant hepatic failure: system modifications and clinical impact, *Transplantation*, 74, 1735, 2002.

[133] van de Kerkhove, M.P. et al., Bridging a patient with acute liver failure to liver transplantation by the AMC-bioartificial liver phase I clinical trial with the AMC-bioartificial liver, *Cell Transplant.*, 12, 563, 2003.

[134] Morsiani, E. et al., Early experiences with a porcine hepatocyte-based bioartificial liver in acute hepatic failure patients, *Int. J. Artif. Organs*, 25, 192, 2002.

[135] Sielaff, T.D. et al., Characterization of the three-compartment gel-entrapment porcine hepatocyte bioartificial liver, *Cell Biol. Toxicol.*, 13, 357, 1997.

[136] Ding, Y.-T. et al., The development of a new bioartificial liver and its application in 12 acute liver failure patients, *World J. Gastroenterol.*, 9, 829, 2003.

31

Tissue Engineering of Renal Replacement Therapy

William H. Fissell
H. David Humes
University of Michigan

31.1 Background

The kidney is unique among body organs in that it is the first organ for which maintenance replacement therapy has become available and widespread. At the time of this writing, approximately a quarter million people in the United States receive maintenance dialysis as a lifesaving treatment for end-stage renal disease (ESRD) [1]. Despite its successes in preventing immediate death from volume overload, hyperkalemia, and uremia, one should not confuse the *ex vivo* application of a technology developed for chemical purification with complete organ replacement.

Epidemiologic data suggest that patients with chronic renal insufficiency who do not yet require renal replacement therapy have worse operative mortality than those with better renal function, suggesting that adequate clearance and metabolic function of the kidney is necessary for recovery from injury [2,3]. Wolfe, Ashby, and Port published a landmark study comparing the survival of ESRD patients on dialysis awaiting kidney transplant to the survival of recipients of a first deceased donor kidney transplant [4]. This study, the first to directly compare survival of otherwise similar groups treated with dialysis or transplant, showed an annual death rate 1.7-fold higher in patients remaining on the wait-list and receiving dialysis compared with the death rate of those receiving a renal transplant. This suggests a survival benefit associated with renal function over and above waste removal. It has been suggested that the poor outcomes seen in U.S. dialysis patients are related to underdosing of dialysis. Multiple observational studies have noted a correlation between delivered dialysis dose and patient survival, yet the data are subject to attack on several grounds: first, dialytic dosing is measured by clearance of a marker molecule, urea, which itself is not particularly toxic; second, measurement of clearance of urea is easily confounded by the volume of distribution of urea, which is related to patient body mass and obesity; and last, delivered dose of dialysis may be a surrogate for another factor, such as patient adherence or physician attentiveness. Prospective trials in ESRD and in acute renal failure (ARF) have failed to show a conclusive cause and effect relationship between dialysis dose and survival [5–7].

Given that mortality in ESRD is high (18 deaths per 100 patient-years at risk) compared with mortality in patients with a functioning kidney, it is instructive to examine the causes of death in dialysis patients. The leading cause of death is cardiovascular disease, accounting for almost half of all deaths in ESRD patients, and this affects diabetics and nondiabetics alike. The second most common cause of death is infection, and even after controlling for infections related to dialysis access, dialysis patients die from pulmonary infections at a rate 16 times that of the general population and 8 times that of the renal transplant population — a group of patients treated with immunosuppressive drugs [1,8].

The molecular mechanisms underlying accelerated cardiovascular disease and susceptibility to infection in the ESRD population remain unclear, and this is a major concern to the tissue engineer who contemplates design considerations for renal replacement therapy. Putative mechanisms of vascular disease in ESRD include the observation that chronic kidney disease before and after initiation of dialysis is associated with increased plasma markers of oxidative stress, which have in turn been proposed as accelerants of vascular disease [9–11]. Similarly, differences in serum and stimulated peripheral blood mononuclear cell cytokine levels between control subjects and subjects with acute and chronic renal failure have been identified in humans and in animal models, and that has been hypothesized to play a role in the diminished immunologic competence observed in ESRD [12,13].

31.2 Renal Physiology

With the explicit understanding that the high mortality observed in ESRD cannot yet be attributed to lack of specific physiologic processes of the native kidney, let us begin to consider the engineering of a renal replacement system. A diagram of the kidney and its functional units, nephrons, are shown in Figure 31.1. The glomerulus is a specialized tuft of capillaries which permit high-flux transudation of fluid out of the bloodstream and into a receptacle, Bowman's capsule. Bowman's capsule drains into

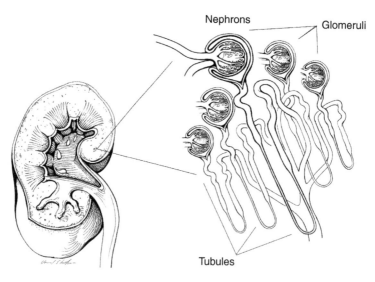

FIGURE 31.1 Diagram of kidney and nephrons.

TABLE 31.1 Renal Functions in the Native Kidney and in Hemodialysis

Function	Renal physiology	Dialysis therapy
Volume homeostasis	Direct control of filtered volume by glomerular capillaries, with passive regulation of reabsorption in the proximal tubule via glomerulo-tubular balance and active control of sodium reabsorption in the cortical collecting system	Patient is weighed at each treatment and technician enters an ultrafiltration volume into the dialysis machine
Control of blood tonicity	Generation of medullary concentration gradient by loop of Henle; ADH-regulated insertion of aquaporin channels into luminal membrane of collecting duct cells; ADH-mediated thirst	Concentration of electrolytes in dialysate prescribed by physician; water and sodium diffuse freely across dialysis membrane; ADH-mediated thirst
Toxin clearance	Bulk filtration by glomerular capillaries, with resorptive concentration of ultrafiltrate into urine, including reabsorption of important metabolites, such glutathione, glucose, amino acids	Passive diffusion of small solutes from blood into dialysate
Acid-base homeostasis	pH-dependent ammoniagenesis in the proximal tubule; active pH-dependent proton excretion in collecting duct	Bicarbonate concentration in dialysate controlled by prescription
Potassium metabolism	Potassium ion excretion by principal cells of the collecting duct under active regulation by aldosterone-mediated Na–K ATPase activity	Serum potassium measured periodically and dialysate potassium concentration adjusted
Calcium–phosphorus metabolism	Hydroxylation of 25-(OH) Vit D_3 in proximal tubule cells; PTH-regulated calcium reabsorption and phosphorus excretion	Calcium present in dialysate; dietary phosphorus absorption blocked with phosphorus binders, oral or intravenous vitamin D analogs prescribed
Regulation of red blood cell mass	Oxygen-dependent synthesis and release of erythropoetin from capillary endothelium	Erythropoetin analogs injected at time of dialysis

a multisegmented tube, the renal tubule, which progressively reabsorbs the majority of the fluid which enters it, but does not reabsorb toxins to the same degree, thus concentrating toxins in the remaining fluid, urine. An understanding of the known physiologic processes of the kidney is a first step towards outlining the task list of a hypothetical artificial kidney, and a comparison to existing dialytic therapy may be helpful.

The major filtration, excretion, and metabolic functions of the healthy kidney are contrasted with dialysis therapy in Table 31.1.

The kidney's algorithm for waste excretion is at first confusing: why does the kidney in effect discard everything and then struggle to reabsorb its losses? In contrast, the body's other major toxin clearinghouse, the liver, has evolved a complex web of enzymes to degrade and detoxify bloodborne toxins. Each normal kidney's filters produce approximately 60 ml/min of an ultrafiltrate of blood, and the kidney's tubules progressively extract salt and water from that ultrafiltrate, concentrating toxins until that ultrafiltrate is urine. A unique and underappreciated feature of the kidney's approach is that the organism has an opportunity via the kidney to excrete novel toxins for which it has not been evolutionarily prepared. The kidney has evolved a narrow set of transport proteins to scavenge from the ultrafiltrate stream the salt, water, glucose, and amino acids necessary for life, and with that toolkit, to discard all that is not needed, whether the body recognizes a particular molecule as a toxin or not. Curiously, despite a half century of research, the identities of the molecules responsible for the uremic state have proven elusive, which argues against an engineering approach that targets individual toxins. This suggests that adsorbent or catalytic systems are not the approach of choice for a tissue-engineered artifical kidney.

The kidney also employs a combination of active and passive negative feedback systems to confer redundancy in the event of failure. In advanced renal failure, when glomerular filtration has dropped to 10% of its original value, despite dilution of the medullary concentrating gradient and increased single-nephron glomerular filtration rate (GFR), the kidney can still maintain volume homeostasis, albeit over a narrower range of intake and output. A successful implantable artificial kidney will need closed-loop feedback algorithms integrating sensors of blood pressure and flow, tonicity, and potassium concentration to autonomously replace the failed organ.

It is worthwhile to include some order-of-magnitude estimates of the fluxes and volumes, and working pressures necessary for renal function. Patients with chronic renal insufficiency develop symptoms insidiously with wide variability in actual GFR at the time of clinical illness, but it is generally accepted that a GFR of 10 ml/min is enough for some patients to survive, and almost all will be manageable with a GFR of 20 ml/min. This corresponds to just under 29 l/day of blood clearance. Typical dietary consumption in the United States includes 1.5 to 2 l of fluid intake, 100 to 200 mEq of sodium, and 90 mEq of potassium, although both of the latter vary widely with dietary intake. This same diet also includes approximately 50 to 100 mM of acid. An ultrafiltrate stream of 20 ml/min thus bears some 4000 mEq of sodium and 120 mEq of potassium over 24 h. Of this 2000 to 4000 Eq of sodium filtered each day, all but 100 to 200 mEq must be returned to the blood stream, illustrating the efficient reclamation of sodium performed by the kidney. Of the 29 l of water, all but one to two l need to be reabsorbed. All of this filtration and reabsorption occur at hydrostatic pressures approximating capillary perfusion pressure of 25 mmHg, or about 0.1 PSI.

31.3 Engineering Renal Replacement

Design of an artifical kidney begins with the enumeration of a set of tasks for the proposed device. At a minimum, to replace present dialytic therapy, an engineered kidney will need to remove approximately 100 to 200 mM of sodium, 100 mM of potassium, 1 to 2 l of water, and 50 to 100 mM of acid from the bloodstream, while also providing 20 to 30 l a day of small-molecule clearance. In addition, calcium-phosphorus balance, in particular, avoidance of hyperphosphatemia, must be maintained, as well as regulation of serum tonicity.

Several of the functions of the native kidney may not need replacing in a tissue-engineered artificial kidney. Regulation of red blood cell mass, (1,25)-OH Vitamin D_3, and buffering of dietary acids are solved problems from an engineering viewpoint; weekly injections and well-tolerated oral medications appear sufficient to replace these metabolic and endocrine functions of the kidney. Serum tonicity is controlled by two redundant pathways in the healthy organism: vasopressin is a short peptide hormone synthesized and released by the periventricular and supraoptic nuclei of the hypothalamus in response to increased plasma tonicity. Elevated levels of vasopressin, also called antidiuretic hormone, stimulate water reabsorption in the kidney and the sensation of thirst. Hemodialysis patients appear to be able to regulate plasma tonicity solely through the latter mechanism, as well as patients with advanced renal insufficiency who have little urinary diluting or concentrating capacity. It is when these patients either imbibe excess water or are deprived of it that they lose osmoregulation. This suggests that although tight control of plasma tonicity is important, it can be achieved in the absence of a renal regulatory mechanism.

Renal potassium excretion varies with dietary input under control of a serum steroid hormone, aldosterone. Longterm success of an implantable artifical kidney will be predicated in part on sensing and responding to serum potassium levels, as it is difficult to tailor a varied diet to contain identical amounts of potassium each day. As failure of potassium control may be promptly fatal, development of online noninvasive potassium monitors is a major challenge in renal tissue engineering. However, given information regarding potassium levels, external control of potassium levels may be accomplished via oral medicines.

The single largest challenge facing renal tissue engineering is providing 30 to 50 l each day of small-molecule clearance. In present clinical practice, this clearance is provided by hemodialysis, peritoneal dialysis, or hemofiltration. Dialysis is an unattractive strategy for providing clearance in a totally implanted device, given the large volumes of electrolytically pure nonpyrogenic fluid necessary. Patients receiving peritoneal dialysis wash their peritoneal cavities with 10 to 15 l of electrolyte each day. Scarring of the peritoneal membrane, caloric load from the hypertonic glucose, and infection related to the dialysis catheter limit the effectiveness of this approach, as well as the labor of maintaining a stock of dialysate near the patient. Dialysate regeneration, wherein toxins are adsorbed or catabolized in the dialysate in essence simply transfers the clearance problem from the blood to the dialysate. In the absence of detailed knowledge of specific molecules to be cleared, selection of reagents for dialysate regeneration is problematic.

A second strategy for clearance is already used to clear excess potassium from patients with chronic hyperkalemia and block dietary phosphorus absorption from patients with hyperphosphatemia. Oral administration of binding resins with high affinity for the solute of choice (Kayexelate® Renegel®) can alter serum electrolyte levels promptly and, if used regularly, can maintain neutral balance for months to years. This strategy in essence uses the enteric epithelium to confer biocompatibility to the chemicals used to adsorb the solutes. The broader applicability of absorbent technologies hinges on the absorption of the toxins responsible for uremia. They are at present unknown and by extension not only is the design of a sorbent difficult, but the transport of such toxins into the intestinal lumen remains speculative.

A third strategy for providing clearance is high-volume ultrafiltration and selective reabsorption of that ultrafiltrate, as occurs in the native nephron. This poses the problem as a two-stage filtration problem: one filter generating an ultrafiltrate and a second reabsorbing part of the ultrafiltrate back into the feed solution of the first membrane. The remaining of the chapter will address the nature of these passive and active membranes and the forces driving mass transport across them in the nephron and present tissue-engineering constructs.

31.4 Hemofiltration

31.4.1 The Renal Glomerulus

The fundamental structure of the glomerulus, and fundamental physiologic determinants of glomerular filtration have been well described. Three distinct layers, the endothelium, the basement membrane — glomerular basement membrane (GBM), and a specialized cell junction, the podocyte slit diaphragm, comprise the filter (Figure 31.2). The glomerulus itself is a tuft of capillaries tethered by smooth muscle-like cells continuous with the arteriolar wall called mesangial cells. The outer surface of the glomerular capillary tube is covered with a mesh-like array of interdigitated epithelial cells called podocytes, and the inner surface of the capillary tube is a specialized endothelium with regular pores called fenestrae. The capillary tuft is enclosed in a small receptacle for ultrafiltrate called Bowman's space, which drains ultrafiltrate into the renal tubule. The glomerulus permits low-molecular weight substances, such as

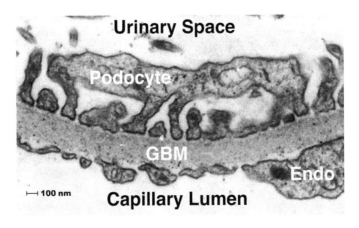

FIGURE 31.2   Transmission electron micrograph of the glomerular filtration barrier, GBM.

electrolytes, urea, creatinine, glucose, and amino acids, to pass with water as ultrafiltrate into Bowman's space for processing by the renal tubule. Larger molecules, such as proteins, which are energetically expensive for the body to synthesize, are retained in the bloodstream. The exact contribution of each component of the glomerular capillary to the filtration barrier is unknown.

The glomerular endothelium is a specialized structure featuring unusual endothelial cells with fenestrations, considered to be the loci at which water enters the glomerular basement membrane. Other vessels in the kidney subject to high-volume convective transport, such as peritubular capillaries, also have fenestrations. The individual glomerular fenestrae are not bridged by proteinaceous diaphragms, although such diaphragms are described in other fenestrated endothelia. Endothelial proteins which localize to other fenestrated endothelia in the kidney and the lung do not localize in the glomerular endothelial fenestrae [14]. Glomerular endothelial cells also form intracapillary ridges, which are thought to impair laminar flow and promote mixing at the capillary walls. The role of the fenestrae in glomerular filtration may solely lie in this mixing and in directing site of entry of fluid into the GBM.

The GBM is a 200 to 300 nm thick gel of extracellular matrix proteins, primarily collagen IV, laminin, fibronectin, and heparan sulfate proteoglycan. The GBM is composed of two separate and superimposed collagen IV networks, one of $\alpha(1,2)$ collagen IV chains adjacent to and synthesized by endothelial cells, and one of $\alpha(3,4,5)$ collagen IV chains adjacent to and synthesized by the podocytes. Laminin, fibronectin, and entactin are thought to crosslink the collagen IV molecules and add mechanical rigidity to the basement membrane. A heparan sulfate proteoglycan, perlecan, is the major proteglycan constituent of the GBM and forms a hydrated meshwork within the GBM.

The glomerular slit diaphragm is a unique intercellular junction whose fundamental ultrastructure was elucidated by Karnovsky et al. 30 years ago [15]. The protein constituents of the glomerular slit diaphragm have been the subject of intense recent study. Nephrin, a 185 D protein product of NPHS1, a gene mutated in congenital nephropathy of the Finnish type, has been localized to the glomerular slit diaphragm [16]. The slit diaphragm is a somewhat delicate structure, in that neutralization of fixed anion charges with polycations such as protamine can induce reversible podocyte rearrangement and obliteration of the slit diaphragm structure. Detailed analysis of the functional properties of the glomerular slit diaphragm have been hampered by the terminally differentiated phenotype of the glomerular podocyte. The podocyte does not easily replicate in cell culture, and has not yet been observed to form the slit diaphragm structure *in vitro*. Strategies for understanding glomerular filtration to date have centered on ultrastructural imaging and mathematical modeling of the glomerular barrier as an array of pores [15,17–21].

31.4.2 Toward a Synthetic Glomerulus

The modern hollow-fiber hemodialyzer is an excellent glomerular analog, providing passage of electrolytes, water, and low-molecular weight toxins while retarding passage of medium and large serum proteins such as albumin. As such, it provides life-saving clearance to a quarter-million dialysis patients in the United States. However, present technology poses several challenges to integration into a durable or implantable artificial organ. The first and most obvious is the fairly low hydraulic permeability of existing membranes. To achieve meaningful filtration rates in clinical practice, transmembrane pressures of 200 to 400 mmHg are applied to the membrane by peristaltic pumps. Incorporation of these technologies into an implantable or wearable organ will require either pumps, a very large membrane area, or significant advances in existing polymer technology. Second, membrane fouling presently limits service lifetime to around 100 h with present technology. Last, and possibly most significant, the glomerulus appears to have a fairly sharp cutoff for molecular weights around 60 kDa, probably attributable to the extraordinary uniformity of the glomerular slit diaphragm. Polymer filters have a normally distributed spectrum of pore sizes arising from the characteristics of the polymer melt or cast. Thus, a sharp transition from passage to retardation with increasing molecular weight is difficult to acheive in practice. A durable hemofilter with a sharp molecular weight cutoff and very high hydraulic permeability is clearly a highly desireable goal of tissue engineering an artifical kidney.

Two recent approaches are worthy of a brief introduction. If, as evidence suggests, the podocyte slit diaphragm is the locus of glomerular permselectivity, growth of the podocyte in tissue or organ culture may promise a glomerular replacement. Unfortunately, podocytes are terminally differentiated epithelial cells and resist expansion in cell or tissue culture. Cultured podocytes harvested from organs do not easily form the octopus-like structures of the podocyte foot processes. In an attempt to expand understanding of podocyte biology, Mundel and colleagues have engineered a conditionally transformed podocyte cell line using a temperature-sensitive SV40 construct [22]. This cell line appears to display many cellular proteins characteristic of differentiated podocytes, and is a promising area for tissue engineering. Live human podocytes are shed in the urine in patients with glomerular disease, and the potential to harvest, transform and expand these cells is an exciting prospect in renal tissue engineering.

Our laboratory has begun testing mechanical analogs of the glomerular slit diaphragm. Using nano-fabrication technology, silicon membranes comprised of slit shaped pores 6 to 100 nm in smallest dimension have been tested for hydraulic permeability. This approach allows extremely narrow pore size dispersions, unprecedented control over pore size and shape, as well as batch fabrication using conventional silicon micromachining techniques. Silicon surfaces are easily functionalized with polymer and peptide moeities giving rise to the possibility of independent control of electrostatic and steric hindrance of molecules by selection of pore size and surface treatment. Most promising, slit-shaped pores, as are found in the glomerular slit diaphragm, potentially have higher hydraulic permeability than an equivalent area of cylindrical pores, while retaining steric hindrance to macromolecules. Initial characterization of these membranes suggests that prototype slit-pore membranes with low porosity ($\sim 10^{-3}$) have hydraulic permeabilities on a par with polymer membranes with porosities two orders of magnitude higher [23]. Further work characterizing protein permselectivity of these membranes is underway.

31.5   Reabsorption of Filtered Fluid

31.5.1   The Renal Tubule

The renal tubule affects the controlled reabsorption of salt, water, glucose, and amino acids from the ultrafiltrate stream. It permits the body to excrete a urine stream that may vary in tonicity from about 100 to about 1000 mOsm under the control of a single hormone, vasopressin. The tubule is loosely described as having, in sequence, eight segments: the proximal convoluted tubule, the descending limb of the loop of Henle, the thick and thin ascending limbs of the loop of Henle, the distal convoluted and straight tubules, and the cortical and medullary collecting ducts. Each segment is composed of one or more individual cell types, which may be interspersed, as in the principal and intercalated cells of the collecting ducts, or may transition from one type to the next along the length of the segment, as occurs in the proximal tubule. Postglomerular blood accompanies the tubule in peritubular capillaries. In concert with the spatial distribution of cell types, the anatomy of the tubule gives rise to a corticomedullary gradient in tonicity, with the renal cortex approximating systemic tonicity, and the inner medulla of the kidney approaching 1200 mOsm. It is this progressive concentration and dilution of the ultrafiltrate stream that permits the kidney to regulate serum tonicity.

A slight majority of tubular reabsorption of salt, water, electrolytes, glucose, and amino acids takes place in the proximal tubule, with the balance occurring more distally. The maximum fractional reabsorption of salt and water by the tubule remains unclear. The driving force for sodium reabsorption in all nephron segments in which it occurs is the basolateral localization of the sodium–potassium ATPase transporter. This plasma membrane protein moves three sodium ions from the interior of the cell to the exterior, while transporting two potassium ions into the cell.

Both movements are against and in fact give rise to the prevailing concentration gradient, and so require energy in the conversion of adenosine triphosphate (ATP) to adenosine diphosphate (ADP). The resulting very low intracellular sodium concentration allows passive transport of sodium into the interior of the cell from the ultrafiltrate stream via apical sodium channels. Intracellular sodium is then transported across the basolateral membrane by the Na–K ATPase, accomplishing bulk transport of sodium from ultrafiltrate to pericapillary interstitium. The transport of ions results in an increase in basolateral tonicity, and water moves from the tubular lumen to the peritubular interstitium under osmotic pressure. In the proximal tubule, this takes place by passive, unregulated paracellular means, but in more distal segments this occurs via highly regulated transcellular pathways.

Broadly speaking, there are two approaches to engineering tubular reabsorption: employing living cells to mimic the function of their native counterparts, or manufacturing a second filtration membrane which permits the passage of salt, water, glucose, and sodium bicarbonate but retards the passage of urea, creatinine, and other uremic toxins. The advantage of the former lies in the simplicity of the approach; there is no need to separately implement each of the many transporters on the apical surface of the cell; supply the cell and the cell will supply not only the transporters but in addition the driving force for reabsorption. The disadvantage of the cellular approach is that it is subject to the same supply pressures as renal transplantation: the cells need to come from somewhere. Despite these pressures, a tissue-engineered bioartificial tubule containing human cells has entered clinical trials in the United States, and the design and engineering of that device will be described.

31.5.2 Isolation and Culture of Proximal Tubule Cells

Our laboratory developed experience in the isolation and culture of porcine proximal tubule cells until we could reproducibly harvest cells and maintain them stably in culture [24,25]. In brief, Yorkshire breed pigs were sacrificed at 4 to 6 weeks of age and their kidneys harvested. The renal cortices were dissected, minced, digested with collagenase, and the resulting mixture was separated on a Percoll density gradient. Renal proximal tubule fragments were isolated and grown in a serum-free hormonally derived medium. After third passage and reaching confluence on 100 mm culture dishes, cells were mobilized with trypsin into a suspension and seeded into polysulfone single hollow-fiber bioreactors for *in vitro* assessment of cell viability and metabolic activity.

Cellular attachment, stability, and confluence on the interior lumen of the bioreactor is of paramount importance. To promote attachment of the cells, the luminal surface of the polysulfone membrane was coated with ProNectin-L, a synthetic protein sharing the intercellular attachment domains of laminin, a protein found in the renal glomerular and tubular basement membrane. Laminin and collagen type IV, key components of the tubular basement membrane, also provide an effective biomatrix for cell attachment and growth. After seeding of the hollow fiber with tubule cells, the hollow fibers were perfused with culture media. As newly seeded cells need time to attach, perfusion was initially performed via diffusion from the exterior through the polysulfone membrane, and after time for attachment, convective flow through the interior of the fiber was initiated. A graduated increase in flow (and thus shear forces) was used to condition the cells and minimize cellular detachment. Studies demonstrated confluence was reached in 7 to 10 days. After 14 days in culture the hollow fiber bioreactors were assessed for cellular confluency and viability. Light microscopy of fixed sections showed evidence of a confluent monolayer formed on the inside of the hollow fibers, and intercellular tight junction formation was verified by measuring low inulin leak rates across the monolayer [26].

31.5.3 Transport and Metabolic Characteristics of Hollow-Fiber Bioreactors

As initial experiments using the single hollow-fiber model were promising, the design was scaled up to use commercially available polysulfone hollow fiber dialysis cartridges from the manufacturers of the single hollow fibers. Single hollow-fiber measurements of transport and metabolic activity were repeated with 97 cm^2 and 0.4 m^2 surface area cartridges.

We further explored the metabolic characteristics of the cultured proximal tubule cells. We examined the transport of glucose, bicarbonate, and glutathione and expressed the data in terms of fractional reabsorption accomplished by the bioreactor. For each of the molecules listed, fractional excretion was measured in the absence and presence of a known inhibitor of an enzyme essential for the reabsorption. In each case, there was evidence of active transport and specific inhibition [26].

The synthesis and secretion of ammonia into the tubule is essential for renal excretion of an acid load, as it buffers secreted protons. Proximal tubule cells are able to increase their ammoniagenesis in response to a decline in pH, and the proximal tubule cells in the bioreactor demonstrated a stepwise increase in ammonia production with changes in pH [26].

The experiments detailed above were performed with porcine tubule cells, and our laboratory has demonstrated similar results in culture, attachment, and activity with human proximal tubule cells from cadaveric organs. The final selection of cell type for use in a renal tubule device rests not only on supply and safety of cells, but also depends on the ability of xenotransplanted cells to participate in the homeostasis of the host.

The above data suggest that our laboratory has successfully isolated and cultured renal proximal tubule cells, established stable confluent monolayers within hollow fiber bioreactors, and scaled the initial construct to a level approximating the number of proximal tubule cells in a single kidney.

31.5.4 Preclinical Characterization of the Renal Assist Device (RAD)

In keeping with its role as a metabolically active replacement for the renal proximal tubule, an extracorporeal circuit was devised that recapitulated nephron anatomy (Figure 31.3). A conventional hollow-fiber dialyser and a hollow fiber bioreactor were connected in series, so that a portion of the ultrafiltrate exiting the hemofilter was directed into the luminal spaces of the bioreactor, and thus presented to the apical aspect of the cultured proximal tubule cells. Concentrated blood exiting the hemofilter was directed to the extraluminal space (conventionally the dialysate compartment) of the bioreactor, just as post glomerular blood surrounds the renal tubules via the peritubular capillaries. In order to allow independent control of the subject's volume status and clearance parameters during experiments, the balance of the ultrafiltrate was discarded and the subject infused with a balanced electrolyte solution, as in a conventional hemofiltration circuit.

The bioartificial kidney setup consists of a filtration device (a conventional hemofilter) followed in series by the tubule RAD unit. Specifically, blood is pumped out of a large animal using a peristaltic pump. The blood then enters the fibers of a hemofilter, where ultrafiltrate is formed and delivered into the fibers of the tubule lumens within the RAD downstream to the hemofilter. Processed ultrafiltrate exiting the RAD is collected and discarded as "urine." The filtered blood exiting the hemofilter enters the RAD through the extracapillary space port and disperses among the fibers of the device. Upon exiting the RAD, the processed blood travels through a third pump and is delivered back to the animal. This additional pump is required to maintain appropriate hydraulic pressures within the RAD. Heparin is delivered continuously into the blood before entering the RAD to diminish clotting within the device. The RAD is oriented horizontally and placed into a temperature-controlled environment. The temperature of the cell compartment of the RAD must be maintained at 37°C throughout its operation to ensure optimal functionality of the cells. Maintenance of a physiologic temperature is a critical factor in the functionality of the RAD. The tubule unit is able to maintain viability because metabolic substrates and low-molecular weight growth factors are delivered to the tubule cells from the ultrafiltration unit and the blood in the extracapillary space. Furthermore, immunoprotection of the cells grown within the hollow fiber is achieved because of the impenetrability of immunoglobulins and immunologically competent cells through the hollow fibers. Rejection of the cells, therefore, does not occur. This arrangement thereby allows the filtrate to enter the internal compartments of the hollow fiber network, lined with confluent monolayers of renal tubule cells for regulated transport and metabolic function.

Large animal studies have been completed with the use of this extracorporeal circuit. Dogs were made uremic by performing bilateral nephrectomies. A double lumen catheter was placed into the internal jugular vein, extending into the heart. After 24 h of postoperative recovery, the dogs were treated either with hemofiltration and the RAD or with hemofilter and a sham control cartridge containing no cells. The blood flow rate to the hemofiltrator was maintained at 80 ml/min, with a controlled ultrafiltration rate of 5 to 7 ml/min. Dogs were treated daily for either 7 or 9 h or for 24 h continuously. The dogs in these experiments developed ARF with average blood urea nitrogen (BUN) and plasma creatinine levels of 68 and 6.6 mg/dl,

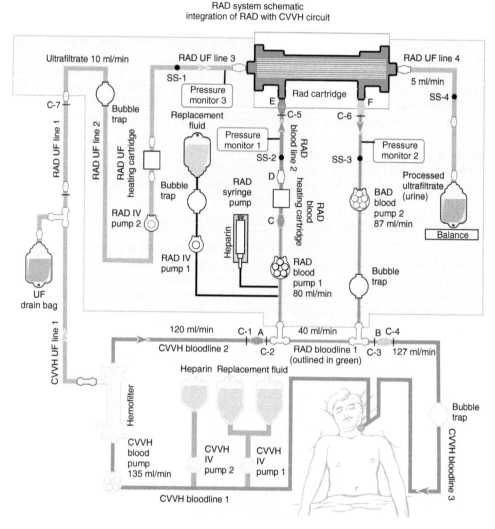

FIGURE 31.3   RAD circuit diagram. (Reprinted with permission from RenaMed Biologics, Inc.)

respectively. The RADs maintained viability and functionality when connected in series to a hemofiltration cartridge within an extracorporeal perfusion circuit in an acutely uremic animal. During a 24-h perfusion period, fewer than 10^5 cells were lost from the RAD, which contained more than 1.4×10^8 cells. Treatment with the RAD and hemofiltration maintained BUN and plasma creatinine levels similar to those of sham controls. In addition, plasma HCO_3, P_i, and K^+ levels were more readily maintained near normal values in RAD treatment than in sham treatment. The RADs were able to reabsorb 40 to 50% of ultrafiltrate volume presented to the devices. Furthermore, active transport of K^+, HCO_3, and glucose was accomplished by the RAD in this *ex vivo* situation. Metabolic activity of the RAD was also shown in these experiments. Virtually no ammonia excretion occurred in the processed ultrafiltrate of the sham control group, in contrast with an ammonia excretion level as high as 100 μM/h in the RAD-treated group. Glutathione processing by the RAD was also shown, with greater than 50% glutathione removal from the ultrafiltrate presented to the RAD. Finally, uremic animals treated with the RAD attained 1,25-$(OH)_2$-D_3 levels of 19.5 ± 0.5 pg/ml, a value no different from the normal levels of the prenephrectomy condition. In contrast, sham treatment resulted in a further fall of 4.0 ± 2.4 pg/ml from the already low plasma levels of 1,25-$(OH)_2$-D_3 in the acutely uremic animals. Thus, these experiments clearly showed that the combination of

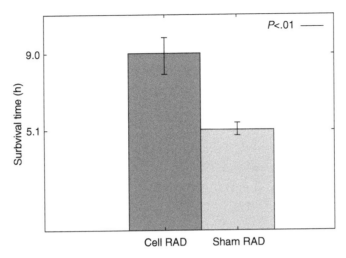

FIGURE 31.4 Survival times of RAD- and sham-treated animals. (Reprinted from Fissell and Humer, Cell therapy of renal failure, Transplantation Proceedings, 35, 2837–2842, 2003, with permission from Elsevier.)

a synthetic hemofiltration cartridge and a RAD in an extracorporeal circuit successfully replaced filtration, transport, and metabolic and endocrinologic functions of the kidney in acutely uremic dogs.

After a series of experiments demonstrating bioactivity, longevity, and systemic activity of the proximal tubule cells in a large animal model, further experiments were conducted to examine the impact of cell therapy on the course of sepsis complicated by renal failure. Septic shock with ARF was chosen as an experimental model for several reasons. First, there is no established animal model of chronic renal failure, largely due to the cost involved in maintaining a herd of animals on dialysis. Second, the time course of well-dialyzed chronic renal failure is months to years of subject survival, which would be prohibitively expensive. Lastly, animal models of septic shock have been well established and can serve as a starting point for understanding the disease physiology.

After two initial studies supported a systemic effect and hemodynamic benefit from cell therapy in large animal models of sepsis [27,28], we pursued further evidence that cell therapy with renal proximal tubule cells altered the physiologic response to sepsis. A porcine model of septic shock was developed from the previous work. It was noted in nonnephrectomized animals that in response to bacteria, the animals rapidly became oligoanuric. We hypothesized that septic shock rapidly induced tubule cell injury, and that replacement of the function of injured tubule cells would confer benefit upon the animals.

Purpose-bred pigs were anesthetized and administered an intraperitoneal dose of bacteria, causing shock and renal failure. An hour later CVVH was initiated with either cell or sham RAD. Urine output and mean arterial pressure declined within the first few hours after insult. Cell-treated animals survived 9.0 ± 0.83 h vs. 5.1 ± 0.4 h ($P < .005$) for sham-treated animals (Figure 31.4).

Serum cytokines were similar between the two groups, with the striking exception of IL-6 and IFN-γ. Treatment with the cell RAD resulted in significantly lower plasma levels of both IL-6 ($P < .04$) and IFN-γ ($P < .02$) throughout the experimental time course compared to sham RAD exposure (Figure 31.5).

This controlled trial of cell therapy of renal failure in a realistic animal model of sepsis has several findings not immediately expected from a priori assumptions regarding renal function. Heretofore, although renal failure has been strongly associated with poor outcome in hospitalized patients, and chronic renal failure is associated with specific defects in humoral and cellular immunity, a direct immunomodulatory effect of the kidney had not been accepted. In this trial, clear differences in survival and clear differences in a serum cytokine associated with mortality in sepsis were found between animals treated with cells and with sham cartridges. This hearkens back to statements made earlier in the chapter: that the increased mortality in renal failure has not been conclusively attributed to inadequate clearance, but may arise from other bioactivity of the kidney. With a series of preclinical experiments demonstrating the extracorporeal circuit

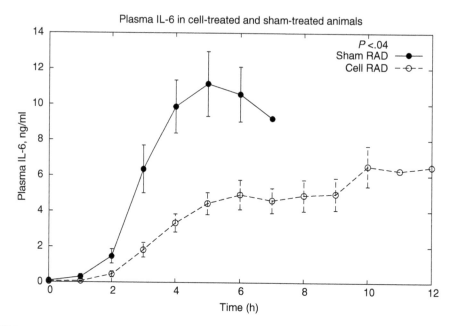

FIGURE 31.5 Serum cytokine levels in septic animals treated with a bioartificial kidney. (Reprinted with modifications from Humer, H.D., Buffington, D.A., Lou, L., Abrishami, S., Wang, M., Xia, J., and Fissell, W.H. *Critical Care Med.*, 31: 2421–2428, 2003, with permission from Lippincott Williams & Wilkins.)

containing the bioreactor, the U.S. Food and Drug Administration granted Investigational New Drug approval to the bioreactor, and a multicenter Phase I/II clinical trial of the bioartificial kidney was begun.

31.6 Clinical Trials of the RAD

With the suggestive preclinical data from the canine studies, the FDA approved an investigational new drug (IND) application to study the RAD containing human cells in patients with acute tubular necrosis and multisystem organ failure who were receiving continuous renal replacement therapy. Human kidney cells were isolated from kidneys donated for cadaveric transplantation but which could not be used for this end due to anatomic or fibrotic defects.

At the time of this writing, ten human patients at two investigational centers have been treated with the RAD in this Phase I trial. The initial clinical experience with the RAD is detailed in Reference 29. The data collected to date suggests that the RAD remains functional with proximal tubules remaining viable and continuing to display differentiated functions of the proximal tubule. The cells demonstrated glutathione degradation and 25-OH Vitamin D_3 hydroxylation out to 24 h of use, the longest time tested to date. This represents an important milestone in medical therapeutics. We have shown the practicability and safety of treating an acute illness characterized by damage to and loss of function of a cell type with human cells grown in tissue culture and delivered to the patient in a bedside bioreactor.

References

[1] U.S. Renal Data System. *USRDS 2003 Annual Data Report*. Atlas of End-Stage Renal Disease in the United States. National Institutes of Health, National Institute of Diabetes and Digestive and Kidney Diseases, 2003.

[2] O'Brien, M.M., Gonzales, R., Shroyer, A.L., Grunwald, G.K., Daley, J., Henderson, W.G., Khuri, S.F., and Anderson, R.J. Modest serum creatinine elevation affects adverse outcome after general surgery. *Kidney Int.*, 62: 585–592, 2002.

[3] Anderson, R.J., O'Brien, M.M., MaWhinney, S., VillaNueva, C.B., Moritz, T.E., Sethi, G.K., Henderson, W.G., Hammermeister, K.E., Grover, F.L., and Shroyer, A.L. Renal failure predisposes patients to adverse outcome after coronary artery bypass surgery. *Kidney Int.*, 55: 1057–1062, 1999.

[4] Wolfe, R.A., Ashby, V.B., Milford, E.L., Ojo, A.O., Ettenger, R.E., Agodoa, L.Y.C., Held, P.J., and Port, F.K. Comparison of mortality in all patients on dialysis, patients on dialysis awaiting transplantation, and recipients of a first cadaveric transplant. *NEJM*, 341: 1725–1730, 1999.

[5] Schiffl, H., Lang, S.M., and Fischer, R. Daily hemodialysis and the outcome of acute renal failure. *NEJM*, 346: 305–310, 2002.

[6] Ronco, C., Bellomo, R., Homel, P., Brendolan, A., Dan, M., Piccinni, P., and La Greca, G. Effects of different doses in continuous veno-venous haemofiltration on outcomes of acute renal failure: a prospective randomised trial. *Lancet*, 356: 26–30, 2000.

[7] Eknoyan, G., Beck, G.J., Cheung, A.K., Daugirdas, J.T., Greene, T., Kusek, J.W., Allon, M., Bailey, J., Delmez, J.A., Depner, T.A., Dwyer, J.T., Levey, A.S., Levin, N.W., Milford, E., Ornt, D.B., Rocco, M.V., Schulman, G., Schwab, S.J., Teehan, B.P., Toto, R., and the Hemodialysis (HEMO) Study Group. Effect of dialysis dose and membrane flux in maintenance hemodialysis. *NEJM*, 347: 2010–2019, 2002.

[8] Sarnak, M.J. and Jaber, B.L. Mortality caused by sepsis in patients with end-stage renal disease compared to the general population. *Kidney Int.*, 58: 1758–1764, 2000.

[9] Himmelfarb, J., Stenvinkel, P., Ikizler, T.A., and Hakim, R.M. The elephant in uremia: oxidant stress as a unifying concept of cardiovascular disease in uremia. *Kidney Int.*, 62: 1524–1538, 2002.

[10] Himmelfarb, J., McMenamin, E., and McMonagle, E. Plasma aminothiol oxidation in chronic hemodialysis patients. *Kidney Int.*, 62: 705–716, 2002.

[11] Himmelfarb, J., McMonagle, E., and McMenamin, E. Plasma protein thiol oxidation and carbonyl formation in chronic renal failure. *Kidney Int.*, 58: 2571–2578, 2000.

[12] Girndt, M., Kohler, H., and Schiedhelm-Weick, E. Production of interleukin-6, tumor necrosis factor alpha and interleukin-10 *in vitro* correlates with the clinical immune defect in chronic hemodialysis patients. *Kidney Int.*, 47: 559–565, 1995.

[13] Girndt, M., Sester, U., Sester, M., Kaul, H., and Kohler, H. Impaired cellular immune function in patients with endstage renal failure. *Nephrol. Dial. Transplant.*, 14: 2807–2830, 1999.

[14] Stan, R.V., Kubitza, M., and Palade, G.E. Pv-1 is a component of the fenestral and stomatal diaphragms in fenestrated endothelia. *Proc. Natl Acad. Sci. USA*, 96: 13203–13207, 1999.

[15] Rodewald, R. and Karnovsky, M.J. Porous substructure of the glomerular slit diaphragm in the rat and the mouse. *J. Cell Biol.*, 60: 423, 1974.

[16] Holzman, L.B., St. John, P.L., Kovari, I.A., Verma, R., Holthofer, H., and Abrahamson, R. Nephrin localizes to the slit pore of the glomerular slit diaphragm. *Kidney Int.*, 56: 1481–1491, 1999.

[17] Ohlson, M., Sorensson, J., and Haraldsson, B. Glomerular size and charge selectivity in the rat as revealed by fitc-ficoll and albumin. *Am. J. Physiol.*, 279: F84–F91, 2000.

[18] Sorensson, J., Ohlson, M., and Haraldsson, B. A quantitative analysis of the glomerular charge barrier in the rat. *Am. J. Physiol. Renal Physiol.*, 280: F646–F656, 2001.

[19] Ohlson, M., Sorensson, J., and Haraldsson, B. A gel-membrane model of glomerular charge and size selectivity in series. *Am. J. Physiol. Renal Physiol.*, 280: F396–F405, 2001.

[20] Drumond, M.C. and Deen, W.M. Structural determinants of glomerular hydraulic permeability. *Am. J. Physiol. Renal Physiol.*, 266: F1–F12, 1994.

[21] Deen, W.M., Satvat, B., and Jamieson, J.M. Theoretical model for glomerular filtration of charged solutes. *Am. J. Physiol. Renal Physiol.*, 238: F126–F139, 1980.

[22] Saleem, M.A., O'Hare, M.J., Reiser, J., Coward, R.J., Inward, C.D., Farren, T., Xing, C.Y., Ni, L., Mathieson, P.W., and Mundel, P. A conditionally immortalized podocyte cell line demonstrating nephrin and podocin expression. *J. Am. Soc. Nephrol.*, 13: 630–638, 2002.

[23] Fissell, W.H., Humes, H.D., Roy, S., and Fleischman, A. Initial characterization of a nanoengineered ultrafiltration membrane. *J. Am. Soc. Nephrol.*, 13: 602A, 2002.

[24] Humes, H.D. and Cieslinski, D.A. Interaction between growth factors and retinoic acid in the induction of kidney tubulogenesis in tissue culture. *Exp. Cell Res.*, 201: 8–15, 1992.

[25] Humes, H.D., Krauss, J.C., Cieslinski, D.A., and Funke, A.J. Tubulogenesis from isolated single cells of adult mammalian kidney: clonal analysis with a recombinant retrovirus. *Am. J. Physiol.*, 271: F42–49, 1996.

[26] Humes, H.D., MacKay, S.M., Funke, A.J., and Buffington, D.A. Tissue engineering of a bioartificial renal tubule assist device: *in vitro* transport and metabolic characteristics. *Kidney Int.*, 55: 2502–2514, 1999.

[27] Fissell, W.H., Dyke, D.B., Weitzel, W.F., Buffington, D.A., Westover, A.J., Mackay, S.M., Gutierrez, J.M., and Humer, H.D., Bioartificial kidney alters cytokine response and hemodynamics in endotoxin- challenged uremic animals. Blood Purif., 20: 55–60, 2002.

[28] Fissell, W.H., Lou, L., Abrishami, S., Buffington, D.A., and Humer, H.D., Bioartificial kidney ameliorates gram-negative bacteria-induced septic shock in uremic animals. *J. Am. Soc. Nephrol.* 14: 456–461, 2003.

[29] Weitzel, W.F., Fissell, W.H., and Humes, H.D. Initial clinical experience with a human proximal tubule cell renal assist device (RAD). *J. Am. Soc. Nephrol.*, 12: 279a, 2001.

32

The Bioengineering of Dental Tissues

Rena N. D'Souza
University of Texas Health Science Center at Houston

Songtao Shi
National Institutes of Health
National Institute of Dental and Craniofacial Research

32.1 Introduction

The proper size, shape, color, and alignment of teeth influences the nature of our smile and determines our uniqueness as individual humans. In addition to their esthetic value, teeth are important for the mastication of food and for proper speech. Despite these critical functions, the importance and uniqueness of teeth are frequently overlooked by health professionals. The loss of dentition to common diseases like caries, periodontal disease and to trauma imposes significant emotional and financial burdens on patients and their families. Despite the overall success of osseo-integrated titanium implants, tooth forms that are bioengineered from natural tissues/cells represent the next wave of dental regenerative medicine. The calcified tooth matrices of enamel, dentin, and cementum each possesses unique biomechanical, structural and biochemical properties. When consideration is given to the bioengineering of whole tooth forms, several challenges exist that relate to the restoration of specific shapes and sizes as well as the (re)generation of these highly specialized mineralized matrices. In order to provide an appreciation for the complexity of the tooth as a whole, this chapter first discusses the components of a mature tooth and its surrounding structures. Next, the basic principles of tooth development that lend the molecular and genetic bases for modern bioengineering strategies is presented. Important contributions from mouse and human genetic studies is also briefly overviewed. Finally, recent data from successful tooth engineering initiatives involving somatic and stem cell approaches along with whole tooth organ strategies is discussed. In projecting future research directions, this chapter concludes with a brief discussion of the challenges and opportunities that exist for bioengineering one of the most complex of all vertebrate organ systems.

32.2 The Tooth and Its Supporting Structures

The crowns of teeth that are exposed in the oral cavity are covered by enamel. Under the enamel is a thick layer of dentin and a soft central core, the pulp chamber (Figure 32.1). Enamel is the hardest calcified structure as it is about 99.5% mineralized. It varies from 2 to 3 mm in thickness at the height of cusps and narrows to a knife-edge thickness at the cementoenamel junction. Enamel is deposited by ameloblasts, cells that are believed to undergo programmed cell death. Because enamel is acellular and nonvital it cannot regenerate itself. Underlying enamel is dentin, a specialized mineralized matrix that shares several biochemical characteristics with bone. In contrast to enamel, dentin is a vital tissue that harbors odontoblastic processes and some nerve endings. The formation of dentin follows the same principles that guide the formation of other hard connective tissues in the body, namely, cementum and bone.

As described by Linde and Goldberg [1] and Butler and Ritchie [2], the composition of dentin matrix and the process of dentinogenesis are highly complex. The organic phase of dentin is composed of proteins, proteoglycans, lipids, various growth factors, and water. Among the proteins, collagen is the most abundant and offers a fibrous matrix for the deposition of carbonate apatite crystals. The collagens that are found in dentin are primarily type I collagen with trace amounts of type V collagen and some type I collagen trimer. An important class of dentin matrix proteins is the noncollagenous proteins or NCPs [2]. The dentin-specific NCPs are dentin phosphoproteins (DPP) or phosphophoryns and dentin sialoprotein (DSP). After type I collagen, DPP is the most abundant of dentin matrix proteins and represents almost 50% of the dentin extracellular matrix. DPP is a polyionic macromolecule that is rich in phosphoserine and aspartic acid. Its high affinity for type I collagen as well as calcium makes it a strong candidate for the initiation of dentin mineralization. DSP accounts for 5 to 8% of the dentin matrix and has a relatively high sialic acid and carbohydrate content. Its role in dentin mineralization is unclear at the present time. For several years it was believed that DSP and DPP were two independent proteins that were encoded by individual genes. DPP and DSP are specific cleavage products of a larger precursor protein that was translated from one large transcript [3]. This single gene encoding for DSP and DPP is named dentin sialophosphoprotein or Dspp.

A second category of NCPs with Ca-binding properties are classified as mineralized tissue-specific as they are found in all the calcified connective tissues, namely, dentin, bone, and cementum. These

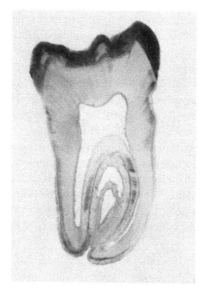

FIGURE 32.1   Component parts of a tooth.

include osteocalcin (OC) and bone sialoprotein (BSP). A serine-rich phosphoprotein called dentin matrix protein 1, Dmp-1, whose expression was first described as being restricted to odontoblasts [4], was later shown to be expressed by osteoblasts and cementoblasts [5]. Other NCPs include osteopontin (OP) and osteonectin (secreted protein, acidic, cycteine-rich; SPARC). The fourth category of dentin NCPs are not expressed in odontoblasts but are primarily synthesized in the liver and are released into the circulation. An example of a serum borne protein is α2HS-glycoprotein. Diffusible growth factors that appear to be sequestered within dentin matrix constitute the fifth group of dentin NCPs. This group includes the BMPs, IGFs, and TGF-βs [6].

The central chamber of the tooth is occupied by a soft connective tissue called the dental pulp that is comprised of a heterogeneous cell population of fibroblasts, undifferentiated mesenchymal cells, nerves, blood vessels, and lymphatics. The regenerative capacity of dental pulp is well documented in the literature and best illustrated by the formation of a layer of reparative dentin beneath a carious lesion or a cavity base. As will be discussed later, somatic stem cells from the dental pulp are capable of regenerating several tissues when transplanted *in vivo*. Cementum is another calcified tissue of mesodermal origin. The cementum covering the apical third of the root is cellular (contains cementocytes), whereas that of the remaining two-thirds is acellular. Since the fibers of the periodontal ligament are anchored within cementum, the regeneration of this complex is important when bioengineering of whole tooth structures is considered.

32.3 Genetic Control of Tooth Development

32.3.1 Stages of Tooth Development

Teeth develop in distinct stages that are easily recognizable at the microscopic level. Hence, stages in odontogenesis are described in classic terms by the histologic appearance of the tooth organ. From early to late, these stages are described as the lamina, bud, cap, and bell (early and late) stages of tooth development [7,8]. Recent advances made in the understanding of the molecular control of tooth development have led to the development of new terminology to describe tooth development as occurring in four phases: initiation, morphogenesis, cell or cyto-differentiation, and matrix apposition (Figure 32.2). The appearance of the dental lamina marks the first visible sign of tooth initiation that is seen about five weeks of human

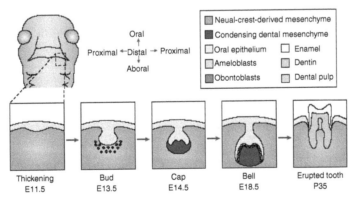

FIGURE 32.2 Stages of tooth development. A schematic frontal view of an embryo head at embryonic day (E)11.5 is shown with a dashed box to indicate the site where the lower (mandibular) molars will form. Below, the stages of tooth development are laid out from the first signs of thickening at E11.5 to eruption of the tooth at around 5 weeks after birth. The tooth germ is formed from the oral epithelium and neural-crest-derived mesenchyme. At the bell stage of development, the ameloblasts and odontoblasts form in adjacent layers at the site of interaction between the epithelium and mesenchyme. These layers produce the enamel and dentin of the fully formed tooth. (Reproduced from Tucker, A. and Sharpe, P. *Nat. Rev. Genet.* 5: 499–508, 2004. With permission.)

development. The inductive influence of the dental lamina to dictate the fate of the underlying ectomesenchyme has been confirmed by several researchers [9]. The table below summarizes the molecules expressed in the epithelium and mesenchyme at this inductive phase. The bud stage is characterized by the continual growth of cells of the dental lamina and ectomesenchyme. The latter is condensed and termed the dental papilla. At this stage, the inductive or tooth-forming potential is transferred from the dental epithelium to the dental papilla. The transition from the bud to the cap stage is an important step in tooth development as it marks the onset of crown formation. The tooth bud assumes the shape of a cap that is surrounded by the dental papilla. The ectodermal compartment of the tooth organ is referred to as the dental or enamel organ. The enamel organ and dental papilla become encapsulated by a sac called the dental follicle that separate the tooth organ papilla from the other connective tissues of the jaws. A cluster of cells called the enamel knot is an important organizing center within the dental organ and is important for the formation of cusps [10,11]. The enamel knot expresses a unique set of signaling molecules that influence the shape of the crown as well as the development of the dental papilla. Similar to the fate of signaling centers in other organizing tissues like the developing limb bud, the enamel knot undergoes programmed cell death or apoptosis, after cuspal patterning is completed at the onset of the early bell stage. As the dental organ assumes the shape of a bell, several layers of cells continue to divide but at differential rates. A single layer of cuboidal cells called the external or outer dental epithelium lines the periphery of the dental organ while cells that border the dental papilla and are columnar in appearance form the internal or inner dental epithelium. The latter gives rise to the ameloblasts, cells responsible for enamel formation. Cells located in the center of the dental organ produce high levels of glycosaminoglycans that are able to sequester fluids as well as growth factors that lead to its expansion. This network of star-shaped cells is named the stellate reticulum. Interposed between the stellate reticulum and the internal dental epithelium is a narrow layer of flattened cells termed the stratum intermedium that express high levels of alkaline phosphatase. The stratum intermedium is believed to influence the biomineralization of enamel. In the region of the apical end of the tooth organ, the internal and external dental epithelial layers meet at a junction called the cervical loop.

At the early bell stage, each layer of the dental organ has assumed special functions and exchanges molecular information that leads to cell differentiation at the late bell stage. The dental lamina that connects the tooth organ to the oral epithelium gradually disintegrates at the late bell stage. At the future cusp tips, cells of the internal dental epithelium stop dividing and assume a columnar shape. The most peripheral cells of the dental papilla organize along the basement membrane and differentiate into odontoblasts, the dentin-forming cells. At this time, the dental papilla is termed the dental pulp. After the first layer of predentin matrix is deposited, cells of the internal dental epithelium differentiate into ameloblasts or enamel-producing cells. As enamel is deposited over dentin matrix, ameloblasts retreat to the external surface of the crown and are believed to undergo programmed cell death. In contrast, odontoblasts line the inner surface of dentin and remain metabolically active throughout the life of a tooth. Root formation then proceeds as epithelial cells proliferate apically and influence the differentiation of odontoblasts from the dental papilla as well as cementoblasts from follicle mesenchyme. This leads to the deposition of root dentin and cementum respectively. The dental follicle that gives rise to components of the periodontium, namely the periodontal ligament fibroblasts, alveolar bone of the tooth socket and the cementum also plays a role during tooth eruption which marks the end phase of odontogenesis.

32.3.2 Molecular Mechanisms That Determine Tooth Shape, Size, and Structure

Similar to other organs like the limb bud, kidney, lung, and hair follicles, tooth development is regulated by temporally and spatially restricted interactions between epithelial and mesenchymal compartments. Molecular approaches used in expression analyses as well as functional *in vivo* and *in vitro* tooth recombinations and bead implantation assays have greatly increased our understanding about the molecular control of tooth development. In addition, the use of genetic approaches involving transgenic mice with targeted inactivation of various genes have provided a powerful means to delineate the *in vivo* functions of individual molecules [12,13].

In vivo and *in vitro* recombination studies have shown that during the formation of the epithelial bud (E12), the inductive potential shifts to the dental mesenchyme that later influences the fate of the enamel organ and its morphogenesis from the bud stage to the early bell stage (E16) [9,14–16]. Reciprocal interactions between the morphologically distinct enamel organ and papilla mesenchyme at the late bell stage (E18) then leads to the differentiation of dentin-forming odontoblasts and enamel-forming ameloblasts. As morphogenesis advances, the matrices of dentin and enamel are deposited in an organized manner and root formation begins. Interactions between the apical extension of the enamel organ (epithelial root sheath) and papilla/follicle mesenchyme lead to the patterning of roots, the differentiation of cementoblasts and the formation of cementum. Hence, during crown and root development, morphogenesis, and cytodifferentiation are controlled by epithelial-mesenchymal interactions.

As depicted in Table 32.1 [13], molecular changes in dental mesenchyme involve proteins in the bone morphogenetic protein (BMP), fibroblast growth factor (FGF) and wingless-type MMTV integration site family WNT families; *sonic hedgehog* (*Shh*) as well as transcriptional molecules like the *Msx-1, -2* homeobox genes; lymphoid enhancer-binding factor 1 (Lef-1) and Pax-9, a member of the paired-box-containing transcription factor gene family. The actions and interactions of these molecules are complex and described eloquently in recent reviews [12,17].

The BMPs are among the best-characterized signals in tooth development. In addition to directly influencing morphogenesis of the enamel organ (see discussion on enamel knot later), epithelial BMP-2 and -4 are able to induce expression of *Msx1, Msx2,* and *Lef-1* in dental mesenchyme as shown in bead implantation assays [18–20]. The shift in *Bmp-4* expression from epithelium to mesenchyme occurs around E12 and is coincident with the transfer of inductive potential from dental epithelium to mesenchyme [18]. In mesenchyme, *Bmp-4* in turn, requires *Msx-1* to induce its own expression [19]. The FGFs, in general, are potent stimulators of cell proliferation and division both in dental mesenchyme and epithelium. *Fgf-2, -4,-8,* and *-9* expression are each restricted to dental epithelium and can stimulate *Msx-1* but not *Msx-2* expression in underlying mesenchyme. *Fgf-8* is expressed early in odontogenesis (E10.5 to E11.5), in presumptive dental epithelium, and can induce the expression of *Pax-9* in underlying mesenchyme. Interestingly, BMP-4 prevents this induction and may share an antagonistic relationship with the FGFs similar to what is observed in limb development [21]. Recent studies by Hardcastle et al., 1998, have shown that Shh in beads cannot induce *Pax9, Msx-1* or *Bmp-4* expression in dental mesenchyme but is able to stimulate other genes encoding the transmembrane protein *patched* (*Ptc)* and *Gli1,* a zinc finger transcription factor [22–24]. Since neither FGF-8 nor BMP-4 can stimulate *Ptc* or *Gli1,* it can be assumed at the present time that the Shh signaling pathway is independent of the BMP and FGF pathways during tooth development [24]. Several *Wnt* genes are expressed during tooth development and may be required for the formation of the tooth bud [12]. These genes are believed to play a role in activating the intracellular pathway involving frizzled receptors, β-catenin and nuclear transport of Lef-1. Other signaling molecules including the Notch genes, epidermal growth factor (EGF), hepatocyte growth factor (HGF) and, platelet derived growth factor (PDGF) families may also influence tooth development, though the exact nature of their involvement remains to be elucidated.

The enamel knot is a transient epithelial structure that appears at the onset of cusp formation. For years, it was thought that the enamel knot controlled the folding of the dental epithelium and hence cuspal morphogenesis. Recently, the morphological, cellular and molecular events leading to the formation and disappearance of the enamel knot have been described, thus linking its role as an organizing center for tooth morphogenesis [11,25,26]. Interestingly, cells of the enamel knot are the only cells within the enamel organ that stop proliferating [10] and that undergo apoptosis [27]. Another intriguing finding linked p21, a cyclin-dependent kinase inhibitor associated with terminal differentiation events, to apoptosis of the enamel knot [11].

The enamel knot cells express several signaling molecule genes including *Bmp-2, -4, -7; Fgf-4, -9; Msx-2* and *Shh* [22,25,28–30]. Although the precise function of each morphogen is not known at the present time, a model for the relationship of inductive signaling molecules involved has been proposed by integrating morphological and molecular data [11]. Since the instructive signaling influence lies with the dental mesenchyme prior to the development of the primary enamel knot, it is likely that this tissue

TABLE 32.1 Genes Expressed During Tooth Development in Mouse[a]

Stage of development	Expressed in epithelium	References	Expressed in mesenchyme	References
Up to epithelial thickening				
(E10–E11)	*Fgf8,9*	[21,60,63,64]	*Activin*	[70]
	Bmp4	[21,35,60,75]	*Pax9*	[21]
	Shh	[24,69]	*Barx1*	[60]
	Islet1	[65]	*Msx1 & Msx2*	[19,83,84]
	Pitx2	[66,67]	*Dlx1,2,3,5,6*	[61,62]
	Wnt 10b	[76]	*Ptc*	[24]
	Follistatin	[70]	*Gli1,2,3*	[24]
	Lef1	[20]	*Lhx6,7*	[63]
	Eda, Edar	[72]		
Bud stage				
(E12–E13)	*Eda, Edar*	[72]	*Runx2*	[71]
	Pitx2	[66,67]	*Bmp4*	[35,75]
			Msx1	[35]
			Lef1	[20]
			Fgf3,10	[73]
			Dlx1,2	[62]
			Lhx6,7	[63]
Cap stage				
(E14–E15)	Enamel knot		Non-EK epithelium	[74]
	p21	[11]	*FgfR*	[72]
	Shh	[24]	*Eda*	
	Edar	[72]		
	Edaradd	[68]		
	Bmp 2,4,7	[75]		
	Wnt10a	[76]		
	Msx2	[11,29]		
	Fgf3,4	[10,73]		
Bell stage				
(E16 Onward)	*Amelogenin*	[77]	*Dspp*	[78]
	Bmp5	[75]		

[a] This list indicates the expression pattern of several genes that are thought to be important in tooth development in the mouse. A more comprehensive list of genes and their expression patterns can be found at the Gene Expression in Tooth web site (http://bite-it.helsinki.fi/) [79]. *Barx*, BarH-like homeobox; *Bmp*, bone morphogenetic protein; *Dlx*, distal-less homeobox; *Dspp*, dentin sialophosphoprotein; E, embryonic stage; *Eda*, ectodysplasin-A; *Edar*, *Eda* receptor; *Edaradd*, EDAR (ectodysplasin-A receptor)-associated death domain; *EK*, enamel knot; *Fgf*, fibroblast growth factor; *FgfR*, fibroblast growth factor receptor; *Gli*, GLI-Kruppel family member; *Lef*, lymphoid enhancer binding factor; *Lhx*, LIM homeodomain genes; *Msx*, homeobox, msh-like; *p21* (*CDKN1A*), cyclin-dependent kinase inhibitor 1A; *Pax*, pairedbox gene; *Pitx*, paired-related homeobox gene; *Ptc*, patched; *Runx*, runt homologue; *Shh*, sonic hedgehog; *Wnt*, wingless-related protein.

Source: Reproduced from Tucker, A. and Sharpe, P. *Nat. Rev. Genet.* 5: 499–508, 2004. With permission.

influences enamel knot formation. In this regard, BMP-4 in condensing dental mesenchyme, functions as a paracrine molecule that can upregulate *Msx2* and *p21* expression within the enamel knot [11,26]. It is hypothesized that *p21* then prevents proliferation within the enamel knot allowing for the growth stimulatory *Fgf-4* to be expressed exclusively in this region [10]. FGF-4 in turn, may act singly or in concert with *Fgf-9* to influence patterning or to regulate expression of downstream genes like *Msx1* in underlying papilla mesenchyme [19,28]. Intriguingly, later in development, BMP-4 participates in the regulation of apoptosis perhaps in an autocrine fashion by involving genes like *Msx-2*.

Mice genetically engineered with targeted mutations in transcription factor genes like *Msx-1*, *Lef-1*, and *Pax9* as well as activin-βA, a member of the TGF-β superfamily, have revealed important information.

Knockouts of *Bmp-2, -4,* and *Shh* have proven less informative largely due to death that occurs *in utero* prior to the onset of tooth development. In *Msx-1, Lef-1, Pax9,* and activin-βA mutant strains, tooth development fails to advance beyond the bud stage. Thus, these molecules are important in directing the fate of the dental mesenchyme and its ability to influence the progress of epithelial morphogenesis to the cap stage [31–34]. Curiously, *Msx2* molars develop fully but show abnormal cuspal patterning, a poorly differentiated stellate reticulum and enamel matrix defects, suggesting that this homeobox gene is involved in the patterning and differentiation of the enamel organ [17; Maas, personal communication]. As reviewed by Tucker and Sharpe [13], molecular information on tooth development can be used to alter the shape and size of teeth. For example, when beads soaked with Bmp4 are placed on mesenchyme within the presumptive incisor region, *Msx1* expression is downregulated and expression of the transcription factor *Barx1* is upregulated. Since *Barx1* is normally restricted to the molar region, its misexpression within the incisor region results in the formation of a molar tooth organ instead of an incisor [35].

In addition to use of the mouse tooth organ model, the legacy of inheritable anomalies of human dentition involving the failure of teeth to develop offers a powerful system for studying the genetic pathways that control the development of human dentition. Familial tooth agenesis is the most common dental anomaly that affects up to 25% of the population. It is transmitted either as an autosomal-dominant, autosomal-recessive or X-linked trait, and presents in syndromic and nonsyndromic forms. Genes involved in epithelial-mesenchymal interactions as shown by studies in the mouse are strong candidates for human tooth agenesis. Until recently, mutations in two genes that encode for the key transcription factors *PAX9* and *MSX1* were associated with agenesis of molars and premolars [36]. Importantly, PAX9 and MSX1 have each been excluded in other families with autosomal dominant forms of tooth agenesis. Recently, tooth agenesis has been linked to a mutation in AXIN2, a molecule known to regulate cell homeostasis [37]. Several members of this four-generation family are affected by or are at risk for colon cancer suggesting a broader role for this molecule in cell proliferation.

Taken together, the data from mouse and human studies have provided valuable insights into the molecular and genetic control of tooth development. As illustrated later, such basic information provided the rationale for tooth bioengineering initiatives for the regeneration of dentin matrix and whole tooth forms.

32.4   Tooth Regenerative Strategies

Over the years, the research on the use of stem cells for clinical therapies has been growing, especially after researchers have found that hematopoietic stem cells, a well-characterized population of postnatal stem cells, have been successfully utilized in clinics to treat hematopoietic diseases [38] autoimmune diseases [39] and solid tumors [40]. Stem cells are defined as cells that have clonogenic and self-renewing capabilities and that differentiate into multiple-cell lineages. In general, there are two kinds of stem cells: embryonic and postnatal stem cells. Embryonic stem cells are derived from mammalian embryos in the blastocyst stage and have the ability to generate any terminally differentiated cell in the body; postnatal stem cells are part of tissue-specific cells of the postnatal organism into which they are committed to differentiate. Stem cell-based tissue regeneration has great clinical potential to regain physiological functions that have been damaged by various diseases.

32.4.1   Human Dental Pulp Stem Cells

Isolation and identification of stem cells are the first step in studying the potential of stem cell-mediated therapy. Postnatal stem cells have been isolated from a variety of tissues including but not limited to skin, liver, brain, bone marrow, and peripheral blood. Recently, dental pulp stem cells (DPSCs) have been successfully isolated from adult dental pulp in extracted human teeth [41]. Similar to the other mesenchymal stem cells, DPSCs are able to generate clonogenic cell colonies *in vitro*. The majority of the individual colonies (67%) failed to proliferate beyond 20 population doublings in the culture, suggesting that only a small portion of cells maintain high proliferation potential *in vitro* [42]. Mixed multi-colony DPSCs show a higher proliferation rate than bone marrow stromal stem cells (BMSSCs) in culture. cDNA microarray analysis demonstrated that highly expressed cyclin-dependent kinase 6 (cdk6) and IGF-2 in DPSCs might be, at least partially, responsible for the promoted progression of cells through G1 to the start of DNA synthesis [43–45], leading to an elevated replicative proliferation. Most postnatal stem cells reside in a specific niche microenvironment to maintain their stemness. To elucidate the DPSCs' niche environment, DPSCs were first found to express various markers associated with endothelial and/or smooth muscle cells such as STRO-1, 3G5, VCAM-1, MUC-18 and α-smooth muscle actin [41,46]. Then, immunohistochemical staining and magnetic beads sorting were applied to confirm that DPSCs, similar to BMSSCs, reside in a perivascular niche microenvironment [46]. Taken together with their clonogenic nature, higher proliferation rate, and specific niche microenvironment, DPSCs satisfy three criteria characteristic of human postnatal somatic stem cells.

32.4.2 Dentin Tissue Regeneration

One of the most important characteristics of DPSCs is their capability to form a dentin/pulp-like complex upon *in vivo* transplantation in conjunction with hydroxyapatite/tricalcium phosphate as a carrier (Figure 32.3). Backscatter EM analysis demonstrated that the dentin-like material formed in the

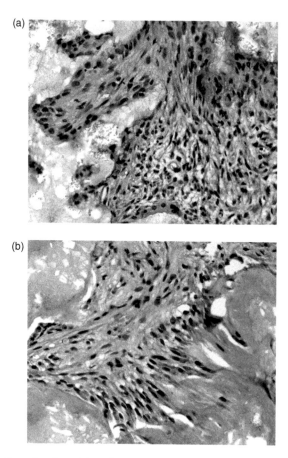

FIGURE 32.3 Hematoxylin and eosin staining of representative DPSC transplants. (a) After one week posttransplantation, DPSC transplants contain connective tissue (*CT*) around HA/TCP carrier (*HA*), without any sign of dentin formation. (b) After six week posttransplantation, DPSCs differentiate into odontoblasts (arrows) that are responsible for the dentin formation on the surface of HA/TCP (*HA*). Original magnification: 40×.

transplants had a globular appearance consistent with the structure of dentin *in situ* [42]. DPSC-mediated odontogenesis is differentiable from BMSSC-mediated osteogenesis by regenerating different organ-like structures and involving different regulating molecules [47]. This implies that critical factor(s) may regulate mineralized matrix forming stem cells to generate defined mineralized tissue along with associated soft tissues. The property of multipotential differentiation of DPSCs has been demonstrated by the findings that under the proper culture conditions, DPSCs are capable of differentiating into osteo/odontogenic cells, adipocytes, and neural cells [42]. However, the assets of multipotential differentiation of DPSCs at functional levels remain to be confirmed.

The findings on DPSCs may provide a potential for utilizing them for dentin and pulp tissues regeneration. Human teeth do not undergo the type of remodeling that is seen in other mineralized tissues such as bone, which remodels to maintain organ integrity. Once a tooth has erupted, dentinal damage caused by mechanical trauma, exposure to chemicals or by infectious processes, induces the formation of reparative dentin that is even though structurally poorly organized, but serves as a protective barrier to the dental pulp with limited capacity [48–52]. It was reported that bone morphogenetic protein-7 is capable of stimulating tertiary dentin formation when applied to freshly cut dentin both *in vitro* and *in vivo* [53,54]. This probably occurs through an osteo/odontogenic induction property of BMP-7, since BMP-7 transfected human fibroblasts were able to express osteogenic characteristic and form bone tissue *in vivo* [55]. DPSCs were also capable of forming reparative dentin structure on the surfaces of regular human dentin [47]. However, it seems that DPSCs exhibit a decreased and altered *in vivo* odontogenic capacity when loaded on the surface of human dentin. Although the reason is not known, it may be associated with the microenvironment that accommodates *in vivo* differentiation of DPSCs [47]. Recently, it was demonstrated that autogenous transplantation of BMP2-treated DPSCs was able to stimulate reparative dentin formation on the amputated pulp [56]. This finding suggests that combination therapy using stem cells and growth factors may improve stem cell-mediated dentin regeneration.

32.4.3 Tooth Regeneration

Recently, whole tooth regeneration *in vivo* has become a hot topic in dental research. Tooth development involves a mutual signaling interaction between epithelial and mesenchymal cells of neural ectodermal origin. It was demonstrated that tooth crown structures including dentin, odontoblasts, pulp chamber, putative Hertwig's root sheath epithelia, putative cementoblasts, and enamel organ could be regenerated using dissociated cells from pig tooth bud tissues (Figure 32.4) [57]. Further, the same research group identified that cultured cells from rat tooth bud were also able to regenerate tooth structure when loaded on the PGA or PLGA scaffolds [58]. These studies demonstrate for the first time that mammalian tooth structure can be regenerated in a system consisting of tooth bud progenitors and the proper scaffold. Moreover, Sharpe's group conducted a promising study to demonstrate that mice embryonic oral epithelium along with nondental stem cells can induce an odontogenic response, showing the expression of odontogenic mesenchymal cell associated genes such as Msx1, Lhx7, and Pax9 [59]. After being transplanted into adult renal capsules, the recombination of embryonic oral epithelium with nondental stem cells (embryonic, neural, and bone marrow stem cells) gave rise to both tooth structure and bone tissue (Figure 32.5) [59]. Also, transplanted embryonic tooth primordial were able to maintain their tooth development potential within an adult environment [59]. This study clearly indicates that the inductive function of embryonic oral epithelium may be an important driving force for future prospects of achieving entire tooth regeneration *in vivo*.

Human DPSCs have been successfully isolated and characterized, which open the door for using these cells for potential tooth structure regeneration. In addition, stem cell-mediated whole tooth structure regeneration implies a great potential for regenerating functional entire tooth *in vivo*. However, substantial experimentation is still required for translating these technologies into clinical applications.

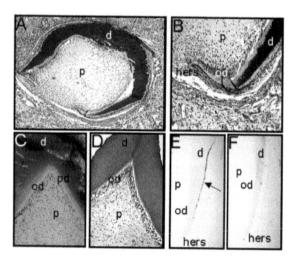

FIGURE 32.4   Histology and immunohistochemistry of a 20-week implant. (a) Von Kossa stain for calcified mineralization in bioengineered tooth crown (50× magnification). Dark brown stain is positive for mineralized tissues. (b) A high-magnification (400×) photomicrograph of the Hertwig's epithelial root sheath is shown, stained by the Von Kossa method to detect calcified mineralization. (c) High-magnification (200×) photomicrograph of cuspal region in bioengineered tooth crown. The tissue was stained by the Von Kossa method. (d) Hematoxylin and eosin (H&E) stain of a positive control porcine third molar cuspal region demonstrates morphology similar to that of the bioengineered tooth structure (200×). (e) BSP immunostain of 20-week bioengineered tooth crown (100×). Positive BSP expression is indicated by the arrow. (f) Negative preimmune control immunostain for BSP in bioengineered tooth crown (100×). Abbreviations: d, dentin; od, odontoblasts; p, pulp; pd, predentin; hers, Hertwig's epithelial root sheath.

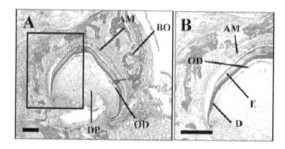

FIGURE 32.5   Recombinant explant between bone-marrow-derived cells and oral epithelium following 12 days of development in a renal capsule. All the tissues visible are donor-derived, since the host kidney makes no cellular contribution to the tissue. Where epithelium in the recombinations was from GFP mice, *in situ* hybridization of sections of these tissues confirmed that all mesenchyme-derived cells were of wildtype origin (not shown). BO, bone; Am, ameloblasts; DP, dental pulp; OD, odontoblasts; E, enamel; D, dentin. Scale bar: 80 μm. (Reproduced from Ohazama, et al. *J. Dent. Res.* 83: 518–522. With permission.)

32.5   Conclusions

The last decade has witnessed an explosion of scientific and technological advances that will undoubtedly propel the field of tooth bioengineering forward. This chapter was limited in scope in as much as only a few tooth regenerative strategies were discussed. Therefore, readers should be mindful of several other dimensions of research that exist. As presented in the current literature, there is much interest in understanding the structural, biomechanical, and bioregulatory features of dentin and bone matrices as well as the complex process of enamel biomineralization and remineralization. Such basic knowledge is essential for the development of tooth-specific biological substitutes that will best restore, maintain, or improve the functions of normal dentition. The legacy of inheritable anomalies that involve tooth patterning and extracellular matrices will continue to provide a powerful means of identifying new molecular pathways that influence normal and abnormal development. Clearly, advances in the field of tooth bioengineering will depend on the clever integration of basic science knowledge from animal and human developmental and genetic studies with emerging technologies in the field of stem cell biology, autologous cell therapy, gene therapy, materials sciences, and nanotechnology. Although the clinical applications for the use of bioengineered tooth forms and matrices remain limitless, several challenges must be surmounted prior to successful therapeutic interventions. As important as the timely diagnosis, accurate prognosis and proper treatment of diseases affecting dentition will be the preparation of host sites within the oral cavity to receive bioengineered materials. In every respect, the field of tooth bioengineering encompasses broad strategies and multidisciplinary approaches directed at restoring one of the most complex organs in vertebrates.

Acknowledgments

The authors acknowledge the support of the National Institute of Dental and Craniofacial Research (NIDCR), National Institutes of Health (NIH). The research program of RDS has been funded through NIH grants DE10517; DE07252; DE12269; DE11663 and DE13368. STS is supported by the Division of Intramural Research at the NIDCR.

References

[1] Linde, A. and Goldberg, M. Dentinogenesis. *Crit. Rev. Oral. Biol. Med.* 4: 679–728, 1993.

[2] Butler, W.T. and Ritchie, H. The nature and functional significance of dentin extracellular matrix proteins. *Int. J. Dev. Biol.* 39: 169–179, 1995.

[3] MacDougall, M. et al. Dentin phosphoprotein and dentin sialoprotein are cleavage products expressed from a single transcript coded by a gene on human chromosome 4. Dentin phosphoprotein DNA sequence determination. *J. Biol. Chem.* 272: 835–842, 1997.

[4] George, A. et al. Characterization of a novel dentin matrix acidic phosphoprotein. Implications for induction of biomineralization. *J. Biol. Chem.* 268: 12624–12630, 1993.

[5] D'Souza, R.N. et al. Gene expression patterns of murine dentin matrix protein 1 (Dmp1) and dentin sialophosphoprotein (DSPP) suggest distinct developmental functions *in vivo. J. Bone Miner. Res.* 12: 2040–2049, 1997.

[6] Smith, A.J. and Lesot, H. Induction and regulation of crown dentinogenesis: embryonic events as a template for dental tissue repair? *Crit. Rev. Oral Biol. Med.* 12: 425–437, 2001.

[7] Ten Cate, A.R. *Oral Histology: Development, Structure, and Function*, 5th ed. St. Louis, MO, Mosby, Inc. 1998.

[8] Avery, J.K. *Oral Development and Histology*. Pine, J.W. (ed.) Baltimore, MD, Waverly Press, 1987.

[9] Mina, M. and Kollar, E.J. The induction of odontogenesis in non-dental mesenchyme combined with early murine mandibular arch epithelium. *Arch. Oral Biol.* 32: 123–127, 1987.

[10] Jernvall, J. et al. Evidence for the role of the enamel knot as a control center in mammalian tooth cusp formation: non-dividing cells express growth stimulating Fgf-4 gene. *Int. J. Dev. Biol.* 38: 463–469, 1994.

[11] Jernvall, J. et al. The life history of an embryonic signaling center: BMP-4 induces p21 and is associated with apoptosis in the mouse tooth enamel knot. *Development* 125: 161–169, 1998.

[12] Thesleff, I. and Sharpe, P. Signalling networks regulating dental development. *Mech. Dev.* 67: 111–123, 1997.

[13] Tucker, A. and Sharpe, P. The cutting-edge of mammalian development; how the embryo makes teeth. *Nat. Rev. Genet.* 5: 499–508, 2004.

[14] Lumsden, A.G.S. Spatial organization of the epithelium and the role of neural crest cells in the initiation of the mammalian tooth. *Development* 103: 155–169, 1988.

[15] Kollar, E.J. and Baird, G.R. The influence of the dental papilla on the development of tooth shape in embryonic mouse tooth germs. *J. Embryol. Exp. Morphol.* 21: 131–148, 1969.

[16] Kollar, E.J. and Baird, G.R. Tissue interactions in embryonic mouse tooth germs. I. Reorganization of the dental epithelium during tooth-germ reconstruction. *J. Embryol. Exp. Morphol.* 24: 159–171, 1970.

[17] Maas, R. and Bei, M. The genetic control of early tooth development. *Crit. Rev. Oral Biol. Med.* 8: 4–39, 1997.

[18] Vainio, S. et al. Identification of BMP-4 as a signal mediating secondary induction between epithelial and mesenchymal tissues during early tooth development. *Cell* 75: 45–58, 1993.

[19] Chen, Y. et al. Msx1 controls inductive signaling in mammalian tooth morphogenesis. *Development* 122: 3035–3044, 1996.

[20] Kratochwil, K. et al. Lef1 expression is activated by BMP-4 and regulates inductive tissue interactions in tooth and hair development. *Genes Dev.* 10: 1382–1394, 1996.

[21] Neubüser, A. et al. Antagonistic interactions between FGF and BMP signaling pathways: a mechanism for positioning the sites of tooth formation. *Cell* 90: 247–255, 1997.

[22] Bitgood, M.J. and McMahon, A.P. Hedgehog and Bmp genes are coexpressed at many diverse sites of cell-cell interaction in the mouse embryo. *Dev. Biol.* 172: 126–138, 1995.

[23] Koyama, E. et al. Polarizing activity, Sonic hedgehog, and tooth development in embryonic and postnatal mouse. *Dev. Dyn.* 206: 59–72, 1996.

[24] Hardcastle, Z. et al. The Shh signalling pathway in tooth development: defects in Gli2 and Gli3 mutants. *Development* 125: 2803–2811, 1998.

[25] Vaahtokari, A. et al. The enamel knot as a signaling center in the developing mouse tooth. *Mech. Dev.* 54: 39–43, 1996a.

[26] Thesleff, I. and Jernvall, J. The enamel knot: a putative signaling center regulating tooth development. *Cold spring harbour Symp. Quant. Biol.* 62: 257–267, 1997.

[27] Vaahtokari, A., Åberg, T., and Thesleff, I. Apoptosis in the developing tooth: association with an embryonic signaling center and suppression by EGF and FGF-4. *Development* 122: 121–129, 1996b.

[28] Kettunen, P. and Thesleff, I. Expression and function of FGFs-4, -8, and -9 suggest functional redundancy and repetitive use as epithelial signals during tooth morphogenesis. *Dev. Dyn.* 211: 256–268, 1998.

[29] MacKenzie, A., Ferguson, M.W., and Sharpe, P.T. Expression patterns of the homeobox gene, Hox-8, in the mouse embryo suggest a role in specifying tooth initiation and shape. *Development* 115: 403–420, 1992.

[30] Iseki, S. et al. Sonic hedgehog is expressed in epithelial cells during development of whisker, hair, and tooth. *Biochem. Biophys. Res. Commun.* 218: 688–693, 1996.

[31] Satokata, I. and Maas, R. Msx1 deficient mice exhibit cleft palate and abnormalities of craniofacial and tooth development. *Nat. Genet.* 6: 348–356, 1994.

[32] van Genderen, C. et al. Development of several organs that require inductive epithelial-mesenchymal interactions is impaired in LEF-1-deficient mice. *Genes Dev.* 8: 2691–2703, 1994.

[33] Peters, H., Neubüser, A., and Balling, R. Pax genes and organogenesis: Pax9 meets tooth development. *Eur. J. Oral Sci.* 106: 38–43, 1998.

[34] Matzuk, M.M., Kumar, T.R., and Bradley, A. Different phenotypes for mice deficient in either activins or activin receptor type II. *Nature* 374: 356–360, 1995.

[35] Tucker, A.S., Al Khamis, A., and Sharpe, P.T. Interactions between Bmp-4 and Msx-1 act to restrict gene expression to odontogenic mesenchyme. *Dev. Dyn.* 212, 533–539, 1998.

[36] Vieira, A.R. Oral clefts and syndromic forms of tooth agenesis as models for genetics of isolated tooth agenesis. *J. Dent. Res.* 82: 162–165, 2003.

[37] Lammi, L. et al. Mutations in AXIN2 cause familial tooth agenesis and predispose to colorectal cancer. *Am. J. Hum. Genet.* 74: 1043–1050, 2004. Epub 2004 Mar 23.

[38] Thomas, E.D. Bone marrow transplantation from bench to bedside. *Ann. NY Acad. Sci.* 770: 34–41, 1995.

[39] Snowden, J.A. et al. Autologous hemopoietic stem cell transplantation in severe RA: a report from the EBMT and ABMTR. *J. Rheumatol.* 31: 482–488, 2004.

[40] Rini, B.I. et al. Allogeneic stem-cell transplantation of renal cell cancer after nonmyeloablative chemotherapy: feasibility, engraftment, and clinical results. *J. Clin. Oncol.* 20: 2017–2024, 2002.

[41] Gronthos, S. et al. Postnatal human dental pulp stem cells (DPSCs) *in vitro* and *in vivo*. *Proc. Natl Acad. Sci. USA* 97: 13625–30136, 2000.

[42] Gronthos, S. et al. Stem cell properties of human dental pulp stem cells. *J. Dental. Res.* 81: 531–535, 2002.

[43] Shi, S., Robey, P.G., and Gronthos, S. Comparison of gene expression profiles for human, dental pulp and bone marrow stromal stem cells by cDNA microarray analysis. *Bone* 29: 532–539, 2001.

[44] Ekholm, S.V. and Reed, S.I. Regulation of G(1) cyclin-dependent kinases in the mammalian cell cycle. *Curr. Opin. Cell Biol.* 12: 676–684, 2000.

[45] Grossel, M.J., Baker, G.L., and Hinds, P.W. cdk6 can shorten G(1) phase dependent upon the N-terminal INK4 interaction domain. *J. Biol. Chem.* 274: 29960–29967, 1999.

[46] Shi, S. and Gronthos, S. Perivascular niche of postnatal mesenchymal stem cells identified in human bone marrow and dental pulp. *J. Bone Miner. Res.* 18: 696–704, 2003.

[47] Batouli, S. et al. Comparison of stem cell-mediated osteogenesis and dentinogenesis. *J. Dental. Res.* 82: 975–980, 2003.

[48] Levin, L.G. Pulpal regeneration. *Pract. Periodont. Aesthet. Dental.* 10: 621–624, 1998.

[49] About, I. et al. Pulpal inflammatory responses following non-carious class V restorations. *Oper. Dental.* 26: 336–342, 2001.

[50] About, I. et al. The effect of cavity restoration variables on odontoblast cell numbers and dental repair. *J. Dental.* 29: 109–117, 2001.

[51] Murray, P.E. et al. Restorative pulpal and repair responses. *J. Am. Dental. Assoc.* 132: 482–491, 2001.

[52] Murray, P.E. et al. Postoperative pulpal and repair responses. *J. Am. Dental. Assoc.* 131: 321–329, 2000.

[53] Rutherford, R.B. and Gu, K. Treatment of inflamed ferret dental pulps with recombinant bone morphogenetic protein-7. *Eur. J. Oral Sci.* 108: 202–206, 2000.

[54] Sloan, A.J., Rutherford, R.B., and Smith, A.J. Stimulation of the rat dentine-pulp complex by bone morphogenetic protein-7 *in vitro*. *Arch. Oral Biol.* 45: 173–177, 2000.

[55] Rutherford, R.B. et al. Bone morphogenetic protein-transduced human fibroblasts convert to osteoblasts and form bone *in vivo*. *Tissue Eng.* 8: 441–52, 2002.

[56] Iohara, K. et al. Dentin regeneration by dental pulp stem cell therapy with recombinant human bone morphogenetic protein 2. *J. Dental. Res.* 83: 590–595, 2004.

[57] Young, C.S. et al. Tissue engineering of complex tooth structures on biodegradable polymer scaffolds. *J. Dental. Res.* 81: 695–700, 2002.

[58] Duailibi, M.T. et al. Bioengineered teeth from cultured rat tooth bud cells. *J. Dental. Res.* 83: 523–528, 2004.

[59] Ohazama, A. et al. Stem-cell-based tissue engineering of murine teeth. *J. Dental. Res.* 83: 518–522, 2004.

[60] Tucker, A.S., Matthews, K.L., and Sharpe, P.T. Transformation of tooth type induced by inhibition of BMP signalling. *Science* 282: 1136–1138, 1998.

[61] Ferguson, C.A., Tucker, A.S., and Sharpe, P.T. Temporospatial cell interactions regulating mandibular and maxillary arch patterning. *Development* 127: 403–412, 2000.

[62] Thomas, B.L. et al. Role of *Dlx-1* and *Dlx-2* genes in patterning of the murine dentition. *Development* 124: 4811–4818, 1997.

[63] Grigoriou, M. et al. Expression of *Lhx6* and *Lhx7*, a novel subfamily of LIM homeodomain genes, suggests a role in mammalian head development. *Development* 125: 2063–2074, 1998.

[64] Tucker, A.S. et al. Fgf-8 determines rostral–caudal polarity in the first branchial arch. *Development* 126: 51–61, 1999.

[65] Mitsiadis, T.A. et al. Role of Islet1 in the patterning of murine dentition. *Development* 130: 4451–4460, 2003.

[66] Mucchielli, M.L. et al. *Otlx2/RIEG* expression in the odontogenic epithelium precedes tooth initiation and requires mesenchyme-derived signals for its maintenance. *Dev. Biol.* 189: 275–284, 1997.

[67] St. Amand, T.R. et al. Antagonistic signals between BMP4 and FGF8 define the expression of *Pitx1* and *Pitx2* in mouse tooth-forming anlage. *Dev. Biol.* 217: 323–332, 2000.

[68] Headon, D.J. et al. Gene defect in ectodermal dysplasia implicates a death domain adaper in development. *Nature* 414: 913–916, 2002.

[69] Sarkar, L. et al. Wnt/Shh interactions regulate ectodermal boundary formation during mammalian tooth development. *Proc. Natl Acad. Sci. USA* 97: 4520–4524, 2000.

[70] Ferguson, C.A. et al. Activin is an essential early mesenchymal signal in tooth development that is required for patterning of the murine dentition. *Genes Dev.* 12: 2636–2649, 1998.

[71] D'Souza, R.N. et al. *Cbfa1* is required for epithelial-mesenchymal interactions regulating tooth development in mice. *Development* 126: 2911–2920, 1999.

[72] Tucker, A.S. et al. Edar/Eda interactions regulate enamel knot formation in tooth morphogenesis. *Development* 127: 4691–4700, 2000.

[73] Kettunen, P. et al. Associations of FGF-3 and FGF-10 with signaling networks regulating tooth morphogenesis. *Dev. Dyn.* 219: 322–332, 2000.

[74] Kettunen, P., Karavanova, I., and Thesleff, I. Responsiveness of developing dental tissues to fibroblast growth factors: expression of splicing alternatives of FGFR1,-2,-3, and of FGFR4; and stimulation of cell proliferation by FGF-2,-4,-8, and -9. *Dev. Genet.* 22: 374–385, 1998.

[75] Åberg, T., Wozney, J., and Thesleff, I. Expression patterns of bone morphogenic proteins (bmps) in the developing mouse tooth suggest poles in morphogenesis and cell differentiation. *Dev. Dyn.* 210: 383–396, 1997.

[76] Sarkar, L. and Sharpe, P.T. Expression of wnt signaling pathway genes during tooth development. *Mech. Dev.* 85: 197–200, 1999.

[77] Snead, M.L., Luo, W., Lau, E.C., and Slavkin, H.C. Spatial and temporal-restricted pattern for amelogenin gene expression during mouse molar tooth organogenesis. *Development* 104: 77–85, 1988.

[78] Bègue-Kirn, C. et al. Dentin sialoprotein, dentin phosphoprotein, enamelysin and ameloblastin: tooth-specific molecules that are distinctively expressed during murine dental differentiation. *Eur. J. Oral Sci.* 106: 963–970, 1998.

[79] Developmental biology programme of the University of Helsinki. *Gene Expression in Tooth* [online], http://biteit.helsinki.fi, 1996.

33

Tracheal Tissue Engineering

Brian Dunham
Paul Flint
Sunil Singhal
Catherine Le Visage
Kam Leong
Johns Hopkins School of Medicine

33.1 Introduction

A seemingly simple, single-lumen structure, the trachea is the sole conduit between the supraglottic airway and the lungs. Humidified and warmed air inspired through the nose travels to the lungs through the relatively thin-walled trachea, which widens slightly at its distal end. At birth, its diameter is approximately 0.5 cm. Tracheal size grows proportionally with the height and weight of the child [1,2]. In a male human adult, the trachea is approximately 12-cm long and 1.5- to 2-cm wide. In an adult female it is approximately 11-cm long and narrower. At its distal end, the carina, it bifurcates into the two mainstem bronchi (Figure 33.1).

Mechanically, the trachea has several functions. As an air conduit one of its most important structural functions is to maintain patency; any significant obstruction of its lumen can result in rapid asphyxiation. It must also be flexible enough to accommodate cervical rotation, flexion, and extension. Furthermore, it has to withstand both negative and positive intraluminal pressures encountered in the respiratory cycle. Approximately 16 to 20 hyaline cartilage rings provide the necessary rigidity; the intervening soft tissue provides the necessary flexibility and compliance to respond to cervical motion and varying intraluminal pressure. The first and most superior ring, the cricoid cartilage, is a complete ring (Figure 33.1). The remaining cartilage rings beneath the cricoid are C-shaped and open posteriorly. The pars membrana spans the open ends of the cartilage rings and is composed of a fibroelastic ligament and longitudinally oriented smooth muscle (Figure 33.2). The ligament prevents overdistention, while contraction of the muscle reduces the size of the lumen. The latter occurs during the cough reflex; the decreased luminal size increases the velocity of the expired air, facilitating airway clearance.

The trachea is lined with a pseudostratified columnar respiratory epithelium that consists of a heterogeneous population of cells that form tight junctions; in the submucosal space are numerous mixed

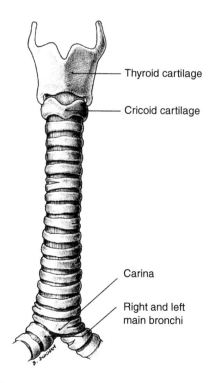

FIGURE 33.1   Anterior view of a human trachea.

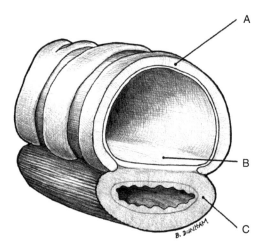

FIGURE 33.2   Cross sectional view of a human trachea segment. (A) C-shaped cartilaginous ring. (B) Pars membrana of the trachea, which is composed of a fibroelastic ligament and smooth muscle. (C) Esophagus.

sero-mucous glands, which decrease in numbers in the distal aspect of the trachea (Figures 33.3a,b). Airway epithelial cells, as well as dendritic cells found in airway epithelium, express major histocompatibility complex class I and II molecules, which endow the epithelium with the properties of an immunologic barrier [3]. The epithelium's major function was once thought to be that of a physical barrier; it is now thought to be far more complex. The airway surface epithelium does indeed possess a variety of intercellular junctional complexes that create a tight and efficient barrier against inhaled pathogens and other noxious agents [4,5]. In addition, airway epithelial cells act together to ensure mucosal defense

(a)

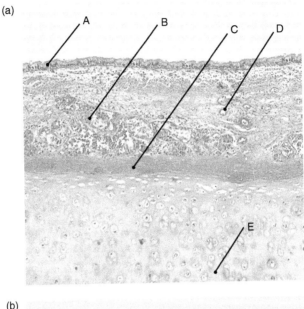

(b)

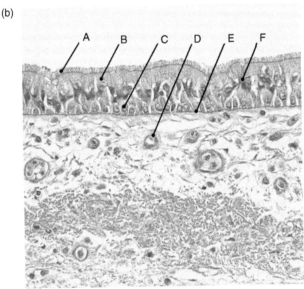

FIGURE 33.3 (a) Photomicrograph of tracheal tissue, hematoxylin and eosin stain ×200. (A) Respiratory epithelium. (B) Mixed sero-mucous glands residing in the lamina propria. (C) Perichondrium. (D) Blood vessel. The lamina propria underlying the epithelium is richly vascularized, which helps to warm the inspired air. (E) Cartilage. (b) Photomicrograph of tracheal respiratory epithelium, hematoxylin and eosin stain ×400. (A) Cilia arising from (B), the columnar ciliated epithelial cells. (C) Basal cell. (D) Blood vessel within the lamina propria. (E) Basement membrane. (F) Goblet cell.

through a variety of mechanisms such as mucociliary clearance, active secretion of ions and regulation of water balance, regulation of airway smooth muscle function, and the release of antibacterial, antioxidant, and anti-inflammatory molecules in the airway surface liquid. Airway epithelium constitutes the interface between the internal milieu and the external environment, and responds to changes in the external environment by secreting a large number of mediators that interact with cells of the immune system and underlying mesenchyme [5]. These mediators include arachidonic acid metabolites, nonprostanoid inhibitory factors, nitric oxide, endothelin, cytokines, and growth factors [3].

Since the epithelium is in direct and permanent contact with the external environment, it is frequently injured. It is capable of rapid restitution if it is denuded [6,7]. After an injury, epithelial cells dedifferentiate, flatten, and migrate rapidly beneath a fibrin–fibronectin plasma-derived gel that contains both adhesive plasma proteins and leukocytes [8]. The response to injury, however, appears to partly depend on the depth of injury. Deep injuries violating the lamina propria and reaching the perichondrial tissues are associated with excessive granulation tissue [9,10]. Bacterial and viral infections, inhaled pollutants and toxic agents, and mechanical stress can severely alter the integrity of the epithelial barrier. The response of the airway surface epithelium to an acute injury includes a succession of cellular events varying from loss of surface epithelial impermeability to partial shedding of the epithelium or even to complete denudation of the basement membrane. In response to chronic injury, the airway epithelial cells can also transdifferentiate, with a shift from serous to mucous cells, from ciliated to secretory cells, or from secretory to squamous cells. Such a remodeling illustrates the marked plasticity and capacity of the airway epithelium to regenerate [4,11]. Given its regenerative capacity, characterization of airway stem cells may eventually lead to clinical, therapeutic benefit [12].

There are at least eight morphologically distinct cells types in human respiratory epithelium. These include columnar ciliated epithelial cells, mucous goblet cells, serous cells, basal cells, Clara cells, pulmonary endocrine cells, as well as intraepithelial nerve cells, and a variety of immune cells. The latter group of cells is comprised of mast cells, intraepithelial lymphocytes, dendritic cells, and macrophages. Serous cells and Clara cells are found beyond the trachea in the more distant airway conduits. The most abundant of the tracheal epithelial cells are the ciliated columnar cells, accounting for approximately 50% of all epithelial cells. Ciliated cells, which arise from either basal or secretory cells, are no longer thought to be terminally differentiated [5,13]. In the adult human trachea, each of these ciliated columnar cells host approximately 300 cilia that beat in an organized fashion to sweep respiratory secretions upward into the larynx and oral cavity.

The second most common cell in the human trachea is the mucous goblet cell, which is characterized by acidic-mucin granules. Secretion into the airway lumen of the correct amount of mucin, a glycoprotein, and the viscoelasticity of the resulting mucus are important parameters for an efficient mucociliary clearance of mucus-entrapped foreign bodies. It is thought that the acidity, due to the sialic acid content of the glycoprotein, determines the viscoelastic profile and hence the relative ease of transport across cilia [5]. These goblet cells are thought to be capable of self-renewal and may differentiate into ciliated cells [14,15], as do the basal cells [16]. The basal cells are short, rounded cells that lie on the basal lamina without extension to the apical surface. They are the only cells in the epithelium that are firmly attached to the basement membrane and, as such, aid in the attachment of more superficial cells to the basement membrane via hemidesmosomal complexes [15,17]. The basal cell is thought to be able to function as a primary stem cell, giving rise to mucous and ciliated epithelial cells [5,18–25]. Pulmonary endocrine cells are found throughout the airway as solitary cells or in clusters. These cells secrete a variety of biogenic amines and peptides, which appear to play an important role in fetal lung development and airway function including the regulation of epithelial cell growth and regeneration.

The trachea's rich arterial blood supply is derived from fine branches of the superior and inferior thyroid arteries, of the internal thoracic arteries, and of the bronchial arteries. Returning blood from tracheal veins eventually travels into the inferior thyroid veins. The incompletion of the C-shaped rings allows the trachea to be in close apposition to the esophagus throughout its length and to share vascular supply. While it does receive its blood supply from named vessels, its vasculature is composed of a rich network of thin vessels. The profuse system of microvessels that immediately underlie the epithelium is of particular importance in the maintenance and regeneration of airway epithelium. There is thought to be a dynamic interplay between plasma-derived molecules, their receptors, airway epithelial cells, and their secretions *in vivo*, which either promotes airway defense or induces disease [26].

The smooth muscle and glands of the trachea are parasympathetically innervated by the vagus nerve, either directly or by the recurrent laryngeal nerves. Sympathetic innervation comes directly from the sympathetic trunk. Tracheal mucosa itself is richly innervated from subepithelial plexuses. The trachea is remarkably sensitive to touch and has a low threshold to elicit a reflexive cough in the presence of foreign material. The subepithelial nerves penetrate the basement membrane at focal points where they branch and spread along the basement membrane with terminal ends extending between epithelial cells and terminating in the airway lumen. The nerves have an obvious sensory role but the full breadth of their exact function is unknown; there is, however, evidence that they might be in direct apposition to pulmonary endocrine cells, with the suggestion of a bi-directional communication between these two cell types [5].

In summary, the trachea is a simple, yet elegant, structure that very effectively resists collapse from negative intraluminal pressures. It is flexible enough to accommodate distension and adapt to cervical rotation, and is lined with a metabolically active and physiologically complex respiratory epithelium that is intimately linked to the underlying mesenchyme and the immune system.

33.2 Tracheal Reconstruction: Previous Attempts

Tracheal reconstruction dates back to at least 1881 when Gluck and Zeller re-anastomosed a transected dog trachea [27]. Over 80 years later in his classic 1964 paper, Grillo described anatomic studies on human cadavers establishing the upper limits of tracheal resection that would allow a direct end-to-end anastomosis without undue anastomotic tension [28]. Subsequently, he and others refined the techniques of tracheal resection with primary anastomosis [29–37]. Today, approximately half of the human adult's and one third of the small child's trachea can safely be resected and primarily anastomosed. Even long segment tracheal stenosis can now often be handled by an operation known as a slide tracheoplasty. The need for more extensive resections is clinically rare. In the adult population, the need for replacement of greater than a half of the trachea usually arises in the setting of a low-grade neoplasm, such as adenoid cystic carcinoma of the trachea, or in the setting of unresectable diseases (such as tracheopathia osteoplastica, relapsing polychondritis, Wegener's granulomatosis, and trauma). In the neonate, it arises in the setting of tracheal agenesis, a congenital absence of tracheal tissue [38,39]. Whenever a significant length of trachea is compromised by disease, it presents a true surgical dilemma as no truly dependable and reliable replacement yet exists. Given the infrequent clinical demand, it is amazing that the literature is rich in attempts to find suitable materials with which to replace tracheal tissue. None of these have been particularly successful and none have found consistent and widely accepted clinical use. A limited review of some these attempts offers valuable insight into the physiology and pathophysiology of the trachea. They fall into several categories: implantation of foreign materials, reconstruction with autogenous tissues, reconstruction, and transplantation of autografts and allografts. The newest category is tissue engineering [38].

A wide variety of materials has been used for solid prostheses, including but not limited to stainless steel, Vitallium, glass, polyethylene, Lucite, silicone, Teflon, Ivalon, polyvinyl chloride, and polyurethane [40–54]. These materials were used as single constituents or in combination; some prostheses used cuffs draped over tracheal ends to encourage fixation of the prosthesis and prevent obstruction with granulation tissue at the anastomotic sites. The solid prostheses have been prone to migrate and dislodge, to obstruct with granulation tissue at the anastomoses, and to develop infections. In addition, they have tended to yield poor epithelialization. No solid prosthesis has ever proven reliable over time. Some may work for an unpredictable amount of time but all eventually fail. In response to the failure of solid prostheses, some groups turned to porous synthetic prostheses to allow tissue ingrowth and promote a greater prosthetic incorporation. A variety of porous prostheses have been attempted; some of these were used in conjunction with tissues such as pericardium, omentum, dermis, pleura, and fascia as well as fibrin and collagen [50,51,55–75]. The porous prostheses have also yielded unsatisfactory results. They have regularly failed to become fully epithelialized, especially in the center, which promoted central granulation, cicatrisation (scar formation), and stenosis. The porous prostheses have also shown a propensity for bacterial colonization.

There have been many attempts to replace trachea with autogenous tissues, including skin, fascia, pericardium, periosteum, buccal mucosa, aortic tissue, esophageal tissue, bronchial tissue, cartilage, bone, and bladder epithelium [38,49,57,58,70,76–92]. The implanted cartilage, despite its autogenous source, often resorbed. In addition to devascularized autogenous tissues, several authors have reported using vascularly pedicled tissue transfers. Although some yielded temporary success, none of these have become commonplace in clinical practice. Some heroic efforts have been documented in the clinical realm. There are several clinical reports of tracheal reconstruction with cutaneous troughs, as well as esophageal transfers. These are difficult procedures whose failure rates, not surprisingly, rise with their increasing complexity.

Experiments using devitalized tissues including cadaveric tracheas have also failed to produce robust results [93,94]. Cadaveric tracheal grafts have been treated in a variety of ways: irradiation, freeze-drying, and chemical treatment [58,95–102]. None has resulted in clinically reliable solutions. Devitalized tracheal tissue is inherently problematic as the cartilage is inevitably doomed to resorption, leading to tracheomalacia, a degenerative softening of the trachea. The degree of reported epithelialization is somewhat variable, but it is doubtful that any was truly and thoroughly effective. Our own laboratory has investigated the use of chemically decellularized cadaveric grafts in a rabbit model, both as anterior window grafts and circumferential grafts (unpublished data; Figures 33.4a,b). All experimental animals that underwent an anterior window replacement survived until their appointed time of sacrifice without suffering from clinically significant stenoses. The circumferential replacement was far more problematic. All grafts, if given sufficient time, gradually stenosed. Gross evaluation confirmed central malacia of the grafts, as well as centrally located mucosal stenoses. Histologic evaluation revealed a significant inflammatory response that showed a predilection for the center of the grafts; epithelial denudation also appeared to coincide with excessive granulation tissue, as is often reported in the literature.

Early experimental use of true transplants, in other words, fresh tracheal allografts in animal models has yielded uniformly poor results [57,58,78,99,103]. Even fresh autografts have been problematic. Smaller

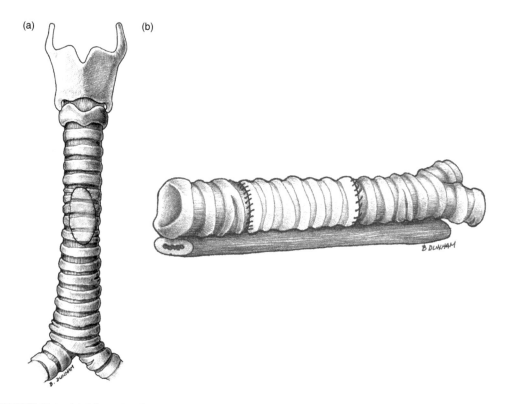

FIGURE 33.4 (a) Schematic of an anterior window replacement. (b) Schematic of a full circumferential tracheal replacment.

segments and patches tend to have a higher success rate. Fibrous degeneration of the autograft is common. Presumably, the failure partly occurs because the fine vasculature of the grafts has not reestablished rapidly enough to support the respiratory mucosa and the underlying mesenchyme before exposure to the external environment engenders a chronic inflammatory response with devolution of the mucosa. One potential solution would be the use of vascular anastomoses to allow for near immediate reconnection of a graft's vascular network. Unfortunately, although vascular anastomoses are technically feasible as recent work by Genden [104,105] has shown, they significantly increase the complexity of the surgical procedure and can increase the risk of failure in the already unforgiving milieu of the airway. Other groups have addressed this issue by incorporating thyroid tissue along with the trachea during transplantation attempts, using the thyroid's larger blood vessels to perform the anastomoses [106,107].

Without immune suppression, fresh allografts have elicited an immune rejection, resulting in ischemic necrosis of the implanted tissues. These allografts, not surprisingly, suffered from resorptive collapse of their tracheal rings and poor reepithelization leading to a short postoperative survival of the animals [57,58,103]. Some improvement in graft viability has occurred with indirect vascularization with omental flaps. This has been especially true for small grafts; longer grafts, however, still have not fared well. Allografts that have undergone cryopreservation and have been supported with an omental flap have shown improved survival, even without immunosuppression[108–110]. However, recipient chondrocytes have not repopulated the grafts' cartilage. Cryopreservation, presumably, has reduced epithelial antigenicity; this phenomenon is not yet fully understood. Bujia [111] showed evidence that the predominant locus of antigenicity is likely to reside in the epithelium and not the cartilage. Liu denuded tracheal epithelium with detergent and reported that these allografts, which were supported with an omental flap, remained viable [112,113]. Not surprisingly, given airway mucosa regenerative capacity, recipient epithelial cells have eventually repopulated the grafts' epithelium.

In an effort to induce immunotolerance, Genden and colleagues pretreated rats with a single portal-vein injection of ultraviolet-B irradiated donor splenocytes seven days prior to circumferential tracheal allograft placement. The pretreatment induced a donor-specific immune hyporesponsiveness and prevented rejection of the grafts [114]. Whether or not immune tolerance can be induced in higher animal models remains to be seen.

Tracheal obstruction is seen time and time again in tracheal reconstruction efforts. Provided that a neotrachea would be able to resist collapse, it still faces a tremendous challenge within its lumen: that of establishing a healthy mucosa, free of chronic inflammation. The luminal compromise is consistently associated with a detrimental soft tissue reaction, in which airway mucosa undergoes a significant reactive thickening with a progressive diminution of the airway lumen and a disruption of the structural integrity of the cartilaginous rings of the trachea. This reaction appears to be superficially analogous to exuberant scarring found in skin tissue. It is thought that prompt and thorough reepithelialization can prevent this complication and maintain luminal patency [115].

Our present ability to intelligently address the infrequent need for tracheal tissue replacement is hampered by our lack of understanding airway mucosal of wound healing. Significant strides have been made in the tissue engineering of cartilage; the ability to engineer a structure that is able to resist collapse and that meets the rigors of clinical application is likely to be imminent. However, a great deal remains to be understood about airway mucosal physiology and healing before true clinical utility will be possible.

Although the need for full circumferential tracheal replacement is limited, the need to address airway mucosal disease is far greater. The incidence of tracheal stenosis after an ischemic mucosal injury is far more prevalent than those etiologies requiring replacement of more than half of the trachea. Since the advent of cuffed endotracheal tubes, the incidence of tracheal stenosis has risen sharply. The presence of a foreign body within the airway, especially one that applies mucosal pressure denuding epithelium and compromising blood flow, can trigger a very poorly understood chronic mucosal inflammation, which often results in mucosal hypertrophy and luminal obstruction. The problem is compounded by the fact that the airway is constantly exposed to the bacteria-laden external environment. So, whereas the literature seems to have focused on circumferential tracheal tissue replacement, clinically there is a far greater need to address focal mucosal disease, underscoring the importance of furthering our understanding of airway mucosal physiology and pathophysiology.

33.3 Tracheal Tissue Engineering

Tracheal tissue engineering is in its infancy. The first true tissue engineering effort appeared only in 1994 [116]. As with any field in its early development, the current work is limited by a lack of detailed understanding of the physiology and pathophysiology of tracheal tissue. Current work is premised on rather crude hypotheses. Efforts in this field will undoubtedly evolve as functional replacement models begin to emerge and the critical issues become more apparent. Most efforts have centered on the construction of a cartilaginous structure that can resist collapse. Recent work has focused on finding the best cellular source and method to recreate the cartilaginous infrastructure, and attempting to introduce an epithelial lining within the lumen of a tissue-engineered construct. Some investigators have made use of mesenchymal stem cells to create cartilage as well as a respiratory epithelium. Other investigators have explored fetal surgery techniques allowing their constructs to partly mature *in utero*. There are also efforts that do not neatly fall into the category of tissue engineering but are worthy of mention.

Whereas there are many reports of attempted tracheal replacement, it was not until 1994 that the first true tissue-engineering attempt appears in the literature when Vacanti [116] and colleagues reported on the circumferential replacement of rat tracheas with tissue-engineered cartilaginous tubes. They seeded chondrocytes harvested from the shoulder of newborn calves onto sheets of nonwoven polyglycolic acid (PGA) mesh. All experimental animals died within one week of the operation. Presumably, the animals succumbed to respiratory distress. There was no discussion of postmortem histology; what happened within the lumen of the airway is not clear. It is likely that a straight cartilaginous tubular structure is inadequate for two important reasons. First, a rigid tube would place a significant mechanical demand on the anastomoses, encouraging failure at those sites. Second, it would be hard for blood vessels to penetrate directly through the cartilage to establish the robust vasculature required to meet the high metabolic demands of airway epithelium. In essence, this design requires that the vascular network arise exclusively from an intraluminal source and travel from mechanically challenged anastomoses over a considerable distance, without support from extratracheal vascular ingrowth. Nonetheless, this is a landmark paper in the evolution of tracheal engineering as it is arguably the first true translational effort to tissue engineer a trachea.

In 2002, Kojima [117] and colleagues addressed the mechanical inadequacies inherent in a straight cartilaginous tube. They isolated chondrocytes and fibroblasts from the nasal septum of 2-month-old sheep. Chondrocyte-seeded PGA mesh was placed in the grooves of a 20 by 50-mm-long helical Silastic template; the entire construct was then wrapped with a fibroblast-laden PGA mesh. In the first arm of the study, the cultured constructs were implanted into dorsal subcutaneous pockets in athymic rats and explanted after eight weeks. In this refined attempt, the rat-incubated construct yielded cartilage biochemically similar to native tissue as measured by sulfated glycosaminoglycan (GAG) content, hydroxyproline content, and cellularity. In the second arm of the study, autologous constructs were implanted underneath the sternocleidomastoid muscle of the donor sheep, allowed to mature *in vivo* for eight weeks before being harvested without preservation of a vascular pedicle and used to replace a surgically created 5-cm circumferential defect in the cervical trachea of the now four-month-old donor sheep. By contrast with the results obtained in the rat model, the sheep-incubated constructs were significantly more cellular and poorer in GAG content than native tissue. As a result the sheep-incubated constructs performed poorly once rotated into the airway, without an attached vascular supply. Postoperatively, none of the sheep survived for more than seven days. They all suffered from either stenosis or clinically significant tracheomalacia. The difference in the results obtained between the two species was tentatively attributed to an inflammatory reaction engendered by the PGA in sheep, leading to hypercellular cartilage poor in GAG content, the cartilaginous component thought to be responsible for establishing stiffness. The authors stated that their future attempts would explore less inflammatory scaffolding material and also preserve the constructs' vascular connection to the surrounding muscle. It will be interesting to see if a helical construct has the appropriate mechanical properties to allow for successful long-term integration of a tissue-engineered trachea. By its very design, a helix precludes the development of a pars membrana.

Searching for an optimal source of chondrocytes, Fuchs and colleagues [118] experimented in 2002 with the use of fetal chondrocytes to tissue engineer cartilage. Specimens of fetal ovine elastic and hyaline cartilage were harvested from ears and tracheas, respectively. Adult aural elastic cartilage served as a comparison. Isolated chondrocytes were seeded onto nonwoven biodegradable scaffolds composed of poly-L-lactic acid (PLLA)-treated PGA polymer and maintained *in vitro* for 6 to 8 weeks. The chondrocytes isolated from fetal elastic and hyaline cartilage grew significantly faster than those harvested from adult elastic cartilage. All fetal constructs showed characteristics of hyaline cartilage, both on gross and microscopic inspection, regardless of whether the cells had originated from hyaline or elastic cartilage. In the fetal constructs, the histologic pattern seen with stains for glycosaminoglycans was comparable to that seen in native hyaline cartilage. The levels of GAG and collagen type II were higher in fetal versus adult constructs. There were, however, no significant differences between GAG, collagen type II, or elastin in fetal constructs derived from hyaline or elastic cartilage. When compared to their native tissue, fetal constructs derived from elastic cartilage contained similar levels of GAG and type II collagen, but had lower levels of elastin. Fetal hyaline constructs did not differ significantly in the levels of GAG, collagen type II, or elastin when compared to their native source. Adult elastic constructs, however, were significantly lower in collagen type II and elastin than the native elastic tissue. It is interesting that chondrocytes derived from elastic cartilage eventually produced hyaline cartilage and grew faster. What governs the formation of hyaline versus elastic cartilage is unknown at this point. The authors concluded that compared with adult constructs, matrix deposition is enhanced in engineered fetal cartilage, closely mimicking native hyaline tissue.

Another study searching for an optimal source of chondrocytes was reported by Kojima and colleagues in [119] 2003 in which they compared ovine nasal and tracheal chondrocytes seeded onto PGA matrices and implanted in nude mice. Chondrocytes from the nasal septal cartilage as well as tracheal rings of 2-month old sheep were seeded onto nonwoven mesh of PGA fibers. The cell-constructs were eventually wrapped around a 7-mm diameter silicone tube and implanted subcutaneously in nude mice for 8 weeks. They evaluated the gross and histologic appearances, the GAG and the hydroxyproline contents, and the biomechanical profile of the constructs. The gross appearances from both groups were found to resemble native tracheal cartilage. The histologic evaluation revealed a mature-appearing hyaline cartilage in both cases with abundant proteoglycan production. There was no evidence of an inflammatory reaction to the implants. The GAG contents of the tracheal-derived and the nasal septum-derived constructs were 70 and 81% of the native tracheal cartilage respectively. The hydroxyproline contents were 91 and 94% for the tracheal-derived and the nasal septum-derived constructs respectively. The tensile modulus for both construct types was, however, only about 15% that of the native trachea. Nonetheless, the constructs were stiff to the touch and grossly resisted collapse. The authors suggested that other compositional features of the extracellular matrix, such as collagen crosslinking, might play an important role in the mechanical profile of tracheal tissue. This study's findings support the use of nasal septal cartilage, which is attractive since it requires a far less invasive procedure for harvest than do most other sources of cartilage.

Later in the same year, Sakata and colleagues [120] reported an attempt to line these cartilaginous tubes with a respiratory epithelium. They constructed ten tubes as previously described [116] and implanted them into the flanks of nude mice, this time for six weeks. Epithelial cells were aseptically harvested from newborn lambs' cervical tracheas and were immediately injected in the implanted cartilaginous tubes after which both ends of the cartilaginous tube were sutured shut. The epithelial cell-lined tubes were then collected at intervals of up to three weeks. Out of ten tubes, six were found to be infected and had not developed an epithelial lining as observed on hematoxylin and eosin sections. The remaining four were lined with a respiratory mucosa, including a submucosal connective tissue, which had "abundant" small blood vessels. Some of the cells had developed cilia three weeks following inoculation into the implanted cartilaginous construct. Although promising, this brief report did not provide histologic details. For example the uniformity of the epithelial lining was not commented upon, and the composition of the epithelium was not reported. The high infection rate highlights one of the challenges of engineering respiratory tissue. The airway is inherently contaminated with a significant bacterial load. Future designs might want to consider a preliminary *in vitro* incubation period of the freshly harvested epithelial cells with antibiotics to reduce the incidence of infection. It might also have been beneficial to attach silastic tubing at either end of the tubes to allow them to be flushed on a regular interval, as the epithelial lining will begin to produce mucin and the nonattached dead cells would be trapped within the lumen increasing the risk of infection. Most importantly, however, this experiment does show that it is possible to line a tissue-engineered cartilaginous lumen with a respiratory epithelium.

There is ample evidence outside the tissue engineering literature that epithelial cells can be successfully introduced into a living luminal cavity. Several investigators have made use of a heterotopic tracheal graft model developed by Terzaghi [121]. These experiments have repeatedly demonstrated that epithelium-denuded tracheal tissue that is subcutaneously implanted in nude animals can easily be repopulated with heterologous cultured rat, rabbit, and human epithelial tracheal cells [16,20,121–124]. More recently, Dupuit and colleagues used human respiratory cells to reconstitute epithelium-denuded rat tracheas, which were implanted in the flanks of nude mice. Four to five weeks postimplantation, a fully differentiated and tightly sealed pseudostratified functional epithelial barrier was documented [125]. Rainer and colleagues successfully transplanted cultured epithelial cells onto fibrotic capsules surgically created in the anterior rectus sheath of inbred Wistar Furth rats. They did so by co-culturing native rat tracheal epithelial cells with irradiated fibroblasts, then suspending them in fibrin glue, which was used as the *in vivo* delivery vehicle [126].

A subsequent study by Kojima went one step further by investigating the co-culture of epithelial cells and chondrocytes. They used ovine nasal septum chondrocytes again, but this time, added nasal septum respiratory cells suspended in a hydrogel into the lumen of the chondrocyte-seeded constructs *in vivo*. The subcutaneously implanted chondrocyte constructs were allowed to mature for six weeks. Histologic examination, four weeks post instillation of the epithelial cells, revealed that only approximately 60% of the lumen was covered with an epithelium. Although the epithelial reconstitution was not complete, this study highlighted the practicality of using a single nasal septal harvest to obtain both cell types [127].

Congenital deformities, such as long-segment stenosis and atresia, in which the tracheal lumen does not form, are relatively infrequent but bear very high mortality rates. Fuchs and colleagues [128] have begun to explore the possibility of early *in utero* intervention using harvested cells seeded onto polymer scaffolding in an ovine model. The intrauterine environment, bathed in sterile amniotic fluid, offers the advantage of avoiding the contaminated milieu of the postnatal airway. In their experiments they performed *in utero* anterior tracheal replacement using native elastic cartilage (group II) or tissue-engineered constructs seeded with heterologous fetal chondrocytes (group I).

For group I, fetal elastic or hyaline cartilage was harvested, respectively, from the ear or tracheal rings. Chondrocyte-seeded PLLA-treated nonwoven PGA mesh was maintained in a rotating oven for 6 to 8 weeks. The constructs were then trimmed into a diamond shape and used to repair a surgically created longitudinal tracheostomy spanning 4 to 5 cartilaginous rings.

In the second group, fetal auricular cartilage was used as autologous free grafts to repair the same anterior defect as created in group I. The animals were sacrificed prenatally and postnatally. The fetal and postnatal survival rates ranged from 80 to 100%. Stridor was, however, present in more than half of the surviving animals. Again, the histologic evaluation of the tissue-engineered grafts revealed hyaline cartilage formation regardless of whether the cell source was hyaline or elastic cartilage. There was mild to moderate deformation of the tracheal lumen in some of the tissue-engineered group. In the native cartilage group, however, the tracheal lumens were severely deformed. The engineered constructs displayed a time-dependent epithelialization of their lumens. Of note, there were no serous or mucinous acini that are normally found in native tissue. All the implants of group II retained their elastic cartilage characteristics, but they also developed a thinner epithelium, which lacked both ciliated cells, as well as serous and mucinous glands. It is interesting to note that the heterologous chondrocytes engrafted without obvious signs of rejection. The authors attributed that phenomenon to the immaturity of the fetal immune system and the low histocompatibility antigen expression of fetal chondrocytes. This study demonstrated the feasibility of using heterologous chondrocytes *in utero*. The authors did not comment on why group II seemed to have poorer epithelialization; it would be interesting to know if lack of rigidity and possibly central mobility of the free grafts impaired proper epithelialization and vascularization.

One important note of caution must be mentioned when evaluating anterior tracheal repair studies. The native trachea has an impressive ability to reform an epithelial mucosa when only a limited amount of tracheal tissue is removed. This phenomenon is routinely seen after tracheostomy. An epithelial membrane will quickly form after the removal of a tracheostomy tube. Studies that perform only anterior reconstruction therefore offer a limited understanding of the potential benefits or drawbacks of the graft materials used.

In 2003, Fuchs and colleagues explored the use of mesenchymal stem cells (MSCs) in tracheal tissue engineering. MSCs are attractive in tracheal tissue engineering because of their ability to form not only cartilage but also muscle with the appropriate biochemical cues. In addition, harvesting MSCs is less invasive than obtaining autologous cartilage and would obviate the need for an additional general anesthesic and its attendant risks. MSCs can be harvested from a number of sites. Umbilical cord blood has successfully been used [129,130]; adipose tissue has yielded a cell population that can undergo chondrogenesis [131]. Though there is at least one report claiming the MSCs can be collected from the peripheral circulation of an adult [132], traditionally, a bone marrow aspirate is the best means of obtaining a sufficient number of cells [133].

Fuchs and colleagues showed the feasibility of using MSCs *in* utero by comparing constructs seeded with bone marrow-derived heterologous ovine MSCs with constructs seeded with chondrocytes derived from ovine elastic cartilage. The stem cells were encouraged to differentiate into chondrocytes by exposure to carefully defined media that contained transforming growth factor as well as ascorbic acid. After 12 weeks of incubation in a rotating bioreactor, the pretreated PGA mesh-cell constructs were then implanted in heterologous fashion into a 4 to 5 ring longitudinal tracheostomy. Normal delivery of the animals was allowed. Tracheal specimens were harvested 4 days after birth. There were no significant differences found in the survival rates prenatally or postnatally. There was again a significant amount of stridor observed, this time in both groups. The engineered cartilage implants form both sources were found to engraft despite their heterologous origin; there were, however, some mononuclear infiltrates seen in some specimens. Both groups maintained their cartilaginous phenotype *in vivo* and both epithelialized their luminal surfaces. The MSC-derived cartilage had a higher GAG content *in vitro*, but after remodeling *in vivo*, both groups had comparable GAG contents.

In 2004, Kojima explored the use of augmenting an MSC-based construct with transforming growth factor β2 (TGF-β2) released from gelatin microspheres. Bone marrow-derived ovine MSCs were seeded onto a PGA nonwoven mesh and cultured for a week with exposure to both TGF-β2 and insulin growth factor 1. The cell-polymer constructs were wrapped around a silicone helical template. Constructs were then coated with glutaraldehyde-cross-linked gelatin microspheres containing TGF-β2. The entire construct was subcutaneously implanted in nude rats for six weeks. The proteoglycan content of the tissue-engineered construct was approximately 79% of that of the native tracheal cartilage. Additionally, the collagen content of the tissue-engineered product was 71% of that of the native's content. They found no statistically significant differences in glycosaminoglycan and hydroxyproline content between native and tissue-engineered tissue. In this instance, the ability to add controlled-release growth factors via microspheres is of significant value as the addition of transforming growth factor has repeatedly been shown to significantly augment chrondogenesis in MSC culture *in vitro* [134,135].

MSC are attractive not only because of their chondrogenic capacity, but also for their potential ability to support the respiratory epithelial cells. We recently described a novel *in vitro* reconstitution system for tracheal epithelium that could be useful for investigating the cellular and molecular interaction of epithelial and mesenchymal cells [136]. In this system, a porous Transwell insert was used as a basement membrane on which MSCs were cultured on the lower side whereas normal human bronchial epithelial cells were cultured on the opposite upper side (Figure 33.5). When co-cultured with MSCs, respiratory epithelial cells maintained their capacity to progressively differentiate and form a functional epithelium, leading to the differentiation of mucin-producing cells.

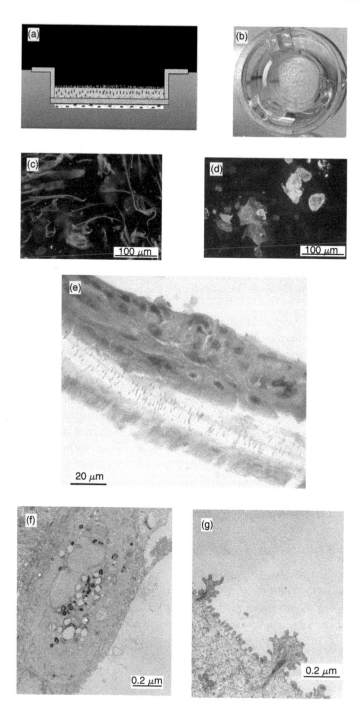

FIGURE 33.5 (a) Schematic representation of the Transwell membrane co-culture of MSCs and respiratory epithelial cells. (b) The macroscopic observation of the upper side of a Transwell membrane with a characteristic accumulation of respiratory epithelial cells. (c) Identification of MSCs during co-culture on a Transwell membrane with immunostaining for vimentin, a mesenchymal cell marker. (d) Identification of respiratory epithelial cells during co-culture on a Transwell membrane with immunostaining for cytokeratin, an epithelial cell marker. (e) Hematoxylin and eosin paraffin section showing a monolayer of MSCs on the bottom, separated from the epithelial cells by a membrane, which is evidenced by visible pores. (f) Transmission electron microscopy (TEM) revealing the presence of granules in some epithelial cells. (g) TEM confirming the development of microvilli and actin filaments the formation of cilia.

So far most of the work has focused on the infrastructure of tracheal tissue: tracheal cartilage. There appears to be momentum gathering in this area, and it is likely that a reliable and functional infrastructure will soon be available. As of today, however, there are no standard endpoints that are commonly accepted to evaluate the overall functionality and health of the constructs. Presumptively, the ideal properties of a tissue-engineered trachea are an ability to resist negative and positive pressures, an airtight lining that promotes a healthy epithelium, torsional flexibility, good vascularization and innervation, and nonimmunogenicity and biocompatibility to ensure bioincorporation and permanence. As the work matures, endpoints are likely to routinely include thorough histological evaluation of the airway epithelium, vascular and neural networks, as well evaluation of biomechanical and biochemical function.

A major challenge in tracheal tissue engineering is the avoidance of the mucosal inflammatory response that leads to its cicatrization and an eventual reduction and obstruction of the lumen. Though recent work has begun to elucidate some of the details, the inflammatory cascade involved in this phenomenon is far from being well understood. The use of microchip arrays may help to shed some light on some of these biochemical events and guide our eventual designs. The ultimate success of a tissue-engineered trachea is likely to be dependent on future basic science efforts in airway mucosal physiology.

It is likely that the promotion of thorough vascularization and innervation will be integral to the health of the construct. So far, very little work has been done to evaluate vascularization. It is presumed, but not known, that complete coverage with a healthy respiratory epithelium at the time of airway implantation will reduce, if not eliminate, the inflammation so often seen. Considering the high metabolic nature of epithelial cells, an adequate network of blood vessels is likely to be a prerequisite. It is feasible that as our understanding of mucosal physiology matures, we might potentially avoid a two-stage operation, but, in the short term, this is unlikely to be the case. Little or no work has focused on the innervation of the tissue-engineered constructs. Though poorly understood, it is likely that neural input is integral to the healthy maintenance of airway epithelium [6]. How much time will be needed to establish robust circulation and innervation? Will growth factors be necessary? Will the development of an airway bioreactor be advantageous?

At this point, it is not known if a respiratory epithelium is an absolute must; at the very least, the surface must be airtight to prevent bacterial colonization. The presumption is that a respiratory epithelium is the ultimate lining, but could an interim lining also work until the native cells reline the lumen [137]? Respiratory epithelial restitution takes place rapidly in native tissue. Would a respiratory epithelial cell-friendly surface, such as a basement membrane, suffice? Over what length could we expect the native epithelial cells to migrate? Will tissue-engineered epithelium be functional and display coordinated movement of its cilia?

Tracheal tissue engineering is currently in the very early stages of its development. As with any new effort, there is great hope that it will produce significant advances in our ability to treat tracheal disease. Its potential, however, is inescapably restrained by our limited understanding of airway physiology and will only flourish with persistently diligent efforts in both biomaterials engineering and basic science.

References

[1] Griscom, N.T. and Wohl, M.E., Dimensions of the growing trachea related to body height. Length, anteroposterior and transverse diameters, cross-sectional area, and volume in subjects younger than 20 years of age, *Am. Rev. Respir. Dis.*, 131, 840, 1985.

[2] Chen, S.J. et al., Measurement of tracheal size in children with congenital heart disease by computed tomography, *Ann. Thorac. Surg.*, 77, 1216, 2004.

[3] Spina, D., Epithelium smooth muscle regulation and interactions, *Am. J. Respir. Crit. Care Med.*, 158, S141, 1998.

[4] Puchelle, E. and Peault, B., Human airway xenograft models of epithelial cell regeneration, *Respir. Res.*, 1, 125, 2000.

[5] Knight, D.A. and Holgate, S.T., The airway epithelium: structural and functional properties in health and disease, *Respirology*, 8, 432, 2003.

[6] Erjefalt, J.S. et al., *In vivo* restitution of airway epithelium, *Cell Tissue Res.*, 281, 305, 1995.

[7] Erjefalt, J.S. and Persson, C.G., Airway epithelial repair: breathtakingly quick and multipotentially pathogenic, *Thorax*, 52, 1010, 1997.

[8] Erjefalt, J.S. et al., Microcirculation-derived factors in airway epithelial repair *in vivo*, *Microvasc. Res.*, 48, 161, 1994.

[9] Dohar, J.E. et al., Acquired subglottic stenosis — depth and not extent of the insult is key, *Int. J. Pediatr. Otorhinolaryngol.*, 46, 159, 1998.

[10] Jung Kwon, O. et al., Tracheal stenosis depends on the extent of cartilaginous injury in experimental canine model, *Exp. Lung Res.*, 29, 329, 2003.

[11] Basbaum, C. and Jany, B., Plasticity in the airway epithelium, *Am. J. Physiol.*, 259, L38, 1990.

[12] Engelhardt, J.F. et al., Progenitor cells of the adult human airway involved in submucosal gland development, *Development*, 121, 2031, 1995.

[13] Ayers, M.M. and Jeffery, P.K., Proliferation and differentiation in mammalian airway epithelium, *Eur. Respir. J.*, 1, 58, 1988.

[14] Evans, M.J. and Plopper, C.G., The role of basal cells in adhesion of columnar epithelium to airway basement membrane, *Am. Rev. Respir. Dis.*, 138, 481, 1988.

[15] Evans, M.J. et al., The role of basal cells in attachment of columnar cells to the basal lamina of the trachea, *Am. J. Respir. Cell Mol. Biol.*, 1, 463, 1989.

[16] Liu, J.Y., Nettesheim, P., and Randell, S.H., Growth and differentiation of tracheal epithelial progenitor cells, *Am. J. Physiol.*, 266, L296, 1994.

[17] Evans, M.J. et al., Junctional adhesion mechanisms in airway basal cells, *Am. J. Respir. Cell Mol. Biol.*, 3, 341, 1990.

[18] Borthwick, D.W. et al., Evidence for stem-cell niches in the tracheal epithelium, *Am. J. Respir. Cell Mol. Biol.*, 24, 662, 2001.

[19] Nettesheim, P. et al., Pathways of differentiation of airway epithelial cells, *Environ. Health Perspect.*, 85, 317, 1990.

[20] Inayama, Y. et al., The differentiation potential of tracheal basal cells, *Lab. Invest.*, 58, 706, 1988.

[21] Inayama, Y. et al., *In vitro* and *in vivo* growth and differentiation of clones of tracheal basal cells, *Am. J. Pathol.*, 134, 539, 1989.

[22] Erjefalt, J.S., Sundler, F., and Persson, C.G., Epithelial barrier formation by airway basal cells, *Thorax*, 52, 213, 1997.

[23] Schoch, K.G. et al., A subset of mouse tracheal epithelial basal cells generates large colonies *in vitro*, *Am. J. Physiol Lung Cell Mol. Physiol.*, 286, L631, 2004.

[24] Hong, K.U. et al., Basal cells are a multipotent progenitor capable of renewing the bronchial epithelium, *Am. J. Pathol.*, 164, 577, 2004.

[25] Hong, K.U. et al., *In vivo* differentiation potential of tracheal basal cells: evidence for multipotent and unipotent subpopulations, *Am. J. Physiol. Lung Cell Mol. Physiol.*, 286, L643, 2004.

[26] Persson, C.G. et al., Plasma-derived proteins in airway defence, disease and repair of epithelial injury, *Eur. Respir. J.*, 11, 958, 1998.

[27] Gluck, T.H. and Zeller, A., Die prophylaktische resektion der trachea, *Arch. Klin. Chir.*, 26, 427, 1881.

[28] Grillo, H.C., Dignan, E.F., and Miura, T., Extensive resection and reconstruction of mediastinal trachea without prosthesis or graft: an anatomical study in man, *J. Thorac. Cardiovasc. Surg.*, 48, 741, 1964.

[29] Grillo, H.C., Circumferential resection and reconstruction of the mediastinal and cervical trachea, *Ann. Surg.*, 162, 374, 1965.

[30] Grillo, H.C., Surgical treatment of postintubation tracheal injuries, *J. Thorac. Cardiovasc. Surg.*, 78, 860, 1979.

[31] Grillo, H.C., Primary reconstruction of airway after resection of subglottic laryngeal and upper tracheal stenosis, *Ann. Thorac. Surg.*, 33, 3, 1982.

[32] Grillo, H.C., Reconstructive techniques for extensive post-intubation tracheal stenosis, *Int. Surg.*, 67, 215, 1982.

[33] Dedo, H.H. and Fishman, N.H., Laryngeal release and sleeve resection for tracheal stenosis, *Ann. Otol. Rhinol. Laryngol.*, 78, 285, 1969.

[34] Montgomery, W.W., Suprahyoid release for tracheal anastomosis, *Arch. Otolaryngol.*, 99, 255, 1974.

[35] Pearson, F.G. et al., Primary tracheal anastomosis after resection of the cricoid cartilage with preservation of recurrent laryngeal nerves, *J. Thorac. Cardiovasc. Surg.*, 70, 806, 1975.

[36] Maddaus, M.A. et al., Subglottic tracheal resection and synchronous laryngeal reconstruction, *J. Thorac. Cardiovasc. Surg.*, 104, 1443, 1992.

[37] Pearson, F.G. and Gullane, P., Subglottic resection with primary tracheal anastomosis: including synchronous laryngotracheal reconstructions, *Semin. Thorac. Cardiovasc. Surg.*, 8, 381, 1996.

[38] Grillo, H.C., Tracheal replacement: a critical review, *Ann. Thorac. Surg.*, 73, 1995, 2002.

[39] Macchiarini, P., Trachea-guided generation: deja vu all over again? *J. Thorac. Cardiovasc. Surg.*, 128, 14, 2004.

[40] Cheng, W.F., Takagi, H., and Akutsu, T., Prosthetic reconstruction of the trachea, *Surgery*, 65, 462, 1969.

[41] Borrie, J. and Redshaw, N.R., Prosthetic tracheal replacement, *J. Thorac. Cardiovasc. Surg.*, 60, 829, 1970.

[42] Demos, N.J. et al., Tracheal regeneration in long-term survivors with silicone prosthesis, *Ann. Thorac. Surg.*, 16, 293, 1973.

[43] Neville, W.E., Bolanowski, P.J., and Soltanzadeh, H., Prosthetic reconstruction of the trachea and carina, *J. Thorac. Cardiovasc. Surg.*, 72, 525, 1976.

[44] Neville, W.E., Prosthetic reconstruction of trachea, *Rev. Laryngol. Otol. Rhinol.*, 103, 153, 1982.

[45] Nelson, R.J. et al., Neovascularity of a tracheal prosthesis/tissue complex, *J. Thorac. Cardiovasc. Surg.*, 86, 800, 1983.

[46] Nelson, R.J. et al., Development of a microporous tracheal prosthesis, *Trans. Am. Soc. Artif. Intern. Organs*, 25, 8, 1979.

[47] Wykoff, T.W., A preliminary report on segmental tracheal prosthetic replacement in dogs, *Laryngoscope*, 83, 1072, 1973.

[48] Michelson, E. et al., Experiments in tracheal reconstruction, *J. Thorac. Cardiovasc. Surg.*, 41, 748, 1961.

[49] Daniel, R.A., Jr., The regeneration of defects of the trachea and bronchi. An experimental study, *J. Thorac. Surg.*, 17, 335, 1948.

[50] Shaw, R.R., Aslami, A., and Webb, W.R., Circumferential replacement of the trachea in experimental animals, *Ann. Thorac. Surg.*, 5, 30, 1968.

[51] Keshishian, J.M., Blades, B., and Beattie, E.J., Tracheal reconstruction, *J. Thorac. Surg.*, 32, 707, 1956.

[52] Spinazzola, A.J., Graziano, J.L., and Neville, W.E., Experimental reconstruction of the tracheal carina, *J. Thorac. Cardiovasc. Surg.*, 58, 1, 1969.

[53] Moncrief, W.H., Jr. and Salvatore, J.E., An improved tracheal prosthesis, *Surg. Forum*, 9, 350, 1958.

[54] Pagliero, K.M. and Shepherd, M.P., Use of stainless steel wire coil prosthesis in treatment of anastomotic dehiscence after cervical tracheal resection, *J. Thorac. Cardiovasc. Surg.*, 67, 932, 1974.

[55] Bottema, J.R. and Wildevuur, C.H., Incorporation of microporous Teflon tracheal prostheses in rabbits: evaluation of surgical aspects, *J. Surg. Res.*, 41, 16, 1986.

[56] Bottema, J.R. et al., Microporous tracheal prosthesis: incorporation and prevention of infection, *Trans. Am. Soc. Artif. Intern. Organs*, 26, 412, 1980.

[57] Bailey, B.J. and Kosoy, J., Observations in the development of tracheal prostheses and tracheal transplantation, *Laryngoscope*, 80, 1553, 1970.

[58] Greenberg, S.D., Tracheal reconstruction: an experimental study, *Arch. Otolaryngol.*, 72, 565, 1960.

[59] Poticha, S.M. and Lewis, F.J., Experimental replacement of the trachea, *J. Thorac. Cardiovasc. Surg.*, 52, 61, 1966.

[60] Beall, A.C., Jr. et al., Circumferential replacement of the trachea with marlex mesh: preliminary report, *Surg. Forum*, 11, 40, 1960.

[61] Beall, A.C., Jr. et al., Tracheal replacement with heavy Marlex mesh. Circumferential replacement of the cervical trachea, *Arch. Surg.*, 84, 390, 1962.

[62] Harrington, O.B. et al., Circumferential replacement of the trachea with Marlex mesh, *Am. Surg.*, 28, 217, 1962.

[63] Ellis, P.R. et al., The use of heavy Marlex mesh for tracheal reconstruction following resection for malignancy, *J. Thorac. Cardiovasc. Surg.*, 44, 520, 1962.

[64] Kosoy, J. et al., Proplast tracheal prosthesis: a preliminary report, *Ann. Otol. Rhinol. Laryngol.*, 86, 392, 1977.

[65] Podoshin, L. and Fradis, M., Reconstruction of the anterior wall of the cervical trachea using knitted dacron, *Ear Nose Throat J.*, 55, 42, 1976.

[66] Ike, O. et al., Experimental studies on an artificial trachea of collagen-coated poly(L-lactic acid) mesh or unwoven cloth combined with a periosteal graft, *ASAIO Trans.*, 37, 24, 1991.

[67] Mendak, S.H., Jr. et al., The evaluation of various bioabsorbable materials on the titanium fiber metal tracheal prosthesis, *Ann. Thorac. Surg.*, 38, 488, 1984.

[68] Takahama, T. et al., A new improved biodegradable tracheal prosthesis using hydroxy apatite and carbon fiber, *ASAIO Trans.*, 35, 291, 1989.

[69] Kaiser, D., Alloplastic replacement of canine trachea with Dacron, *Thorac. Cardiovasc. Surg.*, 33, 239, 1985.

[70] Suh, S.W. et al., Replacement of a tracheal defect with autogenous mucosa lined tracheal prosthesis made from polypropylene mesh, *ASAIO J.*, 47, 496, 2001.

[71] Okumura, N. et al., A new tracheal prosthesis made from collagen grafted mesh, *ASAIO J.*, 39, M475, 1993.

[72] Okumura, N. et al., Experimental reconstruction of the intrathoracic trachea using a new prosthesis made from collagen grafted mesh, *ASAIO J.*, 40, M834, 1994.

[73] Pearson, F.G. et al., The reconstruction of circumferential tracheal defects with a porous prosthesis. An experimental and clinical study using heavy Marlex mesh, *J. Thorac. Cardiovasc. Surg.*, 55, 605, 1968.

[74] Teramachi, M. et al., Porous-type tracheal prosthesis sealed with collagen sponge, *Ann. Thorac. Surg.*, 64, 965, 1997.

[75] Suh, S.W. et al., Development of new tracheal prosthesis: autogenous mucosa-lined prosthesis made from polypropylene mesh, *Int. J. Artif. Organs*, 23, 261, 2000.

[76] Glatz, F. et al., A tissue-engineering technique for vascularized laryngotracheal reconstruction, *Arch. Otolaryngol. Head Neck Surg.*, 129, 201, 2003.

[77] Rush, B.F. and Cliffton, E.E., Experimental reconstruction of the trachea with bladder mucosa, *Surgery*, 40, 1105, 1956.

[78] Barker, W.S. and Litton, W.B., Bladder osteogenesis aids tracheal reconstruction, *Arch. Otolaryngol.*, 98, 422, 1973.

[79] Ohlsen, L. and Nordin, U., Tracheal reconstruction with perichondrial grafts, *Scand. J. Plast. Reconstr. Surg.*, 10, 135, 1976.

[80] Toohill, R.J., Martinelli, D.L., and Janowak, M.C., Repair of laryngeal stenosis with nasal septal grafts, *Ann. Otol. Rhinol. Laryngol.*, 85, 600, 1976.

[81] Kato, R. et al., Tracheal reconstruction by esophageal interposition: an experimental study [see comments], *Ann. Thorac. Surg.*, 49, 951, 1990.

[82] Murakami, S. et al., An experimental study of tracheal reconstruction using a freed piece of the right bronchus, *Thorac. Cardiovasc. Surg.*, 42, 76, 1994.

[83] Carbognani, P. et al., Experimental tracheal transplantation using a cryopreserved aortic allograft, *Eur. Surg. Res.*, 31, 210, 1999.

[84] Bryant, L.R., Replacement of tracheobronchial defects with autogenous pericardium, *J. Thorac. Cardiovasc. Surg.*, 48, 733, 1958.

[85] Cohen, R.C. et al., A new model of tracheal stenosis and its repair with free periosteal grafts, *J. Thorac. Cardiovasc. Surg.*, 92, 296, 1986.

[86] Eckersberger, F., Moritz, E., and Wolner, E., Circumferential tracheal replacement with costal cartilage, *J. Thorac. Cardiovasc. Surg.*, 94, 175, 1987.

[87] Drettner, B. and Lindholm, C.E., Experimental tracheal reconstruction with composite graft from nasal septum, *Acta Otolaryngol.*, 70, 401, 1970.

[88] Martinod, E. et al., Metaplasia of aortic tissue into tracheal tissue. Surgical perspectives, *C. R. Acad. Sci. III*, 323, 455, 2000.

[89] Hubbell, R.N. et al., Irradiated costal cartilage graft in experimental laryngotracheal reconstruction, *Int. J. Pediatr. Otorhinolaryngol.*, 15, 67, 1988.

[90] Zalzal, G.H., Cotton, R.T., and McAdams, A.J., The survival of costal cartilage graft in laryngotracheal reconstruction, *Otolaryngol. Head Neck Surg.*, 94, 204, 1986.

[91] Zalzal, G.H., Cotton, R.T., and McAdams, A.J., Cartilage grafts — present status, *Head Neck Surg.*, 8, 363, 1986.

[92] Lobe, T.E. et al., The application of solvent-processed human dura in experimental tracheal reconstruction, *J. Pediatr. Surg.*, 26, 1104, 1991.

[93] Chahine, A.A., Tam, V., and Ricketts, R.R., Use of the aortic homograft in the reconstruction of complex tracheobronchial tree injuries, *J. Pediatr. Surg.*, 34, 891, 1999.

[94] Dal, T. and Demirhan, B., Reconstruction of tracheal defects with dehydrated human costal cartilage: an experimental study in rats [In Process Citation], *Otolaryngol. Head Neck Surg.*, 123, 607, 2000.

[95] Bjork, V.O. and Rodriguez, L.E., Reconstruction of the trachea and its bifurcation; an experimental study, *J. Thorac. Surg.*, 35, 596, 1958.

[96] Elliott, M.J. et al., Tracheal reconstruction in children using cadaveric homograft trachea, *Eur. J. Cardiothorac. Surg.*, 10, 707, 1996.

[97] Jacobs, J.P. et al., Pediatric tracheal homograft reconstruction: a novel approach to complex tracheal stenoses in children, *J. Thorac. Cardiovasc. Surg.*, 112, 1549, 1996.

[98] Keskin, I.G. et al., Tracheal reconstruction using alcohol-stored homologous cartilage and autologous cartilage in the rabbit model, *Int. J. Pediatr. Otorhinolaryngol.*, 56, 161, 2000.

[99] Scherer, M.A. et al., Experimental bioprosthetic reconstruction of the trachea, *Arch. Otorhinolaryngol.*, 243, 215, 1986.

[100] Jacobs, J.P. et al., Tracheal allograft reconstruction: the total North American and worldwide pediatric experiences, *Ann. Thorac. Surg.*, 68, 1043, 1999.

[101] Yokomise, H. et al., High-dose irradiation prevents rejection of canine tracheal allografts, *J. Thorac. Cardiovasc. Surg.*, 107, 1391, 1994.

[102] Yokomise, H. et al., The infeasibility of using ten-ring irradiated grafts for tracheal allotransplantation even with omentopexy, *Surg. Today*, 26, 427, 1996.

[103] Farrington, W.T., Hung, W.C., and Binns, P.M., Experimental tracheal homografting, *J. Laryngol. Otol.*, 91, 101, 1977.

[104] Genden, E.M. et al., Microvascular transplantation of tracheal allografts model in the canine, *Ann. Otol. Rhinol. Laryngol.*, 112, 307, 2003.

[105] Genden, E.M. et al., Microvascular transfer of long tracheal autograft segments in the canine model, *Laryngoscope*, 112, 439, 2002.

[106] Delaere, P.R. et al., Experimental tracheal allograft revascularization and transplantation, *J. Thorac. Cardiovasc. Surg.*, 110, 728, 1995.

[107] Khalil-Marzouk, J.F., Allograft replacement of the trachea. Experimental synchronous revascularization of composite thyrotracheal transplant, *J. Thorac. Cardiovasc. Surg.*, 105, 242, 1993.

[108] Yokomise, H. et al., Long-term cryopreservation can prevent rejection of canine tracheal allografts with preservation of graft viability, *J. Thorac. Cardiovasc. Surg.*, 111, 930, 1996.

[109] Tojo, T. et al., Tracheal replacement with cryopreserved tracheal allograft: experiment in dogs, *Ann. Thorac. Surg.*, 66, 209, 1998.

[110] Mukaida, T. et al., Experimental study of tracheal allotransplantation with cryopreserved grafts, *J. Thorac. Cardiovasc. Surg.*, 116, 262, 1998.

[111] Bujia, J. et al., Tracheal transplantation: demonstration of HLA class II subregion gene products on human trachea, *Acta Otolaryngol.*, 110, 149, 1990.

[112] Liu, Y. et al., Immunosuppressant-free allotransplantation of the trachea: the antigenicity of tracheal grafts can be reduced by removing the epithelium and mixed glands from the graft by detergent treatment, *J. Thorac. Cardiovasc. Surg.*, 120, 108, 2000.

[113] Liu, Y. et al., A new tracheal bioartificial organ: evaluation of a tracheal allograft with minimal antigenicity after treatment by detergent, *ASAIO J.*, 46, 536, 2000.

[114] Genden, E.M. et al., Portal venous ultraviolet B-irradiated donor alloantigen prevents rejection in circumferential rat tracheal allografts, *Otolaryngol. Head Neck Surg.*, 124, 481, 2001.

[115] Genden, E.M. et al., Orthotopic tracheal allografts undergo reepithelialization with recipient-derived epithelium, *Arch Otolaryngol. Head Neck Surg.*, 129, 118, 2003.

[116] Vacanti, C.A. et al., Experimental tracheal replacement using tissue-engineered cartilage, *J. Pediatr. Surg.*, 29, 201, 1994.

[117] Kojima, K. et al., Autologous tissue-engineered trachea with sheep nasal chondrocytes, *J. Thorac. Cardiovasc. Surg.*, 123, 1177, 2002.

[118] Fuchs, J.R. et al., Engineered fetal cartilage: structural and functional analysis *in vitro*, *J. Pediatr. Surg.*, 37, 1720, 2002.

[119] Kojima, K. et al., Comparison of tracheal and nasal chondrocytes for tissue engineering of the trachea, *Ann. Thorac. Surg.*, 76, 1884, 2003.

[120] Sakata, J. et al., Tracheal composites tissue engineered from chondrocytes, tracheal epithelial cells, and synthetic degradable scaffolding, *Transplant Proc.*, 26, 3309, 1994.

[121] Terzaghi, M., Nettesheim, P., and Williams, M.L., Repopulation of denuded tracheal grafts with normal, preneoplastic, and neoplastic epithelial cell populations, *Cancer Res.*, 38, 4546, 1978.

[122] Engelhardt, J.F., Allen, E.D., and Wilson, J.M., Reconstitution of tracheal grafts with a genetically modified epithelium, *Proc. Natl Acad. Sci. USA*, 88, 11192, 1991.

[123] Johnson, N.F. and Hubbs, A.F., Epithelial progenitor cells in the rat trachea, *Am. J. Respir. Cell Mol. Biol.*, 3, 579, 1990.

[124] Engelhardt, J.F., Yankaskas, J.R., and Wilson, J.M., *In vivo* retroviral gene transfer into human bronchial epithelia of xenografts, *J. Clin. Invest.*, 90, 2598, 1992.

[125] Dupuit, F. et al., Differentiated and functional human airway epithelium regeneration in tracheal xenografts, *Am. J. Physiol. Lung Cell Mol. Physiol.*, 278, L165, 2000.

[126] Rainer, C. et al., Transplantation of tracheal epithelial cells onto a prefabricated capsule pouch with fibrin glue as a delivery vehicle, *J. Thorac. Cardiovasc. Surg.*, 121, 1187, 2001.

[127] Kojima, K. et al., A composite tissue-engineered trachea using sheep nasal chondrocyte and epithelial cells, *FASEB J.*, 17, 823, 2003.

[128] Fuchs, J.R. et al., Fetal tissue engineering: in utero tracheal augmentation in an ovine model, *J. Pediatr. Surg.*, 37, 1000, 2002.

[129] Erices, A., Conget, P., and Minguell, J.J., Mesenchymal progenitor cells in human umbilical cord blood, *Br. J. Haematol.*, 109, 235, 2000.

[130] Erices, A.A. et al., Human cord blood-derived mesenchymal stem cells home and survive in the marrow of immunodeficient mice after systemic infusion, *Cell Transplant*, 12, 555, 2003.

[131] Erickson, G.R. et al., Chondrogenic potential of adipose tissue-derived stromal cells *in vitro* and *in vivo*, *Biochem. Biophys. Res. Commun.*, 290, 763, 2002.

[132] Zvaifler, N.J. et al., Mesenchymal precursor cells in the blood of normal individuals, *Arthr. Res.*, 2, 477, 2000.

[133] Wexler, S.A. et al., Adult bone marrow is a rich source of human mesenchymal "stem" cells but umbilical cord and mobilized adult blood are not, *Br. J. Haematol.*, 121, 368, 2003.

[134] Pittenger, M.F. et al., Multilineage potential of adult human mesenchymal stem cells, *Science*, 284, 143, 1999.

[135] Barry, F. et al., Chondrogenic differentiation of mesenchymal stem cells from bone marrow: differentiation-dependent gene expression of matrix components, *Exp. Cell Res.*, 268, 189, 2001.

[136] Le Visage, C., Dunham, B., Flint, P., and Leong, K.W., Co-culture of mesenchymal stem cells and respiratory epithelial cells to engineer a human composite respiratory mucosa, *Tissue Eng.*, 10, 1426–35, 2004.

[137] Kim, J. et al., Replacement of a tracheal defect with a tissue-engineered prosthesis: early results from animal experiments, *J. Thorac. Cardiovasc. Surg.*, 128, 124, 2004.

Index

Note: Page numbers in *italics* refer to figures and tables

Milton Keynes UK
Ingram Content Group UK Ltd.
UKHW052028071024
449327UK00027B/2475